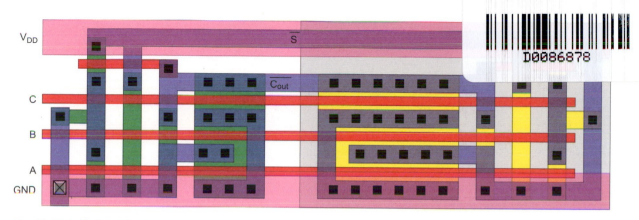

Fig 10.5(b) Full Adder Layout

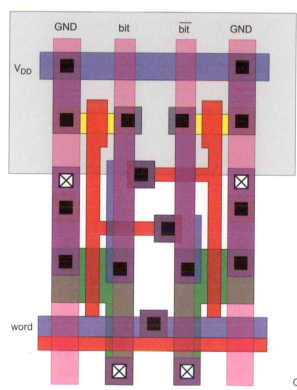

Fig 11.6(a) 6T SRAM Layout

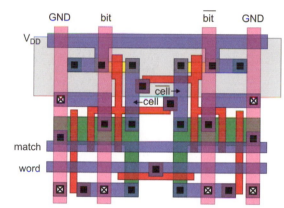

Fig 11.50 CAM Layout

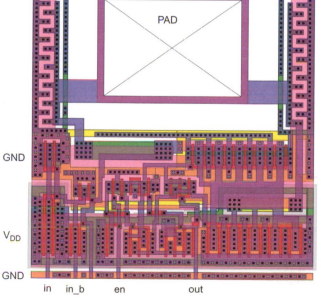

Fig 12.23 I/O Pad Layout

CMOS VLSI Design

A Circuits and Systems Perspective

Third Edition

Neil H. E. Weste
Macquarie University and
The University of Adelaide

David Harris
Harvey Mudd College

PEARSON

Addison
Wesley

Boston San Francisco New York
London Toronto Sydney Tokyo Singapore Madrid
Mexico City Munich Paris Cape Town Hong Kong Montreal

Sponsoring Editor: Matt Goldstein
Project Editor: Maite Suarez-Rivas
Senior Production Supervisor: Juliet Silveri
Production Services and Copyediting: Kathy Smith
Proofreading: Holly McLean-Aldis
Composition and Art: Gillian Hall, The Aardvark Group
Cover Design Supervisor: Joyce Cosentino Wells
Cover Designer: Alison R. Paddock
Text Designer: Delgado and Company, Inc.
Marketing Manager: Michelle Brown
Print Buyer: Caroline Fell

The photograph used on the cover and chapter openers is copyright ©2004 by Cisco Systems Inc., Wireless Networking Business Unit. Used with permission.

Access the latest information about Addison-Wesley titles from our World Wide Web site: http://www.aw-bc.com/computing

Many of the designations used by manufacturers and sellers to distinguish their products are claimed as trademarks. Where those designations appear in this book, and Addison-Wesley was aware of a trademark claim, the designations have been printed in initial caps or all caps.

The programs and applications presented in this book have been included for their instructional value. They have been tested with care, but are not guaranteed for any particular purpose. The publisher does not offer any warranties or representations, nor does it accept any liabilities with respect to the programs or applications.

Library of Congress Cataloging-in-Publication Data
Weste, Neil H. E.
   CMOS VLSI design : a circuits and systems perspective / Neil H.E. Weste,
David Harris.-- 3rd ed.
       p. cm.
   Includes bibliographical references and index.
   ISBN 0-321-14901-7
   1. Rev. ed. of: Principles of CMOS VLSI design. 1993. 2. Integrated
circuits--Very large scale integration--design and construction. 3. Metal
oxide semiconductors, Complementary. I. Harris, David (David M.) II. Weste,
Neil H. E. Principles of CMOS VLSI design. III. Title.

 TK7874.W45 2005
 621.39'5--dc22
                                           2004007092

For information on obtaining permission for the use of material from this work, please submit a written request to Pearson Education, Inc., Rights and Contracts Department, 75 Arlington St., Suite 300, Boston, MA 02116 or fax your request to 617-848-7047.

3 4 5 6 7 8 9 10—CRW—09 08 07

Contents

CHAPTER 2   MOS Transistor Theory

CHAPTER 3   CMOS Processing Technology

CHAPTER 4 Circuit Characterization and Performance Estimation

---

CHAPTER 5 Circuit Simulation

CHAPTER 6 **Combinational Circuit Design**

CHAPTER 7   Sequential Circuit Design

CHAPTER 8 Design Methodology and Tools

CHAPTER 9   **Testing and Verification**

CHAPTER 10 Datapath Subsystems

CHAPTER 11   Array Subsystems

CHAPTER 12 Special-purpose Subsystems

Preface

In the two decades since the first edition of this book was published, CMOS technology has claimed the preeminent position in modern electrical system design and enabled the widespread use of personal computers. In the decade since the second edition was published, continued advances have led to the explosion of the Internet and wireless communications. The transistor counts and clock frequencies of state-of-the-art chips have grown by orders of magnitude.

	1st Edition	2nd Edition	3rd Edition
Year	1985	1993	2004
Transistor Counts	10^5–10^6	10^6–10^7	10^8–10^9
Clock Frequencies	10^7	10^8	10^9
Worldwide Market	\$25B	\$60B	\$170B

This book has been extensively rewritten to reflect the enormous advances in integrated circuit design over the past decade. While the basic principles are largely the same, the practices have changed enormously because of the increases in transistor budgets and clock speeds, the growing challenges of power consumption, and the improvements in productivity and CAD tools.

How to Use This Book

This book intentionally covers more breadth and depth than any course would cover in a semester. It is accessible for a first undergraduate course in VLSI, yet detailed enough for advanced graduate courses and is useful as a reference to the practicing engineer. You are encouraged to pick and choose topics according to your interest. Chapter 1 previews the entire field, while subsequent chapters elaborate on specific topics. Sections are marked as optional (see example icon in margin) if they are not needed to understand subsequent sections. You may skip them on a first reading and return when they are relevant to you.

We have endeavored to include figures whenever possible ("a picture is worth a thousand words") to trigger your thinking. As you encounter examples throughout the text, we urge you to think about them before reading on to the solutions. We have also provided extensive references for those who need to delve deeper into topics introduced in this text. We have emphasized the best practices that are used in industry and warned of pitfalls and fallacies. Our judgments about merits of circuits may become incorrect as technology and applications change, but we believe it is the responsibility of a writer to attempt to separate the good from the bad.

Supplements

Updating and expanding on the supplements available was a major goal of this edition. There is now a volume of support materials for students and professors. All these are available on the Companion Web site for the book, www.aw-bc.com/weste. Supplements to help students with the course include:

- A lab manual with laboratory exercises involving the design of an 8-bit microprocessor covered in Chapter 1.
- A collection of links to VLSI resources including open-source CAD tools and process parameters.
- A student solutions manual that includes answers to selected problems.

Supplements to help instructors with the course include:

- A sample syllabus.
- Lecture slides for an introductory VLSI course.
- An instructors' manual with solutions.

These materials have been prepared exclusively for professors using the book in a course. Please contact your local Addison-Wesley sales representative, or send email to aw.cse@aw.com, for information on how to access them.

Acknowledgments

I (Neil) would like to first acknowledge my coauthor David Harris, who convinced me to come out of "book retirement" to do this edition. Without his tireless efforts, this book would not exist. Next, my long-suffering spouse, Avril, who relented on "Never another book!"—one more time.

Over the past ten or so years, along with my colleagues at Radiata Communications / Cisco Systems, I have learned a lot about getting mixed signal and RF CMOS circuits from conception into production. Much of that experience is embedded in this new edition. Gordon Foyster wrote the software that allowed us to convert mask layouts to postscript, a task that hasn't eased in arcaneness in ten years. Steve Avery helped out with suggestions on how to deal with the mask artwork and various other "book savvy" comments. Geoff Smith did the synthesis of the place and route examples that appear in Chapter 1, and along with Gordon often provided design methodology suggestions. Jared Anderson did the MATLAB and synthesis of the NCO in Chapter 8. Phil Ryan, Greg Zyner, and Mike Webb provided background on digital methodologies and design management. Andrew Adams, Jeffrey Harrison and John Olip guided me in RF issues. Rodney Chandler provided ADC theory and practice. Brian Hart contributed the INL/DNL plots in Chapter 12. Tom McDermott often provided software guidance and other eclectic information. Chris Corcoran often got me out of a muddle with dead laptops or other seemingly insurmountable software or network problems. John O'Sullivan provided useful feedback and was a good person to bounce ideas off. George Bouchaya brought an extensive background in manufacturing. John Haddy provided some of the photographs in this edition.

Bronwyn Forde provided administrative assistance. Dave Leonard and Bill Rossi provided support from afar. Finally, I would like to thank my long-time business partners and friends, Dave Skellern, Chris Beare, and Don MacLennan for a fruitful and productive association. I would like to acknowledge Cisco Systems in general for its support of this revision.

In contrast to the second edition, which was completed on an Apple Macintosh and a Symbolics Ivory processor in a basement in Massachusetts, many of my contributions to this edition were completed using the very technology that has become pervasive in the last ten or so years. Wireless hotspots and ADSL at home and in hotels and airports all over Australia and the world played their part in this edition. And both PC and Mac laptop technology played a huge part in the revision—providing the ability to work anywhere, anytime.

I (David) owe a great deal to the outstanding circuit designers with whom I have worked over the years. Mark Horowitz, Jonathan Allen, Bill Dally, Ivan Sutherland, Jason Stinson, Sam Naffziger, Tom Fletcher, and the Horowitz group at Stanford have shaped the way I think about circuits. I hope this book passes on some of the insights they have taught me. I would also like to thank Peter Cheung of Imperial College for hosting me during a very productive summer of writing in London.

We are indebted to many people for their reviews and suggestions. These people include: Bharadwaj "Birdy" Amrutur, Jacob Baker, Kerry Bernstein, Neil Burgess, Krishnendu Chakrabarty, C. K. Chen, Bill Dally, Nana Dankwa, Azita Emami-Neyestanak, Scott Fairbanks, Tom Fletcher, Jim Frenzel, Claude Gauthier, Ron Ho, David Hopkins, Nan "Ted" Jiang, Marcie Karty, Stephen Keckler, Fabian Klass, Torsten Lehmann, Rich Lethin, Michael Linderman, Dean Liu, Wagdy Mahmoud, Ziyad Mansour, Simon Moore, Alice Parker, Braden Phillips, Parameswaran Ramanathan, Justin Schauer, Ashok Srivastava, James Stine, Gu Wei, Ken Yang, and Evelina Yeung. Jaeha Kim, Tom Grutkowski, and Cecilia Krasuk provided particularly thorough technical reviews of large portions of the manuscript. We apologize in advance to anyone we overlooked.

TSMC kindly provided permission to use the 180 nm SPICE models for many examples. The MOSIS Service has provided measured SPICE parameters for numerous other processes. Artisan contributed data sheet pages from a TSMC 180 nm cell library.

Harvey Mudd College provided chip photographs taken by Kevin Mapp. Steve Rubin developed the open-source Electric Editor used to produce many layouts.

Designers from several companies contributed "war stories" of chip design errors for Section 9.12 under the condition of anonymity. We welcome your war stories as candidates for the next edition. An ideal story would be presented like a detective mystery starting with the symptoms, then the process by which the bug was traced, and concluding with a schematic of the incorrect circuit and the method by which it was corrected.

The E158 CMOS VLSI Design classes of Spring 2002, 2003, and 2004 at Harvey Mudd College and classes of engineers at Qualcomm and Sun Microsystems tested drafts of the manuscript. Some of the engineers who helped improve the manuscript include Matt Aldrich, Kevin Alley, Chi Bui, Ayoob Dooply, Trevor Gile, Brad Greer, Shamit Grover, Eric Henderson, Nick Hertl, Nicole Kang, Clark Korb, Karen Lee, Li-Jen Lin, Michael Linderman, Mark Locascio, Renee Logan, Dimitrios Lymberopoulos, Khurram Malik, Charles Matlack, Joe Petolino, Geoff Shippee, Joshua Smallman, Keith Stevens, Aaron Stratton, Yushi Tian, Daniel Woo, and Amy Yang.

Genevieve Breed, Matthew Erler, Tommy Leung, and David Diaz from Harvey Mudd College developed many of the simulations and figures throughout the text. David Diaz, Sean Kao, and Daniel Lee helped develop the MIPS processor example at Harvey Mudd College. Max Yi ported the MIPS example from Appendix B to Appendix A.

Addison-Wesley has done an admirable job with the grueling editorial and production process. We would particularly like to thank our editors Maite Suarez-Rivas, Matt Goldstein, and Juliet Silveri, our copyeditor Kathy Smith, and our compositor and artist Gillian Hall.

Sally Harris has been editing family books since David was an infant on her lap. She read the page proofs with amazing attention to detail and unearthed hundreds of errors. She (with help from Daniel Harris) also volunteered for the thankless task of checking the bibliography.

We have become painfully aware of the ease with which mistakes creep into a book of this size. The remaining errors are our own fault. We will pay a bounty of $1 to the first person to report each error, paid at the next printing. Please check the errata sheet at `www.aw-bc.com/weste` to see if the bug has already been reported. Send your reports with your name and address to `bugs@cmosvlsi.com`.

N.W.
D.H.
April 2004

Introduction
1

1.1 A Brief History

In 1958, Jack Kilby built the first integrated circuit flip-flop with two transistors at Texas Instruments. In 2003, the Intel Pentium 4 microprocessor contained 55 million transistors and a 512-Mbit dynamic random access memory (DRAM) contained more than half a billion transistors. This corresponds to a compound annual growth rate of 53% over 45 years. No other technology in history has sustained such a high growth rate for so long.

This incredible growth has come from steady miniaturization of transistors and improvements in manufacturing processes. Most other fields of engineering involve tradeoffs between performance, power, and price. However, as transistors become smaller, they also become faster, dissipate less power, and are cheaper to manufacture. This synergy has revolutionized not only electronics, but also society at large.

The processing performance once exclusive to Cray supercomputers is now available in hand-held personal digital assistants. Processing capability that was once necessary for secret military spread-spectrum communications is now available in disposable cellular telephones. Improvements in integrated circuits have enabled space exploration, made automobiles more efficient, revolutionized the nature of warfare, brought vast libraries of information to our Web browsers, and made the world a more interdependent place.

Figure 1.1 shows annual sales in the worldwide semiconductor market. Integrated circuits became a $100B/year business in 1994. The blip in 2000 is associated with the surge in sales for Y2K upgrades followed by the worldwide recession. In 2003, the industry manufactured more than one quintillion (10^{18}) transistors, or more than 100 million for every human being on the planet. Thousands of engineers have made their fortunes in the field. New fortunes lie ahead for those with innovative ideas and the talent to bring their ideas to reality.

During the first half of the 20th century, electronic circuits used large, expensive, power-hungry, and unreliable vacuum tubes. In 1947, John Bardeen and Walter Brattain built the first functioning point contact transistor at Bell Laboratories, shown in Figure 1.2(a) [Riordan97]. It was nearly classified as a military secret, but Bell Labs publicly announced the device in the following year.

We have called it the Transistor, T–R–A–N–S–I–S–T–O–R, because it is a resistor or semiconductor device which can amplify electrical signals as they are transferred

1

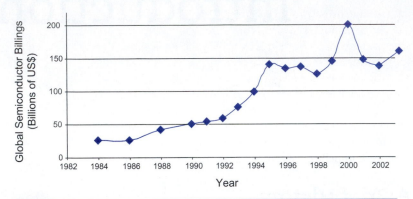

FIG 1.1 Size of worldwide semiconductor market

Source: Semiconductor Industry Association.

through it from input to output terminals. It is, if you will, the electrical equivalent of a vacuum tube amplifier. But there the similarity ceases. It has no vacuum, no fil-ament, no glass tube. It is composed entirely of cold, solid substances.

Ten years later, Jack Kilby at Texas Instruments realized the potential for miniaturiza-tion if multiple transistors could be built on a single piece of silicon. Figure 1.2(b) shows his first prototype of an integrated circuit, constructed from a germanium slice and gold wires.

The invention of the transistor earned the Nobel Prize in Physics in 1956 for Bardeen, Brattain, and their co-worker William Shockley. Kilby received the Nobel Prize in Physics in 2000 for the invention of the integrated circuit.

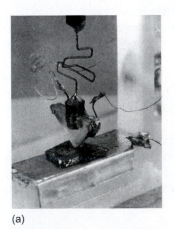

(a)   (b)

FIG 1.2 (a) First transistor (Property of AT&T Archives. Reprinted with permission of AT&T.) and (b) first integrated circuit (Courtesy of Texas Instruments.)

Soon after inventing the point contact transistor, Bell Labs developed the bipolar junction transistor. Bipolar transistors were more reliable, less noisy, and more power-efficient. Early integrated circuits primarily used bipolar transistors. Transistors can be viewed as electrically controlled switches with a control terminal and two other terminals that are connected or disconnected depending on the voltage applied to the control. Bipolar transistors require a small current into the control (base) terminal to switch much larger currents between the other two (emitter and collector) terminals. The quiescent power dissipated by these base currents limits the maximum number of transistors that can be integrated onto a single die. Metal Oxide Semiconductor Field Effect Transistors (MOSFETs) offer the compelling advantage that they draw almost zero control current while idle. They come in two flavors: nMOS and pMOS, using n-type and p-type dopants, respectively. The original idea of field effect transistors dated back to the German scientist Julius Lilienfeld in 1925 [US patent 1,745,175] and a structure closely resembling the MOSFET was proposed in 1935 by Oskar Heil [British patent 439,457], but materials problems foiled early attempts to make functioning devices.

Frank Wanlass at Fairchild described the first logic gates using MOSFETs in 1963 [Wanlass63]. His gates used both nMOS and pMOS transistors, earning the name Complementary Metal Oxide Semiconductor, or CMOS. The circuits used discrete transistors but consumed only nanowatts of power, six orders of magnitude less than their bipolar counterparts. With the development of the silicon planar process, MOS integrated circuits became attractive for their low cost because each transistor occupied less area and the fabrication process was simpler [Vadasz69]. Early processes used only pMOS transistors and suffered from poor performance, yield, and reliability. Processes using nMOS transistors became dominant in the 1970s [Mead80]. Intel pioneered nMOS technology with its 1101 256-bit static random access memory and 4004 4-bit microprocessor shown in Figure 1.3. While the nMOS process was less expensive than CMOS, nMOS logic gates still consumed power while idle. Power consumption became a major issue in the 1980s as

(a)

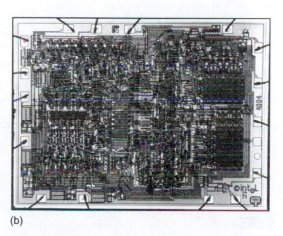
(b)

FIG 1.3 (a) Intel 1101 SRAM (© IEEE 1967 [Vadasz69]) and (b) 4004 microprocessor (Reprinted with permission of Intel Corporation.)

hundreds of thousands of transistors were integrated onto a single die. CMOS processes were widely adopted and have essentially replaced nMOS and bipolar processes for nearly all digital logic applications.

Gordon Moore observed in 1965 that plotting the number of transistors that can be most economically manufactured on a chip gives a straight line on a semilogarithmic scale [Moore65]. At the time he found transistor count doubling every 18 months. This observation has been called *Moore's Law* and has become a self-fulfilling prophecy. Figure 1.4 shows that the number of transistors in Intel microprocessors has doubled every 26 months since the invention of the 4004.

The level of integration of chips has been classified as small-scale, medium-scale, large-scale, and very large-scale. *Small-scale integration* (SSI) circuits such as the 7404 inverter have fewer than 10 gates, with a conversion of roughly half a dozen transistors per gate. *Medium-scale integration* (MSI) circuits such as the 74161 counter have up to 1000 gates. *Large-scale integration* (LSI) circuits such as simple 8-bit microprocessors have up to 10,000 gates. It soon became apparent that new names would have to be created every five years if this naming trend continued and thus the term *very large-scale integration* (VLSI) is used to describe most integrated circuits from the 1980s onward.

A corollary of Moore's law is that transistors become faster, consume less power, and are cheaper to manufacture each year. Figure 1.5 shows that Intel microprocessor clock frequencies have doubled roughly every 34 months. Remarkably, the improvements have accelerated in recent years. Computer performance has grown even more than raw clock speed. Even though an individual CMOS transistor uses very little energy each time it switches, the enormous numbers of transistors switching at very high rates of speed have made power consumption a major design consideration again.

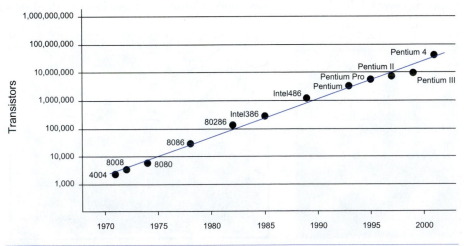

FIG 1.4 Transistors in Intel microprocessors [Intel03]

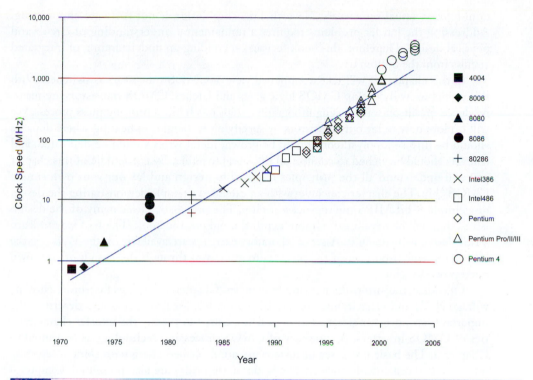

FIG 1.5 Clock frequencies of Intel microprocessors

The 4004 used transistors with minimum dimensions of 10 µm in 1971. The Pentium 4 uses transistors with minimum dimensions of 130 nm in 2003, corresponding to two orders of magnitude in improvement over three decades. Obviously this *scaling* cannot go on forever because transistors cannot be smaller than atoms. However, many predictions of fundamental limits to scaling have already proven wrong. Creative engineers and material scientists have billions of dollars to gain by getting ahead of their competitors. In the early 1990s, experts agreed that scaling would continue for at least a decade but that beyond that point the future was murky. In 2003, we still believe that scaling will continue for at least another decade. The future is yours to invent.

1.2 Book Summary

As VLSI transistor budgets have grown exponentially, designers have come to rely on increasing levels of automation to seek corresponding productivity gains. Many designers spend much of their effort specifying circuits with hardware description languages and seldom look at actual transistors. Nevertheless, chip design is not software engineering. Addressing the harder problems requires a fundamental understanding of circuit and physical design. Therefore, this book focuses on building an understanding of integrated circuits from the bottom up.

In this chapter we will take a simplified view of CMOS transistors as switches. With this model we will develop CMOS logic gates and latches. CMOS transistors are mass-produced on silicon wafers using lithographic steps much like a printing press process. We will explore how to lay out transistors by specifying rectangles indicating where dopants should be diffused, polysilicon should be grown, metal wires should be deposited, and contacts should be etched to connect all the layers together. By the middle of this chapter you will understand all the principles required to design and lay out your own simple CMOS chip. The chapter concludes with an extended example demonstrating the design of a simple 8-bit MIPS microprocessor chip. The processor raises many of the design issues that will be developed in more depth throughout the book. The best way to learn VLSI design is by doing it. A set of laboratory exercises are available on the Web (see the Web Supplement portion of the preface) to guide you through the design of your own microprocessor chip.

Of course, transistors are not simply switches. Chapter 2 develops first-order current-voltage (I-V) and capacitance-voltage (C-V) models for transistors and describes the important second-order effects. These models are used to predict the transfer characteristics of CMOS inverters. A summary of CMOS processing technology is presented in Chapter 3. The basic processes in current use are described along with some interesting process enhancements. Representative layout design rules are also presented. Chapter 4 addresses performance estimation for circuits. The first-order I-V characteristics are both too complicated and too inaccurate to apply by hand to interesting circuits. Fortunately, transistors can be modeled as having an effective resistance and capacitance for the purpose of estimating delay. The relative performance of different gates can be quantified through their logical effort, a technique that we will revisit throughout the book. Wires are also addressed because they are as important as the transistors to overall performance. Simulation is discussed in Chapter 5 and is used to obtain more accurate performance predictions as well as to verify the correctness of circuits and logic. Chapter 6 addresses combinational circuit design. A whole kit of circuit families are available with different tradeoffs in speed, power, complexity, and robustness. Chapter 7 continues with sequential circuit design, including clocking and latching techniques.

The remainder of this book presents a subsystem view of CMOS design. Chapter 8 focuses on a range of current design methods, identifying the issues peculiar to CMOS. Testing and design-for-test techniques are discussed in Chapter 9. Chapter 10 catalogs designs for a host of datapath subsystems including adders, shifters, multipliers, counters, and others. Chapter 11 similarly describes memory subsystems including SRAMs, DRAMs, CAMs, ROMs, and PLAs. Finally, Chapter 12 addresses special-purpose subsystems including clocking, I/O, and mixed-signal blocks. Hardware description languages (HDLs) are used in the design of nearly all digital integrated circuits today. Appendices A and B provide tutorials in Verilog and VHDL, the two dominant HDLs.

A number of sections are marked with an "optional" icon. These sections describe particular subjects in greater detail. You may skip over these sections on a first reading and return to them when they are of practical relevance.

1.3   MOS Transistors

Silicon (Si), a semiconductor, forms the basic starting material for a large class of integrated circuits [Tsividis99]. Pure silicon consists of a three-dimensional *lattice* of atoms. Silicon is a Group IV element, so it forms covalent bonds with four adjacent atoms, as shown in Figure 1.6(a). The lattice is shown in the plane for ease of drawing, but it actually forms a cubic crystal. As all of its valence electrons are involved in chemical bonds, it is a poor conductor. The conductivity can be raised by introducing small amounts of impurities into the silicon lattice. These impurities are called *dopants*. A Group V dopant such as arsenic has five valence electrons. It replaces a silicon atom in the lattice and still bonds to four neighbors, so the fifth valence electron is loosely bound to the arsenic atom, as shown in Figure 1.6(b). Thermal vibration of the lattice at room temperature is enough to set the electron free to move, leaving a positively charged As^+ ion and a free electron. The free electron can carry current so the conductivity is higher. We call this an *n*-type semiconductor because the free carriers are negatively charged electrons. Similarly, a Group III dopant such as boron has three valence electrons, as shown in Figure 1.6(c). The dopant atom can borrow an electron from a neighboring silicon atom, which in turn becomes short by one electron. That atom in turn can borrow an electron, and so forth, so the missing electron, or *hole*, can propagate about the lattice. The hole acts as a positive carrier so we call this a *p*-type semiconductor.

A junction between p-type and n-type silicon is called a *diode*, shown in Figure 1.7. When the voltage on the p-type semiconductor, called the *anode*, is raised above the n-type *cathode*, we say the diode is *forward biased* and current will flow. When the anode voltage is less than or equal to the cathode voltage, the diode is *reverse biased* and almost zero current flows.

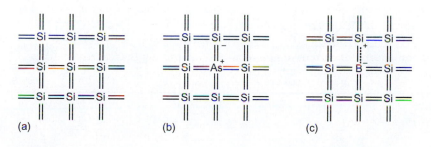

(a)   (b)   (c)

FIG 1.6 Silicon lattice and dopant atoms

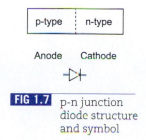

FIG 1.7 p-n junction diode structure and symbol

An *MOS* (Metal-Oxide-Semiconductor) structure is created by superimposing several layers of conducting and insulating materials to form a sandwich-like structure. These structures are manufactured using a series of chemical processing steps involving oxidation of the silicon, the diffusion of impurities into the silicon to give it certain conduction characteristics, and the deposition and etching of aluminum or other metals to provide interconnection in the same way that a printed wiring board is constructed. This is carried out on a single crystal of silicon, which is available as thin flat circular wafers around 15–30 cm in diameter. CMOS technology provides two types of transistors (also called *devices* in this text): an n-type transistor (nMOS) and a p-type transistor (pMOS). Transistor operation is based on electric fields so the devices are also called Metal Oxide Semiconductor Field Effect Transistors (*MOSFETs*) or simply *FETs*. Cross-sections and symbols of these transistors are shown in Figure 1.8. The n+ and p+ regions indicate heavily doped n- or p-type silicon.

Each transistor consists of a stack of the conducting *gate*, an insulating layer of silicon dioxide (SiO_2, better known as glass), and the silicon wafer, also called the *substrate*, *body*, or *bulk*. Gates of early transistors were built from metal, so the stack was called metal-oxide-semiconductor, or MOS. Now the gate is typically formed from polycrystalline silicon (*polysilicon*), but the name stuck. An nMOS transistor is built with a p-type body and has regions of n-type semiconductor adjacent to the gate called the *source* and *drain*. They are physically equivalent and for now we will regard them as interchangeable. The body is typically grounded. A pMOS transistor is just the opposite, consisting of p-type source and drain regions with an n-type body. In a CMOS technology with both flavors of transistors, the substrate is either n-type or p-type. The other flavor of transistor must be built in a special well in which dopant atoms have been locally added to form the body of the opposite type.

The gate is a control input: It affects the flow of electrical current between the source and drain. Consider an nMOS transistor. The body is generally grounded so the p–n junc-

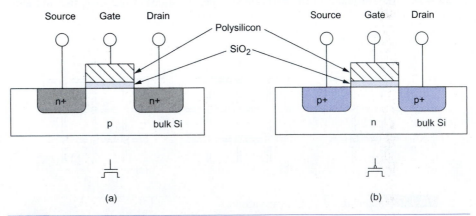

FIG 1.8 nMOS transistor (a) and pMOS transistor (b)

tions of the source and drain to body are reverse-biased. If the gate is also grounded, no current flows through the reverse-biased junctions. Hence, we say the transistor is OFF. If the gate voltage is raised, it creates an electric field that starts to attract free electrons to the underside of the Si–SiO$_2$ interface. If the voltage is raised enough, the electrons outnumber the holes and a thin region under the gate called the *channel* is inverted to act as an n-type semiconductor. Hence, a conducting path of electron carriers is formed from source to drain and current can flow. We say the transistor is ON.

For a pMOS transistor, the situation is again reversed. The body is held at a high potential. When the gate is also at a high potential, the source and drain junctions are reverse-biased and no current flows so the transistor is OFF. When the gate voltage is lowered, positive charges are attracted to the underside of the Si–SiO$_2$ interface. A sufficiently low gate voltage inverts the channel and a conducting path of positive carriers is formed from source to drain, so the transistor is ON. Notice that the symbol for the pMOS transistor has a bubble on the gate, indicating that the transistor behavior is the opposite of the nMOS.

Throughout this book, we will call the high potential V_{DD} or POWER and assume it represents a logic '1' value in digital circuits. In many logic families of the 1970s and 1980s, V_{DD} was set to 5 volts or occasionally higher. Smaller, more recent transistors are unable to withstand such high voltages and have used supplies of 3.3 V, 2.5 V, 1.8 V, 1.5 V, and so forth. The low potential is called GROUND (GND) or V_{SS} and represents a logic '0.' It is normally 0 volts.

In summary, the gate of an MOS transistor controls the flow of current between the source and drain. Simplifying this to the extreme allows the MOS transistors to be viewed as simple on/off switches. When the gate of an nMOS transistor is '1,' the transistor is ON and there is a conducting path from source to drain. When the gate is low, the nMOS transistor is OFF and almost zero current flows from source to drain. A pMOS transistor is just the opposite, being ON when the gate is low and OFF when the gate is high. This switch model is illustrated in Figure 1.9, where *g*, *s*, and *d* indicate gate, source, and drain. It is so useful that it will be our most common model for thinking about circuit behavior.

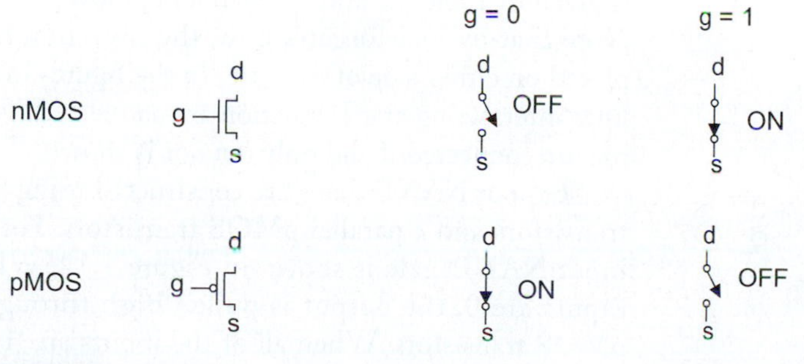

FIG 1.9 Transistor symbols and switch-level models

1.4 CMOS Logic

1.4.1 The Inverter

Figure 1.10(a) shows a CMOS inverter or NOT gate using one nMOS transistor and one pMOS transistor. The horizontal bar at the top indicates V_{DD} and the triangle at the bottom indicates GND. When the input A is '0,' the nMOS transistor is OFF and the pMOS transistor is ON. Thus the output Y is pulled up to '1' because it is connected to V_{DD} but not to GND. Conversely, when A is '1,' the nMOS is ON, the pMOS is OFF, and the Y is pulled down to '0.' This is summarized in the truth table of Table 1.1. The symbol is given in Figure 1.10(b).

Table 1.1	Inverter truth table
A	Y
0	1
1	0

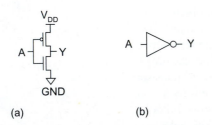

(a)   (b)

FIG 1.10 Inverter schematic (a) and symbol (b) $Y = \overline{A}$

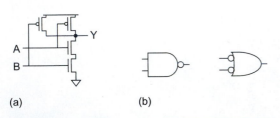

(a)   (b)

FIG 1.11 2-input NAND gate schematic (a) and symbol (b) $Y = \overline{A \cdot B}$

1.4.2 The NAND Gate

Figure 1.11(a) shows a 2-input CMOS NAND gate. It consists of two series nMOS transistors between Y and GND and two parallel pMOS transistors between Y and V_{DD}. If either input A or B is '0,' at least one of the nMOS transistors will be OFF, breaking the path from Y to GND. But at least one of the pMOS transistors will be ON, creating a path from Y to V_{DD}. Hence, the output Y will be '1.' If both inputs are '1,' both of the nMOS transistors will be ON and both of the pMOS transistors will be OFF. Hence, the output will be '0.' The truth table is given in Table 1.2 and the symbol is shown in Figure 1.11(b). Note that by DeMorgan's Law, the inversion bubble may be placed on either side of the gate. In the figures in this book, two lines intersecting at a T-junction are connected. Two lines crossing are connected if and only if a dot is shown.

k-input NAND gates are constructed using k series nMOS transistors and k parallel pMOS transistors. For example, a 3-input NAND gate is shown in Figure 1.12. When any of the inputs are 0, the output is pulled high through the parallel pMOS transistors. When all of the inputs are '1,' the output is pulled low through the series nMOS transistors.

Table 1.2	NAND gate truth table			
A	**B**	**pull-down network**	**pull-up network**	**Y**
0	0	OFF	ON	1
0	1	OFF	ON	1
1	0	OFF	ON	1
1	1	ON	OFF	0

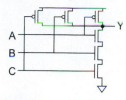

FIG 1.12 3-input NAND gate schematic $Y = \overline{A \bullet B \bullet C}$

1.4.3 Combinational Logic

The inverter and NAND gates are examples of *complementary CMOS logic gates*, also called *static CMOS gates*. In general, a fully complementary CMOS gate has an nMOS *pull-down network* to connect the output to '0' (GND) and pMOS *pull-up network* to connect the output to '1' (V_{DD}), as shown in Figure 1.13. The networks are arranged such that one is ON and the other OFF for any input pattern.

The pull-up and pull-down networks in the inverter each consisted of a single transistor. The NAND gate used a series pull-down network and a parallel pull-up network. More elaborate networks are used for more complex gates. Two or more transistors in series are ON only if all of the series transistors are ON. Two or more transistors in parallel are ON if any of the parallel transistors are ON. This is illustrated in Figure 1.14 for nMOS and pMOS transistor pairs. By using combinations of these constructions, CMOS combinational gates can be constructed.

In general when we join a pull-up network to a pull-down network to form a logic gate as shown in Figure 1.13, they both will attempt to exert a logic level at the output. The possible levels at the output are shown in Table 1.3. From this table it can be seen that the output of a CMOS logic gate can be in four states. The '1' and '0' levels have been encountered with the inverter and NAND gates, where either the pull-up or pull-down is OFF and the other structure is ON. When both pull-up and pull-down are OFF, the *high-impedance* or *floating* Z output state results. This is of importance in multiplexers, memory elements, and bus drivers. The *crowbarred* X level exists when both pull-up and pull-down are simultaneously turned ON. This causes an indeterminate level and also static power to be dissipated. It is usually an unwanted condition in any CMOS digital circuit.

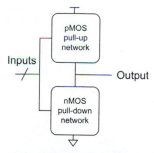

FIG 1.13 General logic gate using pull-up and pull-down networks

Table 1.3	Output states of CMOS logic gate	
	pull-up OFF	**pull-up ON**
pull-down OFF	Z	1
pull-down ON	0	crowbarred (X)

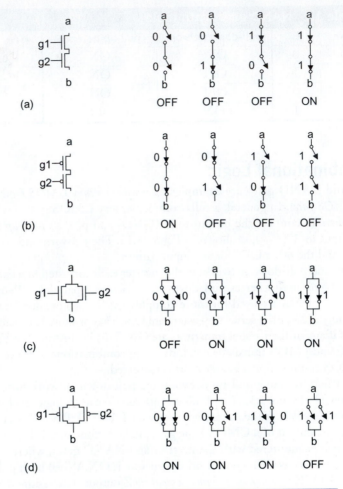

FIG 1.14 Connection and behavior of series and parallel transistors

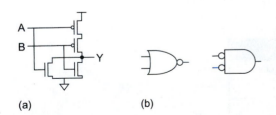

(a)   (b)

FIG 1.15 2-input NOR gate schematic (a) and symbol (b) $Y = \overline{A + B}$

1.4.4 The NOR Gate

A 2-input NOR gate is shown in Figure 1.15. The nMOS transistors are in parallel to pull the output low when either input is high. The pMOS transistors are in series to pull the output high when both inputs are low, as indicated by the truth table of Table 1.4. As with the NAND gate, there is never a case in which the output is crowbarred or left floating.

Table 1.4	NOR gate truth table	
A	**B**	**Y**
0	0	1
0	1	0
1	0	0
1	1	0

Example

Sketch a 3-input CMOS NOR gate.

Solution: Figure 1.16 shows such a gate. If any input is high, the output is pulled low through the parallel nMOS transistors. If all inputs are low, the output is pulled high through the series pMOS transistors.

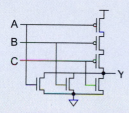

FIG 1.16 3-input NOR gate schematic
$$Y = \overline{A + B + C}$$

1.4.5 Compound Gates

A compound gate is formed by using a combination of series and parallel switch structures. For example, the derivation of the switch connection diagram for the function $Y = \overline{(A \cdot B) + (C \cdot D)}$ is shown in Figure 1.17. This function is sometimes called AND-OR-INVERT-22, or AOI22 because it performs the NOR of a pair of 2-input ANDs. For the nMOS pull-down network, take the uninverted expression $((A \cdot B) + (C \cdot D))$ indicating when the output should be pulled to '0.' The AND expressions $(A \cdot B)$ and $(C \cdot D)$ may be implemented by series connections of switches as shown in Figure 1.17(a). Now taking these as subswitches and ORing the result requires the parallel connection of these two structures, which is shown in Figure 1.17(b). For the pMOS pull-up network we must compute the complementary expression using switches that turn on with inverted polarity. By DeMorgan's Law, this is equivalent to interchanging AND and OR operations. Hence, transistors that appear in series in the pull-down network must appear in parallel in the pull-up network. Transistors that appear in parallel in the pull-down network must appear in series in the pull-up network. This principle is called *conduction complements* and has already appeared in the design of the NAND and NOR gates. In the pull-up network, the parallel combination of A and B is placed in series with the parallel combination of C and D. This progression is evident in Figure 1.17(c) and Figure 1.17(d). Putting the networks together yields the connection diagram (Figure 1.17(e)). The schematic icon is shown in Figure 1.17(f), which shows that this gate can be used in a 2-input inverting multiplexer. If $C = \overline{B}$, then $Y = \overline{A}$ if B is true, while $Y = \overline{D}$ if B is false.

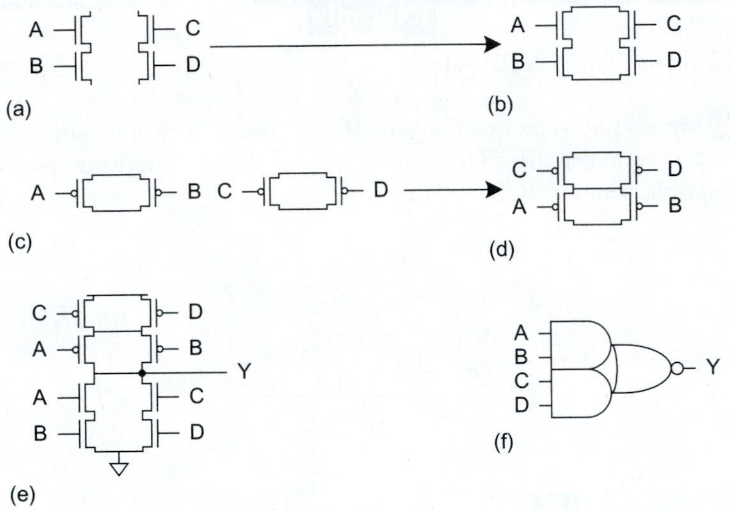

FIG 1.17 CMOS compound gate for function $Y = \overline{(A \bullet B) + (C \bullet D)}$

1.4.6 Pass Transistors and Transmission Gates

The *strength* of a signal is measured by how closely it approximates an ideal voltage source. In general, the stronger a signal, the more current it can source or sink. The power supplies, or *rails*, (V_{DD} and GND) are the source of the strongest '1's and '0's.

An nMOS transistor is an almost perfect switch when passing a '0' and thus we say it passes a *strong* '0.' However, the nMOS transistor is imperfect at passing a '1.' The high voltage level is somewhat less than V_{DD}, as will be explained in Section 2.5.5. We say it passes a *degraded* or *weak* '1.' A pMOS transistor again has the opposite behavior, passing strong '1's but degraded '0's. The transistor symbols and behaviors are summarized in Figure 1.19 with g, s, and d indicating gate, source, and drain. The symbols with arrows represent switches in the OFF and ON positions.

Example

Sketch a complementary CMOS gate computing $Y = \overline{(A + B + C) \cdot D}$.

Solution: Figure 1.18 shows such an OR-AND-INVERT-3-1 (OAI31) gate. The nMOS pull-down network pulls the output low if D is '1' and either A or B or C are '1', so D is in series with the parallel combination of A, B, and C. The pMOS pull-up network is the conduction complement, so D must be in parallel with the series combination of A, B, and C.

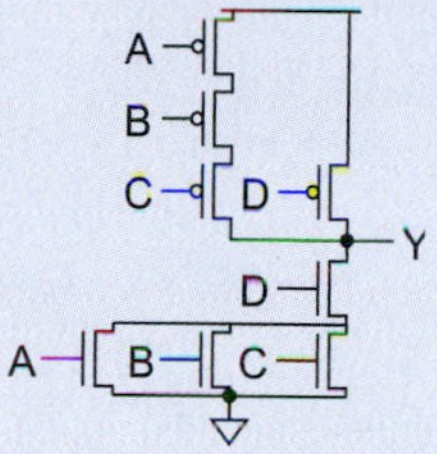

FIG 1.18 CMOS compound gate for function $Y = \overline{(A + B + C) \cdot D}$

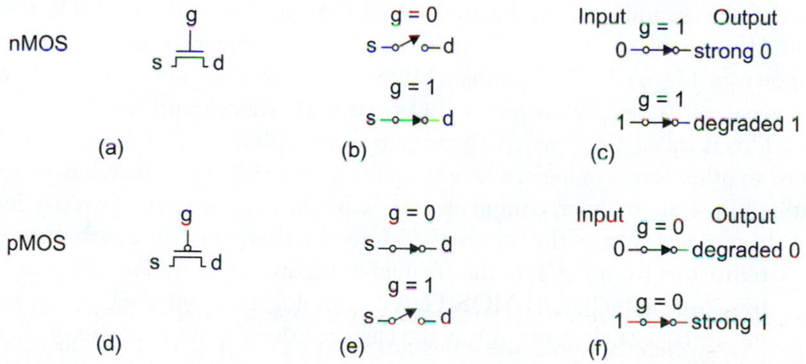

FIG 1.19 Pass transistor strong and degraded outputs

When an nMOS or pMOS is used alone as an imperfect switch, we sometimes call it a *pass transistor.* By combining an nMOS and a pMOS transistor in parallel (Figure 1.20(a)), we obtain a switch that turns on when a '1' is applied to g (Figure 1.20(b)) in which '0's and '1's are both passed in an acceptable fashion (Figure 1.20(c)). We term this a *transmission gate* or *pass gate*. In a circuit where only a '0' or a '1' has to be passed, the appropriate transistor (n or p) can be deleted, reverting to a single nMOS or pMOS device. Note that both the control input and its complement are required by the transmission gate. This is called *double rail* logic. Some circuit symbols for the transmission gate are shown in Figure 1.20(d).[1] None are easier to draw than the simple schematic, so we will use the schematic to represent a transmission gate in this book.

Input	Output

g = 0, gb = 1
a ⟶ b

g = 1, gb = 0
a ⟶ b

g = 1, gb = 0
0 ⟶ strong 0

g = 1, gb = 0
1 ⟶ strong 1

g
a ▭ b
gb

(a)   (b)   (c)

g
a ⬕ b
gb

g
a ⊗ b
gb

g
a ▷ b
gb

(d)

FIG 1.20 Transmission gate

In all of our examples so far, the inputs drive the gate terminals of nMOS transistors in the pull-down network and pMOS transistors in the complementary pull-up network, as was shown in Figure 1.13. Thus the nMOS transistors only need to pass 0's and the pMOS only pass 1's, so the output is always strongly driven and the levels are never degraded. This is called a *fully restored* logic gate and simplifies circuit design considerably. In contrast to other forms of logic, where the pull-up and pull-down switch networks have to be ratioed in some manner, complementary CMOS gates operate correctly independently of the physical sizes of the transistors. Moreover, there is never a path through 'ON' transistors from the '1' to the '0' supplies for any combination of inputs (in contrast to single-channel MOS, GaAs technologies, or bipolar). As we will learn in subsequent chapters, this is the basis for the low static power dissipation in CMOS.

A consequence of the design of complementary CMOS gates is that they must be inverting. The nMOS pull-down network turns ON when inputs are '1,' leading to '0' at the output. We might be tempted to turn the transistors upside down to build a noninverting gate. For example, Figure 1.21 shows a noninverting buffer. Unfortunately, now both the nMOS and pMOS transistors produce degraded outputs, so the technique should be avoided. Instead, we

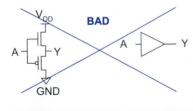

FIG 1.21 Bad noninverting buffer

[1] We call the left and right terminals *a* and *b* because each is technically the source of one of the transistors and the drain of the other.

can build noninverting functions from multiple stages of inverting gates. Figure 1.22 shows a 4-input AND gate built from two levels of inverting complementary CMOS gates. In isolation, the NAND design is simpler. In the context of a larger system, one can optimize the gates depending on the speed and density required.

Similarly, the compound gate of Figure 1.17 could be built with two AND gates, an OR gate, and an inverter. The AND and OR gates in turn could be constructed from NAND/NOR gates and inverters, shown in Figure 1.23, using a total of 20 transistors, as compared to 8 in Figure 1.17. CMOS logic designers must learn to take advantage of the efficiencies of compound gates rather than using large numbers of AND/OR gates.

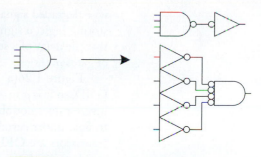

FIG 1.22 Various implementations of a CMOS 4-input AND gate

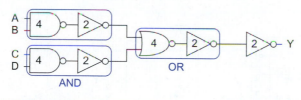

FIG 1.23 Inefficient discrete gate implementation of AOI22 indicating transistor counts

1.4.7 Tristates

Figure 1.24 shows symbols for a *tristate buffer*. When the enable input EN is '1,' the output Y equals the input A, just as in an ordinary buffer. When the enable is '0,' Y is left floating (a 'Z' value). This is summarized in Table 1.5. Sometimes both true and complementary enable signals EN and $\overline{EN}$ are drawn explicitly, while sometimes only EN is shown.

FIG 1.24 Tristate buffer symbol

Table 1.5	Truth table for tristate	
EN / $\overline{EN}$	A	Y
0/1	0	Z
0/1	1	Z
1/0	0	0
1/0	1	1

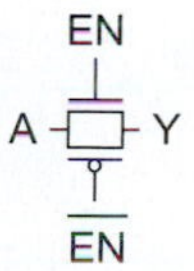

The transmission gate in Figure 1.25 has the same truth table as a tristate buffer. It only requires two transistors but it is a *nonrestoring* circuit. If the input is a noisy or other-

FIG 1.25 Transmission gate

wise degraded signal, the output will receive the same noise. After several stages of nonre-storing logic, a signal can become too degraded to recognize. We will see in Section 4.2 that the delay of a series of nonrestoring gates also increases quadratically with the number of gates in series.

Figure 1.26(a) shows a *tristate inverter*. The output is actively driven from V_{DD} or GND, so it is a restoring logic gate. Unlike any of the gates considered so far, the tristate inverter does not obey the conduction complements rule because it must allow the output to float under certain input combinations. When EN is '0' (Figure 1.26(b)), both enable transistors are OFF, leaving the output floating. When EN is '1' (Figure 1.26(c)), both enable transistors are ON. They are conceptually removed from the circuit, leaving a simple inverter. Figure 1.26(d) shows symbols for the tristate inverter. The complementary enable signal can be generated internally or can be routed to the cell explicitly. A tristate buffer can be built as a tristate inverter following an ordinary inverter.

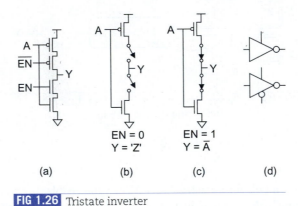

(a)   (b)   (c)   (d)

FIG 1.26 Tristate inverter

Tristates were once commonly used to allow multiple units to drive a common bus, as long as more than one are not simultaneously enabled. Distributing mutually exclusive enable signals in a timely fashion across a large chip is becoming more difficult, so multi-plexers are now preferred.

1.4.8 Multiplexers

Multiplexers are key components in CMOS memory elements and data manipulation structures. A multiplexer chooses the output to be one of several inputs based on a select signal. A two-input, or 2:1 multiplexer, chooses input *D0* when the select is '0' and input *D1* when the select is '1.' The truth table is given in Table 1.6; the logic function is $Y = \overline{S} \cdot D0 + S \cdot D1$.

S / $\overline{S}$	D1	D0	Y
0/1	X	0	0
0/1	X	1	1
1/0	0	X	0
1/0	1	X	1

Table 1.6 Multiplexer truth table

Two transmission gates can be tied together to form a compact 2-input multiplexer, as shown in Figure 1.27(a). The select and its complement enable exactly one of the two transmission gates at any given time. The complementary select $\overline{S}$ is often not drawn in the symbol, as in Figure 1.27(b).

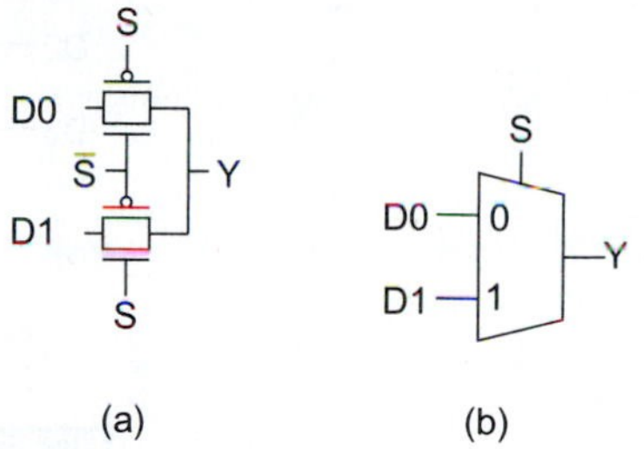

Again, the transmission gates produce a nonrestoring multiplexer. We could build a restoring, inverting multiplexer out of gates in several ways. One is the compound gate of Figure 1.17, connected as shown in Figure 1.28(a). Another is to gang together two tristate inverters, as shown in Figure 1.28(b). Notice that the schematics of these two approaches are nearly identical, save that the pull-up network has been slightly simplified and permuted in Figure 1.28(b). This is possible because the select and its complement are mutually exclusive. The tristate approach is slightly more compact and faster because it requires less internal wire. Again, if the complementary select is generated within the cell, it is omitted from the symbol (Figure 1.28(c)).

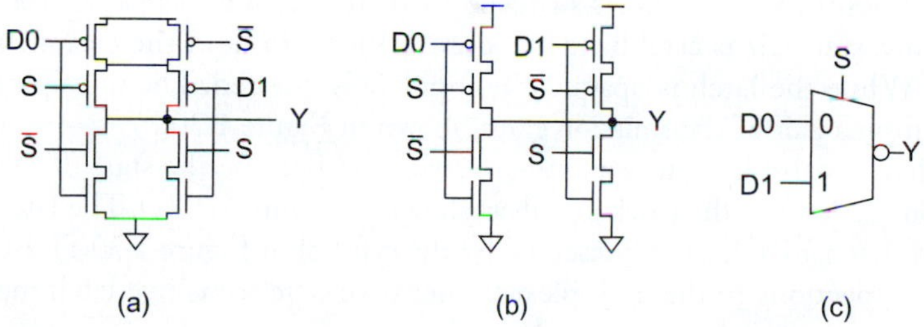

(a)   (b)

FIG 1.27 Transmission gate multiplexer

(a)   (b)   (c)

FIG 1.28 Inverting multiplexer

Larger multiplexers can be built from multiple 2-input multiplexers or by directly ganging together several tristates. The latter approach requires decoded select signals for each tristate. 4-input (4:1) multiplexers using each of these approaches are shown in Figure 1.29. In practice, both inverting and noninverting multiplexers are simply called multiplexers or muxes.

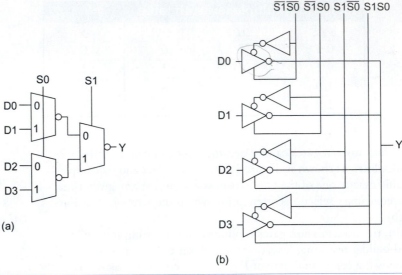

FIG 1.29 4:1 multiplexer

1.4.9 Latches and Flip-Flops

Using the combinational circuits developed so far, we can now build sequential circuits such as latches and flip-flops. A D *latch* using one 2-input multiplexer and two inverters is shown in Figure 1.30(a). It consists of a data input, D, a clock input, CLK, and true and complementary outputs Q and $\overline{Q}$. When CLK = '1,' the latch is *transparent*. $Q = D$ and $\overline{Q}$ = $\overline{D}$ (Figure 1.30(c)). When CLK is switched to '0,' the latch is *opaque*. A feedback path around the inverter pair is established (Figure 1.30(d)) to hold the current state of Q indefinitely. While the latch is opaque, the input D is ignored. The multiplexer can be constructed from a pair of transmission gates, shown in Figure 1.30(b).

The D latch is also known as a *level-sensitive latch* because the state of the output is dependent on the level of the clock signal, as shown in Figure 1.30(e). The latch shown is a positive-level-sensitive latch, represented by the symbol in Figure 1.30(f). By inverting the control connections to the multiplexer, a negative-level-sensitive latch may be constructed.

By combining two level-sensitive latches, one positive-sensitive and one negative-sensitive, we construct an *edge-triggered flip-flop* as shown in Figure 1.31(a-b). By convention, the first latch stage is called the *master* and the second is called the *slave*.

While CLK is low, the master negative-level-sensitive latch output ($\overline{QM}$) follows the D input while the slave positive-level-sensitive latch holds the previous value (Figure 1.31(c)). When the clock transitions from 0 to 1, the master latch ceases to sample the input and holds the D value at the time of the clock transition. The slave latch opens, passing the stored master value ($\overline{QM}$) to the output of the slave latch (Q). The D input is blocked from affecting the output because the master is disconnected from the D input

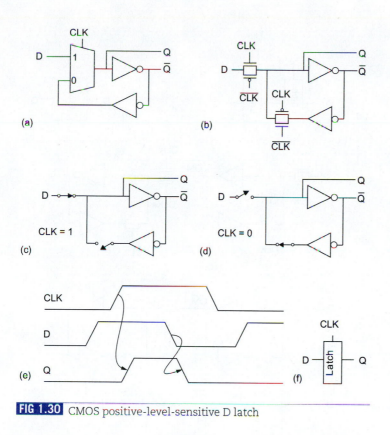

FIG 1.30   CMOS positive-level-sensitive D latch

(Figure 1.31(d)). When the clock transitions from 1 to 0, the slave latch holds its value and the master starts sampling the input again.

In summary, this flip-flop copies D to Q on the rising edge of the clock, as shown in Figure 1.31(e). Thus this device is called a positive-edge triggered flip-flop (also called a *D flip-flop, D register*, or *master–slave flip-flop*). Figure 1.31(f) shows the circuit symbol for the flip-flop. By reversing the latch polarities, a negative edge triggered flip-flop may be constructed. A collection of two or more D flip-flops sharing a common clock input is called a *register*. A register is often drawn as a flip-flop with multi-bit D and Q busses.

In Section 7.2.3 we will see that flip-flops may experience hold-time failures if the system has too much *clock skew*, i.e., if one flip-flop triggers early and another triggers late because of variations in clock arrival times. In industrial designs, a great deal of effort is devoted to timing simulations to catch hold-time problems. When design time is more important (e.g., in academic class projects), hold time problems can be avoided altogether by distributing a two-phase nonoverlapping clock. Figure 1.32 shows the flip-flop clocked with two nonoverlapping phases. As long as the phases do not overlap even with worst-case skews, at least one latch will be opaque at any given time and hold-time problems will never occur.

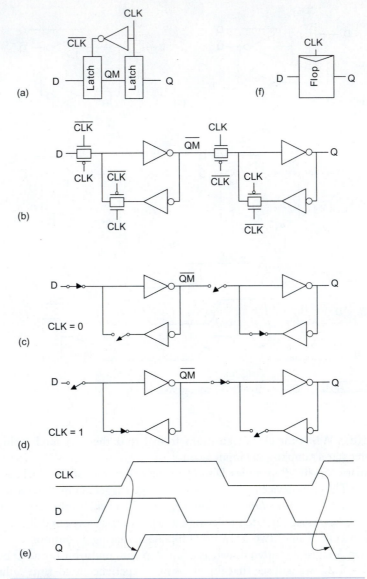

FIG 1.31 CMOS positive-edge-triggered D flip-flop

It is often useful to provide reset and/or enable signals to flip-flops and latches. Such modifications are straightforward and are discussed in Section 7.3.

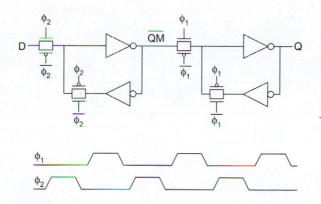

FIG 1.32 CMOS flip-flop with two-phase nonoverlapping clocks

1.5 CMOS Fabrication and Layout

Now that we can design logic gates and latches from transistors, let us consider how the transistors are built. Transistors are fabricated on thin silicon wafers that serve as both a mechanical support and an electrical common point called the *substrate*. We can understand the physical layout of transistors from two perspectives. One is the top view, obtained by looking down on a wafer. The other is the cross-section, obtained by slicing the wafer through the middle of a transistor and looking at it edgewise. We begin by looking at the cross-section of a complete CMOS inverter. We then look at the top view of the same inverter and define a set of masks used to manufacture the different parts of the inverter. The size of the transistors and wires is set by the mask dimensions and is limited by the resolution of the manufacturing process. Continual advancements in this resolution have fueled the exponential growth of the semiconductor industry.

1.5.1 Inverter Cross-section

Figure 1.33 shows a cross-section of the inverter from Section 1.4.1. In this diagram, the inverter is built on a p-type substrate. The pMOS transistor requires an n-type body region, so an n-well is diffused into the substrate in its vicinity. Note that it is also possible to design a CMOS process with an n-type substrate and p-wells to contain the nMOS transistors. As described in Section 1.3, the nMOS transistor has n-type source and drain regions and a polysilicon gate over a thin layer of silicon dioxide (SiO_2, also called *gate oxide*). The pMOS transistor is a similar structure with p-type source and drain regions. The polysilicon gates of the two transistors are tied together somewhere off the page and form the input A. The source of the nMOS transistor is connected to a metal ground line and the source of the pMOS transistor is connected to a metal V_{DD} line. The drains of the

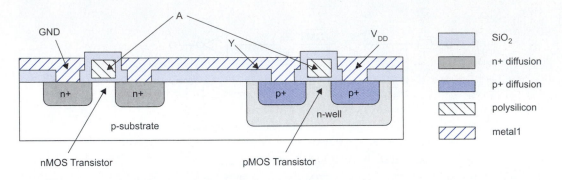

FIG 1.33 Inverter cross-section

two transistors are connected with metal to form the output Y. A thicker layer of SiO$_2$ called *field oxide* prevents metal from shorting to other layers except where contacts are explicitly etched.

The substrate must be tied to a low potential to avoid forward-biasing the p-n junction between the p-type substrate and the n+ nMOS source or drain. Likewise, the n-well must be tied to a high potential. This is generally done by adding heavily doped substrate and well contacts, or *taps*, to connect GND and V_{DD} to the substrate and n-well, respectively, as shown in Figure 1.34 and the inside front cover. The heavy doping is required to establish a good *ohmic contact* that provides low resistance for bidirectional current flow; a metal to lightly doped semiconductor junction forms a *Schottky diode*.

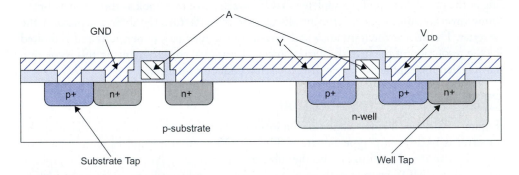

FIG 1.34 Inverter cross-section with well and substrate contacts. Color version on inside front cover.

1.5.2 Fabrication Process

For all their complexity, chips are amazingly inexpensive because all the transistors and wires can be printed in much the same way as books. The fabrication sequence consists of a series of steps in which layers of the chip are defined through a process called *photolithogra-phy*. Because many entire chips are printed at once, the cost of the chip is proportional to the chip area, rather than the number of transistors. As manufacturing advances allow engineers to build smaller transistors and place more transistors in the same area, each transistor gets cheaper. Smaller transistors are also faster because electrons don't have to travel as far to get from the source to the drain! This explains the remarkable trend for computers and electronics to become both cheaper and more capable with each generation.

The inverter could be defined by a hypothetical set of six masks: n-well, polysilicon, n+ diffusion, p+ diffusion, contacts, and metal (for fabrication reasons discussed in Chapter 3, the actual mask set is usually different). Masks specify where the components will be manufactured on the chip. Figure 1.35(a) shows a top view of the six masks. (See also the inside front cover for a color picture.) The cross-section of the inverter from Figure 1.34 was taken along the dashed line.

Consider a very simple fabrication process to illustrate the fundamental ideas. The process begins with the creation of an n-well on a bare p-type silicon wafer. Figure 1.36 shows cross-sections of the wafer after each processing step involved in forming the n-well; Figure 1.36(a) illustrates the bare substrate before processing. Forming the n-well requires adding enough Group V dopants into the silicon substrate to change the substrate from p-type to n-type in the region of the well. To define what regions receive n-wells, we grow a protective layer of oxide over the entire wafer, then remove it where we want the wells. We then add the n-type dopants; the dopants are blocked by the oxide, but enter the substrate and form the wells where there is no oxide. The next paragraph describes these steps in more detail.

The wafer is first *oxidized* in a high-temperature (typically 900°–1200°C) furnace that causes the Si and O_2 to react and become SiO_2 on the wafer surface (Figure 1.36(b)). The oxide must be patterned to define the n-well. An organic photoresist[2] that softens where exposed to light is spun onto the wafer (Figure 1.36(c)). The photoresist is exposed through the n-well mask (Figure 1.35(b)) that allows light to pass through only where the well should be. The softened photoresist is removed to expose the oxide (Figure 1.36(d)). The oxide is etched with hydrofluoric acid (HF) where it is not protected by the photoresist (Figure 1.36(e)), then the remaining photoresist is stripped away using a mixture of acids called *piranha etch* (Figure 1.36(f)). The well is formed where the substrate is not covered with oxide. Two ways to add dopants are diffusion and ion implantation. In the *diffusion* process, the wafer is placed in a furnace with a gas containing the dopants. When heated, dopant atoms diffuse into the substrate. Notice how the well is larger than the hole in the oxide on account of *lateral* diffusion (Figure 1.36(g)). With *ion implantation*, dopant ions are accelerated through an electric field and blasted into the substrate. In either method, the oxide layer prevents dopant atoms from entering the substrate where no well is intended. Finally, the remaining oxide is stripped with HF to leave the bare wafer with wells in the appropriate places.

---

[2] Engineers have experimented with many organic polymers. Brumford and Walker reported in 1958 that Jello™ could be used for masking. They did extensive testing, observing that "various Jellos™ were evaluated with lemon giving the best result."

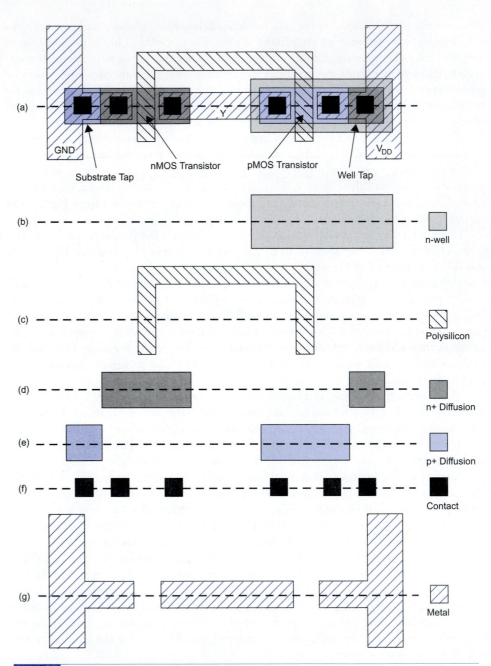

FIG 1.35 Inverter mask set. Color version on inside front cover.

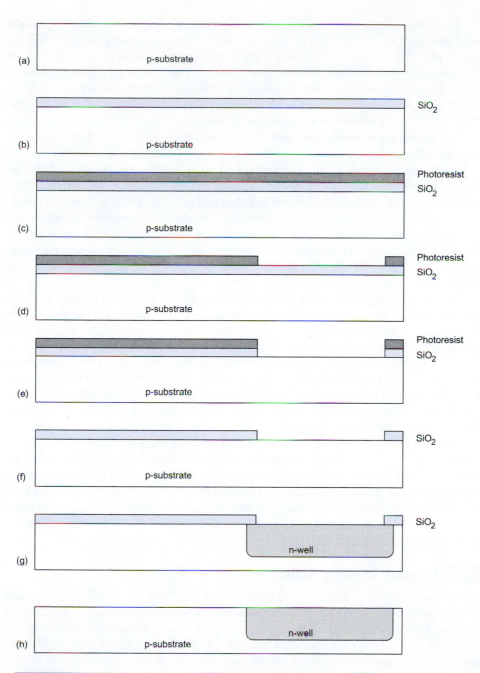

FIG 1.36 Cross-sections while manufacturing the n-well

The transistor gates are formed next. These consist of polycrystalline silicon, generally called *polysilicon*, over a thin layer of oxide. The thin oxide is grown in a furnace. Then the wafer is placed in a reactor with silane gas (SiH_4) and heated again to grow the polysilicon layer through a process called *chemical vapor deposition*. The polysilicon is heavily doped to form a reasonably good conductor. The resulting cross-section is shown in Figure 1.37(a). As before, the wafer is patterned with photoresist and the polysilicon mask (Figure 1.35(c)), leaving the polysilicon gates (Figure 1.37(b)).

The n+ regions are diffused to create the transistor active area and the well contact. As with the well, a protective layer of oxide is formed (Figure 1.37(c)) and patterned with the n-diffusion mask (Figure 1.35(d)) to expose the areas where the dopants are needed (Figure 1.37(d)). Although the n+ regions are typically formed with ion implantation (Figure 1.37(e)), they were historically diffused and thus still are often called *n-diffusion*. Notice that the polysilicon gate over the nMOS transistor blocks the diffusion so the source and drain are separated by a channel under the gate. This is called a *self-aligned* process because the source and drain of the transistor are automatically formed adjacent to the gate without the need to precisely align the masks. Finally, the protective oxide is stripped (Figure 1.37(f)).

The process is repeated for the p-diffusion mask (Figure 1.35(e)) to give the structure of Figure 1.38(a). Oxide is used for masking in the same way, and thus is not shown. The field oxide is grown to insulate the wafer from metal and patterned with the contact mask (Figure 1.35(f)) to leave contact cuts where metal should attach to diffusion or polysilicon (Figure 1.38(b)). Finally, aluminum is sputtered over the entire wafer, filling the contact cuts as well. Sputtering involves blasting aluminum into a vapor that evenly coats the wafer. The metal is patterned with the metal mask (Figure 1.35(g)) and plasma etched to remove metal everywhere except where wires should remain (Figure 1.38(c)). This completes the simple fabrication process.

Modern fabrication sequences are somewhat more elaborate because they must create complex doping profiles around the channel of the transistor and print features that are smaller than the wavelength of the light being used in lithography. However, masks for these elaborations can be automatically generated from the simple set of masks we have just examined. Modern processes may also have five or more layers of metal, so the metal and contact steps must be repeated for each layer. Chip manufacturing has become a commodity, and many different vendors will build designs from a basic set of masks.

1.5.3 Layout Design Rules

Layout design rules describe how small features can be and how closely they can be packed in a particular manufacturing process. Industrial design rules are usually specified in microns. This makes migrating from one process to a more advanced process difficult because not all rules scale in the same way. Mead and Conway [Mead80] popularized lambda-based design rules based on a single parameter, λ, which characterizes the resolution of the process. λ is generally half of the minimum drawn transistor channel length.

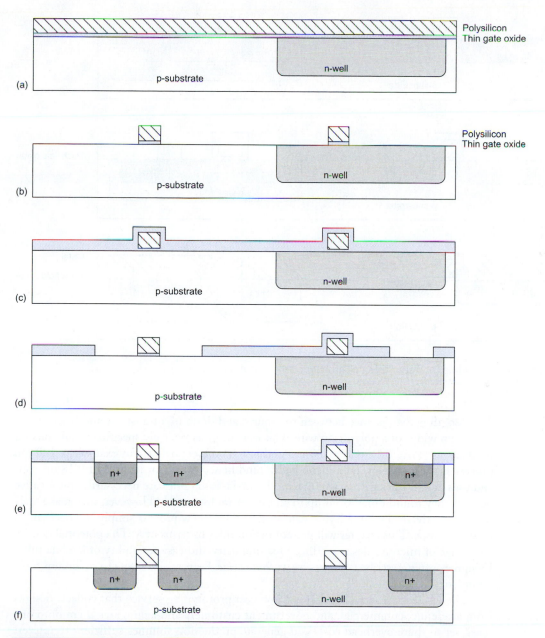

Polysilicon
Thin gate oxide

Polysilicon
Thin gate oxide

FIG 1.37 Cross-sections while manufacturing polysilicon and n-diffusion

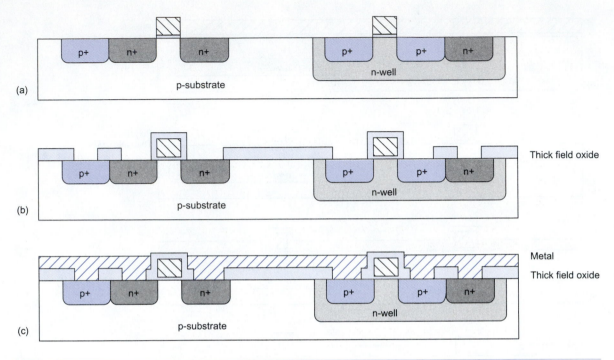

FIG 1.38 Cross-sections while manufacturing p-diffusion, contacts, and metal

This length is the distance between the source and drain of a transistor and is set by the minimum width of a polysilicon wire. This dimension is typically specified in microns for dimensions above 0.18 μm = 180 nm and in nanometers below. For example, a 180 nm process has a minimum polysilicon width (and hence transistor length) of 0.18 microns and uses design rules with λ = 0.09 μm[3]. Lambda-based rules are necessarily conservative because they round dimensions up to an integer multiple of λ. However, they make scaling layout trivial; the same layout can be moved to a new process simply by specifying a new value of λ. This chapter will present design rules in terms of λ. The potential density advantage of micron rules is sacrificed for simplicity and easy scalability of lambda rules. Designers often describe a process by its *feature size*. Feature size refers to minimum transistor length, so λ is half the feature size.

The MOSIS service [Piña02] is a low-cost prototyping service that collects designs from academic, commercial, and government customers and aggregates them onto one mask set to share overhead costs and generate production volumes sufficient to interest fabrication companies. MOSIS has developed a set of scalable lambda-based design rules

---

[3]Some 180 nm lambda-based rules actually set λ = 0.10 μm, then shrink the gate by 20 nm while generating masks. This keeps 180 nm gate lengths but makes all other features slightly larger.

that cover a wide range of manufacturing processes. The rules have become slightly more conservative for more advanced submicron processes, but are fundamentally very similar to those proposed by Mead and Conway in 1980 for a 6-micron process. The rules describe the minimum width to avoid breaks in a line, minimum spacing to avoid shorts between lines, and minimum overlap to ensure two layers completely overlap.

A conservative but easy-to-use set of design rules for layouts with two metal layers in an n-well process is as follows.

- Metal and diffusion have minimum width and spacing of 4 λ.

- Contacts are 2 λ × 2 λ and must be surrounded by 1 λ on the layers above and below.

- Polysilicon uses a width of 2 λ.

- Polysilicon overlaps diffusion by 2 λ where a transistor is desired and has a spacing of 1 λ away where no transistor is desired.

- Polysilicon and contacts have a spacing of 3 λ from other polysilicon or contacts.

- N-well surrounds pMOS transistors by 6 λ and avoids nMOS transistors by 6 λ.

Figure 1.39 shows the basic MOSIS design rules for a process with two metal layers. Section 3.3 elaborates on these rules and compares them with more aggressive industrial design rules.

Transistor dimensions are often specified by their Width/Length (W/L) ratio. For example, the nMOS transistor in Figure 1.39 formed where polysilicon crosses n-diffusion

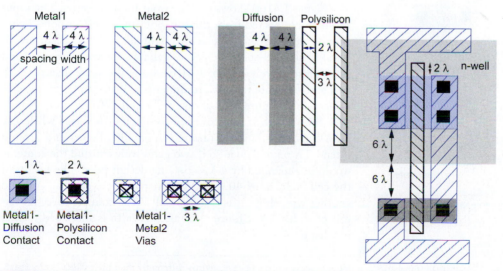

FIG 1.39 Simplified λ-based design rules

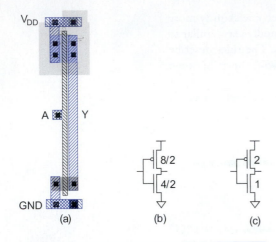

FIG 1.40 Inverter with dimensions labeled

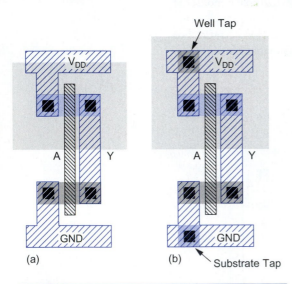

FIG 1.41 Inverter standard cell layout

has a W/L of 4/2. In a 0.6 μm process, this corresponds to an actual width of 1.2 μm and a length of 0.6 μm. Such a minimum-width contacted transistor is often called a unit transistor.[4] pMOS transistors are often wider than nMOS transistors because holes move more slowly than electrons so the transistor has to be wider to deliver the same current. Figure 1.40(a) shows a unit inverter layout with a unit nMOS transistor and a double-sized pMOS transistor. Figure 1.40(b) shows a schematic for the inverter annotated with Width/Length for each transistor. Figure 1.40(c) shows a shorthand we will often use, specifying multiples of unit width and assuming minimum length.

1.5.4 Gate Layout

A good deal of ingenuity can be exercised and a vast amount of time wasted exploring layout topologies to minimize the size of a gate or other *cell* such as an adder or memory element. For many applications, a straightforward layout is good enough and can be automatically generated or rapidly built by hand. This section presents a simple layout style based on a "line of diffusion" rule that is commonly used for standard cells in automated layout systems. This style consists of four horizontal strips: metal ground at the bottom of the cell, n-diffusion, p-diffusion, and metal power at the top. The power and ground lines are often called *supply rails*. Polysilicon lines run vertically to form transistor gates. Metal wires within the cell connect the transistors appropriately.

Figure 1.41(a) shows such a layout for an inverter. The input *A* can be connected from the top, bottom, or left in polysilicon. The output *Y* is available at the right side of the cell in metal. Recall that the p-substrate and n-well must be tied to ground and power, respectively. Figure 1.41(b) shows the same inverter with well and substrate taps placed under the power and ground rails, respectively. Figure 1.42 shows a 3-input NAND gate. Notice how the nMOS transistors are connected in series while the pMOS transistors are connected in parallel. Power and ground extend 2 λ on each side so if two gates were abutted the contents would be separated by 4 λ, satisfying design rules. The height of the cell is 36 λ, or 40 λ if the 4 λ space between the cell and another wire above it is counted. All these examples use transistors of width 4 λ. Choice of transistor width is addressed further in Chapter 4.

---

[4]Such small transistors in modern processes often behave slightly differently than their wider counterparts. Moreover, the transistor will not operate if either contact is damaged. Industrial designers often use a transistor wide enough for two contacts (9 λ) as the unit transistor to avoid these problems.

These cells were designed such that the gate connections are made from the top or bottom in polysilicon. In contemporary standard cells, polysilicon is generally not used as a routing layer so the cell must allow metal2 to metal1 and metal1 to polysilicon contacts to each gate. While this increases the size of the cell, it allows free access to all terminals on metal routing layers.

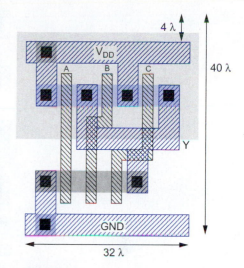

FIG 1.42 3-input NAND standard cell gate layouts

1.5.5 Stick Diagrams

As layout is time-consuming, it is important to have fast ways to plan layout and estimate area before committing to a full layout. *Stick diagrams* are easy to draw because they do not need to be drawn to scale. Figure 1.43 and the inside front cover show stick diagrams for an inverter and a 3-input NAND gate. While this book uses stipple patterns, layout designers use colored pencils or dry-erase markers.

With practice it is easy to estimate the area of a layout from the corresponding stick diagram even though the diagram is not to scale. Although schematics focus on transistors, layout area is usually determined by the metal wires. Transistors are merely small widgets that fit under the wires. We define a *routing track* as enough space to place a wire and the required spacing to the next wire. If our wires have a width of 4 λ and a spacing of 4 λ to the next wire, the track *pitch* is 8 λ, as shown in Figure 1.44(a). This pitch also leaves room for a transistor to be placed between the wires (Figure 1.44(b)). Therefore, as a rule of thumb, it is reasonable to estimate the height and width of a cell by counting the number of metal tracks and multiplying by 8 λ. A slight complication is the required spacing of 12 λ between nMOS and pMOS transistors set by the well, as shown in Figure 1.45(a). This space can be occupied by an additional track of wire, shown in Figure 1.45(b). Therefore, an extra track must be allocated between nMOS and pMOS transistors regardless of whether wire is actually used in that track. Figure 1.46 shows how to count tracks to estimate the size of a 3-input NAND. There are four vertical wire tracks, multiplied by 8 λ per track to give a cell width of 32 λ. There are five horizontal tracks, giving a cell height

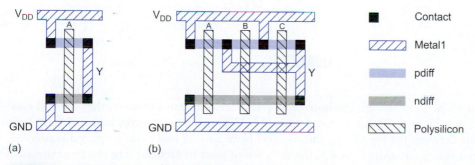

(a)   (b)

■	Contact
▨	Metal1
▨	pdiff
▨	ndiff
▨	Polysilicon

FIG 1.43 Stick diagrams of inverter and 3-input NAND gate. Color version on inside front cover.

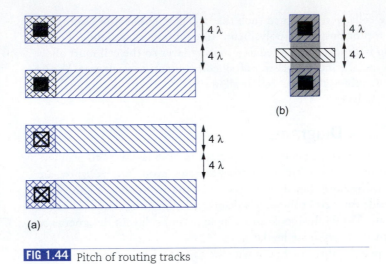

FIG 1.44 Pitch of routing tracks

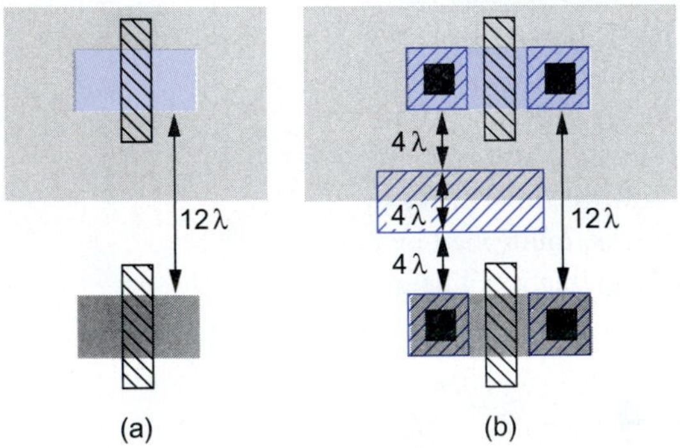

FIG 1.45 Spacing between nMOS and pMOS transistors

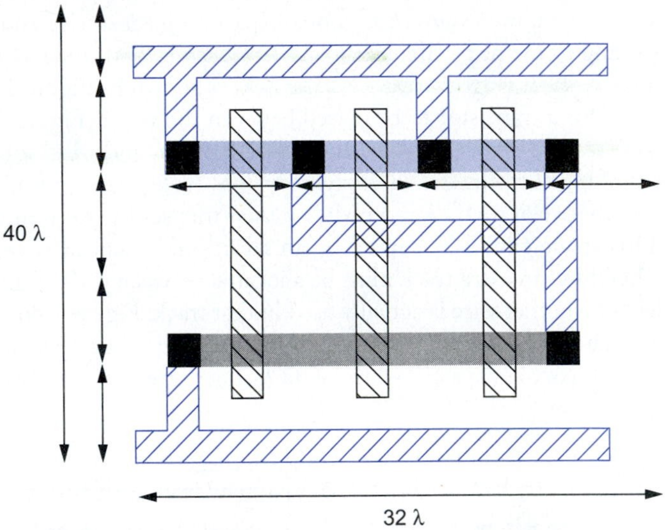

FIG 1.46 3-input NAND gate area estimation

of 40 λ. Even though the horizontal tracks are not drawn to scale, they are still easy to count. Figure 1.42 shows that the actual NAND gate layout agrees in size if the 4 λ spacing between the top of V_{DD} and the next place a metal line could be placed is counted. If transistors are wider than 4 λ, the extra width must be factored into the area estimate. Of course, these estimates are oversimplifications of the complete design rules and a trial layout should be performed for truly critical cells.

Example

Sketch a stick diagram for a CMOS gate computing $Y = \overline{(A + B + C) \cdot D}$ (see Figure 1.18) and estimate the cell width and height.

Solution: Figure 1.47 shows a stick diagram. Counting horizontal and vertical pitches gives an estimated cell size of 40 by 48 λ.

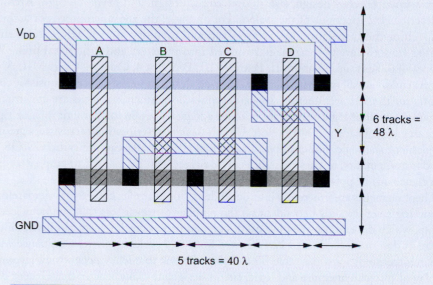

FIG 1.47 CMOS compound gate for function $Y = \overline{(A + B + C) \cdot D}$

1.6   Design Partitioning

By this point, you know that MOS transistors behave as voltage-controlled switches. You know how to build logic gates out of transistors. And you know how transistors are fabricated and how to draw a layout that specifies how transistors should be placed and connected together. You know enough to start building your own simple chips.

The greatest challenge in modern VLSI design is not in designing the individual transistors but rather in managing system complexity. Modern *System-On-Chip* (SOC) designs combine memories, processors, high speed I/O interfaces, and dedicated application-specific logic on a single chip. They use hundreds of millions (soon billions) of transistors. The implementation must be divided among large teams of engineers and each engineer must be highly productive. If the implementation is too rigidly partitioned, each block can be optimized without regard to its neighbors, leading to poor system results. Conversely, if every task is interdependent with every other task, design will progress too slowly. Design managers face the challenge of choosing a suitable tradeoff between these extremes. There is no substitute for practical experience in making these choices, and talented engineers who have experience with multiple designs are very important to the success of a large project. The notion of structured design, which is also used in large software projects, will be introduced in Chapter 8. Structured design uses the principles of hierarchy, regularity, modularity, and locality to manage the complexity.

Digital VLSI design is often partitioned into five interrelated tasks: *architecture* design, *microarchitecture* design, *logic* design, *circuit* design, and *physical* design. Architecture describes the functions of the system. For example, the x86 microprocessor architecture specifies the instruction set, register set, and memory model. Microarchitecture describes how the architecture is partitioned into registers and functional units. The 80386, 80486, Pentium, Pentium II, Pentium III, Pentium 4, Celeron, Cyrix MII, AMD K5, and Athlon are all microarchitectures offering different performance / transistor count tradeoffs for the x86 architecture. Logic describes how functional units are constructed. For example, various logic designs for a 32-bit adder in the x86 integer unit include ripple carry, carry lookahead, and carry select. Circuit design describes how transistors are used to implement the logic. For example, a carry lookahead adder can use static CMOS circuits, domino circuits, or pass transistors. The circuits can be tailored to emphasize high performance or low power. Physical design describes the layout of the chip.

These elements are inherently interdependent. For example, choices of microarchitecture and logic are strongly dependent on the number of transistors that can be placed on the chip, which depends on the physical design and process technology. Similarly, innovative circuit design that reduces a cache access from two cycles to one can influence which microarchitecture is most desirable. The choice of clock frequency depends on a complex interplay of microarchitecture and logic, circuit design, and physical design. Deeper pipelines allow higher frequencies but lead to greater performance penalties when operations early in the pipeline are dependent on those late in the pipeline. Many functions have various logic and circuit designs trading speed for area, power, and design effort. Custom physical design allows more compact, faster circuits and lower manufacturing costs, but involves an enormous labor cost. Automatic layout with CAD systems reduces the labor and achieves faster times to market.

To deal with these interdependencies, microarchitecture, logic, circuit, and physical design must occur, at least in part, in parallel. Microarchitects depend on circuit and physical design studies to understand the cost of proposed microarchitectural features. Engineers are sometimes categorized as "short and fat" or "tall and skinny." Tall, skinny engineers understand something about a broad range of topics. Short, fat engineers understand a large amount about a narrow field. Digital VLSI design favors the tall, skinny engineer who can evaluate how choices in one part of the system impact other parts of the system.

A critical tool for managing complex designs is *hierarchy*. A large system can be partitioned into many *units*. Each unit in turn is composed of multiple *functional blocks*[5]. These blocks in turn are built from *cells*, which ultimately are constructed from transistors. The

---

[5]Some designers refer to both units and functional blocks as *modules*.

system can be more easily understood at the top level by viewing units as black boxes with well-defined interfaces and functions rather than looking at each individual transistor. Hierarchy also facilitates design reuse; a block can be designed and verified once, then used in many places. Logic, circuit, and physical views of the design should share the same hierarchy for ease of verification. A design hierarchy can be viewed as a tree structure with the overall chip as the *root* and the primitive cells as *leafs*.

An alternative way of viewing design partitioning is shown with the Y-chart in Figure 1.48 [Gajski83, Kang03]. The radial lines on the Y-chart represent three distinct design domains: behavioral, structural, and physical. These domains can be used to describe the design of almost any artifact and thus form a very general taxonomy for describing the design process. Within each domain there are a number of levels of design abstraction that start at a very high level and descend eventually to the individual elements that need to be aggregated to yield the top level function (i.e., in the case of chip design and transistors).

The behavioral domain describes what a particular system does. For instance, at the highest level we might state that we desire to build a chip that can generate audio tones of specific frequencies (i.e., a touch-tone generator for a telephone). This behavior can be successively refined to more precisely describe what needs to be done in order to build the tone generator (i.e., the frequencies desired, output levels, distortion allowed, etc.).

At each abstraction level, a corresponding structural description can be described. The structural domain describes the interconnection of modules necessary to achieve a particular behavior. For instance, at the highest level, the touch-tone generator might consist of a keyboard, a tone generator, an audio amplifier, a battery, and a speaker. Eventually at lower levels of abstraction, the individual gate and then transistor connections required to build the tone generator are described.

For each level of abstraction, the physical domain description explains how to physically construct that level of abstraction. At high levels this might consist of an engineering drawing showing how to put together the keyboard, tone generator chip, battery, and speaker in the associated housing. At the top chip level, this might consist of a floorplan, and at lower levels, the actual geometry of individual transistors.

The design process can be viewed as making transformations from one domain to another while maintaining the equivalency of the domains. Behavioral descriptions are transformed to structural descriptions, which in turn are transformed to physical descriptions. These transformations can be manual or automatic. In either case, it is normal design practice to verify the transformation of one domain to the other by some checking process. This ensures that the design intent is carried across the domain boundaries. Hierarchically specifying each domain at successively detailed levels of abstraction allows us to design very large systems.

The reason for strictly describing the domains and levels of abstraction is to define a precise design process in which the final function of the system can be traced all the way back to the initial behavioral description. There should be no opportunity to produce an incorrect design. If anomalies arise, the design process is corrected so that those anomalies will not reoccur in the future. A designer should acquire a rigid discipline with respect to the design process, and be aware of each transformation and how and why it is failproof.

Normally, these steps are fully automated in a modern design process, but it is important to be aware of the basis for these steps in order to debug them if they go astray.

The Y diagram can be used to illustrate each domain and the transformations between domains at varying levels of design abstraction. As the design process winds its way from the outer to inner rings, it proceeds from higher to lower levels of abstraction and hierarchy.

Most of the remainder of this chapter is a case study in the design of a simple microprocessor to illustrate the various aspects of VLSI design applied to a nontrivial system. We begin by describing the architecture and microarchitecture of the processor. We then consider logic design and discuss hardware description languages. The processor is built with static CMOS circuits, which have been examined in Section 1.4 already; transistor level design and netlist formats are discussed. We continue exploring the physical design of the processor including floorplanning and area estimation. Design verification is very important and happens at each level of the hierarchy for each element of the design. Finally, the layout is converted into masks so the chip can be manufactured, packaged, and tested.

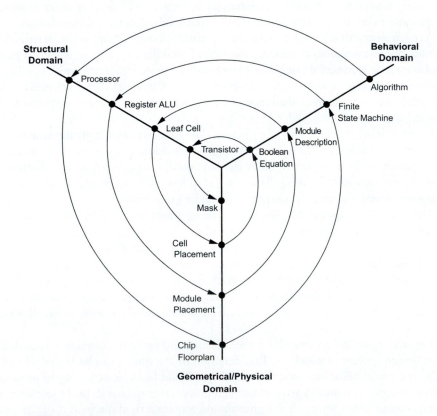

FIG 1.48 Y Diagram. Reproduced from [Kang03] with permission of The McGraw-Hill Companies.

1.7   Example: A Simple MIPS Microprocessor

We consider an 8-bit subset of the Patterson & Hennessy MIPS microprocessor architecture [Patterson04] because it is widely studied and is relatively simple, while still being large enough to illustrate hierarchical design. This section describes the architecture and the multicycle microarchitecture we will be implementing. If you are not familiar with computer architecture, you can regard the MIPS processor as a black box and skip to Section 1.8.

A set of laboratory exercises are available online in which you can learn VLSI design by building the microprocessor yourself using a free open-source CAD tool called *Electric*.

1.7.1   MIPS Architecture

The MIPS32 architecture is a simple 32-bit RISC architecture with relatively few idiosyncrasies. Our subset of the architecture uses 32-bit instruction encodings but only eight 8-bit general-purpose registers named $0–$7. We also use an 8-bit program counter (PC). Register $0 is hardwired to contain the number 0. The instructions are ADD, SUB, AND, OR, SLT, ADDI, BEQ, J, LB, and SB.

The function and encoding of each instruction is given in Table 1.7. Each instruction is encoded using one of three templates: R, I, and J. R-type instructions (*register*-based) are used for arithmetic and specify two source registers and a destination register. I-type

Table 1.7	MIPS instruction set (subset supported)				
Instruction	**Function**		**Encoding**	**op**	**funct**
add $1, $2, $3	addition:	$1 <- $2 + $3	R	000000	100000
sub $1, $2, $3	subtraction:	$1 <- $2 − $3	R	000000	100010
and $1, $2, $3	bitwise and:	$1 <- $2 and $3	R	000000	100100
or $1, $2, $3	bitwise or:	$1 <- $2 or $3	R	000000	100101
slt $1, $2, $3	set less than:	$1 <- 1 if $2 < $3 $1 <- 0 otherwise	R	000000	101010
addi $1, $2, imm	add immediate:	$1 <- $2 + imm	I	001000	n/a
beq $1, $2, imm	branch if equal:	PC <- PC + imm[a]	I	000100	n/a
j destination	jump:	PC <- destination[a]	J	000010	n/a
lb $1, imm($2)	load byte:	$1 <- mem[$2 + imm]	I	100000	n/a
sb $1, imm($2)	store byte:	mem[$2 + imm] <- $1	I	101000	n/a

a. Technically, MIPS addresses specify bytes. Instructions require a four-byte word and must begin at addresses that are a multiple of four. To most effectively use instruction bits in the full 32-bit MIPS architecture, branch and jump constants are specified in words and must be multiplied by four (shifted left two bits) to be converted to byte addresses.

instructions are used when a 16-bit constant (also known as an *immediate*) and two registers must be specified. J-type instructions (*jumps*) dedicate most of the instruction word to a 26-bit jump destination. The format of each encoding is defined in Figure 1.49. The six most significant bits of all formats are the operation code (op). R-type instructions all share op = 000000 and use six more funct bits to differentiate the functions.

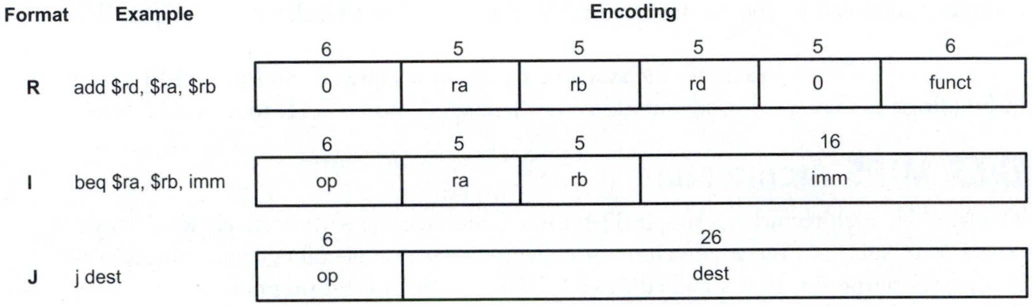

Format	Example			Encoding			

R   add $rd, $ra, $rb

6	5	5	5	5	6
0	ra	rb	rd	0	funct

I   beq $ra, $rb, imm

6	5	5	16
op	ra	rb	imm

J   j dest

6	26
op	dest

FIG 1.49 Instruction encoding formats

We can write programs for the MIPS processor in *assembly language*, where each line of the program contains one instruction such as ADD or BEQ. However, the MIPS hardware ultimately must read the program as a series of 32-bit numbers called *machine language*. An *assembler* automates the tedious process of translating from assembly language to machine language using the encodings defined in Table 1.7 and Figure 1.49. Writing nontrivial programs in assembly language is also tedious, so programmers usually work in a *high-level language* such as C or FORTRAN. A *compiler* translates a program from high-level language *source code* into the appropriate machine language *object code*.

Example

Figure 1.50 shows a simple C program that computes the nth Fibonacci number f_n defined recursively for $n > 0$ as $f_n = f_{n-1} + f_{n-2}, f_{-1} = -1, f_0 = 1$. Translate the program into MIPS assembly language and machine language.

Solution: Figure 1.51 gives a commented assembly language program. Figure 1.52 translates the assembly language to machine language.

continues

```c
int fib(void)
{
    int n = 8;              /* compute nth Fibonacci number */
    int f1 = 1, f2 = -1;    /* last two Fibonacci numbers */

    while (n != 0) {        /* count down to n = 0 */
      f1 = f1 + f2;
      f2 = f1 - f2;
      n = n - 1;
    }
    return f1;
}
```

FIG 1.50 C code for Fibonacci program

```
# fib.asm
# Register usage: $3: n $4: f1 $5: f2
# return value written to address 255
fib:  addi $3, $0, 8      # initialize n=8
      addi $4, $0, 1      # initialize f1 = 1
      addi $5, $0, -1     # initialize f2 = -1
loop: beq $3, $0, end     # Done with loop if n = 0
      add $4, $4, $5      # f1 = f1 + f2
      sub $5, $4, $5      # f2 = f1 - f2
      addi $3, $3, -1     # n = n - 1
      j loop             # repeat until done
end:  sb $4, 255($0)     # store result in address 255
```

FIG 1.51 Assembly language code for Fibonacci program

Instruction	Binary Encoding		Hexadecimal Encoding
addi $3, $0, 8	001000 00000 00011	0000000000001000	20030008
addi $4, $0, 1	001000 00000 00100	0000000000000001	20040001
addi $5, $0, -1	001000 00000 00101	1111111111111111	2005ffff
beq $3, $0, end	000100 00011 00000	0000000000000101	10600005
add $4, $4, $5	000000 00100 00101 00100 00000 100000		00852020
sub $5, $4, $5	000000 00100 00101 00101 00000 100010		00852822
addi $3, $3, -1	001000 00011 00011	1111111111111111	2063ffff
j loop	000010	00000000000000000000000011	08000003
sb $4, 255($0)	110000 00000 00100	0000000011111111	a00400ff

FIG 1.52 Machine language code for Fibonacci program

1.7.2 Multicycle MIPS Microarchitecture

We will implement the multicycle MIPS microarchitecture given in Chapter 5 of [Patterson04] modified to process 8-bit data. The microarchitecture is illustrated in Figure 1.53. The rectangles represent registers or memory. The rounded rectangles represent multiplexers. The ovals represent control logic. Light lines indicate individual signals while heavy lines indicate busses. The control logic and signals are highlighted in blue while the datapath is shown in black. Control signals generally drive multiplexer select signals and register enables to tell the datapath how to execute an instruction.

Instruction execution generally flows from left to right. The program counter (PC) specifies the address of the instruction. The instruction is loaded one byte at a time over four cycles from an off-chip memory into the 32-bit instruction register (IR). The op field (bits 31:26 of the instruction) is sent to the controller, which sequences the datapath through the correct operations to execute the instruction. For example, in an ADD instruction, the two source registers are read from the register file into temporary registers a and b. On the next cycle, the alucontrol unit commands the Arithmetic/Logic Unit (ALU) to add the inputs. The result is captured in the aluout register. On the third cycle, the result is written back to the appropriate destination register in the register file.

The controller is a finite state machine that generates multiplexer select signals and register enables to sequence the datapath. A state transition diagram for the FSM is shown in Figure 1.54. As discussed, the first four states fetch the instruction from memory. The FSM then is dispatched based on op to execute the particular instruction. The FSM states for ADDI are missing and left as an exercise for the reader.

Observe that the controller produces a 2-bit aluop output. The alucontrol unit uses combinational logic to compute a 3-bit alucontrol signal from the aluop and funct fields, as specified in Table 1.8. alucontrol drives multiplexers in the ALU to select the appropriate computation.

Table 1.8	ALUControl determination		
aluop	funct	alucontrol	Meaning
00	x	010	ADD
01	x	110	SUB
10	100000	010	ADD
10	100010	110	SUB
10	100100	000	AND
10	100101	001	OR
10	101010	111	SLT
11	x	x	undefined

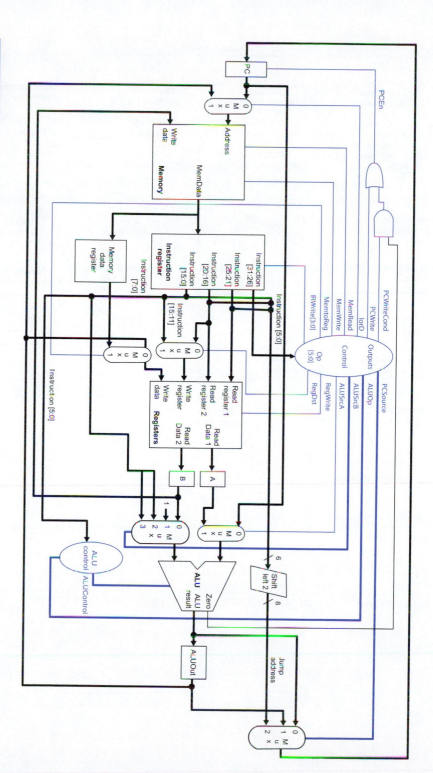

FIG 1.53 Multicycle MIPS microarchitecture. Reprinted from [Patterson04] with permission from Elsevier.

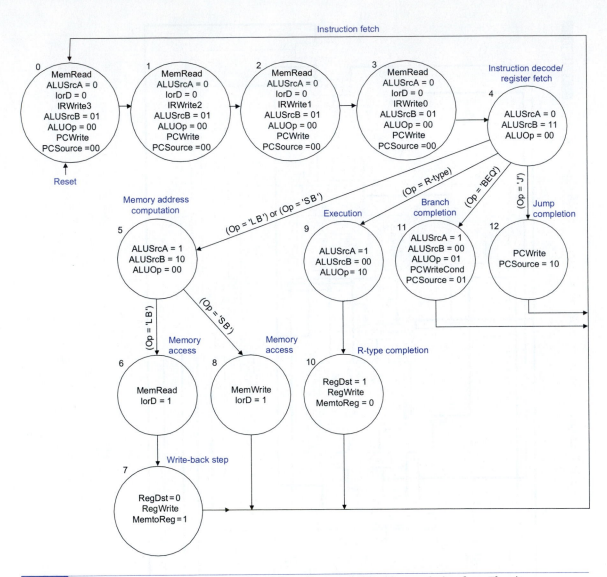

FIG 1.54 Multicycle MIPS control FSM. Reprinted from [Patterson04] with permission from Elsevier.

Example

Referring to Figure 1.53 and Figure 1.54, explain how the MIPS processor fetches and executes the SUB instruction.

Solution: The first step is to fetch the 32-bit instruction. This takes four cycles because the instruction must come over an 8-bit memory interface. On each cycle, we want to fetch a byte from the address in memory specified by the program counter, then increment the program counter by one to point to the next byte.

The fetch is performed by states 0–3 of the FSM in Figure 1.54. Let us start with state 0. The program counter (PC) contains the address of the first byte of the instruction. The controller must select iord = 0 so that the multiplexer sends this address to the memory. memread must also be asserted so the memory reads the byte onto the memdata bus. Finally, irwrite3 should be asserted to enable writing memdata into the most significant byte of the instruction register (IR).

Meanwhile, we need to increment the program counter. We can do this with the ALU by specifying PC as one input, '1' as the other input, and ADD as the operation. To select PC as the first input, alusrca = 0. To select '1' as the other input, alusrcb = 01. To perform an addition, aluop = 00, according to Table 1.8. To write this result back into the program counter at the end of the cycle, pcsource = 00 and pcen = 1 (done by setting pcwritecond = 1).

All of these control signals are indicated in state 0 of Figure 1.54. The other register enables are assumed to be 0 if not explicitly asserted and the other multiplexer selects are don't cares. The next three states are identical except that they write bytes 2, 1, and 0 of the IR, respectively.

The next step is to read the source registers, done in state 4. The two source registers are specified in bits 25:21 and 20:16 of the IR. The register file reads these registers and puts the values into the A and B registers. No control signals are necessary for SUB (although state 4 performs a branch address computation in case the instruction is BEQ).

The next step is to perform the subtraction. Based on the op field (IR bits 31:26), the FSM jumps to state 9 because SUB is an R-type instruction. The two source registers are selected as input to the ALU by setting alusrca = 1 and alusrcb = 00. Choosing aluop = 10 directs the ALU Control decoder to select the alucontrol signal as 110, subtraction. Other R-type instructions are executed identically except that the decoder receives a different funct code (IR bits 5:0) and thus generates a different alucontrol signal. The result is placed in the ALUOut register.

Finally, the result must be written back to the register file in state 10. The data comes from the ALUOut register so memtoreg = 0. The destination register is specified in bits 15:11 of the instruction so regdst = 1. regwrite must be asserted to perform the write. Then the control FSM returns to state 0 to fetch the next instruction.

1.8 Logic Design

We begin the logic design by defining the top-level chip interface and block diagram. We then hierarchically decompose the units until we reach leaf cells. We specify the logic with a Hardware Description Language (HDL), which provides a higher level of abstraction than schematics or layout.

1.8.1 Top-level Interface

The top-level inputs and outputs are listed in Table 1.9. This example uses a two-phase clocking system to avoid hold-time problems. Reset initializes the PC to 0 and the control FSM to the start state. The remainder of the signals are used for an asynchronous 8-bit memory interface (assuming the memory is located off chip). The processor sends an 8-bit address adr and either asserts memread or memwrite. On a read cycle, the memory returns a value on the memdata lines while on a write cycle, the memory accepts input from writedata. In many systems, memdata and writedata can be combined onto a single bidirectional bus, but for this example we preserve the interface of Figure 1.53. Figure 1.55 shows a very simple computer system built from the MIPS processor, external memory, reset switch, and clock generator.

Table 1.9	Top-level inputs and outputs
Inputs	**Outputs**
ph1	adr[7:0]
ph2	writedata[7:0]
reset	memread
memdata[7:0]	memwrite

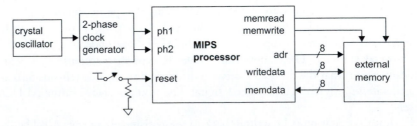

FIG 1.55 MIPS computer system

1.8.2 Block Diagram

The chip is partitioned into three top-level units: the controller, alucontrol, and datapath, as shown in the block diagram in Figure 1.56. The controller comprises the control FSM and the two gates used to compute pcen. The alucontrol consists of combinational logic to drive the ALU. The datapath contains the remainder of the chip, organized as eight identical *bitslices*. This partitioning is influenced by the intended physical design. The datapath contains most of the transistors and is very regular in structure. We can achieve high density with moderate design effort by handcrafting a single bitslice of the datapath, then replicating that bitslice eight times. The controller has much less structure. It is tedious to translate an FSM into gates by hand, and in a new design, the controller is the most likely portion to have bugs and last-minute changes. Therefore, we will specify the controller more abstractly with a hardware description language and automatically generate it using synthesis and place & route tools.

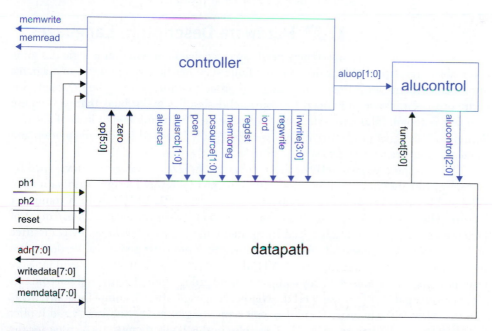

FIG 1.56 Top-level MIPS block diagram

1.8.3 Hierarchy

The best way to design complex systems is to decompose them into simpler pieces. Figure 1.57 shows the design hierarchy for the MIPS processor. The controller and alucontrol are built from a library of standard cells such as NANDs, NORs, and latches. The datapath is

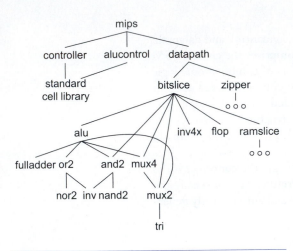

FIG 1.57 MIPS design hierarchy

composed of eight bitslices and a *zipper* that ties the datapath together by driving control signals and register enables to the various bits. The bitslice in turn is composed of the ALU, register file ramslice, flip-flops, and gates. Some of these gates are reused in multiple places. The pieces composing the zipper and ramslice are left out for brevity.

The design hierarchy does not necessarily have to be identical in the logic, circuit, and physical designs. For example, in the logic view, a memory may be best treated as a black box, while in the circuit implementation, it may have a decoder, cell array, column multiplexers, and so forth. Different hierarchies complicate verification, however, because they must be *flattened* until the point that they agree. As a matter of practice, it is best to make logic, circuit, and physical design hierarchies agree as far as possible.

1.8.4 Hardware Description Languages

Designers need rapid feedback on whether a logic design is reasonable. Translating a block diagram and finite state machine state transition diagrams into circuit schematics is time-consuming and prone to error; before going through this entire process it is wise to know if the top-level design has major bugs that will require complete redesign. HDLs provide a way to specify the design at a higher level of abstraction to raise designer productivity. They were originally intended for documentation and simulation, but are now used to synthesize gates directly from the HDL.

The two most popular HDLs are *Verilog* and *VHDL*. Verilog was developed by Advanced Integrated Design Systems (later renamed Gateway Design Automation) in 1984 and became a *de facto* industry open standard by 1991. VHDL, which stands for VHSIC Hardware Description Language, where VHSIC in turn was a Department of Defense project on Very High Speed Integrated Circuits, was developed by committee under government sponsorship. As one might expect from their pedigrees, Verilog is less verbose and closer in syntax to C, while VHDL supports some abstractions useful for large team projects. Many Silicon Valley companies use Verilog while defense and telecommunications companies often use VHDL. Neither language offers a compelling advantage over the other so the industry is saddled with supporting both. Appendices A and B offer brief tutorials on Verilog and VHDL. Examples in this book are given in Verilog for the sake of brevity.

When coding in an HDL, it is important to remember that you are specifying hardware that executes in parallel rather than software that executes sequentially. There are two general coding styles. *Structural* HDL specifies how a cell is composed of other cells or primitive gates and transistors. *Behavioral* HDL specifies what a cell does.

A *logic simulator* simulates both behavioral and structural HDL. A *logic synthesis* tool maps behavioral HDL code onto a *library* of gates called *standard cells* to minimize area while meeting some timing constraints. Only a subset of HDL constructs are synthesizable; this subset is emphasized in the appendices. For example, file I/O commands used in testbenches are obviously not synthesizable. Logic synthesis generally produces circuits that are neither as dense nor as fast as those handcrafted by a skilled designer. Nevertheless, integrated circuit processes are now so advanced that synthesized circuits are good enough for the great majority of application-specific integrated circuits (ASICs) built today. Layout may be automatically generated using *place & route* tools.

Verilog and VHDL models for the MIPS processor are listed in Appendix A.10 and B.7. In Verilog, each cell is called a *module*. The inputs and outputs are declared much as in a C program and bit widths are given for busses. Internal signals must also be declared in a way analogous to local variables. The processor is described hierarchically using structural Verilog at the upper levels and behavioral Verilog for the leaf cells. For example, the controller module shows how a finite state machine is specified in behavioral Verilog and the alucontrol module shows how complex combinational logic is specified. The datapath is specified structurally, containing bitslices, which in turn contain an ALU, which in turn contains a full adder.

The full adder could be expressed structurally as a sum and a carry subcircuit. In turn, the sum and carry subcircuits could be expressed behaviorally. The full adder block is shown in Figure 1.58 while the carry subcircuit is explored further in Section 1.9.

```
module fulladder(input  a, b, c,
                 output s, cout);

   sum s1(a, b, c, s);
   carry c1(a, b, c, cout);
endmodule

module carry (input a, b, c,
              output cout)

   assign cout = (a&b) | (a&c) | (b&c);
endmodule
```

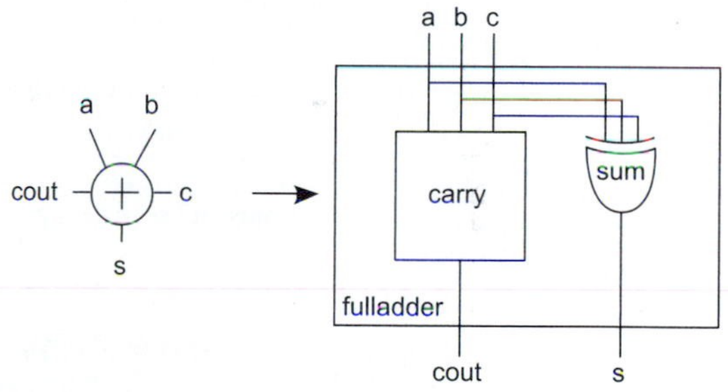

FIG 1.58 Full adder

1.9 Circuit Design

A particular logic function can be implemented in many ways. Should the function be built with ANDs, ORs, NANDs, or NORs? What should the fan-in and fan-out of each gate be? How wide should the transistors be on each gate? These and other choices influence the speed, power, and area of the system and are in the domain of circuit design.

As mentioned earlier, in many design methodologies, logic synthesis tools automatically make these choices, searching through the standard cells for the best implementation. For many applications, synthesis is good enough. When a system has critical requirements of high speed or low power or will be manufactured in large enough volume to justify extra engineering to reduce die area, custom circuit design becomes important for critical portions of the chip.

Circuit designers often draw schematics at the transistor and/or gate level. For example, Figure 1.59 shows two alternative circuit designs for the carry circuit in a full adder. The gate-level design in Figure 1.59(a) requires 26 transistors and four stages of gate delays (recall that ANDs and ORs are built from NANDs and NORs followed by inverters). The transistor-level design in Figure 1.59(b) requires only 12 transistors and two stages of gate delays, illustrating the benefits of optimizing circuit designs to take advantage of CMOS technology.

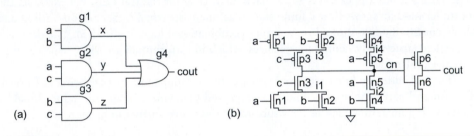

(a)   (b)

FIG 1.59 Carry subcircuit

These schematics are then *netlisted* for simulation and verification. One common netlist format is structural HDL. The gate-level design can be netlisted as shown below.

```
module carry(input   a, b, c,
             output cout)

    wire x, y, z;

    and g1(x, a, b);
    and g2(y, a, c);
    and g3(z, b, c);
    or g4(cout, x, y, z);
endmodule
```

This is a technology-independent structural description, because generic gates have been used and the actual gate implementations have not been specified. The transistor-level netlist is shown below.

```
module carry(input   a, b, c,
             output cout)
```

```
    wire i1, i2, i3, i4, cn;

    tranif1 n1(i1, 0, a);
    tranif1 n2(i1, 0, b);
    tranif1 n3(cn, i1, c);
    tranif1 n4(i2, 0, b);
    tranif1 n5(cn, i2, a);
    tranif0 p1(i3, 1, a);
    tranif0 p2(i3, 1, b);
    tranif0 p3(cn, i3, c);
    tranif0 p4(i4, 1, b);
    tranif0 p5(cn, i4, a);
    tranif1 n6(cout, 0, cn);
    tranif0 p6(cout, 1, cn);
endmodule
```

Transistors are expressed as

```
Transistor-type name(drain, source, gate);
```

`tranif1` corresponds to nMOS transistors that turn ON when the gate is '1' while `tranif0` corresponds to pMOS transistors that turn ON when the gate is '0.'

With the description generated so far, we still do not have the information required to determine the speed of the gate. We need to specify the size of the transistors and the stray capacitance. Because the Verilog language was designed as a switch-level and gate-level language, it is poorly suited to structural descriptions at this level of detail. Hence we turn to another common structural language used by the circuit simulator SPICE. The specification of the transistor-level carry subcircuit at the circuit level might be represented as shown below.

```
.SUBCKT CARRY A B C COUT VDD GND
MN1 I1 A GND GND NMOS W=1U L=0.18U AD=0.3P AS=0.5P
MN2 I1 B GND GND NMOS W=1U L=0.18U AD=0.3P AS=0.5P
MN3 CN C I1 GND NMOS W=1U L=0.18U AD=0.5P AS=0.5P
MN4 I2 B GND GND NMOS W=1U L=0.18U AD=0.15P AS=0.5P
MN5 CN A I2 GND NMOS W=1U L=0.18U AD=0.5P AS=0.15P
MP1 I3 A VDD VDD PMOS W=2U L=0.18U AD=0.6P AS=1 P
MP2 I3 B VDD VDD PMOS W=2U L=0.18U AD=0.6P AS=1P
MP3 CN C I3 VDD PMOS W=2U L=0.18U AD=1P AS=1P
MP4 I4 B VDD VDD PMOS W=2U L=0.18U AD=0.3P AS=1P
MP5 CN A I4 VDD PMOS W=2U L=0.18U AD=1P AS=0.3P
MN6 COUT CN GND GND NMOS W=2U L=0.18U AD=1P AS=1P
MP6 COUT CN VDD VDD PMOS W=4U L=0.18U AD=2P AS=2P
```

```
CI1 I1 GND 2FF
CI3 I3 GND 3FF
CA A GND 4FF
CB B GND 4FF
CC C GND 2FF
CCN CN GND 4FF
CCOUT COUT GND 2FF
.ENDS
```

Transistors are specified by lines beginning with an M as follows:

```
Mname   drain   gate   source   body   type   W=width   L=length
        AD=drain area   AS=source area
```

The body connection is new. Although MOS switches have been masquerading as three terminal devices (gate, source, and drain) until this point, they are in fact four terminal devices with the substrate or well acting as the body terminal. The type specifies whether the transistor is a p-device or n-device. The width, length, and area parameters specify physical dimensions of the actual transistors. Capacitors are specified by lines beginning with C as follows:

```
Cname   node1   node2   value
```

In this description the internal MOS model in SPICE calculates the parasitic capacitances inherent in the MOS transistor using the device dimensions specified. The extra capacitance statements in the above description designate additional routing capacitance not inherent to the device structure. This depends on the physical design of the gate. At the circuit level of structural specification, all connections are given that are necessary to fully characterize the carry gate in terms of speed, power, and connectivity.

1.10 Physical Design

1.10.1 Floorplanning

Physical design begins with a floorplan. The floorplan estimates the area of major units in the chip and defines their relative placements. The floorplan is essential to determine whether a proposed design will fit in the chip area budgeted and to estimate wiring lengths and wiring congestion, so an initial floorplan should be prepared as soon as the logic is loosely defined. As usual, this process involves feedback. The floorplan will often suggest changes to the logic (and microarchitecture), which in turn changes the floorplan. As a complex design begins to stabilize, the floorplan is often hierarchically subdivided to describe the functional blocks within the units.

Figure 1.60 shows the chip floorplan for the MIPS processor including the padframe. The top-level blocks are the controller, alucontrol, and datapath. A wiring channel is located between the controller and datapath to provide room to route the 30 control signals to the datapath. The datapath is further partitioned into the eight bitslices and the zipper. The padframe includes 40 I/O pads, which are wired to the pins on the chip package. There are 29 pads used for signals; the remainder are V_{DD} and GND.

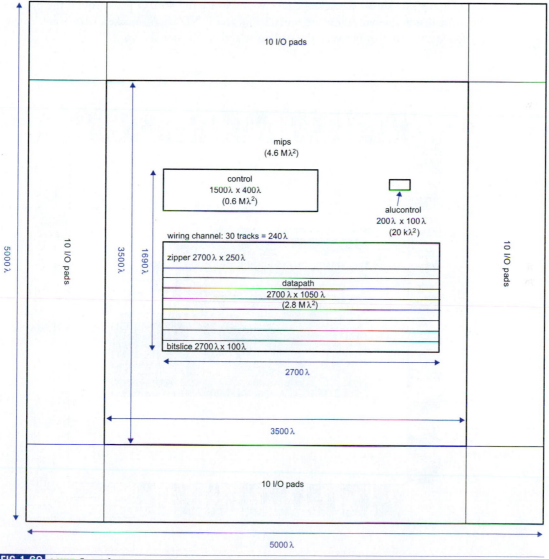

FIG 1.60 MIPS floorplan

The floorplan is drawn to scale and annotated with dimensions. The chip is designed in a 0.6 μm process on a 1.5 × 1.5 mm die so the die is 5000 λ on a side. The maximum possible core area inside the padframe is 3500 λ × 3500 λ = 12.25 Mλ². Due to the wiring channel and wasted space in the upper right corner, the actual core area of 4.6 Mλ² is larger than the sum of the block areas. This design is said to be *pad-limited* because the I/O pads set the chip area. Most commercial chips are *core-limited* because the chip area is set by the logic excluding the pads. In general, blocks in a floorplan should be rectangular because it is difficult for a designer to stuff logic into an odd-shaped region.

Figure 1.61 shows the actual chip layout. Notice the 40 I/O pads around the periphery. Just inside the pad frame are metal2 V_{DD} and GND rings, marked with + and −. For simplicity, the routing between the padframe and core is not shown.

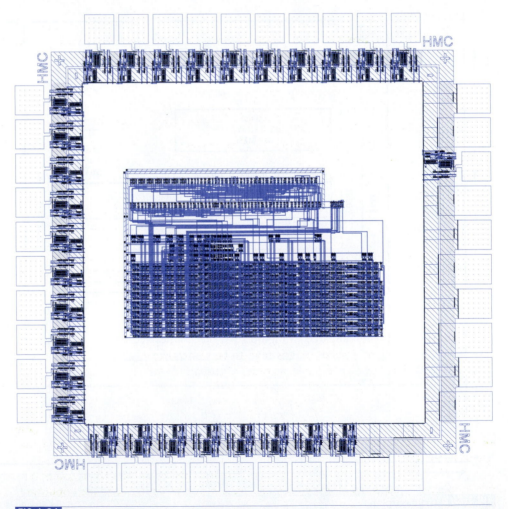

FIG 1.61 MIPS layout

On-chip structures can be categorized as *random logic*, *datapaths*, *arrays*, *analog*, and *input/output* (I/O). Random logic, like the controller, has little structure. Datapaths operate on multi-bit data words and perform roughly the same function on each bit so they consist of multiple identical or nearly identical bitslices repeated in one dimension. Arrays, like RAMs, ROMs, and PLAs, consist of identical cells repeated in two dimensions. Productivity is highest if layout can be reused or automatically generated. Datapaths and arrays are good VLSI building blocks because a single carefully crafted cell is reused in one or two dimensions. Automatic layout generators exist for memory arrays and random logic but are not as mature for datapaths. Therefore, many design methodologies ignore the potential structure of datapaths and instead lay them out with random logic tools except when performance or area are vital. Analog circuits still require careful design and simulation but tend to involve only small amounts of layout. I/O cells are also highly tuned to each fabrication process and are often supplied by the process vendor.

The MIPS layout demonstrates two of the most common layout styles: *standard cells* and *snap-together cells*. Standard cells are used for random logic. Snap-together cells are used in datapaths and arrays. Each is discussed in the following sections.

1.10.2   Standard Cells

A simple standard cell library for logic synthesis is shown in Figure 1.62 and the inside front cover. The standard cells all share the same metal1 V_{DD} and GND locations and 60 λ height so that power and ground can be connected by abutment. Inputs and outputs are provided in metal2.

In a process with two or three metal layers, standard cells are tiled into rows separated by *routing channels*. The number of wires that must be routed sets the height of the routing channels. Layout is often generated with automatic place & route tools. Figure 1.63 shows the controller layout generated by such a tool. In this example, metal1 is used horizontally and metal2 is used vertically. When more layers of metal are available, routing takes place over the cells and routing channels become unnecessary. Note that in this and subsequent layouts, the n-well around the pMOS transistors will usually not be shown.

Automatic synthesis and place & route tools have become good enough to map entire designs onto standard cells. Figure 1.64 shows the entire 8-bit MIPS processor synthesized from the VHDL model given in Appendix B onto a cell library in a 130 nm process with seven metal layers. Compared to Figure 1.61, the synthesized design shows no discernable structure. Synthesized designs tend to be somewhat larger and slower than a good custom design, but they also take an order of magnitude less design time.

1.10.3   Snap-together Cells

The area of the controller in Figure 1.63 is dominated by the routing channels. When the logic is more regular, layout density can be improved by including the wires in cells that "snap together." Snap-together cells require more design and layout effort but lead to smaller area and shorter (i.e., faster) wires. The key issue in designing snap-together cells is *pitch-matching*. Cells that connect together must have the same size in the connecting edge. Figure 1.65 shows several pitch-matched cells. Reducing the size of cell *D* does not

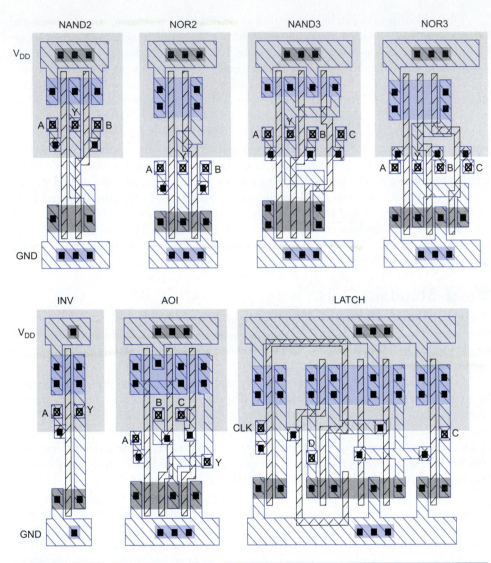

FIG 1.62 Simple standard cell library. Color version on inside front cover.

help the layout area. On the other hand, increasing the size of cell *D* also affects the area of *B* and/or *C*.

Figure 1.66 shows the MIPS datapath in more detail. The eight horizontal bitslices are clearly visible. The zipper at the top of the layout includes a decoder that is pitch–matched to the register file in the datapath.

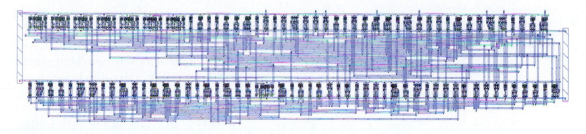

FIG 1.63 MIPS controller layout

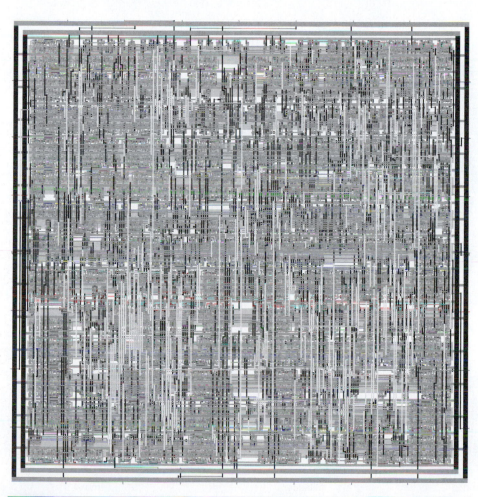

FIG 1.64 Synthesized MIPS processor

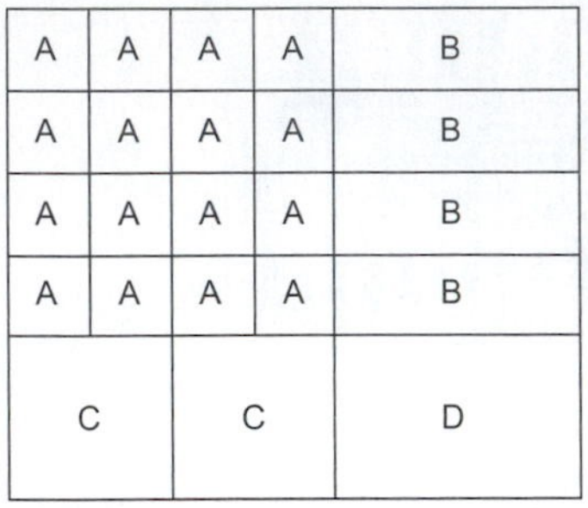

A	A	A	A	B
A	A	A	A	B
A	A	A	A	B
A	A	A	A	B
C		C		D

FIG 1.65 Pitch-matching of snap-together cells

The number of wires that must run over the datapath determines the height of the datapath cells. 80–100 λ are typical heights for relatively simple datapaths. The width of the cell depends on the cell contents. In this layout, metal1 is used for local wiring within a cell. V_{DD} and GND are routed horizontally in metal2. Horizontal metal2 wires also run over the tops of cells to carry data between cells in a bitslice. Vertical metal3 control wires are used for control, including clocks, multiplexer selects, and register enables.

Figure 1.67 shows a blowup of the ALU at the right end of each bitslice. Observe how horizontal metal2 data wires and vertical metal3 multiplexer select lines are routed over the top of the cells so no routing channels are required. The adder is shown as a black box and its design is left as an exercise for the reader.

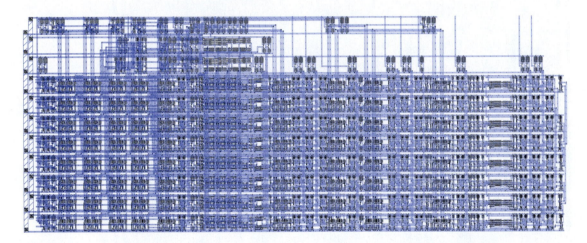

FIG 1.66 MIPS datapath layout

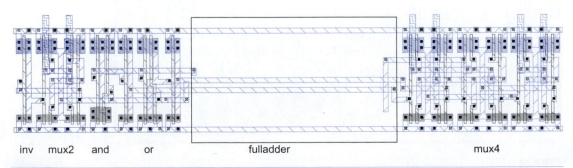

inv   mux2   and   or   fulladder   mux4

FIG 1.67 MIPS ALU layout

1.10.4 Slice Plans

Figure 1.68 shows a *slice plan* for a bitslice of the datapath. The diagram illustrates the ordering of elements in the bitslice and the allocation of wiring tracks. Wires at the left side of a cell are inputs and wires at the right side are outputs. Dots indicate that a wire passes over a cell and is also used in that cell. Each cell is annotated with its type and width. For example, the program counter (pc) is an output of the PC flop and is also used as an input to the srcA and address multiplexers. The slice plan makes it easy to calculate wire lengths and evaluate wiring congestion before laying out the datapath. In this case, it is evident that the greatest congestion takes place over the register file, where eight wiring tracks are required. The slice plan is also useful for estimating area of datapaths.

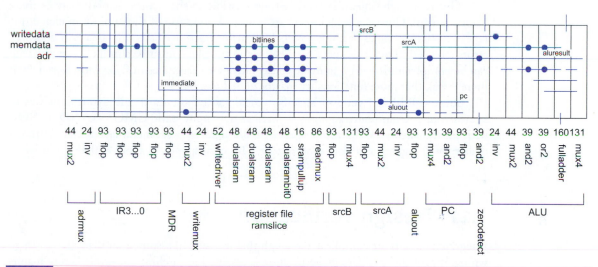

FIG 1.68 Datapath slice plan

1.10.5 Area Estimation

A good floorplan depends on reasonable area estimates, which may be difficult to make before logic is finalized. An experienced designer may be able to estimate block area by comparison to the area of a comparable block drawn in the past. In the absence of data for such comparison, Table 1.10 lists some typical numbers. Be certain to account for large wiring channels at a pitch of 8 λ / track. Larger transistors clearly occupy greater area, so this may be factored into the area estimates as a function of W and L (width and length) [Horowitz02]. Your mileage may vary, but it is clear that datapaths and arrays achieve much higher densities than standard cells. In the MIPS example, the controller random logic has 440 transistors in 0.6 Mλ² for a density of about 1400 λ²/transistor. The datapath has 3461 transistors in a 2.7 Mλ² area for a density of 780 λ²/transistor.

Table 1.10	Typical layout densities
Element	**Area**
random logic (2-level metal process)	$1000 - 1500 \ \lambda^2$ / transistor
datapath	$250 - 750 \ \lambda^2$ / transistor
	or $6 \ WL + 360 \ \lambda^2$ / transistor
SRAM	$1000 \ \lambda^2$ / bit
DRAM (in a DRAM process)	$100 \ \lambda^2$ / bit
ROM	$100 \ \lambda^2$ / bit

Given enough time, it is nearly always possible to shave a few lambda here or there from a design. However, such efforts are seldom a good use of time unless an element is repeated so many times that it accounts for a major fraction of the chip area or if floorplan errors have led to too little space for a block and the block must be shrunk before the chip can be completed. It is wise to make conservative area estimates in floorplans, especially if there is risk that more functionality may be added to a block.

Some cell library vendors specify typical routed standard cell layout densities in kgates/mm^2.[6] Commonly a gate is defined as a 3-input static CMOS NAND or NOR with six transistors. A 180 nm process ($\lambda \sim 0.1 \ \mu$m) with six metal layers may achieve a density of 20 kgates/mm^2 for random logic. This corresponds to about $830 \ \lambda^2$ / transistor.

1.11 Design Verification

Integrated circuits are complicated enough that it is safe to assume that if anything can go wrong, it probably has. Design verification is essential to catching the errors before manufacturing and commonly accounts for one-third to one-half of the effort devoted to a chip.

As design representations become more detailed, verification time increases. It is not practical to simulate an entire chip in a circuit-level simulator such as SPICE for a large number of cycles to prove that the layout is correct. Instead, the design is usually tested for functionality at the architectural level with a model in a language such as C and at the logic level by simulating the HDL description. Then the circuits are checked to ensure they are a faithful representation of the logic and the layout is checked to ensure it is a faithful representation of the circuits, as shown in Figure 1.69. Circuits and layout must meet timing and power specifications as well.

A *testbench* is used to verify that the logic is correct. The testbench instantiates the logic under test. It reads a file of inputs and expected outputs called *test vectors*, applies them to the module under test, and logs mismatches. Appendix A.10 provides an example of a testbench for verifying the MIPS processor logic.

---

[6]1 kgate = 1000 gates

A number of techniques are available for circuit verification. If the logic is synthesized onto a cell library, the postsynthesis gate-level netlist can be expressed in an HDL again and simulated using the same test vectors. Powerful *formal verification* tools are also becoming available that prove that a circuit performs the same Boolean function as the associated logic. These are very useful for handcrafted circuits, but can be slow and difficult to operate. Exotic circuits should be simulated thoroughly to ensure they perform the intended logic function and have adequate noise margins; circuit pitfalls are discussed throughout this book.

Layout vs. Schematic tools (LVS) check to make sure that transistors in a layout are connected in the same way as in the circuit schematic. *Design rule checkers* (DRC) verify that the layout satisfies design rules. *Electrical rule checkers* (ERC) scan for other potential problems such as latch-up, noise problems, or electromigration risks; such problems will also be discussed later in the book.

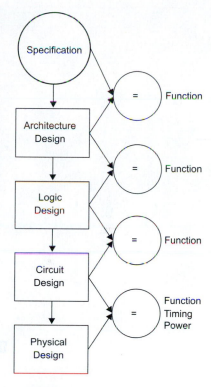

FIG 1.69 Design and verification sequence

1.12 Fabrication, Packaging, and Testing

Once a chip design is complete, it is taped out for manufacturing. *Tapeout* gets its name from the old practice of writing a specification of masks to magnetic tape; now the mask descriptions are usually sent to the manufacturer electronically. Two common formats for mask descriptions are the Caltech Interchange Format (CIF) [Mead80] (mainly used in academia) and the Calma GDS II Stream Format (GDS) [Calma84] (used in industry).

Masks are made by etching a pattern of chrome on glass with an electron beam. A set of masks for a modern process can be very expensive. For example, masks for a large chip in a 180 nm process may cost on the order of a quarter of a million dollars. In a 130 nm process, the mask set may be in the vicinity of a million dollars. The MOSIS service in the United States and its counterparts in Europe and Japan make a single set of masks covering multiple small designs from academia and industry to amortize the cost across many customers.

Integrated circuit fabrication plants (fabs) now cost billions of dollars and become obsolete in a few years. Some large companies still own their own fabs, but an increasing number of fabless semiconductor companies contract out manufacturing to vendors such as TSMC, UMC, and IBM.

Multiple chips are manufactured simultaneously on a single silicon wafer, typically 6″–12″ in diameter. A bare wafer costs about $1000–$5000. Fabrication requires many deposition, masking, etching, and implant steps. Most fabrication plants are optimized for wafer throughput rather than latency, leading to turnaround times of up to 10 weeks, although shorter turnarounds are available for a substantial premium. Figure 1.70 shows a silicon wafer after processing.

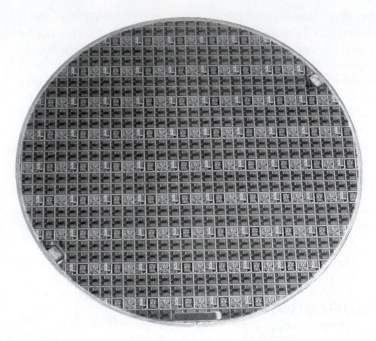

FIG 1.70 Processed 8-inch wafer

FIG 1.71 Chip in a 40-pin dual-inline package

Processed wafers are sliced into dice (chips) and packaged. Figure 1.71 shows a 1.5 × 1.5 mm chip in a 40-pin *dual-inline package* (DIP). This *wire-bonded* package uses thin gold wires to connect the pads on the die to the lead frame in the center cavity of the package. More advanced packages offer different tradeoffs between cost, pin count, pin bandwidth, power handling, and reliability, as will be discussed in Section 12.2. Flip-chip technology places small solder balls directly onto the die, eliminating the bond wire inductance and allowing contacts over the entire chip area rather than just at the periphery.

Even tiny defects in a wafer or dust particles can cause a chip to fail. Chips are tested before being sold. Testers capable of handling high-speed chips cost millions of dollars, so many chips use built-in self-test features to reduce the tester time required. Chapter 9 is devoted to design verification and testing.

Summary

"If the automobile had followed the same development cycle as the computer, a Rolls-Royce would today cost $100, get one million miles to the gallon, and explode once a year . . ."

—Robert X. Cringely

CMOS technology, driven by Moore's Law, has come to dominate the semiconductor industry. This chapter developed the principles of designing a simple CMOS integrated circuit. MOS transistors can be viewed as electrically controlled switches. Complementary CMOS gates are built from pull-down networks of nMOS transistors and pull-up networks of pMOS transistors. Transistors and wires are fabricated on silicon wafers using a series of deposition, lithography, and etch steps. These steps are defined by a set of masks drawn as a chip layout. Design rules specify minimum width and spacing between elements in the layout. The chip design process can be divided into architecture, logic, circuit, and physical design. The performance, area, and power of the chip are influenced by interrelated decisions made at each level. Design verification plays an important role in constructing such complex systems; the reliability requirements for hardware are much greater than those typically imposed on software. The remainder of this book will expand on the material introduced in this chapter.

Exercises

1.1 Extrapolating the data from Figure 1.4, predict the transistor count of a microprocessor in 2010.

1.2 Search the Web for transistor counts of Intel's more recent microprocessors. Make a graph of transistor count vs. year of introduction from the Pentium Processor in 1993 to the present on a semilog scale. How many months pass between doubling of transistor counts?

1.3 Sketch a transistor-level schematic for a CMOS 4-input NOR gate.

1.4 Sketch a transistor-level schematic for a single-stage CMOS logic gate for each of the following functions:

a) $Y = \overline{ABC + D}$

b) $Y = \overline{(AB + C) \cdot D}$

c) $Y = \overline{AB + C \cdot (A + B)}$

1.5 Use a combination of CMOS gates (represented by their symbols) to generate the following functions from A, B, and C.

a) $Y = A$ (buffer)

b) $Y = A\overline{B} + \overline{A}B$ (XOR)

c) $Y = \overline{A}\overline{B} + AB$ (XNOR)

d) $Y = AB + BC + AC$ (majority)

1.6 Sketch a transistor-level schematic of a CMOS 3-input XOR gate. You may assume you have both true and complementary versions of the inputs available.

1.7 Sketch transistor-level schematics for the following logic functions. You may assume you have both true and complementary versions of the inputs available.

a) A 2:4 decoder defined by

$Y0 = \overline{A0} \cdot \overline{A1}$
$Y1 = A0 \cdot \overline{A1}$
$Y2 = \overline{A0} \cdot A1$
$Y3 = A0 \cdot A1$

b) A 3:2 priority encoder defined by

$Y0 = \overline{A0} \cdot (A1 + \overline{A2})$
$Y1 = \overline{A0} \cdot \overline{A1}$

1.8 Sketch a stick diagram for a CMOS 4-input NOR gate from Exercise 1.3.

1.9 Estimate the area of your 4-input NOR gate from Exercise 1.8.

1.10 Using a CAD tool of your choice, layout a 4-input NOR gate. How does its size compare to the prediction from Exercise 1.9?

1.11 Figure 1.72 shows a stick diagram of a 2-input NAND gate. Sketch a side view (cross-section) of the gate from X to X′.

1.12 Figure 1.73 gives a stick diagram for a level-sensitive latch. Estimate the area of the latch.

1.13 Draw a transistor-level schematic for the latch of Figure 1.73. How does the schematic differ from Figure 1.30(b)?

1.14 Consider the design of a CMOS compound OR-AND-INVERT (OAI21) gate computing $F = \overline{(A + B) \cdot C}$.

a) sketch a transistor-level schematic

b) sketch a stick diagram

c) estimate the area from the stick diagram

d) layout your gate with a CAD tool using unit-sized transistors

e) compare the layout size to the estimated area

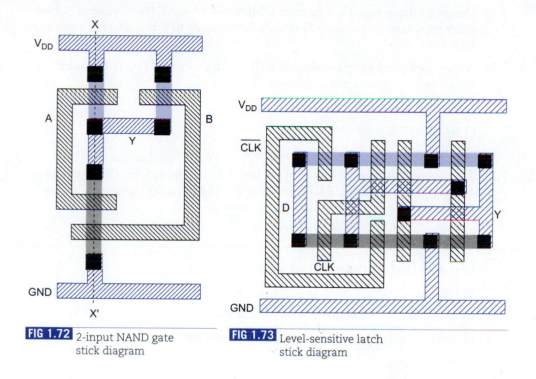

FIG 1.72 2-input NAND gate
stick diagram

FIG 1.73 Level-sensitive latch
stick diagram

1.15 Consider the design of a CMOS compound OR-OR-AND-INVERT (OAI22) gate computing $F = \overline{(A + B) \cdot (C + D)}$.

    a) sketch a transistor-level schematic

    b) sketch a stick diagram

    c) estimate the area from the stick diagram

    d) layout your gate with a CAD tool using unit-sized transistors

    e) compare the layout size to the estimated area

1.16 A 3-input majority gate returns a true output if at least two of the inputs are true. A minority gate is its complement. Design a 3-input CMOS minority gate using a single stage of logic.

    a) sketch a transistor-level schematic

    b) sketch a stick diagram

    c) estimate the area from the stick diagram

1.17 Design a 3-input minority gate using CMOS NANDs, NORs, and inverters. How many transistors are required? How does this compare to a design from Exercise 1.16(a)?

1.18 A carry lookahead adder computes $G = \underline{G_3} + P_3(G_2 + P_2(G_1 + P_1 G_0))$. Consider designing a compound gate to compute $\overline{G}$.

a) sketch a transistor-level schematic

b) sketch a stick diagram

c) estimate the area from the stick diagram

1.19 The course Web page has a series of five labs in which you can learn VLSI design by completing the multicycle MIPS processor described in this chapter. The labs use the open-source Electric CAD tool. They cover:

a) leaf cells: schematic entry, layout, icons, simulation, DRC, ERC, LVS; hierarchical design

b) complex leaf cell design and verification: full adder

c) hierarchical design: ALU assembly, datapath routing, and simulation

d) control design: Verilog, synthesis, place & route

e) chip assembly, pad frame, full-chip verification, tapeout

MOS Transistor Theory

2

2.1   Introduction

In Chapter 1 the Metal-Oxide-Semiconductor (MOS) transistor was introduced in terms of its operation as an ideal switch. In this chapter we will examine the characteristics of MOS transistors in more detail to lay the foundation for predicting their performance. We will also look at second-order effects that are important to power consumption and circuit reliability. Using the transistor models, we will calculate the DC transfer characteristics of logic gates.

Figure 2.1 shows some of the symbols that are commonly used for MOS transistors. The symbols in Figure 2.1(a) will be used where it is only necessary to indicate the switch logic necessary to build a function. If the body (substrate or well) connection needs to be shown, the symbols in Figure 2.1(b) will be used. Figure 2.1(c) shows an example of other symbols that may be encountered in the literature.

The MOS transistor is a *majority-carrier* device in which the current in a conducting channel between the source and drain is controlled by a voltage applied to the gate. In an nMOS transistor, the majority carriers are electrons; in a pMOS transistor, the majority carriers are holes. The behavior of MOS transistors can be understood by first examining an isolated MOS structure with a gate and body but no source or drain. Figure 2.2 shows a simple MOS structure. The top layer of the structure is a good conductor called the *gate*. Early transistors used metal gates, but modern transistors generally use polysilicon, i.e., silicon formed from many small crystals. The middle layer is a very thin insulating film of SiO_2 called the *gate oxide*. The bottom layer is the doped silicon body. The figure shows a p-type body in which the carriers are holes. The body is grounded and a voltage is applied to the gate. The gate oxide is a good insulator so almost zero current flows from the gate to the body.[1]

In Figure 2.2(a), a negative voltage is applied to the gate, so there is negative charge on the gate. The mobile positively charged holes are attracted to the region beneath the gate. This is called the *accumulation* mode. In Figure 2.2(b), a low positive voltage is applied to the gate, resulting in some positive charge on the gate. The holes in the body are repelled from the region directly beneath the gate, resulting in a *depletion* region forming below the gate. In Figure 2.2(c), a higher positive potential exceeding a critical threshold voltage V_t is

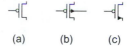

(a)   (b)   (c)

FIG 2.1 MOS transistor symbols

---

[1]Gate oxides are now only a handful of atomic layers thick and electrons sometimes tunnel through the oxide, creating a current from gate to substrate. This effect is explored in Section 2.4.5.

applied, attracting more positive charge to the gate. The holes are repelled further and a small number of free electrons in the body are attracted to the region beneath the gate. This conductive layer of electrons in the p-type body is called the *inversion* layer. The threshold voltage depends on the number of dopants in the body and the thickness t_{ox} of the oxide. It is usually positive, as shown in this example, but can be engineered to be negative.

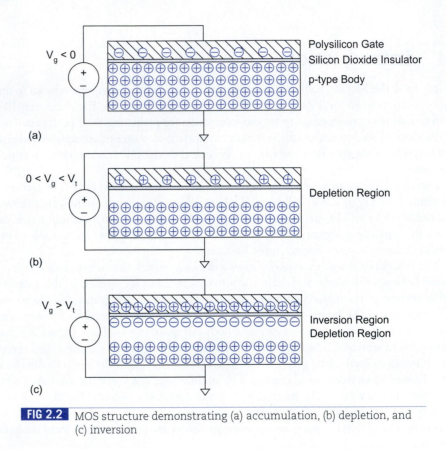

(a)

(b)

(c)

FIG 2.2   MOS structure demonstrating (a) accumulation, (b) depletion, and (c) inversion

Figure 2.3 shows an nMOS transistor with a grounded source and p-type body. The transistor consists of the MOS stack between two n-type regions called the *source* and *drain*. In Figure 2.3(a), the gate-to-source voltage V_{gs} is less than the threshold voltage. The source and drain have free electrons. The body has free holes but no free electrons. The junctions between the body and the source or drain are reverse-biased, so almost zero current flows. This mode of operation is called *cutoff*. In Figure 2.3(b), the gate voltage is greater than the threshold voltage. Now an inversion region of electrons (majority carriers) called the *channel* connects the source and drain, creating a conductive path. The number of carriers and the conductivity increases with the gate voltage. The potential difference

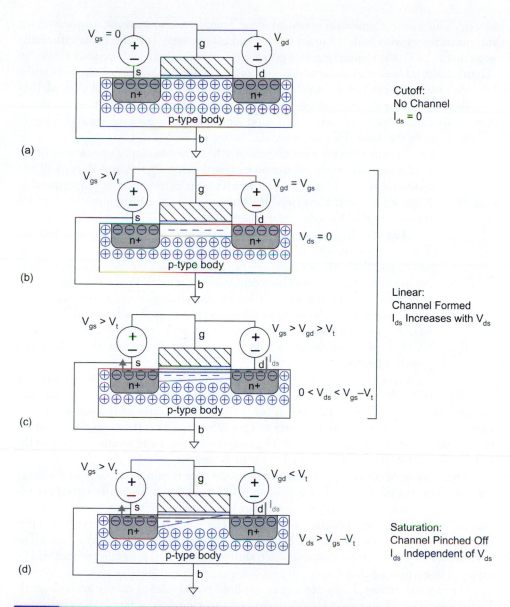

FIG 2.3 nMOS transistor demonstrating cutoff, linear, and saturation regions of operation

between drain and source is $V_{ds} = V_{gs} - V_{gd}$. If $V_{ds} = 0$ (i.e., $V_{gs} = V_{gd}$), there is no electric field tending to push current from drain to source. When a small positive potential V_{ds} is applied to the drain (Figure 2.3(c)), current I_{ds} flows through the channel from drain to source[2]. This mode of operation is termed *linear, resistive, nonsaturated,* or *unsaturated*; the current increases with both the drain voltage and gate voltage. If V_{ds} becomes sufficiently large that $V_{gd} < V_t$, the channel is no longer inverted near the drain and becomes *pinched off* (Figure 2.3(d)). However, conduction is still brought about by the drift of electrons under the influence of the positive drain voltage. As electrons reach the end of the channel, they are injected into the depletion region near the drain and accelerated toward the drain. Above this drain voltage the current I_{ds} is controlled only by the gate voltage and ceases to be influenced by the drain. This mode is called *saturation*.

In summary, the nMOS transistor has three modes of operation. If $V_{gs} < V_t$, the transistor is cut off and no current flows. If $V_{gs} > V_t$ and V_{ds} is small, the transistor acts as a linear resistor in which the current flow is proportional to V_{ds}. If $V_{gs} > V_t$ and V_{ds} is large, the transistor acts as a current source in which the current flow becomes independent of V_{ds}.

The pMOS transistor in Figure 2.4 operates in just the opposite fashion. The n-type body is tied to a high potential so the junctions with the p-type source and drain are normally reverse-biased. When the gate is also at a high potential, no current flows between drain and source. When the gate voltage is lowered by a threshold V_t, holes are attracted to form a p-type channel immediately beneath the gate, allowing current to flow between drain and source. The threshold voltages of the two types of transistors are not necessarily equal, so we use the terms V_{tn} and V_{tp} to distinguish the nMOS and pMOS thresholds.

Although MOS transistors are symmetrical, by convention we say that majority carriers flow from their source to their drain. Because electrons are negatively charged, the source of an nMOS transistor is the more negative of the two terminals. Holes are positively charged so the source of a pMOS transistor is the more positive of the two terminals. In complementary CMOS gates, the source is the terminal closer to the supply rail and the drain is the terminal closer to the output.

The delay of MOS circuits is determined by the time required to charge or discharge the capacitance of the circuits. The gate of an MOS transistor is inherently a good capacitor with a thin dielectric; indeed, its capacitance is responsible for attracting carriers to the channel and thus for the operation of the device. The junctions of the reverse-biased p–n junctions from source or drain to the body contribute additional *parasitic* capacitance. The capacitance of wires interconnecting the transistors is also very significant and will be explored in Section 4.5.2.

We begin in Section 2.2 by deriving an idealized model relating current and voltage (I-V) for a transistor. This model provides a general understanding of transistor behavior but is of limited value. On the one hand, it neglects too many effects that are important in modern transistors with short channel lengths L. Therefore, the model is not sufficient to accurately calculate current. On the other hand, it is still too complicated to use in back-of-

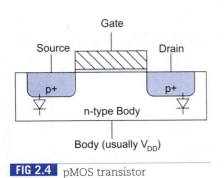

FIG 2.4 pMOS transistor

---

[2]The terminology of source and drain might initially seem backward. Recall that the current in an nMOS transistor is carried by moving electrons with a negative charge. Therefore, positive current from drain to source corresponds to electrons flowing from their source to their drain.

the-envelope calculations when computing the performance of large circuits. Therefore, we will develop even simpler models for performance estimation in Section 2.6. To plausibly simulate transistor behavior, we must consider many second-order effects introduced in Section 2.4 and use circuit simulators based on SPICE [Nagel75] (see Chapter 5) with elaborate models capturing these effects. Nevertheless, the first-order models form a foundation for understanding the second-order effects. In particular, even the smallest transistors still act as voltage-controlled switches with a cutoff region in which the current from source to drain is nearly zero, a linear region in which the transistor acts as a resistor, and a saturation region in which current becomes nearly independent of drain-source voltage.

The remainder of this chapter develops quantitative models of MOS circuit operation. It begins with the ideal Shockley models of transistor I-V characteristics and a study of capacitances. It then explores nonidealities that are important for design. The models are used to predict the DC transfer characteristics of inverters and pass transistors circuits. Finally, a very simple resistor-capacitor (RC) model is introduced for delay estimation.

2.2   Ideal I-V Characteristics

As stated previously, MOS transistors have three regions of operation:

- Cutoff or subthreshold region
- Linear or nonsaturation region
- Saturation region

Let us derive a first-order (ideal Shockley) model [Shockley52, Cobbold70, Sah64] relating the current and voltage (I-V) for an nMOS transistor in each of these regions. In the cutoff region ($V_{gs} < V_t$), there is no channel and almost zero current flows from drain to source. In the other regions, the gate attracts carriers (electrons) to form a channel. The electrons drift from source to drain at a rate proportional to the electric field between these regions. Thus we can compute currents if we know the amount of charge in the channel and the rate at which it moves. We know that the charge on each plate of a capacitor is $Q = CV$. Thus the charge in the channel $Q_{channel}$ is

$$Q_{channel} = C_g \left(V_{gc} - V_t \right) \qquad (2.1)$$

where C_g is the capacitance of the gate to the channel and $V_{gc} - V_t$ is the amount of voltage attracting charge to the channel beyond the minimum required to invert from p to n. The gate voltage is referenced to the channel, which is not grounded. If the source is at V_s and the drain is at V_d, the average is $V_c = (V_s + V_d)/2 = V_s + V_{ds}/2$. Therefore, the mean difference between the gate and channel potentials V_{gc} is $V_{gs} - V_{ds}/2$, as shown in Figure 2.5.

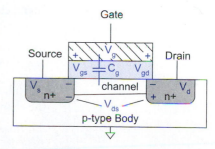

Average gate to channel potential:

$V_{gc} = (V_{gs} + V_{gd})/2 = V_{gs} - V_{ds}/2$

FIG 2.5   Average gate to channel voltage

We can model the gate as a parallel plate capacitor with capacitance proportional to area over thickness. If the gate has length L and width W and the oxide thickness is t_{ox}, as shown in Figure 2.6, the capacitance is

$$C_g = \varepsilon_{ox} \frac{WL}{t_{ox}} \tag{2.2}$$

where the permittivity $\varepsilon_{ox} = 3.9\,\varepsilon_0$ for SiO_2 and ε_0 is the permittivity of free space, $8.85 \cdot 10^{-14}$ F/cm. Often the ε_{ox}/t_{ox} term is called C_{ox}, the capacitance per unit area of the gate oxide.

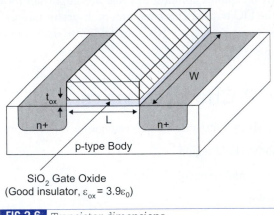

FIG 2.6   Transistor dimensions

Each carrier in the channel is accelerated to an average velocity proportional to the lateral electric field, i.e., the field between source and drain. The constant of proportionality μ is called the *mobility*.

$$v = \mu E \tag{2.3}$$

The electric field E is the voltage difference between drain and source V_{ds} divided by the channel length

$$E = \frac{V_{ds}}{L} \tag{2.4}$$

The time required for carriers to cross the channel is the channel length divided by the carrier velocity: L/v. Therefore, the current between source and drain is the total amount of charge in the channel divided by the time required to cross

$$I_{ds} = \frac{Q_{channel}}{L / v}$$

$$= \mu C_{ox} \frac{W}{L}\left(V_{gs} - V_t - \frac{V_{ds}}{2}\right)V_{ds} \tag{2.5}$$

$$= \beta\left(V_{gs} - V_t - \frac{V_{ds}}{2}\right)V_{ds}$$

where

$$\beta = \mu C_{ox} \frac{W}{L} \tag{2.6}$$

EQ (2.5) describes the linear region of operation, for $V_{gs} > V_t$, but V_{ds} relatively small. It is called *linear* or *resistive* because $V_{ds}/2 << V_{gs} - V_t$, I_{ds} increases almost linearly with V_{ds}, just like an ideal resistor. The geometry and technology-dependent parameters are sometimes merged into a single factor β. Do not confuse this use of β with the same symbol used for the ratio of collector-to-base current in a bipolar transistor. Some texts [Gray01] lump the technology-dependent parameters alone into a constant called "k prime."[3]

$$k' = \mu C_{ox} \tag{2.7}$$

However, if $V_{ds} > V_{dsat} \equiv V_{gs} - V_t$, the channel is no longer inverted in the vicinity of the drain; we say it is pinched off. Beyond this point, called the *drain saturation voltage*, increasing the drain voltage has no further effect on current. Substituting $V_{ds} = V_{dsat}$ at this point of maximum current into EQ (2.5), we find an expression for the saturation current that is independent of V_{ds}. This expression is valid for $V_{gs} > V_t$ and $V_{ds} > V_{dsat}$.

$$I_{ds} = \frac{\beta}{2}\left(V_{gs} - V_t\right)^2 \tag{2.8}$$

It is sometimes convenient to define I_{dsat} as the current of a transistor that is fully ON, i.e., $V_{gs} = V_{ds} = V_{DD}$.

$$I_{dsat} = \frac{\beta}{2}\left(V_{DD} - V_t\right)^2 \tag{2.9}$$

---

[3]Other sources (e.g., MOSIS) define $k' = \frac{\mu C_{ox}}{2}$; check the definition before using quoted data.

EQ (2.10) summarizes the current in the three regions:

$$I_{ds} = \begin{cases} 0 & V_{gs} < V_t & \text{cutoff} \\ \beta\left(V_{gs} - V_t - \dfrac{V_{ds}}{2}\right)V_{ds} & V_{ds} < V_{dsat} & \text{linear} \\ \dfrac{\beta}{2}\left(V_{gs} - V_t\right)^2 & V_{ds} > V_{dsat} & \text{saturation} \end{cases} \tag{2.10}$$

Example

Consider an nMOS transistor in a 180 nm process with W/L = 4/2 λ (i.e., 0.36/0.18 μm). In this process, the gate oxide thickness is 40 Å and the mobility of electrons is 180 cm^2/V • s at 70° C. The threshold voltage is 0.4V. Plot I_{ds} vs. V_{ds} for V_{gs} = 0, 0.3, 0.6, 0.9, 1.2, 1.5, and 1.8 V.

Solution: We first calculate β.

$$\beta = \mu C_{ox} \frac{W}{L} = \left(180 \ \tfrac{\text{cm}^2}{\text{V•s}}\right)\left(\frac{3.9 \cdot 8.85 \cdot 10^{-14} \ \tfrac{\text{F}}{\text{cm}}}{40 \cdot 10^{-8} \ \text{cm}}\right)\left(\frac{W}{L}\right) = 155 \frac{W}{L} \ \tfrac{\mu\text{A}}{\text{V}^2} \tag{2.11}$$

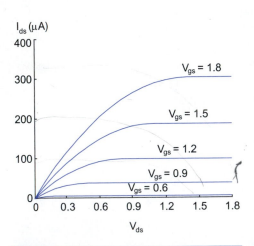

I_{ds} (μA)

V_{gs} = 1.8

V_{gs} = 1.5

V_{gs} = 1.2

V_{gs} = 0.9

V_{gs} = 0.6

V_{ds}

FIG 2.7 I-V characteristics of ideal nMOS transistor

Figure 2.7 shows the I-V characteristics for the transistor. According to the first-order model, the current is zero for gate voltages below V_t. For higher gate voltages, current increases linearly with V_{ds} for small V_{ds}. As V_{ds} reaches the saturation point $V_{gs} - V_t$, current rolls off and eventually becomes independent of V_{ds} when the transistor is saturated.

pMOS transistors behave in much the same way, but with the signs reversed and I-V characteristics in the third quadrant, as shown in Figure 2.8. The mobility of holes in silicon is typically lower than that of electrons. This means that pMOS transistors provide less current than nMOS transistors of comparable size and hence are slower. The symbols μ_n and μ_p are used to distinguish mobility of electrons and of holes in nMOS and pMOS transistors, respectively. The mobility ratio $\mu = \mu_n / \mu_p$ is typically 2–3; we will generally use 2 for examples in this book. The figure reflects a transistor of the same geometry as in Figure 2.7, but with μ_p = 90 cm^2/V • s and V_{tp} = − 0.4 V. Similarly, β_n, β_p, k'_n, and k'_p are sometimes used to distinguish nMOS and pMOS I-V characteristics.

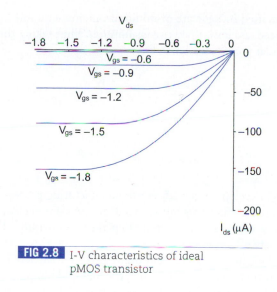

FIG 2.8 I-V characteristics of ideal pMOS transistor

2.3 C-V Characteristics

Each terminal of an MOS transistor has capacitance to the other terminals. In general, these capacitances are nonlinear and voltage dependent (C-V); however, they can be approximated as simple capacitors when their behavior is averaged across the switching voltages of a logic gate. This section first presents simple models of each capacitance suitable for estimating delay and power consumption of transistors. It then explores more detailed models used for circuit simulation. The more detailed models may be skipped on a first reading.

2.3.1 Simple MOS Capacitance Models

The gate of an MOS transistor is a good capacitor. Indeed, its capacitance is necessary to attract charge to invert the channel, so high gate capacitance is required to obtain high I_{ds}. As seen in Section 2.2, the gate capacitor can be viewed as a parallel plate capacitor with the gate on top and channel on bottom with the thin oxide dielectric between. Therefore, the capacitance is

$$C_g = C_{ox}WL \tag{2.12}$$

A capacitor is a two-terminal device. When the transistor is on, the channel extends from the source (and reaches the drain if the transistor is unsaturated, or stops short in saturation). It is not an unreasonable simplification to approximate the gate capacitance as terminating at the source and thus call the capacitance C_{gs}.

Most transistors used in logic are of minimum manufacturable length because this results in greatest speed and lowest power consumption. Thus taking this minimum L as a constant for a particular process, we can define

$$C_g = C_{\text{permicron}} \cdot W \qquad (2.13)$$

where

$$C_{\text{permicron}} = C_{\text{ox}} L = \frac{\varepsilon_{\text{ox}}}{t_{\text{ox}}} L \qquad (2.14)$$

Notice that if we develop a more advanced manufacturing process in which both the channel length and oxide thickness are reduced by the same factor, $C_{\text{permicron}}$ remains unchanged (and has a value of about 1.5–2 fF/μm of gate width). Table 5.5 lists gate capacitance for a variety of processes.

In addition to the gate, the source and drain also have capacitances. These capacitances are not fundamental to operation of the devices, but do impact circuit performance and hence are called *parasitic* capacitors. They arise from the reverse-biased p–n junctions between the source or drain diffusion and the body and hence are also called *diffusion*[4] capacitance C_{sb} and C_{db}. The size of these junctions depends on the area and perimeter of the source and drain diffusion, the depth of the diffusion, the doping levels, and the voltage. As diffusion has both high capacitance and high resistance, it is generally made as small as possible in the layout. There are three types of diffusion regions frequently seen, illustrated with the two series transistors in Figure 2.9. In Figure 2.9(a), each source and drain has its own *isolated* region of contacted diffusion. In Figure 2.9(b), the drain of the bottom transistor and source of the top transistor form a *shared* contacted diffusion region. In Figure 2.9(c), the source and drain are *merged* into an uncontacted region. The average capacitance of each of these types of regions can be calculated or measured from simulation as a transistor switches between V_{DD} and GND. Table 5.5 also lists the capacitance for each scenario for a variety of processes.

For the purposes of hand estimation, you can observe that the diffusion capacitance C_{sb} and C_{db} of contacted source and drain regions is comparable to the gate capacitance (e.g., 1.5-2 fF/μm of gate width). The diffusion capacitance of the uncontacted source or drain is somewhat less because the area is smaller but the difference is usually unimportant for hand calculations. These values of $C_g = C_{sb} = C_{db} \sim 2$fF/$\mu$m will be used in examples throughout the text, but you should obtain the appropriate data for your process using methods to be discussed in Section 5.4.

---

[4]Device engineers more properly call this *depletion* capacitance, but the term *diffusion* capacitance is widely used by circuit designers.

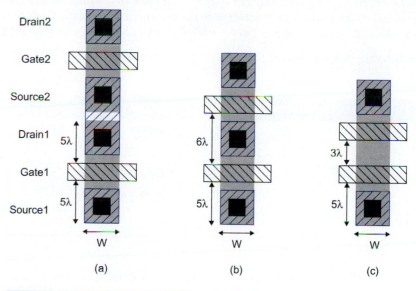

FIG 2.9 Diffusion region geometries

2.3.2 Detailed MOS Gate Capacitance Model

The MOS gate sits above the channel and may partially overlap the source and drain diffusion areas. Therefore, the gate capacitance has two components: the intrinsic capacitance (over the channel) and the overlap capacitances (to the source, drain, and body).

The intrinsic capacitance was approximated as a simple parallel plate in EQ (2.12). Let us call this capacitance $C_0 = WLC_{ox}$. However, the bottom plate of the capacitor depends on the mode of operation of the transistor.

1. *Cutoff.* When the transistor is OFF ($V_{gs} = 0$), the channel is not inverted and charge on the gate is matched with opposite charge from the body. This is called C_{gb}, the gate-to-body capacitance. As V_{gs} increases but remains below a threshold, a depletion region forms at the surface. This effectively moves the bottom plate downward from the oxide, reducing the capacitance.

2. *Linear.* When $V_{gs} > V_t$, the channel inverts and again serves as a good conductive bottom plate. However, the channel is connected to the source and drain, rather than the body. At low values of V_{ds}, the channel charge is roughly shared between source and drain, so $C_{gs} = C_{gd} = C_0/2$. As V_{ds} increases, the region near the drain becomes less inverted, so a greater fraction of the capacitance is attributed to the source and a smaller fraction to the drain.

3. *Saturation.* At $V_{ds} > V_{gs} - V_t$, the transistor saturates and the channel pinches off. At this point, all the intrinsic capacitance is to the source. Because of pinchoff, the capacitance in saturation reduces to $C_{gs} = 2/3\ C_0$ for an ideal transistor [Gray01].

The behavior in these three regions can be approximated as shown in Table 2.1.

Table 2.1	Approximation of intrinsic MOS gate capacitance		
Parameter	**Cutoff**	**Linear**	**Saturation**
C_{gb}	C_0	0	0
C_{gs}	0	$C_0/2$	$2/3\ C_0$
C_{gd}	0	$C_0/2$	0
$C_g = C_{gs} + C_{gd} + C_{gb}$	C_0	C_0	$2/3\ C_0$

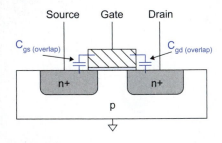

FIG 2.10 Overlap capacitance

The gate overlaps the source and drain by a small amount in a real device and also has fringing fields terminating on the source and drain. This leads to additional overlap capacitances, as shown in Figure 2.10. These capacitances are proportional to the width of the transistor. Typical values are $C_{gsol} = C_{gdol} = 0.2–0.4\ \text{fF}/\mu\text{m}$.

$$C_{gs(\text{overlap})} = C_{gsol}W$$
$$C_{gd(\text{overlap})} = C_{gdol}W$$

(2.15)

The experimentally measured C_{gs} and C_{gd} of a long channel nMOS transistor ($W = 49.2\ \mu\text{m}$, $L = 4.5\ \mu\text{m}$) is shown in Figure 2.11(a) [Sheu87a]. This graph shows the normalized capacitance varying as a function of V_{ds} for a number of $V_{gs} - V_t$ values. Observe that at $V_{ds} = 0$, $C_{gs} = C_{gd} = C_0/2$. As V_{ds} increases, the capacitances approach $C_{gs} = 2/3\ C_0$ and $C_{gd} = 0$, as expected when the transistor is saturated. Figure 2.11(b) shows measured capacitances of a shorter channel transistor ($W = 49.2\ \mu\text{m}$, $L = 0.75\ \mu\text{m}$). Observe that C_{gd} does not go to 0 in saturation because the overlap component $C_{gd(\text{overlap})}$ is significant. Overlap capacitance becomes relatively more important for shorter channel transistors because it is a larger fraction of the total.

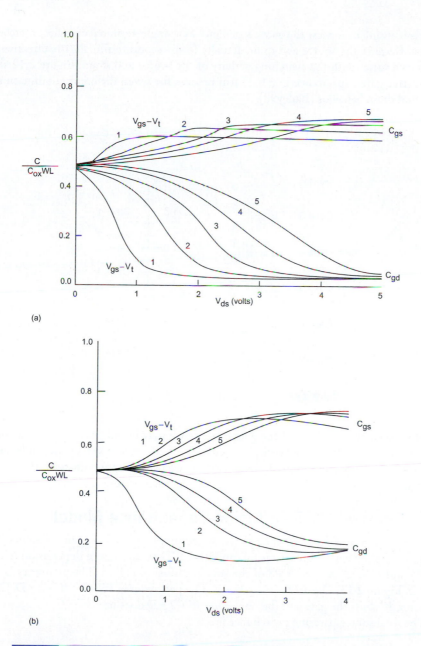

FIG 2.11 Total gate capacitance of an MOS transistor as a function of V_{ds}. ©IEEE 1987.

It is convenient to view the gate capacitance as a single-terminal capacitor attached to the gate. Because the source and drain actually form second terminals, the effective gate capacitance varies with the switching activity of the source and drain. Figure 2.12 shows the effective gate capacitance in a 0.35 μm process for seven different combinations of source and drain behavior [Bailey98].

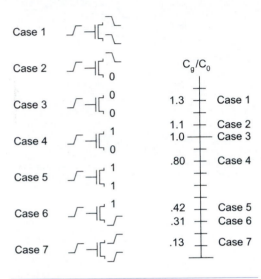

FIG 2.12 Data-dependent gate capacitance

More accurate modeling of the gate capacitance may be achieved by using a charge-based model [Cheng99]. The overlap capacitance also displays a voltage dependence. For the purpose of delay calculation of digital circuits, we usually approximate $C_g = C_{gs} + C_{gd} + C_{gb} \sim C_0$.

2.3.3 Detailed MOS Diffusion Capacitance Model

As mentioned in Section 2.3.1, the reverse-biased p–n junction between the source diffusion and the body contributes parasitic capacitance. The capacitance depends on both the *area AS* and *sidewall perimeter PS* of the source diffusion region. The geometry is illustrated in Figure 2.13. The area is $AS = W \cdot D$. The perimeter is $PS = 2 \cdot W + 2 \cdot D$. Of this perimeter, W abuts the gate and the remaining $W + 2 \cdot D$ does not.

The total source parasitic capacitance is

$$C_{sb} = AS \cdot C_{jbs} + PS \cdot C_{jbssw} \tag{2.16}$$

where C_{jbs} has units of capacitance/area and C_{jbssw} has units of capacitance/length.

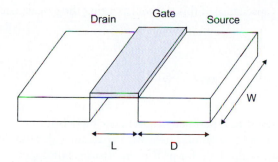

FIG 2.13 Diffusion region geometry

Because the depletion region thickness depends on the reverse bias, these parasitics are nonlinear. The area junction capacitance term is

$$C_{jbs} = C_J \left(1 + \frac{V_{sb}}{\psi_0}\right)^{-M_J} \tag{2.17}$$

C_J is the junction capacitance at zero bias and is highly process-dependent. M_J is the *junction grading coefficient*, typically in the range of 0.5 to 0.33 depending on the abruptness of the diffusion junction. ψ_0 is the *built-in potential* that depends on doping levels.

$$\psi_0 = v_T \ln \frac{N_A N_D}{n_i^2} \tag{2.18}$$

v_T is the *thermal voltage* from thermodynamics, not to be confused with the threshold voltage V_t. It has a value equal to kT/q (26 mV at room temperature), where $k = 1.380 \cdot 10^{-23}$ J/K is Boltzmann's constant, T is absolute temperature (300 K at room temperature), and q $= 1.602 \cdot 10^{-19}$ C is the charge of an electron. N_A and N_D are the doping levels of the body and source diffusion region. n_i is the intrinsic carrier concentration in undoped silicon and has a value of $1.45 \cdot 10^{10}$ cm^{-3} at 300 K.

The sidewall capacitance term is of a similar form but uses different coefficients.

$$C_{jbssw} = C_{JSW} \left(1 + \frac{V_{sb}}{\psi_0}\right)^{-M_{JSW}} \tag{2.19}$$

The capacitance contributed by the sidewall facing the channel can be modified somewhat by the presence of the channel depletion region and the modified doping profiles. In some

SPICE models, the capacitance of this sidewall abutting the gate is specified with another set of parameters:

$$C_{jbsswg} = C_{JSWG}\left(1 + \frac{V_{sb}}{\psi_0}\right)^{-M_{JSWG}} \tag{2.20}$$

Section 5.3.4 discusses SPICE perimeter capacitance models further.

The drain diffusion has a similar parasitic capacitance dependent on AD, PD, and V_{db}. Equivalent relationships hold for pMOS transistors, but doping levels differ. As the capacitances are voltage-dependent, the most useful information to digital designers is the value averaged across a switching transition. This is the C_{sb} or C_{db} value that was presented in Section 2.3.1. Analog designers must minimize the consequences of these variations by using good circuit design.

Example

Calculate the diffusion parasitic C_{db} of the drain of a unit-sized contacted nMOS transistor in a 180 nm process when the drain is at 0 and at V_{DD} = 1.8 V. Assume the substrate is grounded. The transistor characteristics are CJ = 0.98 fF/μm^2, MJ = 0.36, $CJSW$ = 0.22 fF/μm, $CJSWG$ = 0.33 fF/μm, $MJSW$ = $MJSWG$ = 0.10, and ψ_0 = 0.75 V at room temperature.

Solution: From Figure 2.9 we find a unit-size diffusion contact is 4 x 5 λ, or 0.36 x 0.45 μm. The area is 0.162 μm^2 and perimeter is 1.26 μm plus 0.36 μm along the gate. At zero bias, C_{jbs} = 0.98 fF/μm^2, C_{jbssw} = 0.22 fF/μm, and C_{jbsswg} = 0.33 fF/μm. Hence the total capacitance is

$$C_{db}(0\text{ V}) = \left(0.162\ \mu\text{m}^2\right)\left(0.98\ \tfrac{\text{fF}}{\mu\text{m}^2}\right) + $$
$$\left(1.26\ \mu\text{m}\right)\left(0.22\ \tfrac{\text{fF}}{\mu\text{m}}\right) + \left(0.36\ \mu\text{m}\right)\left(0.33\ \tfrac{\text{fF}}{\mu\text{m}}\right) = 0.55\ \text{fF} \tag{2.21}$$

At a drain voltage of V_{DD}, the capacitance reduces to

$$C_{db}(1.8\text{ V}) = \left(0.162\ \mu\text{m}^2\right)\left(0.98\ \tfrac{\text{fF}}{\mu\text{m}^2}\right)\left(1 + \frac{1.8}{0.75}\right)^{-0.36} + $$
$$\left[\left(1.26\ \mu\text{m}\right)\left(0.22\ \tfrac{\text{fF}}{\mu\text{m}}\right) + \left(0.36\ \mu\text{m}\right)\left(0.33\ \tfrac{\text{fF}}{\mu\text{m}}\right)\right]\left(1 + \frac{1.8}{0.75}\right)^{-0.10} = 0.44\ \text{fF} \tag{2.22}$$

For the purpose of manual performance estimation, this nonlinear capacitance is too much effort. An effective capacitance averaged over the switching range is quite satisfactory for digital applications.

Diffusion regions were historically used for short wires called *runners* in processes with limited numbers of metal levels. Diffusion capacitance and resistance are large enough that such practice is now discouraged.

In summary, an MOS transistor can be viewed as a four-terminal device with capacitances between each terminal pair as shown in Figure 2.14. The gate capacitance includes an intrinsic component (to the body, source and drain, or source alone, depending on operating regime) and overlap terms with the source and drain. The source and drain have parasitic diffusion capacitance to the body.

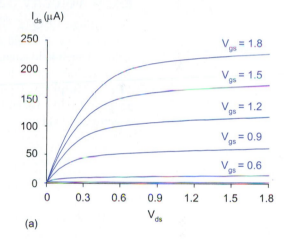

FIG 2.14 Capacitances of an MOS transistor

2.4 Nonideal I-V Effects

The ideal I-V model of EQs (2.5) and (2.8) neglects many effects that are important to modern devices. Figure 2.15 shows the simulated I-V characteristics of a unit nMOS transistor in a 180 nm process.

Compare the characteristics in the linear and saturation regimes (Figure 2.15(a)) to those of the ideal device in Figure 2.7. The saturation current increases less than quadratically with increasing V_{gs}. This is caused by two effects: *velocity saturation* and *mobility degradation*. At high lateral field strengths (V_{ds}/L), carrier velocity ceases to increase linearly with field strength. This is called velocity saturation and results in lower I_{ds} than expected at high V_{ds}. At high vertical field strengths (V_{gs}/t_{ox}), the carriers scatter more often. This mobility degradation effect also leads to less current than expected at high V_{gs}. The saturation current of the nonideal transistor increases slightly with V_{ds}. This is caused by *channel length modulation*, in which higher V_{ds} increases the size of the depletion region around the drain and thus effectively shortens the channel.

There are several sources of leakage resulting in current flow in nominally OFF transistors. Observe in Figure 2.15(b) that at $V_{gs} < V_t$, the current drops off exponentially rather than abruptly becoming zero. This is called *subthreshold conduction*. The threshold voltage itself is influenced by the voltage difference between the source and body; this is called the body effect. The source and drain diffusions are reverse-biased diodes and also experience *junction leakage* into the substrate or well. The current into the gate I_g is ideally 0. However, as the thickness of gate oxides reduces to only a small number of atomic layers, electrons *tunnel* through the gate, causing some gate current.

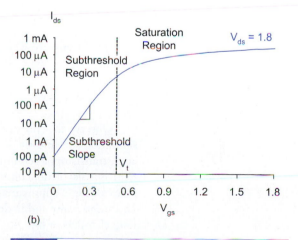

FIG 2.15 Simulated I-V characteristics

Both mobility and threshold voltage decrease with rising temperature. The mobility effect is most important for ON transistors, resulting in lower I_{ds} at high temperature. The threshold effect is most important for OFF transistors, resulting in higher leakage current at high temperature. Clearly, MOS characteristics degrade with temperature.

It is useful to have a qualitative understanding of nonideal effects to predict their impact on circuit behavior and to be able to anticipate how devices will change in future process generations. However, the effects lead to complicated I-V characteristics that are seldom useful for hand calculations. Instead, the effects are built into good transistor models and simulated with SPICE or similar software.

2.4.1 Velocity Saturation and Mobility Degradation

According to EQ (2.3), carrier drift velocity and hence current increase linearly with the lateral electric field $E_{\text{lat}} = V_{ds}/L$ between source and drain. This is only true for weak fields; at high field strength, drift velocity rolls off due to carrier scattering and eventually saturates[5] at v_{sat}, as shown in Figure 2.16. The carrier velocity may be fitted with EQ (2.23) [Murphy80, Cheng99] where E_{sat} is determined empirically. $v_{\text{sat}} = \mu E_{\text{sat}}$ is in the range of $6-10 \cdot 10^6$ cm/s for electrons and $4-8 \cdot 10^6$ cm/s for holes. This corresponds to a saturation field on the order of $2 \cdot 10^4$ V/cm for nMOS transistors.

$$v = \frac{\mu E_{\text{lat}}}{1 + \dfrac{E_{\text{lat}}}{E_{\text{sat}}}} \qquad (2.23)$$

Recall that without velocity saturation, the saturation current is

$$I_{ds} = \mu C_{\text{ox}} \frac{W}{L} \frac{\left(V_{gs} - V_t\right)^2}{2} \qquad (2.24)$$

If the transistor were completely velocity saturated, $v = v_{\text{sat}}$ and the saturation current becomes [Bakoglu90]

$$I_{ds} = C_{\text{ox}} W \left(V_{gs} - V_t\right) v_{\text{sat}} \qquad (2.25)$$

Observe that the drain current is quadratically dependent on voltage without velocity saturation and linearly dependent when fully velocity saturated. For moderate supply voltages, transistors operate in a region where the velocity no longer increases linearly with field, but also is not completely saturated. The α-*power law model* given in EQ (2.26) provides a simple approximation to capture this behavior [Sakurai90]. α is called the *velocity saturation index* and is determined by curve fitting measured I-V data. Transistors with long channels or low V_{DD} display quadratic I-V characteristics in saturation and are mod-

---

[5]Do not confuse the *saturation* region of transistor operation (where $V_{ds} > V_{gs} - V_t$) with *velocity saturation* (where $E_{\text{lat}} = V_{ds}/L$ approaches E_{sat}). In this text, the word "saturation" alone refers to the operating region while "velocity saturation" refers to the limiting of carrier velocity at high field.

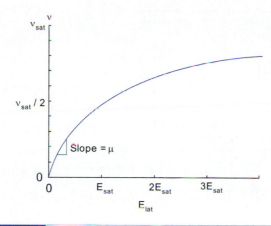

FIG 2.16 Carrier velocity vs. electric field

eled with $\alpha = 2$. As transistors become more velocity saturated, increasing V_{gs} has less effect on current and α decreases, reaching 1 for transistors that are completely velocity saturated. For simplicity, the model uses a straight line in the linear region. Overall, the model is based on three parameters that can be determined empirically from a curve fit of I-V characteristics: α, βP_c, and P_v.

$$I_{ds} = \begin{cases} 0 & V_{gs} < V_t & \text{cutoff} \\ I_{dsat} \dfrac{V_{ds}}{V_{dsat}} & V_{ds} < V_{dsat} & \text{linear} \\ I_{dsat} & V_{ds} > V_{dsat} & \text{saturation} \end{cases}$$ (2.26)

where

$$I_{dsat} = P_c \frac{\beta}{2} \left(V_{gs} - V_t \right)^{\alpha}$$

$$V_{dsat} = P_v \left(V_{gs} - V_t \right)^{\alpha/2}$$ (2.27)

As channel lengths become shorter, the lateral field increases and transistors become more velocity saturated (α closer to 1) if the supply voltage is held constant. For example, a transistor with a 2 μm channel length begins to show the effects of velocity saturation at V_{DD} above 4 V, while a 0.18 μm long transistor begins to experience velocity saturation above 0.36 V. Figure 2.17 compares I_{ds} for a velocity-saturated nMOS transistor with that of an ideal (Shockley model) transistor from Figure 2.17 and with that predicted by the α-power law. The Shockley model grossly overpredicts current at high voltage but the α-power fit is reasonably good. As the transistor becomes severely velocity saturated, there is no performance benefit to raising V_{DD}.

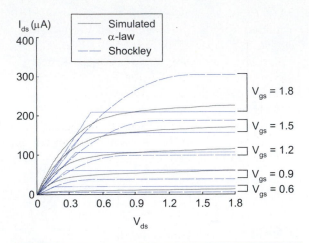

FIG 2.17 I-V characteristics for nMOS transistor with velocity saturation

The low-field mobility of holes is much lower than that of electrons, so pMOS transistors experience less velocity saturation than nMOS for a given V_{DD}. This shows up as a larger value of α for pMOS than for nMOS transistors.

Strong vertical electric fields resulting from large V_{gs} cause the carriers to scatter against the surface and also reduce the carrier mobility μ. This effect is called *mobility degradation*. It can be modeled by replacing μ with a smaller μ_{eff}. The α-power law captures this effect in the parameter α.

2.4.2 Channel Length Modulation

Ideally, I_{ds} is independent of V_{ds} for a transistor in saturation, making the transistor a perfect current source. As discussed in Section 2.3.3, the reverse-biased p–n junction between the drain and body forms a depletion region with a width L_d that increases with V_{db}. The depletion region effectively shortens the channel length to

$$L_{\text{eff}} = L - L_d \tag{2.28}$$

To avoid introducing the body voltage into our calculations, assume the source voltage is close to the body voltage so $V_{db} \sim V_{ds}$. Hence, increasing V_{ds} decreases the effective channel length. Shorter channel length results in higher current; thus I_{ds} increases with V_{ds} in saturation as shown in Figure 2.18. This can be crudely modeled by multiplying EQ (2.10) by a factor of $(1 + \lambda V_{ds})$ [Gray01]. In the saturation region, we find

$$I_{ds} = \beta \frac{\left(V_{gs} - V_t\right)^2}{2}\left(1 + \lambda V_{ds}\right) \tag{2.29}$$

The parameter λ is an empirical channel length modulation factor that should not be confused with the same symbol used in layout design rules. As channel length gets shorter, the effect of the channel length modulation becomes relatively more important. Hence λ is inversely dependent on channel length. This λ channel length modulation model is a gross oversimplification of nonlinear behavior and is more useful for conceptual understanding than for accurate device modeling.

Channel length modulation is very important to analog designers because it reduces the gain of amplifiers. It is generally unimportant for qualitatively understanding the behavior of digital circuits.

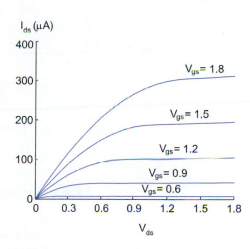

FIG 2.18 I-V characteristics of nMOS transistor with channel length modulation

2.4.3 Body Effect

Until now, we have considered a transistor to be a three-terminal device with gate, source, and drain. However, the body is an implicit fourth terminal. The potential difference between the source and body V_{sb} affects the threshold voltage. The threshold voltage can be modeled as

$$V_t = V_{t0} + \gamma \left(\sqrt{\phi_s + V_{sb}} - \sqrt{\phi_s} \right) \qquad (2.30)$$

where V_{t0} is the threshold voltage when the source is at the body potential, ϕ_s is the *surface potential* at threshold (see a device physics text for further discussion of surface potential), and γ is the *body effect coefficient*, typically in the range 0.4 to 1 $V^{1/2}$. These in turn depend on the doping level N_A. Section 4.4.3.2 will describe how a body bias can intentionally be applied to alter the threshold voltage, permitting tradeoffs between performance and sub-threshold leakage current.

$$\phi_s = 2v_T \ln \frac{N_A}{n_i} \qquad (2.31)$$

$$\gamma = \frac{t_{ox}}{\varepsilon_{ox}} \sqrt{2q\varepsilon_{si}N_A} = \frac{\sqrt{2q\varepsilon_{si}N_A}}{C_{ox}} \qquad (2.32)$$

Example

Consider the nMOS transistor in a 180 nm process with a nominal threshold voltage of 0.4 V and doping level of $8 \cdot 10^{17}$ cm^{-3}. The body is tied to ground with a substrate contact. How much does the threshold change at room temperature if the source is at 1.1 V instead of 0?

Solution: At room temperature the thermal voltage $v_T = kT/q = 26$ mV and $n_i = 1.45 \cdot 10^{10}$ cm^{-3}. The threshold increases by 0.28 V.

$$\phi_s = 2(0.026 \text{ V})\ln\frac{8 \cdot 10^{17} \text{ cm}^{-3}}{1.45 \cdot 10^{10} \text{ cm}^{-3}} = 0.93 \text{ V}$$

$$\gamma = \frac{40 \cdot 10^{-8} \text{ cm}}{3.9 \cdot 8.85 \cdot 10^{-14} \frac{\text{F}}{\text{cm}}}\sqrt{2\left(1.6 \cdot 10^{-19} \text{ C}\right)\left(11.7 \cdot 8.85 \cdot 10^{-14} \frac{\text{F}}{\text{cm}}\right)\left(8 \cdot 10^{17} \text{ cm}^{-3}\right)} = 0.60 \text{ V}^{\frac{1}{2}} \qquad \textbf{(2.33)}$$

$$V_t = 0.4 \text{ V} + \gamma\left(\sqrt{\phi_s + 1.1 \text{ V}} - \sqrt{\phi_s}\right) = 0.68 \text{ V}$$

2.4.4   Subthreshold Conduction

The ideal transistor I-V model assumes current only flows from source to drain when V_{gs} > V_t. In real transistors, current does not abruptly cut off below threshold, but rather drops off exponentially as given in EQ (2.34) [Sheu87b, Cheng99]. This conduction is also known as *leakage* and often results in undesired current when a transistor is nominally OFF. I_{ds0} is the current at threshold and is dependent on process and device geometry; the $e^{1.8}$ term was found empirically. n is a process-dependent term affected by the depletion region characteristics and is typically in the range of 1.4–1.5 for CMOS processes. The final term indicates that leakage is 0 if $V_{ds} = 0$, but increases to its full value when V_{ds} is a few multiples of the thermal voltage v_T (e.g., when V_{ds} > 50 mV). Figure 2.15(b) shows the I-V characteristics on a logarithmic scale illustrating both normal and subthreshold conduction.

$$I_{ds} = I_{ds0}e^{\frac{V_{gs}-V_t}{nv_T}}\left(1 - e^{\frac{-V_{ds}}{v_T}}\right) \qquad \textbf{(2.34)}$$

$$I_{ds0} = \beta v_T^2 e^{1.8} \qquad \textbf{(2.35)}$$

Subthreshold conduction is used to advantage in very low-power analog circuits. It is also particularly important for dynamic circuits and DRAMs, which depend on the storage of charge on a capacitor. Conduction through an OFF transistor discharges the capacitor unless it is periodically refreshed or a trickle of current is available to counter the leakage. Leakage also contributes to power dissipation in idle circuits. Leakage increases exponentially as V_t decreases or as temperature rises, so it is becoming a major problem for chips using low supply and threshold voltages.

Subthreshold conduction is exacerbated by *drain-induced barrier lowering* (DIBL) in which a positive V_{ds} effectively reduces V_t. This effect is especially pronounced in short-channel transistors. It can be modeled as

$$V_t' = V_t - \eta V_{ds} \qquad \textbf{(2.36)}$$

Example

What is the minimum threshold voltage for which the leakage current through an OFF transistor ($V_{gs} = 0$) is 10^3 times less than that of a transistor that is barely ON ($V_{gs} = V_t$) at room temperature if $n = 1.5$? One of the advantages of silicon-on-insulator (SOI) processes is that they have smaller n (see Section 6.7). What threshold is required for SOI if $n = 1.3$?

Solution: $v_T = 26$ mV at room temperature. Assume $V_{ds} \gg v_T$ so leakage is significant. We solve

$$I_{ds}\left(V_{gs} = 0\right) = 10^{-3} I_{ds0} \approx I_{ds0} e^{\frac{-V_t}{nv_T}} \tag{2.37}$$

$$V_t = -nv_T \ln 10^{-3} = 270 \text{ mV}$$

In the CMOS process, leakage rolls off by a factor of 10 for every 90 mV V_{gs} falls below threshold. This is often quoted as a *subthreshold slope* of $S = 90$ mV/decade. In the SOI process, the subthreshold slope S is 78 mV/decade, so a threshold of only 234 mV is required.

where η is the DIBL coefficient, typically in the range of 0.02–0.1.

[Bowman99] extends the α-power law model to a *physical α-power law model* that includes subthreshold conduction as well as velocity saturation effects.

2.4.5 Junction Leakage

The p–n junctions between diffusion and the substrate or well form diodes, as shown in Figure 2.19. The well-to-substrate junction is another diode. The substrate and well are tied to GND or V_{DD} to ensure these diodes remain reverse-biased. However, reverse-biased diodes still conduct a small amount of current I_D.[6]

$$I_D = I_S \left(e^{\frac{V_D}{v_T}} - 1 \right) \tag{2.38}$$

where I_S depends on doping levels and on the area and perimeter of the diffusion region and V_d is the diode voltage (e.g., V_{sb} or V_{db}). When a junction is reverse biased by significantly more than the thermal voltage, the leakage is just $-I_S$, generally in the 0.1–0.01 fA/μm^2 range.

---

[6]Beware that I_D and I_S stand for the diode current and diode reverse-biased saturation currents, respectively. The D and S are not related to drain or source.

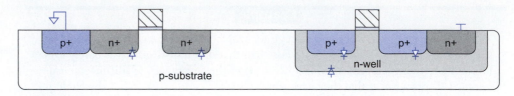

FIG 2.19 Reverse-biased diodes in CMOS circuits

Junction leakage was historically a limiter of storage time on dynamic nodes. In modern transistors with low threshold voltages, subthreshold conduction far exceeds junction leakage.

2.4.6 Tunneling

According to quantum mechanics, there is a finite probability that carriers will tunnel through the gate oxide. This results in *gate leakage* current flowing into the gate. The probability of tunneling drops off exponentially with oxide thickness, and so was negligible until recently.

For oxides thinner than about 15–20 Å, tunneling current becomes a factor and may become comparable to subthreshold leakage in advanced processes. Figure 2.20 plots gate leakage current density J_G against voltage for various oxide thicknesses. Large tunneling currents impact not only dynamic nodes but also quiescent power consumption and thus may limit oxide thicknesses t_{ox} to no less than about 8Å [Song01]. To keep dimensions in perspective, recall that each atomic layer of SiO_2 is about 3 Å, so gate oxides have scaled to only a handful of atomic layers thick. High C_{ox} is important for good transistors, so one research direction is to use an alternative gate insulator with a higher dielectric constant ϵ. A key challenge is finding materials that form a high-quality interface with silicon; one contender is silicon nitride (Si_3N_4) with a dielectric constant of 7.8. Tunneling can purposely be used to create electrically erasable memory devices (see Section 11.3).

Tunneling current is an order of magnitude higher for nMOS than pMOS transistors with SiO_2 gate dielectrics because the electrons tunnel from the conduction band while the holes tunnel from the valence band and see a higher barrier [Hamzaoglu02]. Different dielectrics may have different tunneling properties.

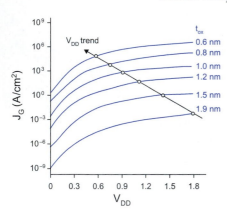

FIG 2.20 Gate leakage current from [Song01]

2.4.7 Temperature Dependence

Transistor characteristics are influenced by temperature [Cobbold66, Vadasz66, Tsividis99, Gutierrez01]. Carrier mobility decreases with temperature. An approximate relation is

$$\mu(T) = \mu(T_r)\left(\frac{T}{T_r}\right)^{-k_\mu} \tag{2.39}$$

where T is the absolute temperature, T_r is room temperature, and k_μ is a fitting parameter generally in the range of 1.2–2.0.

The magnitude of the threshold voltage decreases nearly linearly with temperature and may be approximated by

$$V_t(T) = V_t(T_r) - k_{vt}(T - T_r) \tag{2.40}$$

where k_{vt} is typically in the range of 0.5 to 3.0 mV/K.

Junction leakage also increases with temperature because I_S is strongly temperature dependent.

The combined temperature effects are shown in Figure 2.21, where ON current decreases and OFF current increases with temperature. Similarly, Figure 2.22 shows how the ON current I_{dsat} decreases with temperature. Therefore, circuit performance is generally worst at high temperature. This is called a *negative temperature coefficient*.

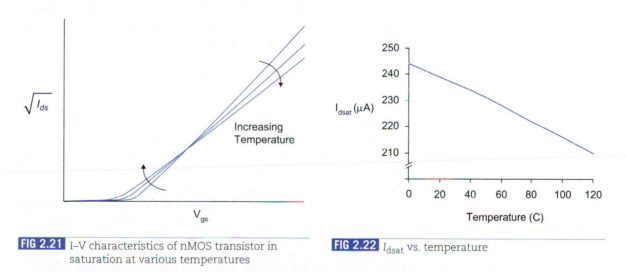

FIG 2.21 I–V characteristics of nMOS transistor in saturation at various temperatures

FIG 2.22 I_{dsat} vs. temperature

Conversely, circuit performance can be improved by cooling. Most systems use natural convection or fans in conjunction with heat sinks, but water cooling, thin-film refrigerators, or even liquid nitrogen can increase performance if the expense is justified. There are many advantages of operating at low temperature [Keyes70, Sun87]. Subthreshold leakage is exponentially dependent on temperature, so lower threshold voltages can be used. Velocity saturation occurs at higher fields, providing more current. As mobility is also higher, these fields are reached at a lower power supply, saving power. Depletion regions become wider, resulting in less junction capacitance. Most wearout mechanisms are temperature-dependent, so transistors are more reliable. However, at low temperatures, transistors break down at lower voltages.

Two popular lab tools for determining temperature dependence in circuits are a can of freeze spray and a heat gun. The former can be used to momentarily "freeze" a chip to see whether performance alters and the other, of course, can be used to heat a chip up. Often these tests are done to quickly determine whether a chip is prone to temperature effects. Be careful—sometimes the sudden temperature change can fracture chips or their packages.

2.4.8 Geometry Dependence

The layout designer draws transistors with width and length W_{drawn} and L_{drawn}. The actual gate dimensions may differ by some factors X_W and X_L. For example, the manufacturer may create masks with narrower polysilicon or may overetch the polysilicon to provide shorter channels (negative X_L) without changing the overall design rules or metal pitch. Moreover, the source and drain tend to diffuse laterally under the gate by L_D, producing a shorter effective channel length that the carriers must traverse between source and drain. Similarly, diffusion of the bulk by W_D decreases the effective channel width. Putting these factors together, we can compute effective transistor lengths and widths that should be used in place of L and W in the current and capacitance equations given elsewhere in the book. The factors of two come from lateral diffusion on both sides of the channel.

$$
\begin{aligned}
L_{\text{eff}} &= L_{\text{drawn}} + X_L - 2L_D \\
W_{\text{eff}} &= W_{\text{drawn}} + X_W - 2W_D
\end{aligned}
\tag{2.41}
$$

Therefore, a transistor drawn twice as long may have an effective length that is more than twice as great. Similarly, two transistors differing in drawn widths by a factor of two may differ in saturation current by more than a factor of two. Threshold voltages also vary somewhat with transistor dimensions [Cheng99].

Recall that long transistors also experience less channel length modulation. Combining the threshold, effective channel length, and channel length modulation effects, a transistor of twice minimum length usually delivers substantially less than half the current of a minimum length device. In general, when currents must be precisely matched (e.g., in sense amplifiers or A/D converters), it is best to use the same width and length for each device. Current ratios can be produced by tying several identical transistors in parallel.

In processes below 0.25 µm, the effective length of the transistor also depends significantly on the orientation of the transistor. Moreover, the amount of nearby polysilicon also affects etch rates during manufacturing and thus channel length. Transistors that must match well should have the same orientation. Dummy polysilicon wires can be placed nearby to improve etch uniformity.

2.4.9 Summary

Although the physics of very small transistors is complicated, the impact of nonideal I-V behavior is fairly easy to understand from the designer's viewpoint.

Threshold drops   Pass transistors suffer a threshold drop when passing the wrong value: nMOS transistors only pull up to $V_{DD} - V_{tn}$, while pMOS transistors only pull down to $|V_{tp}|$. The magnitude of the threshold drop is increased by the body effect. Therefore, pass transistors do not operate very well in modern processes where the threshold voltage is a significant fraction of the supply voltage. Fully complementary transmission gates should be used where both 0's and 1's must be passed well.

Leakage current   Ideally, complementary CMOS gates draw zero current and dissipate zero power when idle. Real gates draw some leakage current. The most important source at this time is subthreshold leakage between source and drain of a transistor that should be cut off. The subthreshold current of an OFF transistor decreases by an order of magnitude for every 60–100 mV that V_{gs} is below V_t. Threshold voltages have been decreasing, so subthreshold leakage has been increasing dramatically. Some processes offer multiple choices of V_t: low-V_t devices are used for high performance in critical circuits, while high-V_t devices are used for low leakage elsewhere.

The transistor gate is a good insulator. However, tunneling current flows through very thin gates. The significance of tunneling is also increasing exponentially and is becoming another important source of leakage current.

Leakage current causes CMOS gates to consume power when idle. It also limits the amount of time that data is retained in dynamic logic, latches, and memory cells. In modern processes, dynamic logic and latches require some sort of feedback to prevent data loss from leakage. Leakage increases at high temperature.

V_{DD}   Velocity saturation and mobility degradation result in less current than expected at high voltage. This means that there is no point in trying to use a high V_{DD} to achieve fast transistors, so V_{DD} has been decreasing with process generation to reduce power consumption. Moreover, the very short channels and thin gate oxides would be damaged by high V_{DD}.

Delay   Transistors in series drop part of the voltage across each transistor and thus experience smaller fields and less velocity saturation than single transistors. Therefore, series transistors tend to be a bit faster than a simple model would predict. For example, two nMOS transistors in series deliver more than half the current of a single nMOS transistor of the same width. This effect is more pronounced for nMOS transistors than pMOS transistors because nMOS transistors have higher mobility to begin with and thus are more velocity saturated.

Matching   If two transistors should behave identically, both should have the same dimensions and orientation and be interdigitated if possible.

2.5   DC Transfer Characteristics

Digital circuits are merely analog circuits used over a special portion of their range. The DC transfer characteristics of a circuit relate the output voltage to the input voltage, assuming the input changes slowly enough that capacitances have plenty of time to charge or discharge. Specific ranges of input and output voltages are defined as valid '0' and '1' logic levels. This section explores the DC transfer characteristics of CMOS gates and pass transistors.

2.5.1   Complementary CMOS Inverter DC Characteristics

Let us derive the DC transfer function (V_{out} vs. V_{in}) for the complementary CMOS inverter shown in Figure 2.23. We begin with Table 2.2, which outlines various regions of operation for the n- and p-transistors. In this table, V_{tn} is the threshold voltage of the n-channel device, and V_{tp} is the threshold voltage of the p-channel device. Note that V_{tp} is negative. The equations are given both in terms of V_{gs}/V_{ds} and V_{in}/V_{out}. As the source of the nMOS transistor is grounded, $V_{gsn} = V_{in}$ and $V_{dsn} = V_{out}$. As the source of the pMOS transistor is tied to V_{DD}, $V_{gsp} = V_{in} - V_{DD}$ and $V_{dsp} = V_{out} - V_{DD}$.

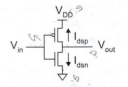

FIG 2.23 A CMOS inverter

Table 2.2	Relationships between voltages for the three regions of operation of a CMOS inverter		
	Cutoff	**Linear**	**Saturated**
nMOS	$V_{gsn} < V_{tn}$	$V_{gsn} > V_{tn}$	$V_{gsn} > V_{tn}$
	$V_{in} < V_{tn}$	$V_{in} > V_{tn}$	$V_{in} > V_{tn}$
		$V_{dsn} < V_{gsn} - V_{tn}$	$V_{dsn} > V_{gsn} - V_{tn}$
		$V_{out} < V_{in} - V_{tn}$	$V_{out} > V_{in} - V_{tn}$
pMOS	$V_{gsp} > V_{tp}$	$V_{gsp} < V_{tp}$	$V_{gsp} < V_{tp}$
	$V_{in} > V_{tp} + V_{DD}$	$V_{in} < V_{tp} + V_{DD}$	$V_{in} < V_{tp} + V_{DD}$
		$V_{dsp} > V_{gsp} - V_{tp}$	$V_{dsp} < V_{gsp} - V_{tp}$
		$V_{out} > V_{in} - V_{tp}$	$V_{out} < V_{in} - V_{tp}$

The objective is to find the variation in output voltage (V_{out}) as a function of the input voltage (V_{in}). This may be done graphically, analytically (see Exercise 2.16), or through simulation [Carr72]. Given V_{in}, we must find V_{out} subject to the constraint that $I_{dsn} = |I_{dsp}|$. For simplicity, we assume $V_{tp} = -V_{tn}$ and that the pMOS transistor is 2–3 times as wide as the nMOS transistor so $\beta_n = \beta_p$. We relax this assumption in Section 2.5.2.

We commence with the graphical representation of the simple algebraic equations described by EQs (2.5) and (2.8) for the two inverter transistors shown in Figure 2.24(a). The plot shows I_{dsn} and I_{dsp} in terms of V_{dsn} and V_{dsp} for various values of V_{gsn} and V_{gsp}. Figure 2.24(b) shows the same plot of I_{dsn} and $|I_{dsp}|$ now in terms of V_{out} for various values of V_{in}. The possible operating points of the inverter, marked with dots, are the values of V_{out} where $I_{dsn} = |I_{dsp}|$ for a given value of V_{in}. These operating points are plotted on V_{out} vs. V_{in} axes in Figure 2.24(c) to show the inverter DC transfer characteristics. The supply current $I_{DD} = I_{dsn} = |I_{dsp}|$ is also plotted against V_{in} in Figure 2.24(d) showing that both transistors are momentarily ON as V_{in} passes through voltages between GND and V_{DD}, resulting in a pulse of current drawn from the power supply.

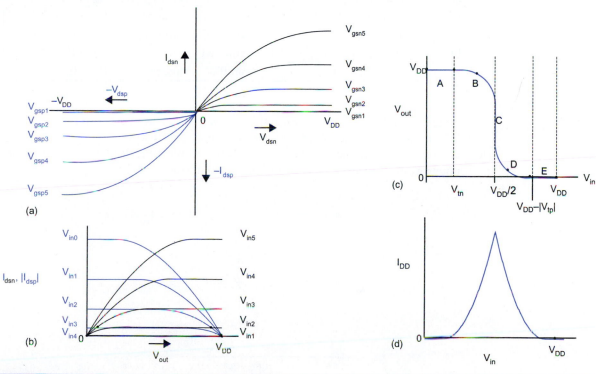

FIG 2.24 Graphical derivation of CMOS inverter DC characteristic

The operation of the CMOS inverter can be divided into five regions indicated on Figure 2.24(c). The state of each transistor in each region is shown in Table 2.3. In region A, the nMOS transistor is OFF so the pMOS transistor pulls the output to V_{DD}. In region B, the nMOS transistor starts to turn ON, pulling the output down. In region C, both transistors are in saturation. Notice that ideal transistors are only in region C for $V_{in} = V_{DD}/2$ and that the slope of the transfer curve in this example is $-\infty$ in this region, corresponding to infinite gain. Real transistors have finite output resistances on account of channel length modulation, described in Section 2.4.2, and thus have finite slopes over a broader region C. In region D, the pMOS transistor is partially ON and in region E, it is completely OFF, leaving the nMOS transistor to pull the output down to GND. Also notice that the inverter's current consumption is zero when the input is within a threshold voltage of the V_{DD} or GND rails. This feature is important for low-power operation.

Table 2.3	Summary of CMOS inverter operation			
Region	Condition	p-device	n-device	Output
A	$0 \leq V_{in} < V_{tn}$	linear	cutoff	$V_{out} = V_{DD}$
B	$V_{tn} \leq V_{in} < V_{DD}/2$	linear	saturated	$V_{out} > V_{DD}/2$
C	$V_{in} = V_{DD}/2$	saturated	saturated	V_{out} drops sharply
D	$V_{DD}/2 < V_{in} \leq V_{DD} - |V_{tp}|$	saturated	linear	$V_{out} < V_{DD}/2$
E	$V_{in} > V_{DD} - |V_{tp}|$	cutoff	linear	$V_{out} = 0$

Figure 2.25 shows simulation results of an inverter from a 180 nm process. The pMOS transistor is twice as wide as the nMOS transistor to achieve approximately equal betas. Simulation matches the simple models reasonably well, although the transition is not quite as steep on account of channel length modulation.

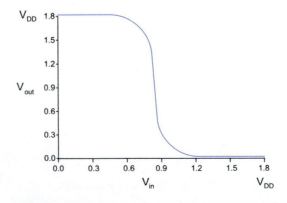

FIG 2.25 Simulated CMOS inverter DC characteristic

The crossover point where $V_{inv} = V_{in} = V_{out}$ is called the *input threshold*. Because both mobility and the magnitude of the threshold voltage decrease with temperature for nMOS and pMOS transistors, the input threshold of the gate is only weakly sensitive to temperature.

2.5.2 Beta Ratio Effects

We have seen that for $\beta_p = \beta_n$, the inverter threshold voltage V_{inv} is $V_{DD}/2$. This may be desirable because it maximizes noise margins (see Section 2.5.3) and allows a capacitive load to charge and discharge in equal times by providing equal current source and sink capabilities (see Section 4.2). Inverters with different beta ratios β_p/β_n are called *skewed* inverters [Sutherland99]. If $\beta_p/\beta_n > 1$ (e.g., 2), the inverter is *HI-skewed*. If $\beta_p/\beta_n < 1$ (e.g., 1/2) the inverter is *LO-skewed*. If $\beta_p/\beta_n = 1$, the inverter has normal skew or is *unskewed*.

A HI-skew inverter has a stronger pMOS transistor. Therefore, if the input is $V_{DD}/2$, we would expect the output will be greater than $V_{DD}/2$. In other words, the input threshold must be higher than for an unskewed inverter. Similarly a LO-skew inverter has a weaker pMOS transistor and thus a lower switching threshold.

Figure 2.26 explores the impact of skewing the beta ratio on the DC transfer characteristics. As the beta ratio is changed, the switching threshold moves. However, the output voltage transition remains sharp. Gates are usually skewed by adjusting the widths of transistors while maintaining minimum length for speed.

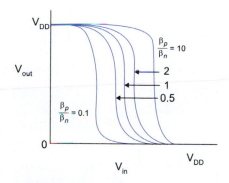

FIG 2.26 Transfer characteristics of skewed inverters

DC transfer characteristics of other complementary CMOS gates can be understood by collapsing the gates into an equivalent inverter. Series transistors can be viewed as a single transistor of greater length. If only one of several parallel transistors is ON, the other transistors can be ignored. If several parallel transistors are ON, the collection can be viewed as a single transistor of greater width.

2.5.3 Noise Margin

Noise margin is closely related to the DC voltage characteristics [Wakerly00]. This parameter allows you to determine the allowable noise voltage on the input of a gate so that the output will not be corrupted. The specification most commonly used to describe noise margin (or *noise immunity*) uses two parameters: the *LOW* noise margin, NM_L, and the *HIGH* noise margin, NM_H. With reference to Figure 2.27, NM_L is defined as the difference in maximum LOW input voltage recognized by the receiving gate and the maximum LOW output voltage produced by the driving gate.

$$NM_L = V_{IL} - V_{OL} \tag{2.42}$$

The value of NM_H is the difference between the minimum HIGH output voltage of the driving gate and the minimum HIGH input voltage recognized by the receiving gate. Thus

$$NM_H = V_{OH} - V_{IH} \tag{2.43}$$

where
V_{IH} = minimum HIGH input voltage
V_{IL} = maximum LOW input voltage
V_{OH} = minimum HIGH output voltage
V_{OL} = maximum LOW output voltage.

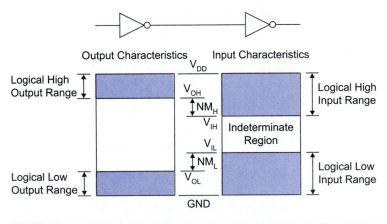

FIG 2.27 Noise margin definitions

Inputs between V_{IL} and V_{IH} are said to be in the *indeterminate region* or *forbidden zone* and do not represent legal digital logic levels. Therefore, it is generally desirable to have V_{IH} as close as possible to V_{IL} and for this value to be midway in the "logic swing," V_{OL} to

V_{OH}. This implies that the transfer characteristic should switch abruptly, that is, there should be high gain in the transition region. For the purpose of calculating noise margins, the transfer characteristic of the inverter and the definition of voltage levels V_{IL}, V_{OL}, V_{IH}, V_{OH} are shown in Figure 2.28. Logic levels are defined at the unity gain point where the slope is −1. This gives a conservative bound on the worst-case static noise margin [Hill68, Lohstroh83, Shepard99]. For the inverter shown, the NM_L is 0.46 V_{DD} while the NM_H is 0.34 V_{DD}. Note that the output is slightly degraded when the input is at its worst legal value; this is called *noise feedthrough* or *propagated noise*. The exercises at the end of the chapter examine graphical and analytical approaches of finding the logic levels and noise margins.

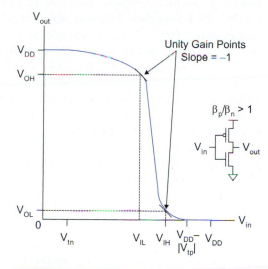

FIG 2.28 CMOS inverter noise margins

Note that if $|V_{tp}| = V_{tn}$, then NM_H and NM_L increase as threshold voltages are increased. If either NM_L or NM_H for a gate are too small (e.g., below about 0.1 V_{DD}), then the gate may be disturbed by noise that occurs on the inputs. Quite often, noise margins are compromised to improve speed. Circuit examples in Chapter 6 will illustrate this tradeoff. Noise sources tend to scale with the supply voltage, so noise margins are best given as a fraction of the supply voltage. A noise margin of 0.4 V is quite comfortable in a 1.8 V process, but marginal in a 5 V process.

DC analysis gives us the *static noise margins* specifying the level of noise that a gate may see for an indefinite duration. Larger noise pulses may be acceptable if they are brief; these are described by *dynamic noise margins* specified by a maximum amplitude as a function of the duration [Lohstroh79, Somasekhar00]. Unfortunately, there is no simple amplitude-duration product that conveniently specifies dynamic noise margins.

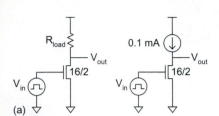

(a)

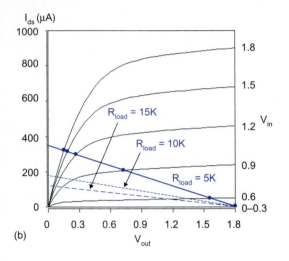

(b)

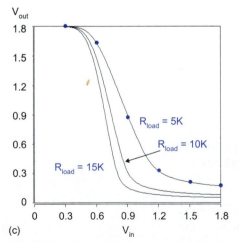

(c)

FIG 2.29 Generic nMOS inverters with resistive or constant current load

2.5.4 Ratioed Inverter Transfer Function

Apart from the complementary CMOS inverter, there are other forms of MOS inverter that can be used to build logic gates. Figure 2.29(a) shows a generic nMOS inverter that uses either a resistive load or a constant current source. For the resistor case, if we superimpose the resistor load line on the I-V characteristics of the pull-down transistor (Figure 2.29(b)), we can see that at $V_{in} = V_{DD}$, the output is some small V_{out} (V_{OL}) (Figure 2.29(c)). When $V_{in} = 0$, V_{out} rises to V_{DD}. As the resistor is made larger, the V_{OL} decreases and the current flowing when the inverter is turned on decreases. Correspondingly, as the load resistor is decreased in value, the V_{OL} rises and the ON current rises. Selection of the resistor value would seek a compromise between V_{OL}, the current drawn, and the pullup delay that increases with the value of the load resistor. Current sources have high output resistance and thus offer sharper transitions.

Neither high-value resistors nor ideal current sources are readily available in most CMOS processes. A more practical circuit called a *pseudo-nMOS inverter* is shown in Figure 2.30(a). It uses a pMOS transistor pullup or *load* that has its gate permanently grounded to approximate a constant current source. Pseudo-nMOS circuits get their name from the early nMOS technology (which preceded CMOS technology as a major systems technology) in which only nMOS transistors were available; the grounded pMOS transistor is reminiscent of a depletion mode nMOS transistor that is always ON.

The transfer characteristics may again be derived by finding V_{out} for which $I_{dsn} = |I_{dsp}|$ for a given V_{in}, as shown in Figure 2.30(b) and Figure 2.30(c). The beta ratio affects the shape of the transfer characteristics and the V_{OL} of the inverter. Larger pMOS transistors offer faster rise times but less sharp transfer characteristics. Figure 2.30(d) shows that when the nMOS transistor is turned on, a constant DC current flows in the circuit.

The gates in this section are called *ratioed* circuits because the transfer function depends on the ratio of the strength of the pull-down transistor to the pullup device. The resistor, current source, or ON transistor is sometimes called a *static load*. It is possible to construct other ratioed circuits such as NAND or NOR gates by replacing the pullup transistors with a single pullup device. Unlike complementary circuits, the ratio must be chosen so the circuit operates correctly despite any variations from nominal component values that may occur during manufacturing. Moreover, ratioed circuits dissipate power continually in certain states (e.g., when the output is low) and have poorer noise margins than complementary circuits.

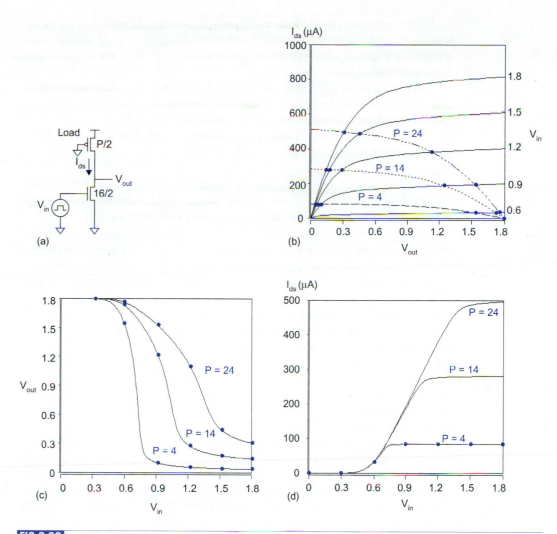

FIG 2.30 Pseudo-nMOS inverter and DC transfer characteristics

Therefore, ratioed circuits tend to be used only in very limited circumstances where they offer critical benefits such as smaller area or reduced input capacitance. We will return to ratioed circuits in Section 6.2.2.

2.5.5   Pass Transistor DC Characteristics

Recall from Section 1.4.6 that nMOS transistors pass '0's well but '1's poorly. We are now ready to better define "poorly." Figure 2.31(a) shows an nMOS transistor with the gate and drain tied to V_{DD}. Imagine that the source is initially at $V_s = 0$. $V_{gs} > V_{tn}$, so the transis-

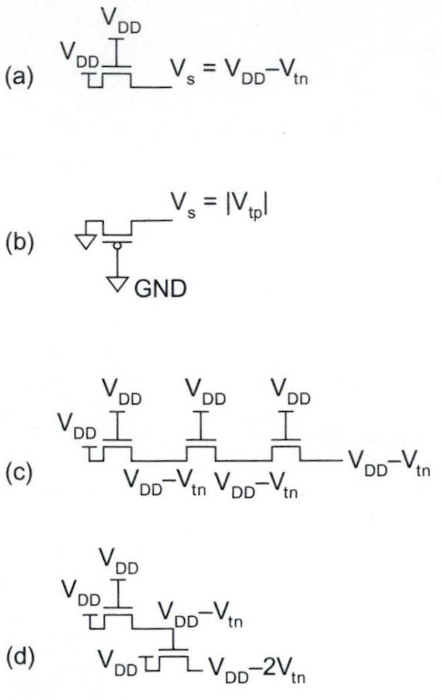

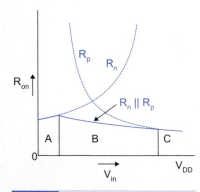

FIG 2.31 Pass transistor threshold drops

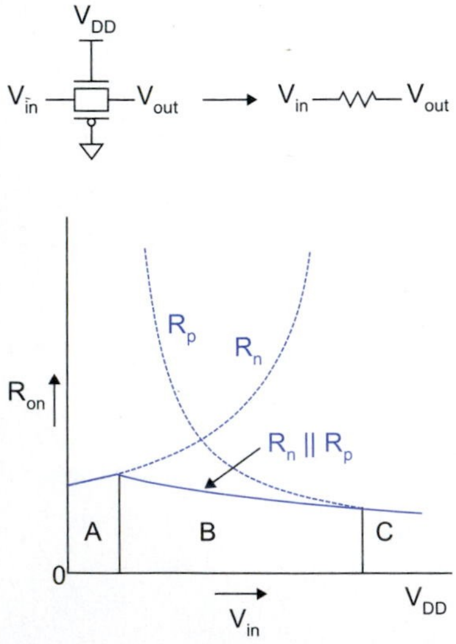

FIG 2.32 Resistance of a transmission gate as a function of input voltage

tor is ON and current flows. If the voltage on the source rises to $V_s = V_{DD} - V_{tn}$, V_{gs} falls to V_{tn} and the transistor cuts itself OFF. Therefore, nMOS transistors attempting to pass a '1' never pull the source above $V_{DD} - V_{tn}$.[7] This loss is sometimes called a *threshold drop*.

Moreover, when the source of the nMOS transistor rises, V_{sb} becomes nonzero. As described in Section 2.4.2, this nonzero source to body potential introduces the body effect that increases the threshold voltage. Using the data from the example in that section, a pass transistor driven with $V_{DD} = 5$ V would produce an output of only 3.34 V, potentially violating the noise margins of the next stage.

Similarly, pMOS transistors pass '1's well but '0's poorly. If the pMOS source drops below $|V_{tp}|$, the transistor cuts off. Hence, pMOS transistors only pull down to within a threshold above GND, as shown in Figure 2.31(b).

As the source can rise to within a threshold voltage of the gate, the output of several transistors in series is no more degraded than that of a single transistor (Figure 2.31(c)). However, if a degraded output drives the gate of another transistor, the second transistor can produce an even further degraded output (Figure 2.31(d)).

Also recall from Section 1.4.6 that a transmission gate consists of an nMOS transistor and a pMOS transistor in parallel with gates controlled by complementary signals. When the transmission gate is ON, at least one of the two transistors is ON for any output voltage and hence the transmission gate passes both '0's and '1's well. The transmission gate is a fundamental and ubiquitous component in MOS logic. It finds use as a multiplexing element, a logic structure, a latch element, and an analog switch. The transmission gate acts as a voltage-controlled resistor connecting the input and the output.

Figure 2.32 plots the transmission gate ON resistance as the input voltage is swept from GND to V_{DD}, assuming the output voltage closely follows. In region *A*, the nMOS transistor is operating linearly and the pMOS is cut off. In region *B*, both transistors are linear. In region *C*, the nMOS transistor is cut off and the pMOS is linear. If both transistors are of equal size, the characteristics are slightly asymmetric because of the better mobility of the nMOS transistor. The effective ON resistance is the parallel combination of the two resistances and is relatively constant across the full range of input voltages.

2.5.6 Tristate Inverter

By cascading a transmission gate with an inverter, the tristate inverter shown in Figure 2.33(a) is constructed. When $EN = 0$ and $ENb = 1$, the output of the inverter is in a tristate condition (the *Y* output is not driven by the *A* input). When $EN = 1$ and $ENb = 0$, the *Y* output is equal to the complement of *A*. The connection between the n- and p-driver transistors can be omitted (Figure

---

[7]Technically, the output can rise higher very slowly on account of subthreshold current.

2.33(b-c)) and the operation remains substantially the same. For the same size n- and p-devices, this tristate inverter is approximately half the speed of a complementary CMOS inverter. The tristate inverter forms the basis for various types of clocked logic, latches, bus drivers, multiplexers, and I/O structures. The circuit in Figure 2.33(d) interchanges the A and enable terminals. It is logically equivalent but electrically inferior because if the output is tristated but A toggles, charge from the internal nodes may disturb the floating output node (see Section 6.3.4 for more discussion of charge sharing).

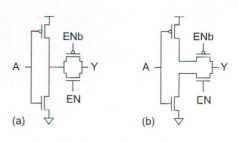

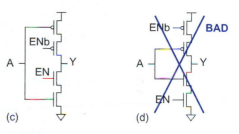

FIG 2.33 Tristate inverter

2.6 Switch-level RC Delay Models

It is very useful to be able to estimate the delay of a gate or compare circuit topologies without resorting to circuit simulation. RC delay models provide a means to make such approximate calculations. They approximate the nonlinear transistor I-V and C-V characteristics with an average resistance and capacitance over the switching range of the gate; this works remarkably well for delay estimation despite its obvious problems in predicting detailed analog behavior.

The RC delay model treats transistors as switches in series with resistors. Define a unit nMOS transistor to have *effective* resistance R. The size of the unit transistor is arbitrary but conventionally refers to a transistor with minimum length and minimum contacted diffusion width (i.e., 4/2 λ). An nMOS transistor of k times unit width has resistance R/k. A unit pMOS transistor has greater resistance, generally in the range of $2R$ – $3R$, because of its lower mobility. Throughout this book we will use $2R$ for examples to keep arithmetic simple.

The resistance at some operating point can be defined as:

$$R = \left(\frac{\partial I_{ds}}{\partial V_{ds}} \right)^{-1} \qquad (2.44)$$

If V_{ds} is small and the transistor is operating in the linear region, the resistance can be approximated by differentiating EQ (2.5) as

$$R = \frac{1}{\beta\left(V_{gs} - V_t\right)} = \frac{1}{\mu C_{ox}} \frac{L}{W} \frac{1}{\left(V_{gs} - V_t\right)} \qquad (2.45)$$

This formula is not particularly useful for calculations because the current rolls off (and resistance increases) as V_{ds} increases; we are more interested in some average current over the switching transition. Moreover, velocity saturation and other effects are important in modern transistors, so estimating resistance from fundamental principles is not very practical. Nevertheless, EQ (2.45) shows that resistance is proportional to L/W and decreases with V_{gs}. A better way to determine resistance is to simulate a transistor driving a known capacitance and measure the time constant (see Section 5.4.5).

Each transistor also has gate and diffusion capacitance. We define C to be the gate capacitance of a unit transistor of either flavor. A transistor of k times unit width has capacitance kC. Diffusion capacitance depends on the size of the source/drain region. Using the approximations from Section 2.3.1, we assume the source or drain of a unit transistor to also have capacitance C. Wider transistors have proportionally greater diffusion capacitance. As discussed in Section 2.3.1, C is typically in the range of 1.5–2 fF/µm of transistor width for a minimum length transistor (2 fF/µm is a reasonable value for rough estimates).

Figure 2.34 shows equivalent RC circuit models for nMOS and pMOS transistors of width k with contacted diffusion on both source and drain. The pMOS capacitors are shown with V_{DD} as their second terminal because the n-well is usually tied high. However, the behavior of the capacitor from a delay perspective is independent of the second terminal voltage so long as it is constant. Hence, we sometimes draw the second terminal as ground for convenience.

The propagation delay of a logic gate can be estimated from the RC models. Figure 2.35 shows how to estimate the delay of a fanout-of-1 inverter. The unit inverter of Figure 2.35(a) is composed from an nMOS transistor of unit size and a pMOS transistor of twice unit width to achieve equal rise and fall resistance. (Remember: If the nMOS mobility were three times that of the pMOS transistor, the pMOS device would have to be made three times wider than the nMOS transistor to achieve symmetric rise and fall times.) Note that if you use a cell library with a larger minimum-sized inverter, it is handy to define the unit transistor as the nMOS transistor in the basic inverter. Figure 2.35(b) gives an equivalent circuit, showing the first inverter driving the second inverter's gate. If the input A rises, the nMOS transistor will be ON and the pMOS OFF. Figure 2.35(c) illustrates this case with the switches removed. The capacitors shorted between two constant supplies are also removed because they are not charged or discharged. The propagation delay of $t_{pd} = R \cdot (6C) = 6RC$ is estimated as the RC time constant of the resistor discharging the diffusion and load capacitances. If an ideal inverter could be constructed with no parasitic diffusion capacitance, the delay would be only $3RC$. This delay of an ideal inverter with no parasitics driving an identical inverter is a figure of merit describing a manufacturing process and is sometimes called τ [Sutherland99].[8] Delay estimates for more complex gates are examined in Section 4.2.

---

[8]Do not confuse this definition of $\tau = 3RC$ with Mead and Conway's definition [Mead80] $\tau = RC$, the delay of an nMOS transistor driving its own gate.

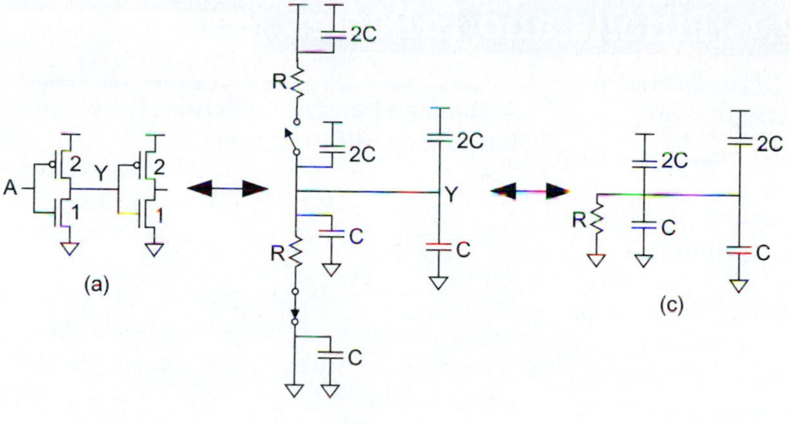

FIG 2.35 Inverter propagation delay

The effective resistance of a transistor passing a value in its poor direction is greater. For example, we can model the effective resistance of a unit nMOS transistor passing a '1' as $2R$ and a unit pMOS transistor passing a '0' as $4R$. The effective resistance of a transmission gate is the parallel combination of the resistances of the two transistors. Figure 2.36 shows that the effective resistance passing a '0' is $R \parallel 4R = (4/5)R$. The effective resistance passing a '1' is $2R \parallel 2R = R$. Hence, a transmission gate made from unit transistors is approximately R in either direction. Note that transmission gates are commonly built using equal-sized nMOS and pMOS transistors. Boosting the size of the pMOS transistor only slightly improves the effective resistance while significantly increasing the capacitance.

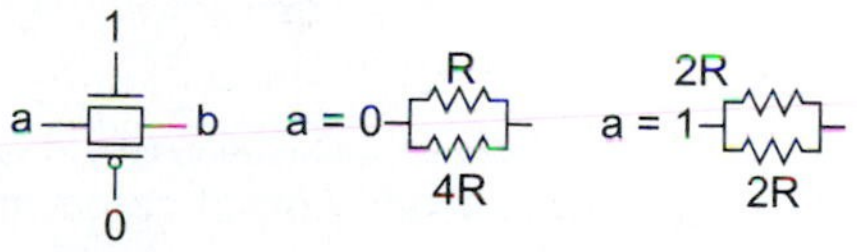

Most digital circuits use minimum-length transistors. If the length is greater than minimum, both resistance and capacitance to first order grow linearly with channel length. Section 5.4 describes how to construct SPICE simulations to extract R and C for a particular fabrication process operating at a particular power supply.

FIG 2.36 Effective resistance of a unit transmission gate

Velocity saturation also impacts the effective resistance. If velocity saturation were not a factor, the effective resistance of two transistors in series would be double that of a single transistor of equal size. If velocity saturation is significant, series transistors each experience a smaller V_{ds} and hence are less velocity saturated. Therefore, the resistance of series transistors is often slightly less than our simple model predicts. The effect is more pronounced for nMOS transistors than pMOS because of the higher mobility and greater degree of velocity saturation.

2.7 Pitfalls and Fallacies

This section lists a number of pitfalls and fallacies that can deceive the novice (or experienced) designer.

Blindly trusting one's models

Models should be viewed as only approximations to reality, not reality itself, and used within their limitations. In particular, simple models like the Shockley or RC models aren't even close to accurate fits for the I-V characteristics of a modern transistor. They are valuable for the insight they give on trends (i.e., making a transistor wider increases its gate capacitance and decreases its ON resistance), not for the absolute values they predict. Cutting-edge projects often target processes that are still under development, so these models should only be viewed as speculative. Finally, processes may not be fully characterized over all operating regimes; for example, don't assume that your models are accurate in the subthreshold region unless your vendor tells you so. Having said this, modern SPICE models do an extremely good job of predicting performance well into the GHz range for well characterized processes and models when using proper design practices (such as accounting for temperature, voltage, and process variation).

Using excessively complicated models for manual calculations

Because models cannot be perfectly accurate, there is little value in using excessively complicated models, particularly for hand calculations. Simpler models give more insight on key tradeoffs and more rapid feedback during design. Moreover, RC models calibrated against simulated data for a fabrication process can estimate delay just as accurately as elaborate models based on a large number of physical parameters but not calibrated to the process.

Assuming a transistor with twice the drawn length has exactly half the current

To first order, current is proportional to W/L. In modern transistors, the effective transistor length is usually shorter than the drawn length, so doubling the drawn length reduces current by more than a factor of two. Moreover, the threshold voltage tends to increase for longer transistors, resulting in less current. Therefore, it is a poor strategy to try to ratio currents by ratioing transistor lengths.

Assuming two transistors in series deliver exactly half the current of a single transistor

To first order, this would be true. However, two transistors in series each see a smaller electric field across the channel and hence are each less velocity saturated. Therefore, two series transistors in a modern process will deliver more than half the current of a single transistor. This is more pronounced for nMOS than pMOS transistors because of the higher mobility and the higher degree of velocity saturation of electrons than holes at a given field. This means that NAND gates perform better than first order estimates might predict.

Using nMOS pass transistors

nMOS pass transistors only pull up to $V_{DD} - V_t$. This voltage may fall below V_{IH} of a receiver, especially as V_{DD} decreases. For example, one author worked with a scan latch containing an nMOS pass transistor that operated correctly in a 250 nm process at 2.5 V. When the latch was ported to a 180 nm process at 1.8 V, the scan chain stopped working. The problem was traced to the pass transistor and the scan chain was made operational in the lab by raising V_{DD} to 2 V. A better solution is to use transmission gates in place of pass transistors.

Summary

In summary, we have seen that MOS transistors are four-terminal devices with a gate, source, drain, and body. In normal operation, the body is tied to GND or V_{DD} so the transistor can be modeled as a three-terminal device. The transistor behaves as a voltage-controlled switch. An nMOS switch is OFF (no path from source to drain) when the gate voltage is below some threshold V_t. The switch turns ON, forming a channel connecting source to drain, when the gate voltage rises above V_t. This chapter has developed more elaborate models to predict the amount of current that flows when the transistor is ON. The transistor operates in three modes depending on the terminal voltage:

- $V_{gs} < V_t$ Cutoff $I_{ds} \sim 0$

- $V_{gs} > V_t, V_{ds} < V_{gs} - V_t$ Linear I_{ds} increases with V_{ds} (like a resistor)

- $V_{gs} > V_t, V_{ds} > V_{gs} - V_t$ Saturation I_{ds} constant (like a current source)

In an ideal transistor, the saturation current depends on $(V_{gs} - V_t)^2$. pMOS transistors are similar to nMOS transistors, but have the signs reversed and deliver about half the current because of lower mobility. From the ideal model, we can derive the DC transfer characteristics and noise margins of logic gates using either the analytical expressions or a graphical load line analysis or simulation.

In a real transistor, the I-V characteristics are more complicated. Modern transistors are extraordinarily small and thus experience enormous electric fields even at low voltage. The high fields cause velocity saturation and mobility degradation that lead to less current than you might otherwise expect. This can be modeled as a saturation current dependent on $(V_{gs} - V_t)^\alpha$, where the velocity saturation index α is less than 2. Moreover, the saturation current does increase slightly with V_{ds} because of channel length modulation. Although simple hand calculations are no longer accurate, the general shape does not change very much and the transfer characteristics can still be derived using the graphical or simulation methods.

Even when the gate voltage is low, the transistor is not completely OFF. Subthreshold current through the channel drops off exponentially for $V_{gs} < V_t$, but is nonnegligible for transistors with low thresholds. Junction leakage currents flow through the reverse-biased p–n junctions. Tunneling current flows through the insulating gate when the oxide becomes thin enough.

Unlike ideal switches, MOS transistors pass some voltage levels better than others. An nMOS transistor passes '0's well, but only pulls up to $V_{DD} - V_{tn}$ when passing '1's. The pMOS passes '1's well, but only pulls down to $|V_{tp}|$ when passing '0's. This threshold drop is exacerbated by the body effect, which increases the threshold voltage when the source is at a different potential than the body.

Transistor speed depends on the ratio of current to capacitance. The two main sources of capacitance in a transistor are the gate capacitance formed by the thin gate oxide and the diffusion capacitance formed by the depletion regions between the source or drain and body. The diffusion capacitance is voltage-dependent, but can be modeled across the digital switching voltages with an average value.

Circuit simulators use very elaborate models to calculate the currents and voltages in a transistor. These models are too complicated to give much insight into the circuit behavior. For the purpose of estimating delay, we can approximate an ON transistor with an effective resistance such that the product of this resistance and the load capacitance matches the gate delay. This approximation is remarkably good and will be developed further in subsequent chapters to explain why some circuits are faster than others and to help optimize gate delay.

[Muller03] offers a comprehensive treatment of device physics at a more advanced level. [Gray01] describes MOSFET models in more detail from the analog designer's point of view.

Exercises

2.1   Consider an nMOS transistor in a 0.6 μm process with $W/L = 4/2$ λ (i.e., 1.2/0.6 μm). In this process, the gate oxide thickness is 100 Å and the mobility of electrons is 350 cm²/V • s. The threshold voltage is 0.7 V. Plot I_{ds} vs. V_{ds} for $V_{gs} = 0, 1, 2, 3, 4$, and 5 V.

2.2   Show that the current through two transistors in series is equal to the current through a single transistor of twice the length if the transistors are well described by the Shockley model. Specifically show that $I_{DS1} = I_{DS2}$ in Figure 2.37 when the transistors are in their linear region: $V_{DS} < V_{DD} - V_t$, $V_{DD} > V_t$ (this is also true in saturation). *Hint:* Express the currents of the series transistors in terms of V_1 and solve for V_1.

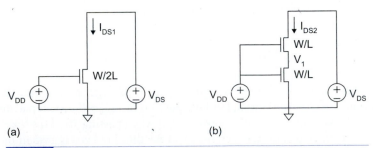

(a)   (b)

FIG 2.37 Current in series transistors

2.3   In Exercise 2.2, the body effect was ignored. If the body effect is considered, will I_{DS2} be equal to, greater than, or less than I_{DS1}? Explain.

2.4   A 90 nm long transistor has a gate oxide thickness of 16 Å. What is its gate capacitance per micron of width?

2.5   Calculate the diffusion parasitic C_{db} of the drain of a unit-sized contacted nMOS transistor in a 0.6 µm process when the drain is at 0 and at V_{DD} = 5 V. Assume the substrate is grounded. The transistor characteristics are CJ = 0.42 fF/µm², MJ = 0.44, $CJSW$ = 0.33 fF/µm, $MJSW$ = 0.12, and ψ_0 = 0.98 V at room temperature.

2.6   Consider the nMOS transistor in a 0.6 µm process with gate oxide thickness of 100 Å. The doping level is N_A = 2 • 10¹⁷ cm⁻³ and the nominal threshold voltage is 0.7 V. The body is tied to ground with a substrate contact. How much does the threshold change at room temperature if the source is at 4 V instead of 0?

2.7   Does the body effect of a process limit the number of transistors that can be placed in series in a CMOS gate at low frequencies?

2.8   Sometimes the substrate is connected to a voltage called the substrate bias to alter the threshold of the nMOS transistors. If the threshold of an nMOS transistor is to be raised, should a positive or negative substrate bias be used?

2.9   An nMOS transistor has a threshold voltage of 0.4 V and a supply voltage of V_{DD} = 1.2 V. A circuit designer is evaluating a proposal to reduce V_t by 100 mV to obtain faster transistors.

   a) By what factor would the saturation current increase (at V_{gs} = V_{ds} = V_{DD}) if the transistor were ideal?

   b) By what factor would the subthreshold leakage current increase at room temperature at V_{gs} = 0? Assume n = 1.4.

   c) By what factor would the subthreshold leakage current increase at 120° C? Assume the threshold voltage is independent of temperature.

2.10   As temperature rises, does the current through an ON transistor increase or decrease? Does current through an OFF transistor increase or decrease? Will a chip operate faster at high temperature or low temperature? Explain.

2.11   Find the subthreshold leakage current of an inverter at room temperature if the input A = 0. Let β_n = $2\beta_p$ = 1 mA/V², n = 1.0, and $|V_t|$ = 0.4 V. Assume the body effect and DIBL coefficients are γ = η = 0.

2.12   Repeat Exercise 2.11 for a NAND gate built from unit transistors with inputs $A = B$ = 0. Show that the subthreshold leakage current through the series transistors is half that of the inverter.

2.13   Repeat Exercises 2.11 and 2.12 when η = 0.04 and V_{DD} = 1.8 V, as in the case of a more realistic transistor. γ has a secondary effect, so assume that it is 0. Did the leakage currents go up or down in each case? Is the leakage through the series transistors more than half, exactly half, or less than half of that through the inverter?

2.14 Peter Pitfall is offering to license to you his patented noninverting buffer circuit shown in Figure 2.38. Graphically derive the transfer characteristics for this buffer. Assume $\beta_n = \beta_p = \beta$ and $V_{tn} = |V_{tp}| = V_t$. Why is it a bad circuit idea?

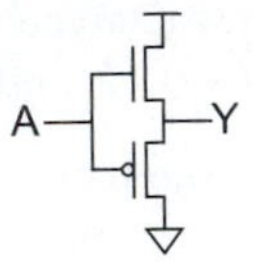

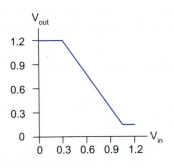

FIG 2.38 Noninverting buffer

2.15 A novel inverter has the transfer characteristics shown in Figure 2.39. What are the values of V_{IL}, V_{IH}, V_{OL}, and V_{OH} that give best noise margins? What are these high and low noise margins?

FIG 2.39 Transfer characteristics

2.16 Section 2.5.1 graphically determined the transfer characteristics of a static CMOS inverter. Derive analytic expressions for V_{out} as a function of V_{in} for regions B and D of the transfer function. Let $|V_{tp}| = V_{tn}$ and $\beta_p = \beta_n$.

2.17 Using the results from Exercise 2.16, calculate the noise margin for a CMOS inverter operating at 1.0 V with $V_{tn} = |V_{tp}| = 0.35$ V, $\beta_p = \beta_n$.

2.18 Repeat Exercise 2.16 if the thresholds and betas of the two transistors are not necessarily equal. Also solve for the value of V_{in} for region C where both transistors are saturated.

2.19 Using the results from Exercise 2.18, calculate the noise margin for a CMOS inverter operating at 1.0 V with $V_{tn} = |V_{tp}| = 0.35$ V, $\beta_p = 0.5\beta_n$.

2.20 Derive V_{out} using the Shockley models for the pseudo-nMOS inverter from Figure 2.30 with $V_{in} = V_{DD}$ as a function of the threshold voltages and beta values of the two transistors. Assume $V_{out} < |V_{tp}|$.

2.21 Give an expression for the output voltage for the pass transistor networks shown in Figure 2.40. Neglect the body effect.

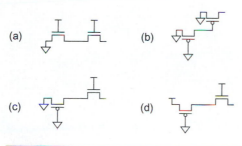

FIG 2.40 Pass transistor networks

2.22 Suppose $V_{DD} = 1.2$ V and $V_t = 0.4$ V. Determine V_{out} in Figure 2.41 for: (a) $V_{in} = 0$ V; (b) $V_{in} = 0.6$ V; (c) $V_{in} = 0.9$ V; and (d) $V_{in} = 1.2$ V. Neglect the body effect.

FIG 2.41 Single pass transistor

CMOS Processing Technology

3

3.1 Introduction

Chapter 1 summarized the steps in a basic CMOS process. These steps are expanded upon in this chapter. Where possible, the processing details are related to the way CMOS circuits and systems are designed. In addition, this chapter investigates some emerging technologies that may become important in the future. Modern CMOS processing is complex, and while coverage of every nuance of modern CMOS processing is not within the scope of this book, we focus on the fundamental concepts.

A fair question from a designer would be "Why do I care how transistors are made?" In many cases, if designers understand the physical process, they will comprehend the reason for the underlying design rules and in turn use this knowledge to create a better design. Understanding the manufacturing steps is also important when debugging some difficult chip failures and improving yield.

3.2 CMOS Technologies

3.2.1 Background

In this section we provide an overview of current CMOS technologies with a simplified treatment of the process steps.

The main CMOS technologies are:

- n-well process
- p-well process
- twin-well process
- triple-well process

By adding bipolar transistors (either silicon or silicon germanium—SiGe), the range of processes can be expanded. Silicon-on-insulator processes are also available through some manufacturers (see Section 3.4.1.2).

Chapter 1 outlined an n-well process. Historically, p-well processes preceded n-well processes. In a p-well process, the nMOS transistors are built in a p-well and the pMOS transistor is placed in the n-type substrate. p-well processes were used to optimize the pMOS transistor performance. Improved techniques allowed good pMOS transistors to be fabricated in an n-well and excellent nMOS transistors to be fabricated in the p-type substrate of an n-well process. Twin-well processes accompanied the emergence of n-well processes. A twin-well process allows the optimization of each transistor type. A third well can be added to create a triple-well process. The triple-well process has emerged to provide good isolation between analog and digital blocks in mixed-signal chips; it is also used to isolate high-density dynamic memory from logic. Most fabrication lines provide a baseline twin-well process that can be upgraded to a triple-well process with the addition of a single mask level. In this section we will present a CMOS process in the 130 or 90 nm generation that is representative of current commercial processes.

To review the material covered in Chapter 1, silicon in its pure or *intrinsic* state is a semiconductor, having bulk electrical resistance somewhere between that of a conductor and an insulator. The conductivity of silicon can be raised by several orders of magnitude by introducing *impurity* atoms into the silicon crystal lattice. These dopants can supply either free electrons or holes. Group III impurity elements such as boron that use up electrons are referred to as *acceptors* because they accept some of the electrons already in the silicon, leaving holes. Similarly, Group V *donor* elements such as arsenic and phosphorous provide electrons. Silicon that contains a majority of donors is known as *n-type*, while silicon that contains a majority of acceptors is known as *p-type*. When n-type and p-type materials are brought together, the region where the silicon changes from n-type to p-type is called a *junction* (junctions also can be formed between two areas with different doping). By arranging junctions in certain physical structures and combining them with other physical structures, various semiconductor devices can be constructed. Over the years, silicon semiconductor processing has evolved sophisticated techniques for building these junctions and other insulating and conducting structures.

3.2.2 Wafer Formation

The basic raw material used in modern semiconductor *fabs* (fabrication facilities) is a *wafer* or disk of silicon, which currently varies from roughly 75 mm to 300 mm (12″—a dinner plate!) in diameter and less than 1 mm thick. Wafers are cut from ingots of single-crystal silicon that have been pulled from a crucible melt of pure molten silicon. This is known as the *Czochralski* method and is currently the most common method for producing single-crystal material. Controlled amounts of impurities are added to the melt to provide the crystal with the required electrical properties. A seed crystal is dipped into the melt to initiate crystal growth. The silicon ingot takes on the same crystal orientation as the seed. A graphite radiator heated by radio-frequency induction surrounds the quartz crucible holding the melt and maintains the temperature a few degrees above the melting point of silicon (1425° C). The atmosphere is typically helium or argon to prevent the silicon from oxidizing.

The seed is gradually withdrawn vertically from the melt while simultaneously being rotated. The molten silicon attaches itself to the seed and recrystallizes as it is withdrawn. The seed withdrawal and rotation rates determine the diameter of the ingot. Growth rates vary from 30 to 180 mm/hour.

3.2.3   Photolithography

Recall that regions of dopants, polysilicon, metal, and contacts are defined using masks. For instance, in places covered by the mask, ion implantation might not occur or the dielectric or metal layer might be left intact. In areas where the mask is absent, the implantation can occur, or dielectric or metal could be etched away. The patterning is achieved by a process called *photolithography*, from the Greek *photo* (light), *lithos* (stone), and *graphe* (picture), which literally means "carving pictures in stone using light." The primary method for defining areas of interest (i.e., where we want material to be present or absent) on a wafer is by the use of *photoresists*. The wafer is coated with the photoresist and subjected to selective illumination through the *photomask*. After the initial patterning of photoresist, other barrier layers such as polycrystalline silicon, silicon dioxide, or silicon nitride can be used as physical masks on the chip. This distinction will become more apparent as this chapter progresses.

A photomask is constructed with chromium (chrome) covered quartz glass. A UV light source is used to expose the photoresist. Figure 3.1 illustrates the lithography process. The photomask has chrome where light should be blocked. The UV light floods the mask from the backside and passes through the clear sections of the mask to expose the organic photoresist (PR) that has been coated on the wafer. A developer solvent is then used to dissolve the soluble unexposed photoresist, leaving islands of insoluble exposed photoresist. This is termed a *negative* photoresist. A *positive* resist is initially insoluble, and when exposed to UV becomes soluble. Positive resists provide for higher resolution than negative resists, but are less sensitive to light. As feature sizes become smaller, the photoresist layers have to be made thinner. This in turn makes them less robust and more subject to failure. In turn, this can impact the overall yield of a process and the cost to produce the chip.

The photomask is commonly called a *reticle* and is usually smaller than the wafer, e.g., 2 cm on a side. A *stepper* moves the reticle to successive locations to completely expose the wafer. *Projection printing* is normally used, in which lenses between the reticle and wafer focus the pattern on the wafer surface. Older techniques include *contact printing*, where the mask and wafer are in contact, and *proximity printing*, where the mask and wafer are close but not touching. The original form of both techniques can cause damage to the mask and are limited in feature definition in the range of 1–2 μm. However, newer approaches to proximity printing are being proposed for future processes. The reticle can be the same size as the area to be patterned (1X) or larger. For instance, 2.5X and 5X steppers with optical reduction have been used in the industry.

The wavelength of the light source influences the minimum feature size that can be printed [Schellenberg03]. In the 1980s, mercury lamps with 436 nm or 365 nm wavelengths were used. At the 0.25 μm process generation, excimer lasers with 248 nm (deep ultraviolet) were adopted and have been used down to the 180 nm node. Currently 193 nm

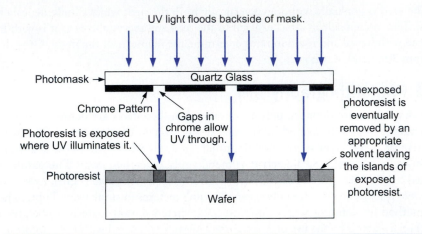

UV light floods backside of mask.

Photomask →

Quartz Glass

Chrome Pattern

Gaps in chrome allow UV through.

Photoresist is exposed where UV illuminates it.

Unexposed photoresist is eventually removed by an appropriate solvent leaving the islands of exposed photoresist.

Photoresist

Wafer

FIG 3.1 Photomasking with a negative resist (lens system between mask and wafer omitted to improve clarity and avoid diffracting the reader ☺)

argon-fluoride lasers are used for the *critical layers* down to the 90 nm node. The critical layers are those that define the device behavior. An example would be the gate (polysilicon), source/drain (diffusion), first metal, and contact masks. The ASML TWINSCAN XT:1250 step and scan lithography tool uses a 193 nm laser to expose up to 150 wafers per hour. It has an alignment tolerance of 8 nm and can process features down to 70 nm. 157 nm fluorine lasers are in development [Nowak02, Miller00] for processes below 90 nm. In the future, 13.4 nm extreme ultraviolet (EUV) light sources may be used, but at present these sources are costly and require prohibitively expensive reflective optics. Each successive UV stepper is more expensive and the throughput of the stepper may decrease. This is just another contributory issue to the spiraling cost of chip manufacturing. The cost of masks is also skyrocketing, forcing chip designers to amortize design and mask expenses across the largest volume possible. This theme will be reinforced in Chapter 8.

Wavelengths comparable to or greater than the feature size cause distortion in the patterns exposed on the photoresist. *Resolution enhancement techniques* precompensate for this distortion so the desired patterns are obtained. These techniques involve modifying the amplitude, phase, or direction of the incoming light. The ends of a line in a layout receive less light than the center, causing nonuniform exposure. *Optical proximity correction* (OPC) makes small changes to the patterns on the masks to compensate for these local distortions. Multiple lines running in parallel on a mask behave as a diffraction grating. *Phase shift masks* (PSM) vary the thickness of the mask to change the phase such that light from adjacent lines are out of phase and cancel where no light is desired. *Off-axis illumination* can also improve contrast for certain types of dense, repetitive patterns. Using these techniques, the resolution can be extended to one-eighth the wavelength of the light.

3.2.4 Well and Channel Formation

Varying proportions of donor and acceptor impurities can be achieved using *epitaxy, deposition,* or *implantation.* Epitaxy involves growing a single-crystal film on the silicon surface (which is already a single crystal) by subjecting the silicon wafer surface to an elevated temperature and a source of dopant material.

Epitaxy can be used to produce a layer of silicon with fewer defects than the native wafer surface and also can help prevent latchup (see Section 4.8.5). Foundries usually provide a choice of *epi* (with epitaxial layer) or non-epi wafers. Microprocessor designers usually prefer to use epi wafers for device uniformity of performance.

Deposition involves placing dopant material onto the silicon surface and then driving it into the bulk using a thermal diffusion step. This can be used to build deep junctions. A step called *chemical vapor deposition* (CVD) can be used for the deposition. As its name suggests, CVD occurs when heated gases react in the vicinity of the wafer and produce a product that is deposited on the silicon surface. CVD is also used to lay down thin films of material later in the CMOS process.

Ion implantation involves subjecting the silicon substrate to highly energized donor or acceptor atoms. When these atoms impinge on the silicon surface, they travel below the surface of the silicon, forming regions with varying doping concentrations. At elevated temperature (> 800° C), diffusion occurs between silicon regions having different densities of impurities, with impurities tending to diffuse from areas of high concentration to areas of low concentration. Therefore, it is important to keep the remaining process steps at as low a temperature as possible once the doped areas have been put into place. However, a high-temperature annealing step is often performed after ion implantation to redistribute dopants more uniformly. Ion implantation is the standard well and source/drain implant method used today.

The first step in most CMOS processes is to define the well regions. In a triple-well process, a deep n-well is first driven into the p-type substrate, usually using high-energy (MeV—Mega electron volt levels) ion implantation as opposed to a thermally diffused operation. This avoids the thermal cycling (i.e., the wafers do not have to be raised significantly in temperature), which improves throughput and reliability. A 2–3 MeV implantation can yield a 2.5–3.5 μm deep n-well. Such a well has a peak dopant concentration just under the surface and for this reason is called a *retrograde* well. This can enhance device performance by providing improved latchup characteristics and reduced susceptibility to vertical punch-through. A thick (3.5–5.5 μm) resist has to be used to block the high energy implantation where no well should be formed. Thick resists and deep implants necessarily lead to fairly coarse feature dimensions for wells, compared to the minimum feature size. Shallower n-well and p-well regions are then implanted. After the wells have been formed, the doping levels can be adjusted (called a *threshold implant*) to set the desired threshold voltages for both nMOS and pMOS transistors. For a given gate and substrate material, the threshold voltage (V_t) depends on the doping level in the substrate (N_A), the oxide thickness (t_{ox}), and the surface state charge (Q_{fc}). The implant can affect both N_A and Q_{fc} and hence V_t. Figure 3.2 shows a typical triple-well structure. As discussed, the nMOS transistors are situated in the p-well located in the deep n-well. The pMOS transistors are located in the shallow (normal) n-well. The figure shows the cross-section of an inverter.

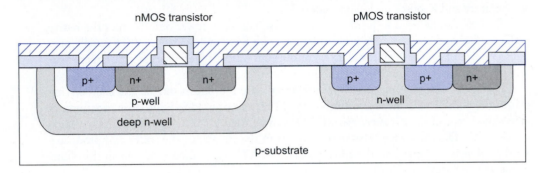

FIG 3.2   Well structure in triple-well process

3.2.5 Silicon Dioxide (SiO$_2$)

Many of the structures and manufacturing techniques used to make silicon integrated circuits rely on the properties of SiO$_2$. Therefore, reliable manufacture of SiO$_2$ is extremely important. In fact, unlike competing materials, silicon has dominated the industry because it has an easily processable oxide (i.e., it can be grown and etched). Various thicknesses of SiO$_2$ may be required, depending on the particular process. Thin oxides are required for transistor gates; thicker oxides might be required for higher voltage devices, while even thicker oxide layers might be required to ensure that transistors are not formed unintentionally in the silicon beneath polysilicon wires (see the next section).

Oxidation of silicon is achieved by heating silicon wafers in an oxidizing atmosphere. Some common approaches are:

- *Wet oxidation*—when the oxidizing atmosphere contains water vapor. The temperature is usually between 900° C and 1000° C. This is also called *pyrogenic* oxidation when a 2:1 mixture of hydrogen and oxygen is used. Wet oxidation is a rapid process.

- *Dry oxidation*—when the oxidizing atmosphere is pure oxygen. Temperatures are in the region of 1200° C to achieve an acceptable growth rate. Dry oxidation forms a better quality oxide than wet oxidation. It is used to form thin, highly controlled gate oxides, while wet oxidation may be used to form thick field oxides.

- *Atomic layer deposition (ALD)*—a process in which a thin chemical layer (material *A*) is attached to a surface and then a chemical (material *B*) is introduced to produce a thin layer of the required layer (i.e., SiO$_2$—this can also be used for other various dielectrics and metals). The process is then repeated and the required layer is built up layer by layer. [George96, Klaus98]. Currently, this is an emerging R&D process.

The oxidation process normally consumes silicon (deposition or ALD does not). Since SiO_2 has approximately twice the volume of silicon, the SiO_2 layer grows almost equally in both vertical directions. Thus, after processing, the SiO_2 projects above and below the original unoxidized silicon surface.

3.2.6   Isolation

Although not immediately obvious, individual devices in a CMOS process need to be isolated from one another so that they do not have unexpected interactions. The source and drain of transistors form reverse-biased p–n junctions with the substrate or well, isolating them from their neighbors. Next, the formation of any parasitic MOS channels must be prevented. This is commonly achieved using a thin *gate oxide* for transistors and a much thicker *field oxide* elsewhere. The thicker oxide increases the threshold voltage to a value above the supply voltage and so prevents a channel from forming in the substrate unless there is an overvoltage condition (actually, these *field devices* are used for I/O protection). In addition to using the thick oxide, the substrate in areas where transistors are not required can be further implanted with dopants to create a *channel-stop* diffusion. The implant increases the impurity concentration in the substrate, which in turn raises the threshold voltage and prevents inversion of an unwanted channel. In early metal gate processes that had a uniform (thin) oxide layer, the channel-stop diffusion surrounded each transistor and was the only method of isolating transistors.

Historically, *LOCOS* or Local Oxidation of Silicon was used to produce varying oxide thicknesses. A common problem with LOCOS-based processes was the transition between thick and thin oxide, which extended some distance laterally because of the way the oxide was grown; this in turn limited the packing density of transistors. The isolation step that is used to achieve isolation between devices in processes at and below the 180 nm node is to form insulating trenches of SiO_2 that surround active areas. Typical trenches in a 90 nm process can be 140 nm wide by 400 nm deep [Kim02]. This is called *shallow trench isolation* (STI). (There are also process options with deep trenches.) STI (Figure 3.3) starts with a *pad oxide* and a silicon nitride layer, which act as the masking layers. Openings in the pad oxide are then used to etch into the well or substrate region (this process can also be used for source/drain diffusion). A liner oxide is then grown to cover the exposed silicon. The trenches are filled with SiO_2 using CVD that does not consume the underlying silicon. The pad oxide and nitride are removed and a *Chemical Mechanical Polishing* (CMP) step is used to planarize the structure. CMP, as its name suggests, combines a mechanical grinding action in which the rotating wafer is contacted by a stationary polishing head while an abrasive mixture is applied. The mixture also reacts chemically with the surface to aid in the polishing action. CMP is used to achieve flat surfaces, which are of central importance in modern processes with many layers.

From the designer's perspective, the presence of a deep n-well and/or trench isolation makes it easier to isolate noise-sensitive (analog or memory) portions of a chip from digital sections. Trench isolation also permits nMOS and pMOS transistors to be placed closer together because the isolation provides a higher source/drain *breakdown voltage—*

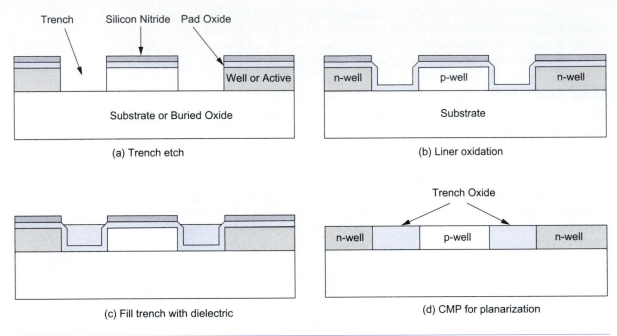

FIG 3.3 Shallow trench isolation

the voltage that a source or drain diodes start to conduct in the reverse-biased condition. The breakdown voltage must exceed the supply voltage (so junctions do not break down during normal operation) and is determined by the junction dimensions and doping levels of the junction formed. Deeper trenches increase the breakdown voltage.

Wells are defined by separate masks. In the case of a twin-well process, only one mask need be defined because the other well is by definition its complement. Triple-well processes have to define at least two masks, one for the deep well and the other for either n-well or p-well.

3.2.7 Gate Oxide

The next step in the process is to form the gate oxide for the transistors. As mentioned, this is most commonly in the form of *silicon dioxide* (SiO_2).

In the case of STI-defined source/drain regions, the gate oxide is grown on top of the planarized structure that occurs at the stage shown in Figure 3.3(d). This is shown in Figure 3.4. The oxide structure is called the *gate stack*. This term arises because current processes seldom use a pure SiO_2 gate oxide, but prefer to produce a stack that consists of a few atomic layers, each 3–4 Å thick, of SiO_2 for reliability, overlaid with a few layers of an *oxynitrided* oxide (one with nitrogen added). The presence of the nitrogen increases the dielectric constant, which decreases the *effective oxide thickness* (EOT); this means that for a given oxide thickness, it performs like a thinner oxide. Being able to use a thicker oxide improves the robustness of the process. This concept is revisited in Section 3.4.1.

Many processes in the 180 nm generation and beyond provide at least two oxide thicknesses (thin for logic transistors and thicker for I/O transistors that must withstand higher voltages). Some processes offer more than one oxide for logic transistors to permit tradeoffs between speed and gate leakage current, as will be discussed in Section 3.4.1. At the 65 nm node, the effective thickness of the thin gate oxide is of the order of 1.5 nm or 15 Å.

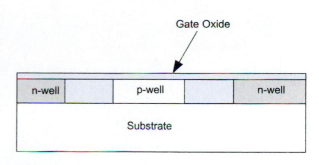

FIG 3.4   Gate oxide formation

3.2.8   Gate and Source/Drain Formation

As a historical note, early metal-gate processes first diffused source and drain regions, and then formed a metal gate. If the gate was misaligned, it could fail to cover the entire channel and lead to a transistor that never turned ON. To prevent this, the metal gate had to overhang the source and drain by more than the alignment tolerance of the process. This created large parasitic gate-to-source and gate-to-drain overlap capacitances that degraded switching speeds.

When silicon is deposited on SiO_2 or other surfaces without crystal orientation, it forms polycrystalline silicon, commonly called *polysilicon* or simply *poly*. An annealing process is used to control the size of the single crystal domains and to improve the quality of the polysilicon. Undoped polysilicon has high resistivity. The resistance can be reduced by implanting it with dopants and/or combining it with a refractory metal. The polysilicon gate serves as a mask to allow precise alignment of the source and drain with the gate. This process is called a *self-aligned polysilicon gate* process. Aluminum could not be used because it would melt during formation of the source and drain.

The steps to define the gate, source, and drain in a self-aligned polysilicon gate are as follows:

- Grow gate oxide wherever transistors are required (area = source + drain + gate)—elsewhere there will be thick oxide (Figure 3.5(a))

- Deposit polysilicon on chip (Figure 3.5(b))

- Pattern polysilicon (both gates and interconnect) (Figure 3.5(c))

- Etch exposed gate oxide—that is, the area of gate oxide that was not covered by polysilicon; at this stage, the chip has windows down to the well or substrate wherever a source/drain diffusion is required (Figure 3.5(d))

- Implant pMOS and nMOS source/drain regions (Figure 3.5(e))

The source/drain implant is relatively low, typically in the range $10^{18}–10^{20}$ cm^{-3} of impurity atoms. Such a *lightly doped drain* (LDD) structure reduces the electric field at the drain junction (the junction with the highest voltage), which improves the immunity of the device to hot electron damage (see Section 4.8.4). The LDD implants are shallow and lightly doped, so they exhibit low capacitance but high resistance. This reduces device performance somewhat because of the resistance in series with the transistor. Consequently,

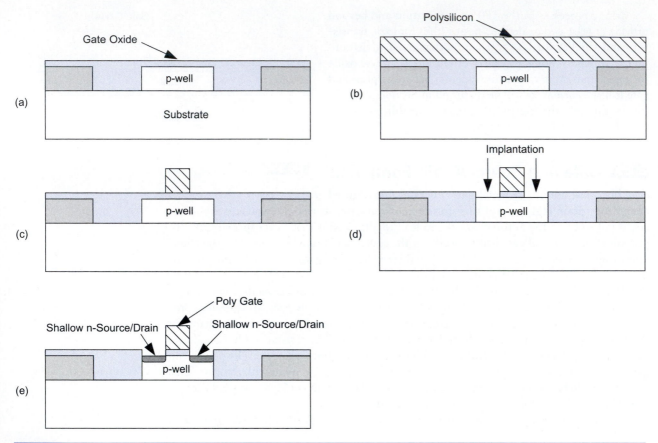

FIG 3.5   Gate and shallow source/drain definition

deeper, more heavily doped source/drain implants are needed to provide devices that combine hot electron suppression with low source/drain resistance. A silicon nitride (Si_3N_4) *spacer* along the edge of the gate serves as a mask to define the location of this deeper diffusion, as shown in Figure 3.6(a). For in-depth coverage of various LDD structures see [Ziegler02].

As mentioned, the polysilicon gate and source/drain diffusion have high resistance due to the resistivity of silicon and their extremely small dimensions. Modern processes form a surface layer of a refractory metal on the silicon to reduce the resistance. A *refractory metal* is one with a high melting point that will not be damaged during subsequent processing. Tantalum, molybdenum, titanium, or cobalt are commonly used. The metal is deposited on the silicon (specifically on the gate polysilicon and/or source/drain regions). A layer of *silicide* is formed when the two substances react at elevated temperatures. In a

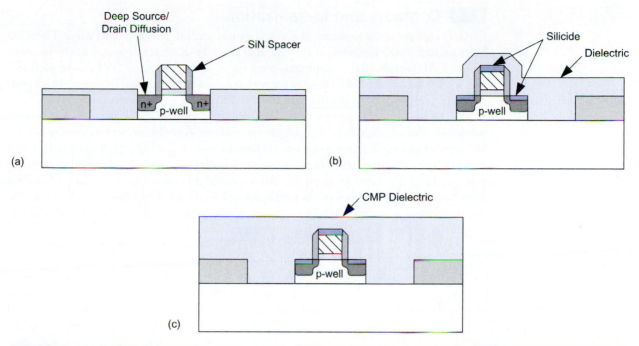

FIG 3.6 Lightly Doped Drain (LDD) structure

polycide process, only the gate polysilicon is silicided. In a silicide process (usually implemented as a self-aligned silicidization—from whence comes the synonymous term *salicide*) both gate polysilicon and source/drain regions are silicided. This process lowers the resistance of the polysilicon interconnect and/or the source and drain.

Figure 3.6(b) shows the resultant structure with gate and source/drain regions silicided. In addition, SiO_2 or an alternative dielectric has been used to cover all areas prior to the next processing steps. The figure shows a resulting structure with some vertical topology typical of older processes.

Figure 3.6(c) shows a structure where CMP has been employed. Achieving a very flat finish allows layers to be stacked vertically without incurring the problems of metal having to traverse rapid transitions in surface height (as shown in Figure 3.6(b)), which can lead to breaks and a plethora of design rules that relate to metal edges.

Polysilicon over diffusion normally forms a transistor gate, so a short metal1 wire is necessary to connect a diffusion output node to a polysilicon input. Some processes add a contact region to the process so that the polysilicon layer can directly connect to the diffusion. Such polysilicon wires are called *local interconnect*. Local interconnect offers denser cell layouts, especially in static RAMs.

3.2.9 Contacts and Metallization

Contact cuts are made to source, drain, and gate according to the contact mask. These are holes etched in the dielectric at the end of the source/drain step covered in the previous section. Aluminum (Al) is commonly used for wires but tungsten (W) can be used as a *plug* to fill the contact holes (to alleviate problems of aluminum conforming to small contacts). In some processes, the tungsten can also be used as a local interconnect layer.

Metallization is the process of building wires to connect the devices. As mentioned previously, conventional metallization uses aluminum. Aluminum can be deposited either by *evaporation* or *sputtering*. Evaporation is performed by passing a high electrical current through a thick aluminum wire in a vacuum chamber. Some of the aluminum atoms are vaporized and deposited on the wafer. An improved form of evaporation that suffers less from contamination focuses an electron beam at a container of aluminum to evaporate the metal. Sputtering is achieved by generating a gas plasma by ionizing an inert gas using an RF or DC electric field. The ions are focused on an aluminum target and the plasma dislodges metal atoms, which are then deposited on the wafer.

Wet or *dry etching* can be used to remove unwanted metal. *Piranha solution* is a 3:1 to 5:1 mix of sulphuric acid and hydrogen peroxide that is used to clean wafers of organic and metal contaminants or photoresist after metal patterning. Plasma etching is a dry etch process with fluorine or chlorine gas used for metallization steps. The plasma charges the etch gas ions, which are attracted to the appropriately charged silicon surface. Very sharp etch profiles can be achieved using plasma etching. The result of the contact and metallization patterning steps is shown in Figure 3.7.

Subsequent intermetal vias and metallization are then applied. Some processes attempt to use a uniform metallization scheme from at least level 2 to $n-1$, where n is the top level of metal. The top level is normally a thicker layer for use in power distribution and as such has relaxed width and spacing constraints. Other processes use successively thicker and wider metal, moving from lower to upper layers, as will be explored in Section 4.5. Such a metallization cross-section is shown in Figure 3.8.

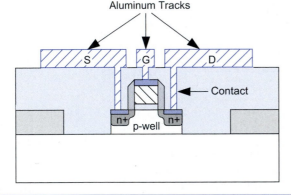

FIG 3.7 Aluminum metallization

3.2.10 Passivation

The final processing step is to add a protective glass layer called *passivation* or *overglass* that prevents the ingress of contaminants. Openings in the passivation layer allow connection to I/O pads and test probe points if needed. After passivation, further steps can be performed such as *bumping*, which allows the chip to be directly connected to a circuit board using plated solder bumps in the pad openings.

3.2.11 Metrology

Metrology is the science of measuring. Everything that is built in a semiconductor process has to be measured to give feedback to the manufacturing process. This ranges from simple optical measurements of line widths to advanced techniques to measure thicknesses of thin films and defects such as voids in copper interconnect. A natural requirement exists for *in situ* real-time measurements so that the manufacturing process can be controlled in a direct feedback manner.

Optical microscopes are used to observe large structures and defects, but are no longer adequate for structures smaller than the wavelength of visible light (~0.5 μm). Scanning electron microscopy (SEM) is used to observe very small features. An SEM raster scans a structure under observation and observes secondary electron emission to produce an image of the surface of the structure. Energy Dispersive Spectroscopy (EDX) bombards a circuit with electrons causing X-ray emission. This can be used for imaging as well. A Transmission Electron Microscope (TEM), which observes the results of passing electrons through a sample (rather than bouncing them off the sample), is sometimes also used to measure structures.

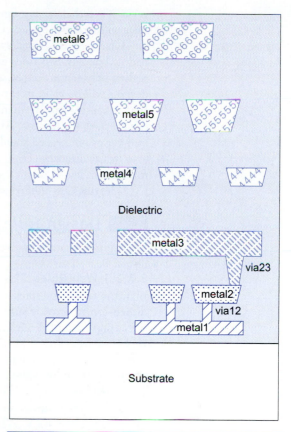

FIG 3.8 Typical metallization cross-section

3.3 Layout Design Rules

Layout rules, also referred to as *design rules* or *ground rules* (IBM), were introduced in Chapter 1 and can be considered a prescription for preparing the photomasks that are used in the fabrication of integrated circuits. The main objective of the layout rules is to build reliably functional circuits in as small an area as possible. In general, design rules represent a compromise between performance and yield. The more conservative the rules are, the more likely it is that the circuit will function. However, the more aggressive the rules are, the greater the opportunity for improvements in circuit performance. This improvement may be the at the expense of yield.

Design rules specify to the designer certain geometric constraints on the layout artwork so that the patterns on the processed wafer will preserve the topology and geometry of the designs. It is important to note that design rules do not represent some hard boundary between correct and incorrect fabrication. Rather, they represent a tolerance that ensures very high probability of correct fabrication and subsequent operation. For example, you may find that a layout that violates design rules can still function correctly and vice versa. Nevertheless, any significant or frequent departure (*design rule waiver*) from design rules will seriously prejudice the success of a design.

Chapter 1 described a version of design rules based on the MOSIS CMOS Scalable Rules. The MOSIS rules are expressed in terms of λ. These rules allow some degree of scaling between processes, as in principle, you only need to reduce the value of λ and the designs will be valid in the next process down in size. Unfortunately, history has shown that processes rarely shrink uniformly. Thus, industry usually uses the actual micron design rules for layouts. At this point in time, custom layout is usually constrained to a number of often-used standard cells or memories, where the effort expended is amortized over many instances. Only for extremely high-volume chips is the cost savings of a smaller full-custom layout worth the labor cost of that layout.

The rules are defined in terms of *feature sizes* (widths), *separations*, and *overlaps*.

3.3.1 Design Rule Background

We begin by examining the reasons for the most important design rules.

3.3.1.1 Well Rules
The n-well is usually a deeper implant (especially a deep n-well) than the transistor source/drain implants, and therefore, it is necessary for the outside dimension to provide sufficient clearance between the n-well edges and the adjacent n^+ diffusions. The inside clearance is determined by the transition of the field oxide across the well boundary. Processes that use STI may permit zero inside clearance. In older LOCOS processes, problems such as the *bird's beak* effect (the lateral space required to transition from thick to thin oxide) usually prevent this. Because the n-well sheet resistance can be several KΩ per square, it is necessary to thoroughly ground the well. This will prevent excessive voltage drops due to well currents. Guidelines on well and substrate taps are given in Section 4.8.5. Where wells are connected to different potentials (say in analog circuits), the spacing rules may differ from *equipotential* wells (all wells at the same voltage—the normal case in digital logic).

Mask Summary: The masks encountered for well specification may include n-well, p-well, and deep n-well. These are used to specify where the various wells are to be placed. Often only one well is specified in a twin-well process (i.e., n-well) and by default the p-well is in areas where the n-well isn't (i.e., p-well equals the logical NOT of the n-well).

3.3.1.2 Transistor Rules
CMOS transistors are generally defined by at least four physical masks. These are active (also called diffusion, diff, or thinox), n-select (also called n-implant, nimp, or nplus), p-select (also called p-implant, pimp, or pplus) and polysilicon (also called poly or polyg). The active mask defines all areas where either n- or p-type diffusion is to be placed *or* where the gates of transistors are to be placed. The gates of transistors are defined by the logical AND of the polysilicon mask and the active mask, i.e., where polysilicon crosses diffusion. The select layers define what type of diffusion is required. n-select surrounds active regions where n-type diffusion is required. p-select surrounds areas where p-type diffusion is required. n-diffusion areas inside p-well regions define

nMOS transistors (or n-diffusion wires). n-diffusion areas inside n-well regions define n-well contacts. p-diffusion areas inside n-wells define pMOS transistors (or p-diffusion wires). p-diffusion areas inside p-wells define substrate contacts (or p-well contacts). Frequently, design systems will only define n-diffusion (ndiff) and p-diffusion (pdiff) to reduce the complexity of the process. The appropriate selects are generated automatically. That is, ndiff will be converted automatically into active with an overlapping rectangle or polygon of n-select.

It is essential for the poly to be completely cross active; otherwise the transistor that has been created will be shorted by a diffusion path between source and drain. Hence, poly is required to extend beyond the edges of the active area. This is often termed the *gate extension*. Active must extend beyond the poly gate so that diffused source and drain regions exist to carry charge into and out of the channel. Poly and active regions that should not form a transistor must be kept separated; this results in a spacing rule from active to polysilicon.

Figure 3.9(a) shows the mask construction for the final structures that appear in Figure 3.9(b).

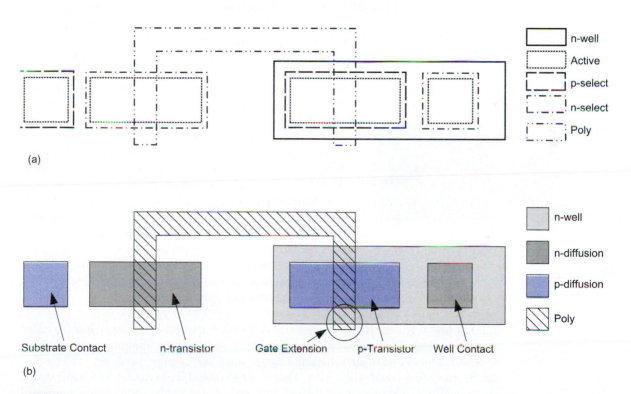

(a)

n-well
Active
p-select
n-select
Poly

(b)

Substrate Contact   n-transistor   Gate Extension   p-Transistor   Well Contact

n-well
n-diffusion
p-diffusion
Poly

FIG 3.9 CMOS n-well process transistor and well/substrate contact construction

Mask Summary: The basic masks (in addition to well masks) used to define transistors, diffusion interconnect (possibly resistors), and gate interconnect are active, n-select, p-select, and polysilicon. These may be called different names in some processes. Sometimes n-diffusion (ndiff) and p-diffusion (pdiff) masks are used to alleviate designer confusion.

3.3.1.3 Contact Rules There are several generally available contacts:

- Metal to p-active (p-diffusion)

- Metal to n-active (n-diffusion)

- Metal to polysilicon

- Metal to well or substrate

Depending on the process, other contacts such as *buried* polysilicon-active contacts may be allowed for local interconnect.

Because the substrate is divided into well regions, each isolated well must be tied to the appropriate supply voltage; that is, the n-well must be tied to V_{DD} and the substrate or p-well must be tied to GND with well or substrate contacts. As mentioned in Section 1.5.1, metal makes a poor connection to the lightly doped substrate or well. Hence, a heavily doped active region is placed beneath the contact. A *split* or *merged* contact is equivalent to two adjacent contacts to n-active and p-active strapped together with metal. This structure is used to tie transistor sources to the substrate or n-well and simultaneously to GND or V_{DD}, as shown at the source of the n-transistor in Figure 3.10. A split contact is shown, which is consistent with modern processes that usually require uniform contact sizes to achieve well-defined etching characteristics. Merged contact structures in older processes may have used an elongated contact rectangle.

Whenever possible, use more than one contact at each connection. This significantly improves yield in many processes because the connection is still made if one of the contacts is malformed.

Mask Summary: The only mask involved with contacts to active or poly is the contact mask, commonly called CONT. Contacts are normally of uniform size.

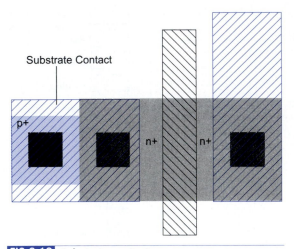

FIG 3.10 Substrate contact

3.3.1.4 Metal Rules Metal spacing may vary with the width of the metal line (so called fat-metal rules). That is, above some metal wire width, the minimum spacing may be increased. This is due to etch characteristics of small versus large metal wires. There may also be maximum metal width rules. That is, single metal wires cannot be greater than a certain width. If wider wires are desired, they are constructed by paralleling a number of smaller wires and adding checkerboard links to tie the wires together. Additionally, there may be spacing rules that are applied to long, closely spaced parallel metal lines.

Older nonplanarized processes required greater width and spacing on upper-level metal wires (e.g., metal3) to prevent breaks or shorts between adjoining wires caused by the vertical topology of the underlying layers. This is no longer a consideration for modern planarized processes. Nevertheless, width and spacing are still greater for thicker metal layers.

Mask Summary: Metal rules may be complicated by varying spacing dependent on width: As the width increases, the spacing increases. Metal overlap over contact might be zero or nonzero. Guidelines will also exist for electromigration, as discussed in Section 4.8.2.

3.3.1.5 Via Rules

Processes may vary in whether they allow *stacked* vias to be placed over polysilicon and diffusion regions. Some processes allow vias to be placed within these areas, but do not allow the vias to straddle the boundary of polysilicon or diffusion. This results from the sudden vertical topology variations that occur at sub-layer boundaries. Most modern planarized processes allow for stacked vias, which reduces the area required to pass from a lower-level metal to a high-level metal.

Mask Summary: Vias are normally of uniform size within a layer. They may increase in size toward the top of a metal stack. For instance, large vias required on power busses are constructed from an array of uniformly sized vias.

3.3.1.6 Other Rules

The passivation or overglass layer is a protective layer of SiO_2 (glass) that covers the final chip. Appropriately sized openings are required at pads and any internal test points.

Some additional rules that might be present in some processes are as follows:

- Extension of polysilicon in the direction that metal wires exit a contact.

- Extension of metal end-of-line region beyond a via.

- Differing pMOS and nMOS gate lengths.

- Differing gate poly extensions depending on the device length or the device construction.

3.3.1.7 Summary

Whereas earlier processes tended to be process driven and frequently had long and involved design rules, processes have become increasingly "designer friendly" or, more specifically, computer friendly (most of the mask geometries for designs are algorithmically produced). Companies sometimes create "generic" rules that span a number of different CMOS foundries that they might use. Some processes have design guidelines that feature structures to be avoided to ensure good yields. Traditionally, engineers followed yield-improvement cycles to determine the causes of defective chips and modify the layout to avoid the most common systematic failures. Time to market and product lifecycles are now so short that yield improvement is only done for the highest volume parts. It is often better to reimplement a successful product in a new, smaller technology rather than worry about improving the yield on the older, larger process.

3.3.2 Scribe Line and Other Structures

The *scribe line* surrounds the completed chip where it is cut with a diamond saw. The construction of the scribe line varies from manufacturer to manufacturer. It is designed to prevent the ingress of contaminants from the side of the chip (as opposed to the top of the chip, which is protected by the overglass).

Several other structures are included on a mask including the *alignment mark*, *critical dimension* structures, *vernier* structures, and *process check* structures [Hess94]. The mask alignment mark is usually placed by the foundry to align one mask to the next. Critical dimension test structures can be measured after processing to check proper etching of narrow polysilicon or metal lines. Vernier structures are used to judge the alignment between layers. A vernier is a set of closely spaced parallel lines on two layers. Misalignment between the two layers can be judged by the alignment of the two verniers. Test structures such as chains of contacts and vias and test transistors are used to evaluate contact resistance and transistor parameters. Often these structures can be placed along the scribe line so they do not consume useful wafer area.

3.3.3 MOSIS Scalable CMOS Design Rules

Academic designs often use the λ-based scalable CMOS design rules from MOSIS because they are simple and freely available, and they allow designs to easily migrate from one process to another. These advantages come at the expense of being conservative because they must work for all manufacturing processes.

MOSIS actually has three sets of rules: SCMOS, SUBM, and DEEP. The SUBM rules are somewhat more conservative than SCMOS rules. DEEP rules are even more conservative. The more conservative rules allow you to use a slightly smaller value of λ while still satisfying all of the micron design rules for a process. Table 3.1 lists some of the foundry processes MOSIS has offered and the associate value of λ for the different rule sets. For example, the AMI 0.5 μm process can use the SCMOS rules with λ = 0.35 μm or the SUBM rules with λ = 0.30 μm. SUBM rules are a good choice for class projects because they are somewhat easier to use than DEEP (no half-λ rules), while still being compatible with most processes. Some processes offer a second polysilicon layer for floating-gate transistors and poly-insulator-poly capacitors used in analog circuits discussed in Section 3.4.3.

For design rules where the minimum drawn gate length exceeds the feature size, MOSIS applies a *polysilicon bias* to shrink the gates by a uniform amount before masks are made. For example, in the SUBM rules for the AMI 0.5 μm process with λ = 0.3 μm, a bias of −0.1 μm is applied to all polysilicon. Thus, a 2 λ transistor gate is 0.5 μm rather than 0.6 μm and a 4 λ gate is 1.1 μm rather than 1.2 μm. When simulating circuits, be sure to use the biased channel lengths to accurately model the transistor behavior. In SPICE, the XL parameter is added to the specified transistor length to find the actual length. For example, a SPICE deck could specify λ = 0.3um for each transistor and include XL = −0.1um in the model file to indicate a biased length of 0.5 μm.

Table 3.1	MOSIS design rule options					
Vendor	Feature Size (μm)	Interconnect Layers	Stacked Vias	SCMOS	SUBM	DEEP
Orbit	2.0	2 metal	No	λ = 1.0 μm		
AMI	1.5	2 metal, 2 poly	No	λ = 0.80 μm	λ = 0.80 μm	
AMI	0.5	3 metal, 1–2 poly	Yes	λ = 0.35 μm	λ = 0.30 μm	
TSMC	0.35	4 metal, 1–2 poly	Yes		λ = 0.20 μm	
TSMC	0.25	5 metal	Yes		λ = 0.15 μm	λ = 0.12 μm
TSMC	0.18	6 metal	Yes		λ = 0.10 μm	λ = 0.09 μm

Section 1.5.3 introduced the basic design rules. Table 3.2 lists in more detail the MOSIS design rules that are generally relevant to designers for a process with N metal layers. Figure 3.11 illustrates many of these rules and they are shown in color on the inside back cover. See www.mosis.org for complete, up-to-date rules with illustrations. Layouts consist of a set of rectangles on various layers such as polysilicon or metal. *Width* is the minimum width of a rectangle on a particular layer. *Spacing* is the minimum spacing between two rectangles on the same or different layers. *Overlap* specifies how much a rectangle must surround another on another layer. Dimensions are all specified in λ except for overglass cuts that do not scale well because they must contact bond wires or probe tips. Select layers are often generated automatically and thus are not shown in the layout. If the active layer satisfies design rules, the select will too.

Contacts and vias must be exactly 2x2 λ. Larger connections are made from arrays of small vias to prevent current crowding at the periphery. The spacing rules of polysilicon or diffusion to arrays of multiple contacts is slightly larger than that to a single contact.

Section 1.5.5 estimated the pitch of lower-level metal to be 8 λ–4 λ for the width, and 4 λ for spacing. Technically, the minimum width and spacing are 3 λ, but the minimum metal contact size is 2x2 λ plus 1 λ surround on each side, for a width of 4 λ. Thus, the pitch for contacted metal lines can be reduced to 7 λ. Moreover, if the lines are drawn at 3 λ and the contacts are staggered so two adjacent lines never have adjacent contacts, the pitch reduces to 6.5 λ. Nevertheless, using a pitch of 8 λ for planning purposes is good practice and leaves a bit of "wiggle room" to solve difficult layout problems.

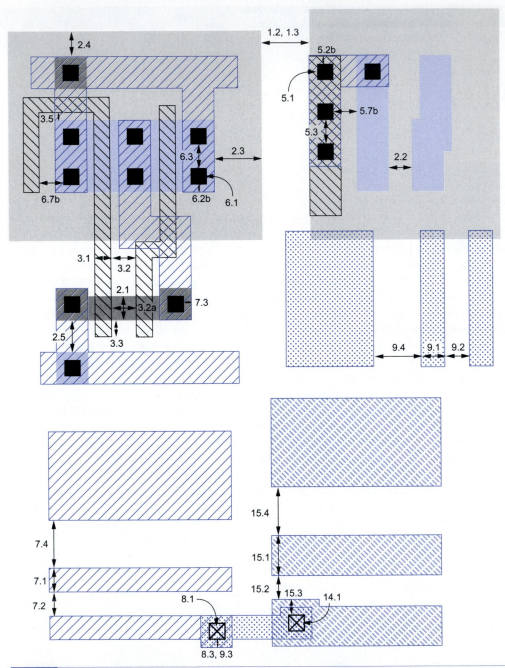

FIG 3.11 Illustrated MOSIS design rules. Color version on inside back cover.

Table 3.2	MOSIS design rules				
Layer	**Rule**	**Description**	**SCMOS**	**SUBM**	**DEEP**
Well	1.1	Width	10	12	12
	1.2	Spacing to well at different potential	9	18	18
	1.3	Spacing to well at same potential	6	6	6
Active (diffusion)	2.1	Width	3	3	3
	2.2	Spacing to active	3	3	3
	2.3	Source/drain surround by well	5	6	6
	2.4	Substrate/well contact surround by well	3	3	3
	2.5	Spacing to active of opposite type	4	4	4
Poly	3.1	Width	2	2	2
	3.2	Spacing to poly over field oxide	2	3	3
	3.2a	Spacing to poly over active	2	3	4
	3.3	Gate extension beyond active	2	2	2.5
	3.4	Active extension beyond poly	3	3	4
	3.5	Spacing of poly to active	1	1	1
Select	4.1	Spacing from substrate/well contact to gate	3	3	3
	4.2	Overlap of active	2	2	2
	4.3	Overlap of substrate/well contact	1	1	1.5
	4.4	Spacing to select	2	2	4
Contact (to poly or active)	5.1, 6.1	Width (exact)	2x2	2x2	2x2
	5.2b, 6.2b	Overlap by poly or active	1	1	1
	5.3, 6.3	Spacing to contact	2	3	4
	5.4, 6.4	Spacing to gate	2	2	2
	5.5b	Spacing of poly contact to other poly	4	5	5
	5.7b, 6.7b	Spacing to active/poly for multiple poly/active contacts	3	3	3
	6.8b	Spacing of active contact to poly contact	4	4	4
Metal1	7.1	Width	3	3	3
	7.2	Spacing to metal1	2	3	3
	7.3, 8.3	Overlap of contact or via	1	1	1
	7.4	Spacing to metal for lines wider than 10 λ	4	6	6

Table 3.2	MOSIS design rules (continued)				
Layer	**Rule**	**Description**	**SCMOS**	**SUBM**	**DEEP**
Via1– Via(N-1)	8.1, 14.1, …	Width (exact)	2x2	2x2	3x3
	8.2, 14.2, …	Spacing to via on same layer	3	3	3
	8.4	Spacing to contacts (if no stacked vias)	2	2	n/a
	8.5	Spacing of via1 to poly or active edge	2	2	n/a
	14.4	Spacing of via2 to via1 (if no stacked vias)	2	2	n/a
Metal2– Metal(N-1)	9.1, …	Width	3	3	3
	9.2, …	Spacing to same layer metal	3	3	4
	9.3, …	Overlap of via	1	1	1
	9.4, …	Spacing to metal for lines wider than 10 λ	6	6	8
Metal3 (3-layer process)	15.1	Width	6	5	n/a
	15.2	Spacing to metal3	4	3	n/a
	15.3	Overlap of via2	2	2	n/a
	15.4	Spacing to metal for lines wider than 10 λ	8	6	n/a
Metal5 (5-layer process)	26.1	Width	n/a	4	4
	26.2	Spacing to metal5	n/a	4	4
	26.3	Overlap of via4	n/a	1	2
	26.4	Spacing to metal for lines wider than 10 λ	n/a	8	8
Metal6 (6-layer process)	30.1	Width	n/a	5	5
	30.2	Spacing to metal6	n/a	5	5
	30.3	Overlap of via5	n/a	1	2
	30.4	Spacing to metal for lines wider than 10 λ	n/a	10	10
Overglass Cut	10.1	Width of bond pad opening	60 μm		
	10.2	Width of probe pad opening	20 μm		
	10.3	Metal overlap of overglass cut	6 μm		
	10.4	Spacing of pad metal to unrelated metal	30 μm		
	10.5	Spacing of pad metal to active or poly	15 μm		

3.3.4 Micron Design Rules

Table 3.3 lists a set of micron design rules for a hypothetical 90 nm process representing an amalgamation of several real processes. Observe that the rules differ slightly but not immensely from lambda-based rules with λ = 0.05 μm.

Table 3.3	Micron design rules for 90nm process		
Layer	Rule	Description	90 nm rule (μm)
Well	1.1	Width	0.75
	1.2	Spacing to well at different potential	1.5
	1.3	Spacing to well at same potential	1.0
Active (diffusion)	2.1	Width	0.15
	2.2	Spacing to active	0.20
	2.3	Source/drain surround by well	0.25
	2.4	Substrate/well contact surround by well	0.25
	2.5	Spacing to active of opposite type	0.30
Poly	3.1	Width	0.09
	3.2	Spacing to poly over field oxide	0.15
	3.2a	Spacing to poly over active	0.15
	3.3	Gate extension beyond active	0.15
	3.4	Active extension beyond poly	0.15
	3.5	Spacing of poly to active	0.10
Select	4.1	Spacing from substrate/well contact to gate	0.25
	4.2	Overlap of active	0.20
	4.3	Overlap of substrate/well contact	0.10
	4.4	Spacing to select	0.30
Contact (to poly or active)	5.1, 6.1	Width (exact)	0.12
	5.2b, 6.2b	Overlap by poly or active	0.01
	5.3, 6.3	Spacing to contact	0.15
	5.4	Spacing to gate	0.10
Metal1	7.1	Width	0.13
	7.2	Spacing to well metal1	0.13
	7.3, 8.3	Overlap of contact or via	0.01
	7.4	Spacing to metal for lines wider than 0.5 μm	0.40
Via1–Via5	8.1, 14.1, …	Width (exact)	0.13
	8.2, 14.2, …	Spacing to via on same layer	0.13

(continued)

Table 3.3	Micron design rules for 90nm process (continued)		
Layer	**Rule**	**Description**	**90 nm rule (μm)**
Metal2–Metal6	9.1, …	Width	0.15
	9.2, …	Spacing to same layer metal	0.15
	9.3, …	Overlap of via	0.01
	9.4, …	Spacing to metal for lines wider than 1.0 μm	0.40
Via6		Width	0.20
		Spacing	0.20
Metal7		Width	0.40
		Spacing to metal7	0.40
		Overlap of via6	0.10
		Spacing to metal7 for lines wider than 1.0 μm	0.50

3.4 CMOS Process Enhancements

3.4.1 Transistors

3.4.1.1 Multiple Threshold Voltages and Oxide Thicknesses It has been mentioned that some processes offer multiple threshold voltages and/or oxide thicknesses. Low-threshold transistors deliver more ON current, but also have greater subthreshold leakage. Providing two or more thresholds permits the designer to use low-V_t devices on critical paths and higher-V_t devices elsewhere to limit leakage power. Multiple masks and implantation steps are used to set the various thresholds.

Thin gate oxides also permit more ON current. However, they break down when exposed to the high voltages needed in I/O circuits. Very thin oxides also contribute to large gate leakage currents. Many processes offer a second, thicker oxide for the I/O transistors (see Section 12.4.3). For example, 3.3 V I/O circuits commonly use 0.35 μm channel lengths and 7 nm gate oxides. An intermediate oxide thickness for low-leakage logic circuits can also be useful. Again, multiple masks are used to define the different oxides.

3.4.1.2 Silicon on Insulator A variant of CMOS that has been available for many years and is currently emerging in prominence is Silicon on Insulator (SOI). As the name suggests, this is a process where the transistors are fabricated on an insulator. Two main insulators are used, SiO_2 and sapphire. One major advantage of an insulating substrate is the elimination of the capacitance between the source/drain regions and body, leading to higher-speed devices. Another major advantage is lower subthreshold leakage. Proponents argue that SOI will be a core technology required to scale devices.

Figure 3.12 shows two common types of SOI. Figure 3.12(a) illustrates a sapphire substrate. In this technology (for example, Peregrine Semiconductor's Ultra Thin Silicon (UTSi), a thin layer of silicon is formed on the sapphire surface. The thin layer of silicon is selectively doped to define different threshold transistors. Gate oxide is grown on top of this and then polysilicon gates are defined. Following this, the nMOS and pMOS transistors are formed by implantation. Figure 3.12(b) shows a silicon-based SOI process. Here, a silicon substrate is used and a *buried oxide* (BOX) is grown on top of the silicon substrate. A thin silicon layer is then grown on top of the buried oxide and this is selectively implanted to form nMOS and pMOS transistor regions. Gate, source, and drain regions are then defined in a similar fashion to a bulk process. Sapphire is optically and RF transparent. As such, it can be of use in optoelectronic areas when merged with III-V based light emitters.

SOI devices and circuits are discussed further in Section 6.7.

3.4.1.3 High-k Gate Dielectrics

MOS transistors need high gate capacitance to attract charge to the channel. This leads to very thin SiO_2 gate dielectrics. Scaling trends indicate the gate leakage will be unacceptably large in such thin gates. Gates could use thicker dielectrics and hence leak less if a material with a higher dielectric constant were available. Materials such as hafnium oxide HfO_2 (dielectric constant $k = 20$), zirconium oxide ZrO_2 ($k = 23$), and silicon nitride Si_3N_4 ($k = 6.5$–7.5) [Ma98] have been proposed. These are called high-k dielectrics in contrast to SiO_2 with $k = 3.9$. They are applied using ALD, metallo-organic chemical vapor deposition (MOCVD) or sputtering [Wong02].

3.4.1.4 Low-leakage Transistors

Another problem with scaling bulk transistors is the subthreshold leakage from drain to source caused by the inability of the gate to turn off the channel. This can be improved by a gate structure where the gate is placed on two, three, or four sides of the channel. A promising candidate solves the problem by forming a vertical channel and constructing the gate in a pincer-like arrangement. These devices have been given the generic name *"finfets"* because the source/drain region forms fins on the silicon surface [Hisamoto98]. Figure 3.13(a) shows a 3D view of a finfet, while Figure 3.13(b) shows the cross-section and Figure 3.13(c) shows the top view. The gate wraps around three sides of the vertical source/drain fins. The width of the device is defined by the height of the fin, so wide devices are constructed by paralleling fins. Various other device structures have been proposed that even try to wrap the gate around the channel.

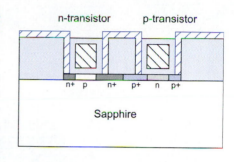

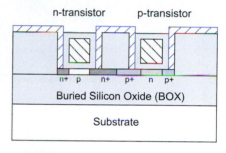

FIG 3.12 SOI types

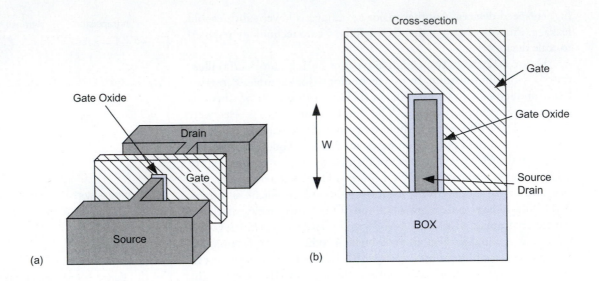

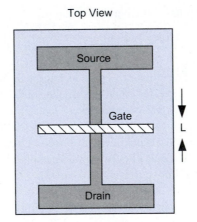

FIG 3.13 Finfet structure

3.4.1.5 Higher Mobility Increasing the mobility (μ) of the semiconductor improves drive current and transistor speed. This has been achieved by using silicon germanium (SiGe) for bipolar transistors in the same process as conventional CMOS devices. A typical SiGe bipolar transistor is shown in Figure 3.14. SiGe transistors can be constructed on conventional CMOS processing by adding a few extra implantation steps. The resulting bipolar transistors have extremely good radio frequency (RF) performance. Advanced SiGe transistors exhibit performance parameters that even III-V compounds such as

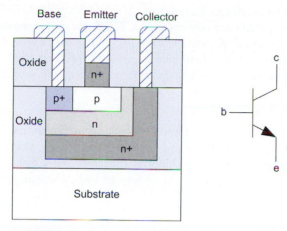

FIG 3.14 SiGe bipolar transistor structure

GaAs and InP find difficult to achieve in production. As a result of the combination of excellent RF performance and high-density digital CMOS, SiGe-based CMOS processes find wide application in communications circuits involving RF and high-speed switching [Hashimoto02, Harame01a, Harame01b].

Silicon germanium can also be used to improve the speed of conventional MOS transistors by creating what is called "*strained silicon*"—silicon into which is implanted germanium atoms that stretch the silicon lattice—as shown in Figure 3.15. This yields an increase in the mobility of the devices over conventional silicon of up to 70% and which corresponds to roughly a 30% increase in performance.

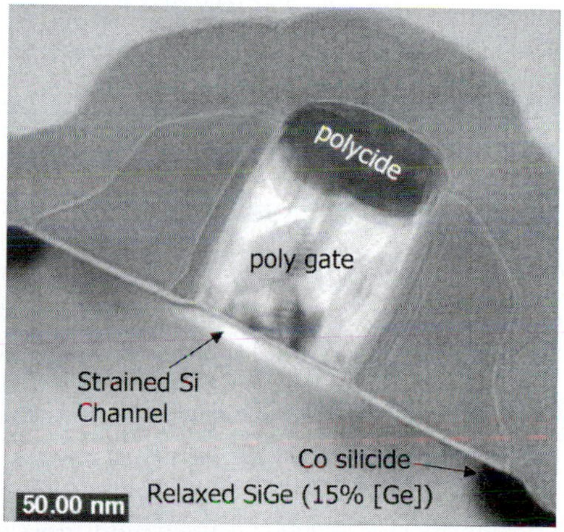

3.4.1.6 Plastic Transistors   MOS transistors can be fabricated with organic chemicals. These transistors show promise in active matrix displays or flexible electronic paper because the devices can be manufactured from an inexpensive chemical solution [Huitema03]. Figure 3.16 shows the structure of a plastic pMOS transistor. The transistor is built "upside down" with the gold gates and interconnect patterned first on the substrate. Then an organic insulator or silicon nitride is laid down, followed by the gold source and drain connections. Finally, the organic semiconductor (pentacene) is laid down. The mobility of the carriers in the plastic pMOS transistor is about 20 cm^2/V · s. This is about one-tenth that of a comparable silicon device. Typical lengths and widths are 5 μm and 400 μm, respectively.

FIG 3.15 IBM strained silicon transistor. Courtesy of International Business Machines Corporation. Unauthorized use not permitted.

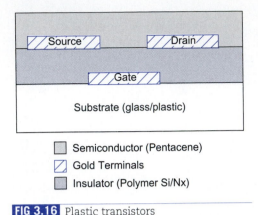

Semiconductor (Pentacene)
Gold Terminals
Insulator (Polymer Si/Nx)

FIG 3.16 Plastic transistors

3.4.1.7 High-voltage Transistors High-voltage MOSFETs can also be integrated onto conventional CMOS processes for switching and high-power applications. Gate oxide thickness and channel length have to be larger than usual to prevent breakdown. Specialized process steps are necessary to achieve very high breakdown voltages.

3.4.2 Interconnect

Interconnect has advanced rapidly. While two or three metal layers were once the norm, CMP has enabled inexpensive processes to include seven or more layers. Copper metal and low-k dielectrics are becoming popular to reduce the resistance and capacitance of these wires.

3.4.2.1 Copper Damascene Process While aluminum has traditionally remained the interconnect metal of choice, a large development effort has centered on using copper as an interconnect metal. This is primarily due to the higher conductivity of copper compared to aluminum. Some challenges of adopting copper include [Merchant01]:

- Copper atoms diffuse into the silicon and dielectrics, destroying transistors.

- The processing required to etch copper wires is tricky.

- Copper oxide forms readily and interferes with good contacts.

- Care has to be taken not to introduce copper into the environment as a pollutant.

Barrier layers have to be used to prevent the copper from entering the silicon surface. A new metallization procedure called the *damascene process* was invented to form this barrier. The process gets its name from the medieval metallurgists of Damascus who crafted fine inlaid swords. In a conventional subtractive aluminum-based metallization step, as we have seen, aluminum is layered on the silicon surface (where vias also have been etched) and then a mask and resist used to define which areas of metal are to be retained. The unneeded metal is etched away. A dielectric (SiO_2 or other) is then placed over the aluminum conductors and the process can be repeated.

A typical copper damascene process is shown in Figure 3.17, which is an adaptation of a dual damascene process flow from Novellus. Figure 3.17(a) shows a barrier layer over the prior metallization layer. This stops the copper from diffusing into the dielectric and silicon. The via dielectric is then laid down (Figure 3.17(b)). A further barrier layer can then be patterned and the line dielectric is layered on top of the structure as shown in Figure 3.17(c). An anti-reflective layer (which helps in the photolithographic process) is added to the top of the sandwich. The two dielectrics are then etched away where the lines and vias are required. A barrier layer such as 10 nm thick Ta or TaN film is then deposited to prevent the copper from diffusing into the dielectrics [Peng02]. As can be seen, a thin layer of the barrier remains at the bottom of the via so the barrier must be conductive. A

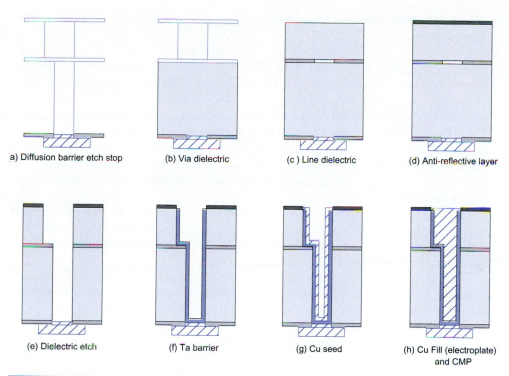

a) Diffusion barrier etch stop (b) Via dielectric (c) Line dielectric (d) Anti-reflective layer

(e) Dielectric etch (f) Ta barrier (g) Cu seed (h) Cu Fill (electroplate) and CMP

FIG 3.17 Copper dual damascene interconnect processing steps

copper seed layer is then coated over the barrier layer (Figure 3.17(g)). The resulting structure is electroplated full of copper and finally the structure is ground flat with CMP as shown in Figure 3.17(h).

3.4.2.2 Low-k Dielectrics SiO_2 has a dielectric constant of k = 3.9–4.2. Low-k dielectrics between wires are attractive because they decrease the wire capacitance [Brown03]. This reduces both wire delay and power consumption. Adding fluorine to the silicon dioxide creates fluorosilicate glass (FSG) with a dielectric constant of 3.6, widely used in 130 nm processes. Adding carbon to the oxide can reduce the dielectric constant to 2.7–3. Alternatively, porous polymer-based dielectrics can deliver even lower dielectric constants. For example, SiLK, from Dow Chemical, has k = 2.6 and may scale to k = 1.6–2.2 by increasing the porosity. Developing low-k dielectrics that can withstand the high temperatures during processing and the forces applied during CMP is a major challenge.

3.4.3 Circuit Elements

While CMOS transistors provide for almost complete digital functionality, the use of CMOS technology as the mixed signal and RF process of choice have driven the addition of special process options to enhance the performance of circuit elements required for these purposes.

3.4.3.1 Capacitors

In a conventional CMOS process, a capacitor can be constructed using the gate and source/drain of an MOS transistor, a diffusion area (to ground or V_{DD}), or a parallel metal plate capacitor (using stacked metal layers). The MOS capacitor has good capacitance per area but is relatively nonlinear if operated over large voltage ranges. The diffusion capacitor cannot be used for a floating capacitor (but is useful as a bypass capacitor). The metal parallel plate capacitor has low capacitance per area. Normally the aim in using a floating capacitor is to have the highest ratio of desired capacitance value to stray capacitance (to ground normally). The bottom metal plate contributes stray capacitance to ground.

Analog circuits frequently require capacitors in the range of 1 to 10 pF. The first method for doing this was to add a second polysilicon layer so that a *poly-insulator-poly* (PIP) capacitor could be constructed. A thin oxide was placed between the two polysilicon layers to achieve capacitance of approximately 1 fF/μm^2.

The most common capacitor used in CMOS processes today is the MIM or metal-insulator-metal capacitor that is normally placed between metal layers n and $n-1$ (where n is normally the top level metal layer) to minimize the stray capacitance of the bottom plate. A typical insulator is an alumina (Al_2O_3)/tantalum pentoxide (Ta_2O_5) sandwich. These capacitors have capacitances of 1–4 fF/μm^2 and provide very area efficient capacitors. A typical MIM capacitor is shown in Figure 3.18(a).

Another type of capacitor that is possible in scaled processes is the fringe (or fractal) capacitor [Samavati98, Sowiati01], which is composed of interdigitated metal fingers as shown in Figure 3.18(b). The original fractal capacitor [Samavati98] has a more involved layout and was necessary in older processes. Successive metal layers can be ganged to achieve more capacitance. If the upper layers of metal are used, a very linear, high-Q and high valued capacitor with low parasitic capacitance to ground can be constructed without any extra process steps at almost the same values as MIM capacitors. This approach works best at 130 nm and below where the metal lines are very closely spaced and thus have high fringing capacitance.

Integrated capacitors have a voltage and temperature dependence. Foundry design guides should be consulted for these parameters.

3.4.3.2 Resistors

In unaugmented processes, resistors can be built from any layer, with the final resistance depending on the resistivity of the layer. Building large resistances in a small area requires layers with high resistivity, particularly polysilicon and diffusion. Diffusion has a large parasitic capacitance to ground, making it unsuitable for high-frequency applications. Polysilicon gates are usually silicided to have low resistivity. The fix for this is to allow for *undoped* high-resistivity polysilicon. This is specified with a mask that blocks the silicide where high-value poly resistors are required. The resistivity can be tuned to around 300–1000 ohms/square, depending on doping levels. Another material used for high-quality resistors is nichrome, although this requires a special processing step.

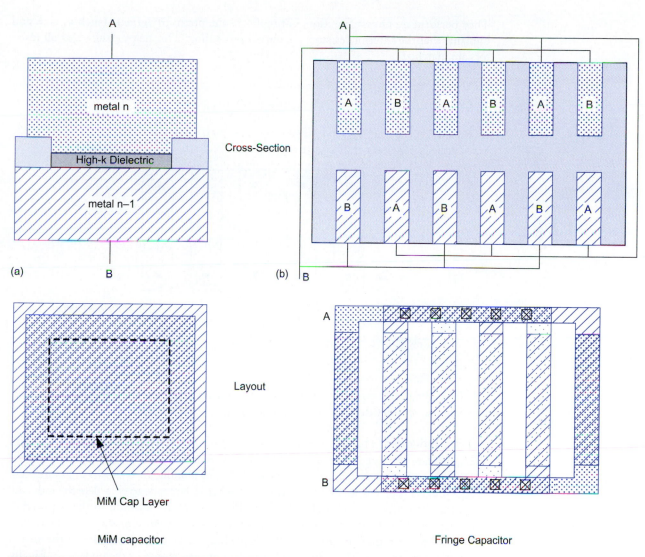

FIG 3.18 MIM and fringe capacitors

A typical resistor layout is shown in Figure 3.19. This geometry is sometimes called a *meander* structure. A number of *unit* resistors have been used so that a variety of matched resistor values can be constructed. For instance, if 20K, 10K, and 15K ohm resistors were required, a unit value of 5K could be used. Then three resistors (as shown) would construct a 15K ohm resistor. The two resistors at the ends are called *dummy resistors* or fingers.

They perform no circuit function, but replicate the proximity effects (such as etch and implant) that the interior resistors see during processing. This helps ensure that all resistors are matched.

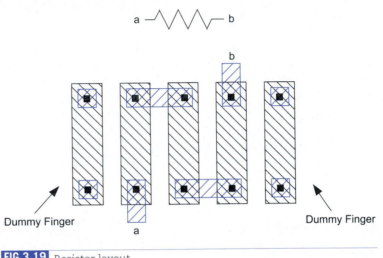

FIG 3.19 Resistor layout

Like integrated capacitors, the various resistor options have temperature and voltage coefficients. Foundry design manuals normally include these values.

3.4.3.3 Inductors

The desire to integrate inductors on chips has increased radically with the upsurge in interest in RF circuits. The most common monolithic inductor is the *spiral* inductor, which is a spiral of upper-level metal. A typical inductor is shown in Figure 3.20. As the process is planar, an underpass connection has to be made to complete the inductor. A typical equivalent model is also shown in Figure 3.20. In addition to the required L, there are several parasitic components. Rs is the series resistance of the metal (and contacts) used to form the inductor. Cp is the parallel capacitance to ground due to the area of the metal wires forming the inductor. Cs is the shunt capacitance of the underpass. Finally, Rp is an element that models the loss incurred in the resistive substrate.

Usually, when considering an inductor, the parameters of interest to a designer are its inductance, the Q of the inductor, and the self-resonant frequency. High Q's are sought to create low phase-noise oscillators, narrow filters, and low-loss circuits in general. Q values for typical planar inductors on a bulk process are in the range from 5 to 10.

The number of turns n required to achieve some inductance L if the wire pitch is $P = W + S$ is [Lee98]

$$n \approx 3\sqrt{\frac{LP}{\mu_0}}$$

(3.1)

where $\mu_0 = 1.2 \cdot 10^{-6}$ H/m is the permeability of free space. Higher-quality inductors can also be manufactured using bond wires between I/O pads. The inductance of a wire of length l and radius r is approximately

$$L \approx \frac{\mu_0 l}{2\pi}\left[\ln\frac{2l}{r} - 0.75\right] \qquad (3.2)$$

or about 1 nH/mm for standard 1 mil (25 μm) bond wires.

Reduction in Q occurs because of the resistive loss in the conductors used to build the inductor (Rs) and the eddy current loss in the resistive silicon substrate (Rp). In an effort to increase Q, designers have resorted to removing the substrate below the inductor using MEMS techniques [Yoon02]. Designers have also used bond wires for high Q inductors. The easiest way to improve the Q of monolithic inductors is to increase the thickness of the top level metal. The Q can also be improved by using a patterned ground shield in polysilicon under the inductor to decrease substrate losses.

3.4.3.4 Transmission Lines

A *transmission line* can be used on a chip to provide a known impedance wire. Two basic kinds of transmission line are commonly used: *microstrip* and *coplanar waveguide*. These are shown in Figure 3.21.

A microstrip transmission line is shown in Figure 3.21(a). It is composed of a wire of width w placed over a ground plane and separated by a dielectric of height h and dielectric constant k. In the chip case, the wire might be the top level of metallization and the ground plane the next metal down.

A coplanar waveguide does not have to have a sublayer ground plane and is shown in Figure 3.21(b). It consists of a wire of width w spaced s on each side from *coplanar* ground wires. The reader is referred to [Wadell01] for detailed design equations.

3.4.3.5 Non-volatile Memory

Non-volatile memory (NVM) retains its state when the power is removed from the circuit. The simplest NVM is a *mask-programmed* ROM cell (see Section 11.5.1). This type of NVM is not reprogrammable or programmable after the device is manufactured. A *one-time programmable* (OTP) memory can be implemented using a fuse constructed of a thin piece of metal through which is passed a current that vaporizes the metal by exceeding the current density in the wire. The first really reprogrammable memories used a stacked polysilicon gate structure and were programmed by applying a high voltage to the device in a manner that caused Fowler-Nordheim tunneling to store a charge on a floating gate. The whole memory could be

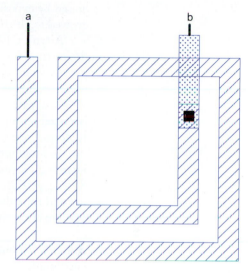

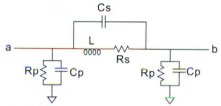

FIG 3.20 Typical spiral inductor and equivalent circuit [Rotella02]

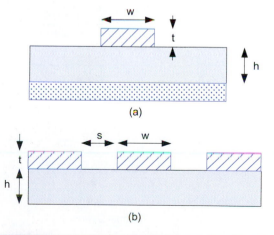

FIG 3.21 Microstrip and coplanar waveguide

erased by exposing it to UV light that knocked the charge off the gate. These memories evolved to *electrically erasable* memories, which are today represented by *Flash* memory.

A typical Flash memory transistor is shown in Figure 3.22 [She02]. The source and drain structures can vary considerably to allow for high-voltage operation, but the dual-gate structure is fairly common. The gate structure is a stacked configuration commencing with a thin tunneling oxide. A floating polysilicon gate sits on top of this oxide and a conventional gate oxide is placed on top of the floating gate. Finally, a polysilicon control gate is placed on top of the gate oxide. The operation of the cell is also shown in Figure 3.22. In normal operation, the floating gate determines whether or not the transistor is conducting. To program the cell, the source is left floating and the control gate is raised to approximately 20 V (using an on-chip voltage multiplier). This causes electrons to tunnel into the floating gate, thus programming it. To deprogram a cell, the drain and source are left floating and the substrate (or well) is connected to 20 V. The electrons stored on the floating gate are attracted away, leaving the gate in an unprogrammed state.

3.4.3.6 Bipolar Transistors   Bipolar transistors were mentioned previously in our discussion of SiGe process options. Both *npn* and *pnp* bipolar transistors can be added to a CMOS process, which is then called a BiCMOS process. These processes tend to be used

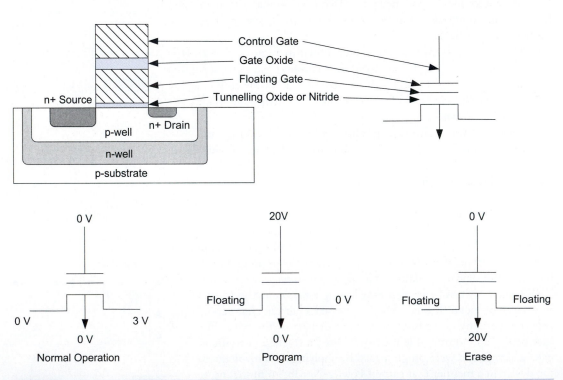

FIG 3.22 Flash memory construction and operation

for specialized analog or high-voltage circuits. In a regular n–well process, a parasitic pnp transistor is present that can be used for circuits such as bandgap voltage references. This transistor is shown in Figure 3.23 with the p-substrate collector, the n-well base, and the p-diffusion emitter. Both process cross-section and layout are shown. This transistor in conjunction with a parasitic npn is the cause of latchup (see Section 4.8.5).

3.4.3.7 Fuses and Antifuses

The use of a fuse was mentioned in the section on non-volatile memory. Fuses can be blown with a high current or zapped by a laser. In the latter case, an area is normally left in the passivation oxide to allow the laser direct access to the metal link that is to be cut. Figure 3.24 shows the layout of a metal fuse.

An *antifuse* is a device that initially has a high resistivity but can become low resistance when a programming voltage is applied. This device requires special processing and is used in programmable logic devices (see Section 8.3.2.2).

3.4.3.8 Micro Electro Mechanical Systems (MEMS)

Semiconductor processes and especially CMOS processes have been used to construct tiny mechanical systems monolithically. A typical device is the well-known air-bag sensor, which is a small accelerometer

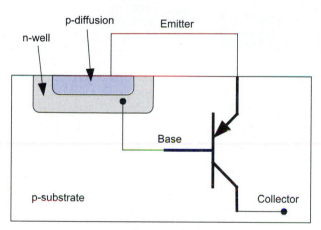

Cross-section

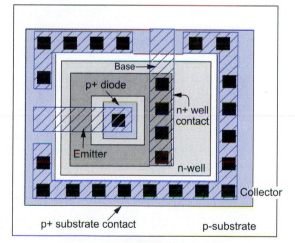

Representative Layout

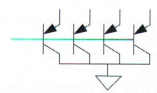

Typical pnp circuit structure as used in voltage reference

FIG 3.23   Parasitic pnp bipolar transistor

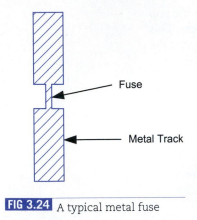

consisting of an air bridge capacitor that can detect sudden changes in acceleration when co-integrated with some conditioning electronics. Structures such as cantilevers, mechanical resonators, and even micromotors have been built. A full discussion of MEMS is beyond the scope of this book, but further material can be found in [Maluf00, Kovacs98].

3.4.4 Beyond Conventional CMOS

Nanotechnology is presently a hot research area seeking alternative structures to replace CMOS when scaling finally runs out of steam. For example, carbon nanotubes have been used to demonstrate transistor behavior and build inverters [Liu01]. They are of interest as the nanotube is smaller than the predicted endpoint for CMOS gate lengths. A nanotube transistor is shown in Figure 3.25 [Collins01]. Presently, the speeds are quite slow and the manufacturing techniques are limited, but they may be of interest in the future.

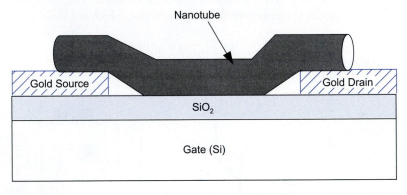

FIG 3.25 Carbon nanotube transistor

3.5 Technology-related CAD Issues

The mask database is the interface between the semiconductor manufacturer and the chip designer. Two basic checks have to be completed to ensure that this description can be turned into a working chip. First, the specified geometric design rules must be obeyed. Second, the interrelationship of the masks must, upon passing through the manufacturing process, produce the correct interconnected set of circuit elements. To check these two requirements, two basic CAD tools are required, namely a *Design Rule Check* (DRC) program and a mask circuit *extraction* program. The most common approach to implementing these tools is to provide a set of subprograms that perform general geometry operations. A particular set of DRC rules or extraction rules for a given CMOS process (or any semiconductor process) is then indicated by a specification of the operations that must be performed on each mask and the inter-mask checks that must be completed. Accompanied by a written specification, these *run sets* are usually the defining specification for a process.

In this section we will examine a hypothetical DRC and extraction system to illustrate the nature of these run sets.

3.5.1 Design Rule Checking (DRC)

Although we can design the physical layout in a certain set of mask layers, the actual masks used in fabrication can be derived from the original specification. Similarly, when we want a program to determine what we have designed by examining the interrelationship of the various mask layers, it may be necessary to determine various logical combinations between masks.

To examine these concepts, let us posit the existence of the following functions (loosely based on the Cadence DRACULA DRC program), which we will apply to a geometric database (i.e., rectangles, polygons, paths):

AND `layer1 layer2 -> layer3`
ANDs layer1 and layer2 together to produce layer3
(i.e., the intersection of the two input mask descriptions).

OR `layer1 layer2 -> layer3`
ORs layer1 and layer2 together to produce layer3
(i.e., the union of the two input mask descriptions).

NOT `layer1 layer2 -> layer3`
Subtracts layer2 from layer1 to produce layer3
(i.e., the difference of the two input mask descriptions).

WIDTH `layer > dimension -> layer3`
Checks that all geometry on layer is larger than dimension.
Any geometry that is not is placed in layer3.

SPACE `layer > dimension -> layer3`
Checks that all geometry on layer is spaced further than dimension.
Any geometry that is not is placed in layer3.

The following layers will be assumed as input:

```
nwell
active
p-select
n-select
poly
poly-contact
active-contact
metal
```

Typically, useful sublayers are first generated. First, the four kinds of active area are isolated. The rule set to accomplish this is as follows:

```
NOT   all nwell -> substrate
AND   nwell active -> nwell-active
NOT   active nwell -> pwell-active
AND   nwell-active p-select -> pdiff
AND   nwell-active n-select -> vddn
AND   pwell-active n-select -> ndiff
AND   pwell-active p-select -> gndp
```

In the above specification, a number of new layers have been specified. For instance, the first rule states that wherever nwell is absent, a layer called substrate exists. The second rule states that all active areas within the nwell are nwell-active. A combination of nwell-active and p-select or n-select yields pdiff (p diffusion) or vddn (well tap).

To find the transistors, the following rule set is used:

```
AND poly ndiff -> ngates
AND poly pdiff -> pgates
```

The first rule states that the combination of polysilicon and ndiff yields the ngates region—all of the n-transistor gates.

Typical design rule checks (DRC) might include the following:

```
WIDTH metal  < 0.13 -> metal-width-error
SPACE metal  < 0.13 -> metal-space-error
```

For instance, the first rule determines if any metal is narrower than 0.13 µm and places the errors in the metal-width-error layer. This layer might be interactively displayed to highlight the errors.

3.5.2 Circuit Extraction

Now imagine that we want to determine the electrical connectivity of a mask database. The following commands are required:

CONNECT layer1 layer2
Electrically connect layer1 and layer2.

MOS name drain-layer gate-layer source-layer substrate-layer
Define an MOS transistor in terms of the component terminal layers. (This is, admittedly, a little bit of magic.)

The connections between layers can be specified as follows:

```
CONNECT active-contact pdiff
CONNECT active-contact ndiff
CONNECT active-contact vddn
CONNECT active-contact gndp
CONNECT active-contact metal
```

```
CONNECT gndp substrate
CONNECT vddn nwell
CONNECT poly-contact poly
CONNECT poly-contact metal
```

The connections between the diffusions and metal are specified by the first seven statements. The last two statements specify how metal is connected to poly.

Finally, the active devices are specified in terms of the layers that we have derived.

```
MOS nmos ndiff ngates ndiff substrate
MOS pmos pdiff pgates pdiff nwell
```

An output statement might then be used to output the extracted transistors in some netlist format (i.e., SPICE format). The extracted netlist is often used to compare the layout against the intended schematic.

It is important to realize that the above run set is manually generated. The data you extract from such a program is only as good as the input. For instance, if parasitic routing capacitances are required, then each layer interaction must be coded. If parasitic resistance is important in determining circuit performance, it also must be specifically included in the extraction run set.

3.6 Manufacturing Issues

As processes have evolved, various design rules have emerged that reflect the complexity of the processing. This section will cover some important areas.

3.6.1 Antenna Rules

When a metal wire contacted to a transistor gate is plasma-etched, it can charge up to a voltage sufficient to break down thin gate oxides. The metal can be contacted to diffusion to provide a path for the charge to bleed away. *Antenna rules* specify the maximum area of metal that can be connected to a gate without a source or drain to act as a discharge element. They are somewhat hard to visualize, but are fixed by placing jumpers to a higher layer of metal to shorten the metal segment or by placing diffusion diodes on wires. The design rule normally defines the maximum ratio of metal area to gate area such that charge on the metal will not damage the gate. The ratios can vary from 100:1 to 5000:1 depending on the thickness of the gate oxide (and hence breakdown voltage) of the transistor in question. Higher ratios apply to thicker gate oxide transistors (i.e., 3.3 V I/O transistors).

Figure 3.26 shows a typical fix to an antenna violation. In the top layout, the wire (L2) exceeds the antenna rule for the process. If this were to be connected to the gate, the gate would possibly break down in subsequent processing steps. The fix is shown in the bottom diagram, where a link to the top metal layer is made so that the length is now L1. At the point the source/drain connection is made, the gate is protected. This occurs at the final metallization step. The link has to be made at the level where a source/drain region

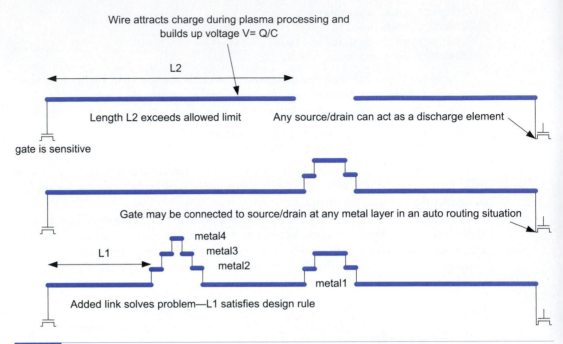

FIG 3.26 Antenna rule violation and fix

will connect to the gate. For safety's sake, in dense layouts it is normally assumed that the source/drain is connected at the final metallization step (in this case metal4). An alternative method is to attach source/drain diodes to problem nets as shown in Figure 3.27. These diodes can be simple junctions of n-diffusion to p-substrate rather than transistor source/drain regions.

3.6.2 Layer Density Rules

Another set of rules that pertain to advanced processes are layer density rules, which specify a minimum and maximum density of a particular layer within a specified area. These are required as a result of the CMP process and the desire to achieve uniform etch rates. For instance, a metal layer might have to have 30% minimum and 70% maximum fill within a 1 mm by 1 mm area. For digital circuits, these density levels are normally reached with normal routing. Analog and RF circuits, on the other hand, are almost by definition, sparse. Thus, gate and metal layers may have to be added manually or by a fill program after design has been completed. In some circumstances, the fill may have to be grounded via n-diffusion diodes. This is especially true for RF circuits where the metal1 fill can be used for a ground plane. Designers must be aware of the fill so that it does not introduce unexpected parasitic capacitance to nearby wires.

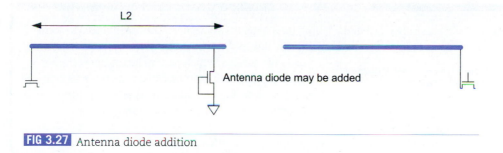

FIG 3.27 Antenna diode addition

3.6.3   Resolution Enhancement Rules

Some resolution enhancement techniques impose further design rules. For example, polysilicon typically uses the narrowest lines and thus needs the most enhancement. This can be simplest if polysilicon gates are only drawn in a single orientation (horizontal or vertical).

3.7   Pitfalls and Fallacies

Targeting a bleeding-edge process
There is a fine balance when you are deciding whether or not to move to a new process for a new design. On the one hand, you are tempted by increased density and speed. On the other hand, support for the new process can initially be expensive (becoming familiar with process rules, CAD tool scripts, porting analog and RF designs, locating logic libraries, etc.). In addition, CMOS foundries frequently tune their processes in the first few months of production, and often, yield improvement steps can reflect back to design rule changes that can impact designs late in their tapeout schedule. For this reason, it is frequently prudent not to jump immediately into a new process when it becomes available. On the other hand, if you are limited in speed or some other attribute that is solved by the new process, then you don't have much choice but to bite the bullet.

Using lambda design rules on commercial designs
Lambda rules have been used in this text for ease of explanation and consistency. They are useable for class designs. However, they are not very useful for production designs for deep sub-micron processes. Of particular concern are the metal width and spacing rules, which are too conservative for most production processes.

Failing to account for the parasitic effects of metal fill
With area density rules, particularly in metal, most design flows include an automatic fill step to achieve the correct metal density. Particularly in analog and RF circuits, it is important to either exclude the automatic fill operation from that area or check circuit performance

(continued)

after the fill by completing a full parasitic extract and rerunning the verification simulation scripts.

Failing to include process calibration test structures

It was mentioned in the discussion on scribe line structures that test structures are frequently inserted here by the silicon manufacturer. Documentation is often unavailable so it is prudent for designers (particularly in academic designs, which receive less support from a foundry) to include their own test structures such as transistors or ring oscillators. This allows designers to calibrate the silicon against simulation models.

3.8 Historical Perspective

In the last 30 years, we (especially those of us who are older in years) have lived through an amazing period of technology evolution in integrated circuits. While some hard limits are being approached in terms of lithography and device dimensions, history shows that the technology takes a path that leads to new devices, applications, and directions.

Summary

Being aware of CMOS process options and directions can greatly influence design decisions. Frequently, the combination of performance and cost possibilities in a new process can provide new product opportunities that were not available previously. Similarly, venerable processes can offer good opportunities with the right product.

One issue that has to be kept in mind is the ever-increasing cost of having a CMOS design fabricated in a leading-edge process. Mask cost for critical layers is in the vicinity of $100K per mask. A full mask set for an advanced process approaches $1M in cost. This in turn is reflected in the types of design and approaches to design that are employed for CMOS chips of the future. For instance, making a design programmable so that it can have a longer product life is a good first start. Chapter 8 covers these approaches in depth.

For more advanced reading on silicon processing, consult textbooks such as [Wolf00].

Exercises

3.1   A 248 nm UV step and scan machine costs $10M and can produce eighty 300 mm diameter, 90 nm node wafers per hour. A 157 nm UV step and scan machine costs $40M and can process twenty 300 mm diameter, 50 nm node wafers per hour. If the machines have a depreciation period of four years, what is the difference in the cost per chip for a chip that occupies 50 square mm at 90 nm resolution if the stepper is used 10 times per process run for the critical layers?

3.2   If the gate oxide thickness in a SiO_2-based structure is 2 nm, what would be the thickness of a HfO_2-based dielectric providing the same capacitance?

3.3   Explain the difference between a polycide and a salicide CMOS process. Which would be likely to have higher performance and why?

3.4   Draw the layout for a pMOS transistor in an n-well process that has active, p-select, n-select, polysilicon, contact, and metal1 masks. Include the well contact to V_{DD}.

3.5   What is the best possible metal for interconnect? Why isn't it used?

3.6   Using Table 3.2, calculate the minimum contacted pitch as shown in Figure 3.28 for metal1, metal3, and metal6 in terms of lambda using the SUBM rules in a 6-layer process. Is there a wiring strategy that can reduce this pitch?

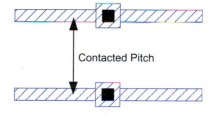

FIG 3.28   Contacted metal pitch

3.7   Using Table 3.2, calculate the minimum uncontacted and contacted transistor pitch as shown in Figure 3.29.

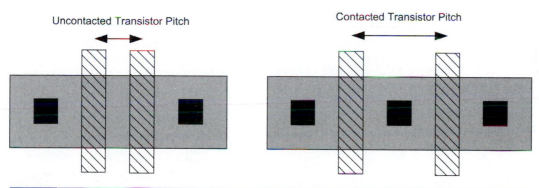

FIG 3.29   Uncontacted and contacted transistor pitch

3.8   Using Figure 3.30 and Table 3.2, calculate the minimum n to p pitch and the minimum inverter height with and without the poly contact to the gate (in). If an SOI process has 2 λ spacing between n and p diffusion, to what are the two pitches reduced?

3.9   Design a metal6 fuse ROM cell in a process where the minimum metal width is 0.5 μm and the maximum current density is 2 mA/μm. A fuse current of less than 10 mA is desired.

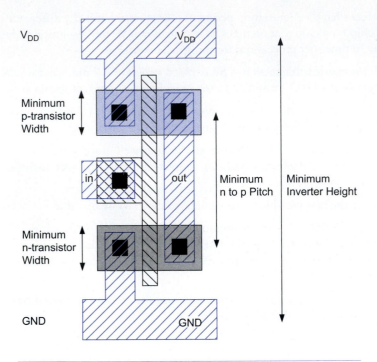

FIG 3.30 Minimum inverter height

Circuit Characterization and Performance Estimation

4

4.1 Introduction

In Chapter 1 we learned how to make chips that work. Now we move on to making chips that work *well*, where "well" can be defined as fast, low in power, inexpensive to manufacture and reliable. Before we can choose which design alternative is best, we must develop ways to estimate the goodness of each option, especially with regard to speed and power consumption.

The most obvious way to characterize a circuit is through simulation, and that will be the topic of Chapter 5. Unfortunately, simulations only inform us how a particular circuit behaves, not how to change the circuit to make it better. Moreover, if we don't know approximately what the result of the simulation should be, we are unlikely to catch bugs in our simulation model. Mediocre engineers rely predominantly on computer tools, but outstanding engineers develop their physical intuition to rapidly estimate the behavior of circuits.

In this chapter we are primarily concerned with the development of simple models that will assist us in the understanding of system performance. In a modern process, interconnect performance can be as important as or even more important than transistor performance. The issues to be considered in this chapter are

- Delay estimation in CMOS gates
- Power dissipation of CMOS logic
- Interconnect delay and signal integrity
- Design margining
- Reliability
- Effects of scaling

4.2 Delay Estimation

In most designs there will be many logic paths that do not require any conscious effort when it comes to speed. However, usually there will be a number of paths, called the *critical paths*, that require attention to timing details. These can be recognized by experience or timing simulation, but most designers use a timing analyzer, which is a design tool that automatically finds the slowest paths in a logic design (see Section 8.4.1.5). The critical paths can be affected at four main levels:

- The architectural/microarchitectural level

- The logic level

- The circuit level

- The layout level

The most leverage is achieved with a good microarchitecture. This requires a broad knowledge of both the algorithms that implement the function and the technology being targeted, such as how many gate delays fit in a clock cycle, how fast addition occurs, how fast memories are accessed, and how long signals take to propagate along a wire. Tradeoffs at the microarchitectural level include the number of pipeline stages, the number of execution units, and the size of memories.

The next level of timing optimization comes at the logic level. Tradeoffs include types of functional blocks (e.g., ripple carry vs. lookahead adders), the number of stages of gates in the cycle, and the fan-in and fan-out of the gates. The transformation from function to gates and registers can be done by experience, by experimentation, or increasingly by logic synthesis. Remember, however, that no amount of skillful logic design can overcome a poor microarchitecture.

Once the logic has been selected, the delay can be tuned at the circuit level by choosing transistor sizes or using other styles of CMOS logic. Finally, delay is dependent on the layout. The floorplan (either manually or automatically generated) is of great importance because it determines the wire lengths that can dominate delay. Tuning of particular cells can also reduce parasitic capacitance.

This section focuses on the logic and circuit optimizations of selecting the number of stages of logic, the types of gates, and the transistor sizes. Chapter 6 addresses other circuit styles and layout techniques. Chapters 10 and 11 examine design of datapath and array functional blocks.

Quick delay estimation is essential to designing critical paths. Although timing analyzers or circuit simulators can compute very detailed switching waveforms and accurately predict delay, good designers cannot be dependent on simulation alone. Simulation or timing analysis only answers how fast a particular circuit operates; they do not resolve the more interesting question of how the circuit could be modified to operate faster. Many novice designers spend countless hours tweaking parameters in a circuit simulator and resimulating only to find tiny improvements. Simple models that can be applied on the back of an envelope are important because they allow us to rapidly estimate delay, understand its origin, and figure out how it can be reduced. This section applies the RC delay model to estimate the delay of logic gates.

We begin with a few definitions:

- *Rise time, t_r* = time for a waveform to rise from 20% to 80% of its steady-state value

- *Fall time, t_f* = time for a waveform to fall from 80% to 20% of its steady-state value

- *Edge rate, t_{rf}* = $(t_r + t_f)/2$

- *Propagation delay time, t_{pd}* = maximum time from the input crossing 50% to the output crossing 50%

- *Contamination delay time, t_{cd}* = minimum time from the input crossing 50% to the output crossing 50%

Intuitively, we know that when an input changes, the output will retain its old value for at least the contamination delay and take on its new value in at most the propagation delay. We sometimes differentiate between the delays for the output rising, t_{pdr}/t_{cdr}, and the output falling, t_{pdf}/t_{cdf}. Rise/fall times are also sometimes called *slopes* or *edge rates*. Propagation and contamination delay times are also called *max-time* and *min-time*, respectively. The gate that charges or discharges a node is called the *driver* and the gates and wire being driven are called the *load*.

4.2.1 RC Delay Models

Section 2.6 developed a lumped RC model for transistors. Although transistors have complex nonlinear current-voltage characteristics, they can be approximated fairly well as a switch in series with a resistor, where the *effective resistance* is chosen to match the average amount of current delivered by the transistor. Transistor gate and diffusion nodes have capacitance.

In this section we apply the model to estimate the delay of logic gates as the RC product of the effective driver resistance and the load capacitance. Usually, logic gates use minimum-length devices for least delay, area, and power consumption. Given this, the delay of a logic gate depends on the widths of the transistors in the gate and the capacitance of the load that must be driven.

4.2.1.1 Effective Resistance and Capacitance

Recall from Section 2.6 that an nMOS transistor with width of one unit is defined to have effective resistance R. The unit-width pMOS has a higher resistance that depends on its mobility relative to the nMOS transistor. For concreteness, let us assume this resistance is $2R$. Wider transistors have lower resistance. For example, a pMOS transistor of double-unit width has effective resistance R. Parallel and series transistors combine like conventional resistors. When multiple transistors are in series, their resistance is the sum of each individual resistance. When multiple transistors are in parallel, the resistance is lower if they are all ON. In many gates, the worst-case delay occurs when only one of several parallel transistors is ON. In that case, the effective resistance is just that of the single transistor.

Example

Sketch a 3-input NAND gate with transistor widths chosen to achieve effective rise and fall resistance equal to that of a unit inverter (R).

Solution: Figure 4.1 shows such a gate. The three nMOS transistors are in series so the resistance is three times that of a unit transistor. Therefore, each must be three times unit width to compensate. In other words, each transistor has resistance $R/3$ and the series combination has resistance R. The three pMOS transistors are in parallel. In the worst case (with one of the three inputs low), only one of the three pMOS transistors is ON. Therefore, each must be twice unit width to have resistance R.

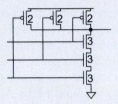

FIG 4.1   3-input NAND gate with unit rise and fall resistance

Also recall that the capacitance consists of gate capacitance and source/drain diffusion capacitance. Let us define the gate capacitance of a unit transistor to be C_g and the diffusion capacitance of its contacted source and drain to each be C_{diff}. In many processes the capacitances are approximately equal and can be labeled $C = C_g = C_{\text{diff}}$ to keep estimation simple. The second terminal of the diffusion capacitor is the body, which is usually tied to ground (for nMOS) or V_{DD} (for pMOS). As the DC voltage on the second terminal is irrelevant to delay, we often draw both capacitances to ground for simplicity. The gate capacitance includes fields terminating on the channel, source, and drain. To make hand analysis tractable, it can be approximated as a single capacitance to the V_{DD} or GND rail. C_g and C_{diff} are proportional to transistor width.

4.2.1.2 Diffusion Capacitance Layout Effects

In a good layout, diffusion nodes are shared wherever possible to reduce the diffusion capacitance. Moreover, the uncontacted diffusion nodes between series transistors are usually smaller than those that must be contacted. Such uncontacted nodes have less capacitance (see Sections 2.3.3 and 5.4.4), although we will neglect the difference for hand calculations.

Example

Annotate the 3-input NAND gate of Figure 4.1 with its gate and diffusion capacitances. Assume all diffusion nodes are contacted.

Solution: Figure 4.2(a) shows the gate and its capacitances. Each input drives five units of gate capacitance. Notice that the capacitors on source diffusions attached to the rails have both terminals shorted together so they are irrelevant to circuit operation. Figure 4.2(b) redraws the gate with these capacitances deleted and the remaining capacitances lumped to ground.

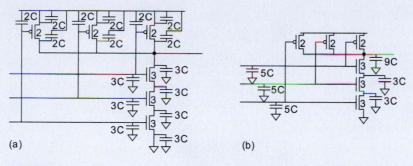

(a)   (b)

FIG 4.2 3-input NAND gate annotated with capacitances

In summary, gate capacitance can be determined directly from the transistor widths in the schematic. Diffusion capacitance depends on the layout. A conservative method of estimating capacitances before layout is to assume uncontacted diffusion between series transistors and contacted diffusion on all other nodes.

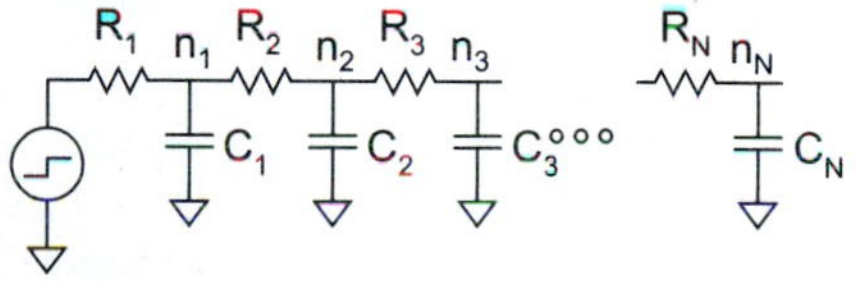

4.2.1.3 Elmore Delay Model Viewing ON transistors as resistors, we see that a chain of transistors can be represented as an RC ladder as shown in Figure 4.3. The Elmore delay [Elmore48] model estimates the delay of an RC ladder as the sum over each node in the ladder of the resistance R_{n-i} between that node and a supply multiplied by the capacitance on the node:

FIG 4.3 RC ladder for Elmore delay

$$t_{pd} = \sum_i R_{n-i} C_i = \sum_{i=1}^{N} C_i \sum_{j=1}^{i} R_j \qquad (4.1)$$

Example

Figure 4.4 shows a layout of the 3-input NAND gate. A single drain diffusion region is shared between two of the pMOS transistors. Estimate the actual diffusion capacitance from the layout.

Solution: Figure 4.5(a) shows the diffusion capacitance on the layout, neglecting those attached to the rails. Figure 4.5(b) redraws the schematic with these capacitances lumped to ground. Observe that two of the pMOS share a single diffusion region so the capacitance is smaller than predicted in Figure 4.2(b).

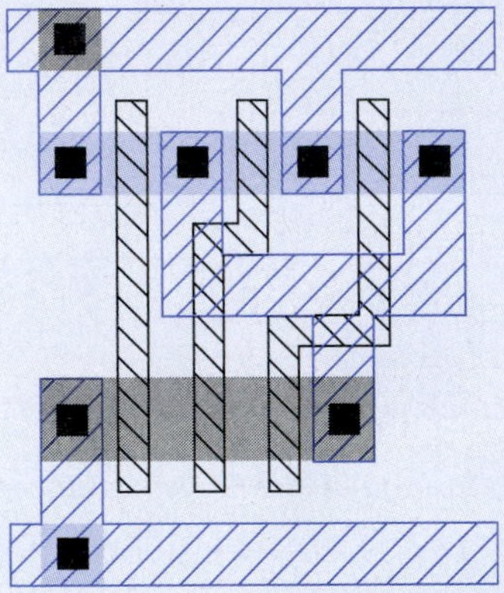

FIG 4.4   3-input NAND gate layout

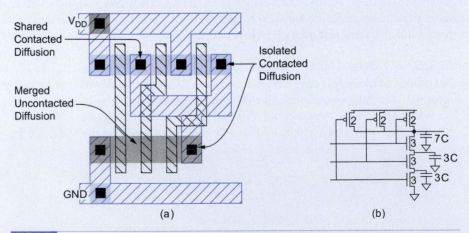

FIG 4.5   3-input NAND annotated with diffusion capacitances extracted from the layout

Example

Sketch a 2-input NAND gate with transistor widths chosen to achieve effective rise and fall resistance equal to a unit inverter. Compute the rising and falling propagation delays (in terms of R and C) of the NAND gate driving h identical NAND gates using the Elmore delay model. If $C = 2$ fF/μm and $R = 2.5$ k$\Omega \cdot \mu$m in a 180 nm process, what is the delay of a fanout-of-3 NAND gate?

Solution: Figure 4.6(a) shows such a NAND gate annotated with diffusion capacitance assuming contacted diffusion on each transistor except between the series nMOS transistors. The two nMOS transistors are in series so each must be made twice as wide to achieve overall unit resistance. The two pMOS transistors are in parallel. In the worst case, only one is ON, so it must be the same width as in the inverter, i.e., twice unit width because pMOS transistors have lower mobility than nMOS transistors. Each input is connected to 4 units of gate width. Hence the output load of h identical NAND gates may be represented as $4h$ units of capacitance. Figure 4.6(b) shows the equivalent circuit for estimating the rising delay. Only one pMOS transistor is ON in the slowest case. The diffusion capacitance of the nMOS transistor is ignored because it is not on the path between the supply rail and the output node Y. This results in a slightly optimistic result. The delay is $t_{pdr} = R \cdot ((6 + 4h)C) = (6 + 4h)RC$. Figure 4.6(c) shows the equivalent circuit for estimating the falling delay. In the worst case, input A is already '1,' so node x is charged up to nearly V_{DD} through the top nMOS transistor. Input B rises, turning on the bottom nMOS transistor and thus discharging both the capacitance on node x and the output capacitance. The Elmore delay is $t_{pdf} = (R/2)(2C) + R \cdot ((6 + 4h)C) = (7 + 4h)RC$. Despite the fact that the rise and fall resistances are equal, the falling propagation delay is slightly longer than the rising delay on account of the time needed to discharge the internal parasitic capacitance.

Observe that the best-case (contamination) delay of the gate could be substantially less. For example, if both inputs fell simultaneously, the output would be pulled up in half the time through the parallel combination of the two pMOS transistors: $t_{cdr} = (R/2)((6 + 4h)C) = (3 + 2h)RC$. If input B were already '1' and input A rises, node x would already have been discharged and thus could be ignored for delay purposes, reducing the falling delay to $t_{cdf} = R((6 + 4h)C) = (6 + 4h)RC$. Hence, for fastest response, the latest input should be connected to the transistor closest to the output node when feasible.

In the 180 nm process, $RC = 5$ ps. Therefore, a fanout-of-3 NAND gate ($h = 3$) has a delay of $(6 + 4 \cdot 3) \cdot 5$ ps = 90 ps.

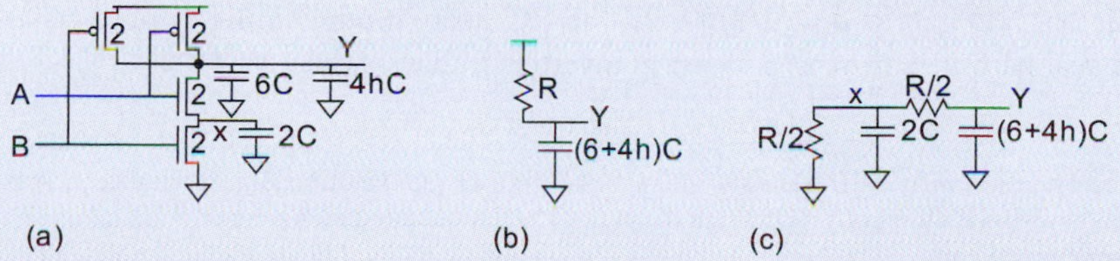

FIG 4.6   NAND gate delay estimation

Example

Suppose the widths of the transistors in the NAND gate are increased by a factor of k but the load is left unchanged at $4h'C$ (where $h' = h$ in the previous example). In other words, the load is equivalent to $h = h'/k$ NAND gates that are also a factor of k larger. Recompute the rising and falling propagation delays using the Elmore delay model.

Solution: Figure 4.7 shows the new equivalent circuits. The Elmore delay model predicts a rising delay of $(6 + 4(h'/k))RC = (6 + 4h)RC$ and a falling delay of $(7 + 4(h'/k))RC = (7 + 4h)RC$.

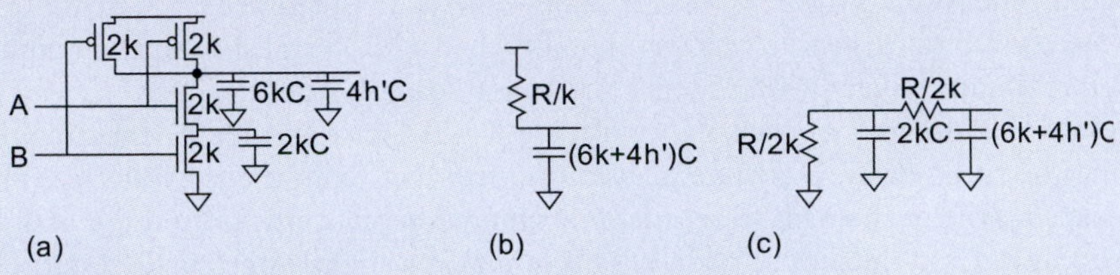

FIG 4.7   Scaled NAND gate delay estimation

Observe that the delay consists of two components. The *parasitic delay* of 6 or 7 is determined by the gate driving its own internal diffusion capacitance. Boosting the width of the transistors decreases the resistance but increases the capacitance so the parasitic delay is ideally independent of the gate size[1]. The *effort delay* of $4(h'/k)C = 4hC$ depends on the ratio (h) of external load capacitance to input capacitance and thus changes with transistor widths. The factor 4 is set by the complexity of the gate. The capacitance ratio is called the *electrical effort* or *fanout* and the term indicating gate complexity is called the *logical effort*. These components will be explored further in the subsequent sections.

It is often helpful to express delay in a process-independent form so that circuits can be compared based on topology rather than speed of the manufacturing process. Moreover, with a process-independent measure for delay, knowledge of circuit speeds gained while working in one process can be carried over to a new process. Recall that the delay of an ideal fanout-of-1 inverter with no parasitic capacitance is $\tau = 3RC$. We denote the normalized delay as multiples of this inverter delay: $d = t_{pd}/\tau$. Hence, the rising delay of the 2-input NAND gate is $d = (4/3)h + 2$. The RC delay model similarly predicts an inverter with real parasitics driving h identical inverters to have a delay of $h + 1$.

---

[1] Of course, gates with wider transistors may use layout tricks so the diffusion capacitance increases less than linearly with width, slightly decreasing the parasitic delay of large gates. Section 8.10.1.3 illustrates folding of wide transistors.

4.2.2   Linear Delay Model

In general the propagation delay of a gate can be written as

$$d = f + p \qquad (4.2)$$

where p is the *parasitic delay* inherent to the gate when no load is attached; f is the *effort delay* or *stage effort* that depends on the complexity and fanout of the gate:

$$f = gh \qquad (4.3)$$

The complexity is represented by the *logical effort*, g [Sutherland99]. An inverter is defined to have a logical effort of 1. More complex gates have greater logical efforts, indicating that they take longer to drive a given fanout. For example, the logical effort of the NAND gate from the previous example is 4/3. A gate driving h identical copies of itself is said to have a *fanout* or *electrical effort* of h. If the load is not identical copies of the gate, the electrical effort can be computed as

$$h = \frac{C_{\text{out}}}{C_{\text{in}}} \qquad (4.4)$$

where C_{out} is the capacitance of the external load being driven and C_{in} is the input capacitance of the gate[2].

Figure 4.8 plots normalized delay vs. electrical effort for an idealized inverter and 2-input NAND gate. The y-intercepts indicate the parasitic delay, i.e., the delay when the gate drives no load. The slope of the lines is the logical effort. The inverter has a slope of 1 by definition. The NAND has a slope of 4/3.

The logical effort and parasitic delay can be estimated using RC models, as will be explored in the next sections, or extracted by curve-fitting simulated data, as discussed in Section 5.5.3. Logic gates fit the linear delay vs. fanout model remarkably well even in advanced processes; for example, Figure 5.28 shows agreement within 0.5 ps in a 180 nm process. A properly calibrated linear delay model is widely used by CAD tools such as logic synthesizers and static timing analyzers, although the notation varies from tool to tool. For example, the popular Synopsys Design Compiler tool uses the following basic model to define delay for a library of gates:

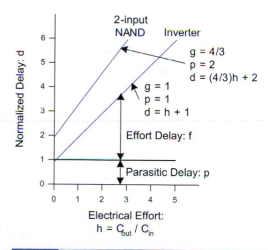

FIG 4.8   Normalized delay vs. fanout

```
delay_rise = intrinsic_rise + rise_resistance * capacitance
delay_fall = intrinsic_fall + fall_resistance * capacitance
```

[2]Some TTL designers say a gate has a fanout of h when it drives h other gates even if the other gates have different capacitances. This definition would not be useful for calculating delay and is best avoided in VLSI design. The term *electrical effort* avoids this potential confusion and emphasizes the parallels with logical effort.

Note that separate rising and falling delays are computed. These parameters are related to the logical effort terms as given in Table 4.1. The effective resistance of a gate increases with the logical effort of the gate but decreases with the gate size (i.e., input capacitance).

Table 4.1	Relationship between Logical Effort and Synopsys Terminology
Logical Effort Term	**Synopsys Term**
d	delay
p	intrinsic
C_{out}	capacitance
g/C_{in}	resistance

Some designers use the term *drive* as the reciprocal of resistance: drive = C_{in}/g. Gates with wider transistors have greater drive. Gates with high logical effort have less drive. Delay can be expressed in terms of drive as

$$d = \frac{C_{out}}{drive} + p \qquad (4.5)$$

4.2.3 Logical Effort

Logical effort of a gate is defined as *the ratio of the input capacitance of the gate to the input capacitance of an inverter that can deliver the same output current.* Equivalently, logical effort indicates how much worse a gate is at producing output current as compared to an inverter, given that each input of the gate may only present as much input capacitance as the inverter.

Logical effort can be measured in simulation from delay vs. fanout plots as the ratio of the slope of the delay of the gate to the slope of the delay of an inverter. Alternatively, it can be estimated by sketching gates. Figure 4.9 shows inverter, NAND, and NOR gates with transistor widths chosen to achieve unit resistance, assuming pMOS transistors have twice the resistance of nMOS transistors[3]. The inverter presents 3 units of input capacitance. The NAND presents 4 units of capacitance on each input, so the logical effort is 4/3. Similarly, the NOR presents 5 units of capacitance, so the logical effort

---

[3]This assumption is made throughout the book. Exercises 4.19–4.20 explore the effects of different relative resistances (see also [Sutherland99]). The overall conclusions do not change very much, so the simple model is good enough for most hand estimates. A simulator or static timing analyzer should be used when more accurate results are required.

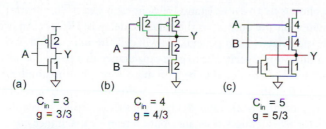

$$C_{in} = 3 \qquad\qquad C_{in} = 4 \qquad\qquad C_{in} = 5$$
$$g = 3/3 \qquad\qquad g = 4/3 \qquad\qquad g = 5/3$$

FIG 4.9 Logic gates sized for unit resistance

is 5/3. This matches our expectation that NANDs are better than NORs because NORs have slow pMOS transistors in series.

Table 4.2 lists the logical effort of common gates. The effort tends to increase with the number of inputs. NAND gates are better than NOR gates because the series transistors are nMOS rather than pMOS. Exclusive-OR gates are particularly costly and have different logical efforts for different inputs. An interesting case is that multiplexers built from ganged tristates, as shown in Figure 1.29(b), have a logical effort of 2 independent of the number of inputs. This might at first seem to imply that very large multiplexers are just as fast as small ones. However, the parasitic delay does increase with multiplexer size; hence, it is generally fastest to construct large multiplexers out of trees of 4-input multiplexers [Sutherland99].

Table 4.2	Logical effort of common gates				
Gate Type	**Number of Inputs**				
	1	**2**	**3**	**4**	**n**
inverter	1				
NAND		4/3	5/3	6/3	$(n + 2)/3$
NOR		5/3	7/3	9/3	$(2n + 1)/3$
tristate, multiplexer	2	2	2	2	2
XOR, XNOR		4, 4	6, 12, 6	8, 16, 16, 8	

4.2.4 Parasitic Delay

The parasitic delay of a gate is the delay of the gate when it drives zero load. It can be estimated with RC delay models. A crude method good for hand calculations is to count only diffusion capacitance on the output node. For example, consider the gates in Figure 4.9, assuming each transistor on the output node has its own drain diffusion contact. Transistor widths were chosen to give a resistance of R in each gate. The inverter has 3 units of diffusion capacitance on the output, so the parasitic delay is $3RC = \tau$. In other words, the normalized parasitic delay is 1. In general, we will call the normalized parasitic delay p_{inv}. p_{inv} is the ratio of diffusion capacitance to gate capacitance in a particular process. It is usually close to 1 and will be considered to be 1 on many examples for simplicity. The NAND and NOR each have 6 units of diffusion capacitance on the output, so the parasitic delay is twice as great ($2p_{inv}$, or simply 2). Table 4.3 estimates the parasitic delay of common gates. Increasing transistor sizes reduces resistance but increases capacitance correspondingly, so parasitic delay is, to first order, independent of gate size. However, wider transistors can be folded and often see less than linear increases in internal wiring parasitic capacitance, so in practice, larger gates tend to have slightly lower parasitic delay.

Table 4.3	Parasitic delay of common gates				
Gate Type	Number of Inputs				
	1	2	3	4	n
inverter	1				
NAND		2	3	4	n
NOR		2	3	4	n
tristate, multiplexer	2	4	6	8	$2n$

This method of estimating parasitic delay is obviously crude. More refined estimates use the Elmore delay counting internal parasitics or extract the delays from simulation. The parasitic delay also depends on the ratio of diffusion capacitance to gate capacitance. For example, in a silicon-on-insulator process in which diffusion capacitance is much less, the parasitic delays will be lower. While knowing the parasitic delay is important for accurately estimating gate delay, we will see in Section 4.3 that the best transistor sizes for a particular circuit are only weakly dependent on parasitic delay. Hence, crude estimates may be acceptable.

Nevertheless, it is important to realize that parasitic delay grows more than linearly with the number of inputs in a real NAND or NOR circuit. For example, Figure 4.10 shows a model of an n-input NAND gate in which the upper inputs were all '1' and the bottom input rises. The gate must discharge the diffusion capacitances of all of the internal nodes as well as the output. The Elmore delay is

$$t_{pd} = R\left(3nC\right) + \sum_{i=1}^{n-1} \left(\frac{iR}{n}\right)\left(nC\right) = \left(\frac{n^2}{2} + \frac{5}{2}n\right)RC \qquad (4.6)$$

FIG 4.10 *n*-input NAND gate parasitic delay

This delay grows quadratically with the number of series transistors *n*, indicating that beyond a certain point it is faster to split a large gate into a cascade of two smaller gates. We will see in 4.2.5.3 that the coefficient of the n^2 term tends to be even larger in real circuits than in this simple model because of gate-source capacitance. In practice, it is rarely advisable to construct a gate with more than four or possibly five series transistors. When building large fan-in gates, trees of NAND gates are better than NOR gates because the nMOS transistors have lower resistance than pMOS transistors of the same size and capacitance.

4.2.5 Limitations to the Linear Delay Model

Logical effort is built on the linear delay model. Although the model works remarkably well for many practical applications, it also has limitations that should be understood when more accuracy is needed.

4.2.5.1 Input and Output Slope The largest source of error in the linear delay model is the input slope effect. Figure 4.11(a) shows a fanout-of-4 inverter driven by ramps with different slopes. Recall that the ON current increases with the gate voltage for an nMOS transistor. We say the transistor is *OFF* for $V_{gs} < V_t$, *fully ON* for $V_{gs} = V_{DD}$, and *partially ON* for intermediate gate voltages. As the rise time of the input increases, the delay also increases because the active transistor is not turned all the way ON at once. Figure 4.11(b) plots average inverter propagation delay vs. input rise time.

Notice that the delay vs. rise time data fits a straight line quite well. [Hedenstierna87] suggests that the line may be modeled as

$$t_{pd} = t_{pd-\text{step}} + t_{\text{edge}} \left(\frac{1 + 2\dfrac{|V_t|}{V_{DD}}}{6} \right) \tag{4.7}$$

where $t_{pd-\text{step}}$ is the propagation delay assuming a step input and t_{edge} is the appropriate edge rate (t_r or t_f).

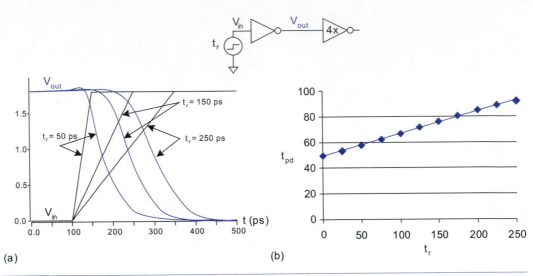

SPICE simulation of slope effect on CMOS inverter delay

A first-order RC circuit has a 20%-80% rise/fall time of $(\ln 0.8 - \ln 0.2)RC = 1.39RC$. The 20%–80% rise/fall time of a gate is roughly 1-1.5x the propagation delay. Hence, the input slope of a gate is related to the propagation delay of the previous gate. The Synopsys delay model supports considering the effect of slope on gate delay by introducing a **slope_rise** / **slope_fall** term. The overall delay of a gate becomes linearly dependent on the delay of the previous gate.

```
delay_rise = intrinsic_rise + rise_resistance · capacitance +
             slope_rise · delay_previous
delay_fall = intrinsic_fall + fall_resistance · capacitance +
             slope_fall · delay_previous
```

Accounting for slopes is important for accurate timing analysis, but is generally more complex than is worthwhile for hand calculations. Fortunately, we will see in Section 4.3 that circuits are fastest when each gate has the same effort delay and when that delay is roughly 4τ. Because slopes are related to edge rate, fast circuits tend to have relatively consistent slopes. If a cell library is characterized with these slopes, it will tend to be used in the regime in which it most accurately models delay.

4.2.5.2 Input Arrival Times Another source of error in the linear delay model is the assumption that one input of a multiple-input gate switches while the others are completely stable. When two inputs to a series stack turn ON simultaneously, the delay will be slightly longer than predicted because both transistors are only partially ON during the initial part of the transition. When two inputs to a parallel stack turn ON simultaneously, the delay will be shorter than predicted because both transistors deliver current to the output.

Figure 4.12 plots the propagation delay of an FO3 2-input NAND gate as a function of the input interarrival time. Input A switches at time 0, while input B switches in the same direction at time t_b. Propagation delay is measured from the latest input rising for falling outputs and from the earliest input falling for rising outputs. When one input arrives well before the other, $|t_b|$ is large and the propagation delay is essentially independent of t_b. This is the case assumed in the linear delay model. The delays are slightly different depending on which input arrives first, as will be explored in Section 5.5.3. When the two inputs arrive at nearly the same time, $|t_b|$ is small. t_{pdf} increases because of the series pulldowns while t_{pdr} decreases because of the parallel pullups.

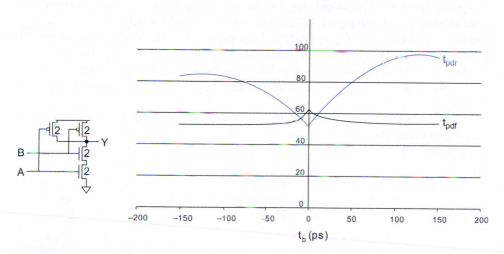

FIG 4.12 Delay sensitivity to input arrival time

4.2.5.3 Gate-Source Capacitance

The examples in Section 4.2.1 assumed that gate capacitance terminates on a fixed supply rail. As discussed in Section 2.3.2, the bottom terminal of the gate oxide capacitor is the channel, which is primarily connected to the source when the transistor is ON. This means that as the source of a transistor changes value, charge is required to change the voltage on C_{gs}. Figure 4.13(a) revisits the 2-input NAND gate example, explicitly showing gate-source capacitances. As node x is discharged on a falling output transition in Figure 4.13(b), C_{gs2} must also be discharged. The delay can now be estimated as $(R/2)(2C + 2C) + R((2 + 2 + 2 + 4h)C) = (8 + 4h)RC$. RC models are more valuable for simplicity than accuracy, so some designers ignore this effect in hand calculations. Note that C_{gs} of the pMOS transistors does not affect delay because it is not on the path between Y and GND. However, the gate capacitance of both nMOS and pMOS would affect the loading and delay of the previous stage.

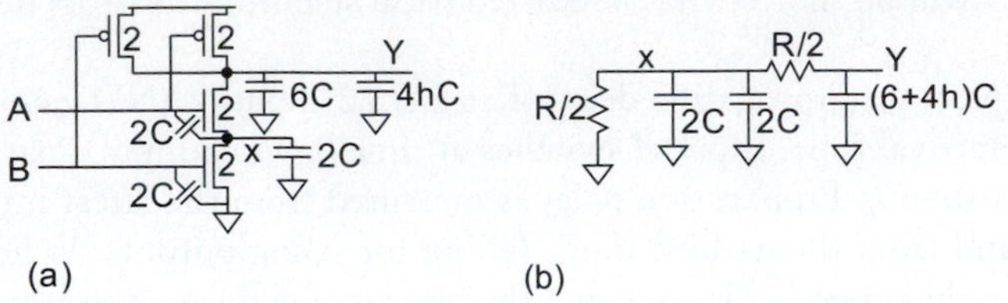

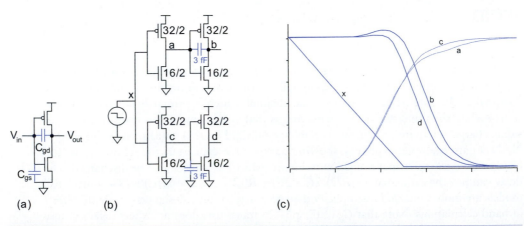

FIG 4.13 NAND gate delay estimation with gate-source capacitance modeled

4.2.5.4 Bootstrapping Transistors also have some capacitance from gate to drain. This capacitance couples the input and output in an effect known as *bootstrapping*, which can be understood by examining Figure 4.14(a). Our models so far have only considered C_{in} (C_{gs}). This figure also considers C_{gd}, the gate to drain capacitance. In the case that the input is rising (the output starts high), the effective input capacitance is $C_{gs} + C_{gd}$. When the output starts to fall, the voltage across C_{gd} changes, requiring the input to supply additional current to charge C_{gd}. In other words, the impact of C_{gd} on gate capacitance is effectively doubled.

To illustrate the effect of the bootstrap capacitance on a circuit, Figure 4.14(b) shows two inverter pairs. The top pair has an extra bit of capacitance between the input and output of the second inverter. The bottom pair has the same amount of extra capacitance from input to ground. When *x* falls, nodes *a* and *c* begin to rise (Figure 4.14(c)). At first, both nodes see approximately the same capacitance, consisting of the two transistors and the extra 3 fF. As node *a* rises, it initially bumps up *b* or "lifts *b* by its own bootstraps." Eventually the nMOS transistors turn ON, pulling down *b* and *d*. As *b* falls, it tugs on *a*

FIG 4.14 The effect of bootstrapping on inverter delay and waveform shape

through the capacitor, leading to the slow final transition visible on node *a*. Also observe that *b* falls later than *d* because of the extra charge that must be supplied to discharge the bootstrap capacitor. In summary, the extra capacitance has a greater effect when connected between input and output as compared to when it is connected between input and ground.

Because C_{gd} is fairly small, bootstrapping is only a mild annoyance in digital circuits. However, if the inverter is biased in its linear region near $V_{DD}/2$, the C_{gd} may appear multiplied by the gain of the inverter. This is known as the *Miller effect* and is of major importance in analog circuits.

4.3 Logical Effort and Transistor Sizing

Designers often need to choose the fastest circuit topology and gate sizes for a particular logic function and to estimate the delay of the design. As has been stated, simulation or timing analysis are poor tools for this task because they only determine how fast a particular implementation will operate, not whether the implementation can be modified for better results and if so, what to change. Inexperienced designers often end up in the "simulate and tweak" loop involving minor changes and many simulations. This is not only tedious but also seldom results in significant improvements. The method of Logical Effort [Sutherland99] provides a simple method "on the back of an envelope" to choose the best topology and number of stages of logic for a function. It allows the designer to quickly estimate the minimum possible delay for the given topology and to choose gate sizes that achieve this delay.

Logical Effort is based on the linear delay model. We first review using the model to estimate the delay of individual logic gates. The method generalizes to predict the delay of multistage logic networks and to choose the best number of stages for a multistage network. This section concludes with an example applying Logical Effort to design a memory decoder and summarizes the key insights from the method. The techniques of Logical Effort will be revisited throughout this text to understand delay of many types of circuits.

4.3.1 Delay in a Logic Gate

The linear delay model of EQs (4.2), (4.3), and (4.4) expresses propagation delay of a logic gate in terms of the complexity of a gate (its logical effort, *g*), the capacitive fanout (electrical effort, *h*), and the parasitic delay, *p*. Let us begin with two examples.

Example

Estimate the delay of the fanout-of-4 (FO4) inverter (i.e., an inverter driving four identical copies) shown in Figure 4.15. Assume the inverter is constructed in a 180 nm process with τ = 15 ps.

Solution: The logical effort of the inverter is g = 1, by definition. The electrical effort is 4 because the load is four gates of equal size. The parasitic delay of an inverter is $p_{inv} \approx 1$. The total delay is $d = gh + p = 1 \cdot 4 + 1 = 5$ in normalized terms, or t_{pd} = 75 ps in absolute terms.

Often path delays are expressed in terms of FO4 inverter delays. While not all designers are familiar with the τ notation, most experienced designers do know the delay of a fanout-of-4 inverter in the process in which they are working. τ can be estimated as 0.2 FO4 inverter delays. Even if the ratio of diffusion capacitance to gate capacitance changes so p_{inv} = 0.8 or 1.2 rather than 1, the FO4 inverter delay only varies from 4.8 to 5.2. Hence, the delay of a gate-dominated logic block expressed in terms of FO4 inverters remains relatively constant from one process to another even if the diffusion capacitance does not.

As a rough rule of thumb, the FO4 delay for a process (in picoseconds) is 1/3 to 1/2 of the channel length (in nanometers). For example, a 180 nm process may have an FO4 delay of 60-90 ps. Delay is highly sensitive to process, voltage, and temperature variations, as will be examined in Section 6.6. The FO4 delay is usually quoted assuming typical process parameters and worst-case environment (low power supply voltage and high temperature).

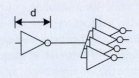

FIG 4.15 Fanout-of-4 (FO4) inverter

4.3.2 Delay in Multistage Logic Networks

Logical Effort generalizes to multistage logic networks. For example, Figure 4.16 shows the logical and electrical efforts of each stage in a multistage path as a function of the sizes of each stage. The path of interest (the only path in this case) is marked with the dashed blue line. Observe that logical effort is independent of size, while electrical effort depends on sizes.

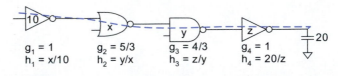

g_1 = 1 g_2 = 5/3 g_3 = 4/3 g_4 = 1
h_1 = x/10 h_2 = y/x h_3 = z/y h_4 = 20/z

FIG 4.16 Multistage logic network

Example

A ring oscillator is constructed from an odd number of inverters, as shown in Figure 4.17. Estimate the frequency of an N-stage ring oscillator.

Solution: The logical effort of the inverter is $g = 1$, by definition. The electrical effort of each inverter is also 1 because it drives a single identical load. The parasitic delay is also 1. The delay of each stage is $d = gh + p = 1 \cdot 1 + 1 = 2$. An N-stage ring oscillator has a period of $2N$ stage delays because a value must propagate twice around the ring to regain the original polarity. Therefore, the period is $T = 2 \cdot 2N$. The frequency is the reciprocal of the period, $1/4N$.

A 31-stage ring oscillator in a 180 nm process has a frequency of $1/(4 \cdot 31 \cdot 15$ ps) = 540 MHz.

Note that ring oscillators are often used as process monitors to judge if a particular chip is faster or slower than nominally expected. One of the inverters should be replaced with a NAND gate to turn the ring off when not in use. The output can be routed to an external pad, possibly through a test multiplexer. The oscillation frequency should be low enough (e.g., 100 MHz) that the path to the outside world is not a limiter.

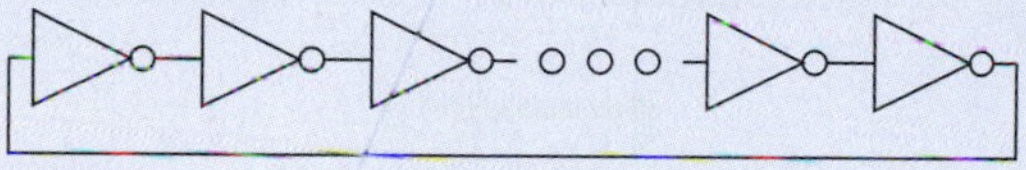

FIG 4.17 Ring oscillator

The *path logical effort G* can be expressed as the products of the logical efforts of each stage along the path.

$$G = \prod g_i \tag{4.8}$$

The *path electrical effort H* can be given as the ratio of the output capacitance the path must drive divided by the input capacitance presented by the path. This is more convenient than defining path electrical effort as the product of stage electrical efforts because we do not know the individual stage electrical efforts until gate sizes are selected.

$$H = \frac{C_{out(path)}}{C_{in(path)}} \tag{4.9}$$

The *path effort F* is the product of the stage efforts of each stage. Can we by analogy state $F = GH$?

$$F = \prod f_i = \prod g_i h_i \tag{4.10}$$

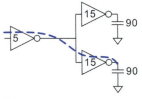

FIG 4.18 Circuit with two-way branch

In paths that branch, $F \neq GH$. This is illustrated in Figure 4.18, a circuit with a two-way branch. Consider a path from the primary input to one of the outputs. The path logical effort is $G = 1 \cdot 1 = 1$. The path electrical effort is $H = 90/5 = 18$. Thus, $GH = 18$. But $F = f_1 f_2 = g_1 h_1 g_2 h_2 = 1 \cdot 6 \cdot 1 \cdot 6 = 36$. In other words, $F = 2GH$ in this path on account of the two-way branch.

We must introduce a new kind of effort to account for branching between stages of a path. This *branching effort b* is the ratio of the total capacitance seen by a stage to the capacitance on the path; in Figure 4.18 it is $(15+15)/15 = 2$.

$$b = \frac{C_{\text{onpath}} + C_{\text{offpath}}}{C_{\text{onpath}}} \tag{4.11}$$

The *path branching effort B* is the product of the branching efforts between stages.

$$B = \prod b_i \tag{4.12}$$

Now we can define the path effort F as the product of the logical, electrical, and branching efforts of the path. Note that the product of the electrical efforts of the stages is actually BH, not just H.

$$F = GBH \tag{4.13}$$

We can now compute the delay of a multistage network. The *path delay D* is the sum of the delays of each stage. It can also be written as the sum of the *path effort delay* D_F and *path parasitic delay* P.

$$
\begin{aligned}
D &= \sum d_i = D_F + P \\
D_F &= \sum f_i \\
P &= \sum p_i
\end{aligned}
\tag{4.14}
$$

The product of the stage efforts is F, independent of gate sizes. The path effort delay is the sum of the stage efforts. The sum of a set of numbers whose product is constant is

Example

Estimate the minimum delay of the path from A to B in Figure 4.19 and choose transistor sizes to achieve this delay. The initial NAND2 gate may present a load of 8λ of transistor width on the input and the output load is equivalent to 45λ of transistor width.

Solution: The path logical effort is $G = (4/3) \cdot (5/3) \cdot (5/3) = 100/27$. The path electrical effort is $H = 45/8$. The path branching effort is $B = 3 \cdot 2 = 6$. The path effort is $F = GBH = 125$. As there are three stages, the best stage effort is $\hat{f} = \sqrt[3]{125} = 5$. The path parasitic delay is $P = 2 + 3 + 2 = 7$. Hence, the minimum path delay is $D = 3 \cdot 5 + 7 = 22$ in units of τ, or 4.4 FO4 inverter delays. The gate sizes are computed with the capacitance transformation: $y = 45 \cdot (5/3)/5 = 15$. $x = (15 + 15) \cdot (5/3)/5 = 10$. We check that the initial 2-input NAND gate should have a size of $(10 + 10 + 10) \cdot (4/3)/5 = 8$, as desired. The transistor sizes in Figure 4.20 are chosen to give the desired amount of input capacitance while achieving equal rise and fall delays.

We can also check that our delay was achieved. The NAND2 gate delay is $d_1 = g_1 h_1 + p_1 = (4/3) \cdot (10 + 10 + 10)/8 + 2 = 7$. The NAND3 gate delay is $d_2 = g_2 h_2 + p_2 = (5/3) \cdot (15 + 15)/10 + 3 = 8$. The NOR2 gate delay is $d_3 = g_3 h_3 + p_3 = (5/3) \cdot 45/15 + 2 = 7$. Hence, the path delay is 22, as predicted.

Many inexperienced designers know that wider transistors offer more current and thus try to make circuits faster by using bigger gates. Increasing the size of any of the gates except the first one only makes the circuit slower. For example, increasing the size of the NAND3 makes the NAND3 faster but makes the NAND2 slower, resulting in a net speed loss. Increasing the size of the initial NAND2 gate does speed up the circuit under consideration. However, it presents a larger load on the path that computes input A, making that path slower. Hence, it is crucial to have a specification of not only the load the path must drive but also the maximum input capacitance the path may present.

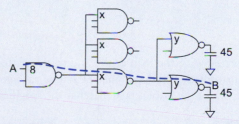

FIG 4.19 Example path

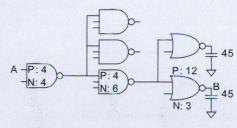

FIG 4.20 Example path annotated with transistor sizes

minimized by choosing all the numbers to be equal. In other words, the path delay is minimized when each stage bears the same effort $\hat{f}$. If a path has N stages and each bears the same effort, that effort must be

$$\hat{f} = g_i h_i = F^{1/N} \qquad (4.15)$$

Thus the minimum possible delay of an N-stage path with path effort F and path parasitic delay P is

$$D = NF^{1/N} + P \qquad (4.16)$$

This is a key result of Logical Effort. It shows that the minimum delay of the path can be estimated knowing only the number of stages, path effort, and parasitic delays without the need to assign transistor sizes. This is superior to simulation, in which delay depends on sizes and you never achieve certainty that the sizes selected are those that offer minimum delay.

It is also straightforward to select gate sizes to achieve this least delay. Combining EQs (4.3) and (4.4) gives us the *capacitance transformation* formula to find the best input capacitance for a gate given the output capacitance it drives.

$$C_{\text{in}_i} = \frac{C_{\text{out}_i} \cdot g_i}{\hat{f}} \qquad (4.17)$$

Starting with the load at the end of the path, work backward applying the capacitance transformation to determine the size of each stage. Check the arithmetic by verifying that the size of the initial stage matches the specification.

4.3.3 Choosing the Best Number of Stages

Given a specific circuit topology, we now know how to estimate delay and choose gate sizes. However, there are many different topologies that implement a particular logic function. Logical Effort tells us that NANDs are better than NORs and that gates with few inputs are better than gates with many. In this section we will also use Logical Effort to predict the best number of stages to use.

Logic designers sometimes estimate delay by counting the number of stages of logic, assuming each stage has a constant "gate delay." This is potentially misleading because it implies that the fastest circuits are those that use the fewest stages of logic. Of course the gate delay actually depends on the electrical effort, so sometimes using fewer stages results in more delay. The following example illustrates this point.

In general, you can always add inverters to the end of a path without changing its function (save possibly for polarity). Let us compute how many should be added for least delay. The logic block shown in Figure 4.22 has n_1 stages and a path effort of F. Consider adding $N - n_1$ inverters to the end to bring the path to N stages. The extra inverters do not

Example

A control unit generates a signal from a unit-sized inverter. The signal must drive unit-sized loads in each bitslice of a 64-bit datapath. The designer can add inverters to buffer the signal to drive the large load. Assuming polarity of the signal does not matter, what is the best number of inverters to add and what delay can be achieved?

Solution: Figure 4.21 shows the cases of adding 0, 1, 2, or 3 inverters. The path electrical effort is $H = 64$. The path logical effort is $G = 1$, independent of the number of inverters. Thus the path effort is $F = 64$. The inverter sizes are chosen to achieve equal stage effort. The total delay is $D = N\sqrt[N]{64} + N$.

The 3-stage design is fastest and is much superior to a single stage. If an even number of inversions were required, the two- or four-stage designs are promising. The four-stage design is slightly faster, but the two-stage design requires significantly less area and power.

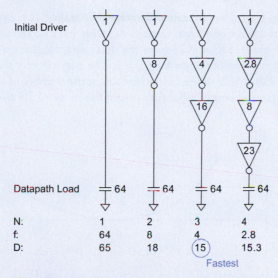

FIG 4.21 Comparison of different number of stages of buffers

change the path logical effort but do add parasitic delay. The delay of the new path is

$$D = NF^{1/N} + \sum_{i=1}^{n_1} p_i + (N - n_1)p_{inv} \qquad (4.18)$$

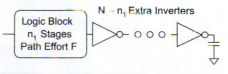

FIG 4.22 Logic block with additional inverters

Differentiating with respect to N and setting to 0 allows us to solve for the best number of stages. The result can be expressed more compactly by defining

$$\rho = F^{1/\hat{N}}$$

to be the best stage effort.

$$\frac{\partial D}{\partial N} = -F^{1/N} \ln F^{1/N} + F^{1/N} + p_{inv} = 0$$

$$\Rightarrow p_{inv} + \rho(1 - \ln\rho) = 0$$

(4.19)

EQ (4.19) has no closed form solution. Neglecting parasitics (i.e., assuming $p_{inv} = 0$), we find the classic result that $\rho = 2.71828$ (e) [Mead80]. In practice, the parasitic delays mean each inverter is somewhat more costly to add. As a result, it is better to use fewer stages, or equivalently a higher stage effort than e. Solving numerically, when $p_{inv} = 1$, we find $\rho = 3.59$.

A path achieves least delay by using $\hat{N} = \log_\rho F$ stages. It is important to understand not only the best stage effort and number of stages but also the sensitivity to using a different number of stages. Figure 4.23 plots the delay increase using a particular number of stages against the number of stages, for $p_{inv} = 1$. The curve is very flat around the optimum. The delay is within 15% of the best achievable if the number of stages is within 2/3 to 1.5 times the theoretical best number (i.e., ρ is in the range of 2.4 to 6).

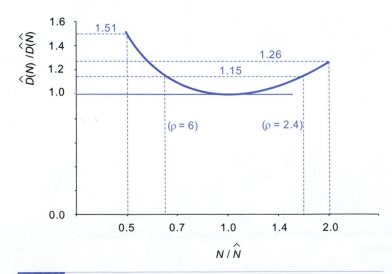

FIG 4.23 Sensitivity of delay to number of stages

Using a stage effort of 4 is a convenient choice and simplifies mentally choosing the best number of stages. This effort gives delays within 2% of minimum for p_{inv} in the range of 0.7 to 2.5. This further explains why a fanout-of-4 inverter has a "representative" logic gate delay.

4.3.4   Example

Consider a larger example to illustrate the application of Logical Effort. Our esteemed colleague Ben Bitdiddle is designing a decoder for a register file in the Motoroil 68W86, an embedded processor for automotive applications. The decoder has the following specifications:

- 16-word register file

- 32-bit words

- Each register bit presents a load of 3 unit-sized transistors on the word line (2 unit-sized access transistors plus some wire capacitance)

- True and complementary versions of the address bits $A[3:0]$ are available

- Each address input can drive 10 unit-sized transistors

As we will see further in Section 11.2.2, a 2^N-word decoder consists of 2^N N-input AND gates. Therefore, the problem is reduced to designing a suitable 4-input AND gate. Let us help Ben determine how many stages to use, how large each gate should be, and how fast the decoder can operate.

The output load on a word line is 32 bits with 3 units of capacitance each, or 96 units. Therefore, the path electrical effort is $H = 96/10 = 9.6$. Each address is used to compute half of the 16 word lines; its complement is used for the other half. Therefore, a $B = 8$-way branch is required somewhere in the path. Now we are faced with a chicken-and-egg dilemma. We need to know the path logical effort to calculate the path effort and best number of stages. However, without knowing the best number of stages, we cannot sketch a path and determine the logical effort for that path. There are two ways to resolve the dilemma. One is to sketch a path with a random number of stages, determine the path logical effort, then use that to compute the path effort and the actual number of stages. The path can be redesigned with this number of stages, refining the path logical effort. If the logical effort changes significantly, the process can be repeated. Alternatively, we know that the logic of a decoder is rather simple so we can ignore the logical effort (assume $G = 1$). Then we can proceed with our design, remembering that the best number of stages is likely slightly higher than predicted because we neglected logical effort.

Taking the second approach, we find the path effort is $F = GBH = (1)(8)(9.6) = 76.8$. Targeting a best stage effort of $\rho = 4$, we find the best number of stages is $N = \log_4 76.8 = 3.1$. Let us select a 3-stage design, recalling that a 4-stage design might be a good choice too when logical effort is considered. Figure 4.24 shows a possible 3-stage design (INV-NAND4-INV).

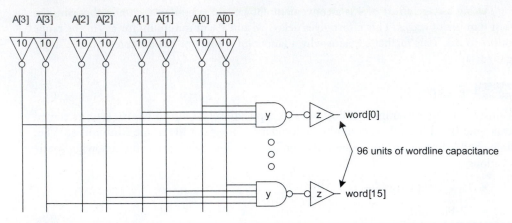

FIG 4.24 3-stage decoder design

The path has a logical effort of $G = 1 \cdot (6/3) \cdot 1 = 2$, so the actual path effort is $F = (2)(8)(9.6) = 154$. The stage effort is $\hat{f} = 154^{1/3} = 5.36$. This is in the reasonable range of 2.4 to 6, so we expect our design to be acceptable. Applying the capacitance transformation, we find gate sizes $z = 96 \cdot 1/5.36 = 18$ and $y = 18 \cdot 2/5.36 = 6.7$. The delay is $3 \cdot 5.36 + 1 + 4 + 1 = 22.1$.

Logical Effort also allows us to rapidly compare alternative designs using a spreadsheet rather than a schematic editor and a large number of simulations. Table 4.4 compares a number of alternative designs. We find a 4-stage design is somewhat faster, as we suspected. The 4-stage NAND2-INV-NAND2-INV design not only has the theoretical best number of stages but also uses simpler 2-input gates to reduce the logical effort and parasitic delay to obtain a 12% speedup over the original design. However, the 3-stage design has a smaller total gate area and dissipates less power.

Table 4.4	Spreadsheet comparing decoder designs			
Design	**Stages N**	**G**	**P**	**D**
NAND4-INV	2	2	5	29.8
NAND2-NOR2	2	20/9	4	30.1
INV-NAND4-INV	**3**	**2**	**6**	**22.1**
NAND4-INV-INV-INV	4	2	7	21.1
NAND2-NOR2-INV-INV	4	20/9	6	20.5
NAND2-INV-NAND2-INV	4	16/9	6	19.7
INV-NAND2-INV-NAND2-INV	5	16/9	7	20.4
NAND2-INV-NAND2-INV-INV-INV	6	16/9	8	21.6

4.3.5 Summary and Observations

Logical Effort provides an easy way to compare and select circuit topologies, choose the best number of stages for a path, and estimate path delay. The notation takes some time to become natural but this author has poured through all the letters in the English and Greek alphabets without finding better notation. It may help to remember d for "**d**elay," p for "**p**arasitic," b for "**b**ranching," f for "e**f**fort," g for "lo**g**ical effort," (or perhaps **g**ain), and h as the next letter after "f" and "g." The notation is summarized in Table 4.5 for both stages and paths.

Table 4.5   Summary of Logical Effort notation

Term	Stage Expression	Path Expression
number of stages	1	N
logical effort	g (see Table 4.2)	$G = \prod g_i$
electrical effort	$h = \dfrac{C_{out}}{C_{in}}$	$H = \dfrac{C_{out(path)}}{C_{in(path)}}$
branching effort	$b = \dfrac{C_{onpath} + C_{offpath}}{C_{onpath}}$	$B = \prod b_i$
effort	$f = gh$	$F = GBH$
effort delay	f	$D_F = \sum f_i$
parasitic delay	p (see Table 4.3)	$P = \sum p_i$
delay	$d = f + p$	$D = \sum d_i = D_F + P$

The method of Logical Effort is applied with the following steps:

1. Compute the path effort: $F = GBH$

2. Estimate the best number of stages: $\hat{N} = \log_4 F$

3. Sketch a path using: $\hat{N}$ stages

4. Estimate the minimum delay: $D = \hat{N}F^{1/\hat{N}} + P$

5. Determine the best stage effort: $\hat{f} = F^{1/\hat{N}}$

6. Starting at the end, work backward to find sizes: $C_{\text{in}_i} = \dfrac{C_{\text{out}_i} \cdot g_i}{\hat{f}}$

CAD tools are very fast and accurate at evaluating complex delay models, so Logical Effort should not be used as a replacement for such tools. Rather, its value arises from "quick and dirty" hand calculations and from the insights it lends to circuit design. Some of the key insights include:

- The idea of a numeric "logical effort" that characterizes the complexity of a logic gate or path allows you to compare alternative circuit topologies and show that some topologies are better than others.

- NAND structures are faster than NOR structures in complementary CMOS circuits.

- Paths are fastest when the effort delays of each stage are about the same and when these delays are close to four.

- Path delay is insensitive to modest deviations from the optimum. Stage efforts of 2.4–6 give designs within 15% of minimum delay. There is no need to make calculations to more than 1–2 significant figures, so many estimations can be made in your head. There is no need to choose transistor sizes exactly according to theory and there is little benefit in tweaking transistor sizes if the design is reasonable.

- Using stage efforts somewhat greater than 4 reduces area and power consumption at a slight cost in speed. Using efforts greater than 6-8 comes at a significant cost in speed.

- Using fewer stages for "less gate delays" does not make a circuit faster. Making gates larger also does not make a circuit faster; it only increases the area and power consumption.

- The delay of a well-designed path is about $\log_4 F$ fanout-of-4 (FO4) inverter delays. Each quadrupling of the load adds about one FO4 inverter delay to the path. Control signals fanning out to a 64-bit datapath therefore incur an amplification delay of about 3 FO4 inverters.

- The logical effort of each input of a gate increases through no fault of its own as the number of inputs grows. Considering both logical effort and parasitic delay, we find a practical limit of about 4 series transistors in logic gates and about 4 inputs to multiplexers. Beyond this fan-in, it is faster to split gates into multiple stages of skinnier gates.

- Inverters or 2-input NAND gates with low logical efforts are best for driving nodes with a large branching effort. Use small gates after the branches to minimize load on the driving gate.

- When a path forks and one leg is more critical than the others, buffer the noncritical legs to reduce the branching effort on the critical path.

4.3.6   Limitations of Logical Effort

Logical Effort is based on the linear delay model and the simple premise that making the effort delays of each stage equal minimizes path delay. This simplicity is the method's greatest strength but also results in a number of limitations:

- The linear delay model fails to capture the effect of input slope. Fortunately, edge rates tend to be about equal in well-designed circuits with equal effort delay per stage.

- The RC delay model neglects the effects of velocity saturation and overestimates the logical effort of NAND structures. It also ignores the body effect. Logical effort may be more accurately characterized through simulation, as shown in Section 5.5.3.

- Logical Effort does not account for interconnect. The effects of nonnegligible wire capacitance and RC delay will be revisited in Section 4.5. Logical Effort is most applicable to high-speed circuits with regular layouts where routing delay does not dominate. Such structures include adders, multipliers, memories, and other datapaths and arrays.

- Logical Effort explains how to design a path for maximum speed but not how to design for minimum area or power given a fixed speed constraint.

- Paths with complex branching are difficult to analyze by hand.

4.3.7   Extracting Logical Effort from Datasheets

When using a standard cell library, you can often extract logical effort of gates directly from the datasheets. For example, Figure 4.25 shows the INV and NAND2 datasheets from the Artisan Components library for the TSMC 180 nm process. The gates in the library come in various drive strengths. INVX1 is the unit inverter; INVX2 has twice the drive. INVXL has the same area as the unit inverter but uses smaller transistors to reduce power consumption on noncritical paths. The X12-X20 inverters are built from three stages of smaller inverters to give high drive strength and low input capacitance at the expense of greater parasitic delay.

From the datasheet, we see the unit inverter has an input capacitance of 3.6 fF. The rising and falling delays are specified separately. We will develop a notation for different delays in Section 6.2.1.5, but will use the average delay for now. The average *intrinsic* or parasitic delay is (25.3 + 14.6)/2 = 20.0 ps. The slope of the delay vs. load capacitance curve is the average of the rising and falling K_{load} values. An inverter driving a fanout of h will thus have a delay of

$$t_{pd} = 20.0 \text{ ps} + \left(3.6 \tfrac{\text{fF}}{\text{gate}}\right)(h \text{ gates})\left(\frac{4.53 + 2.37}{2} \tfrac{\text{ns}}{\text{pF}}\right) = (20.0 + 12.4h) \text{ ps} \qquad \textbf{(4.20)}$$

The slope of the delay vs. fanout curve indicates τ = 12.4 ps and the y-intercept indicates p_{inv} = 20.0 ps, or 1.61 in normalized terms.

By a similar calculation, we find the X1 2-input NAND gate has an average delay from the inner (A) input of

$$t_{pd} = \left(\frac{31.3 + 19.5}{2}\right) \text{ps} + \left(4.2 \tfrac{\text{fF}}{\text{gate}}\right)(h \text{ gates})\left(\frac{4.53 + 2.84}{2} \tfrac{\text{ns}}{\text{pF}}\right) = (25.4 + 15.5h) \text{ ps} \quad \textbf{(4.21)}$$

Thus, the parasitic delay is 2.05 and the logical effort is 1.25. The parasitic delay from the outer (B) input is slightly higher, as expected. The parasitic delay and logical effort of the X2 and X4 gates are similar, confirming our model that logical effort should be independent of gate size for gates of reasonable sizes.

4.4 Power Dissipation

Static CMOS gates are very power-efficient because they dissipate nearly zero power while idle. For much of the history of CMOS design, power was a secondary consideration behind speed and area for many chips. As transistor counts and clock frequencies have increased, power consumption has skyrocketed and now is a primary design constraint.

We begin by reviewing some definitions. The *instantaneous power* $P(t)$ drawn from the power supply is proportional to the supply current $i_{DD}(t)$ and the supply voltage V_{DD}

$$P(t) = i_{DD}(t)V_{DD} \qquad \textbf{(4.22)}$$

The *energy* consumed over some time interval T is the integral of the instantaneous power

$$E = \int_0^T i_{DD}(t)V_{DD}dt \qquad \textbf{(4.23)}$$

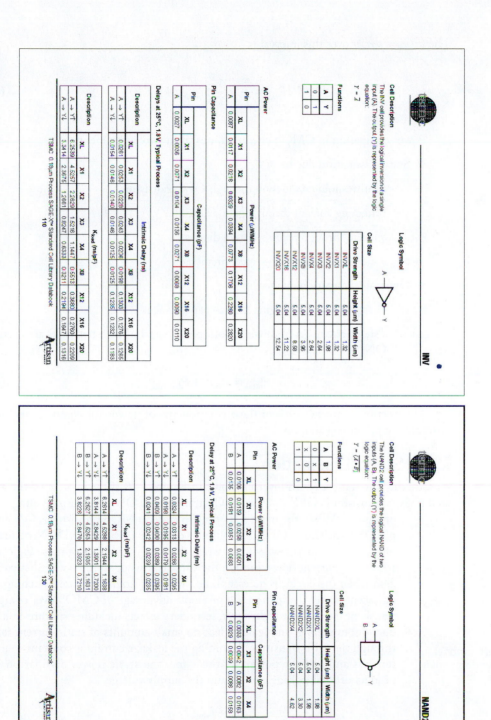

FIG 4.25 Artisan Components cell library datasheets. Reprinted with permission.

The *average power* over this interval is

$$P_{\text{avg}} = \frac{E}{T} = \frac{1}{T}\int_0^T i_{DD}(t)V_{DD}dt \qquad (4.24)$$

Power dissipation in CMOS circuits comes from two components:

- Static dissipation due to
  - subthreshold conduction through OFF transistors
  - tunneling current through gate oxide
  - leakage through reverse-biased diodes
  - contention current in ratioed circuits
- Dynamic dissipation due to
  - charging and discharging of load capacitances
  - "short-circuit" current while both pMOS and nMOS networks are partially ON

$$P_{\text{total}} = P_{\text{static}} + P_{\text{dynamic}} \qquad (4.25)$$

This section quantifies each of these components of power and discusses techniques to minimize power consumption.

4.4.1 Static Dissipation

Considering the static CMOS inverter shown in Figure 4.26, if the input = '0,' the associated nMOS transistor is OFF and the pMOS transistor is ON. The output voltage is V_{DD} or logic '1.' When the input = '1,' the associated nMOS transistor is ON and the pMOS transistor is OFF. The output voltage is 0 volts (GND). Note that one of the transistors is always OFF when the gate is in either of these logic states. Ideally, no current flows through the OFF transistor so the power dissipation is zero when the circuit is quiescent, i.e., when no transistors are switching. Zero quiescent power dissipation is a principle advantage of CMOS over competing transistor technologies. However, secondary effects including subthreshold conduction, tunneling, and leakage lead to small amounts of static current flowing through the OFF transistor. Assuming the leakage current is constant so instantaneous and average power are the same, the static power dissipation is the product of total leakage current and the supply voltage.

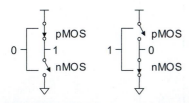

FIG 4.26 CMOS inverter model for static power dissipation evaluation

$$P_{\text{static}} = I_{\text{static}}V_{DD} \qquad (4.26)$$

Example

A digital system in a 1.2 V 100 nm process [Parihar01] has 200 million transistors, of which 20 million are in logic gates and the remainder in memory arrays. The average logic transistor width is 12 λ and the average memory transistor width is 4 λ. The process has two threshold voltages and two oxide thicknesses. Subthreshold leakage for OFF devices is 20 nA/μm for low-threshold devices and 0.02 nA/μm for high-threshold devices. Gate leakage is 3 nA/μm for thin oxides and 0.002 nA/μm for thick oxides. Memories use low-leakage devices everywhere. Logic uses low-leakage devices in all but 20% of the paths that are most critical for performance. Diode leakage is negligible. Estimate the static power consumption. How would the power consumption change if the low-leakage devices were not available?

Solution: There are (20 • 10^6 logic transistors) • (0.2) • (12 λ • (0.05 μm/λ) = 2.4 • 10^6 μm of high-leakage devices and [(20 • 10^6 logic transistors) • (0.8) • (12 λ) + (180 • 10^6 memory transistors) • (4 λ)] • (0.05 μm/λ) = 45.6 • 10^6 μm of low-leakage devices. All devices exhibit gate leakage. On average, half the transistors are OFF and contribute subthreshold leakage. Therefore, the total static current is (2.4 • 10^6 μm) • [(20 nA/μm)/2 + (3 nA/μm)] + (45.6 • 10^6 μm) • [(0.02 nA/μm)/2 + (0.002 nA/μm)] = 32 mA. Static power consumption is (32 mA) • (1.2 V) = 38 mW. This is likely to be small compared to dynamic power consumption, yet large enough to limit the battery life of battery-powered systems on standby.

If low-leakage devices were not available, the total static current would be (2.4 • 10^6 μm + 45.6 • 10^6) • [(20 nA/μm)/2 + (3 nA/μm)] = 624 mA, for standby power of (624 mA) • (1.2 V) = 749 mW.

According to EQ (2.34), OFF transistors still conduct a small amount of subthreshold current. As subthreshold current is exponentially dependent on threshold voltage, it is increasing dramatically as threshold voltages have scaled down.

SiO_2 is a very good insulator, so leakage current through the gate dielectric historically was very low. However, it is possible for electrons to tunnel across very thin insulators; the probability drops off exponentially with oxide thickness. Tunneling current becomes important for transistors around the 130 nm generation with gate oxides of 20Å or thinner.

There is also some small static dissipation due to reverse biased diode leakage between diffusion regions, wells, and the substrate, as shown for an inverter in Figure 2.19. Diode leakage is given by EQ (2.38). In modern processes, diode leakage is generally much smaller than the subthreshold or gate leakage and may be neglected.

In older processes, all three components of static power dissipation were small enough that CMOS was often said to consume "zero" DC power. Leakage power was of concern only to ultra-low-power systems. In 130 nm processes and beyond, the static power is rapidly becoming a primary design issue and vendors now provide leakage data, often in the form of nA/μm of gate length. Hand-held battery-powered devices typically require

Example

For a process with k'_p of 75 $\mu A/V^2$, $V_{tp} = -0.4$ V, and $V_{DD} = 1.8$ V, calculate the static power dissipation of a 32 word x 48-bit ROM that contains a 1:32 pseudo-nMOS row decoder and pMOS pull-ups on the 48-bit lines. The W/L ratio of the pMOS pull-ups is 1. Assume one of the word lines and 50% of the bitlines are high at any given time.

Solution: Each pMOS transistor dissipates power when the output is low.

$$I_{\text{pull-up}} = k'_p \frac{W}{L} \frac{\left(V_{DD} - \left|V_{tp}\right|\right)^2}{2} = 73\mu A$$

$$P_{\text{pull-up}} = V_{DD} I_{\text{pull-up}} = 130\mu W \tag{4.27}$$

We expect to see 31 wordlines and 24 bitlines low, so the total static power is $130\mu W \cdot (31 + 24) = 7.2$ mW.

standby static currents in the 10's to 100's of μA. Eventually, static power dissipation may become comparable to dynamic power even for high-power systems.

Of course, static dissipation can occur in gates such as pseudo-nMOS gates where there is a direct path between power and ground. If such gates are used, this contention current must be factored into the total static power dissipation of the chip.

4.4.2 Dynamic Dissipation

The primary dynamic dissipation component is charging the load capacitance. Suppose a load C is switched between GND and V_{DD} at an average frequency of f_{sw}. Over any given interval of time T, the load will be charged and discharged Tf_{sw} times. Current flows from V_{DD} to the load to charge it. Current then flows from the load to GND during discharge. In one complete charge/discharge cycle, a total charge of $Q = CV_{DD}$ is thus transferred from V_{DD} to GND.

The average dynamic power dissipation is

$$P_{\text{dynamic}} = \frac{1}{T} \int_0^T i_{DD}(t) V_{DD} dt = \frac{V_{DD}}{T} \int_0^T i_{DD}(t) dt \tag{4.28}$$

Taking the integral of the current over some interval T as the total charge delivered during that time, we simplify to

$$P_{\text{dynamic}} = \frac{V_{DD}}{T} \left[Tf_{sw} CV_{DD}\right] = CV_{DD}^2 f_{sw} \tag{4.29}$$

> ### Example
>
> Our 200M transistor digital system from the example on page 189 uses static CMOS for the logic gates with an average activity factor of 0.1 The memory arrays are divided into banks and only the necessary bank is activated so the effective memory activity factor is 0.05. Assume transistors have a gate capacitance of about 2 fF/µm. Neglecting wire capacitance, estimate the dynamic power consumption per MHz of the system.
>
> **Solution:** There are $(20 \cdot 10^6$ logic transistors$) \cdot (12 \lambda) \cdot (0.05 \ \mu m/\lambda) \cdot (2 \ fF/\mu m) =$ 24 nF of logic transistors and $(180 \cdot 10^6$ memory transistors$) \cdot (4 \lambda) \cdot (0.05 \ \mu m/\lambda) \cdot$ 2 fF/µm = 72 nF of memory transistors. The power consumption is $[(0.1) \cdot (24 \cdot 10^{-9}) + (0.05) \cdot (72 \cdot 10^{-9})] \cdot (1.2)^2 = 8.6$ mW/MHz, or 8.6 W at 1 GHz.

Because most gates do not switch every clock cycle, it is often more convenient to express switching frequency f_{sw} as an activity factor α times the clock frequency f. Now the dynamic power dissipation may be rewritten as:

$$P_{\text{dynamic}} = \alpha C V_{DD}^{\ 2} f \qquad (4.30)$$

A clock has an activity factor of $\alpha = 1$ because it rises and falls every cycle. Most data has a maximum activity factor of 0.5 because it transitions only once each cycle. Static CMOS logic has been empirically determined to have activity factors closer to 0.1 because some gates maintain one output state more often than another and because real data inputs to some portions of a system often remain constant from one cycle to the next.

Because the input rise/fall time is greater than zero, both nMOS and pMOS transistors will be ON for a short period of time while the input is between V_{tn} and $V_{DD} - |V_{tp}|$. This results in an additional "short circuit" current pulse from V_{DD} to GND and typically increases power dissipation by about 10% [Veendrick84].

Short circuit power dissipation occurs as both pullup and pulldown networks are partially ON while the input switches [Veendrick84]. It increases as edge rates become slower because both networks are ON for more time. However, it decreases as load capacitance increases because with large loads the output only switches a small amount during the input transition, leading to a small V_{ds} across one of the transistors. Unless the input edge rate is much slower than the output edge rate, short circuit current is a small fraction (< 10%) of current to the load and can be ignored in hand calculations. It is good to use relatively crisp edge rates at the inputs to gates with wide transistors to minimize their short circuit current.

4.4.3   Low-power Design

Total power dissipation is the sum of the static and dynamic dissipation components. Dynamic dissipation has historically been far greater than static power when systems are active, and hence, static power is often ignored, although this will change as gate and subthreshold leakage increase. Many tools are available to assist with power estimation; these are discussed further in Sections 5.5.4 and 8.4.1.7.

Power dissipation has become extremely important to VLSI designers. For high-performance systems such as workstations and servers, dynamic power consumption per chip is often limited to about 150 W by the amount of heat that can be managed with air-cooled systems and cost-effective heatsinks. This number increases slowly with advances in heatsink technology and can be increased significantly with expensive liquid cooling, but has not kept pace with the growing power demands of systems. Therefore, performance may be limited by the inability to cool huge systems with power-hungry circuits operating at high speeds. For battery-based systems such as laptops, cell phones, and PDAs, power consumption sets the battery life of the product. In these systems, most or all of the switching activity may be stopped in an idle or "sleep" mode. Hence, in addition to dynamic power while active, static power consumption may limit the battery life while idle.

Many papers and books have been written on low-power design. Unfortunately, there are no "silver bullets"; low-power consumption is generally achieved by careful design. Power reduction techniques can be divided into those that reduce dynamic power and those that reduce static power.

4.4.3.1 Dynamic Power Reduction

If a process is selected with sufficiently high threshold voltages and oxide thicknesses, static dissipation is small and dynamic dissipation usually dominates while the chip is active. EQ (4.30) shows that dynamic power is reduced by decreasing the activity factors, the switching capacitance, the power supply, or the operating frequency.

Activity factor reduction is very important. Static logic has an inherently low activity factor. Clocked nodes such as the clock network and the clock input to registers have an activity factor of 1 and are very power-hungry. Dynamic circuit families, described in Section 6.2.4, have clocked nodes and a high internal activity factor, so they are also costly in power. Clock gating can be used to stop portions of the chip that are idle; for example, a floating point unit can be turned off when executing integer code and a second level cache can be idled if the data is found in the primary cache. A large fraction of power is dissipated by the clock network itself, so entire portions of the clock network can be turned off where possible. The chip can also sense die temperature and cut back activity if the temperature becomes too high. A drawback of activity factor reduction is that if the system transitions rapidly from an idle mode with little switching to a fully active mode, a large di/dt spike will occur. This leads to inductive noise in the power supply network. Some systems throttle execution, limiting the number of functional units that go from idle to active in each cycle.

Device-switching capacitance is reduced by choosing small transistors. Minimum-sized gates can be used on non-critical paths. Although Logical Effort finds that the best stage effort is about 4, using a larger stage effort increases delay only slightly and greatly reduces transistor sizes. For example, buffers driving I/O pads or long wires may use a stage effort of 8–12 to reduce the buffer size. Interconnect switching capacitance is most effectively reduced through careful floorplanning, placing communicating units near each other to reduce wire lengths.

Voltage has a quadratic effect on dynamic power. Therefore, choosing a lower power supply significantly reduces power consumption. As many transistors are operating in a velocity-saturated regime, the lower power supply may not reduce performance as much as first-order models predict. Voltage can be adjusted based on operating mode; for example, a laptop processor may operate at high voltage and high speed when plugged into an AC adapter, but at lower voltage and speed when on battery power. If the frequency and voltage scale down in proportion, a cubic reduction in power is achieved. For example, the laptop processor may scale back to 2/3 frequency and voltage to save 70% in power when unplugged.

Frequency can also be traded for power. For example, in a digital signal processing system primarily concerned with throughput, two multipliers running at half speed can replace a single multiplier at full speed. At first, this may not appear to be a good idea because it maintains constant power and performance while doubling area. However, if the power supply can also be reduced because the frequency requirement is lowered, overall power consumption goes down.

Commonly used metrics in low-power design are *power*, the *power–delay product*, and the *energy–delay product*. Power alone is a questionable metric because it can be reduced simply by computing more slowly. The power-delay (i.e., energy) product is also suspect because the energy can be reduced by computing more slowly at a lower supply voltage. The energy-delay product (i.e., power • delay^2) is less prone to such gaming.

Overall, the energy-delay product measured in Performance^2/Watt (where performance might be in units of SpecInt) normalized for process only varies by about a factor of two across a wide range of general-purpose microprocessor architectures [Gonzalez96]. This suggests that as long as wasteful practices are avoided, there is little you can do to general-purpose processors except trade the energy consumed by a computation against the delay of the computation. The big power gains are to be made not through tweaking of circuits but by reconsidering algorithms. For example, the Fast Fourier Transform requires far fewer arithmetic operations and hence less power than a Discrete Fourier Transform. Signal-processing systems using datapaths hardwired to a particular operation consume far less power than general-purpose processors delivering the same performance because the datapaths eliminate unnecessary control units.

4.4.3.2 Static Power Reduction   Static power reduction involves minimizing I_{static}. Some circuit techniques such as analog current sources and pseudo-nMOS gates intentionally draw static power. They can be turned off when they are not needed.

Recall that the subthreshold leakage current for $V_{gs} < V_t$ is

$$I_{ds} = I_{ds0}e^{\frac{V_{gs}-V_t}{nv_T}}\left[1 - e^{\frac{-V_{ds}}{v_T}}\right] \qquad (4.31)$$

$$V_t = V_{t0} - \eta V_{ds} + \gamma \left(\sqrt{\phi_s + V_{sb}} - \sqrt{\phi_s} \right) \qquad (4.32)$$

where the η term describes drain-induced barrier lowering and the γ term describes the body effect. For any appreciable V_{ds}, the term in brackets approaches unity and can be discarded. The remaining term can be reduced by increasing the threshold voltage V_{t0}, reducing V_{gs}, reducing V_{ds}, increasing V_{sb}, or lowering the temperature.

Subthreshold leakage power is already a major problem for battery-powered designs in the 180 nm generation and will be growing exponentially as power supplies and thresholds voltages are scaled down in future processes. Many low-power systems need high performance while active and low leakage while idle. The high-performance requirement entails relatively low thresholds, which contribute excessive leakage current in the idle mode. As mentioned in Section 4.4.1, selective application of multiple threshold voltages can maintain performance on critical paths with low-V_t transistors while reducing leakage on other paths with high-V_t transistors.

In low-power battery-operated devices, leakage specifications may be given at 40° C rather than 110° C because battery life is most important in the range of normal ambient temperatures.

Another way to control leakage is through the body voltage using the body effect. For example, low-V_t devices can be used and a *reverse body bias* (RBB) can be applied during idle mode to reduce leakage [Kuroda96]. Alternatively, higher-V_t devices can be used, and then a *forward body bias* (FBB) can be applied during active mode to increase performance [Narendra03]. As we will see in Section 4.7.3, threshold voltages vary from one die to another on account of manufacturing variations. An *adaptive body bias* (ABB) can compensate and achieve more uniform transistor performance despite the variations [Narendra99, Tschanz02]. In any case, the body bias should be kept to less than about 0.5 V. Too much reverse body bias leads to greater junction leakage through a mechanism called *band-to-band tunneling* [Keshavarzi01], while too much forward body bias leads to substantial current through the body to source diodes.

Applying a body bias requires additional power supply rails to distribute the substrate and well voltages. For example, a RBB scheme for a 1.8 V n-well process could bias the p-type substrate at $V_{BBn} = -0.4$ V and the n-well at $V_{BBp} = 2.2$ V. Figure 4.27 shows a schematic and cross-section of an inverter using body bias. In an n-well process, all nMOS transistors share the same p substrate and must use the same V_{BBn}. In a triple-well process, groups of transistors can use different p-wells isolated from the substrate and thus can use different body biases. The well and substrate carry little current, so the bias voltages are relatively easy to generate and distribute.

Alternatively, the source voltage can be raised in sleep mode. This has the double benefit of reducing V_{ds} (to increase V_t through reduced DIBL and to also reduce gate leakage) as well as increasing V_{sb} (to increase V_t through the body effect). However, the source does carry significant current, so generating a stable and adjustable source voltage rail is challenging.

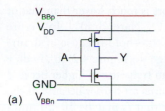

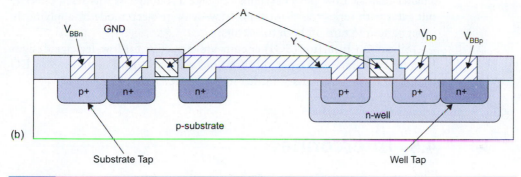

(a)

(b)

FIG 4.27 Body bias

Reducing V_{DD} in standby mode reduces the drain-induced barrier lowering contribution to leakage. It also decreases gate leakage in processes where that component is important. The supply should be maintained at a high enough level to preserve the state of the system [Clark02].

Yet another method of reducing idle leakage current in low-power systems is to turn off the power supply entirely. This could be done externally with the voltage regulator or internally with a series transistor. *Multiple Threshold CMOS* circuits (MTCMOS) use low-V_t transistors for computation and a high-V_t transistor as a switch to disconnect the power supply during idle mode, as shown in Figure 4.28 [Mutoh95][4]. The high-V_t device is connected between the true V_{DD} and the *virtual V_{DDV}* rails connected to the logic gates. The extra transistor increases the impedance between the true and virtual power supply, causing greater power supply noise and gate delay. Bypass capacitance between V_{DDV} and GND stabilizes the supply somewhat, but the capacitance is discharged each time V_{DDV} is disconnected, contributing to the power consumption. Even using a very wide high-V_t transistor, MTCMOS is only suited to systems with small power demands. The pMOS body should be tied to V_{DD} so both V_{DD} and V_{DDV} lines must be routed to all cells. MTCMOS uses carefully designed registers connected to the true supply rails to retain state during idle mode [Kao01].

---

[4]Do not confuse MTCMOS with the use of high-V_t transistors in noncritical gates and low-V_t transistors used in critical gates described earlier.

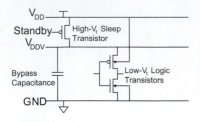

FIG 4.28 MTCMOS

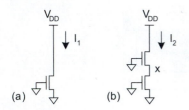

FIG 4.29 Leakage stack effect

The leakage through two series OFF transistors is much lower than that of a single transistor because of the *stack effect* [Ye98, Narendra01]. In Figure 4.29(a), the single transistor has a relatively low threshold because of drain-induced barrier lowering from the high drain voltage. In Figure 4.29(b), node x rises to about 100 mV. The threshold on the bottom transistor is higher because of the small drain voltage. The top transistor also turns off harder because of the negative V_{gs} and the body effect. The net result is that I_2 may be 10–20 times smaller than I_1. Low-power systems can take advantage of this stack effect to put gates with series transistors into a low-leakage sleep mode by applying an input pattern to turn off both transistors.

Silicon on Insulator (SOI) circuits are attractive for low-leakage designs because they have a sharper subthreshold current rolloff (smaller n in EQ (4.31)). SOI will be discussed further in Section 6.7.

4.5 Interconnect

The wires linking transistors together are called *interconnect* and play a major role in the performance of modern systems. Figure 4.30 shows a pair of adjacent wires. The wires have width w, length l, thickness t, and spacing of s from their neighbors and have a dielectric of height h between them and the conducting layer below. The sum of width and spacing is called the wire *pitch*. The thickness to width ratio t/w is called the *aspect ratio*. The dielectric is made of SiO_2 or a low-k material. In the early days of VLSI, transistors were relatively slow. Wires were wide and thick and thus had low resistance. Under those circumstances, wires could be treated as ideal equipotential nodes with lumped capacitance. In modern VLSI processes, transistors switch much faster. Meanwhile, wires have become narrower, driving up their resistance to the point that in many paths the wire RC delay exceeds gate delay. Moreover, the wires are packed very closely together and thus a large fraction of their capacitance is to their neighbors. When one wire switches, it tends to affect its neighbor through capacitive coupling; this effect is called *crosstalk*. On-chip interconnect inductance had been negligible but is now becoming a factor for systems with fast edge rates and closely packed busses. Considering all of these factors, circuit design is now as much about engineering the wires as the transistors that sit underneath.

Early CMOS processes had a single metal layer and for many years only two or three layers were available, but with advances in chemical-mechanical polishing it became far more practical to manufacture many metal layers. A 180 nm process typically has about six to eight metal layers and the layer count has been increasing at a rate of about one per generation. Figure 4.31 shows a cross-section of the Intel 180 nm process metal stack [Yang98]. Metal1 wires are thin and built on a tight pitch to provide dense routing within a cell; observe that the pitch is less than 6 λ, while conservative scalable CMOS rules would dictate an 8 λ pitch. Their resistance is high, but that is acceptable because the wires tend to be short. Top-level metal wires are thicker and built on a wide pitch.

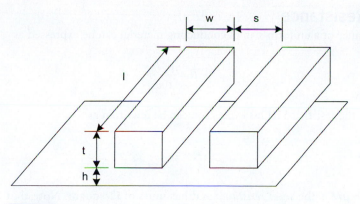

FIG 4.30 Interconnect geometry

Layer	t (nm)	w (nm)	s (nm)	AR
6	1720	860	860	2.0
	1000			
5	1600	800	800	2.0
	1000			
4	1080	540	540	2.0
	700			
3	700	320	320	2.2
	700			
2	700	320	320	2.2
	700			
1	480	250	250	1.9
	800			

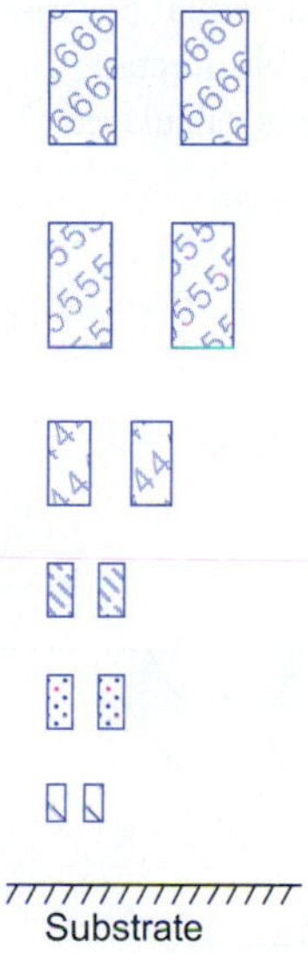

Substrate

FIG 4.31 Layer stack for 6-metal Intel 180 nm process

This low-resistance layer is useful for power, ground, clock, and critical signal routing. In the Intel process, the intermediate layers gradually increase in width and pitch, although in many processes the intermediate layers are uniform. The width and spacing are typically comparable.

4.5.1 Resistance

The resistance of a uniform slab of conducting material can be expressed as

$$R = \frac{\rho}{t}\frac{l}{w} \tag{4.33}$$

where ρ is the resistivity[5]. This expression can be rewritten as

$$R = R_\square \frac{l}{w} \tag{4.34}$$

where $R_\square = \rho/t$ is the *sheet resistance* and has units of Ω/square. Note that a square is a dimensionless quantity corresponding to a slab of equal length and width. This is convenient because resistivity and thickness are characteristics of the process outside the control of the circuit designer and can be abstracted away into the single sheet resistance parameter.

To obtain the resistance of a conductor on a layer, multiply the sheet resistance by the ratio of length to width of the conductor. For example, the resistance of the two shapes in Figure 4.32 are equal because the length-to-width ratio is the same even though the sizes are different. Nonrectangular shapes can be decomposed into simpler regions for which the resistance is calculated [Horowitz83].

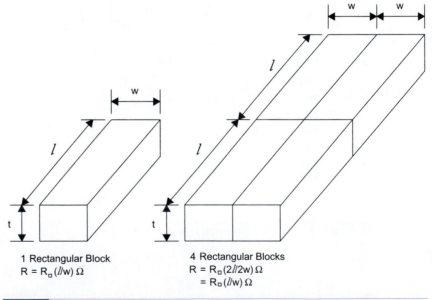

1 Rectangular Block
$R = R_\square (l/w)\ \Omega$

4 Rectangular Blocks
$R = R_\square (2l/2w)\ \Omega$
$= R_\square (l/w)\ \Omega$

FIG 4.32   Two conductors with equal resistance

[5] ρ is used to indicate both resistivity and best stage effort. The meaning should be clear from context.

Table 4.6 shows bulk electrical resistivities of pure metals [Bakoglu90]. The sheet resistance of thin metal films used in wires tends to be slightly higher, e.g., 2.6 $\mu\Omega \cdot$ cm for Cu and 3.5–4.0 $\mu\Omega \cdot$ cm for Al. Most processes prior to the 180 nm generation use aluminum wires. Modern processes often use copper to reduce the resistivity and also to obtain better electromigration characteristics (see Section 4.8.2). Unfortunately, copper must be surrounded by a lower-conductivity diffusion barrier that effectively reduces the wire cross-sectional area and hence raises the resistance. Aluminum does not require such a barrier and thus may actually offer lower resistance for very narrow wires in the future. Electron surface scattering effects in thin conductors also result in somewhat higher resistance for on-chip interconnect than simple bulk resistivity would predict.

Table 4.6	Bulk resistivity of pure metals at 22° C
Metal	Resistivity ($\mu\Omega \cdot$ cm)
Silver (Ag)	1.6
Copper (Cu)	1.7
Gold (Au)	2.2
Aluminum (Al)	2.8
Tungsten (W)	5.3
Molybdenum (Mo)	5.3
Titanium (Ti)	43.0

Table 4.7 shows typical sheet resistances for the 180 nm process with aluminum interconnect. The upper layers of metal have lower resistivity because they are thicker. Metal resistance is determined by the material (usually Al or Cu). The resistivity of polysilicon, diffusion, and wells is significantly influenced by the doping levels. Polysilicon and diffusion are often silicided with $TiSi_2$ (see Section 3.2.8) to reduce the resistance. Interconnect resistance increases with temperature; this effect is particularly pronounced for wells and diffusion.

Contacts and vias also have a resistance associated with them that is dependent on the contacted materials and size of the contact. Typical values are 2–20 Ω. Multiple contacts should be used to form low-resistance connections. Because current crowding tends to occur at the periphery of contacts, design rules dictate multiple small contacts rather than a single large contact, as shown in Figure 4.33. When current turns at a right angle or reverses, a square array of contacts is generally required, while when the flow is in the same direction, fewer contacts can be used.

Table 4.7 Sheet resistances	
Layer	**Sheet Resistance (Ω /□)**
Diffusion (silicided)	3-10
Diffusion (unsilicided)	50-200
Polysilicon (silicided)	3-10
Polysilicon (unsilicided)	50-400
Metal1	0.08
Metal2	0.05
Metal3	0.05
Metal4	0.03
Metal5	0.02
Metal6	0.02

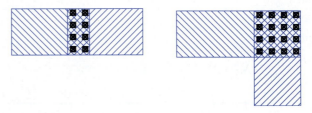

FIG 4.33 Multiple vias for low-resistance connections

4.5.2 Capacitance

An isolated wire over the substrate can be modeled as a conductor over a ground plane. The wire capacitance has two major components: the parallel plate capacitance of the bottom of the wire to ground and the fringing capacitance arising from fringing fields along the edge of a conductor with finite thickness. In addition, a wire adjacent to a second wire on the same layer can exhibit capacitance to that neighbor. These effects are illustrated in Figure 4.34. The classic parallel plate capacitance formula is

$$C = \frac{\varepsilon_{ox}}{h} wl \tag{4.35}$$

Note that oxides are often doped with phosphorous to trap ions before they damage transistors; this oxide has $\varepsilon_{ox} \approx 4.1\varepsilon_0$, as compared to $3.9\varepsilon_0$ for an ideal oxide or lower for low-k dielectrics.

The fringing capacitance is more complicated to compute and requires a numerical field solver for exact results. A number of authors have proposed approximations to this calculation [Barke88, Ruehli73, Yuan82]. One intuitively appealing approximation treats a lone conductor above a ground plane as a rectangular middle section with two hemi-spherical end caps, as shown in Figure 4.35 [Yuan82]. The total capacitance is assumed to be the sum of a parallel plate capacitor of width $w - t/2$ and a cylindrical capacitor of radius $t/2$. This results in an expression for the capacitance that is accurate within 10% for aspect ratios less than 2 and $t \approx h$.

$$C = \varepsilon_{ox} l \left[\frac{w - \frac{t}{2}}{h} + \frac{2\pi}{\ln\left(1 + \frac{2h}{t} + \sqrt{\frac{2h}{t}\left(\frac{2h}{t} + 2\right)}\right)} \right] \qquad (4.36)$$

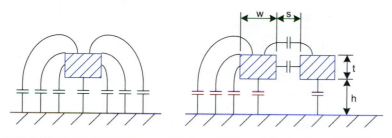

FIG 4.34   Effect of fringing fields on capacitance

An empirical formula that is computationally efficient and relatively accurate is [Meijs84, Barke88]

$$C = \varepsilon_{ox} l \left[\frac{w}{h} + 0.77 + 1.06\left(\frac{w}{h}\right)^{0.25} + 1.06\left(\frac{t}{h}\right)^{0.5} \right] \qquad (4.37)$$

which is good to 6% for aspect ratios less than 3.3.

These formulae do not account for neighbors on the same layer or higher layers. Capacitance interactions between layers can become quite complex in modern multilayer CMOS processes. A conservative upper bound on capacitance can be obtained assuming that the layers above and below the conductor of interest are solid ground planes. Similarly, a lower bound can be obtained assuming there are no other conductors in the sys-tem except the substrate. The upper bound can be used for propagation delay and power estimation while the lower bound can be used for con-tamination delay calculations before layout information is available. A

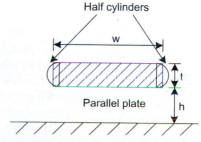

FIG 4.35   Yuan & Trick capacitance model including fringing fields

cross-section of the model used for capacitance upper bound calculations is shown in Figure 4.36. The total capacitance of the conductor of interest is the sum of its capacitance to the layer above, the layer below, and the two adjacent conductors. If the layers above and below are not switching[6], they can be modeled as ground planes and this component of capacitance is called C_{gnd}. Theoretically, wires will have some capacitance to further neighbors, but in practice this capacitance is normally small enough to ignore because most electric fields terminate on the nearest conductors.

$$C_{gnd} = C_{bot} + C_{top}$$
$$C_{total} = C_{gnd} + 2C_{adj}$$

(4.38)

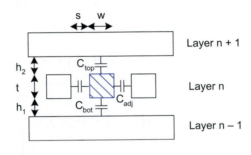

FIG 4.36 Multilayer capacitance model

The capacitances can be computed by generating a table of data with a field solver such as FastCap [Nabors92] or HSPICE. Tables can be generated for different widths and spacings on each layer and you can interpolate between entries if necessary. For example, Table 4.8 tabulates capacitance for the layer stack from Figure 4.31 using a SiOF low-k dielectric with $\varepsilon_{ox} = 3.55\varepsilon_0$ and assuming solid planes of metal above and below. The width w and spacing s indicate multiples of the minimum allowable for the particular metal layer. Three columns for each metal layer indicate the capacitance to each adjacent neighbor, to ground (i.e., the planes above and below, which are assumed on average not to be moving), and total (EQ (4.38)). The capacitance to neighbors accounts for more than 50% of the capacitance for narrow wires. We will see in Section 4.5.4 that coupling to these neighbors has a significant effect on delay and signal integrity and limits the acceptable aspect ratios. The capacitance for metal2 and metal3 lines should be equal because they have the same geometry; the minor variations reflect finite numerical precision in the HSPICE field solver. Table 4.9 shows the same information assuming no metal on other planes. The overall capacitance is slightly smaller because there are no nearby planes for fields to terminate on, but the capacitance to the adjacent neighbors is higher because many of the fringing fields now terminate on the neighbors instead.

---

[6]Or at least consist of a large number of orthogonal conductors that on average cancel each other's switching activities.

Table 4.8 Capacitance table for 180 nm process (aF/µm) with metal planes above and below

w	s	Metal1			Metal2			Metal3			Metal4			Metal5			Metal6		
		C_{adj}	C_{gnd}	C_{tot}	C_{adj}	C_{gnd}	C_{tot}	C_{adj}	C_{gnd}	C_{tot}	C_{adj}	C_{gnd}	C_{tot}	C_{adj}	C_{gnd}	C_{tot}	C_{adj}	C_{gnd}	C_{tot}
1	1	84	43	210	88	57	232	88	57	232	77	78	232	73	94	240	82	64	227
1	1.5	57	52	166	58	68	184	58	68	184	49	93	191	45	112	202	54	76	184
1	2	42	60	144	42	78	162	42	78	162	33	106	173	30	126	186	39	85	164
1	∞	0	112	112	0	132	132	0	132	132	0	154	154	0	167	167	0	137	137
1.5	1	85	53	224	89	70	248	89	70	248	77	99	253	73	119	266	83	80	246
1.5	1.5	58	62	178	59	82	199	59	82	199	49	114	212	45	137	227	56	91	202
1.5	2	43	70	156	42	92	177	42	92	177	34	127	194	30	151	211	41	101	182
1.5	∞	0	123	123	0	147	147	0	147	147	0	174	174	0	192	192	0	154	154
2	1	86	63	236	89	84	263	89	84	263	77	119	274	73	144	291	84	95	264
2	1.5	59	72	190	59	96	214	59	96	214	49	134	232	45	162	252	57	107	220
2	2	44	80	167	43	106	191	43	106	191	34	147	215	30	176	236	41	117	199
2	∞	0	134	134	0	162	162	0	162	162	0	195	195	0	217	217	0	171	171
3	1	87	83	258	90	113	292	90	113	292	77	160	315	73	194	341	86	127	298
3	1.5	60	92	211	59	124	243	59	124	243	49	175	273	45	212	302	58	138	254
3	2	44	101	189	43	135	220	43	135	220	34	188	256	30	226	286	42	148	233
3	∞	0	155	155	0	191	191	0	191	191	0	236	236	0	269	269	0	204	204
4	1	87	104	279	90	141	320	90	141	320	77	201	355	73	245	390	86	159	331
4	1.5	60	113	232	59	153	272	59	153	272	49	216	314	45	262	352	58	170	287
4	2	44	122	210	43	164	249	43	164	249	34	229	296	30	276	336	43	180	266
4	∞	0	176	176	0	219	219	0	219	219	0	278	278	0	321	321	0	237	237
6	1	87	146	320	89	199	377	89	199	377	77	284	437	72	345	490	86	223	396
6	1.5	59	155	274	59	210	328	59	210	328	49	299	396	45	362	451	58	235	351
6	2	44	163	252	42	221	306	42	221	306	33	312	378	30	376	436	43	245	331
6	∞	0	218	218	0	276	276	0	276	276	0	360	360	0	422	422	0	301	301

Figure 4.37 plots total capacitance of a metal2 line from the same process as a function of width for various spacings. For an isolated wire above the substrate, the capacitance is strongly influenced by spacing between conductors. For a wire sandwiched between metal1 and metal3 planes, the capacitance is higher and is more sensitive to the width (determining parallel plate capacitance) but less sensitive to spacing once the spacing is significantly greater than the wire thickness. In either case, the y-intercept is greater than zero so doubling the width of a wire results in less than double the total capacitance. Tight-pitch metal lines have a capacitance of roughly 0.2 fF/µm.

Table 4.9 Capacitance table for 180 nm process (aF/µm) with substrate below and nothing above

w	s	Metal1			Metal2			Metal3			Metal4			Metal5			Metal6		
		C_{adj}	C_{gnd}	C_{tot}	C_{adj}	C_{gnd}	C_{tot}	C_{adj}	C_{gnd}	C_{tot}	C_{adj}	C_{gnd}	C_{tot}	C_{adj}	C_{gnd}	C_{tot}	C_{adj}	C_{gnd}	C_{tot}
1	1	89	28	206	101	19	220	96	16	208	102	15	219	96	15	208	97	14	207
1	1.5	63	33	159	73	22	167	70	18	158	75	17	166	70	17	158	71	15	158
1	2	49	37	136	58	24	140	57	20	133	60	18	138	57	19	132	58	17	132
1	∞	0	87	87	0	70	71	0	60	61	0	59	60	0	60	60	0	55	55
1.5	1	92	33	217	104	21	230	100	17	218	106	16	229	100	17	217	101	15	217
1.5	1.5	66	38	170	76	24	176	74	19	167	78	18	174	74	19	167	75	16	166
1.5	2	52	42	145	61	27	148	60	21	140	63	20	146	60	21	140	61	18	139
1.5	∞	0	94	94	0	75	75	0	64	64	0	62	62	0	63	63	0	58	58
2	1	94	38	227	107	24	238	103	19	225	110	18	237	103	18	225	104	16	224
2	1.5	68	43	178	78	26	183	76	21	174	81	19	181	77	20	173	78	17	173
2	2	53	47	154	63	29	154	62	22	147	66	21	152	62	22	146	63	19	145
2	∞	0	100	100	0	78	79	0	67	67	0	65	65	0	66	66	0	60	60
3	1	98	48	243	111	29	251	108	22	237	114	20	249	108	21	237	109	18	236
3	1.5	71	52	194	82	31	195	81	24	185	85	22	192	81	23	184	82	20	183
3	2	56	57	169	66	34	166	66	25	157	69	24	162	66	25	156	67	21	155
3	∞	0	112	112	0	86	86	0	72	72	0	69	70	0	71	72	0	64	65
4	1	100	58	257	114	33	262	111	25	247	118	23	258	111	24	246	112	20	245
4	1.5	73	62	208	85	36	205	83	27	194	88	25	201	83	26	193	85	22	192
4	2	58	67	182	68	38	175	68	28	165	72	26	171	68	28	165	70	23	163
4	∞	0	124	124	0	92	93	0	77	77	0	74	75	0	76	76	0	68	69
6	1	102	77	282	118	43	279	115	31	262	122	28	272	115	31	260	116	25	258
6	1.5	76	82	233	88	46	222	87	33	208	92	30	215	87	33	207	89	26	204
6	2	61	86	207	72	48	191	72	35	179	76	31	184	72	34	178	73	28	175
6	∞	0	147	147	0	105	105	0	86	86	0	82	83	0	85	85	0	75	76

In practice, the layers above and below the conductor of interest are neither solid planes nor totally empty. One can extract capacitance more accurately by interpolating between these two extremes based on the density of metal on each level. [Chern92] gives formulae for this interpolation accurate to within 10%. However, if the wiring above and below is fairly dense (e.g., a bus on minimum pitch), it is well-approximated as a plane. Dense wire fill is added to many chips for mechanical stability and etch uniformity, making this approximation even more appropriate.

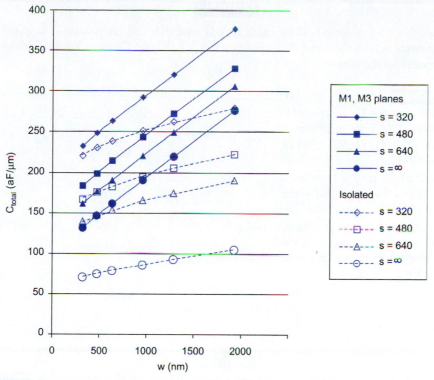

FIG 4.37 Capacitance of metal2 line as a function of width and spacing

4.5.3 Delay

Interconnect increases circuit delay for two reasons. First, the wire capacitance adds loading to each gate. Second, long wires have significant resistance that contributes distributed RC delay or *flight time*. It is straightforward to add wire capacitance to the Elmore delay calculations of Section 4.2.1, so in this section we focus on the RC delay.

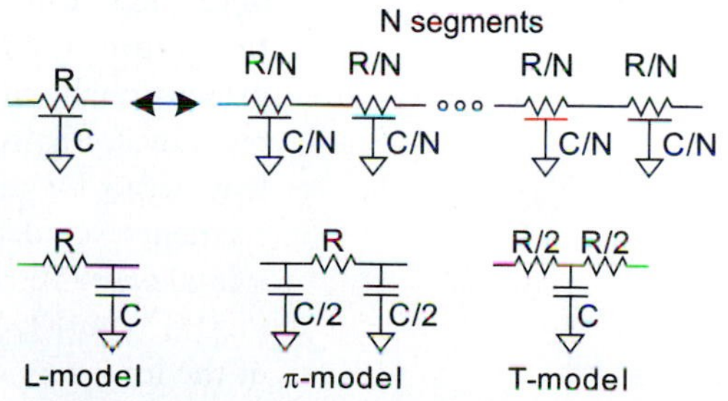

The distributed resistance and capacitance of a wire can be approximated with a number of lumped elements. Three standard approximations are the L-model, π-model, and T-model, so-named because of their shape. Figure 4.38 shows how a distributed RC circuit is equivalent to N distributed RC segments of proportionally less resistance and capacitance, and how these segments

FIG 4.38 Lumped approximation to distributed RC circuit

Example

Consider a 5 mm long, 0.32 μm wide metal2 wire in a 180 nm process. The sheet resistance is 0.05 Ω/□ and the capacitance is 0.2 fF/μm. Construct a 3-segment π-model for the wire.

Solution: The wire is 5000 μm/0.32 μm = 15625 squares in length. The total resistance is (0.05 Ω/□) • (15625□) = 781 Ω. The total capacitance is (0.2 fF/μm) • (5000 μm) = 1 pF. Each π-segment has one-third of this resistance and capacitance. The π-model is shown in Figure 4.39(a); adjacent capacitors can be merged as shown in Figure 4.39(b).

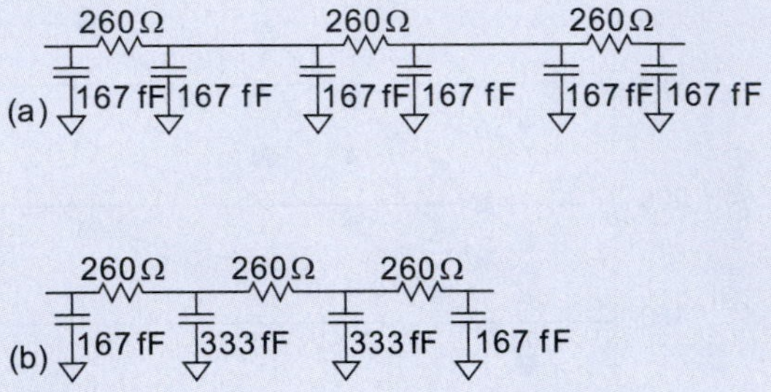

260Ω
260Ω
260Ω
(a)
167 fF
167 fF
167 fF
167 fF
167 fF
167 fF
260Ω
260Ω
260Ω
(b)
167 fF
333 fF
333 fF
167 fF

FIG 4.39 3-segment π-model for wire

Example

A 10x unit-sized inverter drives a 2x inverter at the end of the 5 mm wire from the previous example. The gate capacitance is C = 2 fF/μm and the effective resistance is R = 2.5 kΩ • μm for nMOS transistors. Estimate the propagation delay using the Elmore delay model; neglect diffusion capacitance.

Solution: A unit inverter has a 4 λ = 0.36 μm wide nMOS transistor and an 8 λ = 0.72 μm wide pMOS transistor. Hence, the unit inverter has an effective resistance of (2.5 kΩ • μm)/(0.36 μm) = 6.9 kΩ and a gate capacitance of (0.36 μm + 0.72 μm) • (2 fF/μm) = 2 fF. Larger inverters have proportionally more capacitance and less resistance. Figure 4.40 shows an equivalent circuit for the system using a single-segment π-model. The Elmore delay is t_{pd} = (690 Ω) • (500 fF) + (690 Ω + 781 Ω) • (500 fF + 4 fF) = 1.1 ns. The capacitance of the long wire dominates the delay; the capacitance of the 2x inverter is negligible in comparison.

781Ω
690Ω
500 fF
500 fF
4 fF
Driver
Wire
Load

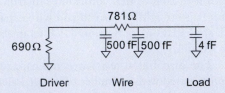

FIG 4.40 Equivalent circuit for example

can be modeled with lumped elements. As the number of segments approaches infinity, the lumped approximation will converge with the true distributed circuit. The L-model is a very poor choice because a large number of segments are required for accurate results. The π-model is much better; three segments are sufficient to give results accurate to 3% [Sakurai83]. The T-model is comparable to the π-model, but produces circuits with more nodes that are slower to solve by hand or with a circuit simulator. Therefore, it is common practice to model long wires with a 3- or 4-segment π-model for simulation.

These results make sense because the Elmore delay of a single-segment L-model is RC while the Elmore delay of a single-segment π- or T-model is $RC/2$. Single-segment π-models are a reasonable approximation for hand calculations.

Because both wire resistance and wire capacitance increase with length, wire delay grows quadratically with length. Using thicker and wider wires, lower-resistance metals such as copper, and lower-dielectric constant insulators helps, but long wires nevertheless often have unacceptable delay. Section 4.6.4 describes how repeaters can be used to break a long wire into multiple segments such that the overall delay becomes a linear function of length.

Polysilicon and diffusion wires (sometimes called *runners*) have high resistance, even if silicided. Diffusion also has very high capacitance. Do not use diffusion. Use polysilicon sparingly, usually in latches and flip-flops.

4.5.4 Crosstalk

As reviewed in Figure 4.41, wires have capacitance to their adjacent neighbors as well as to ground. When wire A switches, it tends to bring its neighbor B along with it on account of capacitive coupling, also called *crosstalk*. If B is supposed to switch simultaneously, this may increase or decrease the switching delay. If B is not supposed to switch, crosstalk causes noise on B. We will see that the impact of crosstalk depends on the ratio of C_{adj} to the total capacitance. Note that the load capacitance is included in the total, so for short wires and large loads, the load capacitance dominates and crosstalk is unimportant. Conversely, for long wires crosstalk is very important.

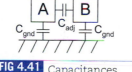

FIG 4.41 Capacitances to adjacent neighbor and to ground

4.5.4.1 Crosstalk Delay Effects
If both a wire and its neighbor are switching, the direction of the switching affects the amount of charge that must be delivered and the delay of the switching. Table 4.10 summarizes this effect. The charge delivered to the coupling capacitor is $Q = C_{adj}\Delta V$, where ΔV is the change in voltage between A and B. If A switches but B does not, $\Delta V = V_{DD}$. The total capacitance effectively seen by A is just the capacitance to ground and to B. If both A and B switch in the same direction, $\Delta V = 0$. Hence, no charge is required and C_{adj} is effectively absent for delay purposes. If A and B switch in the opposite direction, $\Delta V = 2V_{DD}$. Twice as much charge is required. Equivalently, the capacitor can be treated as being effectively twice as large switching through V_{DD}. This is analogous to the Miller effect discussed in Section 4.2.5.4. The *Miller Coupling Factor* (MCF)

describes how the capacitance to adjacent wires is multiplied to find the effective capacitance. Some designers use MCF = 1.5 as a statistical compromise when estimating propagation delays before layout information is available.

Table 4.10	Dependence of effective capacitance on switching direction		
B	**ΔV**	**$C_{\text{eff}(A)}$**	**MCF**
constant	V_{DD}	$C_{\text{gnd}} + C_{\text{adj}}$	1
switching same direction as A	0	C_{gnd}	0
switching opposite to A	$2V_{DD}$	$C_{\text{gnd}} + 2C_{\text{adj}}$	2

Example

Each wire in a pair of 1 mm lines has capacitance of 0.1 fF/µm to ground and 0.1 fF/µm to its neighbor. Each line is driven by an inverter with a 1 kΩ effective resistance. Estimate the contamination and propagation delays of the path. Neglect parasitic capacitance of the inverter and resistance of the wires.

Solution: We find $C_{\text{gnd}} = C_{\text{adj}} = (0.1 \text{ fF/µm}) \cdot (1000 \text{ µm}) = 0.1 \text{ } pF$. The delay is RC_{eff}. The contamination delay is the minimum possible delay, which occurs when both wires switch in the same direction. In that case, $C_{\text{eff}} = C_{\text{gnd}}$ and the delay is $t_{cd} = (1\text{k}\Omega) \cdot (0.1 \text{ pF}) = 100 \text{ ps}$. The propagation delay is the maximum possible delay, which occurs when both wires switch in opposite directions. In this case, $C_{\text{eff}} = C_{\text{gnd}} + 2C_{\text{adj}}$ and the delay is $t_{pd} = (1\text{k}\Omega) \cdot (0.3 \text{ pF}) = 300 \text{ ps}$.

A conservative design methodology assumes neighbors are switching when computing propagation and contamination delays (MCF = 2 and 0, respectively). This leads to a wide variation in the delay of wires. A more aggressive methodology tracks the time window during which each signal can switch. Thus, switching neighbors must be accounted for only if the potential switching windows overlap. Similarly, the direction of switching can be considered. For example, dynamic gates described in Section 6.2.4 precharge high and then fall low during evaluation. Thus, a dynamic bus will never see opposite switching during evaluation.

4.5.4.2 Crosstalk Noise Effects
Suppose wire A switches while B is supposed to remain constant. This introduces noise as B partially switches. We call A the *aggressor* or *perpetrator* and B the *victim*. If the victim is floating, we can model the circuit as a capacitive volt-

age divider to compute the victim noise, as shown in Figure 4.42. $\Delta V_{\text{aggressor}}$ is normally V_{DD}.

$$\Delta V_{\text{victim}} = \frac{C_{\text{adj}}}{C_{\text{gnd}-v} + C_{\text{adj}}} \Delta V_{\text{aggressor}} \qquad (4.39)$$

If the victim is actively driven, the driver will supply current to oppose and reduce the victim noise. We model the drivers as resistors, as shown in Figure 4.43. The peak noise becomes dependent on the time constant ratio k of the aggressor to the victim [Ho01]:

$$\Delta V_{\text{victim}} = \frac{C_{\text{adj}}}{C_{\text{gnd}-v} + C_{\text{adj}}} \frac{1}{1+k} \Delta V_{\text{aggressor}} \qquad (4.40)$$

where

$$k = \frac{\tau_{\text{aggressor}}}{\tau_{\text{victim}}} = \frac{R_{\text{aggressor}} \left(C_{\text{gnd}-a} + C_{\text{adj}} \right)}{R_{\text{victim}} \left(C_{\text{gnd}-v} + C_{\text{adj}} \right)} \qquad (4.41)$$

Figure 4.44 shows simulations of coupling when the aggressor is driven with a unit inverter; the victim is undriven or driven with an inverter of half, equal, or twice the size of the aggressor; and $C_{\text{adj}} = C_{\text{gnd}}$. Observe that when the victim is floating, the noise remains indefinitely. When the victim is driven, the driver restores the victim. Larger (faster) drivers oppose the coupling sooner and result in noise that is a smaller percentage of the supply voltage. Note that during the noise event the victim transistor is in its linear region while the aggressor is in saturation. For equal-sized drivers, this means $R_{\text{aggressor}}$ is two to four times R_{victim}, with greater ratios arising from more velocity saturation [Ho01]. In general, EQ (4.40) is conservative, especially when wire resistance is included [Vittal99]. It is often used to flag nets where coupling can be a problem; then simulations can be performed to calculate the exact coupling noise. Coupling noise is of greatest importance on weakly driven nodes where $k < 1$.

We have only considered the case of a single neighbor switching. When both neighbors switch, the noise will be twice as great. We have also modeled the layers above and below as AC ground planes, but wires on these layers are likely to be switching. For a long line, you can expect about as many

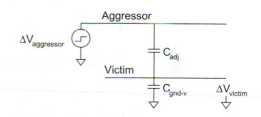

FIG 4.42 Coupling to floating victim

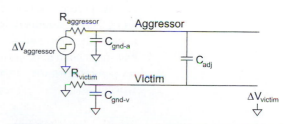

FIG 4.43 Coupling to driven victim

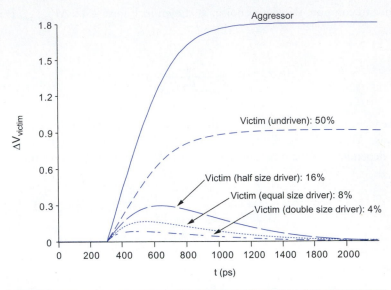

FIG 4.44 Waveforms of coupling noise

lines switching up and switching down, giving no net contribution to delay or noise. However, a short line running over a 64-bit bus in which all 64 bits are simultaneously switching from 0 to 1 will be strongly influenced by this switching.

4.5.5   Inductance

Most design tools consider only interconnect resistance and capacitance. Inductance is beginning to be important for accurately modeling on-chip power grids, clock networks, and wide busses. While the industry still has a very limited understanding of inductive effects, some of the key issues are introduced in this section.

Although we generally discuss current flowing from a gate output to charge or discharge a load capacitance, current really flows in loops. The return path for a current loop is usually the power or ground network; at the frequencies of interest, the power supply is an "AC ground" because the bypass capacitance forms a low-impedance path between V_{DD} and GND. Currents flowing around a loop generate a magnetic field proportional to the area of the loop and the amount of current. Changing the current requires supplying energy to change the magnetic field. This means that changing currents induce a voltage proportional to the rate of change. The constant of proportionality is called the inductance, L[7].

---

[7]L is used to indicate both inductance and transistor channel length. The meaning should be clear from context.

$$V = L\frac{dI}{dt} \qquad (4.42)$$

Inductance and capacitance also set the speed of light in a medium. Even if the resistance of a wire is zero leading to zero RC delay, the speed of light flight-time along a wire of length with inductance and capacitance per unit length of L and C is

$$t_{pd} = l\sqrt{LC} \qquad (4.43)$$

If the current return paths are the same as the conductors on which electric field lines terminate, the signal velocity v is

$$v = \frac{1}{\sqrt{LC}} = \frac{1}{\sqrt{\varepsilon_{ox}\mu_0}} = \frac{c}{\sqrt{3.9}} \qquad (4.44)$$

where μ_0 is the magnetic permeability of free space ($4\pi \cdot 10^{-7}$ H/m) and c is the speed of light in free space ($3 \cdot 10^8$ m/s). In other words, signals travel about half the speed of light. Using low-k (< 3.9) dielectrics raises this velocity. However, many signals have electric fields terminating on nearby neighbors, but currents returning in more distant power supply lines. This raises the inductance and reduces the signal velocity.

Changing magnetic fields in turn produce currents in other loops. This means that signals on one wire can inductively couple onto another; this is called *inductive crosstalk*.

The inductance of a conductor of length l and width w located a height h above a ground plane is approximately

$$L = l\frac{\mu_0}{2\pi}\ln\left(\frac{8h}{w} + \frac{w}{4h}\right) \qquad (4.45)$$

assuming $w < h$ and thickness is negligible. Typical on-chip inductance values are in the range of 0.15–1.5 pH/μm depending on the proximity of the power or ground lines. (Wires near their return path have smaller current loops and lower inductance.)

Current flows along the path of lowest impedance $Z = R + j\omega L$. At high-frequency ω, impedance becomes dominated by inductance. The inductance is minimized if the current flows only near the surface of the conductor closest to the return path. This skin effect can reduce the effective cross-sectional area of thick conductors and raise the effective resistance at high frequency. The skin depth for a conductor is

$$\delta = \sqrt{\frac{2\rho}{\omega\mu_0}} \qquad (4.46)$$

Example

Find the skin depth for signals with 100 ps edge rates on copper interconnect ($\rho = 1.7 \cdot 10^{-8}\,\Omega \cdot$ m).

Solution: 100 ps edge rates correspond to energy at 1.67 GHz.

$$\delta = \sqrt{\frac{2\left(1.7 \times 10^{-8}\,\Omega \cdot m\right)}{\sqrt{\left(2\pi \cdot 1.67 \times 10^{9}\,rad/s\right)\left(4\pi \times 10^{-7}\,H/m\right)}}} = 1.6\ \mu m \qquad (4.47)$$

Skin depth is not a major consideration at the time of writing, but will limit the benefit of thick wires at higher frequencies in the future.

The frequency of importance is the highest frequency with significant power in the Fourier transform of the signal. This is not the chip-operating frequency, but rather is associated with the faster edges. It can be approximated as

$$\omega = \frac{2\pi}{6t_{rf}} \qquad (4.48)$$

Extracting inductance in general is a three-dimensional problem and is extremely time-consuming for complex geometries. Inductance depends on the entire loop and therefore cannot be simply decomposed into sections in the way capacitance is. It is therefore impractical to extract the inductance from a chip layout. Instead, usually inductance is extracted using tools such as FastHenry [Kamon94] for simple test structures intended to capture the worst cases on the chip. This is only possible when the power supply network is highly regular. Power planes are ideal but require a large amount of metal resources. Dense power grids are usually the preferred alternative. Gaps in the power grid force current to flow around the gap, increasing the loop area and greatly increasing inductance. Moreover, large loops couple magnetic fields through other loops formed by conductors at a distance. Therefore, mutual inductive coupling can occur over a long distance, especially when the return path is far from the conductor. Incorporating inductance into simulations is also difficult. Instead, designers usually generate design rules that allow inductance to generally be ignored as long as the rules are followed.

Inductance has always been important for integrated circuit packages where the physical dimensions are large, as will be discussed in Section 12.2.3. On-chip inductance is important for wires where the speed of light flight time is longer than either the rise times of the circuits or the RC delay of the wire. Because speed of light flight time increases linearly and RC delay increases quadratically with length, we can estimate the set of wire lengths for which inductance is relevant [Ismail99].

Example

Consider a metal2 signal line with a sheet resistance of 0.05 Ω/□ and a width of 0.5 μm. The capacitance is 0.2 fF/μm and inductance is 0.5 pH/μm. Compute the velocity of signals on the line and plot the range of lengths over which inductance matters as a function of the rise time.

Solution: The velocity is

$$v = \frac{1}{\sqrt{LC}} = \frac{1}{\sqrt{(0.5 \text{ pH}/\mu m)(0.2 \text{ fF}/\mu m)}} = 10^8 \text{ m/s} = \tfrac{1}{3}c \qquad (4.49)$$

Note that this is 100 mm/ns or 1 mm/10 ps. The resistance is $(0.1 \text{ }\Omega/\square) \cdot (1\square/0.5 \text{ }\mu m) = 0.2$ Ω/μm. Figure 4.45 plots the length of wires for which inductance is relevant against rise times. Above the horizontal line, wires greater than 500 μm are limited by RC delay rather than LC delay. To the right of the diagonal line, rise times are greater than the LC delay. Only in the region between these lines is inductance relevant to delay calculations. This region has very fast edge rates, so inductance is not very important to the delay of highly resistive signal lines at the time of this writing.

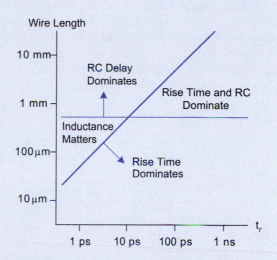

FIG 4.45 Wire lengths and edge rates for which inductance impacts delay

$$\frac{t_r}{2\sqrt{LC}} < l < \frac{2}{R}\sqrt{\frac{L}{C}} \qquad (4.50)$$

As the example illustrated, inductance will only be important to the delay of low-resistance signals such as wide clock lines or the power supply. As edge rates become faster, inductance will become relevant to a larger number of on-chip signals. Inductive crosstalk is also important for wide busses far away from their current return paths.

In power distribution networks, inductance means that if one portion of the chip requires a rapidly increasing amount of current, that charge must be delivered from nearby decoupling capacitors or supply pins; portions of the chip further away are unaware of the changing current needs until a speed-of-light flight time has elapsed and hence will not supply current immediately. Adding inductance to the power grid simulation generally reveals greater supply noise than would otherwise be predicted. Power supplies will be discussed further in Section 12.3.

In wide, thick, upper-level metal lines, resistance and RC delay may be small. This pushes the horizontal line in Figure 4.45 upward, increasing the range of edge rates for which inductance matters. This is especially common for clock signals. Inductance tends to increase the propagation delay and sharpen the edge rate.

To see the effects of inductance, consider a 5 mm metal6 clock line above a metal5 ground plane driving a 2pF clock load. If its width is 4.8 μm (six times minimum), it has resistance of 4 Ω/mm, capacitance of 0.4 pF/mm, and inductance of 0.12 nH/mm. Figure 4.46 presents models of the clock line as a 5-stage π-model without (a) and with (b) inductance. Figure 4.46(c) shows the response of each model to an ideal voltage source with 80 ps rise time. The model including inductance shows a greater delay until the clock begins to rise because of the speed of light flight time. It also overshoots. However, the rising edge is sharper and the rise time is shorter. In some circumstances when the driver impedance is matched to the characteristic impedance of the wire, the sharper rising edge can actually result in a shorter propagation delay measured at the 50% point.

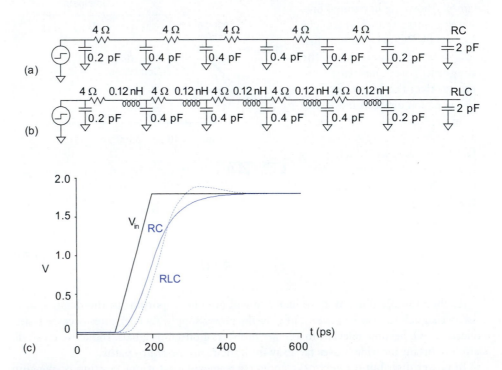

FIG 4.46 Wide clock line modeled with and without inductance

To reduce the inductance and the impact of skin effect when no ground plane is available, it is good practice to split wide wires into thinner sections interdigitated with power and ground lines to serve as return paths. For example, Figure 4.47 shows how a 16 μm wide clock line can be split into four 4 μm lines to reduce the inductance.

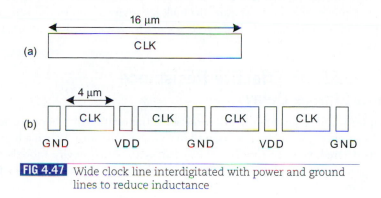

FIG 4.47 Wide clock line interdigitated with power and ground lines to reduce inductance

A bus made of closely spaced wires far above a ground plane is especially susceptible to inductive crosstalk. If all but one wire in the bus rises, each loop induces a magnetic field. These magnetic fields all pass through the loop formed by the nonswitching wire, in turn inducing a current in the victim wire. The noise from each aggressor sums on to the victim in much the same way that multiple primary turns in a transformer couple onto a single secondary turn. Computing the inductive crosstalk requires extracting a mutual inductance matrix for the bus and simulating the system. As this is not yet practical for large chips, designers instead either follow design rules that keep the inductive effects small or ignore inductance and hope for the best. The design rules may be of the form that one power or ground wire must be inserted between every N signal lines on each layer. N is called the signal:return (SR) ratio [Morton99]. $N = 4$ eliminates most inductive effects on noise and delay in a 180 nm process. $N = 2$ means each signal is shielded on one side, also eliminating half the capacitive crosstalk. However, low SR ratios are expensive in terms of metal resources.

In summary, on-chip inductance is difficult to extract. Mutual inductive coupling may occur over a long range, so inductive coupling is difficult to simulate even if accurate values are extracted. Instead, design rules are usually constructed so that inductive effects may be neglected for most structures. A regular power and ground grid or plane with no gaps is essential. Power and ground lines should be interdigitated in wide signal lines such as clocks and between about every four bits in large high-speed busses. Inductance should be incorporated into simulations of the power and clock networks and into the noise and delay calculations for busses with large SR ratios in high-speed designs.

4.5.6 Temperature Dependence

Interconnect capacitance is independent of temperature, but the resistance varies strongly. The temperature coefficients of copper and aluminum are about 0.4%/°C over the normal operating range of circuits; that is, a 100° C increase in temperature leads to 40% higher resistance. At liquid nitrogen temperature, the bulk resistivity of copper drops to 0.22 μΩ-cm, an eight-fold improvement. This suggests great advantages for RC-dominated paths in cooled systems.

4.5.7 An Aside on Effective Resistance and Elmore Delay

In this chapter we have played fast and loose with the relationship between resistance and delay. In practice, the methods work well for hand estimations. When a highly accurate result is desired, you can simulate the circuit with actual device models and wire parameters rather than simplified RC circuits. Nevertheless, rigor compels us to revisit this relationship.

According to the Elmore delay model, a gate with effective resistance R and capacitance C has a propagation delay of RC. A wire with distributed resistance R and capacitance C treated as a single π-segment has propagation delay $RC/2$. Reviewing the properties of RC circuits, we recall that the lumped RC circuit in Figure 4.48(a) has a unit step response of

$$V_{out}(t) = 1 - e^{\frac{-t}{R'C}} \tag{4.51}$$

The propagation delay of this circuit is obtained by solving for t_{pd} when $V_{out}(t_{pd}) = 1/2$:

$$t_{pd} = R'C \ln 2 = 0.69R'C \tag{4.52}$$

The distributed RC circuit in Figure 4.48(b) has no closed form time domain response. Because the capacitance is distributed along the circuit rather than all being at the end, you would expect the capacitance to be charged on average through about half the

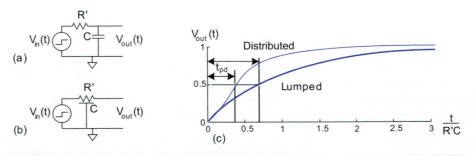

FIG 4.48 Lumped and distributed RC circuit response

resistance and that the propagation delay should thus be about half as great. A numerical analysis finds that the propagation delay is $0.38R'C$.

To reconcile the Elmore model with the true results for a logic gate, recall that logic gates have complex nonlinear I-V characteristics and are approximated as having an effective resistance. If we characterize that effective resistance as $R = R'\ln2$, the propagation delay really becomes the product of the effective resistance and the capacitance: $t_{pd} = RC$. In Section 5.4.5, we will calculate this effective resistance by simulating the delay of a gate driving a capacitive load and measuring the propagation delay.

For distributed circuits, observe that

$$0.38R'C \approx \tfrac{1}{2}R'C\ln 2 = \tfrac{1}{2}RC .$$

Therefore, the Elmore delay model describes distributed delay well if we use an effective wire resistance equal to 69% of that computed with EQ (4.34). This is somewhat inconvenient. The effective resistance is further complicated by the effect of nonzero rise time on propagation delay. Figure 4.49 shows that the propagation delay depends on the rise time of the input and approaches RC for lumped systems and $RC/2$ for distributed systems when the input is a slow ramp. This suggests that when the input is slow, the effective resistance for delay calculations in a distributed RC circuit is equal to the true resistance. Finally, we note that for many analyses such as repeater insertion calculations in Section 4.6.4, the results are only weakly sensitive to wire resistance, so using the true wire resistance does not introduce great error.

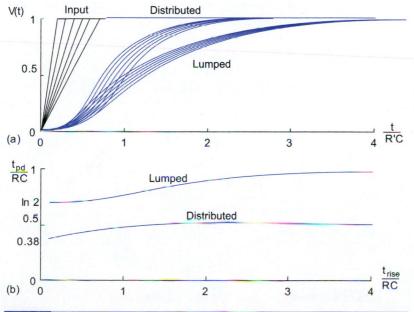

FIG 4.49 Effect of rise time on lumped and distributed RC circuit delays

In summary, it is a reasonable practice to estimate propagation delay of gates using the Elmore delay model as RC where R is the effective resistance of the gate. Similarly, you can estimate the flight time along a wire as $RC/2$ where R is the true resistance of the wire. When more accurate results are needed, it is important to use good transistor models and appropriate input slopes in simulation.

On a separate topic, Section 4.2.1 defined the Elmore delay of an RC ladder. In general, a wire branching to many destinations is usually modeled as an *RC tree* instead. The Elmore delay to node i of an RC tree is

$$T_{D_i} = \sum_{k=1}^{N} R_{ki} C_k \tag{4.53}$$

where N is the number of nodes in the tree, C_k is the capacitance on node k, and R_{ki} is the resistance between the input and node k in common with the path between the input and node i. This simplifies to the simple product of resistance and capacitance for each node in an RC ladder. In trees, the capacitance on branches away from the path to the output is conservatively lumped as if it were at the branch point on the path.

Example

Figure 4.50 models a gate driving wires to two destinations. The gate is represented as a voltage source with effective resistance R_1. The two receivers are located at nodes 3 and 4. The wire to node 3 is long enough that it is represented with a pair of π-segments, while the wire to node 4 is represented with a single segment. Find the Elmore delay from input x to each receiver.

Solution: The Elmore delays are:

$$T_{D_3} = R_1 C_1 + (R_1 + R_2)C_2 + (R_1 + R_2 + R_3)C_3 + R_1 C_4$$
$$T_{D_4} = R_1 C_1 + R_1 C_2 + R_1 C_3 + (R_1 + R_4)C_4 \tag{4.54}$$

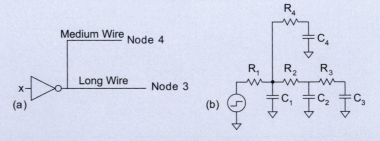

FIG 4.50 Interconnect modeling with RC tree

The Elmore delay can be viewed in terms of the first moment of the impulse response of the circuit. CAD tools can obtain greater accuracy by approximating delay based on higher moments using a technique called *moment matching. Asymptotic Waveform Evaluation* (AWE) uses moment matching to estimate interconnect delay with better accuracy than the Elmore delay model and faster run times than a full circuit simulation [Celik02].

4.6   Wire Engineering

As gate delays continue to improve while long wire delays remain constant or even get slower, wire engineering has become a major part of integrated circuit design. It is necessary to develop a floorplan early in the design cycle, identify the long wires, and plan for them. While floorplanning in such a way that critical communicating units are close to one another has the greatest impact on performance, it is inevitable that long wires will still exist. The designer has a number of techniques to engineer wires for delay and coupling noise. The width, spacing, and layer usage are all under the designer's control. Shielding can be used to further reduce coupling on critical nets. Repeaters inserted along long wires reduce the delay from a quadratic to a linear function of length. Wire capacitance and resistance complicate the use of Logical Effort in selecting gate sizes.

4.6.1   Width and Spacing

The designer selects the wire width, spacing, and layer usage to achieve acceptable delay and noise. By default, minimum pitch wires are preferred for noncritical interconnections for best density. Widening a wire proportionally reduces resistance but increases the capacitance of its top and bottom plates. Table 4.8 showed that this leads to less than a proportional increase in capacitance, so the RC delay product improves, especially for narrow wires. Widening wires also increases the fraction of capacitance of the top and bottom plates, which somewhat reduces coupling noise from adjacent wires. Increasing spacing between wires reduces capacitance to the adjacent wires and leaves resistance unchanged. This improves the RC delay to some extent and significantly reduces coupling noise.

Figure 4.51(a) shows the RC delay of a 10 mm metal2 wire in the process from Section 4.5 sandwiched between metal1 and metal3 planes as a function of wire pitch ($w + s$) for different wire spacings. Figure 4.51(b) shows the coupling capacitance to the two adjacent neighbors as a fraction of the total capacitance. The data shows that for tight pitches, it is better to increase width than spacing to improve delay. However, it is better to increase spacing than width to improve coupling. The best tradeoff clearly depends on the situation.

4.6.2   Layer Selection

Early (1970s) MOS processes offered only a single layer of metal. Polysilicon or diffusion jumpers were required when two signals crossed. Modern processes have six or more metal layers. The lower layers are thin and optimized for a tight routing pitch. Middle layers are

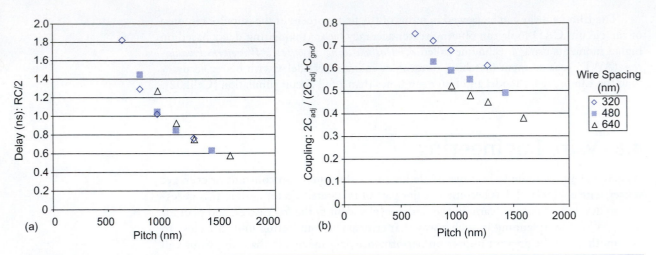

FIG 4.51 Delay and coupling of 10 mm metal2 wire for various wire pitches and spacings

often slightly thicker for lower resistance and better current-handling capability. Upper layers may be even thicker to provide a low-resistance power grid and fast global interconnect. A large number of layers are important to be able to route complex chips and supply power, ground, and the clock. Additional layers are also valuable because they allow a relaxed wire pitch, reducing RC delay and coupling problems. Wiring tracks are a precious resource and are often allocated in the floorplan; the wise designer maintains a reserve of wiring tracks for unanticipated changes late in the design process. A sample allocation of wiring tracks for a six-layer metal process is given in Table 4.11. Critical signals within a unit can be assigned upper-level metal tracks to reduce delay.

Table 4.11	Sample metal layer usage in 6-level process
Layer	**Purpose**
Metal 1	Interconnect within cells
Metal 2/3	Interconnect between cells within units
Metal 4/5	Interconnect between units, critical signals
Metal 6	I/O pads, clock, power, ground

The power grid is usually distributed over multiple layers. Most of the current-handling capability is provided in the upper two layers with lowest resistance. However, the grid must extend down to metal1 or metal2 to provide easy connection to cells.

There is debate over the best use of the growing number of wiring layers to solve inductive and capacitive noise problems. One approach is to dedicate power and ground planes, much like in a printed circuit board. For example, the Alpha 21264 [Gronowski98] used thick metal3 and metal6 planes for ground and power. This was necessary because the I/O pins were all along the periphery of the chip and the planes were needed to provide a low-resistance path for power to the center; better packaging with power and ground bumps across the die reduces the pressure to use planes. Another approach is to shield noise-sensitive nets with power or ground lines; this will be discussed in Section 4.6.3. A third approach is to use differential signaling.

4.6.3   Shielding

As discussed in Section 4.5.4, coupling from adjacent lines impacts both the delay and signal integrity of wires. The coupling can be avoided if the adjacent lines do not switch. It is common practice to shield critical signals with power or ground wires on one or both sides to eliminate coupling. This is costly in area but may be less costly than increasing spacing to the point that the coupling is negligible. For example, clock wires are usually shielded so that switching neighbors do not affect the delay of the clock wire and introduce clock skew. Sensitive analog wires passing near digital signals should also be shielded.

   An alternative to shielding is to interdigitate wires that are guaranteed to switch at different times. For example, if bus *A* switches on the rising edge of the clock and bus *B* switches on the falling edge of the clock, by interleaving the bits of the two busses you can guarantee that both neighbors are constant during a switching event. This avoids the delay impact of coupling; however, you must still ensure that coupling noise does not exceed noise budgets. Figure 4.52 shows wires shielded (a) on one side, (b) on both sides, and (c) interdigitated. Very sensitive signals such as clocks or analog voltages can be shielded above and below as well.

4.6.4   Repeaters

Both resistance and capacitance increase with wire length *l*, so the RC delay of a wire increases with l^2, as shown in Figure 4.53(a). The delay may be reduced by splitting the wire into *N* segments and inserting an inverter or buffer called a *repeater* to actively drive the wire [Glasser85], as shown in Figure 4.53(b). The new wire involves *N* segments with RC flight time of $(l/N)^2$, for a total delay of l^2/N. If the number of segments is proportional to the length, the overall delay increases only linearly with *l*.

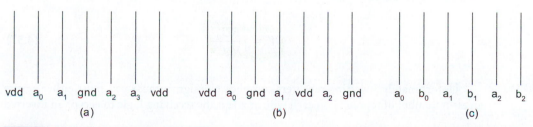

vdd	a$_0$	a$_1$	gnd	a$_2$	a$_3$	vdd		vdd	a$_0$	gnd	a$_1$	vdd	a$_2$	gnd		a$_0$	b$_0$	a$_1$	b$_1$	a$_2$	b$_2$

| (a) | (b) | (c) |

FIG 4.52 Wire shielding topologies

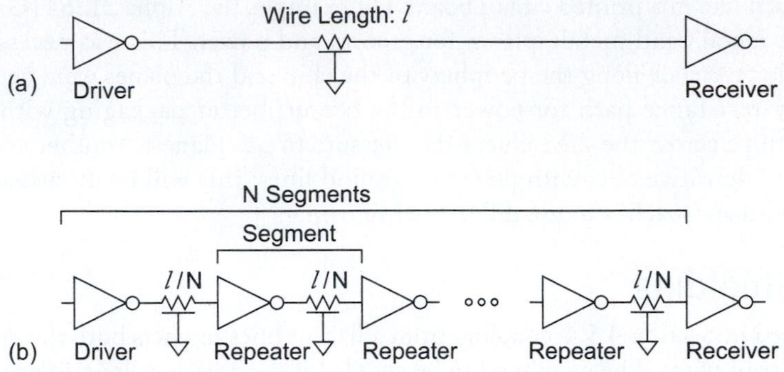

FIG 4.53 Wire with and without repeaters

Using inverters as repeaters gives best performance. Each repeater adds some delay. If the distance is too great between repeaters, the delay will be dominated by the long wires. If the distance is too small, the delay will be dominated by the large number of inverters. As usual, the best distance between repeaters is a compromise between these extremes. Suppose a unit inverter has resistance R and capacitance C' ($C' \approx 3C$ because the inverter is composed of a unit nMOS and double-width pMOS[8]) and a wire has resistance R_w and capacitance C_w per unit length. Consider inserting repeaters of W times unit size. You can show (see Exercise 33) that under the Elmore delay model, neglecting diffusion parasitics, the best length of wire between repeaters is

$$\frac{l}{N} = \sqrt{\frac{2RC'}{R_w C_w}} \qquad (4.55)$$

The delay per unit length of a properly repeated wire is

$$\frac{t_{pd}}{l} = \left(2 + \sqrt{2}\right)\sqrt{RC'R_w C_w} \qquad (4.56)$$

To achieve this delay, the inverters should use an nMOS transistor width of

$$W = \sqrt{\frac{RC_w}{R_w C'}} \qquad (4.57)$$

Unfortunately, inverting repeaters complicate design because you must either ensure an even number of repeaters on each wire or adapt the receiving logic to accept an inverted

---

[8]It is not necessary to use equal rise and fall resistances and Section 6.2.1.6 shows that the delay and area can be slightly improved using smaller pMOS transistors.

input. Some designers use inverter pairs (buffers) rather than single inverters to avoid the polarity problem. The pairs contribute more delay. However, the first inverter size W_1 may be smaller, presenting less load on the wire driving it. The second inverter may be larger, driving the next wire more strongly. You can show that the best size of the second inverter is $W_2 = kW_1$, where $k = 2.06$ if diffusion parasitics are negligible. The distance between repeaters increases to (see Exercise 34)

$$\frac{l}{N} = \sqrt{\frac{2RC'\left(k + \frac{1}{k}\right)}{R_w C_w}} \tag{4.58}$$

The delay per unit length becomes

$$\frac{t_{pd}}{l} = 3.65\sqrt{RC'R_w C_w} \tag{4.59}$$

using transistor widths of

$$W_1 = \frac{W}{\sqrt{k}}, \quad W_2 = W\sqrt{k} \tag{4.60}$$

This typically means that wires driven with noninverting repeaters are only about 7% slower per unit length than those using inverting repeaters. Only about two-thirds as many repeaters are required, simplifying floorplanning. Total repeater area and power increases slightly.

When diffusion parasitics are considered, each repeater becomes slightly more expensive. Therefore, it is better to use fewer repeaters spaced at greater distances, as you would expect from Logical Effort. The best transistor sizes are unchanged. There are no closed-form results, but the problem can be solved numerically or thorough simulation. The overall delay is a weak function of the distance between repeaters, so it is reasonable to increase this distance to reduce the difficulty of finding places in the floorplan for repeaters while only slightly increasing delay. Repeaters impose directionality on a wire. Bidirectional busses and distributed tristate busses cannot use simple repeaters and hence are slower; this favors point-to-point unidirectional communications. Repeaters also draw unusually large crowbar currents because their inputs come from RC lines with slow edge rates, turning both transistors partially ON for a long time.

An alternative to repeaters are *boosters* [Nalamalpu02], which are placed in parallel rather than in series with a wire as shown in Figure 4.54(a). The booster senses when a wire is switching and aids the transition. It relies on the principles of hysteresis and positive feedback to allow bidirectional operation at the expense of reduced noise margins. The

Example

Determine the best distance between repeaters for a minimum pitch metal2 line in a 180 nm process. Assume the metal usage on other layers is dense. The transistor resistance is $3k\Omega \cdot \mu m$ and the gate capacitance is $C = 1.7 \text{ fF}/\mu m$. How far should the repeaters be spaced? How wide should the repeater transistors be? What is the signal velocity on the wire? How do your results change if width and spacing are each increased by 50%? How do they change for a minimum-pitch metal5 line?

Solution: The capacitance of minimum-pitch metal2 lines with planes above and below is $C_w = 0.21 \text{ fF}/\mu m$, according to Table 4.8. The sheet resistance from Table 4.7 is 0.05 Ω/square, or

$$R_w = \frac{0.05 \frac{\Omega}{\text{square}}}{0.32 \frac{\mu m}{\text{square}}} = 0.16 \frac{\Omega}{\mu m} \tag{4.61}$$

$C' = 3C = 5.1 \text{ fF}/\mu m$. According to EQ (4.55), the distance between repeaters should be

$$\sqrt{\frac{2(3000\Omega \cdot \mu m)\left(5.1 \frac{\text{fF}}{\mu m}\right)}{\left(0.16 \frac{\Omega}{\mu m}\right)\left(0.21 \frac{\text{fF}}{\mu m}\right)}} = 950 \ \mu m \tag{4.62}$$

According to EQ (4.57), the nMOS transistor width should be

$$\sqrt{\frac{(3000\Omega \cdot \mu m)\left(0.21 \frac{\text{fF}}{\mu m}\right)}{\left(0.16 \frac{\Omega}{\mu m}\right)\left(5.1 \frac{\text{fF}}{\mu m}\right)}} = 28 \ \mu m \tag{4.63}$$

and the pMOS should be roughly twice that. The repeated signal velocity (ignoring diffusion capacitance) is

$$\left(2 + \sqrt{2}\right)\sqrt{(3000\Omega \cdot \mu m)\left(5.1 \frac{\text{fF}}{\mu m}\right)\left(0.16 \frac{\Omega}{\mu m}\right)\left(0.21 \frac{\text{fF}}{\mu m}\right)} = 77 \frac{\text{ps}}{\text{mm}} \tag{4.64}$$

In comparison, the speed of light in a medium with a relative permittivity of 3.55 is

$$\frac{3 \cdot 10^8 \, \frac{m}{s}}{\sqrt{3.55}} = 0.155 \, \frac{mm}{ps} = 6.3 \, \frac{ps}{mm} \tag{4.65}$$

Even with repeaters, the wire is an order of magnitude slower than the speed of light because it is so resistive.

If the width and spacing increase by 50%, the capacitance is 0.18 fF/μm and the resistance is

$$R_w = \frac{0.05 \, \frac{\Omega}{square}}{1.5 \cdot 0.32 \, \frac{\mu m}{square}} = 0.10 \, \frac{\Omega}{\mu m} \tag{4.66}$$

As the wire is faster, we would expect that the repeaters could be placed further apart but that larger transistors would be needed to drive the longer segments. Reevaluating the same equations gives a distance of 1300 μm between repeaters, transistor widths of 33 μm, and wire delay of 52 ps/mm.

A metal5 line is even faster with a capacitance of 0.24 fF/μm and resistance of

$$R_w = \frac{0.02 \, \frac{\Omega}{square}}{0.8 \, \frac{\mu m}{square}} = 0.025 \, \frac{\Omega}{\mu m} \tag{4.67}$$

The best repeater spacing is 2260 μm with 75 μm transistors, giving a delay of 30 ps/mm. It is clear that the fat, wide upper level metal lines are a precious routing resource.

gate in the center is an *inverting Muller C-element*, defined in Figure 4.54(b), which provides hysteresis by only toggling when both inputs are the same. Skewed inverters with high and low switching points sense the beginning of a transition. Figure 4.54(c) shows simulations of the booster in action on a 6mm long metal2 wire. The wire is initially '0,' so *l* and *h* are initially '1' and *c* is initially '0.' In this state, the booster is OFF and the weak keeper holds the wire low. The input rises and the middle of the wire mid slowly follows. The keeper opposes the transition but is weak enough to have little effect. When the middle reaches the switching point of the LO-skew inverter, node *l* falls. Now the booster fires through both pMOS transistors, serving as positive feedback to strongly pull the middle of the wire high. This is visible as a kink in the mid waveform. The keeper also turns OFF. As the middle reaches the switching point of the HI-skew inverter, node *h* falls. *c* rises and turns OFF the booster because the transition is nearly complete. The keeper turns ON again, holding the middle of the wire high. The waveforms without boosters are also shown; they are slowly rising exponentials. Nalamalpu finds boosters are 20% faster and are inserted at three times the spacing of repeaters. [Dobbalaere95] presents a similar technique using self-timed circuits.

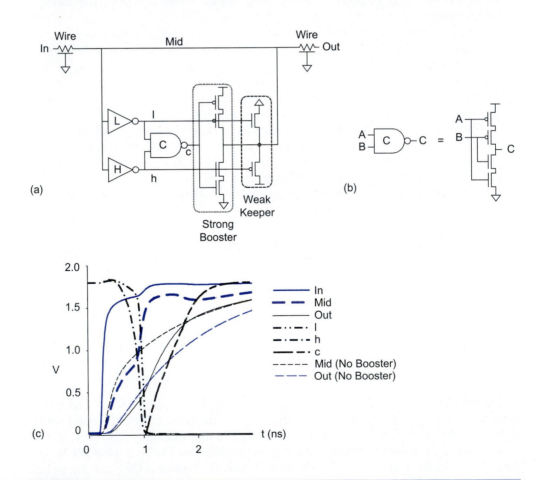

FIG 4.54 Booster

4.6.5 Implications for Logical Effort

Interconnect complicates the application of Logical Effort because the wires have a fixed capacitance. The branching effort at a wire with capacitance C_{wire} driving a gate load of C_{gate} is $(C_{gate} + C_{wire}) / C_{gate}$. This branching effort is not constant; it depends on the size of the gate being driven. The simple rule that circuits are fastest when all stages bear equal effort is no longer true when wire capacitance is introduced. If the wire is very short or very long, there are simple approximations possible, but when the wire and gate loads are comparable, there is no simple method to determine the best stage effort.

Every circuit has some interconnect, but when the interconnect is short ($C_{wire} \ll C_{gate}$), it can be ignored. Alternatively, you can compute the average ratio of wire capacitance to parasitic diffusion capacitance and add this as extra parasitic capacitance when determining parasitic delay. For connections between nearby gates, this generally leads to a best stage effort ρ slightly greater than 4.

Conversely, when the interconnect is long ($C_{wire} \gg C_{gate}$), the gate at the end can be ignored. The path can now be partitioned into two parts. The first part drives the wire while the second receives its input from the wire. The first part is designed to drive the load capacitance of the wire; the extra load of the receiver is negligible. A long wire is often driven by an inverter with a stage effort of 8 to 12 rather than 4 because low stage efforts require large, power-hungry drivers and have little performance advantage when the delay is dominated by the wire RC. The size of the receiver is chosen by practical considerations: Larger receivers may be faster, but they also cost area and power.

The most difficult problems occur when $C_{wire} \approx C_{gate}$. These medium-length wires introduce branching efforts that are a strong function of the size of the gates they drive. Writing a delay equation as a function of the gate sizes along the path and the wire capacitance results in an expression that can be differentiated with respect to gate sizes to compute the best sizes. For the purpose of hand estimation, this is usually too much work. A reasonable alternative is to preserve the stage effort of about four. The initial branching effort of the wire is unknown, so iteration is usually necessary.

4.6.6 Crosstalk Control

Recall from EQ (4.40) that the capacitive crosstalk is proportional to the ratio of coupling capacitance to total capacitance. For modern wires with an aspect ratio (t/w) of 2 or greater, the coupling capacitance can account for two-thirds or more of the total capacitance and crosstalk can create large amounts of noise and huge data-dependent delay variations. Shielding or increasing the spacing or width of wires alleviates crosstalk at the expense of greater area. Three alternative techniques to control crosstalk are *staggered repeaters*, *charge compensation*, and *twisted differential signaling* [Ho03b]. Each technique seeks to cause equal amounts of positive and negative crosstalk on the victim, effectively producing zero net crosstalk.

Figure 4.56(a) shows two wires with staggered repeaters. Each segment of the victim sees half of a rising aggressor segment and half of a falling aggressor segment. Although the cancellation is not perfect because of delays along the segments, staggering is a very

Example

The path in Figure 4.55 contains a medium-length wire modeled as a lumped capacitance. Write an equation for path delay in terms of x and y. How large should the x and y inverters be for shortest path delay? What is the stage effort of each stage?

Solution: From the Logical Effort delay model, we find the path delay is

$$d = \frac{x}{10} + \frac{y+50}{x} + \frac{100}{y} + P \tag{4.68}$$

Differentiating with respect to each size and setting the results to 0 allows us to solve EQ (4.68) for $x = 33$ fF and $y = 57$ fF. The stage efforts are $(33/10) = 3.3$, $(57 + 50)/33 = 3.2$, and $(100/57) = 1.8$. Notice that the first two stage efforts are equal as usual, but the third stage effort is lower. As x already drives a large wire capacitance, y may be rather large (and will bear a small stage effort) before the incremental increase in delay of x driving y equals the incremental decreases in delay of y driving the output.

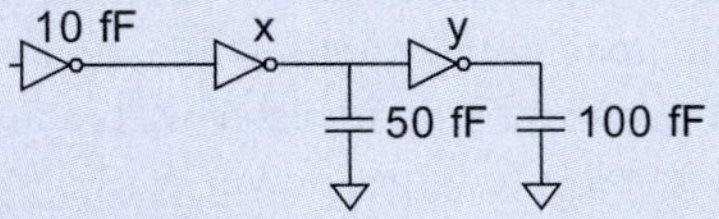

FIG 4.55 Path with medium-length wire

$$\frac{1}{10} - \frac{y+50}{x^2} = 0 \Rightarrow x^2 = 10y + 500$$

$$\frac{1}{x} - \frac{100}{y^2} = 0 \Rightarrow y^2 = 100x \tag{4.69}$$

effective approach. Figure 4.56(b) shows charge compensation in which an inverter and transistor are added between the aggressor and victim. The transistor is connected to behave as a capacitor. When the aggressor rises and couples the victim upward, the inverter falls and couples the victim downward. By choosing an appropriately sized compensation transistor, most of the noise can be cancelled at the expense of the extra circuitry. Figure 4.56(c) shows twisted differential signaling in which each signal is routed differentially. The signals are swapped or *twisted* such that the victim and its complement each see equal coupling from the aggressor and its complement. This approach is expensive in wiring resources, but it very effectively eliminates crosstalk. It is widely used in memory designs, as explored in Section 11.2.3.

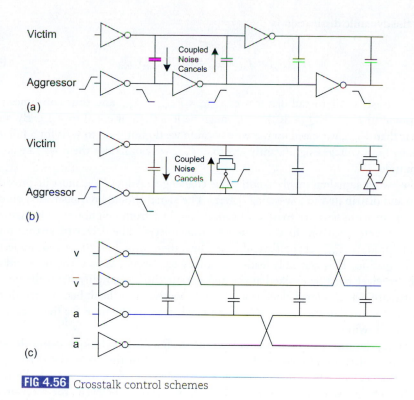

FIG 4.56 Crosstalk control schemes

4.6.7 Low-swing Signaling

Driving long wires is slow because of the RC delay, and expensive in power because of the large capacitance to switch. Low-swing signaling improves performance by sensing when a wire has swung through some small V_{swing} rather than waiting for a full swing. If the driver is turned off after the output has swung sufficiently, the power can be reduced as well. However, the improvements come at the expense of more complicated driver and receiver circuits. Low-swing signaling may also require a twisted differential pair of wires to eliminate common-mode noise that could corrupt the small signal.

The power consumption for low-swing signaling depends on both the driver voltage V_{drive} and the actual voltage swing V_{swing}. Each time the wire is charged and discharged, it consumes $Q = CV_{swing}$. If the effective switching frequency of the wire is αf, the average current is

$$I_{avg} = \frac{1}{T}\int_0^T i_{drive}(t) = \alpha f C V_{swing} \tag{4.70}$$

Hence, the dynamic dissipation is

$$P_{\text{dynamic}} = I_{\text{avg}}V_{\text{drive}} = \alpha f C V_{\text{swing}}V_{\text{drive}} \tag{4.71}$$

In contrast, a rail-to-rail driver uses $V_{\text{drive}} = V_{\text{swing}} = V_{DD}$ and thus consumes power proportional to V_{DD}^2 (EQ (4.30)). V_{swing} must be less than or equal to V_{drive}. By making V_{swing} less than V_{drive}, we speed up the wire because we do not need to wait for a full swing. By making both voltages significantly less than V_{DD}, we can reduce the power by an order of magnitude.

Low-swing signaling involves numerous challenges. A low V_{drive} must be provided to the chip and distributed to low-swing drivers. The signal should be transmitted on differential pairs of wires that are twisted to cancel coupling from neighbors and equalized to prevent interference from the previous data transmitted. The driver must turn on long enough to produce V_{swing} at the far end of the line, then turned off to prevent unnecessary power dissipation. This generally leads to a somewhat larger swing at the near end of the line. The receiver must be clocked at the appropriate time to amplify the differential signal. Distributing a self-timed clock from driver to receiver is difficult because the distances are long so the time to transmit a full-swing clock exceeds the time for the data to complete its small swing.

[Ho03a] describes a synchronous low-swing signaling technique using the system clock for both driver and receiver. During the first half of the cycle, the driver is OFF (high impedance) and the differential wires are equalized to the same voltage. During the second half of the cycle, the drivers turn ON. At the end of the cycle, the receiver senses the differential voltage and amplifies it to full-swing levels. Figure 4.57(a) shows the overall system architecture. Figure 4.57(b) shows the driver for one of the wires. The gates use ordinary V_{DD} while the drive transistors use V_{drive}. Because $V_{\text{drive}} \ll V_{DD}$, nMOS transistors are used for both the pullup and pulldown to deliver low effective resistance in their linear regime. A second driver using the complementary input drives the complementary wire. Figure 4.57(c) shows the differential wires with twisting and equalizing. Figure 4.57(d) shows the clocked sense amplifier based on the SA-F/F that will be described further in Section 7.3.8. The sense amplifier uses pMOS input transistors because the small-swing inputs are close to GND and below the threshold of nMOS transistors.

The design in [Ho03a] used a 1.8 V V_{DD}, 0.65 V V_{drive}, and 0.1 V minimum V_{swing} in a 180 nm process. The clock frequency was 1 GHz, or 10 FO4 inverter delays, to drive a 10 mm wire. Note that the clock period must be long enough to transmit an adequate voltage swing. If the clock period increases, the circuit will actually dissipate more power because the voltage swing will increase to a maximum of V_{drive}.

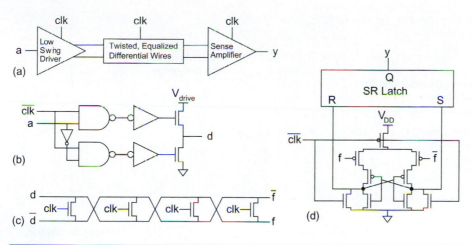

FIG 4.57 Low-swing signaling system

4.7 Design Margin

So far, when considering the various aspects of determining a circuit's behavior, we have only alluded to the variations that might occur in this behavior given different operating conditions. In general, there are three different sources of variation—two environmental and one manufacturing. These are:

- Supply voltage

- Operating temperature

- Process variation

You must aim to design a circuit that will reliably operate over all extremes of these three variables. Failure to do so invites circuit failure, potentially catastrophic system failure, and a rapid decline in reliability (not to mention a loss of customers).

Variations can be modeled with *uniform* or *normal* (Gaussian) statistical distributions. These distributions are shown in Figure 4.58. Uniform distributions are specified with a half-range. For good results, accept variations over the entire half-range. For example, V_{DD} may be specified at 1.2 V +/− 10%. This is a uniform distribution with a 120 mV half-range and all parts should work at any voltage in the range. Normal distributions are specified with a standard deviation σ. Processing variations are usually modeled with normal distributions. Retaining parts with a 3σ distribution will result in 0.26% of parts being rejected. A 2σ retention results in 4.56% of parts being rejected, while 1σ results in a 31.74% rejection rate. Obviously, rejecting parts outside 1σ of nominal would waste a large number of parts. A 3σ or 2σ limit is conventional and a manufacturer with a com-

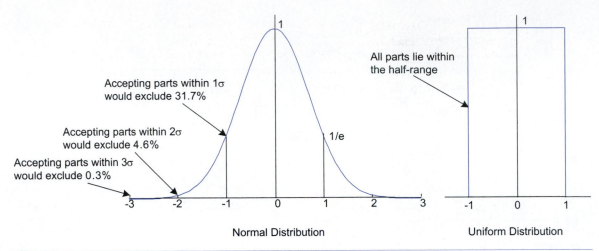

Accepting parts within 1σ
would exclude 31.7%

Accepting parts within 2σ
would exclude 4.6%

Accepting parts within 3σ
would exclude 0.3%

All parts lie within
the half-range

1/e

Normal Distribution

Uniform Distribution

FIG 4.58 Uniform and normal distributions

mercially viable CMOS process should be able to supply a set of device parameters that are guaranteed to yield at this rate. However, as the variations are getting more significant, designers are moving toward statistical rather than worst-case design [Borkar03].

4.7.1 Supply Voltage

Systems are designed to operate at a nominal supply voltage, but this voltage may vary for many reasons including tolerances of the voltage regulator, IR drops along supply rails, and di/dt noise. The system designer may tradeoff power supply noise against resources devoted to power supply regulation and distribution; typically the supply is specified at ±10% around nominal at each logic gate. In other words, the variation has a uniform distribution with a half-range of 10% of V_{DD}. Speed is roughly proportional to V_{DD}, so to first order this leads to ±10% delay variations (check for your process and voltage when this is critical). Power supply variations also appear in noise budgets.

4.7.2 Temperature

Section 2.2.3.7 showed that as temperature increases, drain current decreases. The junction temperature of a transistor is the sum of the ambient temperature and the temperature rise caused by power dissipation in the package. This rise is determined by the power consumption and the package thermal resistance, as discussed in Section 12.2.4.

Table 4.12 lists the ambient temperature ranges for parts specified to commercial, industrial, and military standards. Parts must operate at the bottom end of the ambient range unless they are allowed time to warm up before use. The junction temperature may significantly exceed the maximum ambient temperature. Commonly commercial parts are verified to operate with junction temperatures up to 110° to 125° C.

Table 4.12	Ambient temperature ranges	
Standard	**Minimum**	**Maximum**
Commercial	0° C	70° C
Industrial	−40° C	85° C
Military	−55° C	125° C

4.7.3 Process Variation

Devices and interconnect have variations in film thickness, lateral dimensions, and doping concentrations [Bernstein99]. These variations occur from one wafer to another, between dice on the same wafer, and across an individual die; variation is generally smaller across a die than between wafers. These effects are sometimes called *inter-die* and *intra-die* variations; intra-die variation is also called *process tilt* because certain parameters may slowly and systematically vary across a die. For example, if an ion implanter delivered a greater dose nearer the center of a wafer than near the periphery, the threshold voltages might tilt radially across the wafer.

For devices, the most important variations are channel length L, oxide thickness t_{ox}, and threshold voltage V_t. Channel length variations are caused by photolithography proximity effects, deviations in the optics, and plasma etch dependencies. Oxide thickness is well controlled and generally is only significant between wafers; its effects on performance are often lumped into the channel length variation. Threshold voltages vary because of different doping concentrations and annealing effects, mobile charge in the gate oxide, and discrete dopant variations caused by the small number of dopant atoms in tiny transistors.

For interconnect, the most important variations are line width and spacing, metal and dielectric thickness, and contact resistance. Line width and spacing, like channel length, depend on photolithography and etching proximity effects. Thickness may be influenced by polishing. Contact resistance depends on contact dimensions and the etch and clean steps.

4.7.4 Design Corners

From the designer's point of view, the collective effects of process and environmental variation can be lumped into their effect on transistors: *typical* (also called *nominal*), *fast*, or *slow*. In CMOS, there are two types of transistors with somewhat independent characteristics, so the speed of each can be characterized. Moreover, interconnect speed may vary independently of devices. When these processing variations are combined with the environmental variations, we define *design* or *process corners*. The term *corner* refers to an imaginary box that surrounds the guaranteed performance of the circuits, shown in Figure 4.59. The box is not square because some characteristics such as oxide thickness track between devices, making it impossible to simultaneously find a slow nMOS transistor with thick oxide and a fast pMOS transistor with thin oxide.

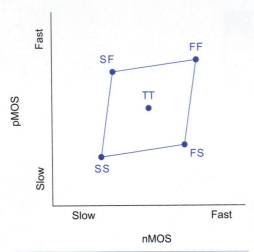

FIG 4.59 Design corners

Table 4.13 lists a number of important design corners. The corners are specified with five letters describing the nMOS, pMOS, interconnect, power supply, and temperature, respectively. The letters are F, T, and S, for *fast*, *typical*, and *slow*. The environmental corners for a 1.8 V commercial process are shown in Table 4.14, illustrating that circuits are fastest at high voltage and low temperature. Circuits are most likely to fail at the corners of the design space, so nonstandard circuits should be simulated at all corners to ensure they operate correctly in all cases. Often integrated circuits are designed to meet a timing specification for typical processing. These parts may be *binned*; faster parts are rated for higher frequency and sold for more money, while slower parts are rated for lower frequency. In any event, the parts must still work in the slowest SSSSS environment. Other integrated circuits are designed to obtain high yield at a relatively low frequency; these parts are simulated for timing in the slow process corner. The fast corner FFFFF has maximum DC power dissipation because threshold voltages are lowest. Other corners are used to check for races and ratio problems where the relative strengths and speeds of different transistors or interconnect are important. The FFFFS corner is important for noise because the edge rates are fast, causing more coupling; the threshold voltages are low; and the leakage is high [Shepard99].

Often the corners are abbreviated to fewer letters. For example, two letters generally refer to nMOS and pMOS. Three refer to nMOS, pMOS, and environment. Four refer to nMOS, pMOS, voltage, and temperature.

Table 4.13	Design corner checks				
Corner					**Purpose**
nMOS	pMOS	Wire	V_{DD}	Temp	
T	T	T	S	S	timing specifications (binned parts)
S	S	S	S	S	timing specifications (conservative)
F	F	F	F	F	DC power dissipation, race conditions, hold time constraints, pulse collapse, noise
F	F	F	F	S	subthreshold leakage noise, overall noise analysis
S	S	F	S	S	races of gates against wires
F	F	S	F	F	races of wires against gates
S	F	T	F	F	pseudo-nMOS & ratioed circuits noise margins, memory read/write, race of pMOS against nMOS
F	S	T	F	F	ratioed circuits, memory read/write, race of nMOS against pMOS

Table 4.14	Environmental corners	
Corner	**Voltage**	**Temperature**
F	1.98	0° C
T	1.8	70° C
S	1.62	125° C

It is important to know the design corner when interpreting delay specifications. For example, the datasheet in Figure 4.25 is specified at the 25° C TTTT corner. The SS corner is 27% slower. The cells are derated at −71% per volt and 0.13%/°C, for additional penalties of 13% each in the low voltage and high temperature corners. These factors are multiplicative, giving SSSS delay of 1.62 times nominal.

[Ho01] and [Chinnery02] find the FO4 inverter delay can be estimated from the effective channel length L_{eff} as:

- $L_{eff} \cdot (0.36 \text{ ps/nm})$ in TTTT corner
- $L_{eff} \cdot (0.50 \text{ ps/nm})$ in TTSS corner
- $L_{eff} \cdot (0.60 \text{ ps/nm})$ in SSSS corner

Note that the effective channel length is aggressively scaled faster than the drawn channel length to improve performance. Typically $L_{eff} = 0.5\text{-}0.7 \ L_{drawn}$. For example, Intel's 180 nm process was originally manufactured with $L_{eff} = 140$ nm and eventually pushed to $L_{eff} = 100$ nm. Our model predicts an FO4 inverter delay of about 70–50 ps in the TTSS corner where design usually takes place.

In addition to working at the standard process corners, chips must function in a very high temperature, high voltage *burn-in* corner (e.g., 125–140° C externally, corresponding to an even higher internal temperature, and 1.3–1.7x nominal V_{DD} [Vollertsen99]), as described in Section 9. While it does not have to run at full speed, it must operate correctly so that all nodes can toggle. The burn-in corner has very high leakage and can dictate the size of keepers and weak feedback on domino gates and static latches.

Processes with multiple threshold voltages and/or multiple oxide thicknesses can see each flavor of transistor independently varying as fast, typical, or slow. This can easily lead to more corners than anyone would care to simulate and raises challenges about identifying what corners must be checked for different types of circuits.

4.7.5 Matching

There are many cases in which the designer cares how well two nominally identical transistors match. For example, in a sense amplifier, the minimum voltage that reliably can be sensed depends on the offset voltage of the amplifier, which in turn depends on mismatch between the input transistors. Differential pairs used in most analog circuits are also very sensitive to mismatch between input transistors. In a clock distribution network, we would like to distribute a clock to all points on a chip at the same time and mismatch leads to clock skew. It is clearly overkill to model one transistor as fast and an adjacent transistor as slow because nearby transistors experience similar processing; such conservative modeling would make design impossible. On the other hand, it is also clear that the two transistors are not identical. Designers should be able to obtain data about process variation from the manufacturer. Unfortunately, this data is often not available. Even if the manufacturer has characterized the process for variation, the data is usually a closely guarded trade secret. Moreover, manufacturers do not like to share the information for fear that it will become part of the process specification and prevent process modifications that lead to faster transistors at the expense of different variabilities. Nevertheless, device matching is a critical parameter for calculating clock skew and is an increasingly important specification for a process.

Mismatches occur from both *systematic variability* and *uncertainty* [Naffziger02]. Systematic variability has a quantitative relationship to a source. For example, an ion implanter can systematically deliver a different dosage to different regions of a wafer. Similarly, polysilicon gates can systematically be etched narrower in regions of high polysilicon density than low density. Uncertainty occurs when the source is unknown, random, or too costly to model. Systematic variability can be modeled and nulled out; for example, in principle, you could examine a layout database and calculate the etching variations as a function of nearby layout, then simulate a circuit with suitably adjusted gate lengths. In practice, systematic variability often must be treated as uncertainty because of limited modeling resources. *Process tilt* describes the tendency for parameters to slowly vary from one corner of the die to another. Adjacent transistors tend to match better than widely separated ones.

Variations in transistor threshold voltage and current have been found experimentally to scale with $1/\sqrt{WL}$ [Pelgrom89]. Device parameters depend on device size, orientation, and nearby polysilicon density. Therefore, for good matching it is best to build identical, relatively large transistors oriented in the same direction. For the same total gate area WL, long-channel transistors match better, so sense amplifiers can be built with longer than minimum devices [Lovett98]. When matching is especially critical, as in clock buffers, you can surround the transistor with a consistent pattern of polysilicon so all transistors see comparable nearby polysilicon densities. Threshold voltage variations are primarily caused by statistical fluctuations in the number of dopant atoms in the channel [Mizuno94]. As transistors shrink, they contain fewer dopant atoms and the statistical fluctuations become more severe.

Matching problems can be characterized as systematic, random, drift, or jitter. Systematic mismatches include factors that can be modeled and simulated at design time, such as wires of different lengths. Random mismatches include most process variations (length, threshold, and interconnect) that are either truly random or too costly to model. These mismatches do not change with time, so they can be nulled out through feedback circuits that detect the variation and compensate. Drift mismatches, notably temperature variation, change slowly with time as compared to the operating frequency of the system. Drift can again be nulled by compensation circuits, but such circuits must sample repeatedly faster than the drift occurs rather than just once at manufacturing or startup. Jitter, often from voltage variations, is the most difficult cause of mismatch. It occurs at frequencies comparable to or faster than the system clock and therefore may not be eliminated through feedback.

As an example, [Harris01b] presents process variation data used to model the clock distribution network of the Itanium 2 processor in the Intel 180 nm process [Yang98]. The primary sources of random mismatch are channel lengths and threshold voltages. The channel length has both a component slowly varying across the die and a random component with no apparent spatial correlation. The slowly varying component varies over a 12.5 nm half-range for transistors separated by 4 mm or more; nearby transistors see less variation and adjacent transistors see no variation. The random component is modeled with a standard deviation of 3.3 nm. The threshold voltage variation has a Gaussian distribution with an inverse area dependence. It is treated as 16.8 mV for small nMOS transistors, 14.6 mV for small pMOS transistors, 7.9 mV for large nMOS transistors, and 6.5 mV for large pMOS transistors. Large transistors are defined as those with width exceeding 12.5 μm.

4.7.6 Delay Tracking

Another common design problem is how to build *matched delays*; for example, clock-delayed domino (see Section 7.5.4.2) needs to provide clocks to gates after their inputs have settled. The clocks must be matched to the gate delay; if they arrive late, the system functions slower, but if they arrive early, the system doesn't work at all. Therefore, it is of great interest to the designer how well two delays can be matched.

The best way to build matched delays is to actually provide replicas of the gates that are being matched. For example, in a static RAM (see Section 11.2.3), replica bitlines are used to determine when the sense amplifier should fire. Any relative variation in wire, diffusion, and gate capacitances happens to both circuits.

In many situations, it is not practical to use replica gates; instead, a chain of inverters can be used. Unfortunately, even if there is no intra-die process variation, the inverter delay may not exactly track the delay it matches across design corners. For example, if the inverter chain were matching a wire delay in the typical corner, it would be faster than the wire in the FFSFF corner and slower than the wire in the SSFSS corner. This variation requires that the designer provide margin in the typical case so that even in the worst case, the matched delay does not arrive too early. How much margin is necessary?

Figure 4.60 shows how gate delays, measured as a multiple of an FO4 inverter delay, vary with process, design corners, temperature, and voltage. The circuits studied include complementary CMOS NAND and NOR gates, domino AND and OR gates, and a 64-bit domino adder with significant wire RC delay. Figure 4.60(a) shows the gate delay of various circuits in different processes. The adder shows the greatest variation because of its wire-limited paths, but all the circuits track to within 20% across processes. This indicates that if a circuit delay is measured in FO4 inverter delays for one process, it will have a comparable delay in a different process. Figure 4.60(b and c) shows gate delay scaling with power supply voltage and temperature. Figure 4.60(d) shows what combination of design corner, voltage, and temperature gives the largest variation in delay normalized to an FO4 inverter in the same combination in the 0.6 μm process. Observe that the variation is

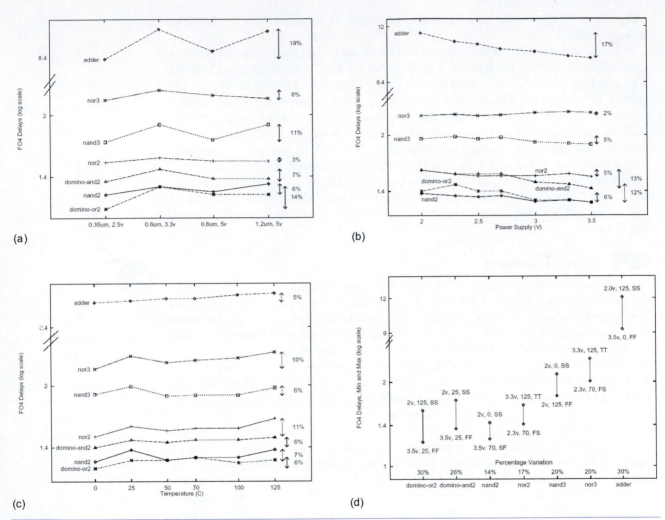

(a)

(b)

(c)

(d)

FIG 4.60 Delay tracking

smallest for simple static CMOS gates that most closely resemble inverters and can reach 30% for some gates. These figures demonstrate that an inverter chain should have a nominal delay about 30% greater than the path it matches so that the inverter output always arrives later than the matched path across all combinations of voltage, temperature, and design corners. This is a hefty margin and discourages the casual use of matched delays.

4.8   Reliability

Designing reliable CMOS chips involves understanding and addressing the potential failure modes [Greenhill02, Bernstein99]. This section addresses reliability problems (*hard errors*) that cause integrated circuits to fail permanently, including:

- Electromigration
- Self-heating
- Hot carriers
- Latchup
- Overvoltage failure

This section also considers transient failures (*soft errors*) that cause the system to crash or lose data. Circuit pitfalls and common design errors are discussed in Section 6.3.

4.8.1   Reliability Terminology

A number of acronyms are commonly used in describing reliability [Tobias95]. MTBF is the *mean time between failures*: (# devices • hours of operation) / # failures. FIT is the *failures in time*, the number of failures that would occur every thousand hours per million devices, or equivalently, 10^9 • (failure rate/hour). 1000 FIT is one failure in 10^6 hours = 114 years. This is good for a single chip. However, if a system contains 100 chips each rated at 1000 FIT and a customer purchases 10 systems, the failure rate is 100 • 1000 • 10 = 10^6 FIT, or one failure every 1000 hours (42 days). Reliability targets of less than 100 FIT are desirable.

Most systems exhibit the *bathtub curve* shown in Figure 4.61. Soon after birth, systems with weak or marginal components tend to fail. This period is called *infant mortality*. Reliable systems then enter their *useful operating life*, in which the failure rate is low. Finally, the failure rate increases at the end of life as the system wears out. It is important to age systems past infant mortality before shipping the products. Aging is accelerated by stressing the part through *burn-in* at higher than normal voltage and temperature. Nodes must toggle during burn-in to stress all parts of the system. Therefore, the circuits must be functional even under burn-in conditions.

Systems are also subjected to accelerated life testing during burn-in conditions to simulate the aging process and evaluate the time to wearout. The results are extrapolated to normal operating conditions to judge the actual useful operating life. This process is time-consuming and comes right at the end of the project. Part of any high-volume chip design will necessarily include designing a reliability assessment program that consists of burn-in boards deliberately stressing a number of chips over an extended period. Designers have tried to develop reliability simulators to predict lifetime [Hu92, Hsu92], but physical testing remains important. For high-volume parts, the source of failures is tracked and common points of failure can be redesigned and rolled into manufacturing.

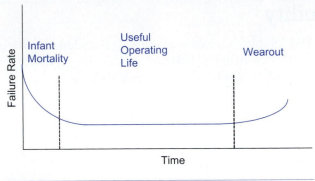

FIG 4.61 Reliability bathtub curve

4.8.2 Electromigration

Electromigration causes wearout of metal interconnect through the formation of voids [Hu95]. High current densities lead to an "electron wind" that causes metal atoms to migrate over time. Remarkable videos taken under a scanning electron microscope show void formation and migration and wire failure [Meier99]. The problem is especially severe for aluminum wires; it is commonly alleviated with an Al-Cu or Al-Si alloy and is much less important for pure copper wires because of the different grain transport properties. The electromigration properties also depend on the grain structure of the metal film.

Electromigration depends on the current density $J = I/wt$. It is more likely to occur for wires carrying a DC current where the electron wind blows in a constant direction than for those with bidirectional currents [Liew90]. Electromigration current limits are usually expressed as a maximum J_{dc}. The mean time to failure (MTTF) also is highly sensitive to operating temperature as given by Black's Equation [Black69]:

$$MTTF \propto \frac{e^{\frac{E_a}{kT}}}{J_{dc}^{\,n}} \qquad (4.72)$$

where E_a is the activation energy that can be experimentally determined by stress testing at high temperatures and n is typically 2. The electromigration DC current limits vary with materials, processing, and desired MTTF and should be obtained from the fabrication vendor. In the absence of better information, a maximum J_{dc} of 1–2 mA/μm^2 is a conservative limit for aluminum wires at 110°C [Rzepka98], although 10 mA/μm^2 or better may be achievable for copper wires [Young00]. Current density may be even more limited in contact cuts.

4.8.3 Self-heating

While bidirectional wires are less prone to electromigration, their current density is still limited by *self-heating*. High currents dissipate power in the wire, which raises its ambient temperature. Hot wires exhibit greater resistance and delay. Electromigration is also highly sensitive to temperature, so self-heating may cause temperature-induced electromigration problems in the bidirectional wires. Brief pulses of high peak currents may even melt the interconnect. Self-heating is dependent on the RMS current density. This can be measured with a circuit simulator or calculated as

$$I_{rms} = \sqrt{\frac{\int_0^T I(t)^2 \, dt}{T}} \qquad \textbf{(4.73)}$$

A reasonable rule to control reliability problems with self-heating is to keep $J_{rms} < 15$ mA/μm^2 for bidirectional aluminum wires [Rzepka98] on a silicon substrate. Self-heating is especially significant for SOI processes because of the poor thermal conductivity of SiO$_2$.

In summary, electromigration from high DC current densities is primarily a problem in power and ground lines. Self-heating limits the RMS current density in bidirectional signal lines. However, do not overlook the significant unidirectional currents that flow through the wires contacting nMOS and pMOS transistors. For example, Figure 4.62 shows which lines in an inverter are limited by DC and RMS currents. Both problems can be addressed by widening the lines or reducing the transistor sizes (and hence the current).

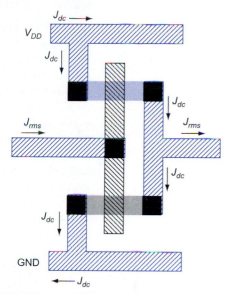

FIG 4.62 Current density limits in an inverter

4.8.4 Hot Carriers

As transistors switch, some high-energy ("hot") carriers may be injected into the gate oxide and become trapped there. The damaged oxide changes the I-V characteristics of the device, reducing current in nMOS transistors and increasing current in pMOS transistors. Damage is maximized when the substrate current I_{sub} is large, which typically occurs when nMOS transistors are in saturation while the input rises. Therefore, the problem is worst for inverters and NOR gates with fast rising inputs and heavily loaded outputs [Sakurai86], and for high power supply voltages.

Hot carriers cause circuit wearout as nMOS transistors become too slow. They can also cause failures of sense amplifiers and other matched circuits if matched components degrade differently [Huh98]. Hot electron degradation can be analyzed with simulators [Hu92, Hsu91, Quader94]. The wear is limited by setting maximum values on input rise-time and stage electrical effort [Leblebici96]. These maximum values depend on the process and operating voltage.

A related aging mechanism is *negative bias temperature instability* (NBTI), which leads to a decrease in pMOS transistor current as transistors wear at high temperature [Doyle91]. NBTI results from trapped holes in the oxide coupled with the creation of interface states. Like hot carriers, it leads to circuit failures from increased delay and poorer matching [Reddy02]. NBTI shifts depend on the electric field seen by the device and can be locked in to the device by high-voltage stress during burn-in; this is good because it allows testing with full NBTI degradation.

4.8.5 Latchup

Early adoption of CMOS processes was slowed by a curious tendency of CMOS chips to develop low-resistance paths between V_{DD} and GND, causing catastrophic meltdown. The phenomenon, called *latchup*, occurs when parasitic bipolar transistors formed by the substrate, well, and diffusion turn ON. With process advances and proper layout procedures, latchup problems can be easily avoided.

The cause of the latchup effect [Estreich82, Troutman86] can be understood by examining the process cross-section of a CMOS inverter, shown in Figure 4.63(a), over which is overlaid an equivalent circuit. In addition to the expected nMOS and pMOS transistors, the schematic depicts a circuit composed of an npn-transistor, a pnp-transistor, and two resistors connected between the power and ground rails (Figure 4.63(b)). The npn-transistor is formed between the grounded n-diffusion source of the nMOS transistor, the p-type substrate, and the n-well. The resistors are due to the resistance through the substrate or well to the nearest substrate and well taps. The cross-coupled transistors form a bistable silicon-controlled rectifier (SCR). Ordinarily, both parasitic bipolar transistors are OFF. Latchup can be triggered when transient currents flow through the substrate during normal chip power-up or when external voltages outside the normal operating range are applied. If substantial current flows in the substrate, V_{sub} will rise, turning ON the npn-transistor. This pulls current through the well resistor, bringing down V_{well} and turning ON the pnp-transistor. The pnp-transistor current in turn raises V_{sub}, initiating a positive feedback loop with a large current flowing between V_{DD} and GND that persists until the power supply is turned off or the power wires melt.

Fortunately, latchup prevention is easily accomplished by minimizing R_{sub} and R_{well}. Some processes use a thin epitaxial layer of lightly doped silicon on top of a heavily doped substrate that offers a low substrate resistance. Most importantly, the designer should place substrate and well taps close to each transistor, as described in Section 1.5.1. A conservative guideline is to place a tap adjacent to every source connected to V_{DD} or GND. If this is not practical, you can obtain more detailed information from the process vendor (they will normally specify a maximum distance for diffusion to substrate/well tap) or try the following guidelines:

- Every well should have at least one tap.

- All substrate and well taps should connect directly to the appropriate supply in metal.

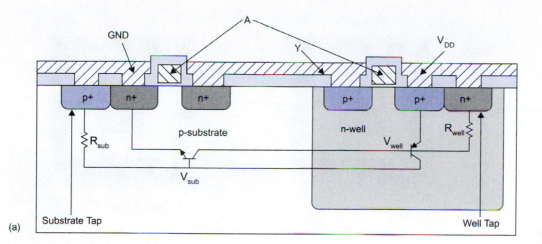

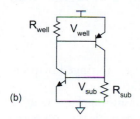

FIG 4.63   Origin and model of CMOS latchup

- A tap should be placed for every 5-10 transistors or every 25-100 µm (this distance is process-dependent).

- nMOS transistors should be clustered together near GND and pMOS transistors should be clustered together near V_{DD}, avoiding convoluted structures that intertwine nMOS and pMOS transistors in checkerboard patterns.

I/O pads are especially susceptible to latchup because external voltages can ring below GND or above V_{DD}, forward biasing the junction between the drain and substrate or well and injecting current into the substrate. In such cases, guard rings should be used to collect the current, as shown in Figure 4.64. Guard rings are simply substrate or well taps tied to the proper supply that completely surround the transistor of concern. For example, the n+ diffusion in Figure 4.64(b) can inject electrons into the substrate if it falls a diode drop below 0 volts. The p+ guard ring tied to ground provides a low-resistance path to collect these electrons before they interfere with the operation of other circuits outside the guard ring. *All* diffusion structures in any circuit connected to the external world must be guard ringed, i.e., n+ diffusion by p+ connected to GND or p+ diffusion by n+ connected to V_{DD}.

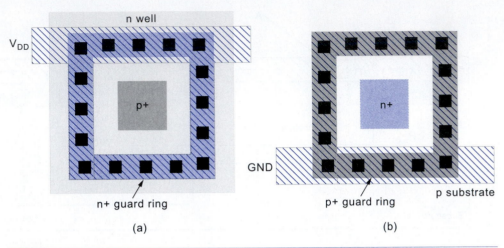

FIG 4.64 Guard rings

For the ultra-paranoid, double guard rings may be employed, i.e., n+ ringed by p+ to GND, then n+ to V_{DD} or p+ ringed by n+ to V_{DD}, then p+ to GND.

SOI processes avoid latchup entirely because they have no parasitic bipolar structures. Also, processes with $V_{DD} < 0.7$ are immune to latchup because the parasitic transistors will never have a large enough base-to-emitter voltage to turn on. In general, low-voltage processes are much less susceptible to latchup problems.

4.8.6 Overvoltage Failure

Tiny transistors can be easily damaged by relatively low voltages. *Overvoltage* reliability problems can arise from electrostatic discharge, oxide breakdown, punchthrough, and time-dependent dielectric breakdown of the gate oxide. *Electrostatic discharge* (ESD) from static electricity entering the I/O pads can cause very large voltage and current transients and is discussed further in Section 12.4.

Undesired voltages applied to the gate can cause *breakdown* and *arcing* across the thin oxide, destroying the device. Higher-than-normal voltages applied between source and drain lead to *punchthrough* when the source/drain depletion regions touch. Both problems lead to a maximum safe voltage that can be applied to transistors. For modern processes, this voltage is often much less than the I/O standard voltage, requiring a second type of transistor with thicker oxides and longer channels to endure the higher I/O voltages.

Gate oxides wear out with time as tunneling currents cause irreversible damage to the oxide; this problem is called *time-dependent dielectric breakdown* (TDDB). The failure rate is exponentially dependent on the temperature and oxide thickness; for a 10-year life at 125° C, the field across the gate $E_{ox} = V_{DD}/t_{ox}$ should be kept below about 7 MV/cm = 0.7 V/nm [Moazzami90]. The problem is greatest when voltage overshoots occur; this can be caused by noisy power supplies or reflections at I/O pads. Reliability is improved by lowering the power supply voltage, minimizing power supply noise, and using thicker oxides on the I/O pads.

4.8.7 Soft Errors

In the 1970s, as dynamic RAMs (DRAMs) replaced core memories, DRAM vendors were puzzled to find DRAM bits occasionally flipping value spontaneously. At first, the errors were attributed to "system noise," "voltage marginality," "sense amplifiers," or "pattern sensitivity," but the errors were found to be random. When the corrupted bit was rewritten with a new value, it was no more likely than any other bit to experience another error. In a classic paper [May79], Intel identified the source of these soft errors as alpha particle collisions that generate electron-hole pairs in the silicon as the particles lose energy. The excess carriers can be collected into the diffusion terminals of transistors. If the charge collected is comparable to the charge on the node, the voltage can be disturbed.

Soft errors are random nonrecurring single bit errors in memory devices, including SRAM, DRAM, registers, and latches. Alpha particles from decaying uranium and thorium impurities in integrated circuit interconnect and packaging is a major source of soft errors at sea level, often causing a *soft error rate* (SER) of 100–2000 FIT/Mb [Hazucha00]. The neutron flux from cosmic rays is low at sea level, but two orders of magnitude larger at aircraft flight altitudes [Ziegler96]. These cosmic rays cause up to 10^6 FIT/Mb at flight altitudes.

Soft errors are minimized by maintaining at least some critical charge Q_{crit} ($Q = CV$) on state nodes. This is difficult as memory cells become smaller and hence have less capacitance, and as power supplies get lower. DRAMs use exotic structures such as trench capacitors to hold sufficient charge. Fortunately, small cells require less Q_{crit} because of their small area to collect carriers.

The error rate is also affected by the proximity of alpha sources. Flip-chip technology with solder bumps bonded directly to the die poses special risks because impurities in the lead bumps emit a high alpha particle flux. Some companies avoid placing solder bumps directly over RAMs or other sensitive circuits. Aged lead (e.g., from Roman pipes) has outlived the contaminant half-lives and is much safer. Similarly, highly purified aluminum interconnect reduces the alpha flux from chip wires [Juhnke95].

Finally, error detecting and correcting codes can be used to tolerate soft errors in memories without data corruption. These codes will be discussed further in Section 10.7.2.

4.9 Scaling

The only constant in VLSI design is constant change. Figure 4.65 shows the year of introduction of Intel microprocessors in each feature size on a logarithmic scale, indicating that feature size reduces by 30% from one generation to the next every 2 to 3 years. As transistors become smaller, they switch faster, dissipate less power, and are cheaper to manufacture! Despite the ever-increasing challenges, process advances have actually accelerated in the past decade. Such scaling is unprecedented in the history of technology. However, scaling also exacerbates noise and reliability issues and introduces new problems. Designers need to be able to predict the effect of this feature size scaling on chip performance to plan future products, ensure existing products will scale gracefully to future processes for cost reduction, and anticipate looming design challenges. This section examines how transistors and interconnect scale, and the implications of scaling for design. The Semiconductor Industry Association prepares and maintains an International Technology Roadmap for Semiconductors predicting future scaling. Section 4.11 gives a case study of how scaling has influenced Intel microprocessors over three decades.

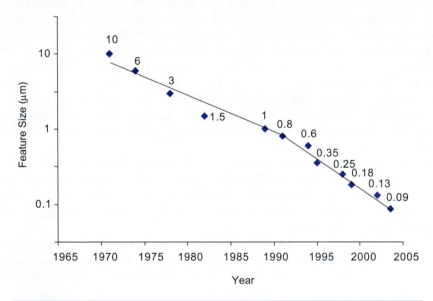

FIG 4.65 Year of introduction of processes

4.9.1 Transistor Scaling

First-order *constant field* MOS scaling theory is based on a model formulated by Dennard [Dennard74]. The characteristics of an MOS device can be maintained and the basic operational characteristics can be preserved if the critical parameters of a device are scaled by a dimensionless factor S. These parameters include

- All dimensions (in the x, y, and z directions)
- Device voltages
- Doping concentration densities

Another approach is *lateral scaling,* in which only the gate length is scaled. This is commonly called a *gate shrink* because it can be done easily to an existing mask database for a design.

The effects of these types of scaling are illustrated in Table 4.15. The industry generally scales process generations with $S = \sqrt{2}$. This doubles the number of transistors per unit area with each generation and doubles transistor performance every two generations under constant field scaling. A 5% gate shrink ($S = 1.05$) is commonly applied as a process becomes mature to boost the speed of components in that process.

Table 4.15   Influence of scaling on MOS device characteristics

Parameter	Sensitivity	Constant Field	Lateral
Scaling Parameters			
Length: L		$1/S$	$1/S$
Width: W		$1/S$	1
Gate oxide thickness: t_{ox}		$1/S$	1
Supply voltage: V_{DD}		$1/S$	1
Threshold voltage: V_{tn}, V_{tp}		$1/S$	1
Substrate doping: N_A		S	1
Device Characteristics			
β	$\dfrac{W}{L}\dfrac{1}{t_{ox}}$	S	S
Current: I_{ds}	$\beta(V_{DD} - V_t)^2$	$1/S$	S
Resistance: R	$\dfrac{V_{DD}}{I_{ds}}$	1	$1/S$
Gate capacitance: C	$\dfrac{WL}{t_{ox}}$	$1/S$	$1/S$
Gate delay: τ	RC	$1/S$	$1/S^2$
Clock frequency: f	$1/\tau$	S	S^2
Dynamic power dissipation (per gate): P	CV^2f	$1/S^2$	S
Chip area: A		$1/S^2$	1
Power density	P/A	1	S
Current density	I_{ds}/A	S	S

For constant field scaling, all device dimensions, including channel length L, width W, and oxide thickness t_{ox}, are reduced by a factor of $1/S$. The supply voltage V_{DD} and the threshold voltages are also reduced by $1/S$. The substrate doping N_A is increased by S. Because both distance and voltage are scaled equally, the electric field remains constant. This has the desirable effect that many nonlinear factors essentially remain unaffected. Because both L and t_{ox} scale at the same rate, the gate capacitance per micron of width $C_{permicron}$ has also remained approximately constant at 1.5–2 fF/μm, as described in Section 2.3.1.

A gate shrink scales only the channel length, leaving other dimensions, voltages, and doping levels unchanged. This offers a quadratic improvement in gate delay according to the first order model. In practice, the gate delay improvement is closer to linear because velocity saturation keeps the current and effective resistance approximately constant.

Historically, feature sizes were shrunk from 6 μm to 1 μm while maintaining a 5 V supply voltage. This was called *constant voltage* scaling and offered quadratic delay improvement as well as cost reduction. It also maintained continuity in I/O voltage standards. Constant voltage scaling increased the electric fields in devices. By the 1 μm generation, velocity saturation was severe enough that decreasing feature size no longer improved device current. Device breakdown from the high field was another risk. Therefore, constant field scaling has been the rule for modern devices.

Example

Most processes have gate capacitance of roughly 2 fF/μm. If the FO4 inverter delay of a process with features size f (in nm) is 1/2 ps • f, estimate the ON resistance of a unit (i.e., 4 λ wide) nMOS transistor.

Solution: An FO4 inverter has a delay of 5 τ = 15 RC. Therefore,

$$RC = \frac{\frac{1}{2}f}{15} = \frac{f}{30} \frac{\text{ps}}{\text{nm}} \tag{4.74}$$

A unit transistor has width $W = 2f$ and thus capacitance of $C = 4f$ fF/μm. Solving for R,

$$R = \left(\frac{f}{30} \frac{\text{ps}}{\text{nm}} \right) \left(\frac{1}{4f} \frac{\text{μm}}{\text{fF}} \right) = 8.33 \text{ k}\Omega \tag{4.75}$$

Note that this is independent of feature size. The resistance of a unit transistor is roughly independent of feature size, while the gate capacitance decreases with feature size. Alternatively, the capacitance per micron is roughly independent of feature size while the resistance • micron decreases with feature size.

The FO4 inverter delay will scale as $1/S$ assuming ideal constant-field scaling. As we saw in Section 4.7.4, this delay is commonly 1/2 ps/nm of the effective channel length for typical processing and worst-case environment. Aggressive processes achieve delays in the short end of the range by building transistors with effective channel lengths somewhat shorter than the feature size would imply.

4.9.2 Interconnect Scaling

Two common approaches to interconnect scaling are to either scale all dimensions or keep the wire height constant. Table 4.16 shows the resistance, capacitance, and delay per unit length for each of these techniques. Wire length decreases for some types of wires, but may increase for others. Local and scaled wires are those that decrease in length during scaling. For example, a wire across a 64-bit ALU is *local* because it becomes shorter as the ALU is migrated to a finer process. A wire across a particular microprocessor is *scaled* because when the microprocessor is shrunk to the new process, the wire will also shrink. However, wires crossing next generation microprocessors are *global* because the next generation microprocessor die is likely to be larger (by a factor of D_c, on the order of 1.1) than the previous generation, so the cross-chip wires get longer rather than shorter. The table also shows scaling of wire delay for interconnect with and without repeaters.

Unrepeated interconnect delay is remaining about constant for local interconnect and increasing for global interconnect. This presents a problem because transistors are getting faster, so the ratio of interconnect to gate delay increases with scaling. An increasing fraction of circuits are limited by wire delay rather than gate delay, and designers must devote greater attention to wire engineering and repeaters.

In older processes where wire width and spacing were much greater than wire thickness, it was advantageous to scale wires by reducing the width and spacing, but not the thickness. This avoids the quadratic increase in wire resistance per unit length and was acceptable because fringing capacitance was a small fraction of the whole. In modern processes with aspect ratios of 1.5–2.2, fringing capacitance accounts for the majority of the total capacitance. Scaling spacing but not height increases the fringing capacitance enough that the extra thickness scarcely improves delay. Moreover, the coupling capacitance to nearby neighbors causes severe crosstalk. Therefore, it is now common to reduce thickness of lower-level metal interconnect with each generation. Of course, process engineers can choose from a continuum of possibilities between linearly scaling wire thickness and keeping wire thickness constant.

Observe that when wire thickness is scaled, the capacitance per unit length remains constant. Hence, a reasonable initial estimate of the capacitance of a minimum-pitch wire is about 0.2 fF/µm, independent of the process. In other words, wire capacitance is roughly 1/10–1/6 of gate capacitance per unit length.

Table 4.16 Influence of scaling on interconnect characteristics

Parameter	Sensitivity	Reduced Thickness	Constant Thickness
Scaling Parameters			
Width: w		$1/S$	
Spacing: s		$1/S$	
Thickness: t		$1/S$	1
Interlayer oxide height: h		$1/S$	
Characteristics Per Unit Length			
Wire resistance per unit length: R_w	$\dfrac{1}{wt}$	S^2	S
Fringing capacitance per unit length: C_{wf}	$\dfrac{t}{s}$	1	S
Parallel plate capacitance per unit length: C_{wp}	$\dfrac{w}{h}$	1	1
Total wire capacitance per unit length: C_w	$C_{wf} + C_{wp}$	1	between $1, S$
Unrepeated RC constant per unit length: t_{wu}	$R_w C_w$	S^2	between S, S^2
Repeated wire RC delay per unit length: t_{wr} (assuming constant field scaling of gates in Table 4.15)	$\sqrt{RCR_w C_w}$	$\sqrt{S}$	between 1, $\sqrt{S}$
Crosstalk noise	$\dfrac{t}{s}$	1	S
Local/Scaled Interconnect Characteristics			
Length: l		$1/S$	
Unrepeated wire RC delay	$l^2 t_{wu}$	1	between $1/S, 1$
Repeated wire delay	$l t_{wr}$	$\sqrt{1/S}$	between $1/S$, $\sqrt{1/S}$
Global Interconnect Characteristics			
Length: l		D_c	
Unrepeated wire RC delay	$l^2 t_{wu}$	$S^2 D_c^2$	between SD_c^2, $S^2 D_c^2$
Repeated wire delay	$l t_{wr}$	$D_c \sqrt{S}$	between D_c, $D_c \sqrt{S}$

4.9.3 International Technology Roadmap for Semiconductors

The incredible pace of scaling requires cooperation among many companies and researchers both to develop compatible process steps and to anticipate and address future challenges before they hold up production. The Semiconductor Industry Association develops and updates the International Technology Roadmap for Semiconductors [SIA02] to forge a consensus so that development efforts are not wasted on incompatible technologies and to predict future needs and direct research efforts. Such an effort to predict the future is inevitably prone to error, and the industry has scaled feature sizes and clock frequencies more rapidly than the roadmap predicted in the late 1990s. Nevertheless, the roadmap offers a more coherent vision than one could obtain by simply interpolating straight lines through historical scaling data.

The ITRS forecasts a major new technology generation, also called *technology node*, approximately every three years. The scaling between generations is traditionally

$$S = \sqrt{2}$$

so the number of transistors per unit area doubles every generation. Table 4.17 summarizes some of the predictions, particularly for high-performance microprocessors. However, serious challenges lie ahead, and major breakthroughs will be necessary in many areas to maintain the scaling on the roadmap. The FO4 delays/cycle figure is extracted assuming the FO4 delay in picoseconds is one-third of the feature size in nanometers. These cycle times appear problematic because it is very difficult for a digital signal to swing rail to rail in less than 6 FO4 delays. Refer to the ITRS for annual updates.

Table 4.17 Predictions from the 2002 ITRS

Year	2001	2004	2007	2010	2013	2016
Feature size (nm)	130	90	65	45	32	22
V_{DD} (V)	1.1–1.2	1–1.2	0.7–1.1	0.6–1.0	0.5–0.9	0.4–0.9
Millions of transistors/die	193	385	773	1564	3092	6184
Wiring levels	8–10	9–13	10–14	10–14	11–15	11–15
Intermediate wire pitch (nm)	450	275	195	135	95	65
Interconnect dielectric constant	3–3.6	2.6–3.1	2.3–2.7	2.1	1.9	1.8
I/O signals	1024	1024	1024	1280	1408	1472
Clock rate (MHz)	1684	3990	6739	11511	19348	28751
FO4 delays/cycle	13.7	8.4	6.8	5.8	4.8	4.7
Maximum power (W)	130	160	190	218	251	288
DRAM capacity (Gbits)	0.5	1	4	8	32	64

4.9.4 Impacts on Design

One of the limitations of first-order scaling is that it gives the wrong impression of being able to scale proportionally to zero dimensions and zero voltage. In reality, a number of factors change significantly with scaling. This section attempts to peer into the crystal ball and predict some of the impacts on design for the future. These predictions are notoriously risky because chip designers have had an astonishing history of inventing ingenious solutions to seemingly insurmountable barriers.

4.9.4.1 Improved Performance and Cost

The most positive impact of scaling is that performance and cost are steadily improving. System architects need to understand the scaling of CMOS technologies and predict the capabilities of the process several years into the future, when a chip will be completed. Because transistors are becoming cheaper each year, architects particularly need creative ideas of how to exploit growing numbers of transistors to deliver more or better functions. When transistors were first invented, the best predictions of the day suggested that they might eventually approach a fifty-cent manufacturing cost. Figure 4.66 plots the number of transistors and average price per transistor shipped by the semiconductor industry over the past three decades [Moore03]. In 2003, you could buy more than 100,000 transistors for a penny.

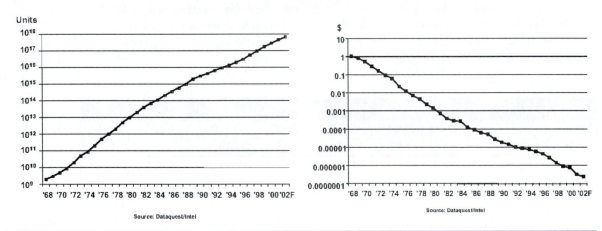

FIG 4.66 Transistor shipments and average price. © IEEE 2003.

4.9.4.2 Interconnect

Scaled transistors are steadily improving in delay, but scaled wires are holding constant or getting worse. Figure 4.67, taken from the 1997 Semiconductor Industry Association Roadmap [SIA97], showed the sum of gate and wire bottoming out at the 250 or 180 nm generation and getting worse thereafter. The wire problem motivated a number of papers predicting the demise of conventional wires. However, the plot

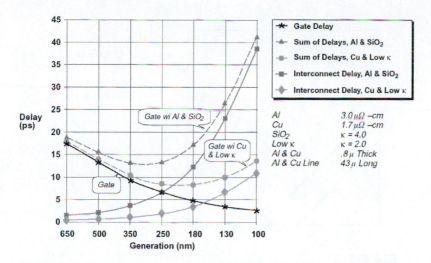

Gate and wire delay scaling. Reprinted from [SIA97] with permission of the Semiconductor Industry Association.

is misleading in two ways. First, the "gate" delay is shown for a single unloaded transistor (delay = RC) rather than a realistically loaded gate (e.g., an FO4 inverter delay = $15RC$). Second, the wire delays shown are for fixed lengths, but as technology scales, most local wires connecting gates within a unit also become shorter [Ho01].

In practice, for short wires, such as those inside a logic gate, the wire RC delay is negligible and will remain so for the foreseeable future. However, the long wires present a considerable challenge. It is no longer possible to send a signal from one side of a large, high-performance chip to another in a single cycle. Also, the "reachable radius" that a signal can travel in a cycle is steadily getting smaller, as shown in Figure 4.68. This requires that microarchitects understand the floorplan and budget multiple pipeline stages for data to travel long distances across the die.

Repeaters help somewhat, but even so, interconnect does not keep up. Moreover, the "repeater farms" must be allocated space in the floorplan. As scaled gates become faster, the delay of a repeater goes down and hence, you should expect it will be better to use more repeaters. This means a greater number of repeater farms are required.

One technique to alleviate the interconnect problem is to use more layers of interconnect. Table 4.18 shows the number of layers of interconnect increasing with each generation in TSMC processes. The lower layers of interconnect are classically scaled to provide high-density short connections. The higher layers are scaled less aggressively, or possibly even reverse-scaled to be thicker and wider to provide low-resistance, high-speed interconnect, good clock distribution networks, and a stiff power grid. Copper and low-k dielectrics were also introduced to reduce resistance and capacitance.

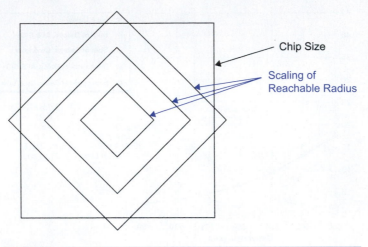

FIG 4.68   Reachable radius scaling

Table 4.18	Scaling of metal layers in TSMC processes
Process (nm)	**Metal Layers**
500	3 (Al)
350	4 (Al)
250	5 (Al)
180	6 (Al, low-k)
150	7 (Cu, low-k)
130	8 (Cu, low-k)
90	9 (Cu, low-k)

Blocks of 50–100 Kgates (1 *Kgate* = 1000 3-input NAND gates or 6000 transistors) will continue to have reasonably short internal wires and acceptably low wire RC delay [Sylvester98]. Therefore, large systems can be partitioned into blocks of roughly this size with repeaters inserted as necessary for communication between blocks.

4.9.4.3 Power   In classical constant field scaling, power density remains constant and overall chip power increases only slowly with die size. In practice, power density has skyrocketed because clock frequencies have increased much faster than classical scaling would predict and V_{DD} is somewhat higher than constant field scaling would demand.

Intel Vice President Patrick Gelsinger gave a keynote speech at the International Solid State Circuits Conference on February 5, 2001 [Gelsinger01]. He showed that microprocessor power consumption had been increasing exponentially (see Figure 4.69)

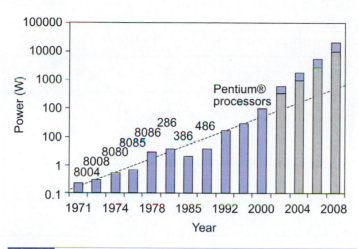

FIG 4.69 Intel processor power consumption. © IEEE 2001.

and was predicted to grow even faster in the coming decade. He predicted "business as usual will not work in the future," and that if scaling continued at this pace, by 2005, high-speed processors would have the power density of a nuclear reactor, by 2010, a rocket nozzle, and by 2015, the surface of the sun! The next day, Intel's stock dropped 8% and Intel has since downplayed the forecast because designers will obviously devote more attention to controlling power.

Dynamic power consumption will not continue to increase at such rates because it will become uneconomical to cool the chips. One reason is that at the time of this writing, high-performance processors are operating at about 12–16 FO4 inverter delays/cycle. It will be very difficult to generate a clock with a period of less than 6 FO4 inverter delays because the clock will look like a sinusoid rather than a square wave [Ho01] and because sequencing overhead becomes excessive; hence, frequency cannot scale faster than raw gate delays too much longer. Another reason is that cache area is likely to become a larger fraction of the die area and caches have low activity factors and lower power dissipation per unit area. Nevertheless, designers will need to budget and plan for power consumption as a factor nearly as important as performance and perhaps more important than area.

Static power consumption will be a growing concern, especially for battery-operated devices. The static power consumption caused by subthreshold leakage was historically negligible but becomes important for threshold voltages below about 0.3–0.4 V, and may become comparable to dynamic power for systems operating below 1 V. Figure 4.70 shows how the static power has historically increased much more rapidly than dynamic power [Moore03]. This is especially problematic because simply turning off clocks in a sleep mode is not sufficient to stop the static power consumption. To slow the increase of leakage, V_t has changed from about $V_{DD}/5$ in older processes to $V_{DD}/3$ in newer processes as

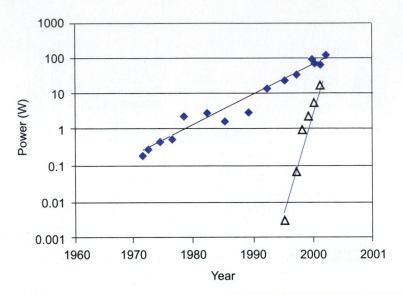

FIG 4.70 Dynamic and static power trends. © IEEE 2003.

V_{DD} has dropped; this has eliminated circuits with threshold drops from the set of viable circuit styles. Systems will use multiple flavors of transistors with different threshold voltages so that gates on the critical path can have fast low-V_t devices while memories and noncritical gates save power with higher-V_t devices. Gate tunneling current is also important for oxides of less than 15–20 Å. Multiple oxide thicknesses are also required. I/O standards of 1.8 or 3.3 V are destructive to the very thin oxides in modern processes, so thicker oxides are used for slower I/O transistors. Thicker oxides may also become an option for memories and noncritical circuits.

Even if power remains constant, lower supply voltage leads to higher current density. This in turn causes higher IR drops and di/dt noise in the supply network (see Sections 6.3.5 and 12.3). These factors lead to more pins and metal resources on a chip being required for the power distribution networks.

4.9.4.4 Productivity The number of transistors that fit on a chip is increasing faster than designer productivity (gates/week). This leads to design teams of increasing size, and difficulty recruiting enough experienced engineers when the economy is good. It has driven a search for design methodologies that maximize productivity, even at the expense of performance and area. Now most chips are designed using synthesis and place and route; the number of situations where custom circuit design is affordable is diminishing. In other words, creativity is shifting from the circuit to the systems level for many designs. On the other hand, performance is still king in the microprocessor world. Design teams in that field are approaching the size of automotive and aerospace teams because the development cost is justified by the size of the market. This drives a need for engineering managers who are skilled in leading such large organizations.

The number of 50–100 Kgate blocks is growing, even in relatively low-end systems. This demands greater attention to floorplanning and placement of the blocks.

One of the best hopes to solve the productivity gap is design reuse. Intellectual property (IP) blocks can be purchased and used as black boxes within a system-on-chip (SoC) in much the same way chips are purchased for a board-level design.

4.9.4.5 Physical Limits

How far will CMOS processes scale? It is clear that scaling cannot continue indefinitely; transistors as we know them today will not work if the oxide is less than an atomic layer thick, the channel less than an atomic layer long, or the charge in the channel less than that of one electron. Numerous papers have been written forecasting the end of silicon scaling. For example, in 1972, the limit was placed at the 0.25 μm generation because of tunneling and fluctuations in dopant distributions [Hoeneisen72, Mead80]; at this generation, chips were predicted to operate at 10–30 MHz! In 1999, IBM predicted that scaling would nearly grind to a halt beyond the 100 nm generation in 2004 [Davari99].

In the authors' experience, seemingly insurmountable barriers have seemed to loom about a decade away. Reasons given for these barriers have included:

- Dynamic power dissipation
- Subthreshold leakage at low V_{DD} and V_t
- Tunneling current through thin oxides
- Poor I-V characteristics due to short channel effects
- Optics for cost-effective manufacturing of small features
- Exponentially increasing costs of fabrication facilities and mask sets
- Electromigration
- Interconnect delay

At the time of writing, it appears that no fundamental barriers exist before the 35 nm generation in 2013 (roughly a decade ahead). Beyond this point, it is difficult to predict the future. Nevertheless, a large number of extremely talented people are continuously pushing the limits and hundreds of billions of dollars are at stake, so we are reluctant to bet against the future of scaling.

4.10   Pitfalls and Fallacies

Defining gate delay for an unloaded gate

When marketing a process, it is common to report gate delay based on an inverter in a ring oscillator (2τ) or even the RC time constant of a transistor charging its own gate capacitance ($1/3\ \tau$). Remember that the delay of a real gate on the critical path should be closer to 5–6τ. When in doubt, ask how "gate delay" is defined or ask for the FO4 inverter delay.

Trying to increase speed by increasing the size of transistors in a path

Most designers know that increasing the size of a transistor decreases its resistance and thus makes it faster at driving a constant load. Novice designers sometimes forget that increasing the size increases input capacitance and makes the previous stage slower, especially when that previous stage belongs to somebody else's timing budget. The authors have seen this lead to lack of convergence in full-chip timing analysis on a large microprocessor because individual engineers boost the size of their own gates until their path meets timing. Only after the weekly full-chip timing roll-up do they discover that their inputs now arrive later because of the greater load on the previous stage. The solution is to include in the specification of each block not only the arrival time but also the resistance of the driver in the previous block.

Trying to increase speed by using as few stages of logic as possible

Logic designers often count "gate delays" in a path. This is a convenient simplification when used properly. In the hands of an inexperienced engineer who believes each gate contributes a gate delay, it suggests that the delay of a path is minimized by using as few stages of logic as possible, which is clearly untrue.

Designing a large chip without considering the floorplan

In the mid-1990s, designers became accustomed to synthesizing a chip from HDL and "tossing the netlist over the wall" to the vendor who would place and route it and manufacture the chip. Many designers were shielded from considering the physical implementation. Now flight times across the chip are a large portion of the cycle time in slow systems and multiple cycles in faster systems. If the chip is synthesized without a floorplan, some paths with long wires will be discovered to be too slow after layout. This requires resynthesis with new timing constraints to shorten the wires. When the new layout is completed, the long wires simply show up in different paths. The solution to this convergence problem is to make a floorplan early and microarchitect around this floorplan, including budgets for wire flight time between blocks. Algorithms termed *timing directed placement* have alleviated this problem, resulting in place & route tools that converge in one or a few iterations.

Not stating process corner or environment when citing circuit performance

Most products must be guaranteed to work at high temperature, yet many papers are written with transistors operating at room temperature (or lower), giving optimistic performance results. For example, at the International Solid State Circuits Conference Intel described a Pentium II processor running at a surprisingly high clock rate [Choudhury97], but when asked, the speaker admitted that the measurements were taken while the processor was "colder than an ice cube."

Similarly, the FFFFF design corner is sometimes called the "published paper" corner because delays are reported under these simulation or manufacturing conditions without bothering to state that fact or report the FO4 inverter delay in the same conditions. Circuits in this corner are about twice as fast as in a manufacturable part.

Providing too little margin in matched delays

We have seen that the delay of a chain of inverters can vary by about 30% as compared to the delay of other circuits across design corners, voltage, and temperature. On top of this, you should expect intra-die process variation and errors in modeling and extraction. If a race condition exists where the circuit will fail when the inverter delay is faster than the gate delay, the experienced designer who wishes to sleep well at night provides generous delay margin under nominal conditions. Remember that the consequences of too little margin can be a million dollars in mask costs for another revision of the chip and far more money in the opportunity cost of arriving late to market.

Failing to plan for process scaling

Many products will migrate through multiple process generations. For example, the Intel Pentium Pro was originally designed and manufactured on a 0.6 micron BiCMOS process. The Pentium II is a closely related derivative manufactured in a 0.35 micron process operating at a lower voltage. In the new process, bipolar transistors ceased to offer performance advantages and were removed at considerable design effort. Further derivatives of the same architecture migrated to 0.25 and 0.18 micron processes in which wire delay did not improve at the same rate as gate delay. Interconnect-dominated paths required further redesign to achieve good performance in the new processes. In contrast, the Pentium 4 was designed with process scaling in mind. Knowing that over the lifetime of the product, device performance would improve but wires would not, designers overengineered the interconnect-dominated paths for the original process so that the paths would not limit performance improvement as the process advanced [Deleganes02].

4.11 Historical Perspective

The incredible history of scaling can be seen in the advancement of the microprocessor. The Intel microprocessor line spans more than three decades. Table 4.19 summarizes the progression from the first 4-bit microprocessor, the 4004, through the Pentium 4, courtesy of the Intel Museum [Intel03]. Over the three decades, feature size has improved nearly one hundred-fold. Transistor budgets multiplied by more than 10,000. Even more remarkably, clock frequencies have also multiplied by almost as much. Even as the challenges have grown in the past decade, scaling has accelerated.

Table 4.19	History of Intel microprocessors over three decades					
Processor	Year	Feature Size (μm)	Transistors	Frequency (MHz)	Word size	Package
4004	1971	10	2.3k	0.75	4	16-pin DIP
8008	1972	10	3.5k	0.5–0.8	8	18-pin DIP
8080	1974	6	6k	2	8	40-pin DIP
8086	1978	3	29k	5–10	16	40-pin DIP
80286	1982	1.5	134k	6–12	16	68-pin PGA
Intel386	1985	1.5–1.0	275k	16–25	32	100-pin PGA
Intel486	1989	1–0.6	1.2M	25–100	32	168-pin PGA
Pentium	1993	0.8–0.35	3.2–4.5M	60–300	32	296-pin PGA
Pentium Pro	1995	0.6–0.35	5.5M	166–200	32	387-pin MCM PGA
Pentium II	1997	0.35–0.25	7.5M	233–450	32	242-pin SECC
Pentium III	1999	0.25–0.18	9.5–28M	450–1000	32	330-pin SECC2
Pentium 4	2001	0.18–0.13	42–55M	1400–3200	32	478-pin PGA

Die photos of the microprocessors also illustrate the remarkable story of scaling. The 4004 [Faggin96] in Figure 4.71 was handcrafted to pack the transistors onto the tiny die. Observe the 4-bit datapaths and register files. Only a single layer of metal was available, so polysilicon jumpers were required when traces had to cross without touching. The masks were designed with colored pencils and were hand-cut from red plastic rubylith. Observe that diagonal lines were used routinely. The 16 I/O pads and bond wires are clearly visible. The processor was used in the Busicom calculator.

The 80286 [Childs84] in Figure 4.72 shows a far more regular appearance. It is partitioned into regular datapaths, random control logic, and several arrays. The arrays include the instruction decoder PLA and memory management hardware. At this scale, individual transistors are no longer visible.

The Intel486™ (originally 80486, but changed because a number cannot be trademarked) integrated an 8KB cache and floating point unit with a pipelined integer datapath, as shown in Figure 4.73. At this scale, individual gates are not visible. The center row is the 32-bit integer datapath. Above is the cache, divided into four 2KB subarrays. Observe that the cache involves a significant amount of logic beside the subarrays. Below are several blocks of synthesized control logic generated with automatic place and route tools. The "more advanced" tools no longer support diagonal interconnect. The wide datapaths in the upper right form the floating point unit.

The Pentium Processor™ in Figure 4.74 provides a superscalar integer execution unit and separate 8KB data and instruction caches. The 32-bit datapath and its associated control logic is again visible in the center of the chip, although at this scale, the individual bitslices of the datapath are difficult to resolve. The instruction cache in the upper left

FIG 4.71 4004 microprocessor. Reprinted with permission of Intel Corporation.

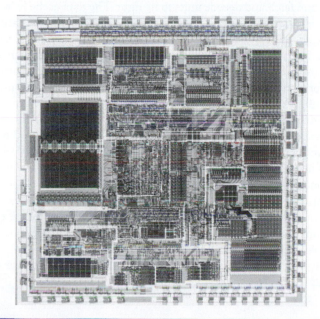

FIG 4.72 80286 microprocessor. Reprinted with permission of Intel Corporation.

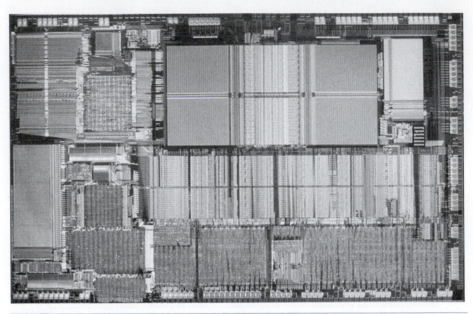

FIG 4.73 Intel486™ microprocessor. Reprinted with permission of Intel Corporation.

feeds the instruction fetch and decode units to its right. The data cache is in the lower left. The bus interface logic sits between the two caches. The pipelined floating point unit, home of the infamous FDIV bug [Price95], is in the lower right. This floorplan is important to minimize wire lengths between units that often communicate, such as the instruction cache and instruction fetch or the data cache and integer datapath. The integer datapath often forms the heart of a microprocessor, and other units surround the datapath to feed it the prodigious quantities of instructions and data that it consumes.

The Pentium™ III Processor, shown in Figure 4.75, offers out-of-order issue of up to three instructions per cycle. The entire left portion of the die is dedicated to 256–512 KB of level 2 cache to supplement the 32KB instruction and data caches. As processor performance outstrips memory system bandwidth, the portion of the die devoted to the cache hierarchy will continue to grow.

The Pentium™ 4 Processor is shown in Figure 4.76. The complexity of a VLSI system is clear from the enormous number of separate blocks that were each uniquely designed by a team of engineers. Indeed, at this scale even major functional units become difficult to resolve. The high operating frequency is achieved with a long pipeline using about 14 FO4 inverter delays per cycle. Remarkably, portions of the integer execution unit are "double-pumped" at twice the regular chip frequency.

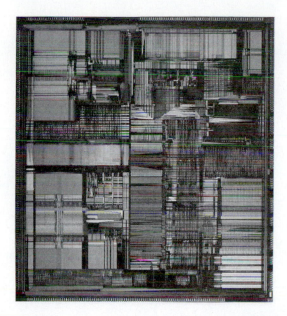

FIG 4.74 Pentium™ microprocessor. Reprinted with permission of Intel Corporation.

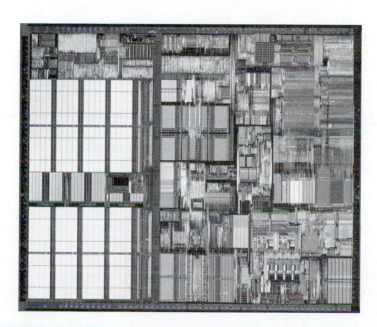

FIG 4.75 Pentium™ III microprocessor. Reprinted with permission of Intel Corporation.

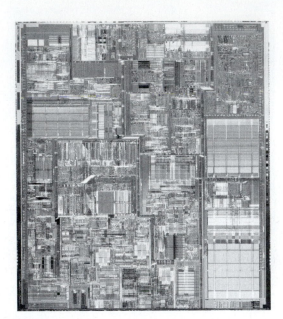

FIG 4.76 Pentium™ 4 microprocessor. Reprinted with permission of Intel Corporation.

Summary

The VLSI designer's challenge is to engineer a system that meets speed requirements while consuming little power or area, operating reliably, and taking little time to design. Circuit simulation is an important tool for calculating delay and will be discussed in depth in Chapter 5, but it takes too long to simulate every possible design; is prone to garbage-in, garbage-out mistakes; and doesn't give insight into why a circuit has a particular delay or how the circuit should be changed to improve delay. The designer must also have simple models to quickly estimate performance by hand and explain why some circuits are better than others.

Although transistors are complicated devices with nonlinear current-voltage and capacitance-voltage relationships, for the purpose of delay estimation in digital circuits, they can be approximated quite well as having constant capacitance and an effective resistance R when ON. Logic gates are thus modeled as RC networks. The Elmore delay model estimates the delay of the network as the sum of each capacitance times the resistance through which it must be charged or discharged. Therefore, the gate delay consists of a parasitic delay (accounting for the gate driving its own internal parasitic capacitance) plus an effort delay (accounting for the gate driving an external load). The effort delay depends on the electrical effort (the ratio of load capacitance to input capacitance, also called *fanout*) and the logical effort (which characterizes the current driving capability of the gate relative to an inverter with equal input capacitance). Even in advanced fabrication processes, the delay vs. electrical effort curve fits a straight line very well. The method of Logical Effort builds on this linear delay model to help us quickly estimate the delay of entire paths based on the effort and parasitic delay of the path. We will use Logical Effort in subsequent chapters to explain what makes circuits fast.

The power consumption of a circuit has both dynamic and static components. The dynamic power comes from charging and discharging the load capacitances and depends on the frequency, voltage, capacitance, and activity factor. The static power comes from circuits that have an intentional path from V_{DD} to GND (like pseudo-nMOS) and from leakage. CMOS circuits have historically consumed relatively low power because complementary CMOS gates dissipate almost zero static power. However, leakage is increasing as feature size decreases, making static power consumption more important.

As feature size decreases, transistors get faster but wires do not. Interconnect delays are now very important. The delay is again estimated using the Elmore delay model based on the resistance and capacitance of the wire and its driver and load. The wire delay grows with the square of its length, so long wires are often broken into shorter segments driven by repeaters. Vast numbers of wires are required to connect all the transistors, so processes provide many layers of interconnect packed closely together. The capacitive coupling between these tightly packed wires can be a major source of noise in a system.

For reliable circuit operation, the designer must ensure that the circuit performs correctly across variations in the operating voltage and temperature, and must develop circuits that are robust for variations in transistor characteristics such as channel length or threshold voltage. Process corners are used to describe the worst-case combination of processing and environment for delay, power consumption, and functionality. The circuits must also be designed to operate correctly even as they age or are subject to cosmic rays and electrostatic discharge.

CMOS processes have been steadily improving for more than 20 years and will continue to do so for at least the next decade. A good designer not only should be familiar with the capabilities of current processes, but also should be able to predict the capabilities of future processes as feature sizes get progressively smaller. In modern constant-field scaling, gate delay improves with channel length. The number of transistors on a chip grows quadratically. The energy for each transistor to switch decreases with the cube of channel length, but the dynamic power density remains about the same because chips have more transistors switching at higher rates. Static power goes up as small transistors have exponentially more leakage. Interconnect capacitance per unit length remains constant, but resistance increases because the wires have a smaller cross-section. Local wires get shorter and have constant delay, while global wires have increasing delay. Architects planning future chips will take advantage of the larger number of faster transistors, but must also rethink existing architectures because of the changing ratio of wire-to-gate delay and the increasing leakage current.

Exercises

4.1 Sketch a 2-input NOR gate with transistor widths chosen to achieve effective rise and fall resistances equal to a unit inverter. Compute the rising and falling propagation delays of the NOR gate driving h identical NOR gates using the Elmore delay model. Assume that every source or drain has fully contacted diffusion when making your estimate of capacitance.

4.2 Sketch a stick diagram for the 2-input NOR. Repeat Exercise 4.1 with better capacitance estimates. In particular, if a diffusion node is shared between two parallel transistors, only budget its capacitance once. If a diffusion node is between two series transistors and requires no contacts, only budget half the capacitance because of the smaller diffusion area.

4.3 Find the rising and falling propagation delays of an unloaded AND-OR-INVERT gate using the Elmore delay model. Estimate the diffusion capacitance based on a stick diagram of the layout.

4.4 Find the worst-case Elmore parasitic delay of an n-input NOR gate.

4.5 Sketch a delay vs. electrical effort graph like that of Figure 4.8 for a 2-input NOR gate using the logical effort and parasitic delay estimated in Section 4.2.3. How does the slope of your graph compare to that of the 2-input NAND? How does the y-intercept compare?

4.6 Let a 4x inverter have transistors four times as wide as those of a unit inverter. If a unit inverter has three units of input capacitance and parasitic delay of p_{inv}, what is the input capacitance of a 4x inverter? What is the logical effort? What is the parasitic delay?

4.7 A three-stage logic path is designed so that the effort borne by each stage is 12, 6, and 9 delay units, respectively. Can this design be improved? Why? What is the best number of stages for this path? What changes do you recommend to the existing design?

4.8 Suppose a unit inverter with three units of input capacitance has unit drive.

a) What is the drive of a 4x inverter?

b) What is the drive of a 2-input NAND gate with 3 units of input capacitance?

4.9 Sketch a 4-input NAND gate with transistor widths chosen to achieve equal rise and fall resistance as a unit inverter. Show why the logical effort is 6/3.

4.10 Consider the two designs of a 2-input AND gate shown in Figure 4.77. Give an intuitive argument about which will be faster. Back up your argument with a calculation of the path effort, delay, and input capacitances x and y to achieve this delay.

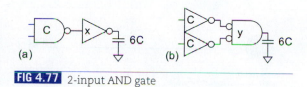

FIG 4.77 2-input AND gate

4.11 Consider four designs of a 6-input AND gate shown in Figure 4.78. Develop an expression for the delay of each path if the path electrical effort is H. What design is fastest for $H = 1$? For $H = 5$? For $H = 20$? Explain your conclusions intuitively.

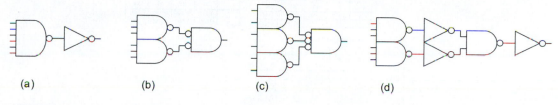

(a)   (b)   (c)   (d)

FIG 4.78 6-input AND gate

4.12 Repeat the decoder design example from Section 4.3.4 for a 32-word register file with 64-bit registers. Determine the fastest decoder design and estimate the delay of the decoder and the transistor widths to achieve this delay.

4.13 Design a circuit at the gate level to compute the following function:

```
if (a == b) y = a;
else y = 0;
```

Let a, b, and y be 16-bit busses. Assume the input and output capacitances are each 10 units. Your goal is to make the circuit as fast as possible. Estimate the delay in FO4 inverter delays using Logical Effort if the best gate sizes were used. What sizes do you need to use to achieve this delay?

4.14 Plot the average delay from input A of an FO3 NAND2 gate from the datasheet in Figure 4.25. Why is the delay larger for the XL drive strength than for the other drive strengths?

4.15 Figure 4.79 shows a datasheet for a 2-input NOR gate in the Artisan Components standard cell library for the TSMC 180 nm process. Find the average parasitic delay and logical effort of the X1 NOR gate A input. Use the value of τ from Section 4.3.7.

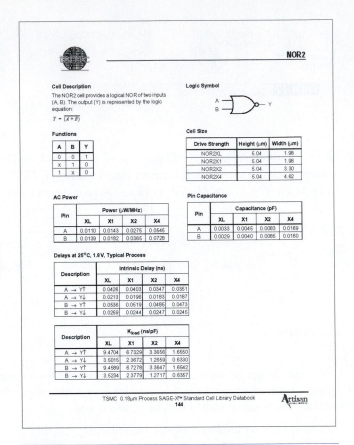

FIG 4.79 2-input NOR datasheet courtesy Artisan Components

4.16   Find the parasitic delay and logical effort of the X2 and X4 NOR gate *A* input. By what percentage do they differ from that of the X1 gate? What does this imply about our model that parasitic delay and logical effort depend only on gate type and not on transistor sizes?

4.17   What are the parasitic delay and logical effort of the X1 NOR gate *B* input? How and why do they differ from the *A* input?

4.18   Parasitic delay estimates in Section 4.2.4 are made assuming contacted diffusion on each transistor on the output node and ignoring internal diffusion. Would parasitic delay increase or decrease if you took into account that some parallel transistors on the output node share a single diffusion contact? If you counted internal diffusion capacitance between series transistors? If you counted wire capacitance within the cell?

4.19 Consider a process in which pMOS transistors have three times the effective resistance as nMOS transistors. A unit inverter with equal rising and falling delays in this process is shown in Figure 4.80. Calculate the logical efforts of a 2-input NAND gate and a 2-input NOR gate if they are designed with equal rising and falling delays.

FIG 4.80 Unit inverter

4.20 Generalize Exercise 4.19 if the pMOS transistors have μ times the effective resistance of nMOS transistors. Find a general expression for the logical efforts of a *k*-input NAND gate and a *k*-input NOR gate. As μ increases, comment on the relative desirability of NANDs vs. NORs.

4.21 Some designers define a "gate delay" to be a fanout-of-3 2-input NAND gate rather than a fanout-of-4 inverter. Using Logical Effort, estimate the delay of a fanout-of-3 2-input NAND gate. Express your result both in τ and in FO4 inverter delays, assuming $p_{inv} = 1$.

4.22 Repeat Exercise 4.21 in a process with a lower ratio of diffusion to gate capacitance in which $p_{inv} = 0.75$. By what percentage does this change the NAND gate delay, as measured in FO4 inverter delays? What if $p_{inv} = 1.25$?

4.23 The 64-bit Naffziger adder [Naffziger96] has a delay of 930 ps in a fast 0.5-μm Hewlett-Packard process with an FO4 inverter delay of about 140 ps. Estimate its delay in a 70 nm process with an FO4 inverter delay of 20 ps.

4.24 An output pad contains a chain of successively larger inverters to drive the (relatively) enormous off-chip capacitance. If the first inverter in the chain has an input capacitance of 20 fF and the off-chip load is 10 pF, how many inverters should be used to drive the load with least delay? Estimate this delay, expressed in FO4 inverter delays.

4.25 The clock buffer in Figure 4.81 can present a maximum input capacitance of 100 fF. Both true and complementary outputs must drive loads of 300 fF. Compute the input capacitance of each inverter to minimize the worst-case delay from input to either output. What is this delay, in τ? Assume the inverter parasitic delay is 1.

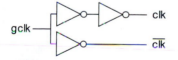

FIG 4.81 Clock buffer

4.26 The clock buffer from Exercise 4.25 is an example of a 1-2 fork. In general, if a 1–2 fork has a maximum input capacitance of C_1 and each of the two legs drives a load of C_2, what should the capacitance of each inverter be and how fast will the circuit operate? Express your answer in terms of p_{inv}.

4.27 A 180 nm standard cell process can have an average switching capacitance of 150 pF/mm². You are synthesizing a chip composed of random logic with an average activity factor of 0.1. Estimate the power consumption of your chip if it has an area of 70 mm² and runs at 450 MHz at $V_{DD} = 0.9$ V.

4.28 You are considering lowering V_{DD} to try to save power in a static CMOS gate. You will also scale V_t proportionally. Will dynamic power consumption go up or down? Will static power consumption go up or down?

4.29   Evaluate the benefits of the stack effect for subthreshold leakage by comparing I_1 and I_2 in Figure 4.29 (this problem is inspired by [Narendra01]). Assume all three transistors are identical and $\gamma = 0$, $n = 1$.

a) If the transistors suffer from no DIBL ($\eta = 0$), prove that $I_2/I_1 = 1/2$, just as you would expect if the transistors behaved as resistors.

b) Does increasing η increase or decrease I_1? By what fraction does I_1 change at room temperature if $\eta = 0.05$ and $V_{DD} = 1.8$ V?

c) Does increasing η increase or decrease I_2? By what fraction does I_2 change at room temperature if $\eta = 0.05$ and $V_{DD} = 1.8$ V?

d) Solve for I_2/I_1 and x as a function of $\Delta = \eta V_{DD}/v_T$ assuming $\Delta \gg 1$.

e) Explain why the stack effect is most important for transistors with significant DIBL.

4.30   Consider a 5 mm long, 4 λ-wide metal2 wire in a 0.6 μm process. The sheet resistance is 0.08 $\Omega/\square$ and the capacitance is 0.2 fF/μm. Construct a 3-segment π-model for the wire.

4.31   A 10x unit-sized inverter drives a 2x inverter at the end of the 5 mm wire from Exercise 30. The gate capacitance is $C = 2$ fF/μm and the effective resistance is $R = 2.5$ k$\Omega \cdot \mu$m for nMOS transistors. Estimate the propagation delay using the Elmore delay model; neglect diffusion capacitance.

4.32   Find the best width and spacing to minimize the RC delay of a metal2 bus in the 180 nm process described in Table 4.8 if the pitch cannot exceed 1000 nm. Minimum width and spacing are 320 nm. First assume that neither adjacent bit is switching. How does your answer change if the adjacent bits may be switching?

4.33   Derive EQ (4.55)–(4.57). Assume the initial driver and final receiver are of the same size as the repeaters so the total delay is N times the delay of a segment. Neglect diffusion parasitics so each segment can be modeled as in Figure 4.82.

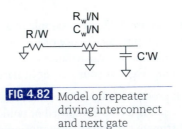

FIG 4.82 Model of repeater driving interconnect and next gate

4.34 Revisit Exercise 33 using pair of inverters (a noninverting buffer) instead of a single inverter. The first inverter in each pair is *W1* times unit width. The second is a factor of *k* larger than the first. Derive EQ (4.58)-(4.60).

4.35 Compute the characteristic velocity (delay per mm) of a repeated wire in the 180 nm process. A unit nMOS transistor has resistance of 2.5 KΩ and capacitance of 0.7 fF, and the pMOS has twice the resistance. Develop a table of results for metal1, metal2, and metal4 on minimum pitch and on double-pitch (twice minimum width and spacing). Use the data and pitches for the Intel 180 nm process. Assume solid metal above and below the wires and that the neighbors are not switching.

4.36 The Pentium 4 operated at 1.4 GHz in Intel's 180 nm process with an FO4 delay of 50 ps in 2001. If a new process is introduced every two years with a scaling factor of

$$S = \sqrt{2},$$

predict microprocessor clock frequencies in 2009, assuming all the improvement comes from ideal constant field scaling and that transistor speed sets clock rate. In practice, the frequency is likely to be higher. Why?

4.37 The path from the data cache to the register file of a microprocessor involves 500 ps of gate delay and 500 ps of wire delay along a repeated wire. The chip is scaled using constant field scaling and reduced height wires to a new generation with $S = 2$. Estimate the gate and wire delays of the path. By how much did the overall delay improve?

Circuit Simulation

5

5.1 Introduction

Fabricating chips is expensive and time-consuming, so designers need simulation tools to explore the design space and verify designs before they are fabricated. Simulators operate at many levels of abstraction, from process through architecture. *Process simulators* such as SUPREME predict how factors in the process recipe such as time and temperature affect device physical and electrical characteristics. *Circuit simulators* such as SPICE use device models and a circuit netlist to predict circuit voltages and currents, which indicate performance and power consumption. *Logic simulators* predict the function of digital circuits and are widely used to verify correct logical operation of designs specified in a hardware description language (HDL). *Architecture simulators* work at the level of instructions and registers to predict throughput and memory access patterns, which influence design decisions such as pipelining and cache memory organization. The various levels of abstraction offer tradeoffs between degree of detail and the size of the system that can be simulated. VLSI designers are primarily concerned with circuit and logic simulation. This chapter focuses on circuit simulation with SPICE. Section 9.3 discusses logic simulation.

Is it better to predict circuit behavior using paper-and-pencil analysis, as has been done in the previous chapters, or with simulation? VLSI circuits are complex and modern transistors have nonlinear, nonideal behavior, so simulation is necessary to accurately predict detailed circuit behavior. Even when closed-form solutions exist for delay or transfer characteristics, they are too time-consuming to apply by hand to large numbers of circuits. On the other hand, circuit simulation is notoriously prone to errors: *garbage in, garbage out* (GIGO). The simulator accepts the model of reality provided by the designer, but it is very easy to create a model that is inaccurate or incomplete. Moreover, the simulator only applies the stimulus provided by the designer and it is very easy to overlook the worst-case stimulus. In the same way that an experienced programmer doesn't expect a program to operate correctly before debugging, an experienced VLSI designer does not expect that the first run of a simulation will reflect reality. Therefore, the circuit designer needs to have a good intuitive understanding of circuit operation and should be able to predict the expected outcome before simulating. Only when expectation and simulation match can there be confidence in the results. In practice, circuit designers depend on both hand analysis and simulation, or as [Glasser85] puts it, "simulation guided through insight gained from analysis."

This chapter presents a brief SPICE tutorial by example. It then discusses models for transistors and diffusion capacitance. The remainder of the chapter is devoted to simulation techniques to characterize a process and to check performance, power, and correctness of circuits and interconnect.

5.2 A SPICE Tutorial

SPICE, a *Simulation Program with Integrated Circuit Emphasis*, was originally developed in the 1970s at Berkeley [Nagel75]. It solves the nonlinear differential equations describing components such as transistors, resistors, capacitors, and voltage sources. SPICE offers many ways to analyze circuits, but digital VLSI designers are primarily interested in *DC* and *transient* analysis that predicts the node voltages given inputs that are fixed or arbitrarily changing in time. SPICE was originally developed in FORTRAN and has some idiosyncrasies, particularly in file formats, related to its heritage. There are free versions of SPICE available on most platforms, but the commercial versions tend to offer more robust numerical convergence. In particular, HSPICE is widely used in industry because it converges well, supports the latest device and interconnect models, and has a large number of enhancements for measuring and optimizing circuits. PSPICE is another commercial version with a free limited student version. The examples throughout this section use HSPICE and generally will not run in ordinary SPICE.

While the details of using SPICE vary with version and platform, all versions of SPICE read an input file and generate a list file with results, warnings, and error messages. The input file is often called a *SPICE deck* and each line a *card* because it was once provided to a mainframe as a deck of punch cards. The input file contains a netlist consisting of components and nodes. It also contains simulation options, analysis commands, and device models. The netlist can be entered by hand or extracted from a circuit schematic or layout in a CAD program.

A good SPICE deck is like a good piece of software. It should be readable, maintainable, and reusable. Comments and white space help make the deck readable. Often the best way to write a SPICE deck is to start with a good deck that does nearly the right thing and then modify it.

The remainder of this section provides a sequence of examples illustrating the key syntax and capabilities of SPICE for digital VLSI circuits. For more detail, consult the Berkeley SPICE manual [Johnson91], the lengthy HSPICE manual, or any number of textbooks on SPICE (such as [Kielkowski95, Foty96]).

5.2.1 Sources and Passive Components

Suppose we would like to find the response of the RC circuit in Figure 5.1(a) given an input rising from 0 to 1.8 V over 50 ps. Because the RC time constant of 100 fF • 2 kΩ = 200 ps is much greater than the input rise time, we intuitively expect the output would look like an exponential asymptotically approaching the final value of 1.8 V with a 200 ps time constant. Figure 5.2 gives a SPICE deck for this simulation and Figure 5.1(b) shows the input and output responses.

Cards beginning with * are comments. The first card of a SPICE deck must be a comment, typically indicating the title of the simulation. It is good practice to treat SPICE input files like computer programs and follow similar procedures for commenting the decks. In particular, giving the author, date, and objective of the simulation at the beginning is helpful when the deck must be revisited in the future (e.g., when a chip is in silicon debug and old simulations are being reviewed to track down potential reasons for failure).

Control cards begin with a dot (.). The `.option post` card instructs HSPICE to write the results to a file for use with a waveform viewer. The last card of a SPICE deck must be `.end`.

Each card in the netlist begins with a letter indicating the type of circuit element. Note that SPICE is case-insensitive. Common elements are given in Table 5.1. In this case the circuit consists of a voltage source named `Vin`, a resistor named `R1`, and a capacitor named `C1`. The nodes in the circuit are named `in`, `out`, and `gnd`. `gnd` is a special node name defined to be the 0 V reference. The units consist of one or two letters. The first character indicates the order of magnitude, as given in Table 5.2. The second letter indicates a unit for human convenience (such as F for farad or s for second) and is ignored by SPICE. For example, the hundred femtofarad capacitor can be expressed as `100fF`, `100f`, or simply `100e−15`.

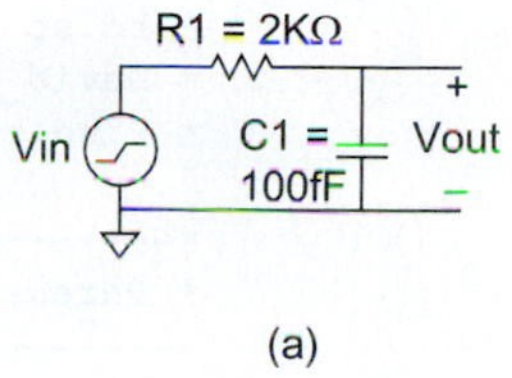

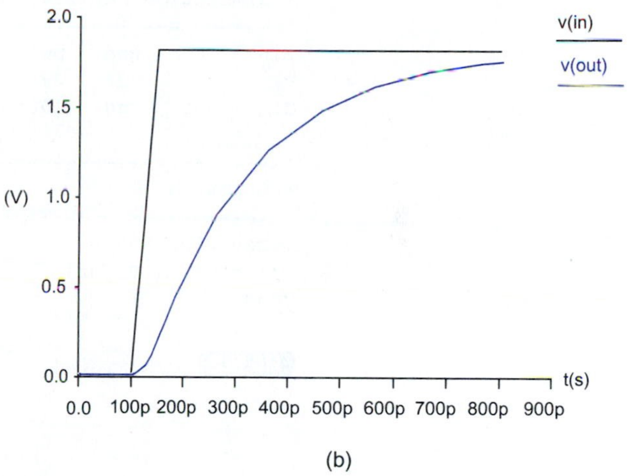

(a)

(b)

FIG 5.1 RC circuit response

The voltage source is defined as a piecewise linear (PWL) source. The waveform is specified with an arbitrary number of (time, voltage) pairs. Other common sources include DC sources and pulse sources. A DC voltage source named `Vdd` that sets node `vdd` to 2.5 V could be expressed as:

```
Vdd    vdd    gnd    2.5
```

Pulse sources are convenient for repetitive signals like clocks. The general form for a pulse source is illustrated in Figure 5.3. For example, a clock with a 1.8 V swing, 800 ps period, 100 ps rise and fall times, and 50% duty cycle (i.e., equal high and low times) would be expressed as

```
Vck    clk    gnd    PULSE 0 1.8 0ps 100ps 100ps 300ps 800ps
```

```
* rc.sp
* David_Harris@hmc.edu 2/2/03
* Find the response of RC circuit to rising input


*-------------------------------------------------------------------
* Parameters and models
*-------------------------------------------------------------------
.option post


*-------------------------------------------------------------------
* Simulation netlist
*-------------------------------------------------------------------
Vin    in     gnd    pwl    0ps 0 100ps 0 150ps 1.8 800ps 1.8
R1     in     out    2k
C1     out    gnd    100f


*-------------------------------------------------------------------
* Stimulus
*-------------------------------------------------------------------
.tran 20ps 800ps
.plot v(in) v(out)
.end
```

FIG 5.2 RC spice deck

PULSE v1 v2 td tr tf pw per

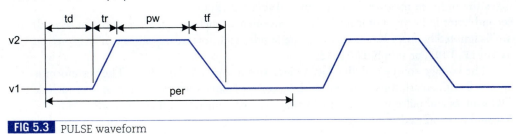

FIG 5.3 PULSE waveform

The stimulus specifies that a transient analysis (.tran) should be performed using a maximum step size of 20 ps for a duration of 800 ps. When plotting node voltages, the step size determines the spacing between points.

The .plot command generates a textual plot of the node variables specified (in this case the voltages at nodes in and out), as shown in Figure 5.4. Similarly, the .print statement prints the results in a multicolumn table. Both commands show the legacy of

Table 5.1	Common SPICE elements
Letter	**Element**
R	Resistor
C	Capacitor
L	Inductor
K	Mutual inductor
V	Independent voltage source
I	Independent current source
M	MOSFET
D	Diode
Q	Bipolar transistor
W	Lossy transmission line
X	Subcircuit
E	Voltage-controlled voltage source
G	Voltage-controlled current source
H	Current-controlled voltage source
F	Current-controlled current source

Table 5.2	SPICE units	
Letter	**Unit**	**Magnitude**
a	atto	10^{-18}
f	femto	10^{-15}
p	pico	10^{-12}
n	nano	10^{-9}
u	micro	10^{-6}
m	milli	10^{-3}
k	kilo	10^{3}
x	mega	10^{6}
g	giga	10^{9}

FORTRAN and line printers. On modern computers with graphical user interfaces, the `.option post` command is usually preferred. It generates a file (in this case, `rc.tr0`) containing the results of the specified (transient) analysis. Then a separate graphical waveform viewer can be used to look at and manipulate the waveforms. AWAVES and COSMOSCOPE are waveform viewers compatible with HSPICE.

```
legend:
  a: v(in)
  b: v(out)

     time        v(in)
  (ab      )       0.              500.0000m        1.0000          1.5000          2.0000
                         +            +              +               +               +
     0.       0.       - 2-------+-------+-------+-------+-------+-------+-------+-------+--
    20.0000p  0.         2        +       +       +       +       +       +       +       +
    40.0000p  0.         2        +       +       +       +       +       +       +       +
    60.0000p  0.         2        +       +       +       +       +       +       +       +
    80.0000p  0.         2        +       +       +       +       +       +       +       +
   100.0000p  0.         2        +       +       +       +       +       +       +       +
   120.0000p  720.000m +b         +       +      a+       +       +       +       +       +
   140.0000p  1.440     +    b    +       +       +       +       +    a  +       +       +
   160.0000p  1.800     +      +b +       +       +       +       +       +a      +       +
   180.0000p  1.800     +      +    b +   +       +       +       +       +a      +       +
   200.0000p  1.800     -+-------+-------b-----+-------+-------+-------+-------+a-----+--
   220.0000p  1.800     +        +       +   b +       +       +       +       +a      +
   240.0000p  1.800     +        +       +  +b   +       +       +       +a      +
   260.0000p  1.800     +        +       +   +  b  +       +       +       +a      +
   280.0000p  1.800     +        +       +   +   b+      +       +       +a      +
   300.0000p  1.800     +        +       +   +   +b      +       +       +a      +
   320.0000p  1.800     +        +       +   +   +   b   +       +       +a      +
   340.0000p  1.800     +        +       +   +   +     b +       +       +a      +
   360.0000p  1.800     +        +       +   +   +      b        +       +a      +
   380.0000p  1.800     +        +       +   +   +      +b       +       +a      +
   400.0000p  1.800     -+-------+-------+-------+-------+---b---+-------+a-----+--
   420.0000p  1.800     +        +       +       +       +       +  b    +       +a      +
   440.0000p  1.800     +        +       +       +       +       +   b   +       +a      +
   460.0000p  1.800     +        +       +       +       +       +  b+    +a      +
   480.0000p  1.800     +        +       +       +       +       +   b    +a      +
   500.0000p  1.800     +        +       +       +       +       +    +b  +a      +
   520.0000p  1.800     +        +       +       +       +       +    +b  +a      +
   540.0000p  1.800     +        +       +       +       +       +    + b +a      +
   560.0000p  1.800     +        +       +       +       +       +    +  b+a      +
   580.0000p  1.800     +        +       +       +       +       +    +  b+a      +
   600.0000p  1.800     -+-------+-------+-------+-------+-------+---b--+-a-----+--
   620.0000p  1.800     +        +       +       +       +       +    +  b +a      +
   640.0000p  1.800     +        +       +       +       +       +    +   b+a      +
   660.0000p  1.800     +        +       +       +       +       +    +   b+a      +
   680.0000p  1.800     +        +       +       +       +       +    +   b+a      +
   700.0000p  1.800     +        +       +       +       +       +    +   b+a      +
   720.0000p  1.800     +        +       +       +       +       +    +   b+a      +
   740.0000p  1.800     +        +       +       +       +       +    +   b+a      +
   760.0000p  1.800     +        +       +       +       +       +    +   b+a      +
   780.0000p  1.800     +        +       +       +       +       +        ba      +
   800.0000p  1.800     -+-------+-------+-------+-------+-------+-------ba-----+--
                         +            +              +               +               +
```

FIG 5.4 Textual plot of RC circuit response

5.2.2 Transistor DC Analysis

One of the first steps in becoming familiar with a new CMOS process is to look at the I-V characteristics of the transistors. Figure 5.5(a) shows test circuits for a unit (4/2 λ) nMOS transistor in a 180 nm process at $V_{DD} = 1.8$ V. The I-V characteristics are plotted in Figure 5.5(b) using the SPICE deck in Figure 5.6.

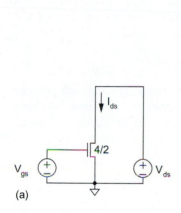

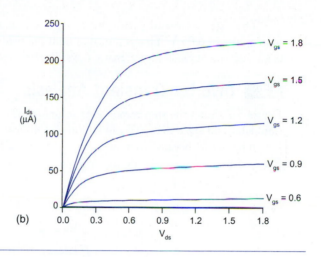

(a)   (b)

FIG 5.5 MOS I-V characteristics

```
* mosiv.sp


*-----------------------------------------------------------------
* Parameters and models
*-----------------------------------------------------------------
.include '../models/tsmc180/models.sp'
.temp 70
.option post


*-----------------------------------------------------------------
* Simulation netlist
*-----------------------------------------------------------------
*nmos
Vgs    g    gnd    0
Vds    d    gnd    0
M1     d    g      gnd    gnd    NMOS    W=0.36u   L=0.18u


*-----------------------------------------------------------------
* Stimulus
*-----------------------------------------------------------------
.dc Vds 0 1.8 0.05 SWEEP Vgs 0 1.8 0.3
.end
```

FIG 5.6 MOSIV SPICE deck

.include reads another SPICE file from disk. In this example, it loads device models that will be discussed further in Section 5.3. The circuit uses two independent voltage sources with default values of 0 V; these voltages will be varied by the .dc command. The nMOS transistor is defined with the MOSFET element M using the syntax

```
Mname   drain   gate   source   body   type   W=<width>   L=<length>
```

The .dc command varies the voltage source Vds DC voltage from 0 to 1.8 V in increments of 0.05 V. This is repeated multiple times as Vgs is swept from 0 to 1.8 V in 0.3 V increments to compute many I_{ds} vs. V_{ds} curves at different values of V_{gs}.

5.2.3 Inverter Transient Analysis

Figure 5.7 shows the step response of an unloaded unit inverter, annotated with propagation delay and 20%–80% rise and fall times. Observe that significant overshoot from bootstrapping occurs because there is no load. The SPICE deck for the simulation is shown in Figure 5.8.

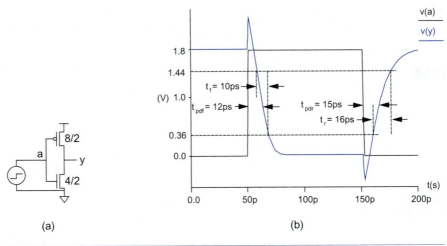

(a)   (b)

FIG 5.7 Unloaded inverter

This deck introduces the use of parameters and scaling. The .param statement defines a parameter named SUPPLY to have a value of 1.8. This is then used to set Vdd and the amplitude of the input pulse. If we wanted to evaluate the response at a different supply voltage, we would simply need to change the .param statement. The .scale sets a scale factor for all dimensions that would by default be measured in meters. In this case, it sets the scale to λ = 90 nm. Now the transistor widths and lengths in the inverter are specified in terms of lambda rather than in meters.

```
* inv.sp

*----------------------------------------------------------------
* Parameters and models
*----------------------------------------------------------------
.param SUPPLY=1.8
.option scale=90n
.include '../models/tsmc180/models.sp'
.temp 70
.option post


*----------------------------------------------------------------
* Simulation netlist
*----------------------------------------------------------------
Vdd    vdd    gnd    'SUPPLY'
Vin    a      gnd    PULSE  0 'SUPPLY' 50ps 0ps 0ps 100ps 200ps
M1     y      a      gnd    gnd    NMOS    W=4    L=2
+ AS=20 PS=18 AD=20 PD=18
M2     y      a      vdd    vdd    PMOS    W=8    L=2
+ AS=40 PS=26 AD=40 PD=26


*----------------------------------------------------------------
* Stimulus
*----------------------------------------------------------------
.tran 1ps 200ps
.end
```

FIG 5.8   INV SPICE deck

Recall that parasitic delay is strongly dependent on diffusion capacitance, which in turn depends on the area and perimeter of the source and drain. As each diffusion region in an inverter must be contacted, the geometry resembles that of Figure 2.9(a). The diffusion width equals the transistor width and the diffusion length is 5 λ. Thus, the area of the source and drain are AS = AD = $5W\lambda^2$ and the perimeters are PS = PD = $(2W + 10)\lambda$. Note that the + sign in the first column of a card indicates that it is a continuation of the previous card. These dimensions are also affected by the scale factor.

5.2.4   Subcircuits and Measurement

One of the simplest measures of a process's inherent speed is the fanout-of-4 inverter delay. Figure 5.9(a) shows a circuit to measure this delay. The nMOS and pMOS transistor sizes (in multiples of a unit 4/2 λ transistor) are listed below and above each gate, respectively. $X3$ is the inverter under test and $X4$ is its load, which is four times larger than $X3$. To first order, these two inverters would be sufficient. However, the delay of $X3$ also depends on the input slope, as discussed in Section 4.2.5.1. One way to obtain a realistic input slope is to drive node c with a pair of FO4 inverters $X1$ and $X2$. Also, as discussed in Section 4.2.5.4, the input capacitance of $X4$ depends not just on its C_{gs} but also on C_{gd}. C_{gd} is Miller-multiplied as node e switches and would be effectively doubled if e switched instantaneously. When e is loaded with *X5*, it switches at a slower, more realistic rate, slightly reducing the effective capacitance presented at node d by *X4*. The waveforms in Figure 5.9(b) are annotated with the rising and falling delays.

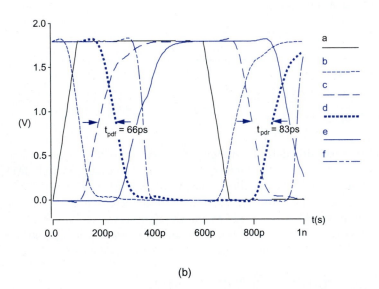

(a)

(b)

FIG 5.9   Fanout-of-4 inverters

SPICE decks are easier to read and maintain when common circuit elements are captured as subcircuits. For example, the deck in Figure 5.10 computes the FO4 inverter delay using an inverter subcircuit.

The `.global` statement defines vdd and gnd as global nodes that can be referenced from within subcircuits. The inverter is declared as a subcircuit with two terminals: a and y. It also accepts two parameters specifying the width of the nMOS and pMOS transis-

```
* fo4.sp
*-----------------------------------------------------------------
* Parameters and models
*-----------------------------------------------------------------
.param SUPPLY=1.8
.param H=4
.option scale=90n
.include '../models/tsmc180/models.sp'
.temp 70
.option post
*-----------------------------------------------------------------
* Subcircuits
*-----------------------------------------------------------------
.global vdd gnd
.subckt inv a y N=4 P=8
M1     y     a     gnd     gnd     NMOS     W='N'     L=2
+ AS='N*5' PS='2*N+10' AD='N*5' PD='2*N+10'
M2     y     a     vdd     vdd     PMOS     W='P'     L=2
+ AS='P*5' PS='2*P+10' AD='P*5' PD='2*P+10'
.ends
*-----------------------------------------------------------------
* Simulation netlist
*-----------------------------------------------------------------
Vdd    vdd    gnd     'SUPPLY'
Vin    a      gnd     PULSE    0 'SUPPLY' 0ps 100ps 100ps 500ps 1000ps
X1     a      b       inv                     * shape input waveform
X2     b      c       inv      M='H'          * reshape input waveform
X3     c      d       inv      M='H**2'       * device under test
X4     d      e       inv      M='H**3'       * load
X5     e      f       inv      M='H**4'       * load on load
*-----------------------------------------------------------------
* Stimulus
*-----------------------------------------------------------------
.tran 1ps 1000ps
.measure tpdr                               * rising prop delay
+     TRIG v(c)     VAL='SUPPLY/2' FALL=1
+     TARG v(d)     VAL='SUPPLY/2' RISE=1
.measure tpdf                               * falling prop delay
+     TRIG v(c)     VAL='SUPPLY/2' RISE=1
+     TARG v(d)     VAL='SUPPLY/2' FALL=1
.measure tpd param='(tpdr+tpdf)/2'          * average prop delay
.measure trise                              * rise time
+     TRIG v(d)     VAL='0.2*SUPPLY' RISE=1
+     TARG v(d)     VAL='0.8*SUPPLY' RISE=1
.measure tfall                              * fall time
+     TRIG v(d)     VAL='0.8*SUPPLY' FALL=1
+     TARG v(d)     VAL='0.2*SUPPLY' FALL=1
.end
```

FIG 5.10 FO4 SPICE deck

tors; these parameters have default values of 4 and 8, respectively. The source and drain area and perimeter are functions of the transistor widths. HSPICE evaluates functions given inside single quotation marks. The functions can include parameters, constants, parentheses, +, -, *, /, and ** (raised to a power).

The simulation netlist contains the power supply, input waveform, and five inverters. Each inverter is a subcircuit (X) element. As N and P are not specified, each uses the default size. The M parameter multiplies all the currents in the subcircuit by the factor given, equivalent to M elements wired in parallel. In this case, the fanouts are expressed in terms of a parameter H. Thus, *X2* has the capacitance and output current of 4 unit inverters, while *X3* is equivalent to 16. Another way to model the inverters would have been to use the N and P parameters:

```
X1    a    b    inv    N=4       P=8       * shape input waveform
X2    b    c    inv    N=16      P=32      * reshape input waveform
X3    c    d    inv    N=64      P=128     * device under test
X4    d    e    inv    N=256     P=512     * load
X5    e    f    inv    N=1024    P=2048    * load on load
```

However, a transistor of four times unit width does not have exactly the same input capacitance or output current as four unit inverters tied in parallel, so the M parameter is usually preferred.

In this example, the subcircuit declaration and simulation netlist are part of the SPICE deck. When working with a standard cell library, it is common to keep subcircuit declarations in their own files and reference them with a .include statement instead. When the simulation netlist is extracted from a schematic or layout CAD system, it is common to put the netlist in a separate file and .include it as well.

The .measure statement measures simulation results and prints them in the listing file. The deck measures the rising propagation delay t_{pdr} as the difference between the time that the input c first falls through $V_{DD}/2$ and the time that the output d first rises through $V_{DD}/2$. TRIG and TARG indicate the trigger and target events between which delay is measured. The .measure statement can also be used to compute functions of other measurements. For example, the average FO4 inverter propagation delay t_{pd} is the mean of t_{pdr} and t_{pdf}, i.e., t_{pd} = 75 ps. The 20%–80% rise time is t_r = 94 ps and the fall time is t_f = 67 ps.

5.2.5 Optimization

In many examples, we have assumed that a *P/N* ratio of 2:1 gives approximately equal rise and fall delays. The FO4 inverter simulation showed that a ratio of 2:1 gives rising delays that are slower than the falling delays because the pMOS mobility is less than half that of the nMOS. You could repeatedly run simulations with different default values of P to find the ratio for equal delay. HSPICE has built-in optimization capabilities that will automatically tweak parameters to achieve some goal and report what parameter value gave the best results. Figure 5.11 shows a modified version of the FO4 inverter simulation using the optimizer.

```
* fo4opt.sp
*-----------------------------------------------------------------
* Parameters and models
*-----------------------------------------------------------------
.param SUPPLY=1.8
.option scale=90n
.include '../models/tsmc180/models.sp'
.temp 70
.option post
*-----------------------------------------------------------------
* Subcircuits
*-----------------------------------------------------------------
.global vdd gnd
.subckt inv a y N=4 P=8
M1   y   a   gnd   gnd   NMOS   W='N'   L=2
+ AS='N*5' PS='2*N+10' AD='N*5' PD='2*N+10'
M2   y   a   vdd   vdd   PMOS   W='P'   L=2
+ AS='P*5' PS='2*P+10' AD='P*5' PD='2*P+10'
.ends
*-----------------------------------------------------------------
* Simulation netlist
*-----------------------------------------------------------------
Vdd   vdd   gnd   'SUPPLY'
Vin   a   gnd   PULSE   0 'SUPPLY' 0ps 100ps 100ps 500ps 1000ps
X1   a   b   inv   P='P1'              * shape input waveform
X2   b   c   inv   P='P1'    M=4       * reshape input waveform
X3   c   d   inv   P='P1'    M=16      * device under test
X4   d   e   inv   P='P1'    M=64      * load
X5   e   f   inv   P='P1'    M=256     * load on load
*-----------------------------------------------------------------
* Optimization setup
*-----------------------------------------------------------------
.param P1=optrange(8,4,16)         * search from 4 to 16, guess 8
.model optmod opt itropt=30        * maximum of 30 iterations
.measure bestratio param='P1/4'    * compute best P/N ratio
*-----------------------------------------------------------------
* Stimulus
*-----------------------------------------------------------------
.tran 1ps 1000ps SWEEP OPTIMIZE=optrange RESULTS=diff MODEL=optmod
.measure tpdr                        * rising propagation delay
+    TRIG v(c)   VAL='SUPPLY/2' FALL=1
+    TARG v(d)   VAL='SUPPLY/2' RISE=1
.measure tpdf                        * falling propagation delay
+    TRIG v(c)   VAL='SUPPLY/2' RISE=1
+    TARG v(d)   VAL='SUPPLY/2' FALL=1
.measure tpd param='(tpdr+tpdf)/2' goal=0  * average prop delay
.measure diff param='tpdr-tpdf' goal = 0   * diff between delays
.end
```

FIG 5.11 FO4OPT SPICE deck

The subcircuits *X1–X4* override their default pMOS widths to use a width of `P1` instead. In the optimization setup, the difference of t_{pdr} and t_{pdf} is measured. The goal of the optimization will be to drive this difference to 0. To do this, `P1` may be varied from 4 to 16, with an initial guess of 8. The optimizer may use up to 30 iterations to find the best value of `P1`. Because the nMOS width is fixed at 4, the best *P/N* ratio is computed as `P1/4`. The transient analysis includes a `SWEEP` statement containing the parameter to vary, the desired result, and the number of iterations.

HSPICE determines that the *P/N* ratio for equal rise and fall delay is 3.6:1, giving a rising and falling delay of 84 ps. This is slower than the 2:1 ratio provides and requires large pMOS transistors that consume area and power, so such a high ratio is seldom used.

A similar scenario is to find the *P/N* ratio that gives lowest average delay. By changing the `.tran` card to use `RESULTS=tpd`, we find a best ratio of 1.4:1 with rising, falling, and average propagation delays of 87, 59, and 73 ps, respectively. Whenever you do an optimization, it is important to consider not only the optimum but also the sensitivity to deviations from this point. Further simulation finds that *P/N* ratios of anywhere from 1.2:1 to 1.7:1 all give an average propagation delay of 73 ps, there is no need to slavishly stick to the 1.4:1 "optimum." The best *P/N* ratio in practice is a compromise between using smaller pMOS devices to save area and power and using larger devices to achieve more nearly equal rise/fall times and avoid the hot electron reliability problems induced by very slow rising edges in circuits with weak pMOS transistors. *P/N* ratios are discussed further in Section 6.2.1.6.

5.2.6 Other HSPICE Commands

The full HSPICE manual fills over 2000 pages and includes many more capabilities than can be described here. A few of the most useful additional commands are covered in this section. Section 5.3 describes transistor models and library calls, while Section 5.6 discusses modeling interconnect with lossy transmission lines.

`.option accurate`

Tighten integration tolerances to obtain more accurate results. Useful for oscillators and high-gain analog circuits or when results seem fishy.

`.option autostop`

Conclude simulation when all `.measure` results are obtained rather than continuing for the full duration of the `.tran` card. This can substantially reduce simulation time.

`.temp 0 70 125`

Repeat the simulation three times at temperatures of 0°, 70°, and 125° C. Device models may contain information about how changing temperature changes device performance.

`.op`

Print the voltages, currents, and transistor bias conditions at the DC operating point.

5.3 Device Models

Most of the examples in Section 5.2 included a file containing transistor models. SPICE provides a wide variety of MOS transistor models with various tradeoffs between complexity and accuracy. Level 1 and Level 3 models were historically important, but they are no longer adequate to accurately model very small modern transistors. BSIM models are more accurate and are presently the most widely used. Some companies use their own proprietary models. This section briefly describes the main features of each of these models. It also describes how to model diffusion capacitance and how to run simulations in various process corners. The model descriptions are intended only as an overview of the capabilities and limitations of the models; refer to a SPICE manual for a much more detailed description if one is necessary.

5.3.1 Level 1 Models

The SPICE Level 1, or Shichman–Hodges Model [Shichman68] is closely related to the Shockley model described in EQ (2.10), enhanced with channel length modulation and the body effect. The basic current model is:

$$I_{ds} = \begin{cases} 0 & V_{gs} < V_t & \text{cutoff} \\ \text{KP}\dfrac{W_{\text{eff}}}{L_{\text{eff}}}\left(1 + \text{LAMBDA} \cdot V_{ds}\right)\left(V_{gs} - V_t - \dfrac{V_{ds}}{2}\right)V_{ds} & V_{ds} < V_{gs} - V_t & \text{linear} \\ \dfrac{\text{KP}}{2}\dfrac{W_{\text{eff}}}{L_{\text{eff}}}\left(1 + \text{LAMBDA} \cdot V_{ds}\right)\left(V_{gs} - V_t\right)^2 & V_{ds} > V_{gs} - V_t & \text{saturation} \end{cases} \tag{5.1}$$

The parameters from the SPICE model are given in ALL CAPS. Notice that β is written instead as KP($W_{\text{eff}}/L_{\text{eff}}$), where KP is a model parameter playing the role of k' from EQ (2.7). W_{eff} and L_{eff} are the effective width and length, as described in Section 2.4.8. The LAMBDA term models channel length modulation (see Section 2.4.2).

The threshold voltage is modulated by the source-to-body voltage V_{sb} through the body effect (see Section 2.4.3). For nonnegative V_{sb}, the threshold voltage is

$$V_t = \text{VTO} + \text{GAMMA}\left(\sqrt{\text{PHI} + V_{sb}} - \sqrt{\text{PHI}}\right) \tag{5.2}$$

Notice that this is identical to EQ (2.30), where VTO is the "zero-bias" threshold voltage V_{t0}, GAMMA is the body effect coefficient γ, and PHI is the surface potential ϕ_s.

The gate capacitance is calculated from the oxide thickness TOX. The default gate capacitance model in HSPICE is adequate for finding the transient response of digital circuits. More elaborate models exist that capture nonreciprocal effects that are important for analog design.

Level 1 models are useful for teaching because they are easy to correlate with hand analysis, but are too simplistic for modern design. Figure 5.12 gives an example of a Level 1 model card illustrating the card syntax. The card also includes terms to compute the diffusion capacitance, as described in Section 5.3.4.

```
.model NMOS NMOS (LEVEL=1 TOX=40e-10 KP=155E-6 LAMBDA=0.2
+                 VTO=0.4 PHI=0.93 GAMMA=0.6
+                 CJ=9.8E-5 PB=0.72 MJ=0.36
+                 CJSW=2.2E-10 PHP=7.5 MJSW=0.1)
```

FIG 5.12   Sample Level 1 MODEL card

5.3.2 Level 2 and 3 Models

The SPICE Level 2 and 3 models add effects of velocity saturation, mobility degradation, subthreshold conduction, and drain-induced barrier lowering. The Level 2 model is based on the Grove-Frohman equations [Frohman69], while the Level 3 model is based on empirical equations that provide similar accuracy and faster simulation times and better convergence. However, these models still do not provide good fits to the measured I-V characteristics of modern transistors.

5.3.3 BSIM Models

The Berkeley Short-Channel IGFET[1] Model (BSIM) is a very elaborate model that is now widely used in circuit simulation. The models are derived from the underlying device physics but use an enormous number of parameters to fit the behavior of modern transistors. BSIM versions 1, 2, 3v3, and 4 are implemented as SPICE levels 13, 39, 49, and 54, respectively.

BSIM version 3v3 requires an entire book [Cheng99] to describe the model. It includes over 100 parameters and the device equations span 27 pages. It is quite good for digital circuit simulation except that it does not model gate leakage. Features of the model include:

- Continuous and differentiable I-V characteristics across subthreshold, linear, and saturation regions for good convergence

- Sensitivity of parameters such as V_t to transistor length and width

- Detailed threshold voltage model including body effect and drain-induced barrier lowering

- Velocity saturation, mobility degradation, and other short-channel effects

- Multiple gate capacitance models

- Diffusion capacitance and resistance models

---

[1]IGFET in turn stands for Insulated-Gate Field Effect Transistor, a synonym for MOSFET.

BSIM version 4 was fairly new at the time this book was being written and adds support for gate leakage and other effects of very thin gates. As these effects are rapidly becoming more important and better characterized, the designer should check with the process engineers before blindly trusting gate leakage models.

Some device parameters such as threshold voltage change significantly with device dimensions. BSIM models can be *binned* with different models covering different ranges of length and width specified by LMIN, LMAX, WMIN, and WMAX parameters. For example, one model might cover transistors with channel lengths from 0.18–0.25 μm, another from 0.25–0.5 μm, and a third from 0.5–5 μm. SPICE will complain if a transistor does not fit in one of the bins.

As the BSIM models are so complicated, it is impractical to derive closed-form equations for propagation delay, switching threshold, noise margins, etc., from the underlying equations. However, it is not difficult to find these properties through circuit simulation. Section 5.4 will show simple simulations to plot the device characteristics over the regions of operation that are interesting to most digital designers and to extract effective capacitance and resistance averaged across the switching transition. The simple RC model continues to give the designer important insight about the characteristics of logic gates.

5.3.4 Diffusion Capacitance Models

The p–n junction between the source or drain diffusion and the body forms a reverse-biased diode. We have seen that the diffusion capacitance determines the parasitic delay of a gate and depends on the area and perimeter of the diffusion. HSPICE provides a number of methods to specify this geometry, controlled by the ACM (Area Calculation Method) parameter, which is part of the transistor model card. The model card must also have values for junction and sidewall diffusion capacitance, as described in Section 2.2.2.3. The diffusion capacitance model is common across most device models including Levels 1–3 and BSIM.

By default, HSPICE models use ACM=0. In this method, the designer must specify the area and perimeter of the source and drain of each transistor. For example, the dimensions of each diffusion region from Figure 2.9 is listed in Table 5.3 (in units of λ^2 for area or λ for perimeter). A SPICE description of the shared contacted diffusion case is shown in Figure 5.13, assuming .option scale is set to the value of λ.

Table 5.3	Diffusion area and perimeter			
	AS1 / AD2	**PS1 / PD2**	**AD1 / AS2**	**PD1 / PS2**
(a) Isolated contacted diffusion	$W \cdot 5$	$2 \cdot W + 10$	$W \cdot 5$	$2 \cdot W + 10$
(b) Shared contacted diffusion	$W \cdot 5$	$2 \cdot W + 10$	$W \cdot 3$	$W + 6$
(c) Merged uncontacted diffusion	$W \cdot 5$	$2 \cdot W + 10$	$W \cdot 1.5$	$W + 3$

```
* (b): Shared contacted diffusion
M1   mid   b   bot   gnd   NMOS   W='w'   L=2
+ AS='w*5'  PS='2*w+10'  AD='w*3'  PD='w+6'
M2   top   a   mid   gnd   NMOS   W='w'   L=2
+ AS='w*3'  PS='w+6'  AD='w*5'  PD='2*w+10'
```

FIG 5.13 SPICE model of transistors with shared contacted diffusion

The SPICE models also should contain parameters CJ, CJSW, PB, PHP, MJ, and MJSW. Assuming the diffusion is reverse-biased and the area and perimeter are specified, the diffusion capacitance between source and body is computed as described in Section 2.3.3.

$$C_{sb} = \text{AS} \cdot \text{CJ} \cdot \left(1 + \frac{V_{sb}}{\text{PB}}\right)^{-MJ} + \text{PS} \cdot \text{CJSW} \cdot \left(1 + \frac{V_{sb}}{\text{PHP}}\right)^{-MJSW} \tag{5.3}$$

The drain equations are analogous, with S replaced by D in the model parameters.

The BSIM3 models offer a similar area calculation model (ACM = 10) that takes into account the different sidewall capacitance on the edge adjacent to the gate. Note that the PHP parameter is renamed to PBSW to be more consistent.

$$C_{sb} = \text{AS} \cdot \text{CJ} \cdot \left(1 + \frac{V_{sb}}{\text{PB}}\right)^{-MJ} + (\text{PS} - W) \cdot \text{CJSW} \cdot \left(1 + \frac{V_{sb}}{\text{PBSW}}\right)^{-MJSW} + $$
$$W \cdot \text{CJSWG} \cdot \left(1 + \frac{V_{sb}}{\text{PBSWG}}\right)^{-MJSWG} \tag{5.4}$$

If the area and perimeter are not specified, they default to 0 in ACM = 0 or 10, grossly underestimating the parasitic delay of the gate. HSPICE also supports ACM = 1, 2, 3, and 12 that provide nonzero default values when the area and perimeter are not specified. Check your models and read the HSPICE documentation carefully.

The diffusion area and perimeter is also used to compute the junction leakage current. However, this current is generally negligible compared to subthreshold leakage in modern devices.

5.3.5 Design Corners

Engineers often simulate circuits in multiple design corners to verify that operation across variations in device characteristics and environment. HSPICE includes the .lib card that makes changing libraries easy. For example, the deck in Figure 5.14 runs three simulations on the step response of an unloaded inverter in the TT, FF, and SS corners.

```
* corner.sp
* Step response of unloaded inverter across process corners


*-------------------------------------------------------------------
* Parameters and models
*-------------------------------------------------------------------
.option scale=90n
.param SUP=1.8 * Must set before calling .lib
.lib '../models/tsmc180/opconditions.lib' TT
.option post


*-------------------------------------------------------------------
* Simulation netlist
*-------------------------------------------------------------------
Vdd    vdd    gnd    'SUPPLY'
Vin    a      gnd    PULSE      0 'SUPPLY' 200ps 0ps 0ps 500ps 1000ps
M1     y      a      gnd        gnd        NMOS    W=4    L=2
+ AS=20 PS=18 AD=20 PD=18
M2     y      a      vdd        vdd        PMOS    W=8    L=2
+ AS=40 PS=26 AD=40 PD=26


*-------------------------------------------------------------------
* Stimulus
*-------------------------------------------------------------------
.tran 1ps 1000ps
.alter
.lib '../models/tsmc180/opconditions.lib' FF
.alter
.lib '../models/tsmc180/opconditions.lib' SS
.end
```

FIG 5.14 CORNER SPICE deck

The deck first sets SUP to the nominal supply voltage of 1.8 V. It then invokes the
.lib card that reads in the library specifying the TT conditions. In the stimulus, the
.alter statement is used to repeat the simulation with changes. In this case, the design
corner is changed. Altogether, three simulations are performed and three sets of waveforms are generated for the three design corners.

The library file is given in Figure 5.15. Depending on what library was specified, the
temperature is set (in degrees Celsius, with .temp) and the V_{DD} value SUPPLY is calculated from the nominal SUP. The library loads the appropriate nMOS and pMOS transistor models. A fast process file might have lower nominal threshold voltages V_{t0}, greater
lateral diffusion L_D, and lower diffusion capacitance values.

```
* opconditions.lib
* For TSMC 180 nm process

* TT: Typical nMOS, pMOS, voltage, temperature
.lib TT
.temp 70
.param SUPPLY='SUP'
.include 'modelsTT.sp'
.endl TT

* SS: Slow nMOS, pMOS, low voltage, high temperature
.lib SS
.temp 125
.param SUPPLY='0.9 * SUP'
.include 'modelsSS.sp'
.endl SS

* FF: Fast nMOS, pMOS, high voltage, low temperature
.lib FF
.temp 0
.param SUPPLY='1.1 * SUP'
.include 'modelsFF.sp'
.endl FF

* FS: Fast nMOS, Slow pMOS, typical voltage and temperature
.lib FS
.temp 70
.param SUPPLY='SUP'
.include 'modelsFS.sp'
.endl FS

* SF: Slow nMOS, Fast pMOS, typical voltage and temperature
.lib SF
.temp 70
.param SUPPLY='SUP'
.include 'modelsSF.sp'
.endl SF
```

FIG 5.15 OPCONDITIONS library

5.4 Device Characterization

Modern SPICE models have so many parameters that the designer cannot easily read key performance characteristics from the model files. A more convenient approach is to run a set of simulations to extract the effective resistance and capacitance, the fanout-of-4 inverter delay, the I-V characteristics, and other interesting data. This section describes these simulations and compares the results across a variety of CMOS processes.

5.4.1 I-V Characteristics

When familiarizing yourself with a new process, a starting point is to plot the current-voltage (I-V) characteristics. Although digital designers seldom make calculations directly from these plots, it is helpful to know the ON current of nMOS and pMOS transistors, how severely velocity-saturated the process is, how the current rolls off below threshold, how the devices are affected by DIBL and body effect, and so forth. These plots are made with DC sweeps, as discussed in Section 5.2.2. Each transistor is 1 μm wide in a representative 180 nm process at 70° C with $V_{DD} = 1.8$ V. The left column shows nMOS behavior and the right column shows pMOS behavior.

Figure 5.16(a) plots I_{ds} vs. V_{ds} at various values of V_{gs}, as was done in Figure 5.5. The saturation current would ideally increase quadratically with $V_{gs} - V_t$, but in this plot shows closer to a linear dependence, indicating that the nMOS transistor is severely velocity-saturated (α closer to 1 than 2 in the α-power model). The increase in saturation current with V_{ds} is caused by channel length modulation. The saturation current for a pMOS transistor is lower than for the nMOS (note the different vertical scales), but the device is not as velocity-saturated. Figure 5.16(b) makes a similar plot for a device with a drawn channel length of 360 nm rather than 180 nm. The current is slightly flatter in saturation, indicating that channel length modulation has less impact at longer channel lengths.

Figure 5.16(c) plots I_{ds} vs. V_{gs} of an nMOS transistor on a semilogarithmic scale for $V_{ds} = 0.1$ V and 1.8 V. The straight line at low V_{gs} indicates that the current rolls off exponentially below threshold. The difference in subthreshold leakage at the varying drain voltage reflects the effects of drain-induced barrier lowering (DIBL) effectively reducing V_t at high V_{ds}. The saturation current I_{dsat} is measured at $V_{gs} = V_{ds} = V_{DD}$, while the OFF current I_{off} is measured at $V_{gs} = 0$ and $V_{ds} = V_{DD}$. The subthreshold slope is about 90 mV/decade and DIBL reduces the effective threshold voltage by about 40 mV over the range of V_{ds}.

Figure 5.16(d) makes a similar plot on a linear scale for $V_{bs} = -0.2$, 0, and 0.2 V. The curves shift horizontally, indicating that the body effect increases the threshold voltage as V_{bs} becomes more negative. V_{ds} is held constant at 0.1 V.

5.4.2 Threshold Voltage

In the Shockley model, the threshold voltage V_t is defined as the value of V_{gs} below which I_{ds} becomes 0. In the real transistor characteristics shown in Figure 5.16(c), subthreshold current continues to flow for $V_{gs} < V_t$, so measuring or even defining the threshold voltage becomes problematic. Moreover, the threshold voltage varies with L, W, V_{ds}, and V_{bs}. At least eleven different methods have been used in the literature to determine the threshold voltage from measured I_{ds}–V_{gs} data [Ortiz-Conde02]. This section will explore two common methods (constant current and linear extrapolation) and a hybrid that combines the advantages of each.

The *constant current* method defines threshold as the gate voltage at a given drain current I_{crit}. This method is easy to use, but depends on an arbitrary choice of critical drain current. A typical choice of I_{crit} is 0.1 μA • (W/L). Figure 5.17 shows how the extracted threshold voltage varies with the choice of $I_{crit} = 0.1$ or 1 μA at $V_{ds} = 100$ mV.

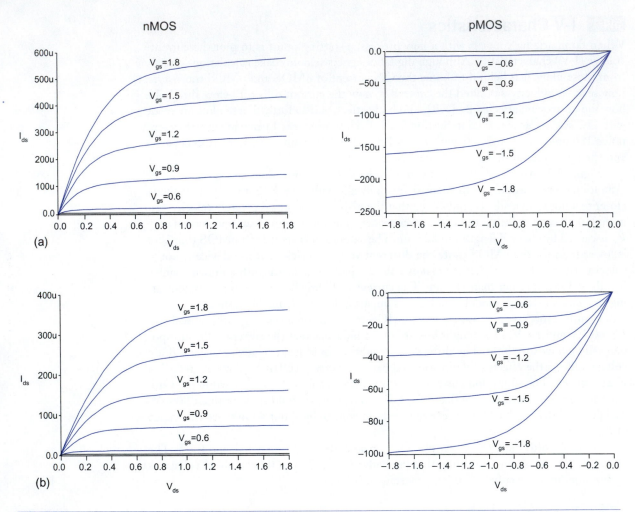

nMOS

pMOS

(a)

(b)

FIG 5.16 MOS I-V characteristics

The *linear extrapolation* (or *maximum-g_m*) method extrapolates the gate voltage from the point of maximum slope on the I_{ds}–V_{gs} characteristics. It is unambiguous but valid only for the linear region of operation (low V_{ds}) because of the series resistance of the source/drain diffusion and because drain-induced barrier lowering effectively reduces the threshold at high V_{ds}. Figure 5.18 shows how the threshold is extracted from measured data using the linear extrapolation method at $V_{ds} = 100$ mV. Observe that this method can give a significantly different threshold voltage and current at threshold, so it is important to check how the threshold voltage was measured when interpreting threshold voltage specifications. I_{crit} is defined to be the value of I_{ds} at $V_{gs} = V_t$.

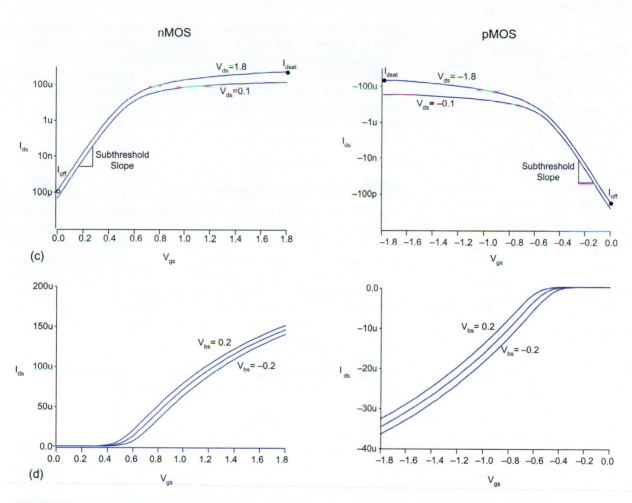

FIG 5.16 MOS I-V characteristics (continued)

[Zhou99] describes a hybrid method of extracting threshold voltage that is valid for all values of V_{ds} and does not depend on an arbitrary choice of critical current. V_t and I_{crit} are found at low V_{ds} (e.g., 100 mV) for a given value of L and W using the linear extrapolation method. For other values of V_{ds}, V_t is defined to be the gate voltage when $I_{ds} = I_{crit}$.

Figure 5.19(a) plots the threshold voltage V_t vs. length for a 16 λ-wide device over a variety of design corners and temperatures. The threshold is extracted using the linear extrapolation method and clearly is not constant. It decreases with temperature and is lower in the FF corner than in the SS corner. In an ideal long-channel transistor, the threshold is independent of width and length. In a real device, the geometry sensitivity

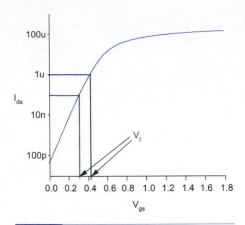

FIG 5.17 Constant current threshold voltage extraction method

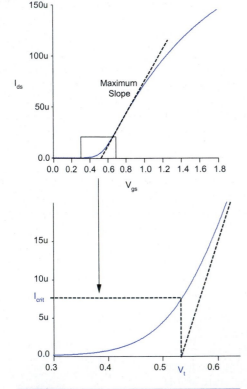

FIG 5.18 Linear extrapolation threshold voltage extraction method

depends on the particular doping profile of the process. Figure 5.19(b) plots V_t against V_{ds} for 16/2 λ transistors using Zhou's method. The threshold voltage decreases with V_{ds} because of DIBL.

The moral of this section is that V_t depends on length, width, temperature, processing, and how you define it. The current does not abruptly drop to zero at threshold and can be significant for leakage-sensitive circuits. When two devices need accurately matched thresholds (e.g., in a sense amplifier), the devices should be adjacent and identical in geometry.

5.4.3 Gate Capacitance

When using RC models to estimate gate delay, we need to know the effective gate capacitance for delay purposes. In Section 2.3.2, we saw that the gate capacitance is voltage-dependent. The gate-to-drain component may be effectively doubled when a gate switches because the gate and drain switch in opposite directions. Nevertheless, we can obtain an effective capacitance averaged across the switching time. We use fanout-of-4 inverters to represent gates with "typical" switching times because we know from logical effort that circuits perform well when the stage effort is approximately 4.

Figure 5.20 shows a circuit for determining the effective gate capacitance of inverter *X4*. The approach is to adjust the capacitance C_{delay} until the average delay from c to g equals the delay from c to d. Because *X6* and *X3* have the same input slope and are the same size, when they have the same delay, C_{delay} must equal the effective gate capacitance of *X4*. *X1* and *X2* are used to produce a reasonable input slope on node c. A single inverter would give reasonable results, but the inverter pair is even better because it provides a slope on c that is essentially independent of the rise time at a. *X5* is the load on *X4* to prevent node e from switching excessively fast, which would overpredict the significance of the gate-to-drain capacitance in *X4*.

Figure 5.21 lists a SPICE deck that uses the optimizer to automatically tune C_{delay} until the delays are equalized. This capacitance is divided by the total gate width (in μm) of *X4* to obtain the capacitance per micron of gate width $C_{permicron}$. This capacitance is listed as C_g (delay) in Table 5.5 for a variety of processes. Note that the deck sets diffusion area and perimeter to 0 to measure only the gate capacitance.

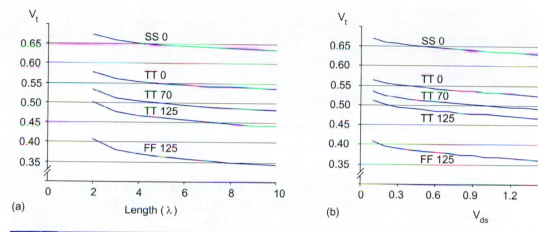

FIG 5.19 MOS threshold voltages

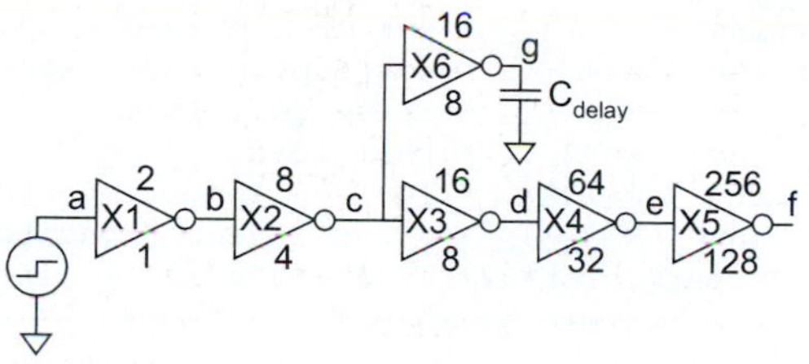

FIG 5.20 Circuit for extracting effective gate capacitance for delay estimation

Gate capacitance is also important for dynamic power consumption, as was given in EQ (4.29). The effective gate capacitance for power is typically somewhat higher than for delay because C_{gd} is effectively doubled from the Miller effect when we wait long enough for the drain to completely switch. Figure 5.22 shows a circuit for measuring gate capacitance for power purposes. A voltage step is applied to the input and the current out of the voltage source is integrated. The effective capacitance for dynamic power consumption is:

$$C_{\text{eff-power}} = \frac{\int i_{\text{in}}(t)\,dt}{V_{DD}} \tag{5.5}$$

Again, this capacitance can be divided by the total transistor width to find the effective capacitance per micron.

```
* capdelay.hsp
* Extract effective gate capacitance for delay estimation.
*-------------------------------------------------------------------
* Parameters and models
*-------------------------------------------------------------------
.option scale=90n
.param SUP=1.8 * Must set before calling .lib
.lib '../models/tsmc180/opconditions.lib' TT
.option post
*-------------------------------------------------------------------
* Subcircuits
*-------------------------------------------------------------------
.global vdd    gnd
.subckt inv    a      y
M1      y      a      gnd    gnd    NMOS   W=16 L=2 AD=0 AS=0 PD=0 PS=0
M2      y      a      vdd    vdd    PMOS   W=32 L=2 AD=0 AS=0 PD=0 PS=0
.ends
*-------------------------------------------------------------------
* Simulation netlist
*-------------------------------------------------------------------
Vdd     vdd    gnd    'SUPPLY'        * SUPPLY is set by .lib call
Vin     a      gnd    pulse 0 'SUPPLY' 1ns 0.5ns 0.5ns 4ns 10ns
X1      a      b      inv            * set appropriate slope
X2      b      c      inv    M=4     * set appropriate slope
X3      c      d      inv    M=8     * drive real load
X4      d      e      inv    M=32    * real load
X5      e      f      inv    M=128   * load on load (important!)
X6      c      g      inv    M=8     * drive linear capacitor
cdelay  g      gnd    'CperMicron*32*(16+32)*90n/1u' * linear capacitor
*-------------------------------------------------------------------
* Optimization setup
*-------------------------------------------------------------------
.measure errorR param='invR - capR' goal=0
.measure errorF param='invF - capF' goal=0
.param CperMicron=optrange(2f, 1f, 3.0f)
.model optmod opt itropt=30
.measure CperMic param = 'CperMicron'
*-------------------------------------------------------------------
* Stimulus
*-------------------------------------------------------------------
.tran .1ns 12ns SWEEP OPTIMIZE = optrange RESULTS=errorR,errorF MODEL=optmod
.measure invR
+       TRIG v(c)   VAL='SUPPLY/2' FALL=1
+       TARG v(d)   VAL='SUPPLY/2' RISE=1
.measure  capR
+       TRIG v(c)   VAL='SUPPLY/2' FALL=1
+       TARG v(g)   VAL='SUPPLY/2' RISE=1
.measure invF
+       TRIG v(c)   VAL='SUPPLY/2' RISE=1
+       TARG v(d)   VAL='SUPPLY/2' FALL=1
.measure  capF
+       TRIG v(c)   VAL='SUPPLY/2' RISE=1
+       TARG v(g)   VAL='SUPPLY/2' FALL=1
.end
```

FIG 5.21 CAPDELAY SPICE deck

5.4.4 Parasitic Capacitance

The parasitic capacitance associated with the source or drain of a transistor includes the gate-to-diffusion overlap capacitance, C_{gol}, and the diffusion area and perimeter capacitance C_{jb} and C_{jbsw}. As discussed in Section 5.3.4, some models assign a different capacitance C_{jbswg} to the perimeter along the gate side. The diffusion capacitance is voltage-dependent, but as with gate capacitance, we can extract an effective capacitance averaged over the switching transition to use for delay estimation.

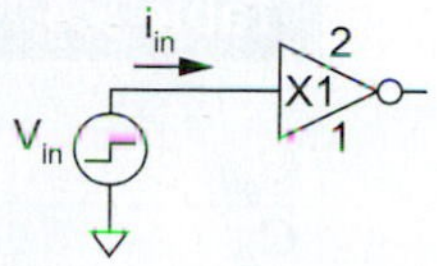

FIG 5.22 Circuit for extracting effective gate capacitance for power estimation

Figure 5.23 shows circuits for extracting these capacitances. They operate in much the same way as the gate capacitance extraction from Section 5.4.3. The first two fanout-of-4 inverters shape the input slope to match a typical gate. $X3$ drives the drain of an OFF transistor $M1$ with specified W, AD, and PD. $X4$ drives a simple capacitor, whose value is optimized so that the delay of $X3$ and $X4$ are equal. This value is the effective capacitance of $M1$'s drain. Similar simulations must be run to find the parasitic capacitances of pMOS transistors.

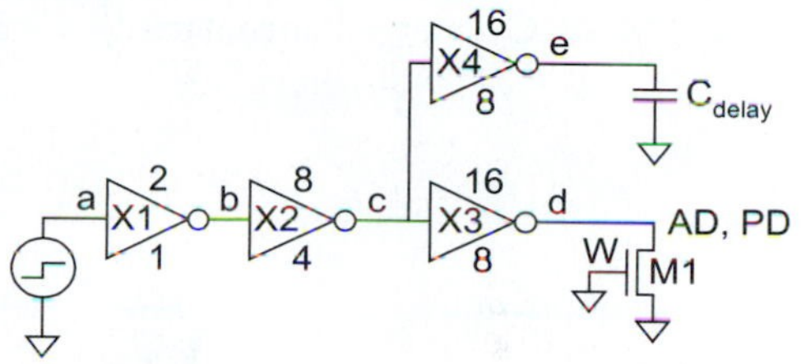

FIG 5.23 Circuit for extracting effective parasitic capacitance for delay estimation

Table 5.4 lists the appropriate values of W, AD, and PD to extract each of the capacitances. The sizes are chosen such that the gate delays and slope on node d are reasonable when a unit transistor is 16 λ wide (as in Figure 5.21). It also gives values to find the effective capacitance C_d of isolated contacted, shared contacted, and merged uncontacted diffusion regions. The capacitance is found, assuming the transistors are wide enough that the perimeter perpendicular to the polysilicon gate is a negligible fraction of the overall capacitance. The AD and PD dimensions are based on the layouts of Figure 2.9; you should substitute your own design rules. The total capacitance of shared and merged regions should be split between the two transistors sharing the diffusion node. The capacitance can be converted to units per micron (or per micron squared) by normalizing for the value of λ. For example, in a 180 nm process with λ = 0.09 μm, if C_{delay} is 53 fF for gate overlap, the capacitance per micron is

$$C_{gol} = \frac{53 \text{ fF}}{(1600\lambda)\left(\frac{0.09\mu m}{\lambda}\right)} = 0.37 \tfrac{\text{fF}}{\mu m} \tag{5.6}$$

5.4.5 Effective Resistance

According to the RC delay model, if a unit transistor has gate capacitance C, parasitic capacitance C_d, and resistance R_n (for nMOS) or R_p (for pMOS), the rising and falling delays of a fanout-of-h inverter with a 2:1 P/N ratio can be found according to Figure 5.24. These delays can readily be measured from the fanout-of-4 inverter simulation in Figure 5.10 by changing h.

Table 5.4   Dimensions for diffusion capacitance extraction

	W (λ)	AD (λ²)	PD (λ)	To find effective C per micron
C_{gol}	1600	0	0	$C_{delay}/1600\lambda$ (per μm)
C_{jb}	0	8000	0	$C_{delay}/8000\lambda^2$ (per μm²)
C_{jbsw}	0	0	1600	$C_{delay}/1600\lambda$ (per μm)
C_{jbswg}	1600	0	1600	$C_{delay}/1600\lambda - C_{gol}$ (per μm)
C_d (isolated contacted)	1600	8000	3200	$C_{delay}/1600\lambda$ (per μm of gate width)
C_d (shared contacted)	3200	9600	3200	$C_{delay}/1600\lambda$ (per μm of gate width)
C_d (merged uncontacted)	3200	4800	3200	$C_{delay}/1600\lambda$ (per μm of gate width)

$$t_{pdr} = \frac{R_p}{2}(3hC + 3C_d) \qquad t_{pdf} = R_n(3hC + 3C_d)$$

(a) Fanout-of-h inverter   (b) Rising delay   (c) Falling delay

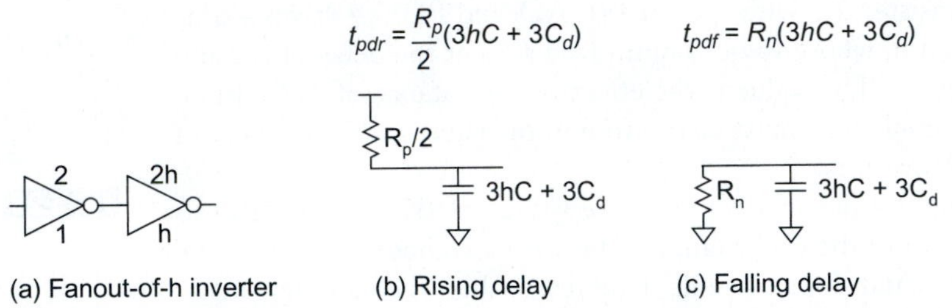

FIG 5.24 RC delay model for fanout-of-h inverter

The dependence on parasitics can be removed by calculating the difference between delays at different fanouts. For example, the difference between delays for $h = 3$ and $h = 4$ are

$$\Delta t_{pdr} = \frac{R_p}{2}(3 \cdot 4 \cdot C + 3C_d) - \frac{R_p}{2}(3 \cdot 3 \cdot C + 3C_d) = \tfrac{3}{2}R_p C$$

$$\Delta t_{pdf} = R_n(3 \cdot 4 \cdot C + 3C_d) - R_n(3 \cdot 3 \cdot C + 3C_d) = 3R_n C \tag{5.7}$$

As C is known from the effective gate capacitance extraction, R_n and R_p are readily calculated. These represent the effective resistance of single nMOS and pMOS transistors for delay estimation.

When two unit transistors are in series, each nominally would have the same effective resistance, giving twice the overall resistance. However, in modern processes where the transistors usually experience some velocity saturation, each transistor sees a smaller V_{ds} and hence less velocity saturation and a lower effective resistance. We can determine this resistance by simulating fanout-of-h tristates in place of inverters, as shown from c to d in Figure 5.25. By a similar reasoning, the difference between delays for $h = 2$ and $h = 3$ is

$$\Delta t_{pdr} = \tfrac{3}{2}\left(2R_{p\text{-series}}\right)C$$

$$\Delta t_{pdf} = 3\left(2R_{n\text{-series}}\right)C \tag{5.8}$$

As C is still known, we can extract the effective resistance of series nMOS and pMOS transistors for delay estimation and should expect this resistance to be slightly smaller than for single transistors.

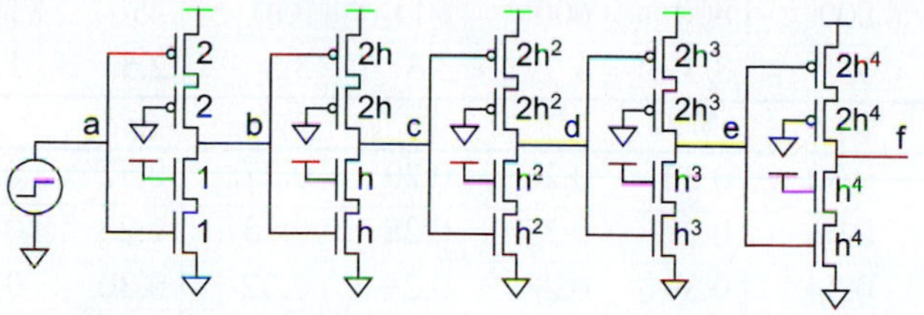

FIG 5.25 Circuit for extracting effective series resistance

It is important to use realistic input slopes when extracting effective resistance because the delay varies with input slope. *Realistic* means that the input and output edge rates should be comparable; if a step input is applied, the output will transition faster and the effective resistance will appear to decrease. h was chosen in this section to give stage efforts close to 4.

5.4.6 Comparison of Processes

Table 5.5 compares the characteristics of a variety of CMOS processes with feature sizes ranging from 2 μm down to 180 nm. Most are based on models published by MOSIS [Piña02] based on wafer test results, but one column uses models provided by TSMC (and indicates the test wafer results were slower than the TT vendor models). The MOSIS models use ACM = 0, so the diffusion sidewall capacitance is treated the same along the gate and the other walls. The 0.8 and 0.6 μm processes operate at either V_{DD} = 5 V (for higher speed) or V_{DD} = 3.3 V (for lower power). All characteristics are extracted for TTTT conditions (70° C).

The table indicates that the diffusion capacitance of an isolated contacted source or drain has been 1–2 fF/μm for both nMOS and pMOS transistors over many generations. The capacitance of a shared contacted diffusion region is slightly higher because it has more area and includes two gate overlaps. The capacitance of the merged diffusion reflects two gate overlaps but a smaller diffusion area.

The effective resistance of a 1 μm wide transistor has decreased with process scaling in proportion to the feature size f. However, the resistance of a unit (4/2 λ) nMOS transistor, $R/2f$, has remained roughly constant around 8 kΩ, as constant field scaling theory would predict. The effective resistance of pMOS transistors is 2–3 times that of nMOS transistors. A pair of nMOS transistors in series each have lower effective resistance than a single device because each has a smaller V_{ds} and thus experiences less velocity saturation. Series pMOS transistors show less pronounced improvement because they were not as velocity-saturated to begin with.

Table 5.5	Device characteristics for a variety of processes										
Vendor		Orbit	AMI	HP	HP	AMI	AMI	TSMC	TSMC	TSMC	TSMC
Model		MOSIS	MOSIS	MOSIS	MOSIS	MOSIS	MOSIS	MOSIS	MOSIS	MOSIS	TSMC
Feature Size f	nm	2000	1600	800	800	600	600	350	250	180	180
V_{DD}	V	5	5	5	3.3	5	3.3	3.3	2.5	1.8	1.8
nMOS											
C_{gol}	fF/μm	0.37	0.18	0.30	0.30	0.20	0.20	0.27	0.57	0.73	0.37
C_{jb}	fF/μm^2	0.05	0.15	0.06	0.09	0.25	0.28	0.73	1.23	0.78	0.77
C_{jbsw}	fF/μm	0.31	0.13	0.34	0.39	0.28	0.29	0.32	0.30	0.22	0.18
C_{jbswg}	fF/μm	n/a	n/a	n/a	n/a	n/a	n/a	n/a	n/a	n/a	0.24
C_d (isolated)	fF/μm	1.19	1.09	1.11	1.27	1.14	1.21	1.63	1.88	1.50	1.12
C_d (shared)	fF/μm	1.62	1.35	1.43	1.60	1.41	1.50	2.04	2.60	2.30	1.62
C_d (merged)	fF/μm	1.48	0.98	1.36	1.49	1.19	1.24	1.60	2.16	2.09	1.41
R_n (single)	kΩ•μm	30.3	25.2	10.1	14.2	9.19	11.9	5.73	4.02	2.96	2.69
R_n (series)	kΩ•μm	22.1	17.1	6.95	10.1	6.28	8.59	4.01	3.10	2.24	2.00
V_{tn} (const. I)	V	0.65	0.53	0.65	0.65	0.70	0.70	0.59	0.48	0.44	0.41
V_{tn} (linear ext.)	V	0.65	0.58	0.75	0.75	0.76	0.76	0.67	0.57	0.55	0.53
I_{dsat}	μA/μm	152	159	380	197	387	216	450	551	560	566
I_{off}	pA/μm	2.26	1.71	9.36	6.59	2.21	1.45	6.57	56.3	78.2	93.9
I_{crit} (linear ext.)	μA/μm	0.05	0.12	0.50	0.50	0.47	0.47	0.89	1.86	4.07	5.12
pMOS											
C_{gol}	fF/μm	0.36	0.23	0.32	0.32	0.29	0.29	0.30	0.66	0.66	0.33
C_{jb}	fF/μm^2	0.13	0.16	0.32	0.37	0.42	0.48	0.91	1.39	0.94	0.85
C_{jbsw}	fF/μm	0.21	0.14	0.11	0.11	0.20	0.21	0.35	0.30	0.15	0.20
C_{jbswg}	fF/μm	n/a	n/a	n/a	n/a	n/a	n/a	n/a	n/a	n/a	0.33
C_d (isolated)	fF/μm	1.42	1.15	1.17	1.26	1.31	1.42	1.89	2.07	1.36	1.24
C_d (shared)	fF/μm	1.92	1.51	1.62	1.72	1.73	1.86	2.37	2.89	2.11	1.79
C_d (merged)	fF/μm	1.52	1.12	1.23	1.29	1.35	1.43	1.83	2.40	1.86	1.56
R_p (single)	kΩ•μm	67.1	62.8	26.7	40.8	19.9	29.6	16.1	8.93	6.55	6.51
R_p (series)	kΩ•μm	53.9	52.5	21.4	33.3	15.4	23.6	13.3	6.91	5.06	5.41
$|V_{tp}|$ (const. I)	V	0.72	0.77	0.91	0.91	0.90	0.90	0.83	0.46	0.45	0.43
$|V_{tp}|$ (linear ext.)	V	0.71	0.76	0.94	0.94	0.93	0.93	0.88	0.52	0.53	0.51
I_{dsat}	μA/μm	70.5	60.6	154	66.6	215.3	99.0	181	245	257	228
I_{off}	pA/μm	2.18	17.9	1.57	1.03	2.08	1.38	2.06	30.1	47.4	25.2
I_{crit} (linear ext.)	μA/μm	0.02	0.03	0.11	0.11	0.16	0.16	0.28	0.59	1.24	1.14
Gates											
C_g (delay)	fF/μm	1.77	1.67	1.67	1.60	1.55	1.48	1.90	2.30	2.13	1.67
C_g (power)	fF/μm	2.24	1.89	1.70	1.90	1.83	1.76	2.20	2.92	2.82	2.06
FO4 inv delay	ps	856	717	297	427	230	312	210	153	99.4	75.6

Threshold voltages are reported at V_{ds} = 100 mV for 16/2 λ devices using both the constant current (at I_{crit} = 0.1(W/L) μA for nMOS and 0.06(W/L) for pMOS) and linear extrapolation methods. Threshold voltages have generally decreased, but not as fast as channel length or supply voltage (because of subthreshold leakage). Therefore, the V_{DD}/V_t ratio is decreasing and pass transistor circuits with threshold drops do not perform well in modern processes.

Saturation current per micron has increased somewhat through aggressive device design as feature size decreases even though constant field scaling would suggest it should remain constant. OFF current was on the order of a few picoamperes per micron in the past, but is now exponentially increasing because of subthreshold conduction through devices with low threshold voltages. The current at threshold using the linear extrapolation method is somewhat higher than the constant current I_{crit}, corresponding to the higher threshold voltages found by the linear extrapolation method.

The gate capacitance for delay has held steady near 2 fF/μm for many generations, as scaling theory would predict. The gate capacitance for power is slightly higher than that for delay as discussed in Section 5.4.3.

The FO4 inverter delay has steadily improved with feature size as constant field scaling predicts. It fits our rule from Section 4.3.1 of 1/3 to 1/2 of the feature size, when delay is measured in picoseconds and feature size in nanometers.

5.4.7 Process and Environmental Sensitivity

Table 5.6 shows how the TSMC180 nm process characteristics vary with process corner, voltage, and temperature. The FO4 inverter delay varies by more than a factor of two between best and worst case. In the TT process, inverter delay varies by about 0.12% / °C and by about 1% for every percent of supply voltage change. These figures agree well with the Artisan library data from Section 4.7.4. Gate and diffusion capacitance change only slightly with process, but effective resistance is inversely proportional to supply voltage and highly sensitive to temperature and device corners. I_{off} subthreshold leakage rises dramatically at high temperature or in the fast corner where threshold voltages are lower.

5.5 Circuit Characterization

The device characterization techniques from the previous section are typically run once by engineers who are familiarizing themselves with a new process. SPICE is used more often to characterize entire circuits. This section gives some pointers on simulating paths and describes how to find the DC transfer characteristics, logical effort, and power consumption of logic gates.

Table 5.6	Process corners of TSMC 180 nm process									
nMOS		T	F	S	F	S	T	T	T	T
pMOS		T	F	S	S	F	T	T	T	T
V_{DD}	V	1.8	1.98	1.62	1.8	1.8	1.98	1.62	1.8	1.8
T	°C	70	0	125	70	70	70	70	0	125
nMOS										
C_{gol}	fF/µm	0.37	0.39	0.35	0.37	0.37	0.37	0.37	0.37	0.37
C_{jb}	fF/µm²	0.77	0.72	0.82	0.72	0.80	0.75	0.78	0.76	0.76
C_{jbsw}	fF/µm	0.18	0.17	0.19	0.17	0.19	0.18	0.18	0.18	0.18
C_{jbswg}	fF/µm	0.24	0.22	0.26	0.23	0.25	0.23	0.25	0.24	0.24
C_d (isolated)	fF/µm	1.12	1.09	1.16	1.08	1.16	1.10	1.13	1.12	1.12
C_d (shared)	fF/µm	1.62	1.58	1.65	1.58	1.66	1.60	1.64	1.62	1.62
C_d (merged)	fF/µm	1.41	1.40	1.43	1.38	1.45	1.40	1.43	1.42	1.41
R_n (single)	kΩ•µm	2.69	1.78	4.23	2.20	3.13	2.43	3.00	2.43	2.91
R_n (series)	kΩ•µm	2.00	1.23	3.42	1.51	2.42	1.78	2.32	1.76	2.17
V_{tn} (const. I)	V	0.41	0.35	0.47	0.30	0.51	0.41	0.41	0.45	0.37
V_{tn} (linear ext.)	V	0.53	0.48	0.60	0.43	0.63	0.53	0.53	0.58	0.50
I_{dsat}	µA/µm	566	834	354	624	508	656	476	623	520
I_{off}	pA/µm	93.9	61.4	84.9	1116	9.9	100	87.8	4.7	689
I_{crit} (linear ext.)	µA/µm	5.12	7.35	3.83	5.29	4.94	5.12	5.12	6.40	4.43
pMOS										
C_{gol}	fF/µm	0.33	0.34	0.31	0.33	0.33	0.33	0.33	0.33	0.33
C_{jb}	fF/µm²	0.85	0.79	0.91	0.90	0.81	0.84	0.87	0.85	0.86
C_{jbsw}	fF/µm	0.20	0.19	0.21	0.21	0.19	0.20	0.20	0.20	0.20
C_{jbswg}	fF/µm	0.33	0.31	0.36	0.35	0.32	0.33	0.34	0.33	0.33
C_d (isolated)	fF/µm	1.24	1.20	1.28	1.28	1.20	1.23	1.26	1.24	1.24
C_d (shared)	fF/µm	1.79	1.75	1.83	1.84	1.73	1.76	1.80	1.78	1.78
C_d (merged)	fF/µm	1.56	1.53	1.58	1.60	1.51	1.54	1.57	1.55	1.56
R_p (single)	kΩ•µm	6.51	4.55	9.60	7.00	6.00	5.93	7.32	6.04	6.86
R_p (series)	kΩ•µm	5.41	3.65	8.03	5.98	4.73	4.84	6.08	4.98	5.68
V_{tn} (const. I)	V	0.43	0.43	0.45	0.50	0.36	0.43	0.43	0.50	0.38
V_{tp} (linear ext.)	V	0.51	0.51	0.52	0.57	0.44	0.51	0.51	0.57	0.46
I_{dsat}	µA/µm	228	345	144	209	248	273	184	238	222
I_{off}	pA/µm	25.2	5.21	748	7.66	127	26.7	23.7	2.76	938
I_{crit} (linear ext.)	µA/µm	1.14	1.40	0.95	1.11	1.16	1.14	1.14	1.20	1.09
Gates										
C_g (delay)	fF/µm	1.67	1.72	1.61	1.71	1.65	1.69	1.65	1.67	1.68
C_g (power)	fF/µm	2.06	2.12	1.98	2.06	2.07	2.09	2.03	2.06	2.06
FO4 inv delay	ps	75.6	53.3	112	74.4	77.6	69.7	83.9	69.4	80.6

5.5.1 Path Simulations

The delay of most static CMOS circuits today is computed with a static timing analyzer (see Section 8.4.1.5). As long as the noise sources (particularly coupling and power supply noise) are controlled, the circuits will operate correctly and will correlate reasonably well with static timing predictions. However, SPICE-level simulation is important for sensitive circuits such as the clock generator and distribution network, custom memory arrays, and novel circuit techniques.

Most experienced designers begin designing paths based on simple models in order to understand what aspects are most important, evaluate design tradeoffs, and obtain a qualitative prediction of the results. The ideal Shockley transistor models, RC delay models, and logical effort are all helpful here because they are simple enough to give insight. When a good first-pass design is ready, the designer simulates the circuit to verify that it operates correctly and meets delay and power specifications. Just as few new software programs run correctly before debugging, the simulation often will be incorrect at first. Unless the designer knows what results to expect, it is tempting to trust the false results that are nicely printed with beguilingly many significant figures. Once the circuit appears to be correct, it should be checked across design corners to verify that it operates in all cases. Section 4.7.4 gave examples of circuits sensitive to various corners.

Simulation is cheap, but silicon revisions are very expensive. Therefore, it is important to construct a circuit model that captures all of the relevant conditions, including real input waveforms, appropriate output loading, and adequate interconnect models. When matching is important, you must consider the effects of mismatches that are not given in the corner files (see Section 5.5.5). However, as SPICE decks get more complicated, they run more slowly, accumulate more mistakes, and are more difficult to debug. A good compromise is to start simple and gradually add complexity, ensuring after each step that the results still make sense.

5.5.2 DC Transfer Characteristics

The `.dc` card is useful for finding the transfer characteristics and noise margins of logic gates. Figure 5.27 shows an example of characterizing static and dynamic inverters (dynamic logic is covered in Section 6.2.4). Figure 5.26(a and b) show the circuit schematics of each gate. Figure 5.26(c) shows the simulation results. The static inverter characteristics are nearly symmetric around $V_{DD}/2$. The dynamic inverter has a lower switching threshold and its output drops abruptly beyond this threshold because positive feedback turns off the keeper.

Note that when the input a is '0' and the dynamic inverter is in evaluation (ϕ = '1'), the output would be stable at either '0' or '1.' To find the transfer characteristics, we initialize the gate with a '1' output using the `.ic` command.

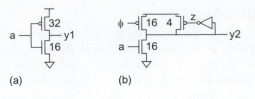

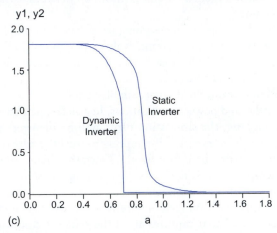

(c)

FIG 5.26 Circuits for DC transfer analysis

5.5.3 Logical Effort

The logical effort and parasitic delay of each input of a gate can be measured by fitting a straight line to delay vs. fanout simulation results. As with the FO4 inverter example, it is important to drive the gate with an appropriate input waveform and to provide two stages of loads. Figure 5.28(a) shows an example of a circuit for characterizing the delay of a 2-input NAND gate *X3* using the M parameter to simulate multiple gates in parallel. Figure 5.28(b) shows that the delay vs. fanout in a 180 nm process fit a straight line of $d_{abs} = 16.7h + 28.9$ ps within 0.5 ps even though the transistors experience all sorts of nonlinear and nonideal effects. This shows that the linear delay model is quite accurate as long as the input and output slopes are consistent.

The SWEEP command is convenient to vary the fanout and repeat the transient simulation multiple times. For example, the following card runs eight simulations varying H from 1 to 8 in steps of 1.

```
.tran 1ps 1000ps SWEEP H 1 8 1
```

To characterize an entire library, you can write a script in a language such as Perl that generates the appropriate SPICE decks, invokes the simulator, and postprocesses the list files to extract the data and do the curve fit.

Recall that τ is the coefficient of h (i.e., the slope) in a delay vs. fanout plot for an inverter. Given that τ = 15 ps in this process, we see the NAND gate has a logical effort of 16.7/15 = 1.11 and a parasitic delay of 28.9/15 = 1.93.

Table 5.7 compares the logical effort and parasitic delay of the different inputs of multi-input NAND gates for rising, falling, and average output transitions in the TSMC 180 nm process. For rising and falling transitions, we still normalize against the value of τ found from the average delay of an inverter. Input A is the outermost (closest to power or ground). As discussed in Section 6.2.1.3, the outer input has higher parasitic delay, but slightly lower logical effort. The rising and falling delays in this process are quite different because pMOS transistors have less than half the mobility of nMOS transistors and because the nMOS transistors are quite velocity-saturated so that series transistors have less resistance than expected.

```
* invdc.sp
* Static and dynamic inverter DC transfer characteristics

*---------------------------------------------------------------
* Parameters and models
*---------------------------------------------------------------
.param SUPPLY=1.8
.option scale=90n
.include '../models/tsmc180/models.sp'
.temp 70
.option post


*---------------------------------------------------------------
* Simulation netlist
*---------------------------------------------------------------
Vdd    vdd    gnd    'SUPPLY'
Va     a      gnd    0
Vclk   clk    gnd    'SUPPLY'
* Static Inverter
M1     y1     a      gnd    gnd    NMOS    W=16    L=2
M2     y1     a      vdd    vdd    PMOS    W=32    L=2
* Dynamic Inverter
M3     y2     a      gnd    gnd    NMOS    W=16    L=2
M4     y2     clk    vdd    vdd    PMOS    W=16    L=2
M5     y2     z      vdd    vdd    PMOS    W=4     L=2
M6     z      y2     gnd    gnd    NMOS    W=4     L=2
M7     z      y2     vdd    vdd    PMOS    W=8     L=2
.ic V(y2) = 'SUPPLY'


*---------------------------------------------------------------
* Stimulus
*---------------------------------------------------------------
.dc Va 0 1.8 0.01
.end
```

FIG 5.27   INVDC SPICE deck for DC transfer analysis

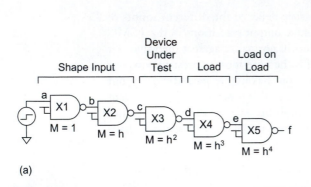

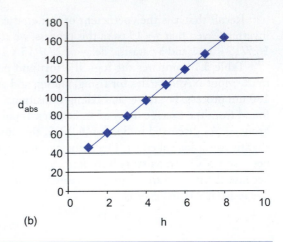

(a)   (b)

FIG 5.28 Logical effort characterization of 2-input NAND gate

Table 5.8 compares the average logical effort and parasitic delay of a variety of gates in many different processes. In each case, the simulations are performed in the TTTT corner for the outer input. For reference, the values of a FO4 inverter delay and τ are given for each process. The logical effort of NAND gates are substantially lower than we predicted in Section 4.2.3 because velocity saturation causes the resistance of N series nMOS transistors to be less than N times that of a single transistor. NOR gate efforts are slightly lower, but the effect is less pronounced because pMOS transistors have lower mobility and thus experience less velocity saturation. The efforts are fairly consistent across process and voltage. The parasitic delays show greater spread because of the variation in the relative capacitances of diffusion and gates.

Table 5.7		Logical effort and parasitic delay of different inputs of multi-input NAND gates					
# of inputs	Input	Rising Logical Effort g_u	Falling Logical Effort g_u	Average Logical Effort g	Rising Parasitic Delay p_u	Falling Parasitic Delay p_d	Average Parasitic Delay p
2	A	1.45	0.82	**1.14**	2.37	1.42	**1.90**
	B	1.36	0.97	**1.17**	1.60	1.22	**1.41**
3	A	1.82	0.80	**1.31**	4.90	2.11	**3.51**
	B	1.73	0.95	**1.34**	3.94	1.86	**2.90**
	C	1.63	1.10	**1.36**	2.60	1.54	**2.07**
4	A	1.96	0.89	**1.42**	6.54	3.08	**4.81**
	B	1.86	1.02	**1.44**	5.71	2.84	**4.28**
	C	1.80	1.14	**1.47**	4.69	2.42	**3.56**
	D	1.71	1.28	**1.49**	3.26	1.97	**2.62**

This data includes more detail than the designer typically wants when doing design by hand; the coarse estimates of logical effort from Table 4.2 are generally sufficient for an initial design. However, the accurate delay vs. fanout information, often augmented with input slope dependence, is very important when characterizing a standard cell library to use with a static timing analyzer

Table 5.8	Logical effort and parasitic delay of gates in various processes										
Vendor		Orbit	AMI	HP	HP	AMI	AMI	TSMC	TSMC	TSMC	TSMC
Model		MOSIS	MOSIS	MOSIS	MOSIS	MOSIS	MOSIS	MOSIS	MOSIS	MOSIS	TSMC
Feature Size f	nm	2000	1600	800	800	600	600	350	250	180	180
V_{DD}	V	5	5	5	3.3	5	3.3	3.3	2.5	1.8	1.8
FO4 inv delay	ps	856	717	297	427	230	312	210	153	99.4	75.6
τ	ps	170	143	59	84	45	60	40	30	20	15
Logical Effort											
Inverter		1.00	1.00	1.00	1.00	1.00	1.00	1.00	1.00	1.00	1.00
NAND2		1.13	1.07	1.07	1.09	1.05	1.08	1.12	1.12	1.17	1.14
NAND3		1.32	1.22	1.21	1.25	1.19	1.24	1.29	1.29	1.36	1.31
NAND4		1.48	1.35	1.33	1.37	1.33	1.38	1.44	1.43	1.49	1.42
NOR2		1.57	1.62	1.59	1.58	1.58	1.60	1.52	1.50	1.46	1.50
NOR3		2.16	2.32	2.23	2.23	2.23	2.30	2.07	2.02	1.94	2.00
NOR4		2.55	2.74	2.61	2.64	2.57	2.68	2.46	2.37	2.27	2.38
Parasitic Delay											
Inverter		1.08	1.05	1.05	1.13	1.18	1.25	1.33	1.18	0.95	1.03
NAND2		1.87	1.79	1.85	2.05	1.92	2.10	2.28	2.07	1.74	1.90
NAND3		3.34	3.22	3.30	3.75	3.40	3.79	4.15	3.65	3.08	3.51
NAND4		4.64	4.42	4.54	5.12	4.70	5.23	5.75	5.01	4.26	4.81
NOR2		2.86	2.97	2.91	3.13	3.29	3.56	3.52	2.95	2.41	2.85
NOR3		5.65	6.22	6.05	6.47	7.02	7.70	6.89	5.61	4.49	5.57
NOR4		7.92	8.71	8.48	9.14	9.68	10.67	9.71	7.93	6.38	7.98

5.5.4   Power and Energy

Recall from Section 4.4 that energy and power are proportional to the supply current. They can be measured based on the current out of the power supply voltage source. For example, the following code uses the INTEGRAL command to measure charge and energy delivered to a circuit during the first 10 ns.

```
.measure charge INTEGRAL I(vdd) FROM=0ns TO=10ns
.measure energy param='charge*SUPPLY'
```

Alternatively, HSPICE allows you to directly measure the instantaneous and average power delivered by a voltage source.

```
.print P(vdd)
.measure pwr AVG P(vdd) FROM=0ns TO=10ns
```

Sometimes it is helpful to measure the power consumed by only one gate in a larger circuit. In that case, you can use a separate voltage source for that gate and measure power only from that source. Unfortunately, this means that vdd cannot be declared as .global.

When the input of a gate switches, it delivers power to the supply through the gate-to-source capacitances. Be careful to differentiate this input power from the power drawn by the gate discharging its internal and load capacitances.

5.5.5 Simulating Mismatches

Many circuits are sensitive to mismatches between nominally identical transistors. For example, the sense amplifiers in Fig 11.17 should respond to a small differential voltage between the inputs. Mismatches between nominally identical transistors add an offset that can significantly increase the required voltage. Merely simulating in different design corners is inadequate because the transistors will still match each other. As discussed in Section 4.7.5, the mismatch between currents in two nominally identical transistors can be primarily attributed to shifts in the threshold voltage and channel length. Figure 5.29 shows an example of simulating this mismatch. Each transistor is replaced by an equivalent circuit with a different channel length and a voltage source modeling the difference in threshold voltage. Note that many binned BSIM models do not allow setting the transistor length shorter than the minimum value supported by the process. Obtaining data on parameter variations can be difficult. Section 4.7.5 gives examples of channel length and threshold variation in a 180 nm process.

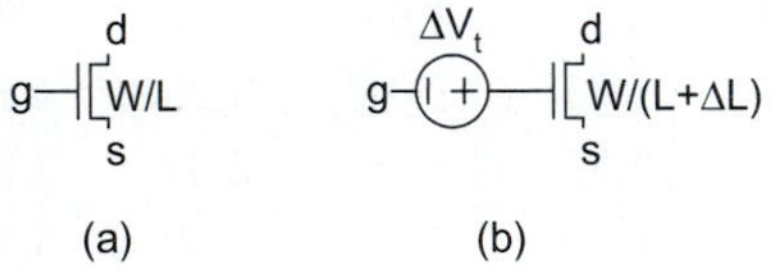

In many cases, the transistors are not adjacent and may see substantial differences in voltage and temperature. For example, two clock buffers in different corners of the chip that see different environments will cause skew between the two clocks. The voltage difference can be modeled with two different voltage sources. The temperature difference is most easily handled through two separate simulations at different temperatures.

FIG 5.29   Modeling mismatch

5.5.6 Monte Carlo Simulation

Monte Carlo simulation can be used to find the effects of random variations on a circuit. It consists of running a simulation repeatedly with different randomly chosen parameter offsets. To use Monte Carlo simulation, the transistor models must include the offset parameters. For example, Figure 5.30 shows the nMOS model card from Figure 5.12 modified to handle length, width, and threshold voltage changes. XL and XW specify offsets to the nominal length and width.

```
.model NMOS NMOS (LEVEL=1 TOX=40e-10 KP=155E-6 LAMBDA=0.2
+                 VTO='0.4+dvthn' PHI=0.93 GAMMA=0.6
+                 CJ=9.8E-5 PB=0.72 MJ=0.36
+                 CJSW=2.2E-10 PHP=7.5 MJSW=0.1
+                 XL='dxl' XW='dxw')
```

FIG 5.30 Sample Level 1 MODEL card with offsets

Consider modifying the FO4 inverter delay simulation from Figure 5.10 to handle the length and threshold voltage mismatches given in Section 4.7.5. The .tran card must be augmented to include the number of times to repeat the simulation with random values. The random distributions of each of the parameters must also be specified. Figure 5.31 shows these changes for 30 Monte Carlo repetitions. The channel length is assigned a uniform offset in the range of 0 +/− 12.5 nm. The nMOS threshold voltage offset has a Gaussian distribution centered around zero with a 1σ variation of 16.8 mV.

```
.param dxl=aunif(0,12.5nm) dvthn=agauss(0,16.8m,1) dvthp=agauss(0,14.6m,1)
.tran 1ps 1000ps SWEEP MONTE=30
```

FIG 5.31 Monte Carlo transient analysis

The .measure cards report average, minimum, maximum, and standard deviation computed from the repeated simulations. In the FO4 example, the average propagation delay is 74.9 ps with a standard deviation of 3.38 ps caused primarily by the channel length variation.

5.6 Interconnect Simulation

Interconnect parasitics can be a very important factor in overall delay. When an actual layout is available, the wire geometry can be directly extracted. If only the schematic is available, the designer may need to estimate wire lengths. For small gates, even the capacitances of the wires inside the gate are important. Therefore, some companies use *parasitic estimator* tools to guess wire parasitics in schematics based on the number and size of the transistors. In any case, the designer must explicitly model long wires based on their estimated lengths in the floorplan.

Once wire length and pitch are known or estimated, they can be converted to a wire resistance R and capacitance C using the methods discussed in Section 4.5. A short wire (where wire resistance is much less than gate resistance) can be modeled as a lumped

FIG 5.32 Four-segment
π model for
interconnect

capacitor. A longer wire can be modeled with a multisegment π-model. A four-segment model such as the one shown in Figure 5.32 is generally quite accurate. The model can be readily extended to include coupling between adjacent lines.

In general, interconnect consists of multiple interacting signal and power/ground lines [Young00]. For example, Figure 5.33(a) shows a pair of parallel signals running between a pair of ground wires. Although it is possible to model the ground lines with a resistance and inductance per unit length, it is usually more practical to treat the supply networks as ideal, then account for power supply noise separately in the noise budget. Figure 5.33(b) shows an equivalent circuit using a single π-segment model. Each line has a series resistance and inductance, a capacitance to ground, and mutual capacitance and inductance. The mutual elements describe how a changing voltage or current in one conductor induce a current or voltage in the other.

HSPICE also supports the w element that models lossy multiconductor transmission lines. This is more convenient than constructing an enormous π-model with resistance, capacitance, inductance, mutual capacitance, and mutual inductance. Moreover, HSPICE has a built-in two-dimensional field solver that can compute all of the terms from a cross-sectional description of the interconnect. Figure 5.34 gives a SPICE deck that uses the field solver to extract the element values and models the lines with the w element.

The deck describes a two-dimensional cross-section of the interconnect that the field solver uses to extract the electrical parameters. The interconnect consists of the two signal traces between two ground wires. Each wire is 2 μm wide and 0.7 μm thick. The copper wires are sandwiched with 0.9 μm of low-k ($\epsilon = 3.55\epsilon_0$) dielectric above and below. The $N = 2$ signal traces are spaced 6 μm from the ground lines and 2 μm from each other and have a length of 6 mm. The HSPICE field solver is quite flexible and is fully documented in the HSPICE manual. It generates the transmission line model and writes it to the

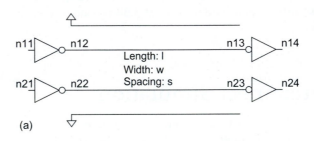

(a)

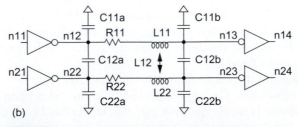

(b)

FIG 5.33 Lossy multiconductor transmission lines

`coplanar.rlgc` file. The file contains resistance, capacitance, and inductance matrices and is shown in Figure 5.35.

```
* interconnect.sp
*-----------------------------------------------------------------
* Parameters and models
*-----------------------------------------------------------------
.param SUPPLY=1.8
.include '../models/tsmc180/models.sp'
.temp 70
.option post
*-----------------------------------------------------------------
* Subcircuits
*-----------------------------------------------------------------
.global vdd gnd
.subckt inv a y N=360nm P=720nm
M1   y    a    gnd   gnd   NMOS   W='N'   L=180nm
+ AS='N*450nm' PS='2*N+900nm' AD='N*450nm' PD='2*N+900nm'
M2   y    a    vdd   vdd   PMOS   W='P'   L=180nm
+ AS='P*450nm' PS='2*P+900nm' AD='P*450nm' PD='2*P+900nm'
.ends
*-----------------------------------------------------------------
* Compute transmission line parameters with field solver
*-----------------------------------------------------------------
.material     oxide        DIELECTRIC          ER=3.55
.material     copper       METAL               CONDUCTIVITY=57.6meg
.layerstack   chipstack    LAYER=(oxide,2.5um)
.fsoptions    opt1         ACCURACY=MEDIUM     PRINTDATA=YES
.shape        widewire     RECTANGLE           WIDTH=2um     HEIGHT=0.7um
.model        coplanar     W                   MODELTYPE=FieldSolver
+ LAYERSTACK=chipstack     FSOPTIONS=opt1      RLGCFILE=coplanar.rlgc
+ CONDUCTOR=(SHAPE=widewire ORIGIN=(0,0.9um)   MATERIAL=copper TYPE=reference)
+ CONDUCTOR=(SHAPE=widewire ORIGIN=(8um,0.9um) MATERIAL=copper)
+ CONDUCTOR=(SHAPE=widewire ORIGIN=(12um,0.9um) MATERIAL=copper)
+ CONDUCTOR=(SHAPE=widewire ORIGIN=(20um,0.9um) MATERIAL=copper TYPE=reference)
*-----------------------------------------------------------------
* Simulation netlist
*-----------------------------------------------------------------
Vdd  vdd   gnd   'SUPPLY'
Vin  n11   gnd   PULSE   0  'SUPPLY' 0ps 100ps 100ps 500ps 1000ps
W1   n12   n22   gnd     n13   n23   gnd     FSmodel=coplanar N=2 l=6mm
X1   n11   n12   inv     M=40
X2   n13   n14   inv     M=20
X3   gnd   n22   inv     M=40
X4   n23   n24   inv     M=20
*-----------------------------------------------------------------
* Stimulus
*-----------------------------------------------------------------
.tran 1ps 1000ps
.end
```

FIG 5.34 SPICE deck for lossy multiconductor transmission line

```
* L(H/m), C(F/m), Ro(Ohm/m), Go(S/m), Rs(Ohm/(m*sqrt(Hz)), Gd(S/(m*Hz))

.MODEL coplanar W MODELTYPE=RLGC, N=2

+ Lo =   6.68161e-007
+         3.67226e-007   6.68161e-007
+ Co =   2.53841e-011
+         -1.36778e-011   2.53841e-011
+ Ro =   12400.8
+         0   12400.8
+ Go =   0
+         0   0
```

FIG 5.35 coplanar.rlgc file

The matrices require a bit of effort to interpret. They are symmetric around the diagonal so only the lower half is printed. The resistances are $R_{11} = R_{22} = 12.4 \ \Omega$/mm. The inductances are $L_{11} = L_{22} = 0.67$ nH/mm and $L_{12} = 0.37$ nH/mm. The capacitance matrix represents coupling capacitances with negative numbers and places the sum of all the capacitances for a trace on the diagonal. Therefore, $C_{11} = C_{22} = 0.117$ pF/mm and $C_{12} = 0.137$ pF/mm. In the π-model, half of each of these capacitances is lumped at each end.

Figure 5.36 shows the voltages along the wires. Observe the *ringing* (i.e., oscillation) caused by the wide wires with high inductance and low resistance driven with sharp edge rates. If the inductance were reduced by moving the ground lines closer to the conductors, the ringing would decrease. The switching line couples onto the quiet victim, causing noise at both ends as well.

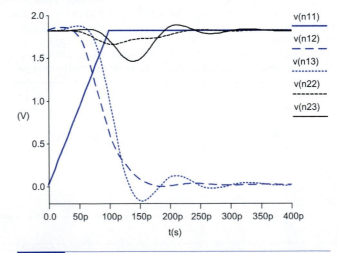

FIG 5.36 Transmission line response

5.7 Pitfalls and Fallacies

Failing to estimate diffusion and interconnect parasitics in simulations

The diffusion capacitance can account for 20% of the delay of an FO4 inverter and more than 50% of the delay of a high fan-in, low fanout gate. Be certain when simulating circuits that the area and perimeter of the source and drain are included in the simulations. Interconnect capacitance is also important, but difficult to estimate. For long wires, the capacitance and RC delay represent most of the path delay. A common error is to ignore wires while doing circuit design at the schematic level, and then discover after layout that the wire delay is important enough to demand major circuit changes and complete change of the layout.

Applying inappropriate input waveforms

Gate delay is strongly dependent on the rise/fall time of the input. For example, the propagation delay of an inverter is substantially shorter when a step input is applied than when an input with a realistic rise time is provided.

Applying inappropriate output loading

Gate delay is even more strongly dependent on the output loading. Some engineers, particularly those in the marketing department, measure gate delay as the delay of an unloaded inverter. This is about one-fifth of the delay of an FO4 inverter or other gate with "typical" loading. When simulating a critical path, it is important to include the estimated load that the final stage must drive.

Choosing inappropriate transistor sizes

Gate delay also depends on transistor widths. Some papers compare a novel design with carefully selected transistor sizes to a conventional design with poorly selected sizes, and arrive at the misleading conclusion that the novel design is superior.

Identifying the incorrect critical path

During preliminary design, it is much more efficient to compare circuits by modeling only the critical paths rather than the entire circuit. However, this requires that the designer correctly identify the path that will be most critical; sometimes this requires much consideration.

Failing to account for hidden scale factors

Many CAD systems introduce scaling factors. For example, a circuit can be drawn with one set of design rules and automatically scaled to the next process generation. The CAD tools may introduce a scaling factor to reflect this change. Specifying the proper transistor sizes reflecting this scaling is notoriously tricky. Simulation results will look good, but mean nothing if scaling is not accounted for properly.

Blindly trusting results from SPICE

Novice SPICE users often trust the results of simulation far too much. This is exacerbated by the fact that SPICE prints results to many significant figures and generates pretty waveforms. As we have seen, there are a multitude of reasons why simulation results may not reflect the behavior of the real circuit.

When first using a new process or tool set, always predict what the results should be for some simple circuits (e.g., an FO4 inverter) and verify that the simulation matches expectation. It doesn't hurt to be a bit paranoid at first. After proving that the flow is correct, lock down all the models and netlist generation scripts with version control if possible. That way, if any changes are made, a valid reason for the change must be evident and the simulations can be revalidated. The motto is: Check and recheck (but don't become compulsive!).

In general, assume SPICE decks are buggy until proven otherwise. If the simulation does not agree with your expectations, look closely for errors or inadequate modeling in the deck.

Using SPICE in place of thinking

A related error, common among perhaps the majority of circuit designers, is to use SPICE too much and one's brain too little. Circuit simulation should be guided by analysis.

Making common SPICE deck errors

Some of the common mistakes in SPICE decks include [Horowitz02]

- Omitting the comment on the first line
- Omitting the new line at the end of the deck
- Omitting the `.option` post command when using a waveform viewer

- Leaving out diffusion parasitics
- Forgetting to set initial values for dynamic logic or sequential circuits

Using incorrect dimensions when `.option scale` is not set

If `.option scale` is not used, a transistor with W = 4, L = 2 would be interpreted as 4 by 2 meters! This often is outside the legal range of sizes in a BSIM model file, causing SPICE to produce error messages. Similarly, a drain diffusion of 3 x 0.5 μm should be specified as PD = 7u AD = 1.5p as opposed to the common mistakes of PD = 7 AD = 1.5 or PD = 7u AD = 1.5u.

Summary

When used properly, SPICE is a powerful tool to characterize the behavior of CMOS circuits. This chapter began with a brief tutorial showing how to perform DC and transient analyses to characterize and optimize simple circuits. SPICE supports many different transistor models. At the time of writing, the BSIM3v3 model is most widely used and describes MOSFET behavior quite well for most digital applications. When specifying the MOSFET connection, you must include not only the terminal connections (drain, gate, source, and body) and width and length, but also the area and perimeter of the source and drain that are used to compute parasitic capacitance.

Modern SPICE models have so many parameters that they are intractable for hand calculations. However, the designer can perform some simple simulations to characterize a process. For example, it is helpful to know the effective gate capacitance and resistance, the diffusion capacitance, and the threshold voltage and leakage current You can also determine the delay of a fanout-of-4 inverter and the logical effort and parasitic delay of a library of gates to make quick estimates of circuit performance.

Most designers use SPICE to characterize real circuits. During preliminary design, you can model the critical path to quickly determine whether a circuit will meet performance requirements. A good model describes not only the circuit itself, but also the input edge rates, the output loading, and parasitics such as diffusion capacitance and interconnect. Most interconnect can be represented with a four-segment π model, although when inductance becomes important, the lossy multiconductor transmission line w element is convenient. Novel and "risky" circuits should be simulated in multiple design corners to ensure they will work correctly across variations in processing and environment. As

SPICE is prone to garbage-in, garbage-out, it is often best to begin with a simple model and debug until it matches expectations. Then more detail can be added and tested incrementally.

Exercises

Note: the book Web page (see preface) contains SPICE models and characterization scripts used to generate the data in this chapter. Unless otherwise stated, try the exercises using the mosistsmc180 model file (extracted by MOSIS from test structures manufactured on the TSMC 180 nm process) in TTTT conditions.

5.1   Find the average propagation delay of a fanout-of-5 inverter by modifying the SPICE deck in Figure 5.10.

5.2   By what percentage does the delay of Exercise 5.1 change if the input is driven by a voltage step rather than a pair of shaping inverters?

5.3   By what percentage does the delay of Exercise 5.1 change if $X5$, the load on the load, is omitted?

5.4   Find the input and output logic levels and high and low noise margins for an inverter with a 3:1 *P/N* ratio.

5.5   What *P/N* ratio maximizes the smaller of the two noise margins for an inverter?

5.6   Generate a set of eight I-V curves like those of Figure 5.16 for nMOS and pMOS transistors in your process.

5.7   The `char.pl` Perl script runs a number of simulations to characterize a process. Look for SPICE models for a newer process on the MOSIS Web site. Use the script to add another column to Table 5.5 for the new process.

5.8   The `charlib.pl` script runs a number of simulations to extract logical effort and parasitic delay of gates in a specified process. Add another column to Table 5.8 for a new process.

5.9   Use the `charlib.pl` script to find the logical effort and parasitic delay of a 5-input NAND gate for the outermost input.

5.10   Exercise 4.10 compares two designs of 2-input AND gates. Simulate each design and compare the average delays. What values of x and y give least delay? How much faster is the delay than that achieved using values of x and y suggested from logical effort calculations? How does the best delay compare to estimates using logical effort? Let $C = 10$ μm of gate capacitance.

5.11   Exercise 4.13 asks you to estimate the delay of a logic function. Simulate your design and compare your results to your estimate. Let one unit of capacitance be a minimum-sized transistor.

Combinational Circuit Design

6.1 Introduction

In Chapter 1, we introduced CMOS logic with the assumption that MOS transistors act as simple switches. *Static CMOS* gates used complementary nMOS and pMOS networks to drive '0' and '1' outputs, respectively. In Chapter 4, we used the RC delay model and logical effort to better understand the sources of delay in static CMOS logic.

In this chapter, we examine alternative CMOS logic configurations, called *circuit families*. While the vast majority of designs synthesize exclusively onto static CMOS libraries and even custom designs use static CMOS for 95% of the logic, high speed, low power, or density restrictions may force another solution. You should always use the circuit that satisfies the application and requires the least design and verification effort. The most commonly used alternative circuit families are ratioed circuits, dynamic circuits, and pass-transistor circuits.

As a grossly oversimplified yet still useful model, the delay of a logic gate depends on its output current I, load capacitance C, and output voltage swing ΔV, as given in EQ (6.1).

$$t \propto \frac{C}{I} \Delta V \tag{6.1}$$

Faster circuit families attempt to reduce one of these three terms. nMOS transistors provide more current than pMOS for the same size and capacitance, so nMOS networks are preferred. Observe that the logical effort is proportional to the C/I term because it is determined by the input capacitance of a gate that can deliver a specified output current.

One drawback of static CMOS is that it requires both nMOS and pMOS transistors on each input. During a falling output transition, the pMOS transistors add significant capacitance without helping the pull-down current; hence, static CMOS has a relatively large logical effort. Many faster circuit families seek to drive only nMOS transistors with the inputs, thus reducing capacitance and logical effort. An alternative mechanism must be provided to pull the output high. Determining when to pull outputs high involves monitoring the inputs, outputs, or some clock signal. Monitoring inputs and outputs inevitably loads the nodes, so clocked circuits are often fastest if the clock can be provided at the ideal time. Another drawback of static CMOS is that all the node voltages must transition between 0 and V_{DD}. Some circuit families use reduced voltage swings to improve propagation delays (and power consumption). This advantage must be weighed against the delay and power of amplifying outputs back to full levels later or the costs of tolerating the reduced swings.

A host of other circuit families have been proposed, but most have never been used in commercial products and are doomed to reside on dusty library shelves. Every transistor contributes capacitance, so most fast structures are simple. Nevertheless, we will describe many of these circuits as a record of ideas that have been explored. A few hold promise for the future, particularly in specialized applications. Many texts simply catalog these circuit families without making judgments. This book attempts to evaluate the circuit families so that designers can concentrate their efforts on the most promising ones, rather than searching for the "gotchas" that were not mentioned in the original papers. Of course, any such evaluation runs the risk of overlooking advantages or becoming incorrect as technology changes, so you should use your own judgment.

Static CMOS logic is particularly popular because of its robustness. Given the correct inputs, it will eventually produce the correct output so long as there were no errors in logic design or manufacturing. Other circuit families are prone to numerous pathologies, including charge sharing, leakage, threshold drops, and ratioing constraints. When using alternative circuit families, it is vital to understand the pathologies and check that the circuits will work correctly in all design corners.

Because we are interested in building performance-optimized chips, both circuit and physical design must be considered. These two phases of design are intimately meshed. The density, behavior, and power dissipation of circuits can have a direct impact on any high-level architectural decision and can allow or preclude options based on the selection of a logic style. For this reason, it is important that the architect and microarchitect have some idea of low-level circuit options.

6.2 Circuit Families

Static CMOS circuits with complementary nMOS pull-down and pMOS pull-up networks are used for the vast majority of logic gates in integrated circuits. They have good noise margins, and are fast, low power, insensitive to device variations, easy to design, widely supported by CAD tools, and readily available in standard cell libraries. When noise does exceed the margins, the gate delay increases because of the glitch, but the gate eventually will settle to the correct answer. Indeed, many ASIC methodologies only allow static CMOS circuits. This section begins with a number of techniques for optimizing static CMOS circuits.

Nevertheless, performance or area constraints occasionally dictate the need for other circuit families. The remainder of this section describes the most popular circuit families, including ratioed and dynamic circuits and pass transistors.

6.2.1 Static CMOS

Designers accustomed to AND and OR functions must learn to think in terms of NAND and NOR to take advantage of static CMOS. In manual circuit design, this is often done through bubble pushing. Compound gates are particularly useful to perform complex functions with relatively low logical efforts. When a particular input is known to be latest, the gate can be optimized to favor that input. Similarly, when either the rising or falling edge is known to be more critical, the gate can be optimized to favor that edge. We have focused on building gates with equal rising and falling delays; however, using smaller pMOS transistors can reduce delay, power, and area. In processes with multiple threshold voltages, multiple flavors of gates can be constructed with different speed/leakage power tradeoffs.

6.2.1.1 Bubble Pushing CMOS stages are inherently inverting, so AND and OR functions must be built from NAND and NOR gates. DeMorgan's Law helps with this conversion:

$$\overline{A \cdot B} = \overline{A} + \overline{B}$$
$$\overline{A + B} = \overline{A} \cdot \overline{B}$$

(6.2)

These relations are illustrated graphically in Figure 6.1. A NAND gate is equivalent to an OR of inverted inputs. A NOR gate is equivalent to an AND of inverted inputs. The same relationship applies to gates with more inputs. Switching between these representations is easy to do on a whiteboard and is often called *bubble pushing*.

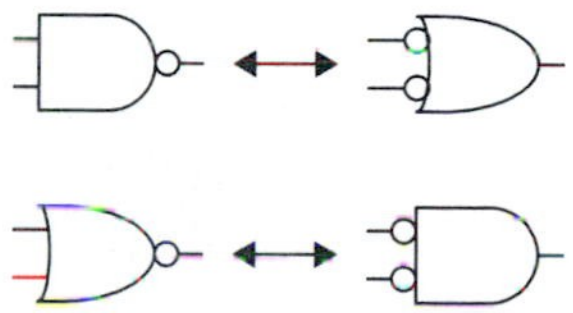

FIG 6.1 Bubble pushing with DeMorgan's law

6.2.1.2 Compound Gates As described in Section 1.4.5, static CMOS also efficiently handles compound gates computing various inverting combinations of AND/OR functions in a single stage. The function $F = AB + CD$ can be computed with an AND-OR-INVERT-22 (AOI22) gate and an inverter, as shown in Figure 6.2.

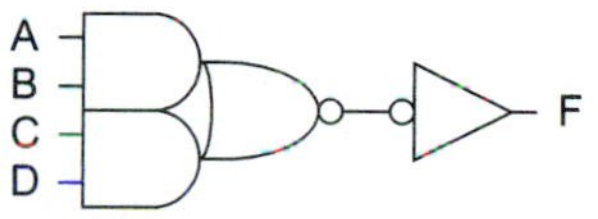

In general, logical effort of compound gates can be different for different inputs. Figure 6.4 shows how logical efforts can be estimated for the AOI21, AOI22, and a more complex compound AOI gate. The transistor widths are chosen to give the same drive as a unit inverter. The logical effort of each input is the ratio of the input capacitance of that input to the input capacitance of the inverter. For the AOI21 gate, this means the logical effort is slightly lower for the OR terminal (C) than for the two AND terminals (A, B). The parasitic delay is crudely estimated from the total diffusion capacitance on the output node by summing the sizes of the transistors attached to the output. The complex AOI will be used for comparison between logic families.

FIG 6.2 Logic using AOI22 gate

Example

Design a circuit to compute $F = AB + CD$ using NANDs and NORs.

Solution: By inspection, the circuit consists of two ANDs and an OR, shown in Figure 6.3(a). In Figure 6.3(b), the ANDs and ORs are converted to basic CMOS stages. In Figure 6.3(c and d), bubble pushing is used to simplify the logic to three NANDs.

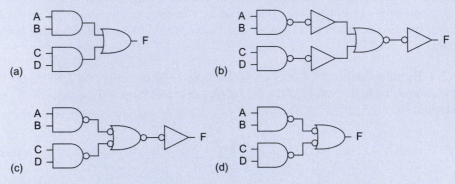

FIG 6.3   Bubble pushing to convert ANDs and ORs to NANDs and NORs

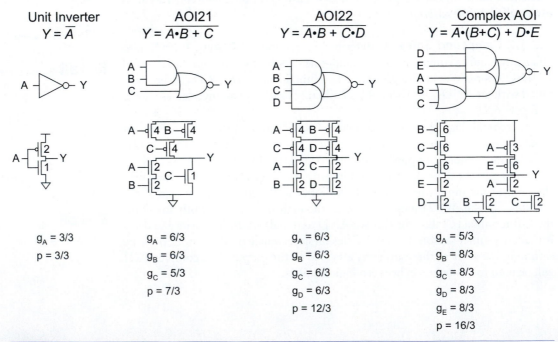

FIG 6.4   Logical efforts and parasitic delays of AOI gates

Example

Calculate the minimum delay, in τ, to compute $F = AB + CD$ using the circuits from Figure 6.2 and Figure 6.3(d). Each input can present a maximum of 20 λ of transistor width. The output must drive a load equivalent to 100 λ of transistor width. Choose transistor sizes to achieve this delay.

Solution: The path electrical effort is $H = 100/20 = 5$ and the branching effort is $B = 1$. The design using NAND gates has a path logical effort of $G = (4/3) \cdot (4/3) = 16/9$ and parasitic delay of $P = (2 + 2) = 4$. The design using the AOI22 and inverter has a path logical effort of $G = (6/3) \cdot 1 = 2$ and a parasitic delay of $P = (12/3 + 1) = 5$. Both designs have $N = 2$ stages. The path efforts $F = GBH$ are 80/9 and 10, respectively. The path delays are $NF^{1/N} + P$, or 10.0 τ and 11.3 τ, respectively. Using compound gates does not always result in faster circuits; simple 2-input NAND gates can be quite fast.

To compute the sizes, we determine the best stage efforts, $\hat{f} = F^{1/N}$, 3.0 and 3.2, respectively. These are in the range of 2.4–6 so we know the efforts are reasonable and the design would not improve too much by adding or removing stages. The input capacitance of the second gate is determined by the capacitance transformation

$$C_{\text{in}_i} = \frac{C_{\text{out}_i} \cdot g_i}{\hat{f}}.$$

For the NAND design,

$$C_{\text{in}} = \frac{100 \, \lambda \cdot (4/3)}{3.0} = 44 \, \lambda.$$

For the AOI22 design,

$$C_{\text{in}} = \frac{100 \, \lambda \cdot (1)}{3.2} = 31 \, \lambda.$$

The paths are shown in Figure 6.5 with transistor widths rounded to integer values.

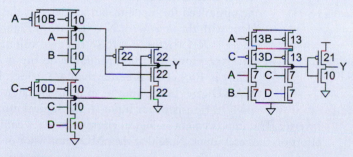

FIG 6.5   Transistor widths for example

6.2.1.3 Input Ordering Delay Effect The logical effort and parasitic delay of different gate inputs is often different. Some logic gates, like the AOI21 in the previous section, are inherently *asymmetric* in that one input sees less capacitance than another. Other gates, like NANDs and NORs, are nominally symmetric but actually have slightly different logical effort and parasitic delays for the different inputs.

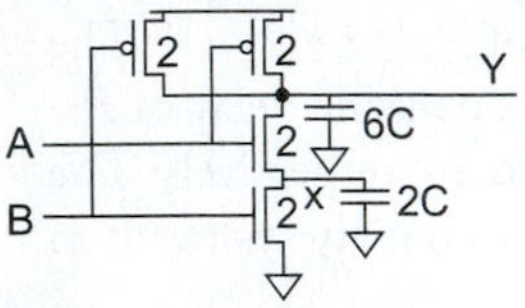

Figure 6.6 shows a 2-input NAND gate annotated with diffusion parasitics. Consider the falling output transition occurring when one input held a stable '1' value and the other rises from '0' to '1.' If input B rises last, node x will initially be at $V_{DD} - V_t \approx V_{DD}$ because it was pulled up through the nMOS transistor on input A. As shown in Section 4.2.1, the Elmore delay is $(R/2)(2C) + R(6C) = 7RC = 2.33\ \tau$[1]. On the other hand, if input A rises last, node x will initially be at 0 V because it was discharged through the nMOS transistor on input B. No charge must be delivered to node x, so the Elmore delay is simply $R(6C) = 6RC = 2\ \tau$.

In general, we define the *outer* input to be the input closer to the supply rail (e.g., B) and the *inner* input to be the input closer to the output (e.g., A). The parasitic delay is smallest when the inner input switches last because the intermediate nodes have already been discharged. Therefore, if one signal is known to arrive later than the others, the gate is fastest when that signal is connected to the inner input.

Table 5.7 listed the logical effort and parasitic delay for each input of various NAND gates, confirming that the inner input has a lower parasitic delay. The logical efforts are lower than initial estimates might predict because of velocity saturation. Interestingly, the inner input has a slightly higher logical effort because the intermediate node x tends to rise and cause negative feedback when the inner input turns ON (see Exercise 6.5) [Sutherland99]. This effect is seldom significant to the designer because the inner input remains faster over the range of fanouts used in reasonable circuits.

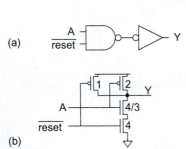

FIG 6.6 NAND gate delay estimation

6.2.1.4 Asymmetric Gates When one input is far less critical than another, even nominally symmetric gates can be made asymmetric to favor the late input at the expense of the early one. For example, consider the path in Figure 6.7(a). Under ordinary conditions, the path acts as a buffer between A and Y. When reset is asserted, the path forces the output low. If reset only occurs under exceptional circumstances and can take place slowly, the circuit should be optimized for input-to-output delay at the expense of reset. This can be done with the *asymmetric* NAND gate in Figure 6.7(b). The pull-down resistance is $R/4 + R/(4/3) = R$, so the gate still offers the same driver as a unit inverter. However, the capacitance on input A is only 10/3, so the logical effort is 10/9. This is better than 4/3, which is normally associated with a NAND gate. In the limit of an infinitely large reset transistor and unit-sized nMOS transistor for input A, the logical effort approaches 1, just like an inverter. The improvement in logical effort of input A comes at the cost of much higher effort on the reset input. Note that the pMOS transistor on the reset input is also shrunk.

FIG 6.7 Resettable buffer optimized for data input

---

[1]Recall that $\tau = 3RC$ is the delay of an inverter driving the gate of an identical inverter.

This reduces its diffusion capacitance and parasitic delay at the expense of slower response to reset.

In other circuits such as arbiters, we may wish to build gates that are perfectly symmetric so neither input is favored. Figure 6.8 shows how to construct a symmetric NAND gate.

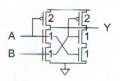

FIG 6.8 Perfectly symmetric 2-input NAND gate

6.2.1.5 Skewed Gates In other cases, one input transition is more important than the other. In Section 2.5.2 we defined *HI-skew* gates to favor the rising output transition and *LO-skew* gates to favor the falling output transition. This favoring can be done by decreasing the size of the noncritical transistor. The logical efforts for the rising (up) and falling (down) transitions are called g_u and g_d, respectively, and are the ratio of the input capacitance of the skewed gate to the input capacitance of an unskewed inverter with equal drive *for that transition*. Figure 6.9(a) shows how a HI-skew inverter is constructed by downsizing the nMOS transistor. This maintains the same effective resistance for the critical transition while reducing the input capacitance relative to the unskewed inverter of Figure 6.9(b), thus reducing the logical effort on that critical transition to $g_u = 2.5/3 = 5/6$. Of course, the improvement comes at the expense of the effort on the noncritical transition. The logical effort for the falling transition is estimated by comparing the inverter to a smaller unskewed inverter with equal pull-down current, shown in Figure 6.9(c), giving a logical effort of $g_d = 2.5/1.5 = 5/3$. The degree of skewing (e.g., the ratio of effective resistance for the fast transition relative to the slow transition) impacts the logical efforts and noise margins; a factor of two is common. Figure 6.10 catalogs HI-skew and LO-skew gates with a skew factor of two. Skewed gates are sometimes denoted with an *H* or an *L* on their symbol in a schematic.

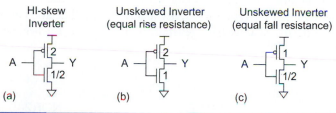

HI-skew Inverter	Unskewed Inverter (equal rise resistance)	Unskewed Inverter (equal fall resistance)
(a)	(b)	(c)

FIG 6.9 Logical effort calculation for HI-skew inverter

Alternating HI-skew and LO-skew gates can be used when only one transition is important [Solomatnikov00]. Skewed gates work particularly well with dynamic circuits, as we shall see in Section 6.2.4.

6.2.1.6 *P/N* Ratios Notice in Figure 6.10 that the average logical effort of the LO-skew NOR2 is actually better than that of the unskewed gate. The pMOS transistors in the unskewed gate are enormous in order to provide equal rise delay. They contribute input capacitance for both transitions, while only helping the rising delay. By accepting a slower

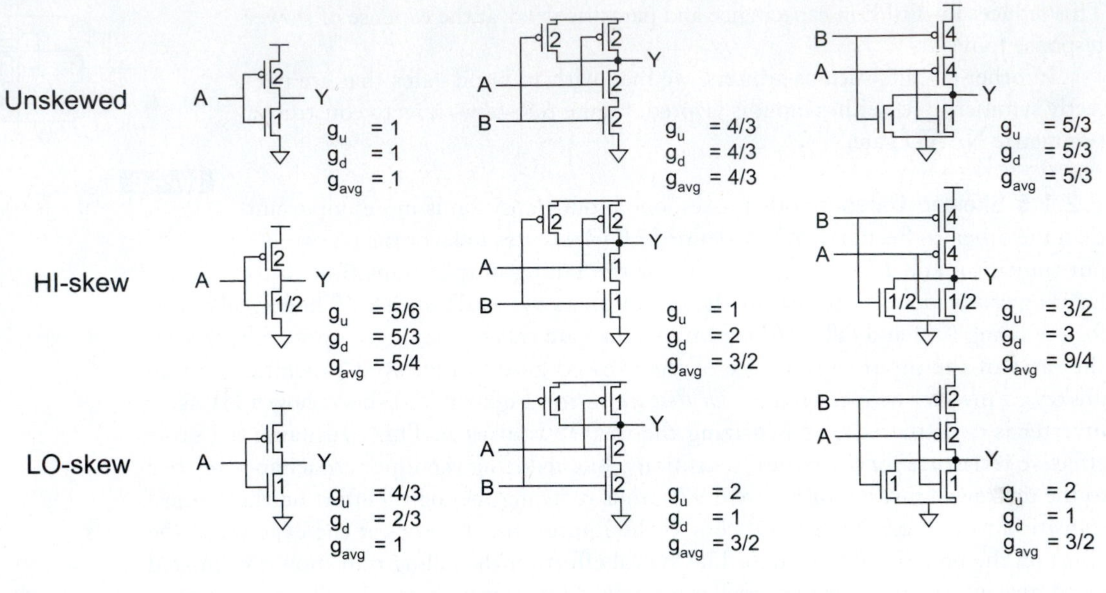

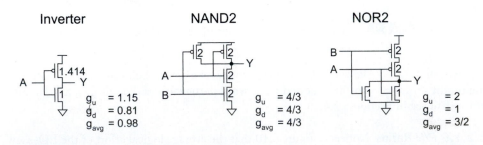

FIG 6.10 Catalog of skewed gates

rise delay, the pMOS transistors can be downsized to significantly reduce input capacitance and average delay.

In general, what is the best *P/N* ratio for logic gates (i.e., the ratio of pMOS to nMOS transistor width)? You will see in Exercise 6.13 that the ratio giving lowest average delay is the square root of the ratio that gives equal rise and fall delays. For processes with a mobility ratio of $\mu_n/\mu_p = 2$ as we have generally been assuming, the best ratios are shown in Figure 6.11.

FIG 6.11 Gates with *P/N* ratios giving least delay

Reducing the pMOS size from 2 to $\sqrt{2} \approx 1.4$ for the inverter gives the theoretical fastest average delay, but this delay improvement is only 2%. However, this significantly reduces the pMOS transistor area. It also reduces input capacitance, which in turn reduces power consumption. Unfortunately, it leads to unequal delay between the outputs. Some paths can be slower than average if they trigger the worst edge of each gate. Excessively slow rising outputs can also cause hot electron degradation. And reducing the pMOS size also moves the switching point lower and reduces the noise margin.

In summary, the P/N ratio of a library of cells should be chosen on the basis of area, power, and reliability, not average delay. For NOR gates, reducing the size of the pMOS transistors significantly improves both delay and area. In most standard cell libraries, the pitch of the cell determines the P/N ratio that can be achieved in any particular gate. Ratios of 1.5–2 are commonly used for inverters.

6.2.1.7 Multiple Threshold Voltages

Some CMOS processes offer two or more threshold voltages. Transistors with lower threshold voltages produce more ON current, but also leak exponentially more OFF current. Libraries can provide both high- and low-threshold versions of gates. The low-threshold gates can be used sparingly to reduce the delay of critical paths [Kumar94, Wei98]. Skewed gates can use low-threshold devices on only the critical network of transistors.

6.2.2 Ratioed Circuits

Ratioed circuits, introduced in Section 2.5.4, use weak pull-up devices and stronger pull-down devices. They reduce the input capacitance and hence improve logical effort by eliminating large pMOS transistors loading the inputs, but depend on the correct ratio of pull-up to pull-down strength. If the pull-up is too strong, V_{OLmax} may be too high; V_{OLmax} is best chosen to be less than V_{tn} so the low output does not turn ON the next stage. If the pull-up is too weak, the rising delay will be too slow. Ratioed circuits also dissipate static power while the output is low, so they must be used in a limited fashion where they provide significant benefits.

6.2.2.1 Pseudo-nMOS

Figure 6.12 shows *pseudo-nMOS* logic gates, which are the most common form of CMOS ratioed logic. The pull-down network is like that of a static gate, but the pull-up network has been replaced with a single pMOS transistor that is grounded so it is always ON. The pMOS transistor width is selected to be about 1/4 the strength (i.e., 1/2 the effective width) of the nMOS pull-down network as a compromise between noise margin and speed; this best size is highly process-dependent, but is usually in the range of 1/3 to 1/6.

To calculate the logical effort of pseudo-nMOS gates, suppose a complementary CMOS unit inverter delivers current I in both rising and falling transitions. For the widths shown, the pMOS transistors produce $I/3$ and the nMOS networks produce $4I/3$. The logical effort for each transition is computed as the ratio of the input capacitance to that of a complementary CMOS inverter with equal current for that transition. For the

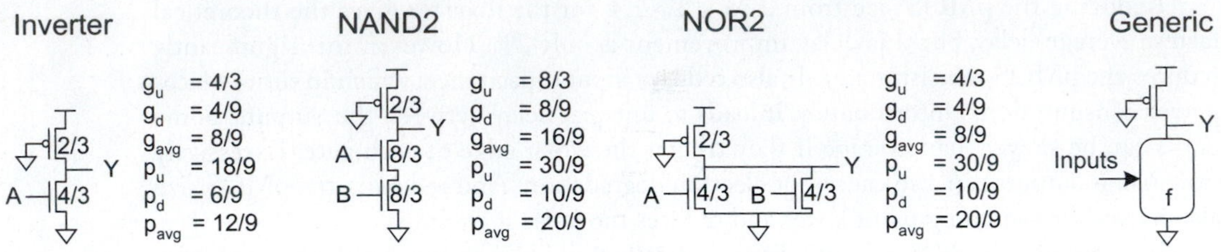

Inverter	NAND2	NOR2	Generic

Inverter

g_u = 4/3
g_d = 4/9
g_{avg} = 8/9
p_u = 18/9
p_d = 6/9
p_{avg} = 12/9

2/3

A — 4/3   Y

NAND2

g_u = 8/3
g_d = 8/9
g_{avg} = 16/9
p_u = 30/9
p_d = 10/9
p_{avg} = 20/9

2/3 Y

A — 8/3
B — 8/3

NOR2

g_u = 4/3
g_d = 4/9
g_{avg} = 8/9
p_u = 30/9
p_d = 10/9
p_{avg} = 20/9

2/3

A — 4/3   B — 4/3   Y

Generic

Inputs

Y

f

FIG 6.12 Pseudo-nMOS logic gates

falling transition, the pMOS transistor effectively fights the nMOS pull-down. The output current is estimated as the pull-down current minus the pull-up current, $(4I/3 - I/3) = I$. Therefore, we will compare each gate to a unit inverter to calculate g_d. For example, the logical effort for a falling transition of the pseudo-nMOS inverter is the ratio of its input capacitance (4/3) to that of a unit complementary CMOS inverter (3), i.e., 4/9. g_u is three times as great because the current is 1/3 as much.

The parasitic delay is also found by counting output capacitance and comparing it to an inverter with equal current. For example, the pseudo-nMOS NOR has 10/3 units of diffusion capacitance as compared to 3 for a unit-sized complementary CMOS inverter, so its parasitic delay pulling down is 10/9. The pull-up current is 1/3 as great, so the parasitic delay pulling up is 10/3.

As can be seen, pseudo-nMOS is slower on average than static CMOS for NAND structures. However, it works well for NOR structures. The logical effort is independent of the number of inputs in wide NORs, so pseudo-nMOS is useful for fast wide NOR gates or NOR-based structures like ROMs and PLAs when power permits.

Pseudo-nMOS gates will not operate correctly if $V_{OL} > V_{IL}$ of the receiving gate. This is most likely in the SF design corner where nMOS transistors are weak and pMOS transistors are strong. Designing for acceptable noise margin in the SF corner forces conservative choice of weak pMOS transistors in the normal corner. A biasing circuit can be used to reduce process sensitivity, as shown in Figure 6.15. The goal of the biasing circuit is to create a V_{bias} that causes P2 to deliver 1/3 the current of N2, independent of the relative mobilities of the pMOS and nMOS transistors. Transistor N2 has width of 3/2 and hence produces current $3I/2$ when ON. Transistor N1 is tied ON to act as a current source with 1/3 the current of N2, i.e., $I/2$. P1 acts as a current mirror using feedback to establish the bias voltage sufficient to provide equal current as N1, $I/2$. The size of P1 is noncritical so long as it is large enough to produce sufficient current and is equal in size to P2. Now, P2 ideally also provides $I/2$. In summary, when A is low, the pseudo-nMOS gate pulls up with a current of $I/2$. When A is high, the pseudo-nMOS gate pulls down with an effective current of $(3I/2 - I/2) = I$. To first order, this biasing technique sets the relative currents strictly by transistor widths, independent of relative pMOS and nMOS mobilities.

Example

Design a k-input AND gate with DeMorgan's Law using static CMOS inverters followed by a k-input pseudo-nMOS NOR, as shown in Figure 6.13. Let each inverter be unit-sized. If the output load is an inverter of size H, determine the best transistor sizes in the NOR gate and estimate the average delay of the path.

Solution: The path electrical effort is H and the branching effort is $B = 1$. The inverter has a logical effort of 1. The pseudo-nMOS NOR has an average logical effort of 8/9 according to Figure 6.12. The path logical effort is $G = 1 \cdot (8/9) = 8/9$, so the path effort is $8H/9$. Each stage should bear an effort of $\hat{f} = \sqrt{8H/9}$. Using the capacitance transformation gives NOR pull-down transistor widths of

$$C_{\text{in}} = \frac{gC_{\text{out}}}{\hat{f}} = \frac{(8/9)H}{\sqrt{8H/9}} = \frac{\sqrt{8H}}{3}$$

unit-sized inverters. As a unit inverter has three units of input capacitance, the NOR transistor nMOS widths should be $\sqrt{8H}$. According to Figure 6.12, the pull-up transistor should be half this width. The complete circuit marked with nMOS and pMOS widths is drawn in Figure 6.14.

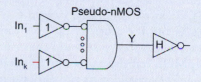

FIG 6.13 k-input AND gate driving load of H

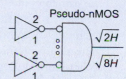

FIG 6.14 k-input AND marked with transistor widths

We estimate the average parasitic delay of a k-input pseudo-nMOS NOR to be $(8k + 4)/9$. The total delay in τ is

$$D = N\hat{f} + P = \frac{4\sqrt{2}}{3}\sqrt{H} + \frac{8k+13}{9}.$$

Increasing the number of inputs only impacts the parasitic delay, not the effort delay.

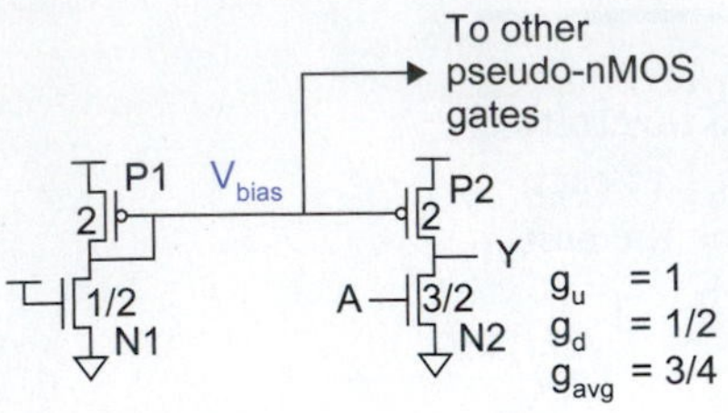

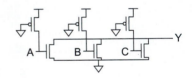

FIG 6.15 Replica biasing of pseudo-nMOS gates

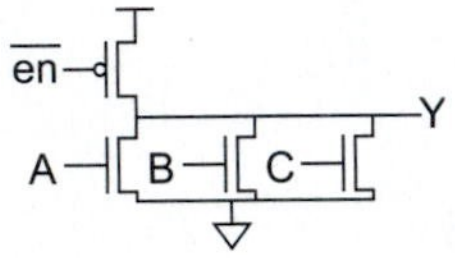

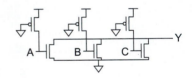

FIG 6.16 Pseudo-nMOS gate with enabled pull-up

FIG 6.17 3-input NOR drawn as multidrain logic

Such replica biasing permits the 1/3 current ratio rather than the conservative 1/4 ratio in the previous circuits, resulting in lower logical effort. The bias voltage V_{bias} can be distributed to multiple pseudo-nMOS gates. Ideally, V_{bias} will adjust itself to keep V_{OL} constant across process corners. Unfortunately, the currents through the two pMOS transistors do not exactly match because their drain voltages are unequal, so this technique still has some process sensitivity. Also note that this bias is relative to V_{DD}, so any noise on either the bias voltage line or the V_{DD} supply rail will impact circuit performance.

Turning off the pMOS transistor can reduce power when the logic is idle or during IDDQ test mode (see Section 9.6.8), as shown in Figure 6.16.

An alternate way to represent the pseudo-nMOS gate is to draw the pull-up at the input and the pull-down transistors with open drains, as shown in Figure 6.17. Multiple gates are tied together in a "wired-OR" fashion. This representation is called *CMOS Multidrain Logic* by the inventors [Wu87], but offers no advantages over normal pseudo-nMOS circuits in modern CAD environments.

6.2.2.2 Ganged CMOS
Figure 6.18 illustrates pairs of CMOS inverters ganged together. The truth table is given in Table 6.1, showing that the pair compute the NOR function. Such a circuit is sometimes called a *symmetric^2 NOR* [Johnson88], or more generally, *ganged CMOS* [Schultz90]. When one input is '0' and the other '1,' the gate can be viewed as a pseudo-nMOS circuit with appropriate ratio constraints. When both inputs are '0,' both pMOS transistors turn on in parallel, pulling the output high faster than they would in an ordinary pseudo-nMOS gate. Moreover, when both inputs are '1,' both pMOS transistors turn OFF, saving static power dissipation. As in pseudo-nMOS, the transistors are sized so the pMOS are about 1/4 the strength of the nMOS and the pull-down current matches that of a unit inverter. Hence,

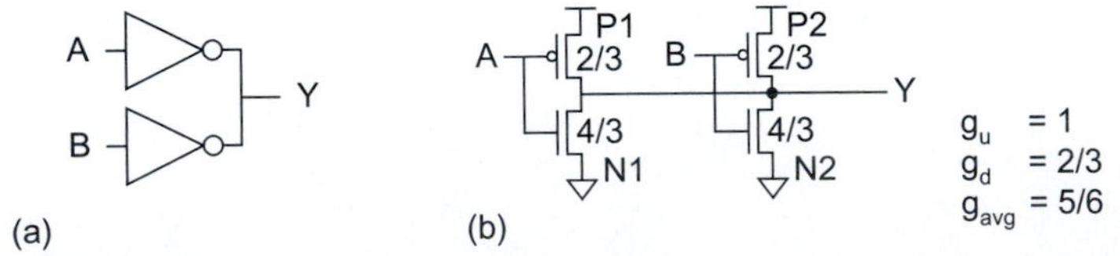

FIG 6.18 Symmetric 2-input NOR gate

---

[2]Do not confuse this use of *symmetric* with the concept of *symmetric* and *asymmetric* gates from Section 6.2.1.4.

Table 6.1		Operation of symmetric NOR				
A	*B*	*N1*	*P1*	*N2*	*P2*	*Y*
0	0	OFF	ON	OFF	ON	1
0	1	OFF	ON	ON	OFF	~ 0
1	0	ON	OFF	OFF	ON	~ 0
1	1	ON	OFF	ON	OFF	0

the symmetric NOR achieves both better performance and lower power dissipation than a 2-input pseudo-nMOS NOR.

Johnson also showed that symmetric structures can be used for NOR gates with more inputs and even for NAND gates (see Exercises 6.23–6.24). The 3-input symmetric NOR also works well, but the logical efforts of the other structures are unattractive.

6.2.2.3 Source Follower Pull-up Logic

Figure 6.19 shows a *Source Follower Pull-up Logic* (SFPL) 4-input NOR gate [Simon92]. It is similar to a pseudo-nMOS gate except that the pull-up is controlled by the inputs. *N6–N9* and *P1* form a pseudo-nMOS NOR function. The gate of the pull-up *P1* is driven by a parallel *source follower* consisting of drive transistors *N1–N4* and load transistor N_{load}. When one input turns on, the source follower pulls node *x* to approximately $V_{DD}/2$. This tends to partially turn off *P1*, which allows smaller nMOS pulldowns *N6–N9* to be used. However, *N1–N4* also load the input, so the overall reduction in input capacitance is not clear. SFPL is primarily applicable to constructing wide NOR gates.

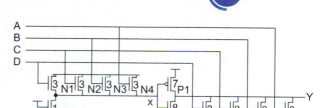

FIG 6.19   SFPL 4-input NOR gate

6.2.3   Cascode Voltage Switch Logic

Cascode Voltage Switch Logic (CVSL[3]) [Heller84] seeks the performance of ratioed circuits without the static power consumption. It uses both true and complementary input signals and computes both true and complementary outputs using a pair of nMOS pull-down networks, as shown in Figure 6.20(a). The pull-down network *f* implements the logic function as in a static CMOS gate, while $\overline{f}$ uses inverted inputs feeding transistors arranged in the conduction complement. For any given input pattern, one of the pull-down networks will be ON and the other OFF. The pull-down network that is ON will pull that output low. This low output turns ON the pMOS transistor to pull the opposite output high. When the opposite output rises, the other pMOS transistor turns OFF so no static power dissipation occurs. Figure 6.20(b) shows a CVSL AND/NAND gate. Observe how the

---

[3] Many authors call this circuit family *Differential Cascode Voltage Switch Logic* (DCVS [Chu86] or DCVSL [Ng96]). The term *cascode* comes from analog circuits where transistors are placed in series.

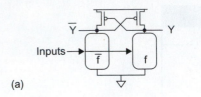

(a)

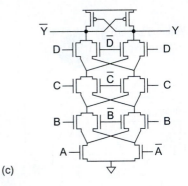

(b)

(c)

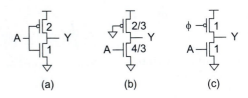

FIG 6.20 CVSL gates

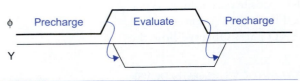

(a)   (b)   (c)

FIG 6.21 Comparison of (a) static CMOS, (b) pseudo-nMOS, and (c) dynamic inverters

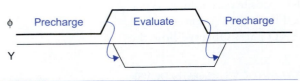

FIG 6.22 Precharge and evaluation of dynamic gates

pull-down networks are complementary, with parallel transistors in one and series in the other. Figure 6.20(c) shows a 4-input XOR gate. The pull-down networks share A and $\overline{A}$ transistors to reduce the transistor count by two. Sharing is often possible in complex functions, and systematic methods exist to design shared networks [Chu86].

CVSL has a potential speed advantage because all of the logic is performed with nMOS transistors, thus reducing the input capacitance. As in pseudo-nMOS, the size of the pMOS transistor is important. It fights the pull-down network, so a large pMOS transistor will slow the falling transition. Unlike pseudo-nMOS, the feedback tends to turn off the pMOS, so the outputs will eventually settle to a legal logic level. A small pMOS transistor is slow at pulling the complementary output high. In addition, the CVSL gate requires both the low- and high-going transitions, adding more delay. Contention current during the switching period also increases power consumption.

Pseudo-nMOS worked well for wide NOR structures. Unfortunately, CVSL also requires the complement, a slow tall NAND structure. Therefore, CVSL is poorly suited to general NAND and NOR logic. Even for symmetric structures like XORs, it tends to be slower than static CMOS, as well as more power-hungry [Chu87, Ng96]. However, the ideas behind CVSL help us understand dual-rail domino and complementary pass-transistor logic discussed in later sections.

6.2.4 Dynamic Circuits

Ratioed circuits reduce the input capacitance by replacing the pMOS transistors connected to the inputs with a single resistive pull-up. The drawbacks of ratioed circuits include slow rising transitions, contention on the falling transitions, static power dissipation, and a non-zero V_{OL}. Dynamic circuits circumvent these drawbacks by using a clocked pull-up transistor rather than a pMOS that is always ON. Figure 6.21 compares (a) static CMOS, (b) pseudo-nMOS, and (c) dynamic inverters. Dynamic circuit operation is divided into two modes, shown in Figure 6.22. During *precharge*, the clock ϕ is '0,' so the clocked pMOS is ON and initializes the output Y high. During *evaluation*, the clock is '1' and the clocked pMOS turns OFF. The output may remain high or may be discharged low through the pull-down network. Dynamic circuits are the fastest commonly used circuit family because they have lower input capacitance and no contention during switching. They also have zero static power dissipation.

However, they require careful clocking, consume significant dynamic power, and are sensitive to noise during evaluation. Clocking of dynamic circuits will be discussed in much more detail in Section 7.5.

In Figure 6.21(c), if the input *A* is '1' during precharge, contention will take place because both the pMOS and nMOS transistors will be ON. When the input cannot be guaranteed to be '0' during precharge, an extra clocked evaluation transistor can be added to the bottom of the nMOS stack to avoid contention as shown in Figure 6.23. The extra transistor is sometimes called a *foot*. Figure 6.24 shows generic *footed* and *unfooted* gates[4].

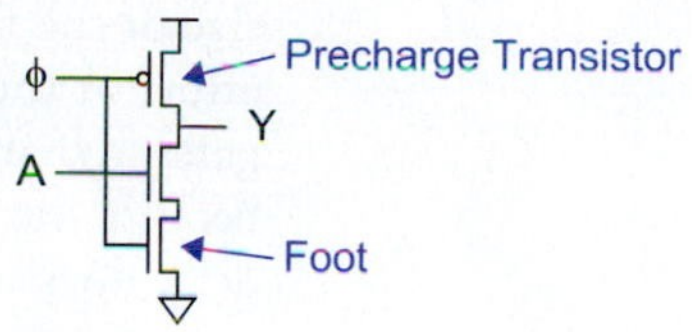

Figure 6.25 estimates the falling logical effort of both footed and unfooted dynamic gates. As usual, the pull-down transistors' widths are chosen to give unit resistance. Precharge occurs while the gate is idle and often may take place more slowly. Therefore, the precharge transistor width is chosen for twice unit resistance. This reduces the capacitive load on the clock and the parasitic capacitance at the expense of greater rising delays. We see that the logical efforts are very low. Footed gates have higher logical effort than their unfooted counterparts but are still an improvement over static logic. In practice, the logical effort of footed gates is better than predicted because velocity saturation means series nMOS transistors have less resistance than we have estimated. Moreover, logical efforts are also slightly better than predicted because there is

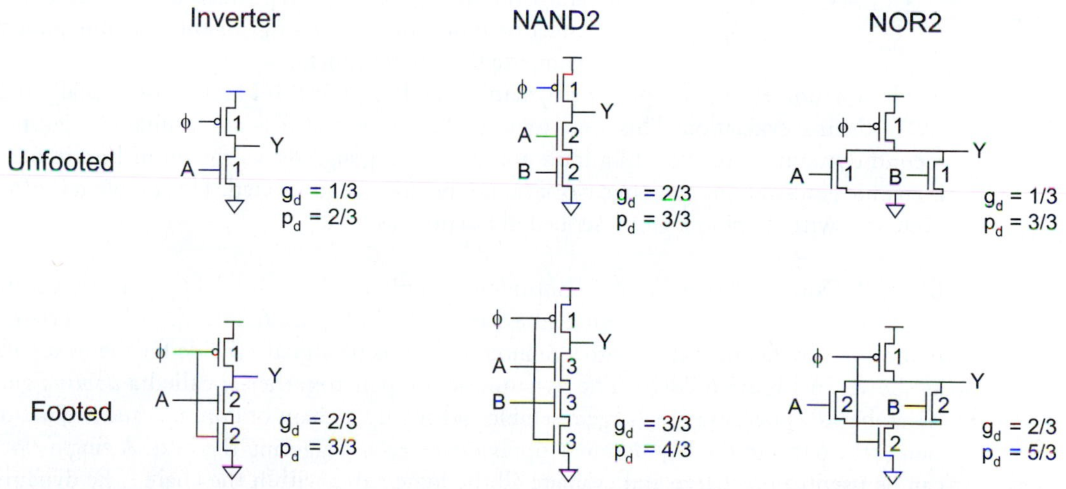

FIG 6.23 Footed dynamic inverter

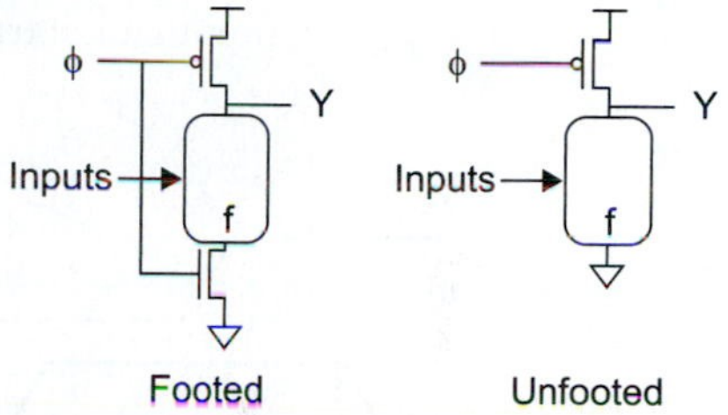

FIG 6.24 Generalized footed and unfooted dynamic gates

FIG 6.25 Catalog of dynamic gates

---

[4]The footed and unfooted terminology is from IBM [Nowka98]. Intel calls these styles D1 and D2, respectively.

no contention between nMOS and pMOS transistors during the input transition. The size of the foot can be increased relative to the other nMOS transistors to reduce logical effort of the other inputs at the expense of greater clock loading. Like pseudo-nMOS gates, dynamic gates are particularly well suited to wide NOR functions or multiplexers because the logical effort is independent of the number of inputs. Of course, the parasitic delay does increase with the number of inputs because there is more diffusion capacitance on the output node. Characterizing the logical effort and parasitic delay of dynamic gates is tricky because the output tends to fall much faster than the input rises, leading to potentially misleading dependence of propagation delay on fanout [Sutherland99].

A fundamental difficulty with dynamic circuits is the *monotonicity* requirement. While a dynamic gate is in evaluation, the inputs must be *monotonically rising*. That is, the input can start LOW and remain LOW, start LOW and rise HIGH, start HIGH and remain HIGH, but not start HIGH and fall LOW. Figure 6.26 shows waveforms for a footed dynamic inverter in which the input violates monotonicity. During precharge, the output is pulled HIGH. When the clock rises, the input is HIGH so the output is discharged LOW through the pull-down network, as you would want to have happen in an inverter. The input later falls LOW, turning off the pull-down network. However, the precharge transistor is also OFF so the output floats, staying LOW rather than rising as it would in a normal inverter. The output will remain low until the next precharge step. In summary, the inputs must be monotonically rising for the dynamic gate to compute the correct function.

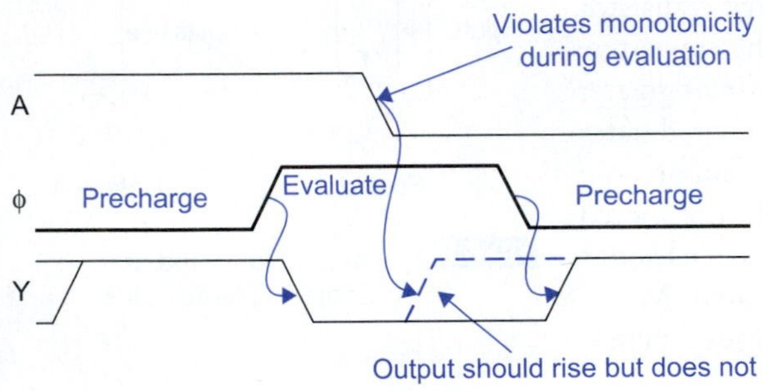

FIG 6.26 Monotonicity problem

Unfortunately, the output of a dynamic gate begins HIGH and monotonically falls LOW during evaluation. This monotonically falling output X is not a suitable input to a second dynamic gate expecting monotonically rising signals, as shown in Figure 6.27. Dynamic gates sharing the same clock cannot be directly connected. This problem is often overcome with domino logic, described in the next section.

6.2.4.1 Domino Logic The monotonicity problem can be solved by placing a static CMOS inverter between dynamic gates, as shown in Figure 6.28(a). This converts the monotonically falling output into a monotonically rising signal suitable for the next gate, as shown in Figure 6.28(b). The dynamic-static pair together is called a *domino* gate [Krambeck82] because precharge resembles setting up a chain of dominos and evaluation causes the gates to fire like dominos tipping over, each triggering the next. A single clock can be used to precharge and evaluate all the logic gates within the chain. The dynamic output is monotonically falling during evaluation, so the static inverter output is monotonically rising. Therefore, the static inverter is usually a HI-skew gate to favor this rising output. Observe that precharge occurs in parallel, but evaluation occurs sequentially. This explains why precharge is usually less critical. The symbols for the dynamic NAND, HI-skew inverter, and domino AND are shown in Figure 6.28(c).

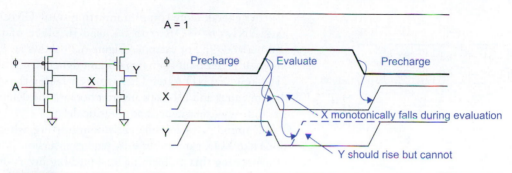

FIG 6.27 Incorrect connection of dynamic gates

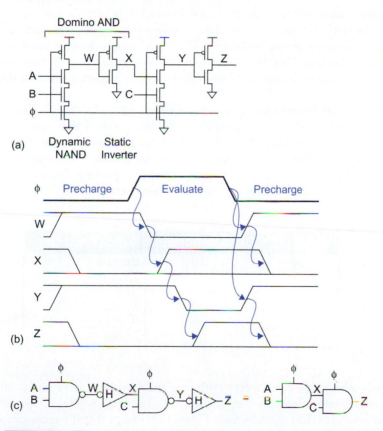

FIG 6.28 Domino gates

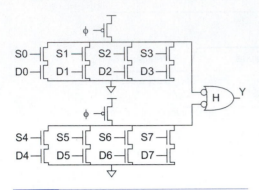

FIG 6.29 Domino gate using logic in static CMOS stage

In general, more complex inverting static CMOS gates such as NANDs or NORs can be used in place of the inverter [Sutherland99]. For example, Figure 6.29 shows an 8-input domino multiplexer built from two 4-input dynamic multiplexers and a HI-skew NAND gate. This is often faster than an 8-input dynamic mux and HI-skew inverter because the dynamic stage has less diffusion capacitance and parasitic delay.

Domino gates are inherently noninverting, while some functions like XOR gates necessarily require inversion. Three methods of addressing this problem include pushing inversions into static logic, delaying clocks, and using dual-rail domino logic. In many circuits such as arithmetic logic units (ALUs), the necessary XOR gate at the end of the path can be built with a conventional static CMOS XOR gate driven by the last domino circuit. However, the XOR output no longer is monotonically rising and thus cannot directly drive more domino logic. A second approach is to directly cascade dynamic gates without the static CMOS inverter, delaying the clock to the later gates to ensure the inputs are monotonic during evaluation. This is commonly done in content-addressable memories (CAMs) and NOR-NOR PLAs and will be discussed in Sections 7.5.4 and 11.7. The third approach, dual-rail domino logic, is discussed in the next section.

6.2.4.2 Dual-rail Domino Logic   *Dual-rail domino* gates encode each signal with a pair of wires. The input and output signal pairs are denoted with _h and _l, respectively. Table 6.2 summarizes the encoding. The _h wire is asserted to indicate that the output of the gate is "high" or '1.' The _l wire is asserted to indicate that the output of the gate is "low" or '0.' When the gate is precharged, neither _h nor _l is asserted. The pair of lines should never be both asserted simultaneously during correct operation.

Table 6.2	Dual-rail domino signal encoding	
sig_h	*sig_l*	Meaning
0	0	precharged
0	1	'0'
1	0	'1'
1	1	invalid

Dual-rail domino gates accept both true and complementary inputs and compute both true and complementary outputs, as shown in Figure 6.30(a). Observe that this is identical to static CVSL circuits from Figure 6.20 except that the cross-coupled pMOS transistors are instead connected to the precharge clock. Therefore, dual-rail domino can be viewed as a dynamic form of CVSL, sometimes called DCVS [Heller84]. Figure

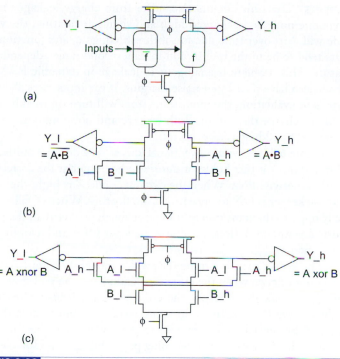

FIG 6.30 Dual-rail domino gates

6.30(b) shows a dual-rail AND/NAND gate and Figure 6.30(c) shows a dual-rail XOR/XNOR gate. The gates are shown with clocked evaluation transistors, but can also be unfooted. Dual-rail domino is a complete logic family in that it can compute all inverting and noninverting logic functions. However, it requires more area, wiring, and power. Dual-rail structures also lose the efficiency of wide dynamic NOR gates because they require complementary tall dynamic NAND stacks.

Dual-rail domino signals not only the result of a computation but also indicates when the computation is done. Before computation completes, both rails are precharged. When the computation completes, one rail will be asserted. A NAND gate can be used for completion detection as shown in Figure 6.31. This is particularly useful for asynchronous circuits [Williams91, Sparsø01].

Coupling can be reduced in dual-rail signal busses by interdigitating the bits of the bus, as shown in Figure 6.32. Each wire will never see more than one aggressor switching at a time because only one of the two rails switches in each cycle.

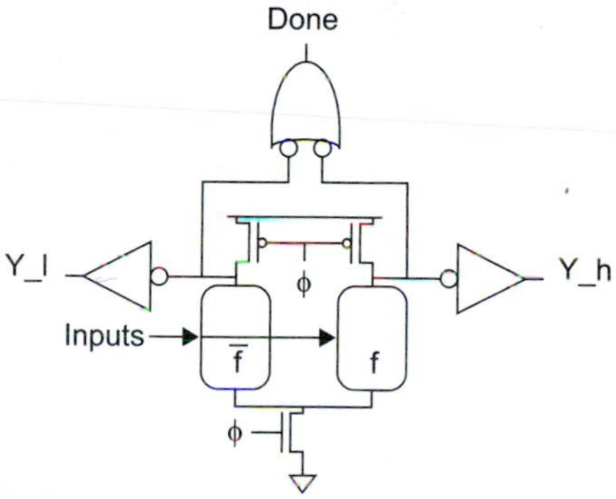

FIG 6.31 Dual-rail domino gate with completion detection

6.2.4.3 Keepers Dynamic circuits also suffer from charge leakage on the dynamic node. If a dynamic node is precharged high and then left floating, the voltage on the dynamic node will drift over time due to subthreshold, gate, and junction leakage. The time constants tend to be in the millisecond to nanosecond range, depending on process and temperature. This problem is analogous to leakage in dynamic RAMs. Moreover, dynamic circuits also have poor input noise margins. If the input rises above V_t while the gate is in evaluation, the input transistors will turn on weakly and can incorrectly discharge the output. Both leakage and noise margin problems can be addressed by adding a *keeper* circuit.

Figure 6.33 shows a conventional keeper on a domino buffer. The keeper is a weak transistor that holds, or *staticizes*, the output at the correct level when it would otherwise float. When the dynamic node X is high, the output Y is low and the keeper is ON to prevent X from floating. When X falls, the keeper initially opposes the transition so it must be much weaker than the pull-down network. Eventually Y rises, turning the keeper OFF and avoiding static power dissipation.

The keeper must be strong (i.e., wide) enough to compensate for any leakage current drawn when the output is floating and the pull-down stack is OFF. Strong keepers also improve the noise margin because when the inputs are slightly above V_t the keeper can supply enough current to hold the output high. Figure 5.26 showed the DC transfer characteristics of a dynamic inverter. As the keeper width k increases, the switching point shifts right. However, strong keepers also increase delay, typically by 5%–10%. Keeper transistors are usually on the order of 1/10 the strength of the pull-down stack, although they may need to be stronger on wide NOR gates or multiplexers in particularly leaky processes. For small dynamic gates, this ratio implies that the keeper must be weaker than a minimum-sized transistor. This is achieved by increasing the keeper length, as shown in Figure 6.34(a). Long keeper transistors increase the capacitive load on the output Y. This can be avoided by splitting the keeper, as shown in Figure 6.34(b).

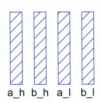

a_h b_h a_l b_l

FIG 6.32 Reducing coupling noise on dual-rail busses

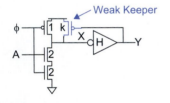

FIG 6.33 Conventional keeper

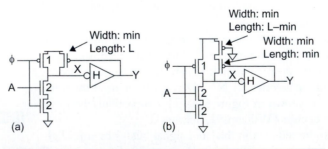

FIG 6.34 Weak keeper implementations

Figure 6.35 shows a *differential keeper* for a dual-rail domino buffer. When the gate is precharged, both keeper transistors are OFF and the dynamic outputs float. However, as soon as one of the rails evaluates low, the opposite keeper turns ON. The differential keeper is fast because it does not oppose the falling rail. As long as one of the rails is guaranteed to fall promptly, the keeper on the other rail will turn on before excessive leakage or noise causes failure. Of course, dual-rail domino can also use a pair of conventional keepers.

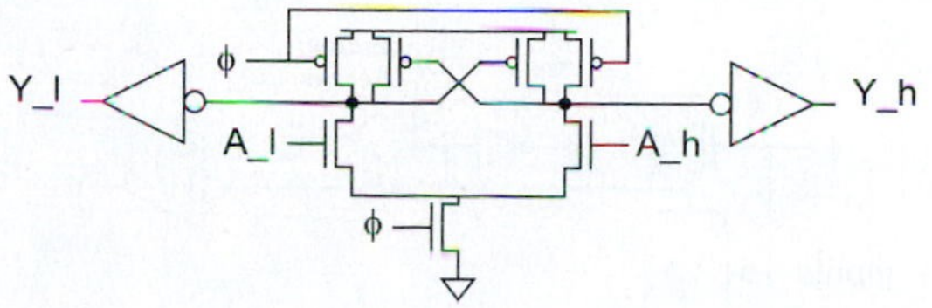

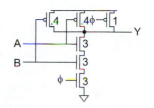

FIG 6.35 Differential keeper

An alternative approach is to build the keeper with a small complementary pull-up stack connected to the inputs rather than a weak pMOS transistor driven by the inverted output, as shown in Figure 6.36 for a 2-input NAND gate. In the terminology of Section 6.2.1.5, such a gate can be viewed as a very LO-skew (e.g., skew factor 5) static CMOS gate augmented with precharge and (possibly) evaluation transistors. Skew factors of 4–8 are reasonable. The approach has been called *noise tolerant precharge* (NTP) [Yamada95, Murabayashi96] or *monotonic static CMOS* [Thorp99]. Such noise-tolerant precharge gates have the advantage that the keeper will eventually recover from a noise event, in contrast to standard keepers that never recover if the input glitches enough to turn them off. Noise-tolerant precharge does not work well for wide NOR structures because it requires many series pMOS transistors. It is also most practical for gates with wide input transistors where it is feasible to construct a comparatively weak complementary network with minimum-sized pMOS transistors. Figure 6.37 plots the static noise margin and delay of dynamic NAND2 gates using conventional keepers and

FIG 6.36 Noise-tolerant precharge

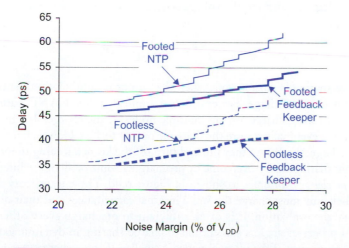

FIG 6.37 Delay vs. noise margin of conventional and noise-tolerant precharge dynamic NAND2s

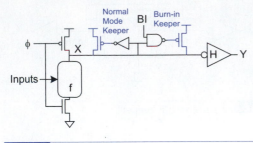

FIG 6.38 Burn-in conditional keeper

NTP [Harris03]. Increasing the keeper size improves noise margin at the expense of delay. For the same noise margin, the conventional keepers are consistently faster.

During *burn-in*, the chip operates at reduced frequency, but at very high temperature and voltage. This causes severe leakage that can overpower the keeper in wide dynamic NOR gates where many nMOS transistors leak in parallel. Figure 6.38 shows a domino gate with a *burn-in conditional keeper* [Alvandpour02]. The *BI* signal is asserted during burn-in to turn on a second keeper in parallel with the primary keeper. The second keeper slows the gate during burn-in, but provides extra current to fight leakage.

Domino circuits with delayed clocks can use full keepers consisting of cross-coupled inverters to hold the output either high or low, as discussed in Section 7.5.2. Many other keeper techniques have been proposed including a "leaker" that is always ON [Krambeck82], input-controlled refresh [Lakshmanan01], the inverter technique [Covino97], pMOS pull-up technique [D'Souza96], mirror technique [Wang00l], and twin-transistor technique [Balamurugan01], but the conventional keeper is generally satisfactory because it is simple and has little cost in speed and power dissipation.

6.2.4.4 Secondary Precharge Devices Dynamic gates are subject to problems with *charge sharing* [Oklobdžija86]. For example, consider the 2-input dynamic NAND gate in Figure 6.39(a). Suppose the output Y is precharged to V_{DD} and inputs A and B are low. Also suppose that the intermediate node x had a low value from a previous cycle. During evaluation, input A rises, but input B remains low so the output Y should remain high. However, charge is shared between C_x and C_Y, shown in Figure 6.39(b). This behaves as a capacitive voltage divider and the voltages equalize at

$$V_x = V_Y = \frac{C_Y}{C_x + C_Y} V_{DD} \qquad (6.3)$$

Charge sharing is most serious when the output is lightly loaded (small C_Y) and the internal capacitance is large. For example, 4-input dynamic NAND gates and complex AOI gates can share charge among multiple nodes. If the charge-sharing noise is small, the keeper will eventually restore the dynamic output to V_{DD}. However, if the charge-sharing noise is large, the output may flip and turn off the keeper, leading to incorrect results.

Charge sharing can be overcome by precharging some or all of the internal nodes with *secondary precharge transistors*, as shown in Figure 6.40. These transistors should be small because they only must charge the small internal capacitances and their diffusion capacitance slows the evaluation. It is often sufficient to precharge every other node in a tall stack. SOI processes are less susceptible to charge sharing in dynamic gates because the diffusion capacitance of the internal nodes is smaller. If some charge sharing is acceptable, a gate can be made faster by predischarging some internal nodes [Ye00].

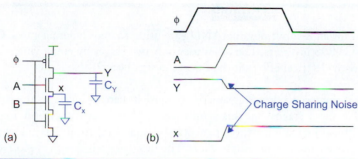

FIG 6.39 Charge-sharing noise

In summary, domino logic was originally proposed as a fast and compact circuit technique. In practice, domino is prized for its speed. However, by the time feet, keepers, and secondary precharge devices are added for robustness, domino is seldom much more compact than static CMOS and it demands a tremendous design effort to ensure robust circuits. When dual-rail domino is required, the area exceeds static CMOS.

6.2.4.5 Logical Effort of Dynamic Paths

In Section 4.3.3, we found the best stage effort by hypothetically appending static CMOS inverters onto the end of the path. The best effort depended on the parasitic delay and was 3.59 for $p_{inv} = 1$. When we employ alternative circuit families, the best stage effort may change. For example, with domino circuits, we may consider appending domino buffers onto the end of the path. Figure 6.41 shows that the logical effort of a domino buffer is G = 5/9 for footed domino and 5/18 for unfooted domino. Therefore, each buffer appended to a path actually decreases the path effort. Hence, it is better to add more buffers, or equivalently, to target a lower stage effort than you would in a static CMOS design.

[Sutherland99] showed that the best stage effort is $\rho = 2.76$ for paths with footed domino and 2.0 for paths with unfooted domino. In paths mixing footed and unfooted domino, the best effort is somewhere between these extremes. As a rule of thumb, just as you target a stage effort of 4 for static CMOS paths, you can target a stage effort of 2–3 for domino paths.

We have also seen that it is possible to push logic into the static CMOS stages between dynamic gates. The following example explores under what circumstances this is beneficial.

In summary, dynamic stages are fast because they build logic using nMOS transistors. Moreover, the low logical efforts suggest that using a relatively large number of stages is beneficial. Pushing logic into the static CMOS stages uses slower

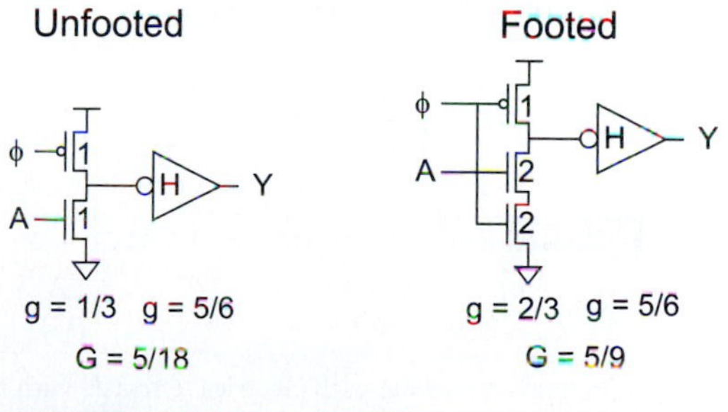

OPTIONAL

Secondary Precharge Transistor

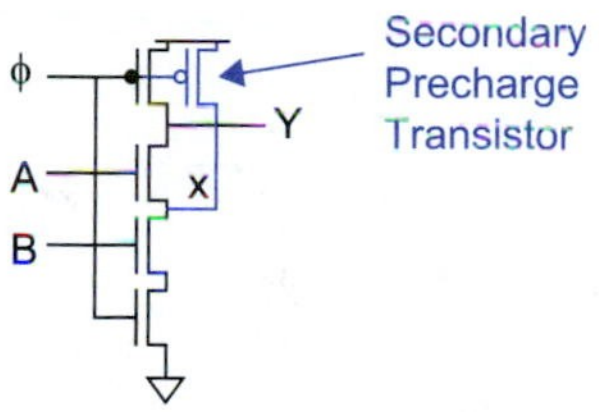

FIG 6.40 Secondary precharge transistor

Unfooted

g = 1/3   g = 5/6

G = 5/18

Footed

g = 2/3   g = 5/6

G = 5/9

FIG 6.41 Logical efforts of domino buffers

Example

Figure 6.42 shows two designs for an 8-input domino AND gate using footed dynamic gates. One uses four stages of logic with static CMOS inverters. The other uses only two stages by employing a HI-skew NOR gate. For what range of path electrical efforts is the 2-stage design faster?

Solution: You might expect that the second design is superior because it scarcely increases the complexity of the static gate and uses half as many stages, but this is only true for low electrical efforts. Figure 6.43 shows the paths annotated with (a) logical effort, (b) parasitic delay, and (c) total delay. The parasitic delays only consider diffusion capacitance on the output node. The delay of each design is plotted against path electrical effort H.[5] For $H > 2.9$, the 4-stage design becomes preferable because the domino gates are effective buffers.

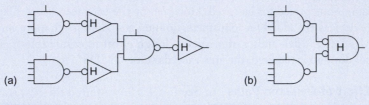

FIG 6.42 8-input domino AND gates

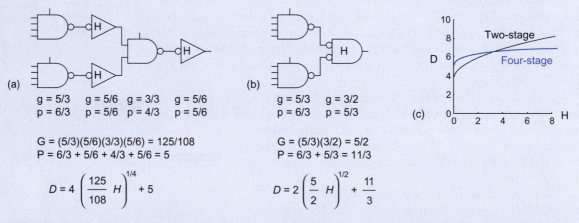

(a)

g = 5/3	g = 5/6	g = 3/3	g = 5/6
p = 6/3	p = 5/6	p = 4/3	p = 5/6

G = (5/3)(5/6)(3/3)(5/6) = 125/108
P = 6/3 + 5/6 + 4/3 + 5/6 = 5

$$D = 4\left(\frac{125}{108}H\right)^{1/4} + 5$$

(b)

g = 5/3	g = 3/2
p = 6/3	p = 5/3

G = (5/3)(3/2) = 5/2
P = 6/3 + 5/3 = 11/3

$$D = 2\left(\frac{5}{2}H\right)^{1/2} + \frac{11}{3}$$

(c)

FIG 6.43 8-input domino AND delays

---

[5]Do not confuse the path electrical effort H with the letter H designating the HI-skew static CMOS gates in the schematic.

pMOS transistors and reduces the number of stages. Thus, it is usually good to use static CMOS gates only on paths with low electrical effort.

6.2.4.6 Multiple-output Domino Logic (MODL)

It is often necessary to compute multiple functions where one is a subfunction of another or shares a subfunction. *Multiple-output domino logic* (MODL) [Hwang89, Wang97] saves area by combining all of the computations into a multiple-output gate.

A popular application is in addition, where the carry-out c_i of each bit of a 4-bit block must be computed, as will be discussed in Section 10.2.2.2. Each bit position i in the block can either propagate the carry (p_i) or generate a carry (g_i). The carry-out logic is

$$
\begin{aligned}
c_1 &= g_1 + p_1 c_0 \\
c_2 &= g_2 + p_2(g_1 + p_1 c_0) \\
c_3 &= g_3 + p_3(g_2 + p_2(g_1 + p_1 c_0)) \\
c_4 &= g_4 + p_4(g_3 + p_3(g_2 + p_2(g_1 + p_1 c_0)))
\end{aligned}
\tag{6.4}
$$

This can be implemented in four compound AOI gates, as shown in Figure 6.44(a). Notice that each output is a function of the less significant outputs. The more compact MODL design is often called a *Manchester carry chain*, shown in Figure 6.44(b). Note that the intermediate outputs require secondary precharge transistors. Also note that care must be taken for certain inputs to be mutually exclusive in order to avoid *sneak paths*. For example, in the adder we must define

$$
\begin{aligned}
g_i &= a_i b_i \\
p_i &= a_i \oplus b_i
\end{aligned}
\tag{6.5}
$$

If p_i were defined as $a_i + b_i$, a sneak path could exist when a_4 and b_4 are '1' and all other inputs are '0.' In that case, $g_4 = p_4 = 1$. c_4 would fire as desired, but c_3 would also fire incorrectly, as shown in Figure 6.45.

6.2.4.7 NP and Zipper Domino

Another variation on domino is shown in Figure 6.46(a). The HI-skew inverting static gates are replaced with predischarged dynamic gates using pMOS logic. For example, a footed dynamic p-logic NAND gate is shown in Figure 6.46(b). When ϕ is 0, the first and third stages precharge high while the second stage predischarges low. When ϕ rises, all the stages evaluate. Domino connections are possible, as shown in Figure 6.46(c). The design style is called *NP Domino* or *NORA Domino* (NO RAce) [Gonclaves83, Friedman84].

NORA has two major drawbacks. The logical effort of footed p-logic gates is generally worse than that of HI-skew gates (e.g., 2 vs. 3/2 for NOR2 and 4/3 vs. 1 for NAND2). Secondly, NORA is extremely susceptible to noise. In an ordinary dynamic

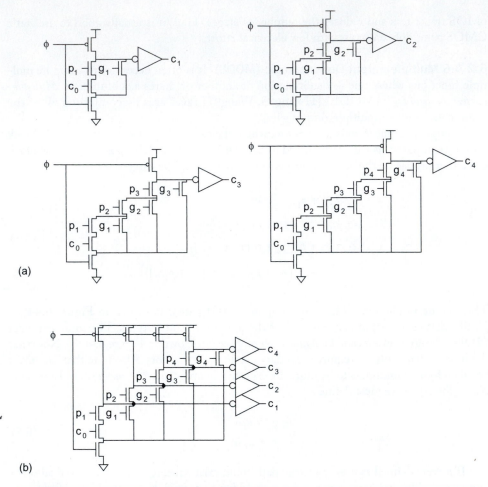

(a)

(b)

FIG 6.44 Conventional and MODL carry chains

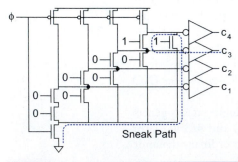

FIG 6.45 Sneak path

gate, the input has a low noise margin (about V_t), but is strongly driven by a static CMOS gate. The floating dynamic output is more prone to noise from coupling and charge sharing, but drives another static CMOS gate with a larger noise margin. In NORA, however, the sensitive dynamic inputs are driven by noise-prone dynamic outputs. Given these drawbacks and the extra clock phase required, there is little reason to use NORA.

Zipper domino [Lee86] is a closely related technique that leaves the precharge transistors slightly ON during evaluation by using precharge clocks that swing between 0 and $V_{DD} - |V_{tp}|$ for the pMOS precharge and V_{tn} and V_{DD} for the nMOS precharge. This

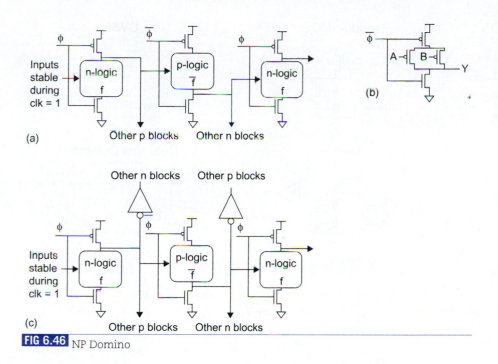

FIG 6.46 NP Domino

plays much the same role as a keeper. Zipper never saw widespread use in the industry [Bernstein99].

6.2.5 Pass-transistor Circuits

In the circuit families we have explored so far, inputs are applied only to the gate terminals of transistors. In pass-transistor circuits, inputs are also applied to the source/drain diffusion terminals. These circuits build switches using either nMOS pass transistors or parallel pairs of nMOS and pMOS transistors called *transmission gates*. Many authors have claimed substantial area, speed, and/or power improvements for pass transistors compared to static CMOS logic. In specialized circumstances this can be true; for example, pass transistors are essential to the design of efficient 6-transistor static RAM cells used in most modern systems (see Section 11.2). Full adders and other circuits rich in XORs also can be efficiently constructed with pass transistors. In certain other cases, we will see that pass-transistor circuits are essentially equivalent ways to draw the fundamental logic structures we have explored before. An independent evaluation finds that for most general-purpose logic, static CMOS is superior in speed, power, and area [Zimmermann97a].

For the purpose of comparison, Figure 6.47 shows a 2-input multiplexer constructed in a wide variety of pass-transistor circuit families along with static CMOS, pseudo-nMOS, CVSL, and single- and dual-rail domino. Some of the circuit families are dual-rail, producing both true and complementary outputs, while others are single-rail and may require an additional inversion if the other polarity of output is needed. U XOR V can be

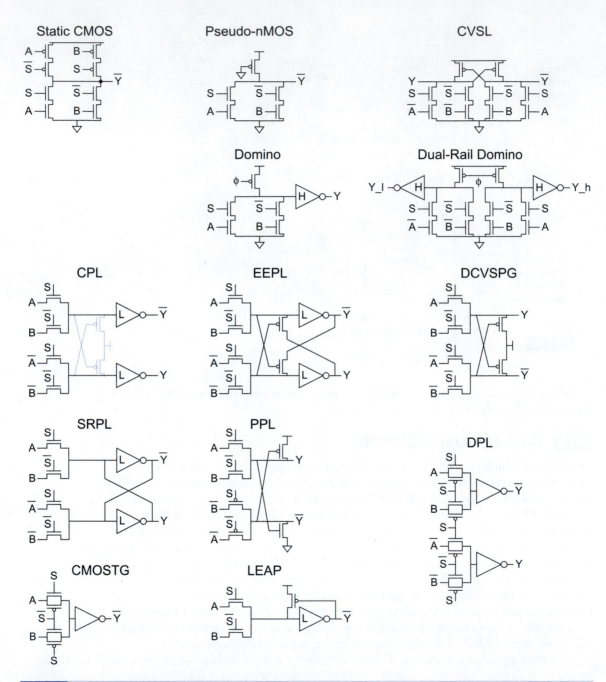

FIG 6.47 Comparison of circuit families for 2-input multiplexers

computed with exactly the same logic using $S = U$, $\overline{S} = \overline{U}$, $A = V$, $B = \overline{V}$. This shows that static CMOS is particularly poorly suited to XOR because the complex gate and two additional inverters are required; hence, pass-transistor circuits become attractive. In comparison, static CMOS NAND and NOR gates are relatively efficient and benefit less from pass transistors.

This section first examines mixing CMOS with transmission gates, as is common in multiplexers and latches. It next examines *Complementary Pass-transistor Logic* (CPL), which can work well for XOR-rich circuits like full adders and *LEAn integration with Pass transistors* (LEAP), which illustrates single-ended pass-transistor design. Finally, it catalogs and compares a wide variety of alternative pass-transistor families.

6.2.5.1 CMOS with Transmission Gates

Structures such as tristates, latches, and multiplexers are often drawn as transmission gates in conjunction with simple static CMOS logic. For example, Figure 1.27 introduced the transmission gate multiplexer using two transmission gates. The circuit was nonrestoring; i.e., the logic levels on the output are no better than those on the input so a cascade of such circuits may accumulate noise. To buffer the output and restore levels, a static CMOS output inverter can be added, as in Figure 6.47 (CMOSTG).

At first, CMOS with transmission gates might appear to offer an entirely new range of circuit constructs. A careful examination shows that the topology is actually almost identical to static CMOS, as was seen in Section 2.5.6. If multiple stages of logic are cascaded, they can be viewed as alternating transmission gates and inverters. Figure 6.48(a) redraws the multiplexer to include the inverters from the previous stage that drive the diffusion inputs but to exclude the output inverter. Figure 6.48(b) shows this multiplexer drawn at the transistor level. Observe that this is identical to the static CMOS multiplexer of Figure 6.47 except that the intermediate nodes in the pull-up and pull-down networks are shorted together as *N1* and *N2*.

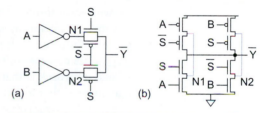

FIG 6.48 Alternate representations of CMOSTG in a 2-input inverting multiplexer

The shorting of the intermediate nodes has two effects on delay. The effective resistance decreases somewhat (especially for rising outputs) because the output is pulled up or down through the parallel combination of both pass transistors rather than through a single transistor. However, the effective capacitance increases slightly because of the extra diffusion and wire capacitance required for this shorting. This is apparent from layouts of the multiplexers; the transmission gate design in Figure 6.49(a) requires contacted diffusion on *N1* and *N2* while the static CMOS gate in Figure 6.49(b) does not. In most processes the improved resistance dominates for gates with moderate fanouts, making shorting generally beneficial.

There are several factors that favor the static CMOS representation over CMOS with transmission gates. If the inverter is on the output rather than the input, the delay of the gate depends on what is driving the input as well as the capacitance driven by the output. This input driver sensitivity makes characterizing the gate more difficult and is incompati-

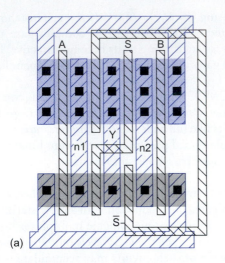 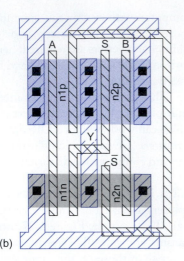

FIG 6.49 Multiplexer layout comparison

ble with most timing analysis tools. Novice designers often erroneously characterize transmission gate circuits by applying a voltage source directly to the diffusion input. This makes transmission gate multiplexers look very fast because they only involve one transistor in series rather than two. For accurate characterization, the driver must also be included. A second drawback is that diffusion inputs to tristate inverters are susceptible to noise that may incorrectly turn on the inverter; this is discussed further in Section 6.3. Finally, the contacts slightly increase area and their capacitance increases power consumption.

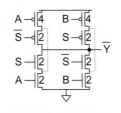

FIG 6.50 Logical effort of transmission gate circuit

If transmission gates are used, it is common practice to make the pMOS and nMOS transistors roughly the same width rather than use a double-width pMOS. This is done because the two transistors pass a signal in parallel. The incremental resistance benefit of a wider pMOS transistor is not worth the extra gate and diffusion capacitance it presents. The effective resistance of a transmission gate built from unit transistors is approximately R in both rising and falling directions, as discussed in Section 2.6. The logical effort of circuits involving transmission gates is computed by drawing stages that begin at gate inputs rather than diffusion inputs, as in Figure 6.50 for a transmission gate multiplexer. The effect of the shorting can be ignored, so the logical effort from either the A or B terminals is 6/3, just as in a static CMOS multiplexer. Note that the parasitic delay of transmission gate circuits with multiple series transmission gates increases rapidly because of the internal diffusion capacitance, so it is seldom beneficial to use more than two transmission gates in series without buffering.

6.2.5.2 Complementary Pass Transistor Logic (CPL) CPL [Yano90] can be understood as an improvement on CVSL. CVSL is slow because one side of the gate pulls down, and then the cross-coupled pMOS transistor pulls the other side up. The size of the cross-coupled device is an inherent compromise between a large transistor that fights the pull-down excessively and a small transistor that is slow pulling up. CPL resolves this problem by making one half of the gate pull up while the other half pulls down.

Figure 6.51(a) shows the CPL multiplexer from Figure 6.47 rotated sideways. If a path consists of a cascade of CPL gates, the inverters can be viewed equally well as being on the output of one stage or the input of the next. Figure 6.51(b) redraws the mux to include the inverters from the previous stage that drives the diffusion input, but to exclude the output inverters. Figure 6.51(c) shows the mux drawn at the transistor level. Observe that this is identical to the CVSL gate from Figure 6.47 except that the internal node of the stack can be pulled up through the weak pMOS transistors in the inverters.

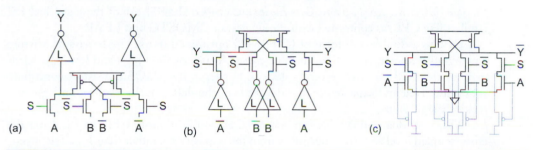

FIG 6.51 Alternate representations of CPL

When the gate switches, one side pulls down well through its nMOS transistors. The other side pulls up. CPL can be constructed without cross-coupled pMOS transistors, but the outputs would only rise to $V_{DD} - V_t$ (or slightly lower because the nMOS transistors experience the body effect). This costs static power because the output inverter will be turned slightly ON. Adding weak cross-coupled devices helps bring the rising output to the supply rail while only slightly slowing the falling output. The output inverters can be LO-skewed to reduce sensitivity to the slowly rising output.

6.2.5.3 Lean Integration with Pass Transistors (LEAP)

Like CPL, LEAP[6] [Yano96] builds logic networks using only fast nMOS transistors, as shown in Figure 6.47. It is a single-ended logic family in that the complementary network is not required, thus saving area and power. The output is buffered with an inverter, which can be LO-skewed to favor the asymmetric response of an nMOS transistor. The nMOS network only pulls up to $V_{DD} - V_t$ so a pMOS feedback transistor is necessary to pull the internal node fully high, avoiding power consumption in the output inverter. The pMOS width is a tradeoff between fighting falling transitions and assisting the last part of a rising transition; it generally should be quite weak and the circuit will fail if it is too strong. LEAP can be a good way to build wide 1-of-N hot multiplexers with many of the advantages of pseudo-nMOS but

---

[6]The LEAP topology was reinvented under the name *Single Ended Swing Restoring Pass Transistor Logic* [Pihl98].

without the static power consumption. It was originally proposed for use in a pass transistor logic synthesis system because the cells are compact.

Unlike most circuit families that can operate down to $V_{DD} \geq \max(V_{tn}, |V_{tp}|)$, LEAP is limited to operating at $V_{DD} \geq 2V_t$ because the inverter must flip even when receiving an input degraded by a threshold voltage.

6.2.5.4 Other Pass Transistor Families There have been a host of pass transistor families proposed in the literature, including *Differential Pass Transistor Logic* (DPTL) [Pasternak87, Pasternak91], *Double Pass Transistor Logic* (DPL) [Suzuki93], *Energy Economized Pass Transistor Logic* (EEPL) [Song96], *Push–Pull Pass Transistor Logic* (PPL) [Paik96], *Swing–Restored Pass Transistor Logic* (SRPL) [Parameswar96], and *Differential Cascode Voltage Switch with Pass Gate Logic* (DCVSPG) [Lai97]. All of these are dual-rail families like CPL, as contrasted with the single-rail CMOSTG and LEAP.

DPL is a double-rail form of CMOSTG optimized to use single pass transistors where only a known '0' or '1' needs to be passed. It passes good high and low logic levels without the need for level-restoring devices. However, the pMOS transistors contribute substantial area and capacitance, but do not help the delay much, resulting in large and relatively slow gates.

The other dual-rail families can be viewed as modifications to CPL. EEPL drives the cross-coupled level restoring transistors from the opposite rail rather than V_{DD}. The inventors claimed this led to shorter delay and lower power dissipation than CPL, but the improvements could not be confirmed [Zimmermann97a]. SRPL cross-couples the inverters instead of using cross-coupled pMOS pull-ups. This leads to a ratio problem in which the nMOS transistors in the inverter must be weak enough to be overcome as the pass transistors try to pull up. This tends to require small inverters, which make poor buffers. DCVSPG eliminates the output inverters from CPL. Without these buffers, the output of a DCVSPG gate makes a poor input to the diffusion terminal of another DCVSPG gate because a long unrestored chain of nMOS transistors would be formed, leading to delay and noise problems. PPL also has unbuffered outputs and associated delay and noise issues. DPTL generalizes the output buffer structure to consider alternatives to the cross-coupled pMOS transistors and LO-skewed inverters of CPL. All of the alternatives are slower and larger than CPL.

6.3 Circuit Pitfalls

Circuit designers tend to use simple circuits because they are robust. Elaborate circuits, especially those with more transistors, tend to add more area, more capacitance, and more things that can go wrong. Static CMOS is the most robust circuit family and should be used whenever possible. This section catalogs a variety of circuit pitfalls that can cause chips to fail. They include:

- threshold drops
- ratio failures

- leakage
- charge sharing
- power supply noise
- coupling
- minority carrier injection
- back-gate coupling
- diffusion input noise sensitivity
- race conditions
- delay matching
- metastability
- hot spots
- soft errors
- process sensitivity

Capacitive and inductive coupling were discussed in Section 4.5. Sneak paths were discussed in Section 6.2.4.6. Reliability issues such as soft errors impacting circuit design were discussed in Section 4.8. Timing-related problems including race conditions, delay matching, and metastability will be examined in Sections 7.2.3, 7.5.4, and 7.6.1. The other pitfalls are described here.

6.3.1 Threshold Drops

Pass transistors are good at pulling in a preferred direction, but only swing to within V_t of the rail in the other direction; this is called a *threshold drop*. For example, Figure 6.52 shows a pass transistor driving a logic '1' into an inverter. The output of the pass transistor only rises to $V_{DD} - V_t$. Worse yet, the body effect increases this threshold voltage because $V_{sb} > 0$ for the pass transistor. The degraded level is insufficient to completely turn off the pMOS transistor in the inverter, resulting in static power dissipation. Indeed, for low V_{DD}, the degraded output can be so poor that the inverter no longer sees a valid input logic level V_{IH}. Finally, the transition becomes lethargic as the output approaches $V_{DD} - V_t$. Threshold drops were sometimes tolerable in older processes where $V_{DD} \approx 5V_t$, but are seldom acceptable in modern processes where the power supply has been scaled down faster than the threshold voltage to $V_{DD} \approx 3V_t$. As a result, pass transistors must be replaced by full transmission gates or may use weak pMOS feedback transistors to pull the output to V_{DD}, as was done in several pass transistor families.

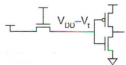

FIG 6.52 Pass transistor with threshold drop

6.3.2 Ratio Failures

Pseudo-nMOS circuits illustrated ratio constraints that occur when a node is simultaneously pulled up and down, typically by strong nMOS transistors and weak pMOS transistors. The weak transistors must be sufficiently small that the output level falls below V_{IL} of the next stage by some noise margin. Ideally, the output should fall below V_t so the next stage does not conduct static power. Ratioed circuits should be checked in the SF and FS corners.

Another example of ratio failures occurs in circuits with feedback. For example, dynamic keepers, level-restoring devices in SRPL and LEAP, and feedback inverters in static latches all have weak feedback transistors that must be ratioed properly.

Ratioing is especially sensitive for diffusion inputs. For example, Figure 6.53(a) shows a static latch with a weak feedback inverter. The feedback inverter must be weak enough to be overcome by the series combination of the pass transistor and the gate driving the D input, as shown in Figure 6.53(b). This cannot be verified by checking the latch alone; it requires a global check of the latch and driver. Worse yet, if the driver is far away, the series wire resistance must also be considered, as shown in Figure 6.53(c).

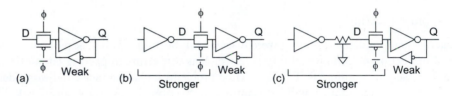

FIG 6.53 Ratio constraint on static latch with diffusion input

6.3.3 Leakage

Leakage current is a growing problem as technology scales, especially for dynamic nodes and wide NOR structures. Recall that leakage arises from subthreshold conduction, gate tunneling, and reverse-biased diode leakage. Subthreshold conduction is presently the most important component because V_t is low and getting lower, but gate tunneling will become profoundly important too as oxide thickness diminishes. Besides causing static power dissipation, leakage can result in incorrect values on dynamic or weakly driven nodes. The time required for leakage to disturb a dynamic node by some voltage ΔV is

$$t = \frac{C_{node}\Delta V}{I_{leak}} \tag{6.6}$$

Subthreshold leakage gradually discharges dynamic nodes through transistors that are nominally OFF. Fully dynamic gates and latches without keepers are not viable in most modern processes. DRAM refresh times are also set by leakage and DRAM processes must minimize leakage to have satisfactory retention times.

Even when a keeper is used, it must be wide enough. This seems trivial because the keeper is fully ON while leakage takes place through transistors that are supposed to be OFF. However, in wide dynamic NOR structures, many parallel nMOS transistors may be leaking simultaneously. Similar problems apply to wide pseudo-nMOS NOR gates and PLAs. Leakage increases exponentially with temperature, so the problem is especially bad at burn-in. For example, a preliminary version of the Sun UltraSparc V had difficulty with burn-in because of excess leakage.

Subthreshold leakage is much lower through two OFF transistors in series than through a single transistor because the outer transistor has a lower drain voltage and sees a much lower effect from DIBL. Multiple threshold voltages are also frequently used to achieve high performance in critical paths and lower leakage in other paths.

6.3.4   Charge Sharing

Charge sharing was introduced in Section 6.2.4.4 in the context of a dynamic gate. Charge sharing can also occur when dynamic gates drive pass transistors. For example, Figure 6.54 shows a dynamic inverter driving a transmission gate. Suppose the dynamic gate has been precharged and the output is floating high. Further suppose the transmission gate is OFF and $Y = 0$. If the transmission gate turns on, charge will be shared between X and Y, disturbing the dynamic output.

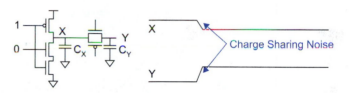

FIG 6.54   Charge sharing on dynamic gate driving pass transistor

6.3.5   Power Supply Noise

V_{DD} and GND are not constant across a large chip. Both are subject to *power supply noise* caused by IR drops and di/dt noise. IR drops occur across the resistance R of the power supply grid between the supply pins and a block drawing a current I, as shown in Figure 6.55. di/dt noise occurs across the power supply inductance L as the current rapidly changes. di/dt noise can be especially important for blocks that are idle for several cycles and then begin switching. Power supply noise hurts performance and can degrade noise margins. Typical targets are for power supply noise on the order of 5%–10% of V_{DD}. Power supply noise causes both noise margin problems and delay variations. The noise margin issues can be managed by placing sensitive circuits near each other and having them share a common low-resistance power wire.

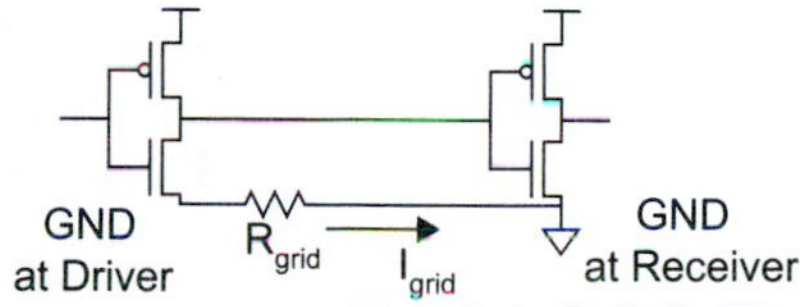

FIG 6.55   Power supply IR drops

Power supply noise can be estimated from simulations of the chip power grid, bypass capacitance, and packaging, as will be discussed in Section 12.3. Figure 6.56 shows a plot of the simulated power supply voltage (V_{DD} – GND) at points across the Itanium 2 [Harris01b].

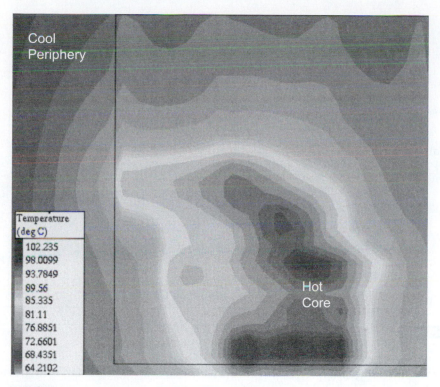

Cool
Periphery

Temperature
(deg C)

102.235
98.0099
93.7849
89.56
85.335
81.11
76.8851
72.6601
68.4351
64.2102

Hot
Core

FIG 6.57 Itanium 2 die temperature. © IEEE 2001.

6.3.7 Minority Carrier Injection

It is sometimes possible to drive a signal momentarily outside the rails, either through capacitive coupling or through inductive ringing on I/O drivers. In such a case, the junctions between drain and body may momentarily become forward-biased, causing current to flow into the substrate. This effect is called *minority carrier injection* [Chandrakasan01]. For example, in Figure 6.58, the drain of an nMOS transistor is driven below GND, injecting electrons into the p-type substrate. These can be collected on a nearby transistor diffusion node (Figure 6.58(a)), disturbing a high voltage on the node. This is a particular problem for dynamic nodes and sensitive analog circuits.

Minority carrier injection problems are avoided by keeping injection sources away from sensitive nodes. In particular, I/O pads should not be located near sensitive nodes. Noise tools can identify potential coupling problems so the layout can be modified to reduce coupling. Alternatively, the sensitive node can be protected by an intermediate substrate or well contact. For example in Figure 6.58(b), most of the injected electrons will be collected into the substrate contact before reaching the dynamic node. In I/O pads, it is

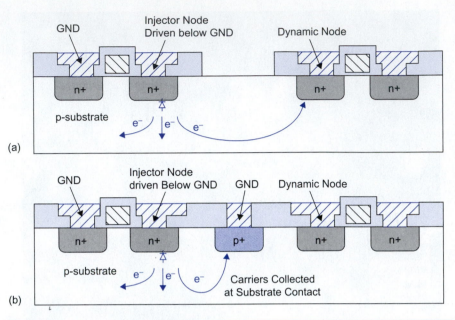

FIG 6.58 Minority carrier injection and collection

common to build *guard rings* of substrate/well contacts around the output transistors. Guard rings were illustrated in Figure 4.64.

6.3.8 Back-gate Coupling

Dynamic gates driving multiple-input static CMOS gates are susceptible to the *back-gate coupling* effect [Chandrakasan01] illustrated in Figure 6.59. In this example, a dynamic NAND gate drives a static NAND gate. The gate-to-source capacitance C_{gs1} of *N1* is shown explicitly. Suppose that the dynamic gate is in evaluation and its output X is floating high. The other input B to the static NAND gate is initially low. Therefore, the NAND output Y is high and the internal node W is charged up to $V_{DD} - V_t$. At some time B rises, discharging Y and W through transistor *N2*. The source of *N1* falls. This tends to bring the gate along for the ride because of the C_{gs1} capacitance, resulting in a droop on the dynamic node X. As with charge sharing, the magnitude of the droop depends on the ratio of C_{gs1} to the total capacitance on node X.

Back-gate coupling is eliminated by driving the input closer to the rail. For example, if X drove *N2* instead of *N1*, the problem would be avoided. Otherwise, the back-gate coupling noise must be included in the dynamic noise budget.

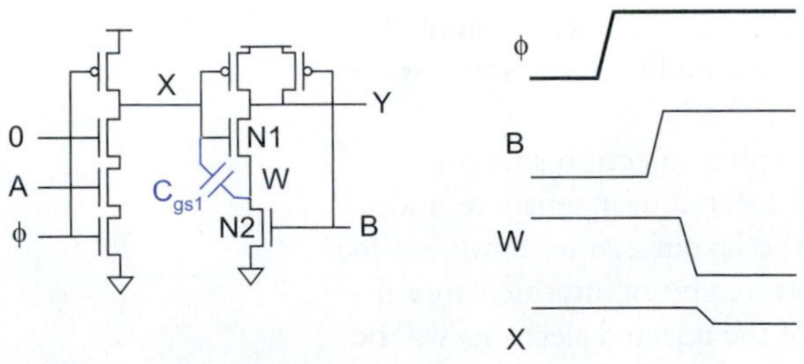

FIG 6.59 Back-gate coupling

6.3.9 Diffusion Input Noise Sensitivity

Figure 6.53(a) showed a static latch with an exposed diffusion input. Such an input is also particularly sensitive to noise. For example, imagine that power supply noise and/or coupling noise drove the input voltage below $-V_t$ relative to GND seen by the transmission gate, as shown in Figure 6.60. V_{gs} now exceeds V_t for the nMOS transistor in the transmission gate, so the transmission gate turns on. If the latch had contained a '1,' it could be incorrectly discharged to '0.' A similar effect can occur for voltage excursions above V_{DD}.

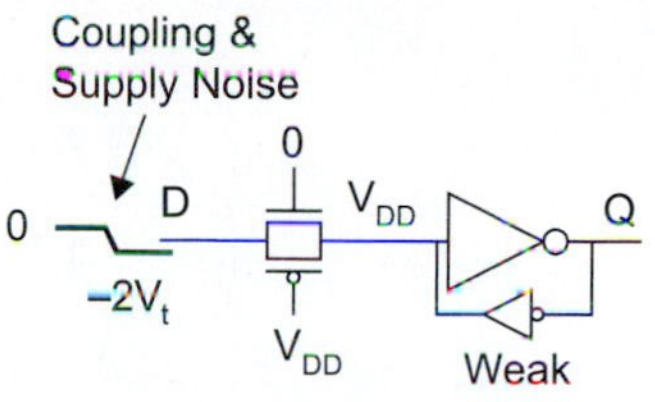

FIG 6.60 Noise on diffusion input of latch

For this reason, along with the ratio issues discussed in Section 6.3.2, standard cell latches are usually built with buffered inputs rather than exposed diffusion nodes. Exposing the diffusion input results in a faster latch and can be used in datapaths where the inputs are carefully controlled and checked.

6.3.10 Process Sensitivity

Marginal circuits can operate under nominal process conditions, but fail in certain process corners or when the circuit is migrated to another process. Novel circuits should be simulated in all process corners and carefully scrutinized for any process sensitivities. They should also be verified to work at all voltages and temperatures, including the elevated voltages and temperatures used during burn-in and the lower voltage that might be used for low-power versions of a part.

When a design is likely to be migrated to another process for cost-reduction, circuits should be designed to facilitate this migration. You can expect that leakage will increase, threshold drops will become a greater fraction of the supply voltage, wire delay will become a greater portion of the cycle time, and coupling may get worse as aspect ratios of wires increase. For example, the Pentium 4 processor was originally fabricated in a 180 nm process. Designers placed repeaters closer than was optimal for that process because they knew the best repeater spacing would become smaller as transistor dimensions were reduced later in the product's life [Kumar01].

6.3.11 Example: Domino Noise Budgets

Domino logic is important in many high-performance microprocessors, but requires careful verification because it is sensitive to noise. Noise in static CMOS gates usually results in greater delay, but noise in domino logic can produce incorrect results. This section reviews the various noise sources that can affect domino gates and presents a sample noise budget.

Dynamic outputs are especially susceptible to noise when they float high, held only by a weak keeper. Dynamic inputs have low noise margins (approximately V_t). Noise issues that should be considered include [Chandrakasan01]:

- **Charge leakage** Subthreshold leakage on the dynamic node is presently most important, but gate leakage will become important, too. Subthreshold leakage is worst for wide NOR structures at high temperature (especially during burn-in). Keepers must be sized appropriately to compensate for leakage.

- **Charge sharing** Charge sharing can take place between the dynamic output node and the nodes within the dynamic gate. Secondary precharge transistors should be added when the charge sharing could be excessive. Do not drive dynamic nodes directly into transmission gates because charge sharing can occur when the transmission gate turns ON.

- **Capacitive coupling** Capacitive coupling can occur on both the input and output. The inputs of dynamic gates have the lowest noise margin, but are actively driven by a static gate, which fights coupling noise. The dynamic outputs have more noise tolerance, but are weakly driven. Coupling is minimized by keeping wires short and increasing the spacing to neighbors or shielding the lines. Coupling can be extremely bad in processes below 250 nm because the wires have such high aspect ratios.

- **Back-gate coupling** Dynamic gates connected to multiple-input CMOS gates should drive the outer input when possible. This is not a factor for dynamic gates driving inverters.

- **Minority carrier injection** Dynamic nodes should be protected from nodes that can inject minority carriers. These include I/O circuits and nodes that can be coupled far outside the supply rails. Substrate/well contacts and guard rings can be added to protect dynamic nodes from potential injectors.

- **Power supply noise** Static gates should be located close to the dynamic gates they drive to minimize the amount of power supply noise seen.

- **Soft errors** Alpha particles and cosmic rays can disturb dynamic nodes. The probability of failure is reduced through large node capacitance and strong keepers.

- **Noise feedthrough** Noise that pushes the input of a previous stage to near its noise margin will cause the output to be slightly degraded, as was shown in Figure 2.28.

- **Process corner effects** Noise margins are degraded in certain process corners. Dynamic gates have the smallest noise margin in the FS corner where the nMOS transistors have a low threshold and the pMOS keepers are weak. HI-skew static gates have the smallest noise margins in the SF corner where the gates are most skewed.

In a domino gate, the noise-prone dynamic output drives a static gate with a reasonable noise margin. The noise-sensitive dynamic gate is strongly driven by a noise-resistant static gate. In an NP domino gate or clock-delayed domino gate, the noise-prone dynamic output directly drives a noise-sensitive dynamic input, making such circuits particularly risky.

Consider a noise budget for a 3.3 V process [Harris01a]. A HI-skew inverter in this process has V_{IH} = 2.08 V, resulting in NM_H = 37% of V_{DD} if V_{OH} = V_{DD} A dynamic gate with a small keeper has V_{IL} = 0.63 V, resulting in NM_L = 19% of V_{DD}. Table 6.3 allocates these margins to the primary noise sources. In a full design methodology, different margins can be used for different gates. For example, wide NOR structures have no charge sharing noise, but may see significant leakage instead. More coupling noise could be tolerated if other noise sources are known to be smaller. Noise analysis tools are discussed further in Section 8.4.2.7.

Table 6.3	Sample domino noise budget	
Source	Dynamic Output	Dynamic Input
Charge sharing	10	n/a
Coupling	17	7
Supply noise	5	5
Feedthrough noise	5	7
Total	37%	19%

6.4   More Circuit Families

Static CMOS is satisfactory for the great majority of logic gates in modern integrated circuits and an assortment of domino, pass-transistor circuits, and pseudo-nMOS accounts for nearly all of the remaining gates. A large number of other circuit families have been proposed in the literature. This section describes some of these circuit families and their strengths and limitations.

6.4.1   Differential Circuits

Several differential circuit families using nMOS pull-down networks are derived from the basic CVSL form, as shown in Figure 6.61.

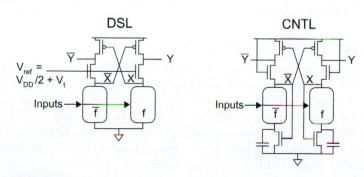

FIG 6.61 Differential circuit families

6.4.1.1 Differential Split-Level (DSL)   *Differential Split-Level* (DSL) [Pfennings85] places nMOS transistors in series with the basic CVSL pull-down networks. By applying a reference voltage of $V_{DD}/2 + V_t$ to these transistors, the swing on the internal nodes (X and $\overline{X}$) are limited to $0 - V_{DD}/2$. This reduces the parasitic delay of the pull-down stacks. The lower internal voltages also lead to lower electric fields across the pull-down transistors. The inventors took advantage of this lower voltage to reduce the channel lengths of the transistors without compromising hot-electron reliability. They claimed a tenfold speedup over a static CMOS full adder; this was attributed to a factor of 2 for the CVSL structure, another factor of 2 from the low-swing signals, and a factor of 2.5 for using shorter transistors.

In a modern process, transistors are generally as short as can be reliably manufactured, so DSL cannot use even shorter transistors. The authors have been unable to reproduce any advantage over static CMOS in a submicron process. The resistance of the extra series transistor does not help. Another disadvantage of DSL is that the voltages on the pMOS gates only swing between 0 and $V_{DD}/2$. Therefore, the pull-up that should be OFF is actually partially ON, resulting in static power dissipation. Finally, generating and distributing the reference voltage requires some effort and the reference may be sensitive to power supply noise and threshold voltage variations.

6.4.1.2 Cascode Nonthreshold Logic (CNTL)   *Cascode Nonthreshold Logic* (CNTL) [Wang89] is derived from DSL by adding a transistor and shunting capacitor to the bottom of each pull-down network and setting the reference voltage to V_{DD} rather than $V_{DD}/2 + V_t$. The series transistors are connected with negative feedback. The internal swing is limited to V_t to $V_{DD} - V_t$ and there is much less quiescent current draw than in DSL because the pMOS transistors are nearly turned OFF. CNTL requires more area than CVSL and the extra series transistors tend to slow it down, although large shunting capacitors partially alleviate this problem.

CNTL is a variant of Nonthreshold Logic (NTL), shown in Figure 6.62, which is essentially a pseudo-nMOS gate with an extra transistor and shunting capacitor in series with the pull-down network. The shunting capacitor is built from the gate of an nMOS transistor. NTL consumes static power and is slower than pseudo-nMOS.

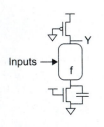

FIG 6.62 Nonthreshold Logic

6.4.2 Sense-amplifier Circuits

Sense amplifiers magnify small differential input voltages into larger output voltages. They are commonly used in memories in which differential bitlines have enormous capacitive loads (see Section 11.2.3). Because of the large load, the bitlines swing slowly. To reduce this delay, the bitline voltages are first equalized. Then, when they are driven apart, the sense amplifier can detect a small swing and bring it up to normal logic levels. This reduces the ΔV term in EQ (6.1); in other words, it reduces the delay by avoiding waiting for a full swing on the bitlines. Sense amplifiers offer potential for reducing delay in heavily loaded logic circuits as well.

Figure 6.63 shows more differential circuit families derived from CVSL. These families add sense amplifiers to dual-rail domino (also repeated in the figure) to detect a small differential voltage and amplify it to a full-rail output. They will be discussed in detail later in this section.

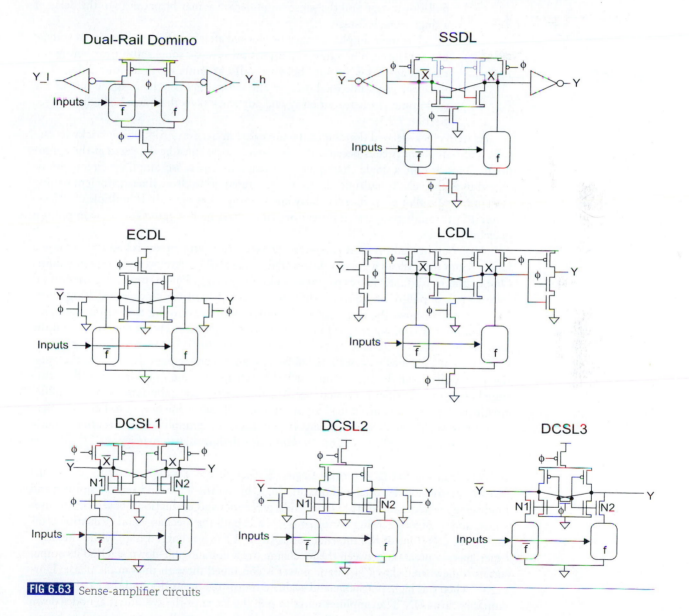

FIG 6.63 Sense-amplifier circuits

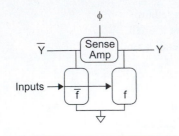

FIG 6.64 Generic sense-amplifier circuit

Figure 6.64 shows a generic sense-amplifier circuit. It works best for complex pull-down networks that would have a large RC delay. The sense amplifier fires after a small ΔV develops. Once fired, it turns on a driver with a low resistance to slew the outputs between rails. The combined delay of the pull-down stage and the sense-amplifier stage may be better than the delay of a single complex stage.

The sense amplifiers must be clocked after a sufficient differential voltage has developed. Therefore, the inputs must settle some setup time before the clock edge. The outputs become valid at some clock-to-out delay after the clock edge. The total delay of the sense-amplifier circuit is the sum of the setup time, clock-to-out delay, and any clock skew that must be budgeted (see Section 7.2.5).

As with clock-delayed domino (to be discussed in Section 7.5.4.2), it is tricky to cascade sense-amplifier circuits because the successive clocks must be delivered at the appropriate times. If only a single clock phase is used, only one sense-amplifier circuit can be placed in each cycle. If multiple clocks are generated using delay lines, sufficient timing margin must be allowed so that the delay line is always slow enough. If multiple clocks are generated through completion detection, time must be budgeted for the completion detection circuits.

An inherent tradeoff exists between the setup time and circuit reliability because a longer setup time allows a greater differential voltage ΔV to develop and overcome noise. One of the important sources of noise is charge sharing. For example, Figure 6.65(a) shows a pair of pull-down networks that are particularly sensitive to charge sharing noise. Figure 6.65(b) shows the response as the inputs arrive, assuming the outputs are precharged, node X carries a residual low voltage from a previous cycle's operation, and the sense amp is inactive. Observe that charge sharing from the large internal diffusion capacitance on node X initially causes Y to fall faster than its complement. Eventually, the resistive path pulls down the correct output Y. This charge sharing noise increases the setup time before the amplifier can safely fire. Yet another risk for unbuffered sense-amplifier circuits is that unequal output loading or coupling will cause one output to fall faster than the other, resulting in incorrect sensing. In summary, sense-amplifier circuits offer promise for special-purpose applications, but present many design risks to manage.

6.4.2.1 Sample Set Differential Logic (SSDL)

Sample Set Differential Logic (SSDL) [Grotjohn86] modifies dual-rail domino logic by adding a clocked sense amplifier and modifying the clocking. Rather than using precharge and evaluation phases, SSDL uses *sample* and *set* phases. During sample, ϕ is low and both the precharge and evaluation transistors are ON. One of the internal nodes (X or $\overline{X}$) is precharged high while the other experiences contention between the precharge transistor and pull-down stack so its output settles somewhere below V_{DD}. Static power is consumed through the sample phase. During set when ϕ is high, precharge and evaluate transistors turn OFF and the clocked sense amplifier turns ON. The amplifier tends to pull the lower of the two internal nodes down to GND. At first, it tends to pull down the other side as well, so it is helpful to have a keeper (shown in blue) to restore the high level.

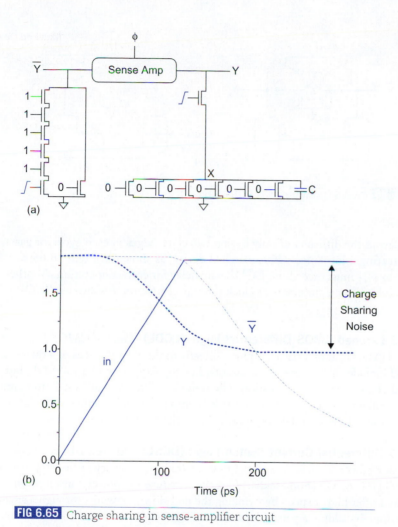

FIG 6.65 Charge sharing in sense-amplifier circuit

6.4.2.2 Enable/Disable CMOS Differential Logic (ECDL) *Enable/Disable CMOS Differential Logic* (ECDL) [Lu88b, Lu91] improves on SSDL by eliminating the static power consumption. The sense amplifier is made from a pair of cross-coupled clocked inverters, as redrawn in Figure 6.66(a) to emphasize the inverters. The cycle is again divided into two phases of operation: *enable* and *disable*. When ϕ is high, the gate is disabled. Both outputs are pulled low and the pull-up stack is turned OFF. When ϕ falls, the gate is enabled. The cross-coupled pMOS transistors are both initially ON and attempt to pull the outputs high. One output will be held down by its pull-down stack and will lag. Positive feedback will pull one output fully high and the other back fully low. The sense amplifier rising delay is somewhat longer than in SSDL because it pulls high through two series pMOS transistors.

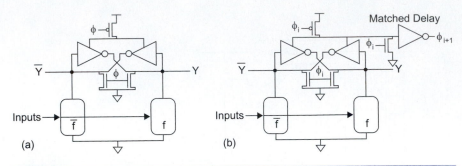

FIG 6.66 Enable/Disable CMOS Differential Logic

To avoid the difficulty of only having two clock edges in each cycle for gates, Lu proposes creating a local clock with matched delays, as shown in Figure 6.66(b). The delay from ϕ_i to ϕ_{i+1} must exceed the ECDL gate delay for correct operation. Another possibility would be to generate the next clock through *completion detection* as the OR of the two outputs.

6.4.2.3 Latched CMOS Differential Logic (LCDL) *Latched CMOS Differential Logic* (LCDL) [Wu91] adds a sense amplifier directly to the output nodes of a dual-rail domino gate and includes n-latches on the outputs. The topology is similar to SSDL, but the non-inverted clock is used for evaluation. The sense amplifier fires at exactly the same time as the dual-rail gate, so there is a serious risk of amplifying noise rather than signal. This can be overcome with a second clock to delay firing the amplifier.

6.4.2.4 Differential Current Switch Logic (DCSL) Differential circuits can consume significant power because one of the outputs transitions every cycle. *Differential Circuit Switch Logic* (DCSL) [Somasekhar96] seeks to reduce the power consumption of internal nodes and offer higher speed by swinging the pull-down networks through a small voltage. This is done by adding a pair of feedback transistors *N1* and *N2* to the SSDL and ECDL structures to cut off the pull-down networks before the internal nodes rise far above 0.

DCSL1 is a "precharge high" circuit related to SSDL and LCDL. When the clock is low, the outputs precharge high. When the clock rises, the circuit begins evaluation. As one side or the other pulls low, the sense amplifier accelerates the transition. *N1* or *N2* turns off to prevent the internal nodes of the pull-down stack on the other side from rising too much.

DCSL2 is a "precharge low" circuit related to ECDL. It again adds *N1* and *N2* to prevent the internal nodes from rising too much. DCSL3 improves on DCSL2 by replacing the two predischarge transistors with a single equalization transistor.

Because the sense amplifiers fire at the same time as the outputs begin to fall, DCSL is sensitive to amplifying noise instead of signal. It also performs poorly for $V_{DD} < 5V_t$. LVDCSL [Somasekhar98] operates better at low voltages, but uses a complex sense amplifier.

6.4.3 BiCMOS Circuits

Bipolar transistors can deliver a much higher output current than can CMOS transistors of equal input capacitance. Therefore, they can be used to build gates with low logical effort and are good for driving large capacitive loads. Gates mixing bipolar and CMOS transistors are called *BiCMOS*.

Figure 6.67 shows a BiCMOS NAND gate using two npn bipolar transistors. An npn transistor behaves as a switch between the collector and the emitter controlled by the base. The base voltage must be about 0.7 V above the emitter to turn the transistor ON. The BiCMOS gate contains an ordinary CMOS NAND gate to compute x. If A or B is '0,' x will be driven to '1.' This turns on $Q2$ and pulls the output Y up. When Y is high, $M1$ turns ON, pulling down w and turning off $Q1$. If A and B are both '1,' $M3$ and $M2$ are both 'ON.' If Y begins at '1,' w will rise to '1' and turn on $Q1$. $Q1$ in turn discharges Y to '0.'

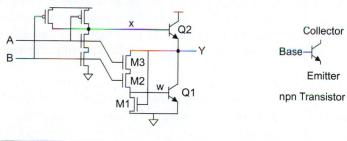

FIG 6.67 BiCMOS NAND gate

Unfortunately, bipolar transistors have an inherent V_{be} drop of about 0.7 V when ON. Hence, Y will never rise above $V_{DD} - V_{be}$. This was acceptable at V_{DD} = 5 V, tolerable at V_{DD} = 3.3 V, and perhaps manageable with elaborate circuit tricks at V_{DD} = 2.5 V. In modern processes with low supply voltages, $V_{DD} - V_{be}$ is too low to form a valid logic level, so BiCMOS circuits are no longer particularly useful for digital logic in processes below 0.35 μm. Moreover, CMOS circuits have been scaled much more aggressively than bipolar transistors, so the short-channel CMOS transistors are now competitive in performance with older, larger bipolar transistors.

6.4.4 Other Circuit Families

Other circuit families in the literature include *Regenerative Push–Pull Differential Logic* (RPPDL) [Partovi94], *Switched Output Differential Structure* (SODS) [Acosta95], *Current Sensing Differential Logic* (CSDL) [Park99], *Dynamic Current Mode Logic* (DyCML) [Allam01], *Subthreshold Logic* [Soeleman01], *Swing Limited Logic* (SLL) [Fahim02], and *Race Logic Architecture* (RALA) [Lee02]. It is left as an exercise to the reader to evaluate the advantages and pitfalls of these and future circuit families.

6.5 Low-power Logic Design

Given the logic families covered in this chapter, what are the particular strategies to achieve low-power dissipation? For the answers, we return to the basics introduced in Section 4.4. Commencing with dynamic power, recall that this component of power dissipation is proportional to α, the activity factor, C, the capacitance switched, f, the frequency of operation and V_{DD}^2, where V_{DD} is the supply voltage.

The first step is to choose a suitable process. Dynamic power is lowest in the most advanced (shortest channel length) manufacturing process available. Capacitance per micron of channel width is approximately 2 fF/μm in most processes, but more advanced processes use minimum-sized transistors with narrower channel widths. Moreover, the processes run at lower V_{DD}. Thus, constant field scaling provides a cubic improvement in dynamic power for a given function. Static power increases with more advanced processes because subthreshold leakage goes up as the threshold voltages decrease. In the past, dynamic power has dominated. In the present and future, low-power processes will be optimized to have higher threshold voltages than their high-performance counterparts. Advanced processes are also more expensive, so tradeoffs between power and cost may be necessary.

The most obvious parameter to adjust for low dynamic power is the supply voltage. Having made the process choice, the supply voltage can usually be further scaled, provided adequate speed and voltage margins are retained. For instance, in a process with a 3.3 V nominal V_{DD} and 0.6 V threshold voltages, V_{DD} can be scaled down to say 1.8 V ($V_{DD\text{min}}$ – $3V_t$). This comes at a substantial reduction in speed, but if the critical path still meets the design target, then this approach can be taken. In particular, smaller processes are faster, offering more headroom to trade away speed for supply voltage. Depending on the workload, the voltage and clock frequency can both be adjusted to conserve power when the system does not need maximum performance. Chips may also sense their own die temperature and throttle the supply voltage or instruction issue rate when the temperature exceeds the operating envelope.

A variation on adjusting power supplies is to divide the logic into high-speed and low-power groups run from separate power supplies. Some designs have been proposed that have dual-supply rails embedded in each logic cell so the decision of what supply to use is left as a late binding routing option. Fast logic is connected to the high supply and slow logic is connected to the low supply. Level converters have to be inserted between the two styles of logic. Given that the low V_{DD} supply transistors can take the high V_{DD} on their gates, the interface between high V_{DD} gate and low V_{DD} gate can be a direct connection. For low V_{DD} to high V_{DD} the CVSL inverter/buffer circuit shown in Figure 6.68(a) can be used. The level converters add delay, power, and area, so they should be used sparingly. [Usami94] proposes *clustered voltage scaling* where the low V_{DD} cells are grouped at the end of each cycle of logic. Then, the level conversion back to high V_{DD} can be

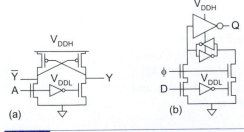

FIG 6.68 Level conversion circuits

built into the register at the end of the cycle. Figure 6.68(b) shows such a latch with integrated level conversion.

The next dynamic power parameter to attack for low-power operation is C, the switched capacitance. This consists of gate capacitance, diffusion capacitance, and wire capacitance. Good floorplanning reduces the number of long wires in a system. Good layout minimizes the size of the diffusion regions for a particular gate width. Thus, gate and diffusion capacitance are both reduced by using a small number of small transistors. Custom layout of datapaths and arrays usually provides shorter wires and lower capacitance than automatic placement and routing.

The activity factor offers more opportunity for power savings. Complementary CMOS logic has a typical activity factor of about 0.1, while dynamic logic has an activity factor of 0.5 and clocks have an activity factor of 1. Gating the clock to registers in unused units saves power not only in the registers but also in all the downstream logic because the activity factor is 0 if the inputs do not switch.

What circuit families are best for low power? Dynamic and pseudo-nMOS gates appear attractive because they eliminate the bulky pMOS transistors that account for 2/3 of the gate width in complementary CMOS logic. However, dynamic gates have a high activity factor that results in high dynamic power consumption. Pseudo-nMOS static power dissipation will dwarf the dynamic power in most applications. Pass transistor circuits have received much hype for their low-power merits. However, complementary CMOS benefits from efficient layout of simple gates, no swing restoration circuitry, single rail logic designs, and better scaling at low V_{DD}/V_t ratios; [Zimmermann97a] demonstrated that pass transistors have a higher power-delay product in most situations. Unless you are really motivated, minimum-sized complementary CMOS with a low P/N ratio is normally the best tradeoff of reliable circuit technology and low power.

Short circuit current is minimized by using sharp edges. Slow, heavily loaded nodes should be avoided. Logical effort teaches us that delay is minimized if all stages have a stage effort of about 4. If a path has slack (time to spare), the power and area can be improved by reducing the size of the largest gates. Efforts exceeding about 8 have diminishing returns and lead to excessive short circuit currents and hot electron reliability problems.

6.6   Comparison of Circuit Families

Table 6.4 summarizes the characteristics of the circuit families described in this chapter. The number of transistors required for k-input gates are listed. Differential circuit families are those that require true and complementary inputs and generate true and complementary outputs. Static power indicates that the gate may consume power while quiescent; this is often not acceptable for battery-operated devices. Circuits with rail-to-rail outputs swing between GND and V_{DD}. Dynamic nodes are those that have been precharged and may float or be only weakly held by a keeper; they are particularly sensitive to noise. Restoring logic families are those whose output logic levels are better than the input logic levels; if families are not restoring, buffers must be periodically placed between gates.

Ratioed circuits are those whose operation depends on the relative strength of nMOS and pMOS transistors; they must be sized properly for correct operation. Circuits are cascadeable if the output of a gate is a legal input to another gate of the same family without any special delayed clocking or self-timing. For example, domino gates sharing a common clock can be cascaded, but dynamic gates cannot be without violating monotonicity. Robustness characterizes the amount of care required to ensure a gate will work. Highly robust circuits like static CMOS will eventually get the right answer independent of sizing and noise, while less robust circuits are more sensitive. Undesirable characteristics are marked in blue.

A large number of circuit families have been presented in this section. A natural question is how to choose the appropriate circuit family for the application.

Static CMOS logic is the best option for the vast majority of CMOS circuits. It is noise-immune, dissipates no static power, and is fast. Highly automated tools and readily available libraries exist to synthesize, place, and route static logic. Don't overlook compound AOI and OAI gates. High fan-in static CMOS gates offer low power but have large logical effort and are best split into multiple stages of simpler gates when speed is essential.

Certain high fan-in functions are implemented much more efficiently with pseudo-nMOS or dynamic NOR gates because the logical effort is independent of the width. Examples include ROMs, PLAs, and CAMs. Pseudo-nMOS static power dissipation can be a problem for battery-operated systems, but sometimes the pMOS pull-up can be turned OFF during idle periods to save power.

Domino logic remains the technique of choice for high-speed applications, especially in high-performance microprocessors. However, it has poor noise margins and is susceptible to noise from charge sharing, coupling, leakage, and alpha particles. If you are not prepared to exhaustively simulate the gates at the circuit level with back-annotated capacitances from the layout, do not consider domino. Remember that the precharge time will rob the speed advantage over static designs in poorly designed clocking schemes (this will be discussed further in Section 7.5.1). Many novices (and pros too!) have been caught by not understanding all the problems that can arise when domino logic is used.

Pass transistors have their vocal advocates, but transmission gate logic can be viewed as an alternative way of drawing static CMOS gates with the driving stage at the output rather than the input. Of the multitude of pass-transistor circuit families that have been proposed, CPL is the most promising.

Other circuit families offer potential for niche applications (i.e., low noise generation in sensitive analog circuits), but one must be wary of pitfalls and consider carefully why so many circuit families have never seen commercial application.

Table 6.4	Comparison of circuit families									
Family	nMOS	pMOS	Differential	Static Power	Rail-to-rail Output	Dynamic Nodes	Restoring	Ratioed	Cascadeable	Robustness
Static CMOS	k	k	NO	NO	YES	NO	YES	NO	YES	HIGH
Pseudo-nMOS	k	1	NO	YES	NO	NO	YES	YES	YES	MEDIUM
SFPL	2k + 2	1	NO	YES	NO	NO	YES	YES	YES	MEDIUM
CVSL	2k	2	YES	NO	YES	NO	YES	NO	YES	HIGH
Dynamic	k + 1	1	NO	NO	YES	YES	YES	NO	NO	LOW
Domino	k + 2	2	NO	NO	YES	YES	YES	NO	YES	LOW
Dual-rail Domino	2k + 3	4	YES	NO	YES	YES	YES	NO	YES	LOW
CMOSTG	k	k	NO	NO	YES	NO	YES	NO	YES	HIGH
LEAP	k	2	NO	NO	YES	NO	YES	YES	YES	MEDIUM
DPL	2k	2k	YES	NO	YES	NO	YES	NO	YES	HIGH
CPL	2k	4	YES	NO	YES	NO	YES	NO	YES	MEDIUM
EEPL	2k	4	YES	NO	YES	NO	YES	NO	YES	MEDIUM
SRPL	2k	2	YES	NO	YES	NO	YES	YES	YES	LOW
DCVSPG	2k − 2	2	YES	NO	YES	NO	NO	NO	YES	MEDIUM
PPL	k	k	YES	NO	YES	NO	NO	NO	YES	LOW
DSL	2k + 2	2	YES	YES	NO	NO	YES	NO	YES	MEDIUM
CNTL	2k + 4	2	YES	YES	NO	NO	YES	NO	YES	MEDIUM
NTL	k + 1	1	NO	YES	NO	NO	YES	YES	YES	MEDIUM
SSDL	2k + 6	6	YES	YES	YES	NO	YES	NO	NO	VERY LOW
EDCL	2k + 4	3	YES	NO	YES	NO	YES	NO	NO	VERY LOW
LCDL	2k + 8	6	YES	NO	YES	NO	YES	NO	NO	VERY LOW
DCSL1	2k + 7	4	YES	NO	YES	NO	YES	NO	NO	VERY LOW
BiCMOS	2k + 1	k	NO	YES	NO	NO	YES	NO	YES	MEDIUM

6.7 Silicon-on-Insulator Circuit Design

Silicon-on-Insulator (SOI) technology has been a subject of research for decades, but has become commercially important since it was adopted by IBM for PowerPC microprocessors in 1998 [Shahidi02]. SOI is attractive because it offers potential for higher performance and lower power consumption, but also has a higher manufacturing cost and some unusual transistor behavior that complicates circuit design.

The fundamental difference between SOI and conventional bulk CMOS technology is that the transistor source, drain, and body are surrounded by insulating oxide rather than the conductive substrate or well (called the *bulk*). Using an insulator eliminates most of the parasitic capacitance of the diffusion regions. However, it means that the body is no longer tied to GND or V_{DD} through the substrate or well. Any change in body voltage modulates V_t, leading to both advantages and complications in design.

Figure 6.69 shows a cross-section of an inverter in a SOI process. The process is similar to standard CMOS, but starts with a wafer containing a thin layer of SiO_2 buried beneath a thin single-crystal silicon layer. Section 3.4.1 discussed several ways to form this buried oxide. Shallow trench isolation is used to surround each transistor by an oxide insulator. Figure 6.70 shows a scanning electron micrograph of a 6-transistor static RAM cell in a 0.22 µm IBM SOI process.

SOI devices are categorized as partially depleted (PD) or fully depleted (FD). A depletion region empty of free carriers forms in the body beneath the gate. In FD SOI, the body is thinner than the channel depletion width, so the body charge is fixed and thus the body voltage does not change. In PD SOI, the body is thicker and its voltage can vary depending on how much charge is present. This varying body voltage in turn changes V_t through the body effect. FD SOI has been difficult to manufacture because of the thin body, so PD SOI appears to be the most promising technology.

Throughout this section we will concentrate on nMOS transistors. pMOS transistors have analogous behaviors.

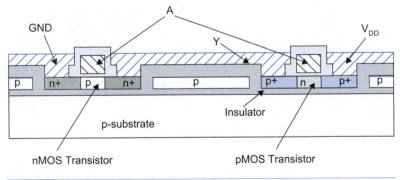

FIG 6.69 SOI inverter cross-sections

6.7.1 Floating Body Voltage

The key to understanding PD SOI is to follow the body voltage. If the body voltage were constant, the threshold voltage would be constant as well and the transistor would behave much like a conventional bulk device except that the diffusion capacitance is lower.

In PD SOI, the floating body voltage varies as it charges or discharges. Figure 6.71 illustrates the mechanisms by which charges enter into or exit from the body [Bernstein00]. There are two paths through which charge can slowly build up in the body:

- Reverse-biased drain-to-body D_{db} and possibly source-to-body D_{sb} junctions carry small diode leakage currents into the body.

- High-energy carriers cause impact ionization, creating electron–hole pairs. Some of these electrons are injected into the gate or gate oxide. (This is the mechanism for hot-electron wearout described in Section 4.8.4.) The corresponding holes accumulate in the body. This effect is most pronounced at V_{DS} above the intended operating point of devices and is relatively unimportant during normal operation. The impact ionization current into the body is modeled as a current source I_{ii}.

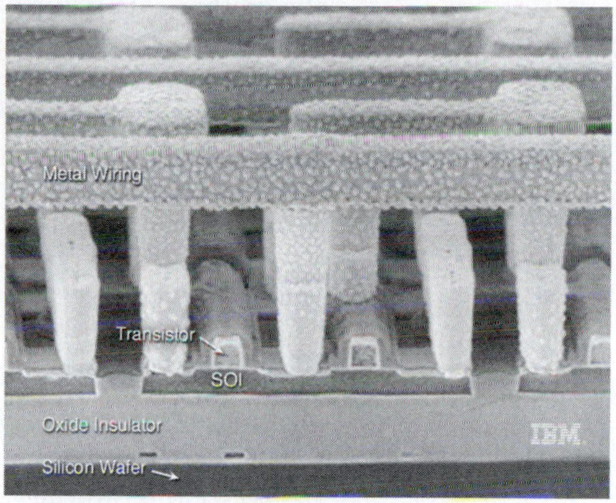

The charge can exit the body through two other paths:

- As the body voltage increases, the source-to-body D_{sb} junction becomes slightly forward-biased. Eventually, the charge exiting from this junction equals the charge leaking in from the drain-to-body D_{db} junction.

- A rising gate or drain capacitively couples the body upward, too. This may strongly forward-bias the source-to-body D_{sb} junction and rapidly spill charge out of the body.

In summary, when a device is idle long enough (on the order of microseconds), the body voltage will reach equilibrium when based on the leakage currents through the source and drain junctions. When the device then begins switching, the charge may spill off the body, shifting the body voltage (and threshold voltage) significantly.

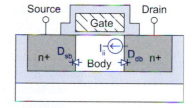

FIG 6.70 IBM SOI process electron micrograph. Courtesy of International Business Machines Corporation. Unauthorized use not permitted.

FIG 6.71 Charge paths to/from floating body

6.7.2 SOI Advantages

A major advantage of SOI is the lower diffusion capacitance. The source and drain abut against oxide on the bottom and sidewalls not facing the channel, essentially eliminating the parasitic capacitance of these sides. This results in a smaller parasitic delay and lower dynamic power consumption.

A more subtle advantage is the potential for lower threshold voltages. In bulk processes, threshold voltage varies with channel length. Hence, variations in polysilicon etching show up as variations in threshold voltage. The threshold voltage must be high enough in the worst (lowest) case to limit subthreshold leakage, so the nominal threshold voltage must be higher. In SOI processes, the threshold variations tend to be smaller. Hence, the nominal V_t can be closer to worst-case. Lower nominal V_t results in faster transistors, especially at low V_{DD}.

As discussed in Section 2.4.4, CMOS devices have a subthreshold slope of about $nv_T\ln 10$, where $v_T = kT/q$ is the thermal voltage (26 mV at room temperature) and n is process-dependent. Bulk CMOS processes typically have $n \approx 1.5$, corresponding to a subthreshold slope of 90 mV/decade. In other words, for each 90 mV decrease in V_{gs} below V_t, the subthreshold leakage current reduces by an order of magnitude. Misleading claims have been made suggesting SOI has $n = 1$ and thus an ideal subthreshold slope of only 60 mV/decade. IBM has found that real SOI devices actually have subthreshold slopes of 75–85 mV/decade. This is better than bulk, but not as good as the hype would suggest. Double-gate MOSFETs and FINFETs discussed in Section 3.4.1 are variations on SOI transistors that offer lower subthreshold slopes because the gate surrounds the channel on more sides and thus turns the transistor off more abruptly.

Finally, SOI is immune to latchup because the insulating oxide eliminates the parasitic bipolar devices that could trigger latchup.

6.7.3 SOI Disadvantages

PD SOI suffers from the *history effect*. Changes in the body voltage modulate the threshold voltage and thus adjust gate delay. The body voltage depends on whether the device has been idle or switching, so gate delay is a function of the switching history. Overall, the elevated body voltage reduces the threshold and makes the gates faster, but the uncertainty makes circuit design more challenging. The history effect can be modeled in a simplified way by assigning different propagation and contamination delays to each gate. IBM found the history effect tends to result in about an 8% variation in gate delay, which is modest compared to the combined effects of manufacturing and environmental variations [Shahidi02].

Unfortunately, the history effect causes significant mismatches between nominally identical transistors. For example, if a sense amplifier has repeatedly read a particular input value, the threshold voltages of the differential pair will be different, introducing an offset voltage in the sense amplifier. This problem can be circumvented by adding a contact to tie the body to ground or to the source for sensitive analog circuits.

Another PD SOI problem is the presence of a parasitic bipolar transistor within each transistor. As shown in Figure 6.72, the source, body, and drain form an emitter, base, and collector of an npn bipolar transistor. In an ordinary transistor, the body is tied to a supply, but in SOI, the body/base floats. If the source and drain are both held high for an extended period of time while the gate is low, the base will float high as well through diode leakage. If the source should then be pulled low, the npn transistor will turn ON. A current I_B flows from body/base to source/emitter. This causes βI_B to flow from the drain/collector to source/emitter. The bipolar transistor gain β depends on the channel length and doping levels but can be greater than 1. Hence, a significant pulse of current can flow from drain to source when the source is pulled low even though the transistor should be OFF.

This pulse of current is sometimes called *pass-gate leakage* because it commonly happens to OFF pass transistors where the source and drain are initially high and then pulled low. It is not a major problem for static circuits because the ON transistors oppose the glitch. However, it can cause malfunctions in dynamic latches and logic. Thus, dynamic nodes should use strong keepers to hold the node steady.

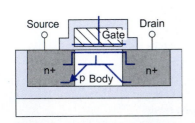

FIG 6.72 Parasitic bipolar transistor in PD SOI

A third problem common to all SOI circuits is self-heating. The oxide is a good thermal insulator as well as an electrical insulator. Thus, heat dissipated in switching transistors tends to accumulate in the transistor rather than spreading rapidly into the substrate. Individual transistors dissipating large amounts of power may become substantially warmer than the die as a whole. At higher temperature they deliver less current and hence are slower. Self-heating can raise the temperature by 10°–15° C for clock buffer and I/O transistors, although the effects tend to be much smaller for logic transistors.

6.7.4 Implications for Circuit Styles

In summary, SOI is attractive for fast CMOS logic. The smaller diffusion capacitance offers a lower parasitic delay. Lower threshold voltages offer better drive current and lower gate delays. Moreover, SOI is also attractive for low-power design. The smaller diffusion capacitance reduces dynamic power consumption. The speed improvements can be traded for lower supply voltage to further reduce dynamic power. Sharper subthreshold slopes offer the opportunity for reduced static leakage current, especially in FINFETs.

Complementary static CMOS gates in PD SOI behave much like their bulk counterparts except for the delay improvement. The history effect also causes pattern-dependent variation in the gate delay.

Circuits with dynamic nodes must cope with a new noise source from pass gate leakage. In particular, dynamic latches and dynamic gates can lose the charge on the dynamic node. Figure 6.73 shows the pass gate leakage mechanism. In each case, the dynamic node X is initially high and the transistor connected to the node is OFF. The source of this transistor starts high and pulls low, turning on the parasitic bipolar transistor and partially discharging X. To overcome pass gate leakage, X should be staticized with a cross-coupled inverter pair for latches or a pMOS keeper for dynamic gates. The staticizing transistors must be relatively strong (e.g., 1/4 as strong as the normal path) to fight the leakage. The gates are slower because they must overcome the strong keepers. Dynamic gates may predischarge the internal nodes to prevent pass gate leakage, but then must deal with charge sharing onto those internal nodes.

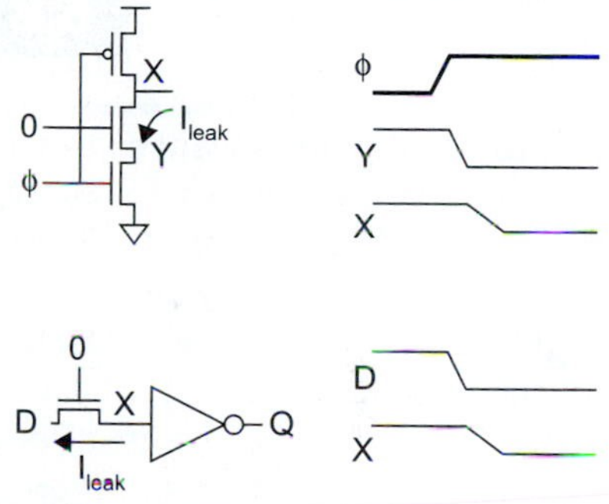

FIG 6.73 Pass gate leakage in dynamic latches and gates

Analog circuits, sense amplifiers, and other circuits that depend on matching between transistors suffer from major threshold voltage mismatches caused by the history of the floating body. They require body contacts to eliminate the mismatches by holding the body at a constant voltage. Gated clocks also have greater clock skew because the history effect makes the clock switch more slowly on the first active cycle after the clock has been disabled for an extended time.

6.7.5 Summary

In summary, Silicon-on-Insulator is attractive because it greatly reduces the source/drain diffusion capacitance, resulting in faster and power-efficient transistors. It also is immune to latchup. Partially depleted SOI is the most practical technology and also boosts drive current because the floating body leads to lower threshold voltages.

SOI design is more challenging because of the floating body effects. Gate delay becomes history-dependent because the voltage of the body depends on the previous state of the device. This complicates device modeling and delay estimation. It also contributes to mismatches between devices. In specialized applications like sense amplifiers, a body contact may be added to create a fully depleted device.

A second challenge with SOI design is pass-gate leakage. Dynamic nodes may be discharged from this leakage even when connected to OFF transistors. Strong keepers can fight the leakage to prevent errors.

Finally, the oxide surrounding SOI devices is a good thermal insulator. This leads to greater self-heating. Thus, the operating temperature of individual transistors may be up to 10°–15° C higher than that of the substrate. Self-heating reduces ON current and makes modeling more difficult.

This section only scratches the surface of a subject worthy of entire books. In particular, SOI static RAMs require special care because of pass gate leakage and floating bodies. [Bernstein00] offers a definitive treatment of partially depleted SOI circuit design and [Kuo01] surveys the literature of SOI circuits.

6.8 Pitfalls and Fallacies

Failing to plan for advances in technology

There are many advances in technology that change the relative merits of different circuit techniques. For example, interconnect delays are not improving as rapidly as gate delays, threshold drops are becoming a greater portion of the supply voltage, and leakage currents are increasing. Failing to anticipate these changes leads to inventions whose usefulness is short-lived.

A salient example is the rise and fall of BiCMOS circuits. Bipolar transistors have a higher current output per unit input capacitance (i.e., a lower logical effort) than CMOS circuits in the 0.8 μm generation, so they became popular, particularly for driving large loads. In the early 1990s, hundreds of papers were written on the subject. The Pentium and Pentium Pro processors were built using BiCMOS processes. Investors poured at least $40 million into a startup company called Exponential, which sought to build a fast PowerPC processor in a BiCMOS process.

Unfortunately, technology scaling works against BiCMOS because of the faster CMOS transistors, lower supply voltages, and larger numbers of transistors on a chip. The relative benefit of bipolar transistors over fine-geometry CMOS decreased. As discussed in Section 6.4.3, the V_{be} drop became an unacceptable fraction of the power supply. Finally, the static power consumption caused by bipolar base currents limits the number of bipolar transistors that can be used.

The Pentium II was based on the Pentium Pro design, but the bipolar transistors had to be removed because they no longer provided advantages in the 0.35 μm generation. Despite a talented engineering team, Exponential failed entirely, ultimately producing a processor that lacked compelling performance advantages and dissipated far more power than anything else on the market [Maier97].

Comparing a well-tuned new circuit to a poor example of existing practice

A time-honored way to make a new invention look good is to tune it as well as possible and compare it to an untuned strawman held up as an example of "existing practice." For example, [Zimmermann97a] points out that most papers finding pass-transistor adders faster than static CMOS adders use 40-transistor static adder cells rather than the faster and smaller 28-transistor cells (Figure 10.4).

Ignoring driver resistance when characterizing pass-transistor circuits

Another way to make pass-transistor circuit families look about twice as fast as they really are is to drive diffusion inputs with a voltage source rather than with the output stage of the previous gate.

Reporting only part of the delay of a circuit

Clocked circuits all have a setup time and a clock-to-output delay. A good way to make clocked circuits look fast is to only report the clock-to-output delay. This is particularly common for the sense-amplifier logic families.

Making outrageous claims about performance

Many published papers have made outrageous performance claims. For example, while comparing full adder designs, some authors have found that DSL and dual-rail domino are 8–10x faster than static CMOS. Neither statement is anywhere close to what designers see in practice; for example, [Ng96] finds that an 8x8 multiplier built from DSL is 1.5x faster and one built from dual-rail domino is 2x faster than static CMOS.

In general, "there ain't no such thing as a free lunch" in circuit design. CMOS design is a fairly mature field and designers are not stupid (or at least not all designers are stupid all the time), so if some new invention seems too good to be true, it probably is. Beware of papers that push the advantages of a new invention without disclosing the inevitable tradeoffs. The tradeoffs may be acceptable, but they must be understood.

Building circuits without adequate verification tools

It is impractical to manually verify circuits on chips that have many millions (soon billions) of transistors. Automated verification tools should check for any pitfalls common to widely used circuit families. If you cannot afford to buy or write appropriate tools, stick with robust static CMOS logic.

6.9 Historical Perspective

Despite rumors of their demise, dynamic circuits remain essential to high-performance digital design. Dynamic circuits predate the widespread use of CMOS. In an nMOS process, pMOS transistors were not available to build complementary gates. One strategy was to build ratioed gates, as shown in Figure 6.74. Conceptually, the ratioed gate consists of an nMOS pull-down network and some pull-up device. A resistor would be simple, but large resistors consume a large layout area in typical MOS processes. Another technique is to use an nMOS transistor with the gate tied to V_{GG}. If $V_{GG} = V_{DD}$, the nMOS transistor will only pull up to $V_{DD} - V_t$. Worse yet, the threshold is increased by the body effect. Thus, using $V_{GG} > V_{DD}$ was attractive. To eliminate this extra supply voltage, some nMOS processes offered *depletion mode* transistors. These transistors, indicated with the thick bar, are identical to

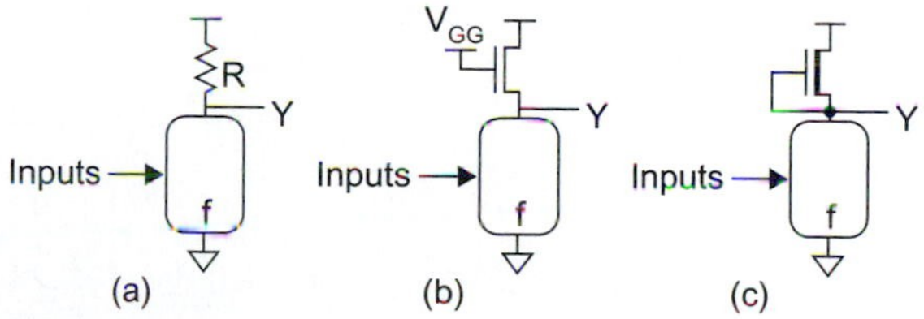

FIG 6.74 nMOS ratioed gates

ordinary *enhancement mode* transistors except that an extra ion implantation was performed to create a negative threshold voltage. The depletion mode pull-ups have their gate wired to the source so $V_{gs} = 0$ and the transistor is always weakly ON.

All the ratioed gates consume static power whenever the outputs are low. The speed is proportional to the RC product, so fast gates need low-resistance pull-ups, exacerbating the power problem. An alternative was to use dynamic gates. The classic MOS textbook of the early 1970s [Penney72] devotes 29 pages to describing a multitude of dynamic gate configurations. Unfortunately, dynamic gates suffer from the monotonicity problem, so each phase of logic may contain only one gate. Phases were separated using nMOS pass transistors that behaved as dynamic latches. Figure 6.75 shows an approach using two-phase nonoverlapping clocks. Each gate precharges in one phase while the subsequent latch is opaque. It then evaluates while making the latch transparent. This approach is prone to charge-sharing noise when the latch opens and precharge only rises to $V_{DD} - V_t$. Numerous four-phase clocking techniques were also developed.

With the advent of CMOS technology, dynamic logic lost its advantage of power consumption. However, chip space was at a premium and dynamic gates could eliminate most of the pMOS transistors to save area. Domino gates were developed at Bell Labs for a 32-bit adder in the BELLMAC-32A microprocessor to solve problems of both area and speed [Krambeck82, Shoji82]. Domino allows multiple noninverting gates to be cascaded in a single phase.

High-performance microprocessors have boosted clock speeds faster than simple process improvement would allow, so the number of gate delays per cycle has shrunk. The DEC Alpha microprocessors pioneered this trend through the 1990s [Gronowski98] and most other CPUs have followed. Domino circuits have become crucial to achieving these fast cycle times. IBM is a notable exception, relying almost exclusively on complementary CMOS logic with fast time to market in cutting-edge processes to deliver adequate mainframe processor performance [Curran02].

Some older domino designs leave out the keeper to save area and gain a slight performance advantage. This has become more difficult as leakage and coupling noise have increased with process scaling. The 0.35 µm Alpha 21164 was one of the last designs to have no keeper (and to use dynamic latches). Its fully dynamic operation gave advantages in both speed and area, but during test it had a minimum operating frequency of 20 MHz

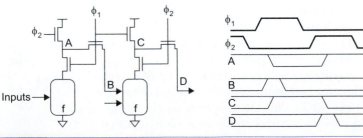

FIG 6.75 nMOS dynamic gates

to retain state. In the Alpha 21264, leakage current had increased to the point that keepers were essential. Modern designs always need keepers. As an interesting aside, the Alpha microprocessors also did not use scan latches because scan cost area and a small amount of performance. This proved unfortunate on the Alpha 21264, which was difficult to debug because of the limited observability into the processor state. Now virtually all design methodologies require scan capability in the latches or registers, as will be discussed in Section 9.6.

Contemporary domino design involves substantial overhead besides the keepers. Dual-rail domino is often necessary for nonmonotonic functions. Secondary precharge devices are used to prevent charge sharing. Multiple clock phases are used to tolerate clock skew and each phase requires a clock buffer and wire. Dynamic inputs and outputs must be shielded or spaced far from aggressors to avoid coupling noise problems. Considering all these factors, domino no longer has significant area advantages over complementary CMOS gates for general-purpose logic. However, static RAMs and register files discussed in Chapter 11 can be viewed as special applications of domino that do achieve high density.

Summary

Circuit delay is related to the $(C/I)\Delta V$ product of gates. This chapter explored alternative combinational circuit structures to improve the C/I ratio or respond to smaller voltage swings. Many of these techniques trade higher power consumption and/or lower noise margins for better delay. While complementary CMOS circuits are quite robust, the alternative circuit families have pitfalls that must be understood and managed. The chapter also examined the layout of cells and layout optimizations to reduce parasitic capacitance.

Three of the commonly used alternatives to complementary CMOS are domino, pseudo-nMOS, and pass transistor logic. Each attempts to reduce the input capacitance by performing logic mostly through nMOS transistors.

Pseudo-nMOS replaces the pMOS pull-up network with a single weak pMOS transistor that is always ON. The pMOS transistor dissipates static power when the output is low. If it is too weak, the rising transition is slow. If it is too strong, V_{OL} is too high and the power consumption increases. When the static power consumption is tolerable, pseudo-nMOS gates work well for wide NOR functions.

Dynamic gates resemble pseudo-nMOS, but use a clocked pMOS transistor in place of the weak pull-up. When the clock is low, the gates precharge high. When the clock rises, the gates evaluate, pulling the output low or leaving it floating high. The input of a dynamic gate must be monotonically rising while the gate is in evaluation, but the output monotonically falls. Domino gates consist of a dynamic gate followed by an inverting static gate and produce monotonically rising outputs. Therefore, domino gates can be cascaded, but only compute noninverting functions. Dual-rail domino accepts true and complementary inputs and produces true and complementary outputs to provide any logic function at the expense of larger gates and twice as many wires. Dynamic gates are also sensitive to noise because V_{IL} is close to the threshold voltage V_t and the output floats.

Major noise sources include charge sharing, leakage, and coupling. Therefore, domino circuits typically use secondary precharge transistors, keepers, and shielded or carefully routed interconnect. The high-activity factors of the clock and dynamic node make domino power hungry. Despite all of these challenges, domino offers a 1.5–2x speedup over static CMOS, giving it a compelling advantage for the critical paths of high-performance systems.

Pass-transistor circuits use inputs that drive the diffusion inputs as well as the gates of transistors. Many pass-transistor techniques have been explored and Complementary Pass Transistor logic has proven to be one of the most effective. This dual-rail technique uses networks of nMOS transistors to compute true and complementary logic functions. The nMOS transistors only pull up to $V_{DD} - V_t$, so cross-coupled pMOS transistors boost the output to full-rail levels. Some designers find that pass-transistor circuits are faster and smaller for functions such as XOR, full adders, and multiplexers that are clumsy to implement in static CMOS. Because of the threshold drop, the circuits do not scale well as V_{DD}/V_t decreases.

As designer productivity, measured in gates per designer per week, must increase exponentially, static CMOS circuits become even more attractive because they are so simple and robust.

Exercises

6.1   Design a fast 6-input OR gate in each of the following circuit families. Sketch an implementation using two stages of logic (e.g., NOR6 + INV, NOR3 + NAND2, etc.). Label each gate with the width of the pMOS and nMOS transistors. Each input can drive no more than 30 λ of transistor width. The output must drive a 60/30 inverter (i.e., an inverter with a 60 λ-wide pMOS and 30 λ-wide nMOS transistor). Use logical effort to choose the topology and size for least average delay. Estimate this delay using logical effort. When estimating parasitic delays, count only the diffusion capacitance on the output node.

a)   static CMOS

b)   pseudo-nMOS with pMOS transistors 1/4 the strength of the pull-down stack

c)   domino (a footed dynamic gate followed by a HI-skew inverter); only optimize the delay from rising input to rising output

6.2   Simulate each gate you designed in Exercise 6.1. Determine the average delay (or rising delay for the domino design). Logical effort is only an approximation. Tweak the transistor sizes to improve the delay. How much improvement can you obtain?

6.3   Sketch a schematic for a 12-input OR gate built from NANDs and NORs of no more than 3 inputs each.

6.4   Design a static CMOS circuit to compute $F = (A + B)(C + D)$ with least delay. Each input can present a maximum of 30 λ of transistor width. The output must drive a load equivalent to 500 λ of transistor width. Choose transistor sizes to achieve least delay and estimate this delay in τ.

6.5   Figure 6.76 shows two series transistors modeling the pull-down network of 2-input NAND gate.

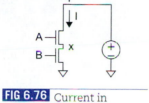

a)   Plot I vs. A using ideal transistor models for $0 \leq A \leq 1$, $B = Y = 1$, $V_t = 0$, $\beta = 1$. On the same axes, plot I vs. B for $0 \leq B \leq 1$, $A = 1$. *Hint*: you will need to solve for x; this can be done best numerically.

b)   Using your results from (a), explain why the inner input of a 2-input NAND gate has a slightly greater logical effort than the outer input.

FIG 6.76 Current in series transistors

6.6   What is the logical effort of an OR-AND-INVERT gate at either of the OR terminals? At the AND terminal? What is the parasitic delay if only diffusion capacitance on the output is counted?

6.7   Simulate a 3-input NOR gate in your process. Determine the logical effort and parasitic delay from each input.

6.8   Using the data sheet from Figure 4.25, find the rising and falling logical effort and parasitic delay of the *X1* 2-input NAND gate from the *A* input.

6.9   Repeat Exercise 6.8 for the *B* input. Explain why the results are different for the different inputs.

6.10   Sketch HI-skew and LO-skew 3-input NAND and NOR gates. What are the logical efforts of each gate on its critical transition?

6.11   Derive a formula for g_u, g_d, and g_{avg} for HI-skew and LO-skew k-input NAND gates with a skew factor of $s < 1$ (i.e., the noncritical transistor is s times normal size) as a function of s and k.

6.12   Design an asymmetric 3-input NOR gate that favors a critical input over the other two. Choose transistor sizes so the logical effort on the critical input is 1.5. What is the logical effort of the noncritical inputs?

6.13   Prove that the P/N ratio that gives lowest average delay in a logic gate is the square root of the ratio that gives equal rise and fall delays.

6.14   Let $\rho(g, p)$ be the best stage effort of a path if one is free to add extra buffers with a parasitic delay p and logical effort g. For example, Section 4.3.3 showed that $\rho(1, 1) = 3.59$. It is easy to make a plot of $\rho(1, p)$ by solving EQ (4.19) numerically; this gives the best stage effort of static CMOS circuits where the inverter has a parasitic delay of p. Prove the following result, which is useful for determining the best stage effort of domino circuits where buffers have lower logical efforts:

$$\rho(g, p) = g\rho(1, \tfrac{p}{g})$$

6.15 Simulate a fanout-of-4 inverter. Use a unit-sized nMOS transistor. How wide must the pMOS transistor be to achieve equal rising and falling delays? What is the delay? How wide must the pMOS transistor be to achieve minimum average delay? What is the delay? How much faster is the average delay?

6.16 Many standard cell libraries choose a *P/N* ratio for an inverter in between that which would give equal rising and falling delays and that which would give minimum average delay. Why is this done?

6.17 A static CMOS NOR gate uses 4 transistors, while a pseudo-nMOS NOR gate uses only 3. Unfortunately, the pseudo-nMOS output does not swing rail to rail. If both the inputs and their complements are available, it is possible to build a 3-transistor NOR that swings rail to rail without using any dynamic nodes. Show how to do it. Explain any drawbacks of your circuit.

6.18 Sketch pseudo-nMOS 3-input NAND and NOR gates. Label the transistor widths. What are the rising, falling, and average logical efforts of each gate?

6.19 Sketch a pseudo-nMOS gate that implements the function

$$F = \overline{A(B+C+D)+E \cdot F \cdot G}.$$

6.20 Design an 8-input AND gate with an electrical effort of 6 using pseudo-nMOS logic. If the parasitic delay of an *n*-input pseudo-nMOS NOR gate is $(4n + 2)/9$, what is the path delay?

6.21 Simulate a pseudo-nMOS inverter in which the pMOS transistor is half the width of the nMOS transistor. What are the rising, falling, and average logical efforts? What is V_{OL}?

6.22 Repeat Exercise 6.21 in the FS and SF process corners.

6.23 Sketch a 3-input symmetric NOR gate. Size the inverters so that the pull-down is four times as strong as the net worst-case pull-up. Label the transistor widths. Estimate the rising, falling, and average logical efforts. How do they compare to a static CMOS 3-input NOR gate?

6.24 Sketch a 2-input symmetric NAND gate. Size the inverters so that the pull-up is four times as strong as the net worst-case pull-down. Label the transistor widths. Estimate the rising, falling, and average logical efforts. How do they compare to a static CMOS 2-input NAND gate?

6.25 Compare the average delays of a 2, 4, 8, and 16-input pseudo-NMOS and SFPL NOR gate driving a fanout of 4 identical gates.

6.26 Sketch a 3-input CVSL OR/NOR gate.

6.27 Sketch dynamic footed and unfooted 3-input NAND and NOR gates. Label the transistor widths. What is the logical effort of each gate?

6.28 Sketch a 3-input dual-rail domino OR/NOR gate.

6.29 Sketch a 3-input dual-rail domino majority/minority gate. This is often used in domino full adder cells. Recall that the majority function is true if more than half of the inputs are true.

6.30 Compare a standard keeper with the noise tolerant precharge device. Larger pMOS transistors result in a higher V_{IL} (and thus better noise margins) but more delay. Simulate a 2-input footed NAND gate and plot V_{IL} vs. delay for various sizes of keepers and noise tolerant precharge transistors.

6.31 Design a 4-input footed dynamic NAND gate driving an electrical effort of 1. Estimate the worst charge sharing noise as a fraction of V_{DD} assuming that diffusion capacitance on uncontacted nodes is about half of gate capacitance and on contacted nodes it equals gate capacitance.

6.32 Repeat Exercise 6.31, generating a graph of charge sharing noise vs. electrical effort for $h = 0, 1, 2, 4$, and 8.

6.33 Repeat Exercise 6.31 if a small secondary precharge transistor is added on one of the internal nodes.

6.34 Perform a simulation of your circuits from Exercise 6.31. Explain any discrepancies.

6.35 Design a domino circuit to compute $F = (A + B)(C + D)$ as fast as possible. Each input may present a maximum of 30 λ of transistor width. The output must drive a load equivalent to 500 λ of transistor width. Choose transistor sizes to achieve least delay and estimate this delay in τ.

6.36 Redesign the memory decoder from Section 4.3.4 using footed domino logic. You can assume you have both true and complementary monotonic inputs available, each capable of driving 10 unit transistors. Label gate sizes and estimate the delay.

6.37 Sketch an NP Domino 8-input AND circuit.

6.38 Sketch a 4:1 multiplexer. You are given four data signals D0, D1, D2, and D3, and two select signals, S0 and S1. How many transistors does each design require?

a) Use only static CMOS logic gates.

b) Use a combination of logic gates and transmission gates.

6.39 Sketch 3-input XOR functions using each of the following circuit techniques:

a) static CMOS

b) pseudo-nMOS

c) dual-rail domino

d) CPL

e) EEPL

f) DCVSPG

      g) SRPL

      h) PPL

      i) DPL

      j) LEAP

6.40 Repeat Exercise 6.39 for a 2-input NAND gate.

6.41 Design sense-amplifier gates using each of the following circuit families to compute an 8-input XOR function in a single gate: SSDL, ECDL, LCDL, DCSL1, DCSL2, DCSL3. Each true or complementary input can drive no more than 24 λ of transistor width. Each output must drive a 32/16 λ inverter. Simulate each circuit to determine the setup time and clock-to-out delays.

6.42 Figure 6.77 shows a Switched Output Differential Structure (SODS) gate. Explain how the gate operates and sketch waveforms for the gate acting as an inverter/buffer. Comment on the strengths and weaknesses of the circuit family.

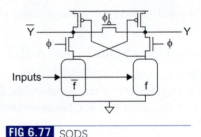

FIG 6.77 SODS

6.43 Choose one of the circuit families (besides SODS, Exercise 6.42) mentioned in Section 6.4.4 or published in a recent paper. Critically evaluate the original paper in which the circuit was proposed. Sketch an inverter or buffer and explain how it operates, including appropriate waveforms. What are the strengths of the circuit family? If you were the circuit manager choosing design styles for a large chip, what concerns might you have about the circuit family?

Sequential Circuit Design

7

7.1 Introduction

The previous chapter addressed *combinational* circuits in which the output is a function of the current inputs. This chapter discusses *sequential* circuits in which the output depends on previous as well as current inputs; such circuits are said to have *state*. Finite state machines and pipelines are two important examples of sequential circuits.

Sequential circuits are usually designed with flip-flops or latches, which are sometimes called *memory elements*, that hold data called *tokens*. The purpose of these elements is not really memory; instead, it is to enforce sequence, to distinguish the *current* token from the *previous* or *next* token. Therefore, we will call them *sequencing elements* [Harris01a]. Without sequencing elements, the next token might catch up with the previous token, garbling both. Sequencing elements delay tokens that arrive too early, preventing them from catching up with previous tokens. Unfortunately, they inevitably add some delay to tokens that are already critical, decreasing the performance of the system. This extra delay is called *sequencing overhead*.

This chapter considers sequencing for both static and dynamic circuits. *Static circuits* refer to gates that have no clock input, such as complementary CMOS, pseudo-nMOS, or pass transistor logic. *Dynamic circuits* refer to gates that have a clock input, especially domino logic. To complicate terminology, sequencing elements themselves can be either static or dynamic. A sequencing element with *static storage* employs some sort of feedback to retain its output value indefinitely. An element with *dynamic storage* generally maintains its value as charge on a capacitor that will leak away if not refreshed for a long period of time. The choices of static or dynamic for gates and for sequencing elements can be independent.

Sections 7.2–7.4 explore sequencing elements for static circuits, particularly flip-flops, 2-phase transparent latches, and pulsed latches. Section 7.5 delves into a variety of ways to sequence dynamic circuits. A periodic clock is commonly used to indicate the timing of a sequence. Section 7.6 describes how external signals can be synchronized to the clock and analyzes the risks of synchronizer failure. Wave pipelining is discussed in Section 7.7. Clock generation and distribution will be examined further in Section 12.5.

The choice of sequencing strategy is intimately tied to the design flow that is being used by an organization. Thus, it is important before departing on a design direction to ensure that all phases of design capture, synthesis, and verification can be accommodated.

This includes such aspects as cell libraries (Are the latch or flip-flop circuits and models available?); tools such as timing analyzers (Can timing closure be achieved easily?); and automatic test generation (Can self-test elements be inserted easily?).

7.2   Sequencing Static Circuits

Recall from Section 1.4.9 that *latches* and *flip-flops* are the two most commonly used sequencing elements. Both have three terminals: data input (D), clock (*clk*), and data output (Q). The latch is transparent when the clock is high and opaque when the clock is low; in other words, when the clock is high, D flows through to Q as if the latch were just a buffer, but when the clock is low, the latch holds its present Q output even if D changes. The flip-flop is an edge-triggered device that copies D to Q on the rising edge of the clock and ignores D at all other times. These are illustrated in Figure 7.1. The unknown state of Q before the first rising clock edge is indicated by the pair of lines at both low and high levels.

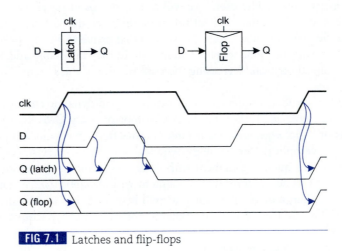

FIG 7.1   Latches and flip-flops

This section explores the three most widely used methods of sequencing static circuits with these elements: flip-flops, 2-phase transparent latches, and pulsed latches [Unger86]. An ideal sequencing methodology would introduce no sequencing overhead, allow sequencing elements back-to-back with no logic in between, grant the designer flexibility in balancing the amount of logic in each clock cycle, tolerate moderate amounts of clock skew without degrading performance, and consume zero area and power. We will compare these methods and explore the tradeoffs they offer. We will also examine a number of transistor-level circuit implementations of each element.

7.2.1 Sequencing Methods

Figure 7.2 illustrates three methods of sequencing blocks of combinational logic. In each case, the clock waveforms, sequencing elements, and combinational logic are shown. The horizontal axis corresponds to the time at which a token reaches a point in the circuit. For example, the token is captured in the first flip-flop on the first rising edge of the clock. It propagates through the combinational logic and reaches the second flip-flop on the second rising edge of the clock. The dashed vertical lines indicate the boundary between one clock cycle and the next. The clock period is T_c. In a 2-phase system, the phases may be separated by $t_{nonoverlap}$. In a pulsed system, the pulse width is t_{pw}.

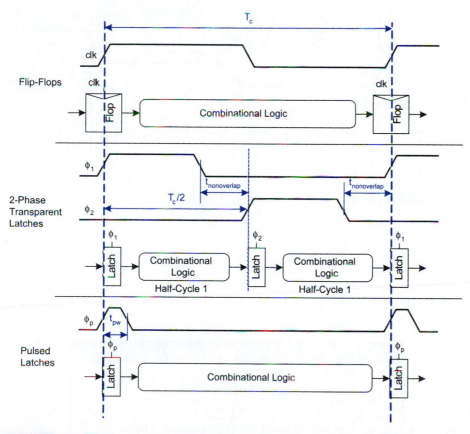

FIG 7.2   Static sequencing methods

Flip-flop-based systems use one flip-flop on each cycle boundary. Tokens advance from one cycle to the next on the rising edge. If a token arrives too early, it waits at the flip-flop until the next cycle. Recall that the flip-flop can be viewed as a pair of back-to-back latches using *clk* and its complement, as shown in Figure 7.3. If we separate the latches, we can divide the full cycle of combinational logic into two phases, sometimes called *half-cycles*. The two latch clocks are often called ϕ_1 and ϕ_2. They may correspond to *clk* and its complement $\overline{clk}$ or may be nonoverlapping ($t_{nonoverlap} > 0$). At any given time, at least one clock is low and the corresponding latch is opaque, preventing one token from catching up with another. The two latches behave in much the same manner as two watertight gates in a canal lock [Mead80]. Pulsed latch systems eliminate one of the latches from each cycle and apply a brief pulse to the remaining latch. If the pulse is shorter than the delay through the combinational logic, we can still expect that a token will only advance through one clock cycle on each pulse.

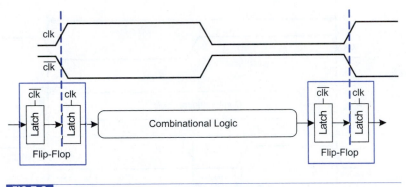

FIG 7.3 Flip-flop viewed as back-to-back latch pair

Table 7.1 defines the delays and timing constraints of the combinational logic and sequencing elements. These delays may differ significantly for rising and falling transitions and can be distinguished with an *r* or *f* suffix. For brevity, we will use the overall maximum and minimum.

Table 7.1	Sequencing element timing notation
Term	**Name**
t_{pd}	Logic Propagation Delay
t_{cd}	Logic Contamination Delay
t_{pcq}	Latch/Flop Clock-to-Q Propagation Delay
t_{ccq}	Latch/Flop Clock-to-Q Contamination Delay
t_{pdq}	Latch D-to-Q Propagation Delay
t_{cdq}	Latch D-to-Q Contamination Delay
t_{setup}	Latch/Flop Setup Time
t_{hold}	Latch/Flop Hold Time

Figure 7.4 illustrates these delays in a *timing diagram*. In a timing diagram, the horizontal axis indicates time and the vertical axis indicates logic level. A single line indicates that a signal is high or low at that time. A pair of lines indicates that a signal is stable but that we don't care about its value. Criss-crossed lines indicate that the signal might change at that time. A pair of lines with cross-hatching indicates that the signal may change once or more over an interval of time.

Figure 7.4(a) shows the response of combinational logic to the input A changing from one arbitrary value to another. The output Y cannot change instantaneously. After the contamination delay t_{cd}, Y may begin to change or *glitch*. After the propagation delay t_{pd}, Y must have settled to a final value. The contamination delay and propagation delay may be very different because of multiple paths through the combinational logic. Figure 7.4(b) shows the response of a flip-flop. The data input must be stable for some window around the rising edge of the flop if it is to be reliably sampled. Specifically, the input D must have settled by some *setup time* t_{setup} before the rising edge of *clk* and should not change again until a *hold time* t_{hold} after the clock edge. The output begins to change after a *clock-to-Q contamination delay* t_{ccq} and completely settles after a *clock-to-Q propagation delay* t_{pcq}. Figure 7.4(c) shows the response of a latch. Now the input D must set up and hold around the falling edge that defines the end of the sampling period. The output initially changes t_{ccq} after the latch becomes transparent on the rising edge of the clock and settles by t_{pcq}. While the latch is transparent, the output will continue to track the input after some D-to-Q delay t_{cdq} and t_{pdq}. Section 7.4.4 discusses how to measure the setup and hold times and propagation delays in simulation.

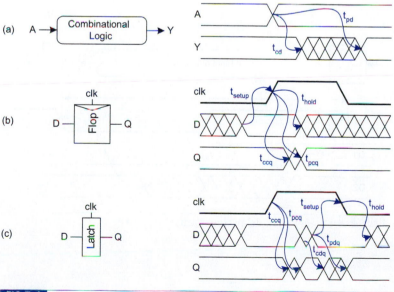

FIG 7.4 Timing diagrams

7.2.2 Max-Delay Constraints

Ideally, the entire clock cycle would be available for computations in the combinational logic. Of course, the sequencing overhead of the latches or flip-flops cuts into this time. If the combinational logic delay is too great, the receiving element will miss its setup time and sample the wrong value. This is called a *setup time failure* or *max-delay failure*. It can be solved by redesigning the logic to be faster or by increasing the clock period. This section computes the actual time available for logic and the sequencing overhead of each of our favorite sequencing elements: flip-flops, two-phase latches, and pulsed latches.

Figure 7.5 shows the max-delay timing constraints on a path from one flip-flop to the next, assuming ideal clocks with no skew. The path begins with the rising edge of the clock triggering *F1*. The data must propagate to the output of the flip-flop *Q1* and through the combinational logic to *D2*, setting up at *F2* before the next rising clock edge. This implies that the clock period must be at least

$$T_c \geq t_{pcq} + t_{pd} + t_{setup} \tag{7.1}$$

Alternatively, we can solve for the maximum allowable logic delay, which is simply the cycle time less the sequencing overhead introduced by the propagation delay and setup time of the flip-flop.

$$t_{pd} \leq T_c - \underbrace{\left(t_{setup} + t_{pcq}\right)}_{\text{sequencing overhead}} \tag{7.2}$$

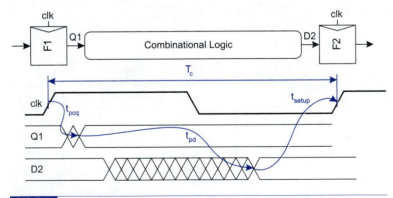

FIG 7.5 Flip-flop max-delay constraint

Example

The Arithmetic/Logic Unit (ALU) *self-bypass path* limits the clock frequency of some pipelined microprocessors. For example, the Integer Execution Unit (IEU) of the Itanium 2 contains self-bypass paths for six separate ALUs, as shown in Figure 7.6(a) [Fetzer02]. The path for one of the ALUs begins at registers containing the inputs to an adder, as shown in Figure 7.6(b). The adder must compute the sum (or difference, for subtraction). A *result multiplexer* chooses between this sum, the output of the logic unit, and the output of the shifter. Then a series of *bypass multiplexers* selects the inputs to the ALU for the next cycle. The early bypass multiplexer chooses among results of ALUs from previous cycles and is not on the critical path. The 8:1 middle bypass multiplexer chooses a result from any of the six ALUs, the early bypass mux, or the register file. The 4:1 late bypass multiplexer chooses a result from either of two results returning from the data cache, the middle bypass mux result, or the immediate operand specified by the next instruction. The late bypass mux output is driven back to the ALU to use on the next cycle. Because the six ALUs and the bypass multiplexers occupy a significant amount of area, the critical path also involves 2 mm wires from the result mux to middle bypass mux and from the middle bypass mux back to the late bypass mux. (Note: In the Itanium 2, the ALU self-bypass path is built from four-phase skew-tolerant domino circuits. For the purposes of these examples, we will hypothesize instead that it is built from static logic and flip-flops or latches.)

For our example, the propagation delays and contamination delays of the path are given in Table 7.2. Suppose the registers are built from flip-flops with a setup time of 62 ps, hold time of −10 ps, propagation delay of 90 ps, and contamination delay of 75 ps. Calculate the minimum cycle time T_c at which the ALU self-bypass path will operate correctly.

Table 7.2	Combinational logic delays	
Element	**Propagation Delay**	**Contamination Delay**
Adder	590 ps	100 ps
Result Mux	60 ps	35 ps
Early Bypass Mux	110 ps	95 ps
Middle Bypass Mux	80 ps	55 ps
Late Bypass Mux	70 ps	45 ps
2-mm wire	100 ps	65 ps

continued

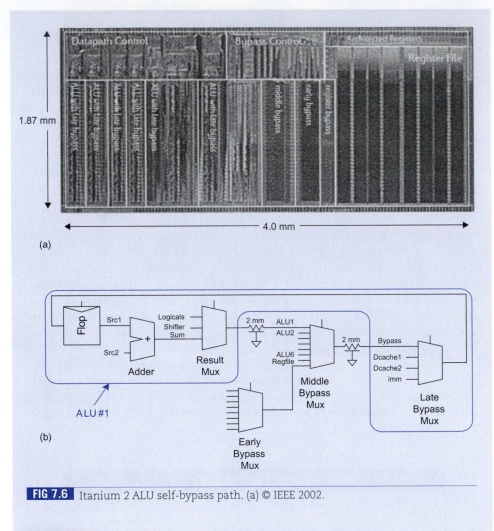

FIG 7.6 Itanium 2 ALU self-bypass path. (a) © IEEE 2002.

Solution: The critical path involves propagation delays through the adder (590 ps), result mux (60 ps), middle bypass mux (80 ps), late bypass mux (70 ps), and two 2-mm wires (100 ps each), for a total of t_{pd} = 1000 ps. According to EQ (7.1), the cycle time T_c must be at least 90 + 1000 + 62 = 1152 ps.

Figure 7.7 shows the analogous constraints on a path using two-phase transparent latches. Let us assume that data *D1* arrives at *L1* while the latch is transparent (ϕ_1 high). The data propagates through *L1*, the first block of combinational logic, *L2*, and the second block of combinational logic. Technically, *D3* could arrive as late as a setup time before the falling edge of ϕ_1 and still be captured correctly by *L3*. To be fair, we will insist that *D3* nominally arrive no more than one clock period after *D1* because, in the long run, it is impossible for every single-cycle path in a design to consume more than a full clock period. Certain paths may take longer if other paths take less time; this technique is called *time borrowing* and will be addressed in Section 7.2.4. Assuming the path takes no more than a cycle, we see the cycle time must be

$$T_c \geq t_{pdq1} + t_{pd1} + t_{pdq2} + t_{pd2} \tag{7.3}$$

Once again, we can solve for the maximum logic delay, which is the sum of the logic delays through each of the two phases. The sequencing overhead is the two latch propagation delays. Notice that the nonoverlap between clocks does not degrade performance in the latch-based system because data continues to propagate through the combinational logic between latches even while both clocks are low. Realizing that a flip-flop can be made from two latches whose delays determine the flop propagation delay and setup time, we see EQ (7.4) is closely analogous to EQ (7.2).

$$t_{pd} = t_{pd1} + t_{pd2} \leq T_c - \underbrace{\left(2t_{pdq}\right)}_{\text{sequencing overhead}} \tag{7.4}$$

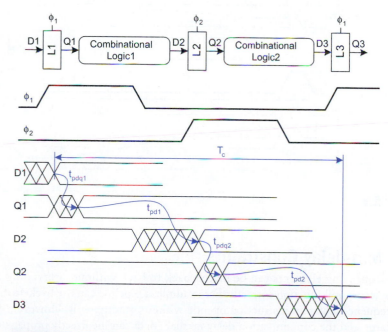

FIG 7.7 Two-phase latch max-delay constraint

If the pulse is wide enough, the max-delay constraint for pulsed latches is similar to that of two-phase latches except that only one latch is in the critical path, as shown in Figure 7.8(a). However, if the pulse is narrower than the setup time, the data must set up before the pulse rises, as shown in Figure 7.8(b). Combining these two cases gives

$$T_c \geq \max\left(t_{pdq} + t_{pd}, \, t_{pcq} + t_{pd} + t_{\text{setup}} - t_{pw}\right) \tag{7.5}$$

Solving for the maximum logic delay shows that the sequencing overhead is just one latch delay if the pulse is wide enough to hide the setup time

$$t_{pd} \leq T_c - \underbrace{\max\left(t_{pdq}, \, t_{pcq} + t_{\text{setup}} - t_{pw}\right)}_{\text{sequencing overhead}} \tag{7.6}$$

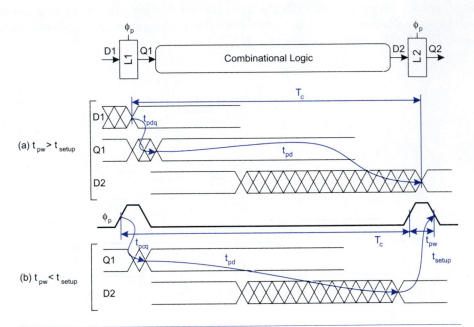

FIG 7.8 Pulsed latch max-delay constraint

7.2.3 Min-delay Constraints

Ideally, sequencing elements can be placed back to back without intervening combinational logic and still function correctly. For example, a pipeline can use back-to-back registers to sequence along an instruction opcode without modifying it. However, if the hold time is large and the contamination delay is small, data can incorrectly propagate through two successive elements on one clock edge, corrupting the state of the system. This is

Example

Recompute the ALU self-bypass path cycle time if the flip-flop is replaced with a pulsed latch. The pulsed latch has a pulse width of 150 ps, a setup time of 40 ps, a hold time of 5 ps, a *clk*-to-*Q* propagation delay of 82 ps and contamination delay of 52 ps, and a *D*-to-*Q* propagation delay of 92 ps.

Solution: t_{pd} is still 1000 ps. According to EQ (7.5), the cycle time must be at least 92 + 1000 = 1092 ps.

called a *race condition*, *hold time failure*, or *min-delay failure*. It can only be fixed by redesigning the logic, not by slowing the clock. Therefore, designers should be very conservative in avoiding such failures because modifying and refabricating a chip is very expensive and time-consuming.

Figure 7.9 shows the min-delay timing constraints on a path from one flip-flop to the next assuming ideal clocks with no skew. The path begins with the rising edge of the clock triggering *F1*. The data may begin to change at *Q1* after a *clk*-to-*Q* contamination delay, and at *D2* after another logic contamination delay. However, it must not reach *D2* until at least the hold time t_{hold} after the clock edge, lest it corrupt the contents of *F2*. Hence we solve for the minimum logic contamination delay:

$$t_{cd} \geq t_{\text{hold}} - t_{ccq} \tag{7.7}$$

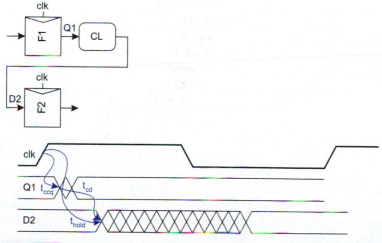

FIG 7.9 Flip-flop latch min-delay constraint

Example

In the ALU self-bypass example with flip-flops from Figure 7.6, the earliest input to the late bypass multiplexer is the *imm* value coming from another flip-flop. Will this path experience any hold time failures?

Solution: No. The late bypass mux has t_{cd} = 45 ps. The flip-flops have t_{hold} = −10 ps and t_{ccq} = 75 ps. Hence, EQ (7.7) is easily satisfied.

If the contamination delay through the flip-flop exceeds the hold time, you can safely use back-to-back flip-flops. If not, you must explicitly add delay between the flip-flops (e.g., with a buffer) or use special slow flip-flops with greater than normal contamination delay on paths that require back-to-back flops. Scan chains are a common example of paths with back-to-back flops.

Figure 7.10 shows the min-delay timing constraints on a path from one transparent latch to the next. The path begins with data passing through *L1* on the rising edge of ϕ_1. It must not reach *L2* until a hold time after the previous falling edge of ϕ_2 because *L2* should have become safely opaque before *L1* becomes transparent. As the edges are separated by $t_{\text{nonoverlap}}$, the minimum logic contamination delay through each phase of logic is

$$t_{cd1}, t_{cd2} \geq t_{\text{hold}} - t_{ccq} - t_{\text{nonoverlap}} \tag{7.8}$$

(Note that our derivation found the minimum delay through the first half-cycle, but that the second half-cycle has the same constraint.)

This result shows that by making $t_{\text{nonoverlap}}$ sufficiently large, hold time failure can be avoided entirely. However, generating and distributing nonoverlapping clocks is very challenging at high speeds. Therefore, most commercial transparent latch-based systems use the clock and its complement. In this case, $t_{\text{nonoverlap}}$ = 0 and the contamination delay constraint is the same between the latches and flip-flops.

This leads to an apparent paradox: The contamination delay constraint applies to each phase of logic for latch-based systems, but to the entire cycle of logic for flip-flops. Therefore, latches seem to require twice the overall logic contamination delay as compared to flip-flops. Yet flip-flops can be built from a pair of latches! The paradox is resolved by observing that a flip-flop has an internal race condition between the two latches. The flip-flop must be carefully designed so that it always operates reliably.

Figure 7.11 shows the min-delay timing constraints on a path from one pulsed latch to the next. Now data departs on the rising edge of the pulse but must hold until after the falling edge of the pulse. Therefore, the pulse width effectively increases the hold time of the pulsed latch as compared to a flip-flop.

$$t_{cd} \geq t_{\text{hold}} - t_{ccq} + t_{pw} \tag{7.9}$$

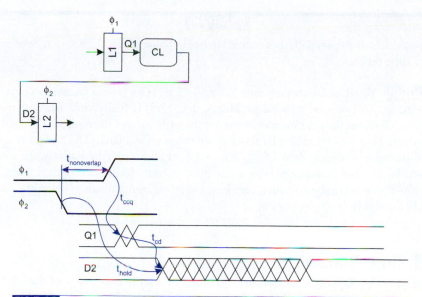

FIG 7.10 Two-phase latch min-delay constraint

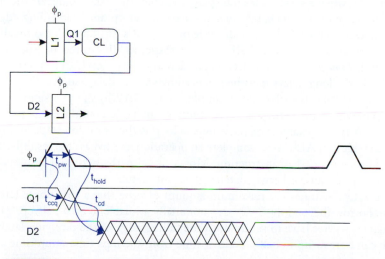

FIG 7.11 Pulsed latch min-delay constraint

Example

If the ALU self-bypass path uses pulsed latches in place of flip-flops, will it have any hold-time problems?

Solution: Yes. The late bypass mux has t_{cd} = 45 ps. The pulsed latches have t_{pw} = 150 ps, t_{hold} = 5 ps, and t_{ccq} = 52 ps. Hence, EQ (7.9) is badly violated. *Src1* may receive *imm* from the next instruction rather than the current instruction. The problem could be solved by adding buffers after the *imm* pulsed latch. The buffers would need to add a minimum delay of $t_{hold} - t_{ccq} + t_{pw} - t_{cd}$ = 58 ps. Alternatively, the *imm* pulsed latch could be replaced with a flip-flop without slowing the critical path. If the flip-flop were designed with a very long (>110 ps) contamination delay, the race would be avoided.

7.2.4 Time Borrowing

In a system using flip-flops, data departs the first flop on the rising edge of the clock and must set up at the second flop before the next rising edge of the clock. If the data arrives late, the circuit produces the wrong result. If the data arrives early, it is blocked until the clock edge, and the remaining time goes unused. Therefore, we say the clock imposes a *hard edge* because it sharply delineates the cycles.

In contrast, when a system uses transparent latches, the data can depart the first latch on the rising edge of the clock, but does not have to set up until the falling edge of the clock on the receiving latch. If one half-cycle or stage of a pipeline has too much logic, it can borrow time into the next half-cycle or stage, as illustrated in Figure 7.12(a) [Bernstein99]. *Time borrowing* can accumulate across multiple cycles. However, in systems with feedback, the long delays must be balanced by shorter delays so that the overall loop completes in the time available. For example, Figure 7.12(b) shows a single-cycle self-bypass loop in which time borrowing occurs across half-cycles, but the entire path must fit in one cycle. A typical example of a self-bypass loop is the execution stage of a pipelined processor in which an ALU must complete an operation and bypass the result back for use in the ALU on a dependent instruction. Most critical paths in digital systems occur in self-bypass loops because otherwise latency does not matter.

Figure 7.13 illustrates the maximum amount of time that a two-phase latch-based system can borrow (beyond the $T_c/2 - t_{pdq}$ nominally available to each half-cycle of logic). Because data does not have to set up until the falling edge of the receiving latch's clock, one phase can borrow up to half a cycle of time from the next (less setup time and nonoverlap):

$$t_{\text{borrow}} \le \frac{T_c}{2} - \left(t_{\text{setup}} + t_{\text{nonoverlap}} \right)$$ (7.10)

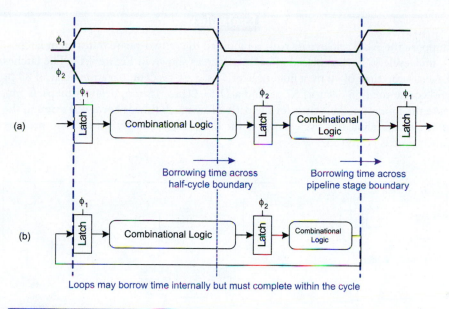

FIG 7.12 Time borrowing

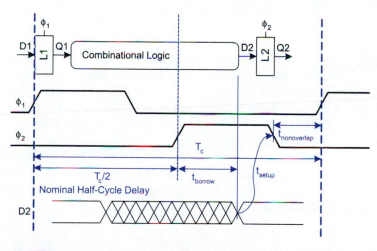

FIG 7.13 Maximum amount of time borrowing

Example

Suppose the ALU self-bypass path is modified to use two-phase transparent latches. A mid-cycle ϕ_2 latch is placed after the adder, as shown in Figure 7.14. The latches have a setup time of 40 ps, a hold time of 5 ps, a *clk*-to-*Q* propagation delay of 82 ps and contamination delay of 52 ps, and a *D*-to-*Q* propagation delay of 82 ps. Compute the minimum cycle time for the path. How much time is borrowed through the mid-cycle latch at this cycle time? If the cycle time is increased to 2000 ps, how much time is borrowed?

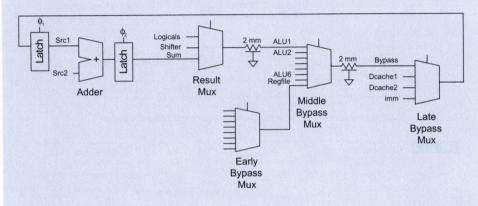

FIG 7.14 ALU self-bypass path with two-phase latches

Solution: According to EQ (7.3), the cycle time is $T_c = 82 + 590 + 82 + 410 = 1164$ ps. The first half of the cycle involves the latch and adder delays and consumes $82 + 590 = 672$ ps. The nominal half-cycle time is $T_c/2 = 582$ ps. Hence, the path borrows 90 ps from the second half-cycle. If the cycle time increases to 2000 ps and the nominal half-cycle time becomes 1000 ps, time borrowing no longer occurs.

Pulsed latches can be viewed as transparent latches with a narrow pulse. If the pulse is wider than the setup time, pulsed latches are also capable of a small amount of time borrowing from one cycle to the next.

$$t_{\text{borrow}} \le t_{pw} - t_{\text{setup}} \tag{7.11}$$

Time borrowing has two benefits for the system designer. The most obvious is *intentional time borrowing*, in which the designer can more easily balance logic between half-cycles and pipeline stages. This leads to potentially shorter design time because the balancing can take place during circuit design rather than requiring changes to the microarchitecture to explicitly move functions from one stage to another. The other is

Example

If the ALU self-bypass path uses pulsed latches, how much time may it borrow from the next cycle?

Solution: None. Because the path is a feedback loop, if its outputs arrive late and borrow time, the path begins later on the next cycle. This in turn causes the outputs to arrive later. Time borrowing can only be used to balance logic within a pipeline but, despite the wishes of many designers, it does not increase the amount of time available in a clock cycle.

opportunistic time borrowing. Even if the designer carefully equalizes the delay in each stage at design time, the delays will differ from one stage to another in the fabricated chip because of process and environmental variations and inaccuracies in the timing model used by the CAD system. In a system with hard edges, the longest cycle sets the minimum clock period. In a system capable of time borrowing, the slow cycles can opportunistically borrow time from faster ones and average out some of the variation.

Some experienced design managers forbid the use of intentional time borrowing until the chip approaches tapeout. Otherwise designers are overly prone to assuming that their pipeline stage can borrow time from adjacent stages. When many designers make this same assumption, all of the paths become excessively long. Worse yet, the problem may be hidden until full chip timing analysis begins, at which time it is too late to redesign so many paths. Another solution is to do full-chip timing analysis starting early in the design process.

7.2.5 Clock Skew

The analysis so far has assumed ideal clocks with zero skew. In reality clocks have some uncertainty in their arrival times that can cut into the time available for useful computation, as shown in Figure 7.15(a). The bold *clk* line indicates the latest possible clock arrival time. The hashed lines show that the clock might arrive over a range of earlier times because of skew. The worst scenario for max delay in a flip-flop-based system is that the launching flop receives its clock late and the receiving flop receives its clock early. In this case, the clock skew is subtracted from the time available for useful computation and appears as sequencing overhead. The worst scenario for min delay is that the launching flop receives its clock early and the receiving clock receives its clock late, as shown in Figure 7.15(b). In this case, the clock skew effectively increases the hold time of the system.

$$t_{pd} \leq T_c - \underbrace{\left(t_{pcq} + t_{setup} + t_{skew} \right)}_{\text{sequencing overhead}}$$

$$(7.12)$$

$$t_{cd} \geq t_{hold} - t_{ccq} + t_{skew}$$

$$(7.13)$$

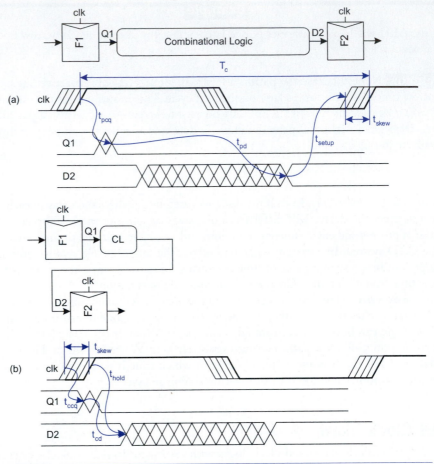

FIG 7.15 Clock skew and flip-flops

In the system using transparent latches, clock skew does not degrade performance. Figure 7.16 shows how the full cycle (less two latch delays) is available for computation even when the clocks are skewed because the data can still arrive at the latches while they are transparent. Therefore, we say that transparent latch-based systems are *skew-tolerant*. However, skew still effectively increases the hold time in each half-cycle. It also cuts into the window available for time borrowing.

$$t_{pd} \leq T_c - \underbrace{\left(2t_{pdq} \right)}_{\text{sequencing overhead}}$$ (7.14)

$$t_{cd1}, t_{cd2} \geq t_{\text{hold}} - t_{ccq} - t_{\text{nonoverlap}} + t_{\text{skew}}$$ (7.15)

Example

If the ALU self-bypass path from Figure 7.6 can experience 50 ps of skew from one cycle to the next between flip-flops in the various ALUs, what is the minimum cycle time of the system? How much clock skew can the system have before hold time failures occur?

Solution: According to EQ (7.12), the cycle time should increase by 50 ps to 1202 ps. The maximum skew for which the system can operate correctly at any cycle time is $t_{cd} - t_{hold} + t_{ccq} = 45 - (-10) + 75 = 130$ ps.

$$t_{borrow} \leq \frac{T_c}{2} - \left(t_{setup} + t_{nonoverlap} + t_{skew} \right) \tag{7.16}$$

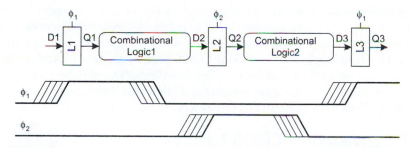

FIG 7.16 Clock skew and transparent latches

Pulsed latches can tolerate an amount of skew proportional to the pulse width. If the pulse is wide enough, the skew will not increase the sequencing overhead because the data can arrive while the latch is transparent. If the pulse is narrow, skew can degrade performance. Again skew effectively increases the hold time and reduces the amount of time available for borrowing (see Exercise 7.7).

$$t_{pd} \leq T_c - \underbrace{\max\left(t_{pdq}, t_{pcq} + t_{setup} - t_{pw} + t_{skew} \right)}_{\text{sequencing overhead}} \tag{7.17}$$

$$t_{cd} \geq t_{hold} + t_{pw} - t_{ccq} + t_{skew} \tag{7.18}$$

$$t_{borrow} \leq t_{pw} - \left(t_{setup} + t_{skew} \right) \tag{7.19}$$

In summary, systems with hard edges (e.g., flip-flops) subtract clock skew from the time available for useful computation. Systems with softer edges (e.g., latches) take advantage of the window of transparency to tolerate some clock skew without increasing the sequencing overhead. Clock skew will be addressed further in Section 12.5. In particular, different amounts of skew can be budgeted for min-delay and max-delay checks. Moreover, nearby sequential elements are likely to see less skew than elements on opposite corners of the chip. Current automated place & route tools spend considerable effort to model clock delays and insert buffer elements to minimize clock skew, but skew is a growing problem for systems with aggressive cycle times.

7.3   Circuit Design of Latches and Flip-flops

Conventional CMOS latches are built using pass transistors or tristate buffers to pass the data while the latch is transparent and feedback to hold the data while the latch is opaque. We begin by exploring circuit designs for basic latches, then build on them to produce flip-flops and pulsed latches. Many latches accept reset and/or enable inputs. It is also possible to build logic functions into the latches to reduce the sequencing overhead.

A number of alternative latch and flip-flop structures have been used in commercial designs. The True Single Phase Clocking (TSPC) technique uses a single clock with no inversions to simplify clock distribution. The Klass Semidynamic Flip-Flop (SDFF) is a fast flip-flop using a domino-style input stage. Differential flip-flops are good for certain applications. Each of these alternatives are described and compared.

7.3.1 Conventional CMOS Latches

Figure 7.17(a) shows a very simple transparent latch built from a single transistor. It is compact and fast but suffers four limitations. The output does not swing from *rail-to-rail* (i.e., from GND to V_{DD}); it never rises above $V_{DD} - V_t$. The output is also *dynamic*; in other words, the output floats when the latch is opaque. If it floats long enough, it can be disturbed by leakage (see Section 6.3.3). *D* drives the *diffusion input* of a pass transistor directly, leading to potential noise issues (see Section 6.3.9) and making the delay harder to model with static timing analyzers. Finally, the state node is *exposed*, so noise on the output can corrupt the state. The remainder of the figures illustrate improved latches using more transistors to achieve more robust operation.

Figure 7.17(b) uses a CMOS transmission gate in place of the single nMOS pass transistor to offer rail-to-rail output swings. It requires a complementary clock $\overline{\phi}$, which can be provided as an additional input or locally generated from ϕ through an inverter. Figure 7.17(c) adds an output inverter so that the state node *X* is isolated from noise on the output. Of course, this creates an inverting latch. Figure 7.17(d) also behaves as an inverting latch with a buffered input but unbuffered output. As discussed in Sections 2.5.6 and 6.2.5.1, the inverter followed by transmission gate is essentially equivalent to a tristate inverter but has a slightly lower logical effort because the output is driven by both transistors of the transmission gate in parallel. Both (c) and (d) are fast dynamic latches.

In modern processes, subthreshold leakage is large enough that dynamic nodes retain their values for only a short time, especially at the high temperature and voltage encountered during burn-in test. Therefore, practical latches need to be staticized, adding feedback to prevent the output from floating, as shown in Figure 7.17(e). When the clock is '1,' the input transmission gate is ON, the feedback tristate is OFF, and the latch is transparent. When the clock is '0,' the input transmission gate turns OFF. However, the feedback tristate turns ON, holding X at the correct level. Figure 7.17(f) adds an input inverter so the input is a transistor gate rather than unbuffered diffusion. Unfortunately, both (e) and (f) reintroduced output noise sensitivity: A large noise spike on the output can propagate backward through the feedback gates and corrupt the state node X. Figure 7.17(g) is a very robust transparent latch that addresses all of the deficiencies mentioned so far: The latch is static, all nodes swing rail-to-rail, the state noise is isolated from output noise, and the input drives transistor gates rather than diffusion. Such a latch is widely used in standard cell applications including the Artisan standard cell library [Artisan02]. It is recommended for all but the most performance- or area-critical designs.

In semicustom datapath applications where input noise can be better controlled, the inverting latch of Figure 7.17(h) may be preferable because it is faster and more compact; for example, Intel uses this as a standard datapath latch [Karnik01]. Figure 7.17(i) shows the *jamb latch*, a variation of (g) that reduces the clock load and saves two transistors by using a weak feedback inverter in place of the tristate. This requires careful circuit design to ensure that the tristate is strong enough to overpower the feedback inverter in all process corners. Figure 7.17(j) shows another jamb latch commonly used in register files and Field Programmable Gate Array (FPGA) cells. Many such latches read out onto a single D_{out} wire and only one is enabled at any given time with its RD signal. The Itanium 2 processor uses the latch shown in Figure 7.17(k) [Naffziger02]. In the static feedback, the pulldown stack is clocked, but the pullup is a weak pMOS transistor. Therefore, the gate driving the input must be strong enough to overcome the feedback. The Itanium 2 cell library also contains a similar latch with an additional input inverter to buffer the input when the previous gate is too weak or far away. With the input inverter, the latch can be viewed as a cross between the designs shown in (g) and (i). Some latches add one more inverter to provide both true and complementary outputs.

Figure 8.53 shows layouts for the latch of Figure 7.17(h) with a built-in clock inverter. The state node X can be shared between the transmission gate and tristate diffusions.

The dynamic latch of Figure 7.17(d) can also be drawn as a clocked tristate, as shown in Figure 7.18(a). Such a form is sometimes called *clocked CMOS* (C^2MOS) [Suzuki73]. The conventional form using the inverter and transmission gate is slightly faster because the output is driven through the nMOS and pMOS working in parallel. C^2MOS is slightly smaller because it eliminates two contacts. Figure 7.18(b) shows another form of the tristate that swaps the data and clock terminals. It is logically equivalent but electrically inferior because toggling D while the latch is opaque can cause charge sharing noise on the output node [Suzuki73].

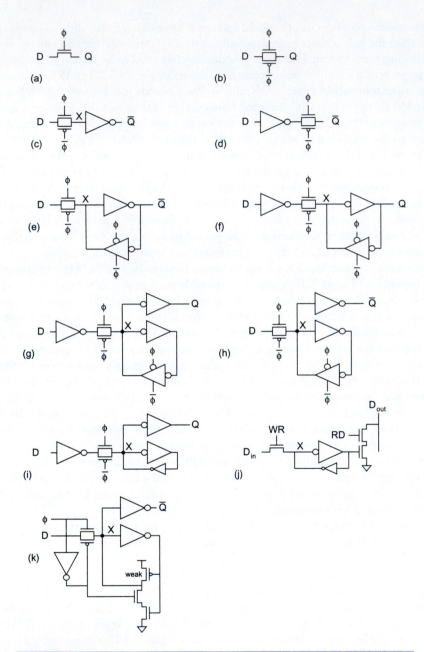

FIG 7.17 Transparent latches

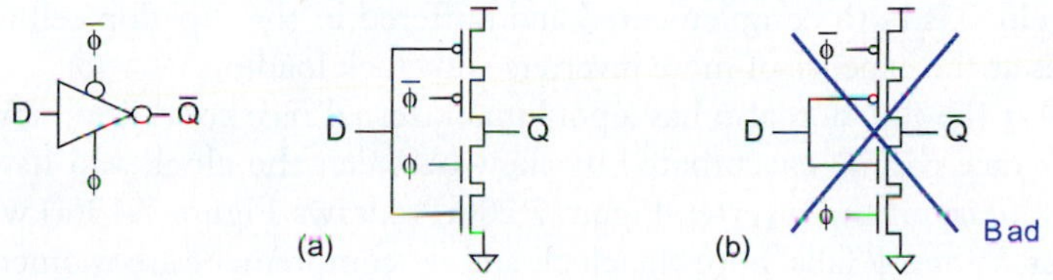

FIG 7.18 C^{2}MOS Latch

All of the latches shown so far are transparent while ϕ is high. They can be converted to active-low latches by swapping ϕ and $\overline{\phi}$.

7.3.2 Conventional CMOS Flip-flops

Figure 7.19(a) shows a dynamic inverting flip-flop built from a pair of back-to-back dynamic latches [Suzuki73]. Either the first or the last inverter can be removed to reduce delay at the expense of greater noise sensitivity on the unbuffered input or output. Figure 7.19(b) adds feedback and another inverter to produce a noninverting static flip-flop. The PowerPC 603 microprocessor datapath used this flip-flop design without the input inverter or $\overline{Q}$ output [Gerosa94].

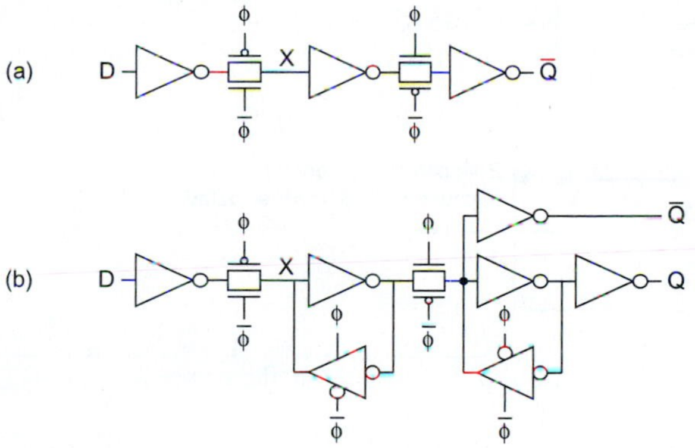

FIG 7.19 Flip-flops

Flip-flops usually take a single clock signal ϕ and locally generate its complement $\overline{\phi}$. If the clock rise/fall time is very slow, it is possible that both the clock and its complement will simultaneously be at intermediate voltages, making both latches transparent and increasing the flip-flop hold time. In ASIC standard cell libraries (such as the Artisan library), the clock is both complemented and buffered in the flip-flop cell to sharpen up the edge rates at the expense of more inverters and clock loading.

Recall that the flip-flop also has a potential internal race condition between the two latches. This race can be exacerbated by skew between the clock and its complement caused by the delay of the inverter. Figure 7.20(a) redraws Figure 7.19(a) with a built-in clock inverter. When ϕ falls, both the clock and its complement are momentarily low as shown in Figure 7.20(b), turning on the clocked pMOS transistors in both transmission gates. If the skew (i.e., inverter delay) is too large, the data can sneak through both latches on the falling clock edge, leading to incorrect operation. Figure 7.20(c) shows a C^{2}MOS dynamic flip-flop built using C^{2}MOS latches rather than inverters and transmission gates [Suzuki73]. Because each stage inverts, data passes through the nMOS stack of one latch and the pMOS of the other, so skew that turns on both clocked pMOS transistors is not a hazard. However, the flip-flop is still susceptible to failure from very slow edge rates that turn both transistors partially ON. The same skew advantages apply even when an even number of inverting logic stages are placed between the latches; this technique is sometimes called *NO RAce* (NORA) [Gonclaves83]. In practice, most flip-flop designs carefully control the delay of the clock inverter so the transmission gate design is safe and slightly faster than C^{2}MOS [Chao89].

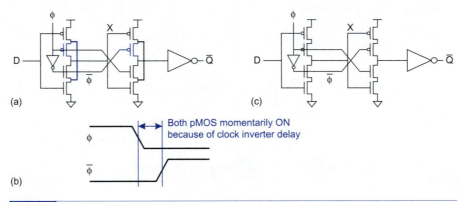

FIG 7.20 Transmission gate and NORA dynamic flip-flops

All of these flip-flop designs still present potential min-delay problems between flip-flops, especially when there is little or no logic between flops and the clock skew is large or poorly analyzed. For VLSI class projects where careful clock skew analysis is too much work and performance is less important, a reasonable alternative is to use a pair of two-phase nonoverlapping clocks instead of the clock and its complement, as shown in Figure 7.21. The flip-flop captures its input on the rising edge of ϕ_1. By making the nonoverlap large enough, the circuit will work despite large skews. However, the nonoverlap time is

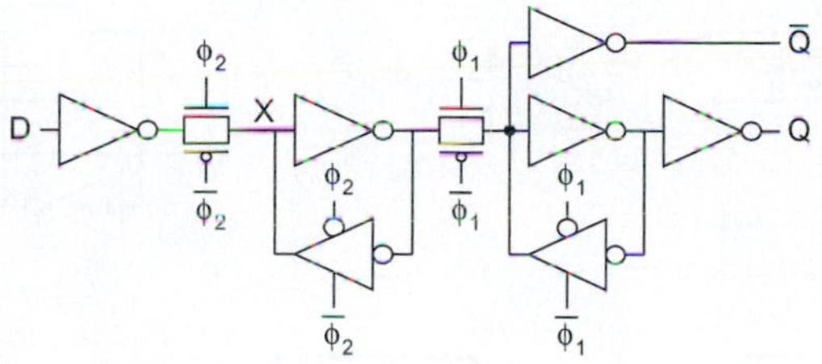

FIG 7.21 Flip-flop with two-phase nonoverlapping clocks

not used by logic, so it directly increases the setup time and sequencing overhead of the flip-flop (see Exercise 7.8).

7.3.3 Pulsed Latches

A pulsed latch can be built from a conventional CMOS transparent latch driven by a brief clock pulse. Figure 7.22(a) shows a simple pulse generator, sometimes called a *clock chopper* or *one-shot* [Harris01a].

The *Naffziger pulsed latch* used on the Itanium 2 processor consists of the latch from Figure 7.17(j) driven by even shorter pulses produced by the generator of Figure 7.22(b) [Naffziger02]. This pulse generator uses a fairly slow (weak) inverter to produce a pulse with a nominal width of about one-sixth of the cycle (125 ps for 1.2 GHz operation). When disabled, the internal node of the pulse generator floats high momentarily, but no keeper is required because the duration is short. Of course, the enable signal has setup and hold requirements around the rising edge of the clock, as shown in Figure 7.22(c).

Figure 7.22(d) shows yet another pulse generator used on an NEC RISC processor [Kozu96] to produce substantially longer pulses. It includes a built-in dynamic transmission-gate latch to prevent the enable from glitching during the pulse.

Many designers consider short pulses risky. The pulse generator should be carefully simulated across process corners and possible RC loads to ensure the pulse is not degraded too badly by process variation or routing. However, the Itanium 2 team found that the pulses could be used just as regular clocks as long as the pulse generator had adequate drive.

The *Partovi pulsed latch* in Figure 7.23 eliminates the need to distribute the pulse by building the pulse generator into the latch itself [Partovi96, Draper97]. The weak cross-coupled inverters in the dashed box staticize the circuit, although the latch is susceptible to back-driven output noise on Q or $\overline{Q}$ unless an extra inverter is used to buffer the output. The Partovi pulsed latch was used on the AMD K6 and Athlon [Golden99], but is slightly slower than a simple latch [Naffziger02]. It was originally called an *Edge Triggered Latch* (ETL), but strictly speaking is a pulsed latch because it has a brief window of transparency.

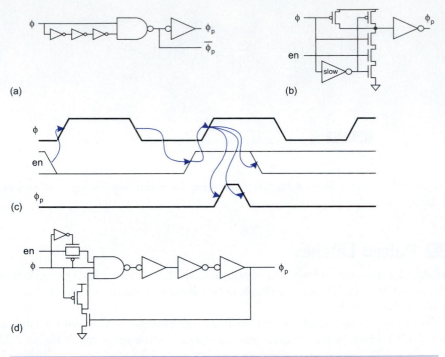

FIG 7.22 Pulse generators

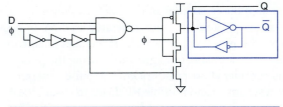

FIG 7.23 Partovi pulsed latch

7.3.4 Resettable Latches and Flip-flops

Most practical sequencing elements require a reset signal to enter a known initial state on startup. Figure 7.24 shows latches and flip-flops with reset inputs. There are two types of reset: *synchronous* and *asynchronous*. Asynchronous reset forces Q low immediately, while synchronous reset waits for the clock. Synchronous reset signals must be stable for a setup and hold time around the clock edge while asynchronous reset is characterized by a propagation delay from reset to output. Synchronous reset simply requires ANDing the input D with $\overline{reset}$. Asynchronous reset requires gating both the data and the feedback to force the reset independent of the clock. The tristate NAND gate can be constructed from a NAND gate in series with a clocked transmission gate.

Settable latches and flip-flops force the output high instead of low. They are similar to resettable elements of Figure 7.24 but replace NAND with NOR and $\overline{reset}$ with *set*. Figure 7.25 shows a flip-flop combining both asynchronous set and reset.

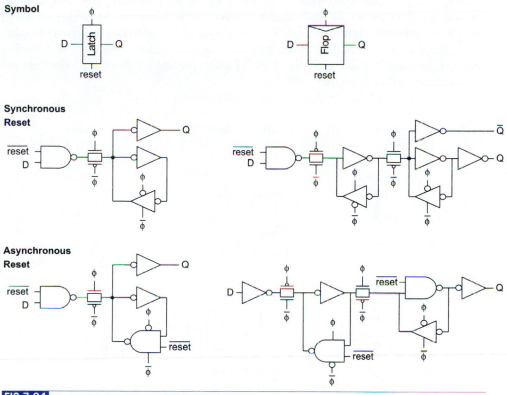

FIG 7.24 Resettable latches and flip-flops

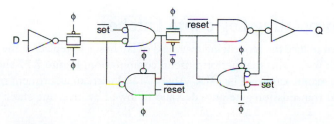

FIG 7.25 Flip-flop with asynchronous set and reset

7.3.5 Enabled Latches and Flip-flops

Sequencing elements also often accept an enable input. When enable *en* is low, the element retains its state independently of the clock. The enable can be performed with an input multiplexer or clock *gating*, as shown in Figure 7.26. The input multiplexer feeds back the old state when the element is disabled. The multiplexer adds area and delay. Clock gating does not affect delay from the data input and the AND gate can be shared among multiple clocked elements. Moreover, it significantly reduces power consumption because the clock on the disabled element does not toggle. However, the AND gate delays the clock, potentially introducing clock skew. Section 12.5.5 addresses techniques to minimize the skew by building the AND gate into the final buffer of the clock distribution network. *en* must be stable while the clock is high to prevent glitches on the clock, as will be discussed further in Section 7.4.3.

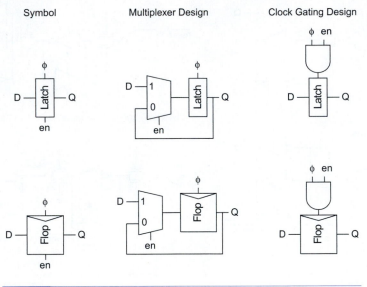

FIG 7.26 Enabled latches and flip-flops

7.3.6 Incorporating Logic into Latches

Another way to reduce the sequencing overhead of latches is to replace some of the inverters in the latch with gates that perform useful computation. Figure 7.27 shows two ways to do this in dynamic latches. The DEC Alpha 21164 used an assortment of latches built from a clocked transmission gate preceded and followed by inverting static CMOS gates such as NANDs, NORs, or inverters [Bowhill95]. This provides the low overhead of the transmission gate latch while preserving the buffered inputs and outputs. The *mux-latch* consists of two transmission gates in parallel controlled by clocks gated with the corresponding select signals. It integrates the multiplexer function with no extra delay from the *D* inputs to the *Q* outputs except the small amount of extra diffusion capacitance on the state node. Note that the setup time on the select inputs is relatively high. The clock gating will introduce skew unless the clocking methodology systematically plans to gate all clocks. The same principles extend to static latches and flip-flops.

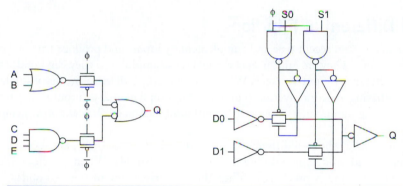

FIG 7.27   Combining logic and latches

7.3.7   Klass Semidynamic Flip-flop (SDFF)

The *Klass semidynamic flip-flop* (SDFF) [Klass99] shown in Figure 7.28 is a cross between a pulsed latch and a flip-flop. Like the Partovi pulsed latch, it operates on the principle of intersecting pulses. However, it uses a dynamic NAND gate in place of the static NAND. While the clock is low, *X* precharges high and *Q* holds its old state. When the clock rises, the dynamic NAND evaluates. If *D* is '0,' *X* remains high and the top nMOS transistor turns OFF. If *D* is '1' and *X* starts to fall low, the transistor remains ON to finish the transition. This allows for a very short pulse and short hold time. The weak cross-coupled inverters staticize the flip-flop and the final inverter buffers the output node.

Like a pulsed latch, the SDFF accepts rising inputs slightly after the rising clock edge. Like a flip-flop, falling inputs must set up before the rising clock edge. It is called *semidynamic* because it combines the dynamic input stage with static operation. The SDFF is slightly faster than the Partovi pulsed latch but loses the skew tolerance and time borrowing capability. The Sun UltraSparc III built logic into the SDFF very efficiently by replacing the single transistor connected to *D* with a collection of transistors performing the OR or multiplexer functions [Heald00].

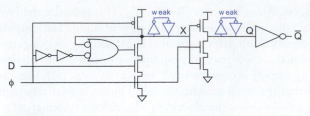

FIG 7.28 Klass semidynamic flip-flop

7.3.8 Differential Flip-flops

Differential flip-flops accept true and complementary inputs and produce true and complementary outputs. They are built from a clocked sense amplifier so they can rapidly respond to small differential input voltages. While they are larger than an ordinary single-ended flip-flop—having an extra inverter to produce the complementary output—they work well with low-swing inputs such as register file bitlines (Section 11.2.3) and low-swing busses (Section 4.6.7).

Figure 7.29(a) shows a differential *sense-amplifier flip-flop* (SA-F/F) receiving differential inputs and producing a differential output [Matsui94]. When the clock is low, the internal nodes X and $\overline{X}$ precharge. When the clock rises, one of the two nodes is pulled down, while the cross-coupled pMOS transistors act as a keeper for the other node. The SR latch formed by the cross-coupled NAND gates behaves as a slave stage, capturing the output and holding it through precharge. The flip-flop can amplify and respond to small differential input voltages, or it can use an inverter to derive the complementary input from D. This flip-flop was used in the Alpha 21264 [Gronowski98]. It has a small clock load and avoids the need for an inverted clock. If the two input transistors are replaced by true and complementary nMOS logic networks, the SA-F/F can also perform logic functions at the expense of greater setup time [Klass99].

The original SA-F/F suffers from the possibility that one of the internal nodes will float low if the inputs switch while the clock is high. The StrongArm 110 processor [Montanaro96] adds the weak nMOS transistor shown in Figure 7.29(a) to fully staticize the flip-flop at the expense of a small amount more internal loading and delay.

Although the sense amplifier stage is fast, the propagation delay through the two cross-coupled NAND gates hurts performance. The NAND gates serve as a slave SR latch and are only necessary to convert the monotonically falling pulsed X signals to static Q outputs; they can be replaced by HI-skew inverters when Q drives domino gates. Alternatively, Figure 7.29(b) shows how to build a faster slave latch to replace the cross-coupled NANDs at the expense of eight more transistors [Nikolić00]. When the X signal falls, it turns on the pMOS transistor to immediately pull the Q output high. It also drives one of the inverters high, which pulls down the $\overline{Q}$ output through the opposite nMOS transistor.

The four blue transistors serve as small cross-coupled tristate keepers that hold the outputs after the master stage precharges, but turn off to avoid contention when the outputs need to switch. This slave latch still involves two gate delays for the falling output, but the delays are faster because the gates avoid crowbar current, have lower logical effort, and are skewed to favor the critical edges.

The AMD K6 used another differential flip-flop shown in Figure 7.29(c) at the interface from static to self-resetting domino logic [Draper97]. The master stage consists of a self-resetting dual-rail domino gate. Assume the internal nodes are initially precharged. On the rising edge of the clock, one of the two will pull down and drive the corresponding output high. The OR gate detects this and produces a *done* signal that precharges the internal nodes and resets the outputs. Therefore, the flip-flop produces pulsed outputs primarily suitable for use in subsequent self-resetting domino gates (see Section 7.5.2.4). The cross-coupled pMOS transistors improve the noise immunity while the cross-coupled inverters staticize the internal nodes.

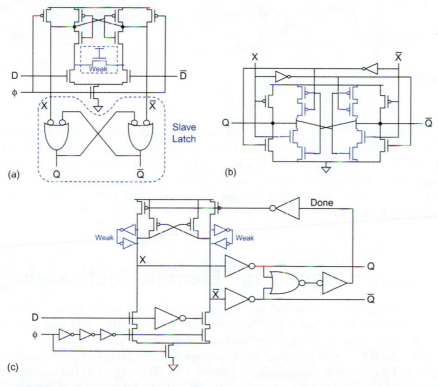

FIG 7.29 Differential flip-flops

7.3.9 True Single-phase-clock (TSPC) Latches and Flip-flops

Conventional latches require both true and complementary clock signals. In modern CMOS systems, the complement is normally generated locally with an inverter in the latch cell. In the late 1980s, some researchers worked to avoid the complementary signal. The *True Single-Phase-Clock* (TSPC) latches and flip-flops replace the inverter-transmission gate or C²MOS stage with a pair of stages requiring only the clock, not its complement [Ji-ren87, Yuan89]. Figure 7.30(a and b) show active high and low TSPC dynamic latches. Figure 7.30(c) shows a TSPC dynamic flip-flop. Note that this flip-flop produces a momentary glitch on $\overline{Q}$ after the rising clock edge when D is low for multiple cycles; this increases the activity factor of downstream circuits and costs power. [Afghahi90] extends the TSPC principle to handle domino, RAMs, and other precharged circuits.

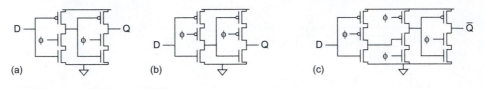

FIG 7.30 TSPC latches and flip-flops

The dynamic TSPC latches were used on the groundbreaking Alpha 21064 microprocessor [Dobberpuhl92]. Logic can be built into the first stage of each latch. The latch is not easy to staticize [Afghahi90]. In any case, the clock must also be reasonably sharp to prevent races when both transistors are partially ON [Larsson94]. The Alpha 21164 reverted to conventional dynamic latches for an estimated 10% speed improvement [Bowhill95]. In summary, TSPC is primarily of historic interest.

7.4 Static Sequencing Element Methodology

This section examines a number of issues designers must address when selecting a sequencing element methodology. We begin with general issues, and then proceed to techniques specific to flip-flops, pulsed latches, and transparent latches.

Until the 0.5 or 0.35 μm generation, leakage was relatively low and thus dynamic latches held their state for acceptably long times. The DEC Alpha 21164 was one of the last major microprocessors to use a dynamic latching methodology in a 0.35 μm process in the mid-1990s. It required a minimum operating frequency of 1/10th full speed to retain state, even during testing. Modern systems generally require static sequencing elements to hold state when clocks are gated or the system is tested at a moderate frequency. Leakage is usually worst during burn-in testing at elevated temperature and voltage, where the chip must still function correctly to ensure good toggle coverage. Static elements are larger and somewhat slower than their dynamic counterparts.

Similarly, the growing difficulty and cost of debugging and testing has forced engineers to build design-for-test (DFT) features into the sequencing elements. The most important feature is *scan*, a special mode in which the latches or flip-flops can be chained together into a large shift register so that they can be read and written under external control during testing. This technique is discussed further in Section 9. Scan has become particularly important because chips have so many metal layers that most internal signals cannot be directly reached with probes. Moreover, some *flip-chips* are mounted upside down, making physical access even more difficult. Scan can dramatically decrease the time required to debug a chip and reduce the cost of testing, so most design methodologies dictate that all sequencing elements must be scannable despite the extra area this entails. The Alpha 21264 did not support full scan and was very difficult to debug, leading to a later-than-desired release.

Clock distribution is another key challenge. As we will see in Section 12.5, it is very difficult to distribute a single clock across a large die in a fashion that gets it to all sequencing elements at nearly the same time. Controlling the clock skew on more than one clock is even more difficult, so almost all modern designs distribute a single high-speed clock. Other signals such as complementary clocks, pulses, and delayed clocks are generated locally where they are needed. The clock edge rates must be relatively sharp to avoid races in which both the master and slave latches are partially on simultaneously. The global clock may have slow edge rates after propagating along long wires, so it is typically buffered locally (either in each sequencing element or in a buffer cell serving a bank of elements) to sharpen the edge rates. Clock power, from the clock distribution network and the clocked loads, typically accounts for one third to one half of the total chip power consumption. Therefore, clocks are often gated with an AND gate in the local clock buffer to turn off the sequencing elements for inactive units of the chip.

All bistable elements are subject to soft errors from alpha particles or cosmic rays striking the circuits and injecting charge onto sensitive nodes (see Section 4.8.7). Sequencing elements require relatively high capacitance on the state node to achieve low soft error rates. This can set a lower bound on the minimum transistor sizes on that node.

7.4.1 Choice of Elements

Flip-flops, pulsed latches, and transparent latches offer tradeoffs in sequencing overhead, skew tolerance, and simplicity.

7.4.1.1 Flip-flops
As we have seen, flip-flops have fairly high sequencing overhead but are popular because they are so simple. Nearly all engineers understand how flip-flops work. Some synthesis tools and timing analyzers handle flip-flops much more gracefully than transparent latches. Many ASIC methodologies use flip-flops exclusively for pipelines and state machines. If performance requirements are not near the cutting edge of a process, flip-flops are clearly the right choice in today's CAD flows.

7.4.1.2 Pulsed Latches Pulsed latches are faster than flip-flops and offer some time-borrowing capability at the expense of greater hold times. They have fewer clocked transistors and hence lower power consumption. If intentional time borrowing is not necessary, you can model a pulsed latch as a flip-flop triggered on the rising edge of the pulse with a lower delay but a lengthy hold time. This makes pulsed latches relatively easy to integrate into flip-flop-based CAD flows. Moreover, the pulsed latches still offer opportunistic time borrowing to compensate for modeling inaccuracies even if the intentional time borrowing is not used.

The long hold times make pulsed latches unsuitable for use in pipelines with no logic between pipeline stages. One solution is to use ordinary flip-flops in place of the pulsed latches in these circumstances where speed is not important. Unfortunately, some pulsed latches fan out to multiple paths, some of which are short and others long. The Itanium 2 processor used the *clocked deracer* in conjunction with Naffziger pulsed latches, as shown in Figure 7.31 [Naffziger02]. These were placed before the receiving latches on short paths and block incoming paths while the receiving latch is transparent. They automatically adapt to pulse with variation and hence have a shorter nominal propagation delay than buffers, but also consume more power than buffers because of the clock loading [Rusu03].

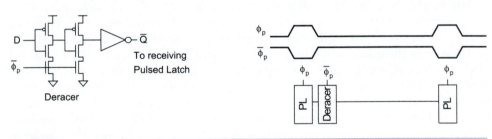

FIG 7.31 Clocked deracer

7.4.1.3 Transparent Latches Transparent latches also have lower sequencing overhead than flip-flops and are attractive because they permit nearly half a cycle of time borrowing. One latch must be placed in each half-cycle. Data can arrive at the latch any time the latch is transparent. A convenient design approach is to nominally place the latch at the beginning of each half-cycle. Then time borrowing occurs when the logic in one half-cycle is longer than nominal and data does not arrive at the next latch until some time into the next half-cycle.

Figure 7.32 illustrates pipeline timing for short and long logic paths between latches. When the path is short (a), the data arrives at the second latch early and is delayed until the rising edge of ϕ_2. Therefore, it is natural to consider latches residing at the beginning of their half-cycle because short paths automatically adjust to operate this way. When the path is longer (b), it borrows time from the first half-cycle into the second. Notice how

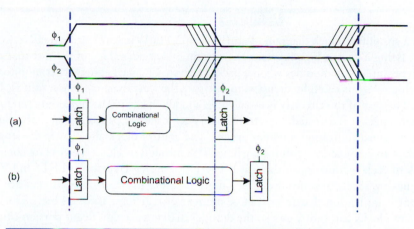

FIG 7.32 Latch placement and time borrowing

clock skew does not slow long paths because the data does not arrive at the latch until after the latest skewed rising edge.

Logic blocks involving multiple signals must ensure that each signal path passes through two latches in each cycle. Signals can be classified as phase 1 or phase 2 and logic gates must receive all their inputs from the same phase. Section 7.4.3 develops a formal notation of timing types to track when signals are safe to use.

7.4.2 Low-power Sequential Design

All of the techniques mentioned in Section 6.5 can be used in common to create low-power libraries for sequential elements. However, because flip-flops and latches are normally clocked every cycle, both the core latch design and the clock distribution network must be carefully scrutinized to achieve low dynamic power. Keep device sizes small inside the core latch and minimize the number of clocked transistors. Pulsed latches are attractive because they have fewer clocked transistors than two-phase latches or flip-flops. The conventional flip-flop of Figure 7.19(b) is also power-efficient because it is simple. [Stojanovic99] presents an extensive study of power and delay in sequential elements.

Scan latches and flip-flops (see Chapter 9) used for testing increase the internal sequential element switched load. Unfortunately, there is a tradeoff between testability and power consumption. The decision is normally toward testability.

Clock gating can be effectively used to turn off sections of circuitry that are not required during certain time intervals. When clock gaters are inserted, the relative delay between blocks must be carefully monitored to ensure that no clock races occur (see Section 12.5.5).

Example

Figure 7.33 shows a simplified block diagram of an IEEE 802.11a Wireless Local Area Network receiver [Ryan01] . Being a packet-based system, the clocking takes advantage of the fact that the clocks only have to be applied to modules when a packet arrives. The packet activity is monitored by the packet detector, which is carefully designed to dissipate the least dynamic power as it is "listening" constantly. The rest of the circuitry is usually idle and all clocks to the remaining modules are turned off. When a packet is detected, a control block issues a start of packet signal to the clock gater that controls the synchronization module (Synch). The control block also figures the length of the packet and thus, in conjunction with the pipeline delay through modules, knows how long to turn the local clock to each module in the data pipe. The Fast Fourier Transform (FFT) and Viterbi Decoder modules are similarly controlled. The Viterbi Decoder uses a register exchange technique, which has the ramification that it has a large number of registers (128 • 64 bits). During clocking each register clocks constantly and so the decoder dissipates a significant portion of the dynamic power in the signal processing chain. Fortunately, the decoder contains very little logic between registers. Thus, it can be operated on a lower supply voltage to reduce dynamic power.

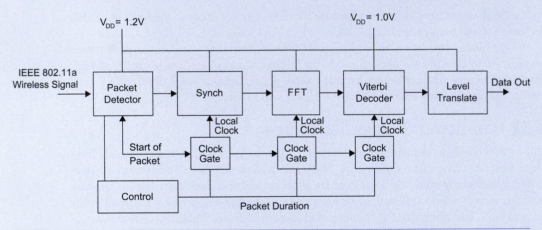

FIG 7.33 Wireless system with clock gating and variable power supply

7.4.3 Two-phase Timing Types

As discussed in Section 7.2.3, latches with two-phase nonoverlapping clocks (ϕ_1 and ϕ_2) are attractive for class projects because with an adequately long clock period and sufficiently great nonoverlap, they are guaranteed to be safe from both setup and hold problems as long as they are used correctly. Logic must be divided into phases 1 and 2. Signals can only interact with other signals in the same phase. Passing through a latch changes the phase of the signal. The situation becomes slightly more complicated when gated clocks and domino circuits are mixed with the latches. [Noice83] describes a method of timing types that can be appended to signal names to keep track of which signals can be safely combined at inputs to gates and latches.

In the two-phase timing discipline, a signal can belong to either phase 1 or phase 2 and be of one of three classes: *stable*, *valid*, or *qualified clock*. A signal is said to be stable during phase 1 (*_s1*) if it settles to a value before ϕ_1 rises and remains constant until after ϕ_1 falls. It is said to be valid during phase 1 (*_v1*) if it settles to a value before ϕ_1 falls and remains at that value until after ϕ_1 falls. It is said to be a phase 1 gated or *qualified clock* (*_q1*) if it either rises and falls like ϕ_1 or remains low for the entire cycle. By definition, ϕ_1 is a *_q1* signal. Phase 2 signals are analogous. Figure 7.34 illustrates the timing of each of these types.

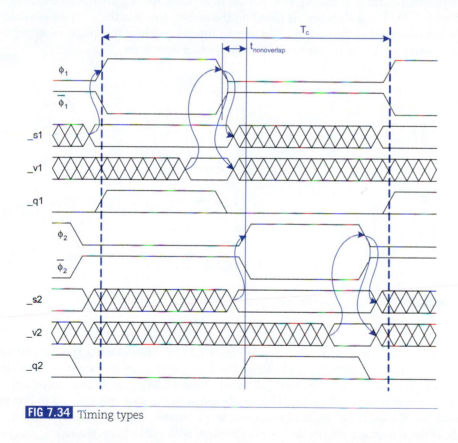

FIG 7.34 Timing types

Latches must take qualified clocks (either *_q1* or *_q2* signals) to their clock inputs. A phase 1 latch requires a *_s1* or *_v1* input (so that the input satisfies setup and hold times around the falling edge of ϕ_1) and produces a *_s2* output because the output settles while

ϕ_1 is high (before ϕ_2 rises) and does not change again until the next time ϕ_1 is high (after ϕ_2 falls). A phase 2 latch requires a _s2 or _v2 input and produces a _s1 output. Qualified clocks are formed as the AND of a clock phase or another qualified clock with a stable signal belonging to the same phase. The qualifying signal must be stable to ensure there are no glitches in the clock. Qualified clocks are only used at the clock terminals of latches or dynamic logic. A block of static CMOS combinational logic requires that all inputs belong to the same phase. If all inputs are stable, the output is also stable. If any are valid, the output is valid. The phase of a domino gate is defined by the clock or qualified clock driving its evaluation transistor. The precharge transistor accepts the complement of the other phase. The inputs must be stable or valid during the evaluation phase and the output is valid during that phase because it settles before the end of the phase and does not change until precharge at the beginning of the next phase. All of these rules are illustrated in Figure 7.35. The definitions are based on the assumption that the propagation delays are short compared to the cycle time so that no time borrowing takes place; however, the connections continue to be safe even if time borrowing does occur.

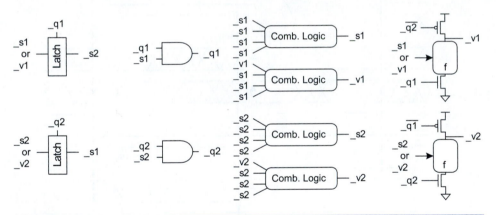

FIG 7.35 Rules for combining timing types

Figure 7.38(a) redraws the flip-flop of Figure 7.21 built from master and slave latches using two-phase nonoverlapping clocking. The flip-flop changes its output on the rising edge of ϕ_1. Both input and output are _s2 signals. Figure 7.38(b) shows an enabled version of the flip-flop using clock gating. The enable signal to the slave must be _s1 to prevent glitches on the qualified clock; in other words, the enable must not change while ϕ_1 is high. If the system is built primarily from flip-flops with _s2 outputs, the enable must be delayed through a phase 2 latch to become _s1. Alternatively, the master (ϕ_2) latch could be enabled, but this requires that the enable set up half a cycle earlier.

Example

Annotate each of the signals in Figure 7.36 with its timing type. If the circuit contains any illegal connections, identify the problems and explain why the connections could cause malfunctions.

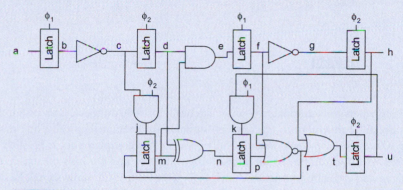

FIG 7.36 Example circuit for timing type checking

Solution: Figure 7.37 shows the timing types of each signal. $t\_??$ is the OR of h_s1 and r_s2. Hence, it might change after the rising edge of ϕ_2 or ϕ_1. Excessive clock skew on ϕ_2 could cause a hold time violation, affecting the result seen at u_s1.

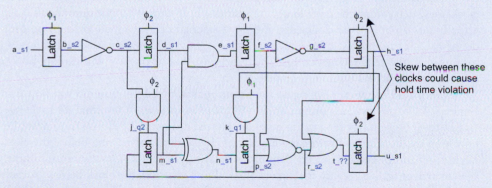

FIG 7.37 Annotated circuit showing timing types

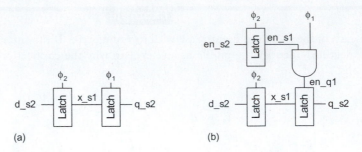

FIG 7.38 Flip-flops using two-phase nonoverlapping clocks

Even when conventional two-phase latches with 50% duty cycles are used, the timing types are still convenient to track which signals can interact. Typically, one distributes a single 50% duty cycle clock *clk* and locally generates its complement $\overline{clk}$. In such a case, *clk* plays the role of ϕ_1 and $\overline{\phi}_2$ while $\overline{clk}$ plays the role of $\overline{\phi}_1$ and ϕ_2. This means that both the precharge and evaluate transistors of dynamic gates receive the same signal. Because there is no nonoverlap, you must analyze each path to ensure no hold problems exist. In particular, be careful to guarantee a stable enable signal for gated clocks.

7.4.4 Characterizing Sequencing Element Delays

Previous sections have derived sequencing element performance in terms of the setup and hold times and propagation and contamination delays. These delays are interrelated and are used for budgeting purposes. For example, a flip-flop might still capture its input properly if the data changes slightly less than a setup time before the clock edge. However, the clock-to-Q delay might be quite long in this situation. If we call t_{DC} the time that the data actually sets up before the clock edge and t_{CQ} the actual delay from clock to Q, we could define t_{setup} as the smallest value of t_{DC} such that $t_{CQ} \le t_{pcq}$. Moreover, we could choose t_{pcq} to minimize the sequencing overhead $t_{\text{setup}} + t_{pcq}$. In this section we will explore how to characterize these delays through simulation.

Figure 7.39 shows the timing of a conventional static edge-triggered flip-flop from Figure 7.19(b). Delays are normalized to a 75 ps FO4 inverter. The actual *clk*-to-Q (t_{CQ}) and D-to-Q (t_{DQ}) delays for a rising input are plotted against the D-to-*clk* (t_{DC}) delay, i.e., how long the data arrived before the clock rises. If the data arrives long before the clock, t_{CQ} is short and essentially independent of t_{DC} delay. $t_{DQ} = t_{DC} + t_{CQ}$, so it increases linearly as data arrives earlier because the data is blocked and waits for the clock before proceeding. As the data arrives closer to the clock, t_{CQ} begins to rise. However, t_{DQ} initially decreases and reaches a minimum when t_{CQ} has a slope of -1 (note the axes are not to scale). Therefore, let us define the setup time t_{setup} as t_{DC} at which this minimum t_{DQ} occurs and the propagation delay t_{pcq} as t_{CQ} at this time. The contamination delay t_{ccq} is the minimum t_{CQ} that occurs when the input arrives early.

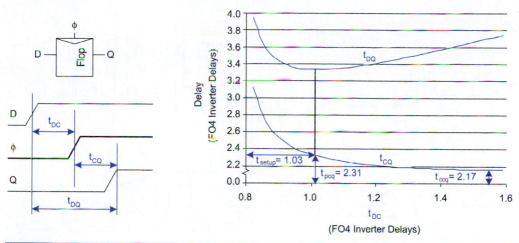

FIG 7.39 Flip-flop delay vs. data arrival time

In general, the delays can differ for inputs of '0' and '1.' Figure 7.40 plots t_{CQ} vs. t_{DC} for the four combinations of rising and falling D and Q. The setup times Δ_{DC0} and Δ_{DC1} are the times that D must fall or rise, respectively, before the clock so that the data is properly captured with the least possible t_{DQ}. Observe that this flip-flop has a longer setup time but shorter propagation delay for low inputs than high inputs. The hold times t_{hold0} and t_{hold1} are the times that D must rise or fall, respectively, after the clock so that the old value of '0' or '1' is captured instead of the new value. Observe that the hold times are typically negative. The contamination delay $t_{ccq0/1}$ again is the lowest possible t_{CQ} and occurs when the input changes well before the clock edge. When only one delay is quoted for a flip-flop timing parameter, it is customarily the worst of the '0' and '1' delays.

The *aperture width* t_a is the width of the window around the clock edge during which the data must not transition if the flip-flop is to produce the correct output with a propagation delay less than t_{pcq}. The aperture times for rising and falling inputs are

$$t_{ar} = t_{setup1} + t_{hold0}$$
$$t_{af} = t_{setup0} + t_{hold1}$$

(7.20)

If the data transitions within the aperture, Q can become metastable and take an unbounded amount of time to settle. Metastability is discussed further in Section 7.6.1.

If D is a very short pulse, the flip-flop may fail to capture it even if D is stable during the setup and hold times around the rising clock edge. Similarly, if the clock pulse is too short, the flip-flop may fail to capture stable data. Well-characterized libraries sometimes specify minimum pulse widths for the clock and/or data as well as setup and hold times.

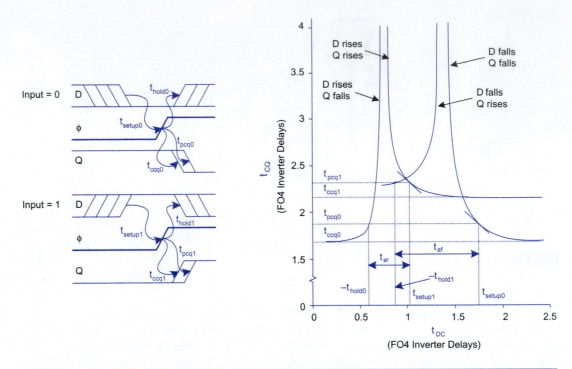

FIG 7.40 Flip-flop setup and hold times

Level-sensitive latches have somewhat different timing constraints because of their transparency, as shown in Figure 7.41 for a conventional static latch from Figure 7.17(g) using a pulse width of 4 FO4 inverter delays. As with an edge-triggered flip-flop, if the data arrives before the clock rises ($t_{DCr} > 0$), it must wait for the clock. In this region the clock-to-Q t_{CrQ} delay is nearly constant and t_{DQ} increases as the data arrives earlier. If the data arrives after the clock rises while the latch is transparent, t_{DQ} is essentially independent of the arrival time. The data must set up before the falling edge of the clock. The second set of labels on the X-axis indicates the D-to-clk fall time t_{DCf} As the data arrives too close to the falling edge, t_{DQ} increases. Now to achieve low t_{DQ}, we choose the setup time before the knee of the curve, e.g., 5% greater than its minimum value. The setup time is measured relative to the falling edge of the clock. If the data changes less than a hold time after the falling edge of the clock, Q may momentarily glitch. Thus, the hold time t_{hold} for a latch is defined to be $-t_{DCf}$ for which Q displays a negligible glitch.

Pulsed latches have setup and hold times measured around the falling edge of the clock. However, designers often wish to treat pulsed latches as edge-triggered flip-flops from the perspective of timing analysis. Therefore, we can define "virtual" setup and hold times relative to the rising clock edge [Stojanovic99]. For example, the pulsed latch in

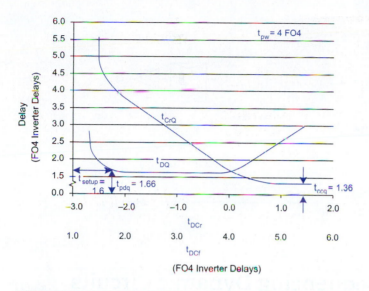

FIG 7.41 Latch delay vs. data arrival time

Figure 7.41 has $t_{setup-virtual} = t_{setup} - t_{pw} = -2.4$ FO4 but $t_{pcq-virtual} = t_{pdq} + (t_{pw} - t_{setup}) = 4.06$ FO4 so the total sequencing overhead of $t_{pdq} = t_{setup-virtual} + t_{pcq-virtual}$ is unaffected by the change of reference or pulse width. The virtual hold time is now $t_{hold-virtual} = t_{hold} + t_{pw} = 2.6$ FO4, which is positive as one should expect because the input must hold long after the rising edge of the clock.

The delays vary with input slope, voltage, and temperature. The contamination delay should be measured in the environment where it is shortest while the setup and hold times and propagation delay should be measured in the environment where it is longest.

The designer can trade off setup time, hold time, and propagation delay. Figure 7.42 shows the effects of adding delay t_{buf} to the clock, D, or Q terminals of a flip-flop. Recall that the sequencing overhead depends on the sum of the setup time and propagation delay while the minimum delay between flip-flops depends on the hold time less the contamination delay. Adding delay on either the input or output eases min-delay at the expense of sequencing overhead. Many standard cell libraries intentionally use slow flip-flops so that logic designers do not have to worry about hold time violations. Adding delay on the clock simply shifts when the flop activates. The sequencing overhead does not change, but the system can accommodate more logic in the previous cycle and less in the next cycle. This is similar to time borrowing in latch-based systems, but must be done intentionally by adjusting the clock rather than opportunistically by taking advantage of transparency. Some authors refer to delaying the clock as *intentional clock skew*. This book reserves the term *clock skew* for uncertainty in the clock arrival times.

$$t'_{setup} = t_{setup} + t_{buf}$$

$$t'_{pcq} = t_{pcq}$$

$$t'_{hold} = t_{hold} - t_{buf}$$

$$t'_{setup} = t_{setup}$$

$$t'_{pcq} = t_{pcq} + t_{buf}$$

$$t'_{hold} = t_{hold}$$

$$t'_{setup} = t_{setup} - t_{buf}$$

$$t'_{pcq} = t_{pcq} + t_{buf}$$

$$t'_{hold} = t_{hold} + t_{buf}$$

FIG 7.42 Delay tradeoffs

7.5 Sequencing Dynamic Circuits

Dynamic and domino circuits operate in two steps: precharge and evaluation. Ideally, the delay of a path should be the sum of the evaluation delays of each gate along the path. This requires some careful sequencing to hide the precharge time. Traditional domino circuits discussed in Section 7.5.1 divide the cycle into two half-cycles. One phase evaluates while the other precharges, then the other evaluates while the first precharges. Transparent latches hold the result of each phase while it precharges. This scheme hides the precharge time but introduces substantial sequencing overhead because of the latch delays and setup time. A variety of skew-tolerant domino circuit schemes described in Section 7.5.2 use overlapping clocks to eliminate the latches and the sequencing overhead. Section 7.5.3 expands on skew-tolerant domino clocking for unfooted dynamic gates.

Recall that dynamic gates require that inputs be monotonically rising during evaluation. They produce monotonically falling outputs. Domino gates consist of dynamic gates followed by inverting static gates to produce monotonically rising outputs. Because of these two levels of inversion, domino gates can only compute noninverting logic functions. We have seen that dual-rail domino gets around this problem by accepting both true and complementary inputs and producing both true and complementary outputs. Dual-rail domino is not always practical. For example, dynamic logic is very efficient for building wide NOR structures because the logical effort is independent of the number of inputs. However, the complementary structure is a tall NAND, which is quite inefficient. When inverting functions are required, an alternative is to use a dynamic gate that produces monotonically falling outputs, but delays the clock to the subsequent dynamic gate so that the inputs are stable by the time the gate enters evaluation. Section 7.5.4 explores a selection of these *nonmonotonic* techniques.

7.5.1 Traditional Domino Circuits

Figure 7.43(a) shows a traditional domino clocking scheme. While the clock is high, the first half-cycle evaluates and the second precharges. While the clock is low, the second evaluates and the first precharges. With this ping-pong approach, the precharge time does not appear in the critical path. The inverting latches hold the result of one half-cycle while that half-cycle precharges and the next evaluates. The data must arrive at the first half-cycle latch a setup time before the clock falls. It propagates through the latch, so the overhead of each latch is the maximum of its setup time and D-to-Q propagation delay [Harris97]. Assuming the propagation delay is longer, the time available for computation in each cycle is

$$t_{pd} = T_c - 2t_{pdq} \tag{7.21}$$

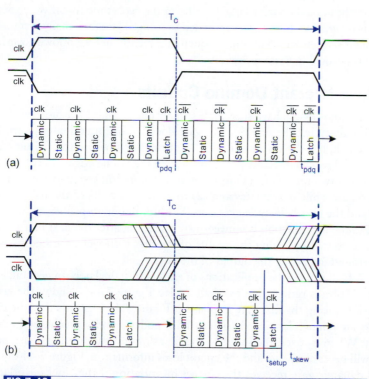

(a)

(b)

FIG 7.43   Traditional domino circuits

Figure 7.43(b) shows the pipeline with clock skew. Data is launched into the first dynamic gate of each cycle on the rising edge of the clock and must set up before the falling edge. Hence, clock skew cuts into the time available for computation in each half-cycle. This is even worse than flip-flops, which pay clock skew once per cycle. Assuming the skew and setup time are greater than the propagation delay, the time for computation becomes

$$t_{pd} = T_c - 2t_{\text{setup}} - 2t_{\text{skew}} \tag{7.22}$$

Moreover, like flip-flops, traditional domino circuits suffer from imbalanced logic. Gates cannot borrow time into the next half-cycle, so a fraction of a gate delay at the end of each half-cycle may be wasted. This penalty is hard to quantify, but clearly the ability to borrow time intentionally or opportunistically would help performance.

In summary, traditional domino circuits have high sequencing overhead from latch delay, clock skew, and imbalanced logic. For heavily pipelined systems with short cycle times, this overhead can be such a large fraction of the cycle time that it wipes out the performance advantage that domino was intended to bring. Therefore, many system designers have developed skew-tolerant domino sequencing techniques with lower overhead. The next section is devoted to these techniques.

7.5.2 Skew-tolerant Domino Circuits

Traditional domino circuits have such high sequencing overhead because they have a hard edge in each half-cycle: The first domino gate does not begin evaluating until the rising edge of the clock, but the result must set up at the latch before the falling edge of the clock. If we could remove the latch, we could soften the falling edge and cut the overhead. The latch serves two functions: (1) to prevent nonmonotonic signals from entering the next domino gate while it evaluates and (2) to hold the results of the half-cycle while it precharges and the next half-cycle evaluates. Within domino pipelines, all the signals are monotonic, so the first function is unnecessary. Moreover, after the next half-cycle has had sufficient time to evaluate using the results of the first half-cycle, the first half-cycle can precharge without impacting the output of the next.

Figure 7.44 illustrates the implications of eliminating the latch. In general, let logic be divided into N phases rather than two half-cycles. Figure 7.44(a) shows the last domino gate in phase 1 driving the first gate in phase 2. Figure 7.44(b) shows that the circuit fails if the clocks are nonoverlapping. When ϕ_1 falls, nodes a and b precharge high and low, respectively. When ϕ_2 rises, the input to the first domino gate in this phase has already fallen, so c will never discharge and the circuit loses information. Figure 7.44(c) shows that the second dynamic gate receives the correct information if the clocks overlap. Now ϕ_2 rises while b still holds its correct value. Therefore, the first phase 2 domino gate can evaluate using the results of phase 1. When ϕ_1 falls and b precharges low, c holds its value. Without a keeper, c can float either high or low. Figure 7.45 shows a *full keeper* consisting of weak cross-coupled inverters to hold the output either high or low. In summary, the latches can be eliminated at phase boundaries as long as the clocks overlap and the first dynamic gate of each phase uses a full keeper.

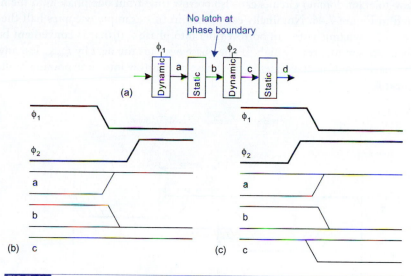

(a)

(b)

(c)

FIG 7.44 Eliminating latches in skew-tolerant domino circuits

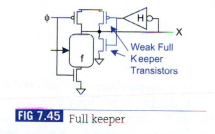

FIG 7.45 Full keeper

In general, as long as the clock overlap is long enough that the second phase can evaluate before the first precharges, the latch between phases is unnecessary. Let us define t_{hold} as the required overlap so that the second phase can evaluate before the first precharges. It is typically a small negative number because the dynamic gate evaluation is fast, but precharge is slow and must ripple through the static stage. The clocks must overlap enough such that they still overlap by t_{hold} even under worst-case clock skew[1]. The sequencing overhead is zero because data propagates from one domino gate to the next without waiting at any sequencing elements. Therefore, we use the generic name *skew-tolerant domino*

[1]Do not confuse this t_{hold}, the amount of time that the clocks must overlap in a skew-tolerant domino pipeline, with t_{hold} on a sequencing element, the time that the data must remain stable after the clock edge.

for domino circuits with overlapping clocks that eliminate the latches between phases [Harris01a]. Using more clock phases also helps spread the power consumption across the cycle rather than drawing large noisy current spikes on the two clock edges.

Skew-tolerant domino circuits can also borrow time from one phase into the next, as illustrated in Figure 7.46. Nominally each phase in this example occupies half the cycle. However, a ϕ_1 dynamic gate can borrow time into phase 2 if that is convenient because both clocks are simultaneously high. If one phase overlaps the next by $t_{overlap}$ less any clock skew, the maximum time that gates in one phase can borrow into time nominally allocated for the next is

$$t_{borrow} = t_{overlap} - t_{hold} - t_{skew} \tag{7.23}$$

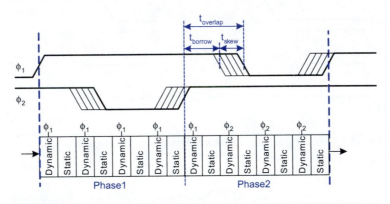

FIG 7.46 Time borrowing in skew-tolerant domino circuits

[Williams91] observed that self-timed pipelines could use overlapping clocks to eliminate latches, but such asynchronous design has not been widely adopted. The Alpha 21164 overlapped clocks in the ALU to eliminate the mid-cycle latch and improve performance [Bowhill95]. Since then, most high-performance synchronous systems using domino have employed some form of skew-tolerant domino to avoid the high sequencing overhead of traditional domino.

There are many ways to produce overlapping clocks. In general, you can use N separate clock phases. Each phase can use 50% duty-cycle waveforms or can stretch the falling edge for even greater overlap. Generating multiple overlapping clocks with low skew is a challenge. The remainder of this section describes a number of methods that have been used successfully.

7.5.2.1 Two-phase Skew-tolerant Domino and OTB Domino
Figure 7.47 shows a clock generator for the two-phase skew-tolerant domino system from Figure 7.46. The generator uses *clock choppers* (also called *clock stretchers*) that delay the falling edge to provide the overlap. A potential problem with two-phase systems is that if a phase of logic

has short contamination delay, the data can race through while both clocks are high.

Opportunistic Time Borrowing (OTB) Domino addresses the race problem by introducing two more clocks (*clk* and *clkb*) with 50% duty cycles that are used on the first gate of each half-cycle, as shown in Figure 7.48. These first gates block data that arrives too early so that it will not race ahead. The delayed clocks *clkd* and *clkbd* play the role of ϕ_1 and ϕ_2. OTB domino was used on the Itanium processor [Rusu00]. However, OTB domino has relatively short overlap and time borrowing capability set by the delay of the clock chopper. The next section describes how to achieve better performance with four phases.

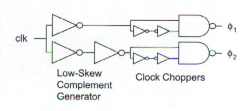

FIG 7.47   Two-phase skew-tolerant domino clock generator

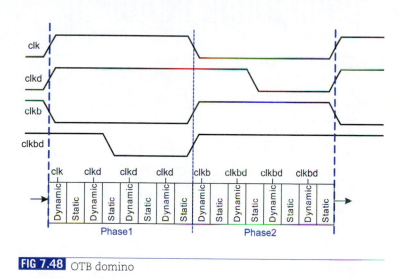

FIG 7.48   OTB domino

7.5.2.2 Four-phase Skew-tolerant Domino   Figure 7.49 shows a four-phase skew-tolerant domino system. Each of the phases has a 50% duty cycle and is spaced a quarter cycle after the previous one, so the nominal overlap is a quarter cycle. The clocks are never all simultaneously high so race problems are solved unless skew approaches a quarter cycle. According to EQ (7.23), the maximum time available for borrowing from one phase to the next is

$$t_{\text{borrow}} = T_c / 4 - t_{\text{hold}} - t_{\text{skew}} \qquad (7.24)$$

Figure 7.50(a) shows a local clock generator producing the four phases. ϕ_1 and ϕ_3 are produced directly from the global clock and its complement. ϕ_2 and ϕ_4 are delayed by buffers with nominal quarter cycle latency. By using both clock edges, each phase is guaranteed to overlap the next phase independent of clock frequency. Variations in these buffer delays with

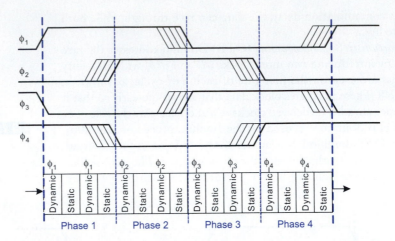

FIG 7.49 Four-phase skew-tolerant domino

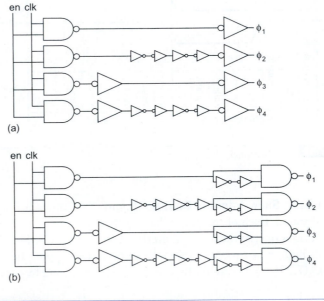

FIG 7.50 Clock generator for four-phase skew-tolerant domino

process, voltage, and temperature can reduce the overlap and available time for borrowing. To avoid excessive pessimism, remember that in the fast corner where overlaps are short, the rest of the gates are also faster. The clock generator also includes a built-in enable.

In general, clock choppers can be used to produce even greater overlap at the expense of greater race concerns. The Itanium II uses four-phase skew-tolerant domino with duty cycles exceeding 50% [Naffziger02]. Figure 7.50(b) shows a four-phase clock generator with clock choppers to provide longer duty cycles. [Harris01a] describes four-phase circuit methodology in much more detail, including testability and a generalization of timing types from Section 7.4.3.

7.5.2.3 *N*-phase Skew-tolerant Domino

Another approach to domino clocking is to use a chain of buffers to produce a unique phase for each level of logic in a cycle. Figure 7.51 shows two ways of producing these phases. In Figure 7.51(a), half the phases are generated off the rising edge of the clock and half off the falling edge. In this way, each phase is guaranteed to overlap the next independent of cycle time. In Figure 7.51(b), all of the phases are generated off the rising edge. If the clock period is long, the final phase must delay its falling edge to guarantee it will still overlap the first phase of the next cycle. The SR latch ensures that the last phase, ϕ_6, will not rise until after *clk* falls (to avoid min-delay problems) and will not fall until after *clk* rises (to ensure overlap of ϕ_1).

A number of design teams have independently developed these techniques. The approach of one phase for each level of logic has been called *Delayed Reset* (IBM [Nowka98]), *Cascaded Reset* (IBM [Silberman98]), and *Delayed Clocking* (Sun [Heald00]). The phase generator for Cascaded Reset domino is well suited to driving footless dynamic gates and will be discussed further in Section 7.5.3.

7.5.2.4 Self-resetting (Postcharge) Domino

In the methods examined so far, the timing of the precharge operation has been controlled by the clock generator. An alternative approach, called *Self-Resetting* or *Postcharge Domino*, is to control the precharge based on the output of the domino gate. Figure 7.52 shows a simple self-resetting domino gate. When the domino gate evaluates and the output rises, a timing chain produces a precharge signal $\overline{reset}$ to precharge the dynamic stage (and possibly assist pulling the HI-skew inverter low, particularly if the inverter is highly skewed). Once the output has fallen, the precharge signal turns off the precharge transistors and the gate is ready to evaluate again. The input must have fallen before the gate reenters evaluation so the gate does not repeatedly pulse on a steady input. Therefore, self-resetting gates accept input pulses and produce output pulses whose duration of five gate delays is determined by the delay of the timing chain. As long as the first inverter in the timing chain is small compared to the rest of the load on node *Y*, its extra loading has negligible impact on performance.

Self-resetting gates save power because they reduce the loading on the clock. Moreover, they only toggle the precharge signal when the gate evaluates low. In Section 11.2.2, we will see that this is particularly useful for RAM decoders. Only one of many word lines in a RAM will rise on each cycle, so a self-resetting decoder saves power by resetting only that line without applying precharge to the other word line drivers. For example, an IBM SRAM [Chappell91], the Intergraph Clipper cache [Heald93], and the Sun UltraSparc I cache [Heald98] use self-resetting gates[2].

---

[2]Confusingly referred to as "delayed reset" by Sun in [Lev95, Heald98].

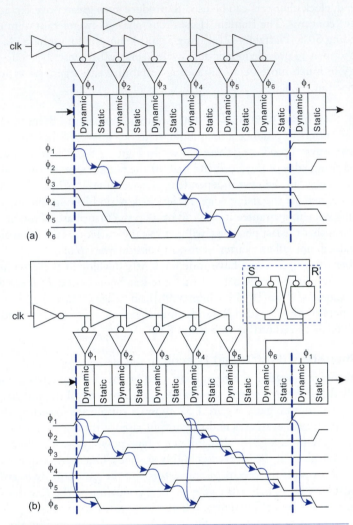

FIG 7.51 N-phase skew-tolerant domino

Self-resetting AND gates in these decoders often receive the address inputs as static levels rather than pulses. *Predicated self-resetting* AND gates [Amrutur01] wait for the input to fall before precharging the output to stretch the pulse width and prevent multiple output pulses when the input is held high, as shown in Figure 7.53. The first inverter in the timing chain is replaced by a *generalized Muller C-element*, shown in blue, whose output does not rise until both Y and one of the inputs have fallen. This only works for functions such as AND or OR-AND where one of the inputs is in series with all of the others.

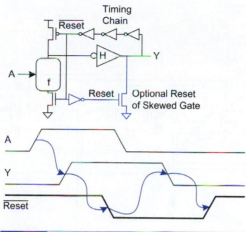

FIG 7.52 Self-resetting gate

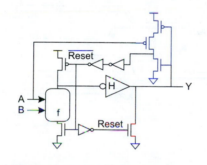

FIG 7.53 Predicated self-resetting gate

[Proebsting91] applies self-resetting techniques to NORA gates for buffers and memory decoders. Figure 7.54 shows an example of a postcharged buffer for a memory chip. It rapidly amplifies the chip select signal CS and provides a series of pulses that serve as clocks for large (multi-pF) loads across the chip. The clock chopper produces a pulse to trigger the first stage of the buffer. The buffer consists of alternating extremely HI- and LO-skew inverters with logical efforts of approximately 2/3 and 1/3, respectively. Each inverter also receives a postcharge signal from a subsequent stage to assist the weak device in resetting the gate. The very small transistor serves as a keeper, so the gates can be viewed as unfooted NTP dynamic nMOS and pMOS inverters. Forward moving pulses trigger each gate. Signals from four stages ahead feed back to postcharge the gate. The buffer is roughly twice as fast as an ordinary chain of inverters because of the lower logical efforts. It also avoids the need for an external clock to precharge the dynamic gates.

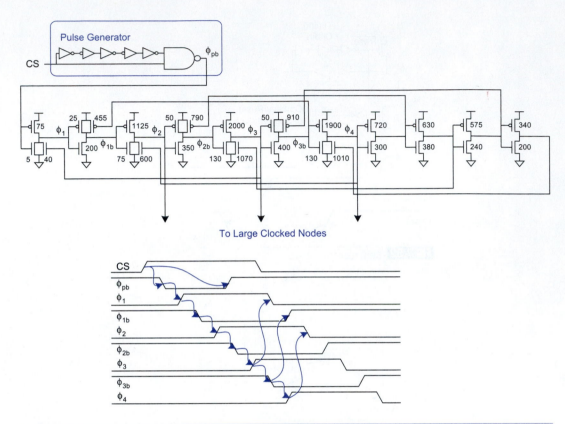

FIG 7.54 Postcharged buffer

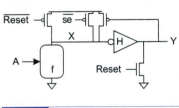

FIG 7.55 SRCMOS

IBM has developed an extensive methodology for self-resetting domino gates called *SRCMOS* [Haring96] that has been applied to circuits including a register file [Hwang99a], 64-bit adder [Hwang99b], and the S/390 G4 CPU cache [Webb97]. SRCMOS gates are typically unfooted dynamic gates followed by highly skewed static inverters, as shown in Figure 7.55. True and complementary reset signals precharge the dynamic stage and help pull the output low. An additional weak *static evaluation* transistor converts the gate into pseudo-nMOS when the global $\overline{se}$ signal is asserted to assist with testing and low-frequency debug. The inputs and outputs are pulses. The reset signals are generated from the gate outputs or from a global reset.

To avoid the overhead and timing constraints of reset circuitry on every gate, the reset signals can be derived from the output of the first gate in a pipeline and delayed through buffers to reset subsequent gates. Figure 7.56 shows an example of an SRCMOS macro adapted from [Hwang99b]. The upper portion represents an abstract datapath. None of the keepers or static evaluation devices are shown. The center is a timing chain that provides reset pulses to each gate. These pulses may be viewed as N-phase skew-tolerant domino clocks. The bottom shows a pulse generator. In normal operation, the power-on reset signal is low and the static evaluation signal $\overline{se}$ high. Assume that all of the gates have been precharged. When the input pulse arrives at A, the datapath will begin evaluating. The first stage must use dual-rail (or in general, 1-of-N hot) encoding so that Y_1_h or Y_1_l will rise when the stage has completed. This triggers the pulse generator, which raises the done signal and initiates a reset. A wave of low-going reset pulses propagates along the timing chain to precharge each gate. One of the reset pulses also precharges the pulse generator, terminating the reset operation. At this point, the datapath can accept a new input pulse. If the data idles low, none of the nodes toggle and the circuit consumes no dynamic power.

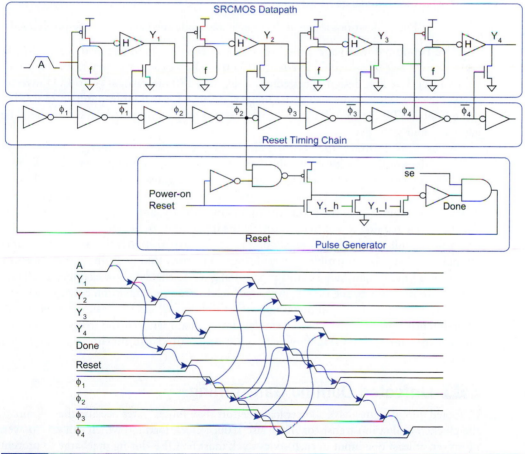

FIG 7.56 SRCMOS macro

The power-on reset forces *done* and *reset* high to initialize the pipeline at startup. When the static evaluation signal is asserted, the reset pulses are inhibited. In this mode, the datapath gates behave as pseudo-nMOS rather than dynamic, permitting low-frequency test and debug.

Self-resetting gates require very careful design because they act on pulses rather than static levels. Some of the timing checks include [Narayanan96]:

- Pulse overlap constraints
  - Pulses arriving at series transistors must overlap so the dynamic gate can pull down through all the transistors
- Pulse width constraints
  - Pulses must be wide enough for a gate to evaluate
- Collision avoidance constraints
  - Pulses must not arrive at dynamic gates while the gates are being precharged

The Pentium 4 uses yet another form of self-resetting domino called *Globally-Reset Domino with Self-Terminating Precharge* (Global STP) to achieve very fast cycle times [Hinton01]. The first design operated at 2 GHz in a 180 nm process (< 16 FO4 inverter delays / cycle). More remarkably, the integer execution was double-pumped to 4 GHz using Global STP domino. Each cycle has time for only eight gate delays: four dynamic gates and four static gates.

Figure 7.57 illustrates the Global STP circuits. A frequency doubler generates pulses off both edges of the clock to drive the datapath. Each stage of the datapath is a domino gate with a keeper (k) and precharge transistor (p). The gates are shown using HI-skew inverters but could use any HI-skew inverting static gate. The small NAND gates save power by only turning on the precharge transistor if the dynamic gate had evaluated low. The first stage requires a foot to only sample the input while ϕ_1 is high. The last stage also uses a foot, a full keeper, and more complex reset circuitry to stretch the width of the output pulse so that it is compatible with static logic. The reset timing chain must be carefully designed to produce precharge clocks properly aligned to the data. For example, ϕ_3 should be timed to rise close to the time Y_1 evaluates high to prevent contention between the precharge transistor and the pulldown network. Global STP circuit design can be a very labor-intensive process. IBM used a similar timing chain without the frequency doubler on an experimental 1 GHz PowerPC chip and called the method *cascaded reset* [Silberman98].

7.5.3 Unfooted Domino Gate Timing

Unfooted domino gates have a lower logical effort than footed gates because they eliminate the clocked evaluation transistor. They also reduce clock loading, which can save power. However, at least one input in each series stack must be OFF during precharge to prevent crowbar current flowing from V_{DD} to GND through the precharge device and ON stack.

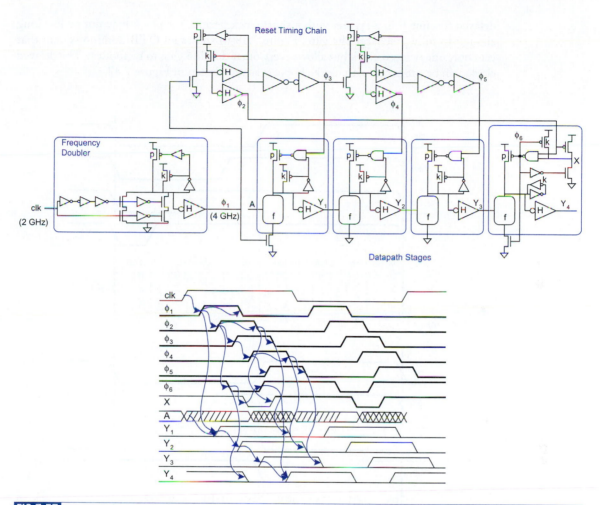

FIG 7.57 Global STP macro

The easiest way to ensure this is to require that the input come from a previous domino gate that has completed precharge before the footless gate begins precharge. Moreover, the previous gate must not output a '1' again until the unfooted gate is in evaluation.

One way to ensure these constraints is to delay the falling edge of clocks to footless gates as shown in Figure 7.58(a). The first domino gate is footed to accept static inputs that might be high during precharge. The subsequent unfooted gates begin evaluating at the same time but have their precharge delayed until the previous gate has precharged. Multiple delayed clocks can be used to allow multiple stages of unfooted gates. For example, the Itanium II processor uses one footed gate followed by four unfooted gates in the first half-cycle of the execution stage for the 64-bit adder [Fetzer02]. If the falling edge is delayed too much in a system with a short clock period, the clock may not be low long enough to fully precharge the gate. Figure 7.58(b) shows an OTB domino system that uses only one delayed clock but allows every other domino gate to be footless. The delayed clocks can be produced with clock choppers as was shown in Figure 7.47.

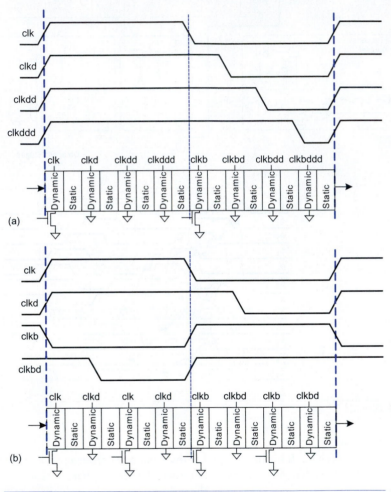

FIG 7.58 Clocking domino pipelines with unfooted gates

The precharge time on each of the delayed phases in Figure 7.58(a) becomes shorter because the falling edge is delayed but the rising edge is not. It is not strictly necessary for all the rising edges to coincide; some delay can be accepted so long as the delayed clock is in evaluation by the time the input arrives at its unfooted gate. Figure 7.59 shows a *delayed precharge* clock buffer [Colwell95] used on the Pentium II. The delayed clocks are produced with skewed buffers that have fast rising edges but slower falling edges.

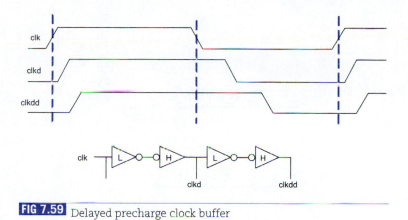

FIG 7.59 Delayed precharge clock buffer

Self-resetting domino also works well with unfooted gates. The inputs are pulses rather than levels. As long as the pulses are only high while the gate is in evaluation, no precharge contention will occur. For example, Figure 7.54, Figure 7.56, and Figure 7.57 illustrate self-resetting circuits with unfooted gates in some or all of the stages.

The consequence of precharging an unfooted gate before its input has fully fallen low is excess power consumption rather than outright circuit failure. Therefore, delays can be set to nominally avoid precharge contention, yet accept that, under worst-case clock skew, contention may occur in a few places.

7.5.4 Nonmonotonic Techniques

The monotonicity requirement forces domino gates to perform only noninverting functions. Dual-rail domino accepts true and complementary inputs and produces true and complementary outputs. This works reasonably well for circuits such as XORs at the expense of twice the hardware. However, domino is particularly poorly suited to wide NOR functions. Figure 7.60 compares a dual-rail domino 4-input OR/NOR gate to a 4-input dynamic NOR. The dual-rail design tends to be very slow because the complementary gate is a tall NAND with a logical effort of 5/3. On the other hand, a dynamic wide NOR is very compact and has a logical effort of only 2/3. The problem is exacerbated for wider gates.

The output of a dynamic gate is monotonically falling so it cannot directly drive another dynamic gate controlled by the same clock, as was shown in Figure 6.27. However, if the rising edge of the clock for the second gate is delayed until the first gate has fully evaluated, the second gate sees a stable input and will work correctly, as shown by Figure 7.61. The primary tradeoff in such *clock-blocked* circuits is the amount of delay: If the delay is too short, the circuit will fail, but as the delay becomes longer, the circuit sacrifices the performance advantages that dynamic logic was supposed to provide. This challenge is exacerbated by process and environmental variations that require margins on the delay in the nominal case so that the circuit continues to operate correctly in the worst case.

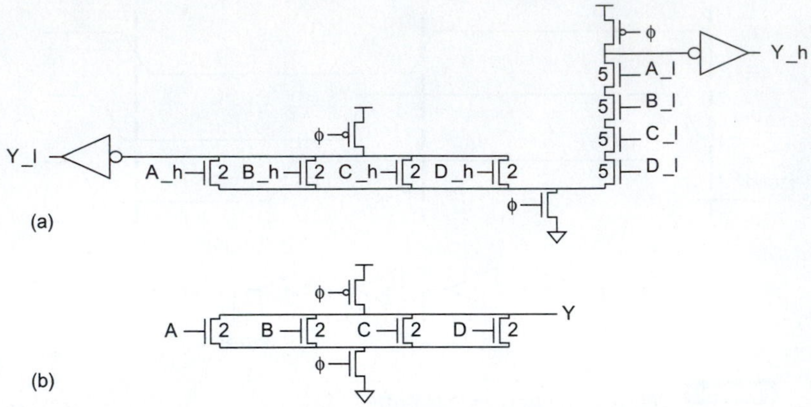

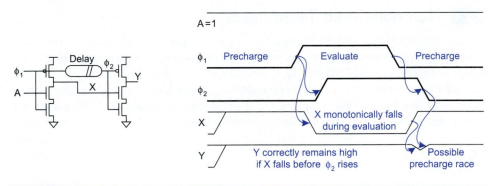

FIG 7.60 Comparison of NOR gates

Figure 7.61 also illustrates the *precharge race* problem. When X precharges while Y is still in evaluation, Y will start to fall. If ϕ_2 falls too late, Y will incorrectly glitch low. We can alleviate this problem by latching Y before X precharges or by delaying the falling edge of ϕ_1.

This section addresses a number of nonmonotonic techniques using delayed clocks to directly cascade dynamic gates and examines the margins required for matched delays.

FIG 7.61 Cascading dynamic gates with a delayed clock

7.5.4.1 Delay Matching Figure 7.62 shows a number of simple delay elements. The buffer delay can be set by adjusting gate widths. The buffer with transmission gates provides flexibility for longer delays. The current-starved inverter and switched capacitance designs use a reference voltage to adjust the delay externally. The digitally controlled current-starved inverter uses several digital signals rather than an analog voltage to adjust delay.

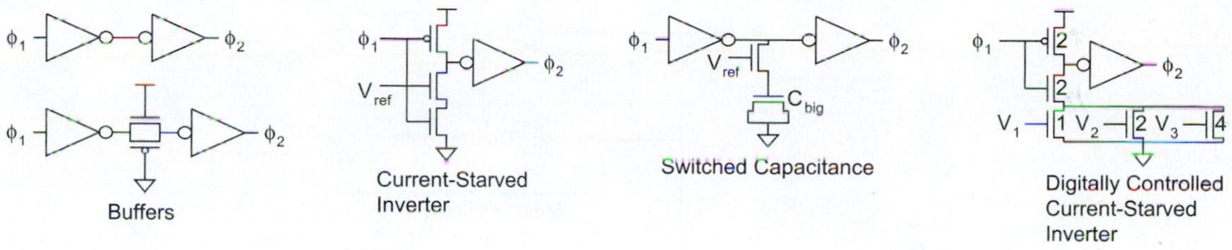

FIG 7.82 Delay elements

Section 4.7.6 showed that the delay of gates can vary by as much as 30% relative to an FO4 inverter across process, voltage, and temperature variations. Therefore, the delay line should provide some margin to guarantee it always is slower than the gate it must match. For example, [Yee00] uses a 20% margin. Many industrial designs use even more margin to ensure the circuit will have good yield in high-volume production. (Who wants to explain to the big boss why he or she wasted millions of dollars for the sake of saving a few picoseconds?) You should always make sure that the circuit works correctly in all process and environmental corners because it is not obvious which corner will cause the worst-case mismatches. Moreover, random device variations and inaccuracies in the parasitic extraction and device models cause further mismatch that cannot be captured through the design corner files. Yet another problem is that matching differs from one process to another, potentially requiring expensive redesign of circuits with matched delays when they are ported to the next process generation. Adjustable delay lines are attractive because the margin can be set more aggressively and increased after fabrication (as was done in [Vangal02]); however, generating and distributing a low-noise reference voltage can be challenging.

The key to good matching is to make the delay circuit behave like the gate it should match as much as possible. A good technique is to use a *dummy gate* in the delay line, as shown in Figure 7.63 for a 2:1 dynamic multiplexer. The dummy gate replicates the gate being matched so that to first order, process and environmental variations will affect both identically. The input pattern is selected for worst-case delay.

You might be tempted to use longer-than-minimum length transistors to create long delays, but this is not good because transistor length variations will affect the delay circuit much differently than the gate it matches.

Despite all of these difficulties, delay matching has been used for decades in specialized circumstances that require wide NOR operation such as CAMs and PLAs (see Sections 11.6 and 11.7). [Yee00] proposes wider use of delay matching in datapath applications and names the practice *Clock-Delayed* (CD) Domino.

7.5.4.2 Clock-delayed Domino   In the simplest CD Domino scheme, logic is *levelized* as shown in Figure 7.64(a). The boxes represent domino gates annotated with their worst-case delay. Delay elements produce clocks tuned to the slowest gate in each level. The overall path delay is the sum of the delays of each element, which may be longer than the

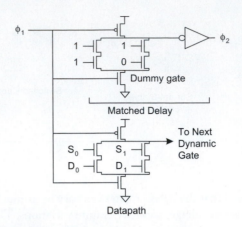

FIG 7.63 Delay matching with dummy gates

actual critical path through logic. An alternative scheme is to clock each gate at a time matched to its latest input, as shown in Figure 7.64(b). This better matches the critical path at the expense of more delay elements and design effort. CD Domino is most effective for functions where high fan-in gates can be converted to wide dynamic NORs.

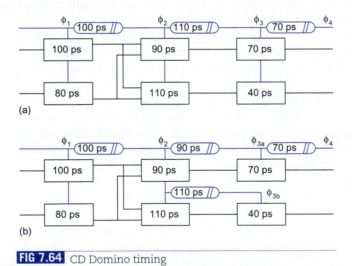

FIG 7.64 CD Domino timing

7.5.4.3 Race-based Nonmonotonic Logic The Itanium II processor uses a specialized nonmonotonic structure called an *annihilation* gate for high fan-in AND functions such as a 6-input decoder [Naffziger02]. An ordinary high fan-in AND gate requires many series transistors. Using DeMorgan's Law, it can be converted to a wide NOR with complementary inputs. The annihilation gate in Figure 7.65 performs this NOR function very rapidly

while generating a monotonically rising output suitable as an input to subsequent domino gates. It can be viewed as a dynamic NOR followed by a domino buffer with no clock delay. This introduces a race condition, but the two stages are carefully sized so the NOR will always win the race.

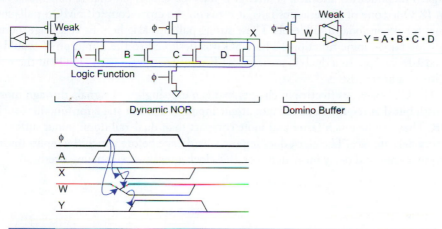

FIG 7.65 Annihilation gate

Initially, both X and W are precharged. The inputs must set up and hold around the rising edge of ϕ. When ϕ rises and the gate evaluates, W begins pulling down. If one or more of the inputs are asserted, X will also pull down, cutting off the transistor that was discharging W. The keeper will restore W back to a high level and the output Y will remain low. If all of the inputs are low, X will remain high, W will discharge, and Y will monotonically rise. The full keepers hold both X and W after evaluation. The gate has a built-in race: X must fall quickly so that W does not droop too much and cause a glitch on Y. The annihilation gate requires very careful design and attention to noise sources, but is fast and compact.

The annihilation gate is a new incarnation of a long-lost circuit called *Latched Domino* [Pretorius86] shown in Figure 7.66. The Latched Domino gate adds a cross-coupled nMOS transistor to help pull down node X. It also replaces the full keepers with ordinary keepers. As long as the glitches on X and W are small enough, Y_h and Y_l are good monotonic dual-rail outputs.

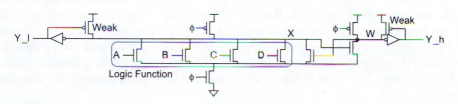

FIG 7.66 Latched domino gate

Intel uses a similar gate called a *Complementary Signal Generator* (CSG) shown in Figure 7.67 to produce dual-rail outputs from single rail inputs in a 5 GHz ALU [Vangal02]. Again, nodes X and W precharge and the inputs must set up before the rising edge of ϕ. When ϕ rises, W begins to discharge. If any of the inputs are true, X also begins to discharge. The pulldown and keeper strengths must be chosen so that X falls much faster than W. Once one of these nodes falls, it turns on the cross-coupled pMOS pull-ups to restore the other node to full levels. These strong pull-ups also help fight leakage, permitting wide fan-in logic functions. The CSG was designed so the glitch on W would not exceed 10% of V_{DD}. In a dual-V_t process, low V_t transistors were used on all but the noise-sensitive input transistors.

The CSG is very effective in circuits that can use single-rail signals through most of the path but that require dual-rail monotonic inputs to the last stage for functions such as XOR. They can be much faster and more compact than dual-rail domino but suffer from the very delicate race. The clock does impose a hard edge before which the inputs must set up so that skew and delay mismatches on this clock appear as sequencing overhead.

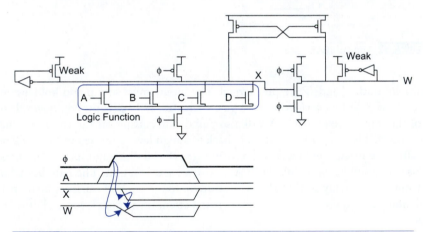

FIG 7.67 Complementary signal generator

7.5.4.4 Output Prediction Logic Clock-delayed and race-based dynamic logic represent two extremes in nonmonotonic logic. Both consist of two cascaded dynamic gates. CD Domino delays the clock to the second gate until the first has had time to fully discharge so the second gate will not glitch. Race-based logic such as annihilation gates and CSGs do not delay the clock but use transistor and keeper sizing to ensure the glitch on the second gate remains acceptably small. *Output Prediction Logic* (OPL) fits between these two extremes, delaying the clock by a moderate amount and accepting modest glitches [McMurchie00]. The delay is chosen as a compromise between performance and glitch size.

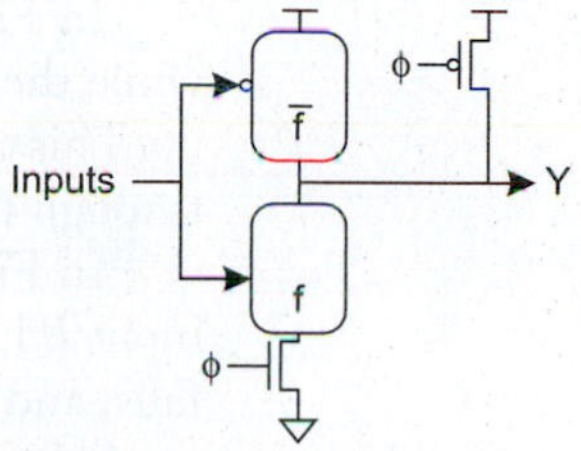

Figure 7.68 shows a basic OPL gate consisting of a Noise-Tolerant Precharge dynamic stage that was discussed in Section 6.2.4.3. You can view it either as a complementary CMOS structure with clocked evaluation and precharge transistors or as a dynamic gate plus a complementary pMOS pullup network. Like an ordinary dynamic gate, the output precharges high while the clock is low, then evaluates low when the clock rises and the appropriate inputs are asserted. However, like a static CMOS gate, the output can pull back high through the pMOS network to recover from output glitches.

Figure 7.69 shows a chain of OPL 2-input NAND gates. Each receives a clock delayed from the previous stage. As the stages are inverting, it resembles a chain of CD Domino gates. The amount of delay is critical to the circuit operation. Suppose A is '1' and all the unnamed outer inputs are also '1' so B, D, and F should pull low and C and E stay high. OPL precharges all the outputs to predict each output will remain high. The gates can be very fast because only half of the outputs have to transition. Figure 7.70 shows three cases of short (a), long (b), and medium (c) clock delays between a pair of OPL inverters. Simulating OPL is tricky because if all the gates are identical, the outputs will tend to settle at a metastable point momentarily, then diverge as the previous gate transitions. To break this misleading symmetry, a small parasitic capacitance C_p was added to node B.

FIG 7.68　OPL gate

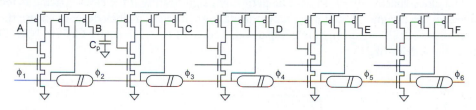

FIG 7.69　Chain of OPL gates

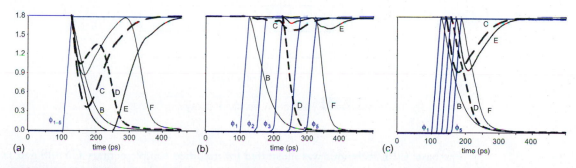

FIG 7.70　OPL waveforms for various clock delays

In Figure 7.70(a), all the clocks rise simultaneously. ϕ_2 arrives at the second stage while the input B is still high so C pulls most of the way low. When B falls, C rises back up. This causes D to fall, E to rise, and F to fall. In this mode of operation, the data ripples through the gates much as in static CMOS and the path delay is rather slow.

In Figure 7.70(b), the clock spacing is 50 ps. ϕ_2 arrives at the second stage after the input B has pulled most of the way low so C remains high. After another delay, ϕ_3 rises, D falls, and so forth. In this mode of operation, the OPL chain behaves in clock-blocked mode just like clock-delayed domino. The path delay is the sum of the clock delays plus the propagation delay of the final stage, which again is rather slow because the clock delay is lengthy.

In Figure 7.70(c), the clock spacing is 15 ps. ϕ_2 arrives at the second stage as the input B is falling so C glitches slightly, then returns to a good high value. After another delay, D falls. Again, the path delay is essentially the sum of the clock delays and final stage delay, but it is now faster because the clock delay is shorter than required for CD domino. The extra speed comes at the expense of some glitching.

A challenge in designing OPL gates is to choose just the right clock spacing. It should be as short as possible but not too short. Figure 7.71 plots the delay from A to F against the spacing between clocks. The nMOS transistors are 2 units wide and the figure compares the performance for pMOS of 1, 3, or 5 units. Wider pMOS transistors have slower evaluation delays but recover better from glitches. The lowest path delay occurs with a clock spacing of 10–15 ps. The path slows significantly if the clock spacing is too short, so the designer should nominally provide some margin in clock delay to ensure the worst case is still long enough. In comparison, a chain of complementary CMOS NAND gates has a delay of 213 ps.

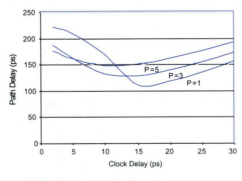

FIG 7.71 Path delay vs. clock delay

The basic OPL technique was illustrated for modified complementary CMOS gates that are relatively slow but recover quickly from large glitches. It also applies to other circuit families that have faster evaluation delays for high fan-in NOR structures such as pseudo-nMOS or dynamic gates, as illustrated in Figure 7.72(a and b). Pseudo-nMOS OPL is faster at evaluating because of the lower logical effort, but slower at recovery if the glitch is large. Dynamic OPL gates evaluate even faster but cannot recover at all if the glitch is large enough to flip the keeper. Using a low-skew feedback inverter improves the glitch tolerance for the keeper. As the best delay between clocks is a function of both evaluation delay and glitch tolerance, pseudo-nMOS and dynamic OPL are comparable in performance. Dynamic gates dissipate less power than pseudo-nMOS but may fail entirely if the clock delay is too short. Figure 7.72(c) shows a differential OPL gate using cross-coupled pMOS keepers that do not fight the initial transition and that can recover from arbitrarily large glitches [Kio01]. The inventors found that this was the fastest family of all, nearly five times faster than static CMOS.

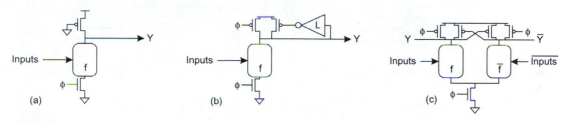

FIG 7.72 Alternative OPL circuit families

7.5.5 Static-to-domino Interface

Static CMOS gates require inputs that are levels and may produce nonmonotonic glitches on the outputs. Domino gates require inputs that are monotonic during evaluation and produce pulses on the outputs. Therefore, interface circuitry is necessary at the static-to-domino interface to avoid glitches and circuitry at the domino-to-static interface to convert the pulses into levels.

7.5.5.1 Static-to-domino Interface
Falling static inputs to domino gates must set up by the time the gate begins evaluation and should not change until evaluation is complete. This imposes a hard edge and the associated clock skew penalties, so the static-to-domino interface is relatively expensive. High-performance skew-tolerant domino pipelines build entire loops out of domino to avoid paying the skew at the static-to-domino interface.

   A simple solution to avoiding glitches at the interface is to latch the static signals as shown in Figure 7.73(a). The latch is opaque while the domino gates evaluate. Figure 7.73(b) shows that the latch does not need to be placed at the end of the previous half-cycle. The static logic must be designed to set up before domino gates enter evaluation. The latch prevents the next token from arriving too early and upsetting the domino input.

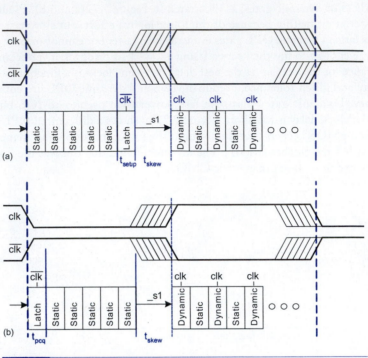

FIG 7.73 Latch at static to domino interface

In systems using flip-flops or pulsed latches, another approach is to capture the input on the clock edge with a flop or latch that produces monotonically rising outputs, as shown in Figure 7.74. The SA/F-F produces dual-rail monotonic outputs if the SR latch is replaced by HI-skew inverters. The K6 differential flip-flop also produces dual-rail monotonic pulsed outputs suitable for self-resetting logic that requires pulsed inputs. In any of these cases, you can build logic into the latch or flip-flop. For example, Figure 7.75 shows a single-rail *pulsed domino flip-flop* or *entry latch* (ELAT) with integrated logic used on UltraSparc and Itanium 2 [Klass99, Naffziger02]. It can be viewed as a fully dynamic version of the Klass SDFF. Falling inputs must set up before the clock edge, but rising inputs can borrow a small amount of time after the edge. The output is a monotonically rising signal suitable as an input to subsequent domino gates. The pulsed domino flip-flop can also use a single pulsed nMOS transistor in place of the two clocked devices [Mehta99].

7.5.5.2 Domino-to-static Interface

Domino outputs are pulses that terminate when the gates precharge. Static logic requires levels that remain stable until they are sampled, independent of the clock period. At the domino-to-static interface, another latch is required as a pulse-to-level converter. The output of this latch can borrow time into subsequent static logic, so the latch does not impose a hard edge.

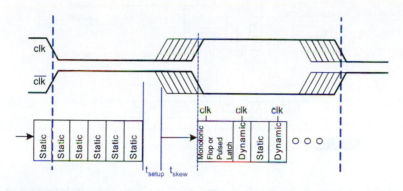

FIG 7.74 Monotonic flip-flop or pulsed latch at static to domino interface

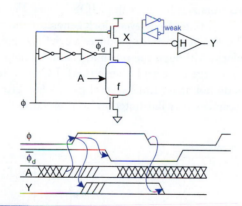

FIG 7.75 Pulsed domino flip-flop with integrated logic

Figure 7.76 shows a domino gate with a simple built-in output latch. The HI-skew inverter is replaced with a clocked inverter. The critical path still passes through only the pMOS transistor, so the latch is nearly as fast as a simple inverter. On the falling edge of the clock, the latch locks out the precharge, holding the result of the domino gate until the next rising edge of the clock. A weak inverter staticizes the Y output. Y should typically be buffered before driving long wires to prevent noise from backdriving the latch. Note that Y does glitch low shortly after the rising edge of the clock. The glitch can cause excess power dissipation in the static logic. Dual-rail domino outputs can avoid the glitch at the cost of greater delay by using the SR latch shown in Figure 7.29(a or b).

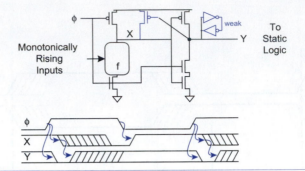

FIG 7.76 Domino gate with pulse-to-level output conversion

The Itanium 2 uses a *dynamic latch converter* (DLC) on the last domino gate in each half-cycle to hold the output by delaying the precharge until the next rising clock edge. This provides greater skew tolerance in domino paths and allows the output to drive static logic. An ordinary dynamic gate receives the same clock for the precharge (*RCLK*) and evaluation (*ECLK*) transistors and has a weak pMOS keeper. Figure 7.77 shows a DLC that is a "bolt-on" block consisting of a delayed clock generator and an extra nMOS keeper to make a full keeper The *RCLK* generator produces a brief low-going precharge pulse on the rising edge of the clock. Although the precharge and evaluate transistors may be on momentarily, this is not a large concern because the DLC operates the last gate of the half-cycle so the inputs do not arrive until several gate delays after the clock edge. The DLC also may include scan circuitry illustrated in Section 9.6.5.

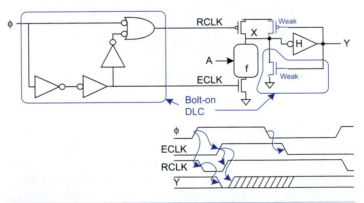

FIG 7.77 Dynamic latch converter

In self-resetting domino, the reset pulse for the last gate can also be delayed so that the domino output is compatible with static logic. For example, Figure 7.57 showed such a pulse generator for Global STP domino.

7.5.6 Delayed Keepers

Dynamic gates with high leakage current will eventually discharge to an invalid logic level unless they have strong keepers. The problem is especially severe when the inputs use many parallel low-V_t transistors. Unfortunately, the strong keeper slows the dynamic gate, reducing the performance advantage it was supposed to provide. As discussed in Section 6.2.4.3 for the burn-in keeper, this problem can be addressed by breaking the keeper into two parts. One part operates in the typical fashion. The second part turns on after some delay when the gate has had adequate time to evaluate. This combines the advantage of fast initial evaluation from the smaller keeper with better long-term leakage immunity from the two keepers in parallel.

Figure 7.78(a) shows such a *conditional keeper* [Alvandpour02]. *P2* is the conventional feedback keeper. *P1* turns on three gate delays after φ rises to help fight leakage. Figure 7.78(b) shows *High-Speed Domino* that leaves *X* floating momentarily until *P1* turns ON [Allam00]. *Skew-Tolerant High-Speed Domino* uses two transistors in series as the second keeper [Jung01], as shown in Figure 7.78(c). The inverting delay logic (IDL) can be an inverter, three inverters in series, or some other inverting structure with greater delay.

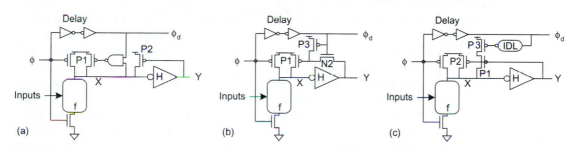

FIG 7.78 Delayed keepers

A challenge with any of these delayed keeper techniques is to ensure that the second part of the keeper turns on at a suitable time after the input arrives, but before too much leakage occurs. They work best for the first gate after a phase boundary, where the inputs are known to set up by the time the clock rises [Alvandpour02].

7.6 Synchronizers

Sequencing elements are characterized by a setup and hold time. If the data input changes before the setup time, the output reflects the new value after a bounded propagation delay. If the data changes after the hold time, the output reflects the old value after a bounded propagation delay. If the data changes during the *aperture* between the setup and hold times, the output may be unpredictable and the time for the output to settle to a good logic level may be unbounded. Properly designed synchronous circuits guarantee the data is stable during the aperture. However, many interesting systems must interface with data coming from sources that are not synchronized to the same clock. For example, the user can press a key at any time and data coming over a network can be aligned with a clock of differing phase or frequency.

A *synchronizer* is a circuit that accepts an input that can change at arbitrary times and produces an output aligned to the synchronizer's clock. Because the input can change during the synchronizer's aperture, the synchronizer has a nonzero probability of producing a *metastable* output [Chaney73]. This section first examines the response of a latch to an analog voltage that can change near the sampling clock edge. The latch can enter a metastable state for some amount of time that is unbounded, although the probability of remaining metastable drops off exponentially with time. Therefore, you can build a simple synchronizer by sampling a signal, waiting until the probability of metastability is acceptably low, then sampling again. In certain circumstances, the relationship of the data and clock timing is more predictable, permitting more reliable synchronizers.

7.6.1 Metastability

A latch is a bistable device; i.e., it has two stable states (0 and 1). Under the right conditions, that latch can enter a metastable state in which the output is at an indeterminate level between 0 and 1. For example, Figure 7.79 shows a simple model for a static latch consisting of two switches (probably transmission gates in practice) and two inverters. While the latch is transparent, the sample switch is closed and the hold switch open (Figure 7.79(a)). When the latch goes opaque, the sample switch opens and the hold switch closes (Figure 7.79(b)). Figure 7.79(c) shows the DC transfer characteristics of the two inverters. Because $A = B$ when the latch is opaque, the stable states are $A = B = 0$ and $A = B = V_{DD}$. The metastable state is $A = B = V_m$, where V_m is not a legal logic level. This point is called *metastable* because the voltages are self-consistent and can remain there indefinitely. However, any noise or other disturbance will cause A and B to switch to one of the two stable states. Figure 7.79(d) shows an analogy of a ball on a hill. The top of the hill is a metastable state. Any disturbance will cause the ball to roll down to one of the two stable states on the left or right side of the hill.

Figure 7.80(a) plots the output of the latch from Figure 7.17(g) as the data transitions near the falling clock edge. If the data changes at just the wrong time t_m within the aperture, the output can remain at the metastable point for some time before settling to a valid logic level. Figure 7.80(b) plots t_{DQ} vs. $t_{DC} - t_m$ on a semilogarithmic scale for a rising input and output. The delay is less than or equal to t_{pdq} for inputs that meet the setup time and increases for inputs that arrive too close to t_m. The points marked on the graph will be used in the example at the end of this section.

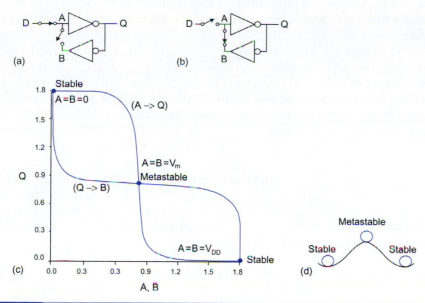

FIG 7.79 Metastable state in static latch

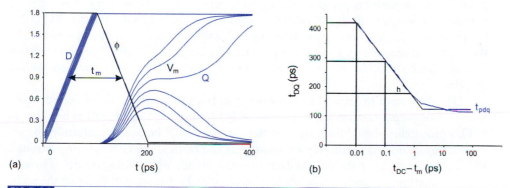

FIG 7.80 Metastable transients and propagation delay

The cross-coupled inverters behave like a linear amplifier with gain G when A is near the metastable voltage V_m. The inverter delay can be modeled with an output resistance R and load capacitance C. We can predict the behavior in metastability by assuming that the initial voltage on node A when the latch becomes opaque at time $t = 0$ is

$$A(0) = V_m + a(0) \qquad\qquad (7.25)$$

where $a(0)$ is a small signal offset from the metastable point. Figure 7.81 shows a small-signal model for $a(t)$. The behavior after time 0 is given by the first-order differential equation

$$\frac{Ga(t) - a(t)}{R} = C\frac{da(t)}{dt} \qquad (7.26)$$

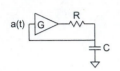

Solving this equation shows that the positive feedback drives $a(t)$ exponentially away from the metastable point with a time constant determined by the gain and RC delay of the cross-coupled inverter loop.

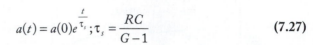

$$a(t) = a(0)e^{\frac{t}{\tau_s}}; \tau_s = \frac{RC}{G-1} \qquad (7.27)$$

FIG 7.81 Small signal model of bistable element in metastability

Suppose the node is defined to reach a legal logic level when $|a(t)|$ exceeds some deviation ΔV. The time to reach this level is

$$t_{DQ} = \tau_s\left[\ln\Delta V - \ln a(0)\right] \qquad (7.28)$$

This shows that the latch propagation delay increases as $A(0)$ approaches the metastable point and $a(0)$ approaches 0. The delay approaches infinity if $a(0)$ is precisely 0, but this can never physically happen because of noise. However, there is no upper bound on the possible waiting time t required for the signal to become valid. If the input $A(t)$ is a ramp that passes through V_m at time t_m, $a(0)$ is proportional to $t_{DC} - t_m$. Observe that EQ (7.28) is a good fit to the log-linear portion of Figure 7.80(b). The time constant τ_s is essentially the reciprocal of the gain-bandwidth product [Flannagan85]. Therefore, the feedback loop in a latch should have a high gain-bandwidth product to resolve from metastability quickly.

Designers need to know the probability that latch propagation delay exceeds some time t'. Longer propagation delays are less likely because they require $a(0)$ to be closer to 0. This probability should decrease with the clock period T_c because a uniformly distributed input change is less likely to occur near the critical time. Projecting through EQ (7.28) shows that it should also decrease exponentially with waiting time t'. Theoretical and experimental studies [Chaney83, Veendrick80, Horstmann89] find that the probability can be expressed as

$$P(t_{DQ} > t') = \frac{T_0}{T_c}e^{-\frac{t'}{\tau_s}} \text{ for } t' > h \qquad (7.29)$$

where T_0 and τ_s can be extracted through simulation [Baghini02] or measurement. Intuitively, T_0 / T_c describes the probability that the input would change during the aperture,

Example

Find τ_s, T_0, and h for the latch using the data in Figure 7.80.

Solution: h is the propagation delay above which the data fits a good straight line on a log-linear scale. In Figure 7.80, this appears to be approximately 175 ps. The probability that the delay exceeds some t' is the chance that the input changing at a random time falls within the small aperture that leads to the high delay. We can choose two points on the linear portion of the plot and solve for the two unknowns. For example, choosing (0.1 ps, 290 ps) and (0.01 ps, 415 ps), we solve

$$P\left(t_{DQ} > 290 \text{ ps}\right) = \frac{0.1\,ps}{T_c} = \frac{T_0}{T_c} e^{-\frac{290 \text{ ps}}{\tau_s}}$$

$$P\left(t_{DQ} > 415 \text{ ps}\right) = \frac{0.01\,ps}{T_c} = \frac{T_0}{T_c} e^{-\frac{415 \text{ ps}}{\tau_s}}$$

$$(7.30)$$

T_c drops out of the equations and we find τ_s = 54 ps and T_0 = 21 ps. Recall that this data was taken for a rising input. A conservative design should also consider the falling input and take data in the slow rather than typical environment.

causing metastability, and the exponential term describes the probability that the output has not resolved after t' if it did enter metastability. The model is only valid for sufficiently long propagation delays (h significantly greater than t_{pdq}).

We have seen that a good synchronizer latch should have a feedback loop with a high-gain-bandwidth product. Conventional latches have data and clock transistors in series, increasing the delay (i.e., reducing the bandwidth). Figure 7.82 shows a synchronizer flip-flop in which the feedback loops simplify to cross-coupled inverter pairs [Dike99]. Furthermore, the flip-flop is reset to 0, and then is only set to 1 if $D = 1$ to minimize loading on the feedback loop.

The flip-flop consists of master and slave jamb latches. Each latch is reset to 0 while $D = 0$. When D rises before ϕ, the master output X is driven high. This in turn drives the slave output Q high when ϕ rises. The pulldown transistors are just large enough to overpower the cross-coupled inverters, but should add as little stray capacitance to the feedback loops as possible. X and Q are buffered with small inverters so they do not load the feedback loops.

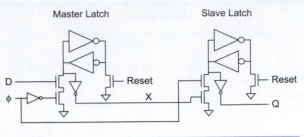

FIG 7.82 Fast synchronizer flip-flop

7.6.2 A Simple Synchronizer

A synchronizer accepts an input D and a clock ϕ. It produces an output Q that ought to be valid some bounded delay after the clock. The synchronizer has an aperture defined by a setup and hold time around the rising edge of the clock. If the data is stable during the aperture, Q should equal D. If the data changes during the aperture, Q can be chosen arbitrarily. Unfortunately, it is impossible to build a perfect synchronizer because the duration of metastability can be unbounded. We define synchronizer failure as occurring if the output has not settled to a valid logic level after some time t'.

Figure 7.83 shows a simple synchronizer built from a pair of flip-flops. *F1* samples the asynchronous input D. The output X may be metastable for some time, but will settle to a good level with high probability if we wait long enough. *F2* samples X and produces an output Q that should be a valid logic level and be aligned with the clock. The synchronizer has a latency of one clock cycle, T_c. It can fail if X has not settled to a valid level by a setup time before the second clock edge.

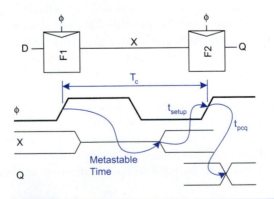

FIG 7.83 Simple synchronizer

Each flip-flop samples on the rising clock edge when the master latch becomes opaque. The slave latch merely passes along the contents of the master and does not sig-

Example

A particular synchronizer flip-flop in a 0.25 μm process has $\tau_s = 20$ ps and $T_0 = 15$ ps [Dike99]. Assuming the input toggles at $N = 50$ MHz and the setup time is negligible, what is the minimum clock period T_c for which the MTBF exceeds one year?

Solution: 1 year $\approx \pi \cdot 10^7$ seconds. Thus, we must solve

$$\pi \cdot 10^7 = \frac{T_c e^{\frac{T_c}{20 \cdot 10^{-12}}}}{\left(5 \cdot 10^7\right)\left(15 \cdot 10^{-12}\right)} \qquad (7.31)$$

numerically for a minimum clock period of 625 ps (1.6 GHz).

Example

How much longer must we wait for a 1000-year MTBF?

Solution: Solving a similar equation gives 760 ps. Increasing the waiting time by 135 ps improved MTBF by a factor of 1000.

nificantly affect the probability of metastability. If the synchronizer receives an average of N asynchronous input changes at D each second, the probability of synchronizer failure in any given second is

$$P(\text{failure}) = N \frac{T_0}{T_c} e^{\frac{-\left(T_c - t_{\text{setup}}\right)}{\tau_s}} \qquad (7.32)$$

and the mean time between failures increases exponentially with cycle time

$$MTBF = \frac{1}{P(\text{failure})} = \frac{T_c e^{\frac{T_c - t_{\text{setup}}}{\tau_s}}}{NT_0} \qquad (7.33)$$

The acceptable MTBF depends on the application. For medical equipment where synchronizer reliability is crucial and latency is relatively unimportant, the MTBF can be chosen to be longer than the life of the universe ($\sim 10^{19}$ seconds) by waiting more than one clock cycle before using the data. For noncritical applications, the MTBF can be chosen to be merely longer than the designer's expected duration of employment at the company!

7.6.3 Communicating Between Asynchronous Clock Domains

A common application of synchronizers is in communication between asynchronous clock domains, i.e., blocks of circuits that do not share a common clock. Suppose System A is controlled by *clkA* that needs to transmit *N*-bit data words to System B, which is controlled by *clkB*, as shown in Figure 7.84. The systems can represent separate chips or separate units within a chip using unrelated clocks. Each word should be received by system B exactly once. System A must guarantee that the data is stable while the flip-flops in System B sample the word. It indicates when new data is valid by using a request signal (*Req*), so System B receives the word exactly once rather than zero or multiple times. System B replies with an acknowledge signal (*Ack*) when it has sampled the data so System A knows when the data can safely be changed. If the relationship between *clkA* and *clkB* is completely unknown, a synchronizer is required at the interface.

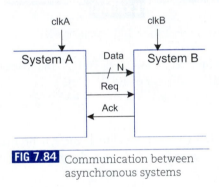

FIG 7.84 Communication between asynchronous systems

The request and acknowledge signals are called *handshaking* lines. Figure 7.85 illustrates two-phase and four-phase handshaking protocols. The four-phase handshake is level-sensitive while the two-phase handshake is edge-triggered. In the four-phase handshake, system A places data on the bus. It then raises *Req* to indicate that the data is valid. System B samples the data when it sees a high value on *Req* and raises *Ack* to indicate that the data has been captured. System A lowers *Req*, then system B lowers *Ack*. This protocol requires four transitions of the handshake lines. In the two-phase handshake, system A places data on the bus. Then it changes *Req* (low to high or high to low) to indicate that the data is valid. System B samples the data when it detects a change in the level of *Req* and toggles *Ack* to indicate that the data has been captured. This protocol uses fewer transitions (and thus possibly less time and energy), but requires circuitry that responds to edges rather than levels.

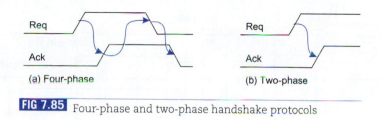

FIG 7.85 Four-phase and two-phase handshake protocols

Req is not synchronized to *clkB*. If it changes at the same time *clkB* rises, System *B* may receive a metastable value. Thus, System *B* needs a synchronizer on the *Req* input. If the synchronizer waits long enough, the request will resolve to a valid logic level with very high probability. The synchronizer may resolve high or low. If it resolves high, the rising request was detected and System *B* can sample the data. If it resolves low, the rising request was just missed. However, it will be detected on the next cycle of *clkB*, just as it would have been if the rising request occurred just slightly later. *Ack* is not synchronized to *clkA*, so it also requires a synchronizer.

Figure 7.86 shows a typical two-phase handshaking system [Crews03]. *clkA* and *clkB* operate at unrelated frequencies and each system may not know the frequency of its counterpart. Each system contains a synchronizer, a level-to-pulse converter, and a pulse-to-level converter. System A asserts *ReqA* for one cycle when *DataA* is ready. We will refer to this as a *pulse*. The XOR and flip-flop form a pulse-to-level converter that toggles the level of *Req*. This level is synchronized to *clkB*. When an edge is detected, the level-to-pulse converter produces a pulse on *ReqB*. This pulse in turn toggles *Ack*. The acknowledge level is synchronized to *clkA* and converted back to a pulse on *AckA*. The synchronizers add significant latency so the throughput of asynchronous communication can be much lower than that of synchronous communication.

7.6.4 Common Synchronizer Mistakes

Although a synchronizer is a simple circuit, it is notoriously easy to misuse. For example, the AMD 9513 system timing controller, AMD 9519 interrupt controller, Zilog Z-80 Serial I/O interface, Intel 8048 microprocessor, and AMD 29000 microprocessor are all said to have suffered from metastability problems [Wakerly00].

One way to build a bad synchronizer is to use a bad latch or flip-flop. The synchronizer depends on positive feedback to drive the output to a good logic level. Therefore, dynamic latches without feedback such as Figure 7.17(a–d) do not work. The probability of failure grows exponentially with the time constant of the feedback loop. Therefore, the loop should be lightly loaded. The latch from Figure 7.17(f) is a poor choice because a large capacitive load on the output will increase the time constant; Figure 7.17(g) is a much better choice.

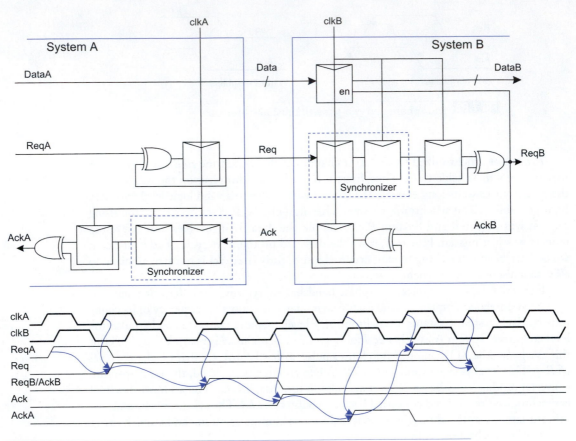

FIG 7.86 Two-phase handshake circuitry with synchronizers

Another error is to capture inconsistent data. For example, Figure 7.87(a) shows a single signal driving two synchronizers (each consisting of a pair of back-to-back flip-flops). If the signal is stable through the aperture, *Q1* and *Q2* will be the same. However, if the signal changes during the aperture, *Q1* and *Q2* might resolve to different values. If the system requires that *Q1* and *Q2* be identical representations of the data input, they must come from a single synchronizer.

Another example is to synchronize a multi-bit word where more than one bit might be changing at a time. For example, if the word in Figure 7.87(b) is transitioning from 0000 to 1111, the synchronizer might produce a value such as 0101 that is neither the old nor the new data word. For this reason, the system in Figure 7.86 synchronized only the *Req/Ack* signals and used them to indicate that data was stable to sample or finished being sampled. *Gray codes* (see Section 10.7.3) are also useful for counters whose outputs must be synchronized because exactly one bit changes on each count so the synchronizer is guaranteed to find either the old or the new data value.

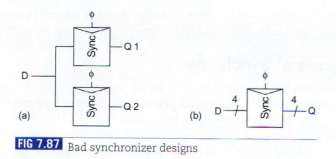

FIG 7.87 Bad synchronizer designs

In general, synchronizer bugs are intermittent and very difficult to locate and diagnose. For this reason, the number of synchronizers in a system should be strictly limited.

7.6.5   Arbiters

The *arbiter* of Figure 7.88(a) is closely related to the synchronizer. It determines which of two inputs arrived first. If the spacing between the inputs exceeds some aperture time, the first input should be acknowledged. If the spacing is smaller, exactly one of the two inputs should be acknowledged, but the choice is arbitrary. For example, in a television game show, two contestants may pound buttons to answer a question. If one presses the button first, she should be acknowledged. If both press the button at times too close to distinguish, the host may choose one of the two contestants arbitrarily.

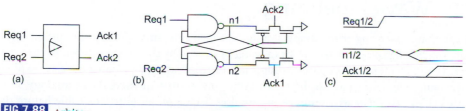

FIG 7.88 Arbiter

Figure 7.88(b) shows an arbiter built from an SR latch and a four-transistor metastability filter. If one of the request inputs arrives well before the other, the latch will respond appropriately. However, if they arrive at nearly the same time, the latch may be driven into metastability, as shown in Figure 7.88(c). The filter keeps both acknowledge signals low until the voltage difference between the internal nodes $n1$ and $n2$ exceeds V_t, indicating that a decision has been made. Such an asynchronous arbiter will never produce metastable outputs. However, the time required to make the decision can be unbounded, so the acknowledge signals must be synchronized before they are used in a clocked system.

Arbiters can be generalized to select 1-of-N or M-of-N inputs. However, such arbiters have multiple metastable states and require careful design [van Berkel99].

7.6.6 Degrees of Synchrony

The simple synchronizer from Section 7.6.2 accepts inputs that can change at any time, but has a nonzero probability of failure. In practice, many inputs may not be aligned to a single system clock, but they may still be predictable. Table 7.3 provides a classification of degrees of synchrony between input signals and the receiver system clock [Messerschmitt90] based on the difference in phase $\Delta\phi$ and frequency Δf.

[Dally98] describes a number of synchronizers that have zero failure probability and possibly lower latency when the input is predictable. They are based on the observation that either the signal or a copy of the signal delayed by t_a will be stable throughout the aperture. Hence, a synchronizer that can predict the input arrival time can choose the signal or its delayed counterpart to safely sample. Mesochronous signals are synchronized by measuring the phase difference and delaying the input enough to ensure it falls outside the aperture. Plesiochronous signals can be synchronized in a similar fashion, but the phase difference slowly varies so the delay must be occasionally adjusted. Because the frequencies differ, the synchronizer requires some control flow to handle the missing or extra data items. Periodic signals also require control flow and use a clock predictor to calculate where the next clock edge will occur and whether the signal must be delayed to avoid falling in the aperture.

7.7 Wave Pipelining

Recall that sequencing elements are used in pipelined systems to prevent the current token from overtaking the next token or from being overtaken by the previous token in the pipeline. If the elements propagate through the pipeline at a fairly constant rate, explicit sequencing elements may not be necessary to maintain sequence. As an analogy, fiber optic cables carry data as a series of light pulses. Many pulses enter the cable before the first one reaches the end, yet the cable does not need internal latches to keep the pulses separated because they propagate along the cable at a well-controlled velocity. The maximum data rate is limited by the dispersion along the line that causes pulses to smear over time and blur into one another if they become too short.

Figure 7.89 compares traditional pipelining with wave pipelining. In both cases, the pipeline contains combinational logic separated by registers (Figure 7.89(a)). The registers *F1* and *F2* receive clocks *clk1* and *clk2* that are nominally identical, but might experience skew. Figure 7.89(b) shows traditional pipelining. The data is launched on the rising edge of *clk1*. Its propagation is indicated by the hashed cone. *D2* becomes stable somewhere between the contamination and propagation delays after the clock edge (neglecting the flip-flop *clk*-to-*Q* delay). *D2* must not change during the setup and hold aperture around *clk2*, marked with the gray box. The figure shows two successive cycles in which tokens *i*

Table 7.3   Degrees of synchrony

Classification	Periodic	Δφ	Δf	Description
Synchronous	Yes	0	0	Signal has same frequency and phase as clock. Safe to sample signal directly with the clock. **Example:** Flip-flop to flip-flop on chip.
Mesochronous	Yes	constant	0	Signal has same frequency, but is out of phase with the clock. Safe to sample signal if it is delayed by a constant amount to fall outside aperture. **Example:** Chip-to-chip where chips use same clock signal, but might have arbitrarily large skews.
Plesiochronous	Yes	varies slowly	small	Signal has nearly the same frequency. Phase drifts slowly over time. Safe to sample signal if it is delayed by a variable but predictable amount. Difference in frequency can lead to dropped or duplicated data. **Example:** Board-to-board where boards use clock crystals with small mismatches in nominally identical rates.
Periodic	Yes	varies rapidly	large	Signal is periodic at an arbitrary frequency. Periodic nature can be exploited to predict and delay accordingly when data will change during aperture. **Example:** Board-to-board where boards use different frequency clocks.
Asynchronous	No	unknown	unknown	Signal may change at arbitrary times. Full synchronizer is required. **Example:** Input from pushbutton switch.

and $i + 1$ move through the pipeline. Each token passes through the combinational logic in a single cycle. Figure 7.89(c) shows wave pipelining with a clock of twice the frequency. Token i enters the combinational logic, but takes two cycles to reach *F2*. Meanwhile, token $i + 1$ enters the logic a cycle later. As long as each token is stable to sample at *F2* and the cones do not overlap, the pipeline will operate correctly with the same latency but twice the throughput.

[Burleson98] gives a tutorial on wave pipelining and derives the timing constraints. In general, a wave pipeline can contain N tokens between each pair of registers. The maximum value of N is limited by the ratio of propagation delay to dispersion of the logic cones:

$$N < \frac{t_{pd}}{t_{pd} - t_{cd}} \qquad (7.34)$$

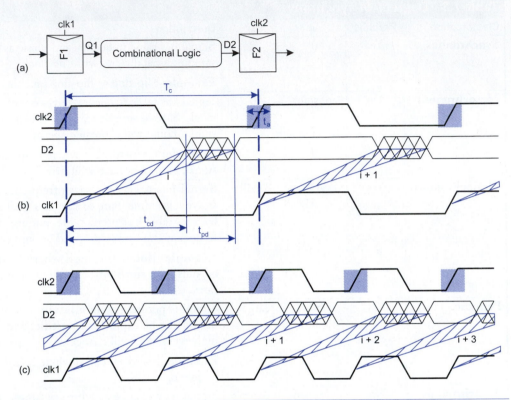

FIG 7.89 Wave pipelining

If the contamination and propagation delays are nearly equal, the combinational logic can contain many tokens simultaneously. In practice, the delays tend to be widely variable because of voltage, temperature, and processing as well as differences in path lengths through the logic. Clock skew and sequencing overhead also eat into the timing budgets. In practice, even achieving $N = 2$ simultaneous tokens can be difficult and wave pipelining has not achieved widespread popularity for general-purpose logic.

7.8   Pitfalls and Fallacies

Incompletely reporting flip-flop delay

The effective delay of a flip-flop is its minimum D-to-Q time. This is the sum of the setup time t_{setup} and the clk-to-Q delay t_{pdq} if these delays are defined to minimize the sum. Some engineers focus on only the clk-to-Q delay or define setup and clk-to-Q delays in a way that does not minimize the sum.

Failing to check hold times

One of the leading reasons that chips fail to operate even though they appear to simulate correctly is hold time violations, especially violations caused by unexpected clock skew. Unless a design uses two-phase nonoverlapping clocks, the clock skew should be carefully modeled and the hold times should be checked with a static timing analyzer. These checks should happen as soon as a block is designed so that errors can be corrected immediately. For example, a large microprocessor used a wide assortment of delayed clocks to solve setup time problems on long paths. Hold times were not checked until shortly before tapeout, leading to a significant schedule slip when many violations were found.

Choosing a sequencing methodology too late in the design cycle

Designers may choose from many sequencing methodologies, each of which has tradeoffs. The best methodology for a particular application is very debatable, and engineers love a good debate. If the sequencing methodology is not settled at the beginning of the project, experience shows that engineers will waste tremendous amounts of time redoing work as the method changes, or supporting and verifying multiple methodologies. Projects need a strong technical manager to demand that a team choose one method at the beginning and stick with it.

Failing to synchronize asynchronous inputs

Unsynchronized inputs can cause strange and wonderful sporadic system failures that are very difficult to locate. For example, a finite state machine running off one clock received a READY input from a UART running on another clock when the UART had data available, as shown in Figure 7.90. The designer reasoned that synchronizing the READY signal was unimportant because if it changed near the clock edge of the FSM, she did not care whether it was detected in one cycle or the next. Moreover, the clock was so slow that metastability would have time to resolve. However, the FSM occasionally failed by jumping to seemingly random states that could never legally occur. After two months of debugging, she realized that the problem was triggered if the asynchronous READY signal was asserted a few gate delays before the FSM clock edge. The propagation delay through the combinational logic was different for various bits of the next state logic. Some bits had changed to their new values while others were still at their old values, so the FSM could jump to an undefined state. Registering the READY signal with the FSM clock before it drove the combinational logic solved the problem.

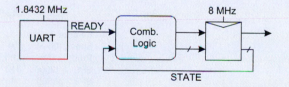

FIG 7.90 Unsynchronized input

Building faulty synchronizers

Designers have found many ways to build faulty synchronizers. For example, if an asynchronous input drives more than one synchronizer, the two synchroniz-

continued

ers can resolve to different values. If they must produce consistent outputs, only one synchronizer should be used. In another example, synchronizers must not accept multi-bit inputs where more than one of the bits can change simultaneously. This would pose the risk that some of the bits resolve as changed while others resolve in their old state, resulting in an invalid pattern that is neither the old nor the new input word. In yet another example, synchronizers with poorly designed feedback loops can be much slower than expected and can have exponentially worse mean time between failures.

7.9   Case Study: Pentium 4 and Itanium 2 Sequencing Methodologies

The Pentium 4 and Itanium 2 represent two philosophies of high-performance microprocessor design sometimes called *Speed Demon* and *Braniac*, respectively. The Pentium 4 was designed by Intel for server and desktop applications and has migrated into laptop computers as well. The Itanium 2 was jointly designed by Hewlett-Packard and Intel for high-end server applications. Figure 7.91 shows the date of introduction and the performance of several generations of these processors.

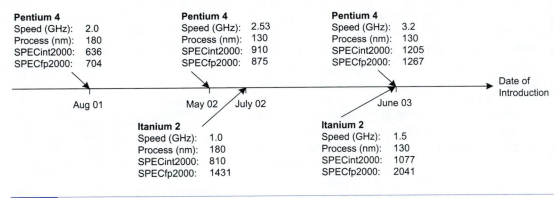

FIG 7.91 Microprocessor timeline

The Pentium 4 uses a very long (20+ stage) pipeline with few stages of logic per cycle to achieve extremely high frequencies. It issues up to three instructions per cycle, but the long pipeline causes severe penalties for branch mispredictions and cache misses, so the overall average number of instructions executed per cycle is relatively low. Figure 4.76 showed a die photo of the 42-million transistor Pentium 4. The chip consumes around 55 watts. A top-of-the-line Pentium 4 sold in 1000-unit quantities for around $400–$600 (depending on price pressure from competitor AMD). The chip has aggressively migrated into Intel's most advanced processes both to achieve high performance and to reduce the die size and manufacturing cost. The Speed Demon approach also gives Intel bragging rights to the highest clock frequency microprocessors, which is important because many consumers compare processors on clock frequency rather than benchmark performance. [Hrishikesh02] argues that the best logic depth is only 6 to 8 FO4 inverter delays per cycle.

In contrast, the Itanium 2 focuses on executing many instructions per cycle at a lower clock rate. It uses an 8-stage integer pipeline clocked at about half the rate of the Pentium 4 in the same process, so each cycle accommodates about twice as many gate delays (roughly 20–24 FO4 inverter delays, compared to roughly 10–12 for the Pentium 4). However, it issues up to six instructions per cycle and has a very high-bandwidth memory and I/O system to deliver these instructions and their data. As a result, it achieves nearly the same integer performance and much better floating-point benchmark results than the Pentium 4. Moreover, it also performs well on multiprocessor and transaction processing tasks typical of high-end servers. Figure 7.92 shows a die photo of the Itanium 2 with a 3MB level 3 (L3) cache; notice that the three levels of cache occupy most of the die area and most of the 221 million transistors. The 1.5 GHz model with 6MB cache bumps the transistor count to 410 million and further dwarfs the processor core. The chip consumes about 130 watts, limited by the cost of cooling multiprocessor server boxes. A high-end Itanium 2 sold for more than $4000 because the server market is much less price-sensitive. The chip has lagged a year behind the Pentium 4 in process technology.

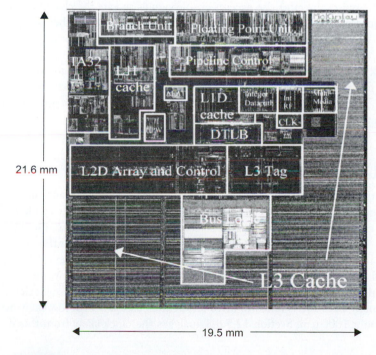

FIG 7.92 Itanium II die photo. © IEEE 2002.

7.9.1 Pentium 4 Sequencing

The Pentium 4 actually operates at three different internal clock rates [Hinton01, Kurd01]. In addition to the *core clock* that drives most of the logic, it has a double-speed *fast clock* for the ALU core and a half-speed *slow clock* for noncritical portions of the chip. The core clock is distributed across the chip using a triple spine, as will be shown in Section 12.5.4.3. These clocks drive pulsed latches, flip-flops, and self-resetting domino gates.

The ALU runs at a remarkable rate of twice the core clock frequency (about 6 FO4 inverter delays). To achieve this speed, it is stripped down to just the essential functions of the bypass multiplexer and the 16-bit add/subtract unit. Other less commonly used blocks such as the shifter and multiplier operate at core frequency. The ALU uses unfooted domino gates. The gates produce pulsed outputs and precharge in a self-timed fashion using the Global STP approach described in Section 7.5.2.4. These circuits demanded extensive verification by expert circuit designers to ensure the domino gates function reliably.

The Pentium 4 uses pulsed latches operating at all three clock speeds. Figure 7.93 shows pulse generators that receive the core clock and produce the appropriate output pulses. The medium-speed pulse generator produces a pulse on the rising edge of the core clock. The pulse width can be shaped by the adjustable delay buffer to provide both long pulses (offering more time borrowing) and short pulses (to prevent hold-time problems). The buffer is built from a digitally controlled current-starved inverter with four discrete settings. The pulse generator also accepts enable signals to gate the clock or save power on unused blocks. The slow pulse generator produces a pulse on every other rising edge of the core clock. To do this, it receives a sync signal that is asserted every other cycle. While the sync signal must be distributed globally, it is more convenient than distributing a half-speed clock because it can accept substantial skew while still being stable around the clock edge. The fast pulse generator produces pulses on both the rising and falling edges of the core clock. Therefore, the core clock should have nearly equal high and low times, i.e., 50% duty cycle, so the pulses are equally spaced.

7.9.2 Itanium 2 Sequencing

The Itanium 2 operates at a single primary clock speed, but also makes use of extensive domino logic and pulsed latches [Naffziger02, Fetzer02, Rusu03]. The clock is distributed across the chip using an H-tree, as will be shown in Section 12.5.4.2. The H-tree drives 33 second-level clock buffers distributed across the chip. These buffer outputs, called SLCBOs, in turn drive local clock gaters that serve banks of sequencing elements within functional blocks. There are 24 different types of clock gaters producing inverted, stretched, delayed, and pulsed clocks. Figure 7.94 shows some of these clocks. Each gater comes in many sizes and is tuned to drive different clock loads with low skew over regions of up to about 1000 μm. Section 12.5.6.3 analyzes the clock skew from this distribution network.

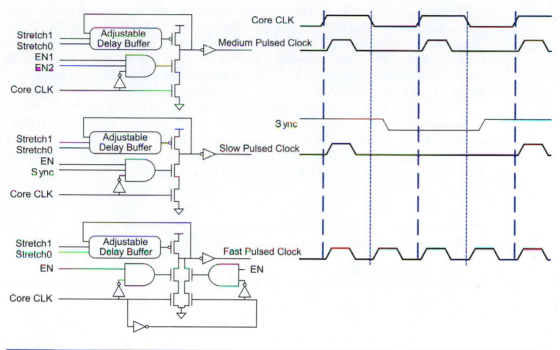

FIG 7.93 Pulse generators

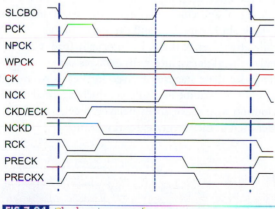

FIG 7.94 Clock gater waveforms

In the Itanium 2, 95% of the static logic blocks use Naffziger-pulsed latches with 125-ps wide pulses called *PCK*, as were described in Section 7.3.3. The pulsed latches are fast, permit a small amount of time borrowing, and present a small load to the clock. In situations where more time borrowing is needed, the gater may produce a wider pulsed clock *WPCK*. As discussed in Section 7.4.1.2, *clocked deracers* using *NPCK* can be inserted between back-to-back pulsed latches to prevent hold time violations.

The Itanium 2 uses extensive amounts of domino logic to achieve high performance at the expense of power consumption and careful design. Figure 7.95 shows a typical four-phase skew-tolerant domino pipeline from the Itanium 2. *CK* and *NCK* are clocks with a duty cycle slightly higher than 50% that are playing the roles of ϕ_1 and ϕ_3. They are delayed with buffers to produce *CKD* and *NCKD* (ϕ_2 and ϕ_4).

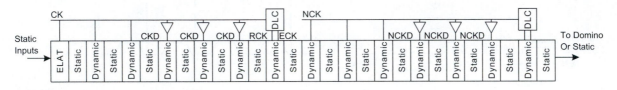

FIG 7.95 Four-phase skew-tolerant domino pipeline

The last gate in each phase uses a dynamic latch converter (DLC) to hold the output so that it can drive static logic and retain its state when the clock stops, as was discussed in Section 7.5.5.2. The DLC also provides scan capability at each half-cycle boundary to help with test.

At static-to-dynamic interfaces, inputs pass through pulsed entry latches (ELATs) that capture the static signal and convert it into a single-rail monotonic dynamic signal. These ELATs were shown in Figure 7.75 and can perform logic as well as latching. Some ELATs use *PCK*, while others derive the pulse internally from *CK*.

In some especially critical paths, alternating stages use unfooted domino gates. The falling edge of the clocks for these stages is delayed further to avoid contention during precharge. Figure 7.96 shows an extreme example in which a footed gate is followed by three stages of unfooted domino with successively delayed precharge edges, as was done in the 64-bit Naffziger adder used in the integer execution units.

FIG 7.96 Clocking for footless domino

Summary

This chapter has examined the tradeoffs of sequencing with flip-flops, two-phase transparent latches, and pulsed latches. The ITRS forecasts cycle times dropping well below 10 FO4 delays (see Table 4.17). Minimizing sequencing overhead will be very important in these high-performance systems. Flip-flops are the simplest, but have the greatest sequencing overhead. Transparent latches are most tolerant of skew and allow the most time borrowing, but require greater design effort to partition logic into half-cycles instead of cycles. Pulsed latches have the lowest sequencing overhead, but are most susceptible to min-delay problems. Table 7.4 compares the sequencing overhead, minimum delay constraint, and time borrowing capability of each technique. All of the techniques are used in commercial products, and the designer's choice depends on the design constraints and CAD tools.

Table 7.4	**Comparison of sequencing elements**		
	Sequencing overhead $(T_c - t_{pd})$	Minimum logic delay t_{cd}	Time borrowing t_{borrow}
Flip-Flops	$t_{pcq} + t_{setup} + t_{skew}$	$t_{hold} - t_{ccq} + t_{skew}$	0
Two-Phase Transparent Latches	$2t_{pdq}$	$t_{hold} - t_{ccq} - t_{nonoverlap} + t_{skew}$ in each half-cycle	$\dfrac{T_c}{2} - \left(t_{setup} + t_{nonoverlap} + t_{skew}\right)$
Pulsed Latches	$\max\left(t_{pdq}, t_{pcq} + t_{setup} - t_{pw} + t_{skew}\right)$	$t_{hold} - t_{ccq} + t_{pw} + t_{skew}$	$t_{pw} - \left(t_{setup} + t_{skew}\right)$

In class projects for introductory VLSI classes, timing analysis is often rudimentary or nonexistent. Using two-phase nonoverlapping clocks generated off chip is attractive because you can guarantee the chip will have no max-delay or min-delay failures if the clock period and nonoverlap are sufficiently large. However, it is not practical to generate and distribute two nonoverlapping phases on a large, high-performance commercial chip.

The great majority of low- and mid-performance designs and some high-speed designs use flip-flops. Flip-flops are very easy to use and are well understood by most designers. Even more importantly, they are handled well by synthesis tools and timing analyzers. Unfortunately, in systems with few gate delays per cycle, the sequencing overhead can consume a large fraction of the cycle. Moreover, many standard cell flip-flops are intentionally rather slow to prevent hold time violations at the expense of greater sequencing overhead.

Most two-phase latch systems distribute a single clock and locally invert it to drive the second latch. These systems tolerate significant amounts of clock skew without loss of performance and can borrow time to balance delay intentionally or opportunistically. However, the systems require more effort to understand because time borrowing distributes the timing constraints across many stages of a pipeline rather than isolating them at each stage. Not all timing analyzers handle latches gracefully, especially when there are different amounts of clock skew between different clocks [Harris99]. Two-phase latches have been used in the Alpha 21064 and 21164 [Gronowski98], PowerPC 603 [Gerosa94], and many other IBM designs.

Pulsed latches have low sequencing overhead. They present a tradeoff when choosing pulse width: A wide pulse permits more time borrowing and skew tolerance, but makes min-delay constraints harder to meet. Pulsed latches are also popular because they can be modeled as fast flip-flops with a lousy hold time from the point of view of a timing analyzer (or novice designer) if intentional time borrowing is not permitted. The min-delay problems can be largely overcome by mixing pulsed latches for long paths and flip-flops for short paths. Unfortunately, many real designs have paths in which the propagation delay is very long but the contamination delay is very short, making robust design more challenging. Pulsed latches have been used on Itanium 2 [Naffziger02], Pentium 4 [Kord01], Athlon [Draper97], and CRAY 1 [Unger86]. However, they can wreak havoc with conventional commercially available design flows and are best avoided unless the performance requirements are extreme.

Domino circuits are widely used in high-performance systems because they are 1.5-2x faster than static CMOS. Traditional domino circuits with latches have high sequencing overhead that wastes much of the potential speedup, so most designers have moved to skew-tolerant techniques. Static-to-domino interfaces impose hard edges and the associated sequencing overhead, motivating the use of domino throughout critical loops. Single-rail domino only computes noninverting functions, so most loops require dual-rail domino that consumes more area, wiring, and power and is ill-suited to wide NORs. An alternative is to push the inverting functions to the end of the pipeline, using single-rail domino through most of the pipeline and nonmonotonic static logic at the end. The area savings comes at the cost of one hard edge in the cycle.

Four-phase or delayed reset skew-tolerant domino circuits work well in datapaths because the clock generation is relatively simple. Self-resetting domino is ideally suited to memories where the decoder power consumption is greatly reduced by only precharging the output that switched and where the number of unique circuits to design is relatively small. It was also used on the Pentium 4, but was costly in terms of designer effort because so many pulse constraints must be satisfied.

Clock-delayed domino is used in wide dynamic NOR functions where the power consumption of pseudo-nMOS is unacceptable. For example, it is an important technique for CAMs and PLAs. The delay matching raises an unpleasant tradeoff between speed and correct operation, requiring significant margin for safe operation. The risk of race conditions deters many designers from using it more widely. Annihilation gates and complementary signal generators are interesting special cases in which no clock gate delay at all is required. Output prediction logic is also interesting, but has yet to be proven in a large application.

When inputs to a system arrive asynchronously, they cannot be guaranteed to meet setup or hold times at clocked elements. Even if we do not care whether an input arrived in one cycle or the next, we must ensure that the clocked element produces a valid logic level. Unfortunately, if the element samples a changing input at just the wrong time, it may produce a metastable output that remains invalid for an unbounded amount of time. The probability of metastability drops off exponentially with time. Systems use synchronizers to sample the asynchronous input and hold it long enough to resolve to a valid logic level with very high probability before passing it onward.

Most synchronous VLSI systems use opaque sequencing elements to separate one token from the next. In contrast, many optical systems transmit data as pulses separated in time. As long as the propagation medium does not disperse the pulses too badly, they can be recovered at a receiver. Similarly, if a VLSI system has low dispersion, i.e., nearly equal contamination and propagation delays, it can send more than one wave of data without explicit latching. Such wave pipelining offers the potential of high throughput and low sequencing overhead. However, it is difficult to perform in practice because of the variability of data delay.

Exercises

Use the following timing parameters for the questions in this section.

Table 7.5	Sequencing element parameters				
	Setup Time	clk-to-Q Delay	D-to-Q Delay	Contamination Delay	Hold Time
Flip-flops	65 ps	50 ps	n/a	35 ps	30 ps
Latches	25 ps	50 ps	40 ps	35 ps	30 ps

7.1   For each of the following sequencing styles, determine the maximum logic propagation delay available within a 500 ps clock cycle. Assume there is zero clock skew and no time borrowing takes place.

a) Flip-flops

b) Two-phase transparent latches

c) Pulsed latches with 80 ps pulse width

7.2   Repeat Exercise 7.1 if the clock skew between any two elements can be up to 50 ps.

7.3   For each of the following sequencing styles, determine the minimum logic contamination delay in each clock cycle (or half-cycle, for two-phase latches). Assume there is zero clock skew.

a) Flip-flops

b) Two-phase transparent latches with 50% duty cycle clocks

c) Two-phase transparent latches with 60 ps of nonoverlap between phases

d) Pulsed latches with 80 ps pulse width

7.4   Repeat Exercise 7.3 if the clock skew between any two elements can be up to 50 ps.

7.5   Suppose one cycle of logic is particularly critical and the next cycle is nearly empty. Determine the maximum amount of time the first cycle can borrow into the second for each of the following sequencing styles. Assume there is zero clock skew and that the cycle time is 500 ps.

a) Flip-flops

b) Two-phase transparent latches with 50% duty cycle clocks

c) Two-phase transparent latches with 60 ps of nonoverlap between phases

d) Pulsed latches with 80 ps pulse width

7.6   Repeat Exercise 7.5 if the clock skew between any two elements can be up to 50 ps.

7.7   Prove EQ (7.17).

7.8   Consider a flip-flop built from a pair of transparent latches using nonoverlapping clocks. Express the setup time, hold time, and clock-to-Q delay of the flip-flop in terms of the latch timing parameters and $t_{nonoverlap}$.

7.9   For the path in Figure 7.97, determine which latches borrow time and if any setup time violations occur. Repeat for cycle times of 1200, 1000, and 800 ps. Assume there is zero clock skew and that the latch delays are accounted for in the propagation delay Δ's.

a) $\Delta 1 = 550$ ps; $\Delta 2 = 580$ ps; $\Delta 3 = 450$ ps; $\Delta 4 = 200$ ps

b) $\Delta 1 = 300$ ps; $\Delta 2 = 600$ ps; $\Delta 3 = 400$ ps; $\Delta 4 = 550$ ps

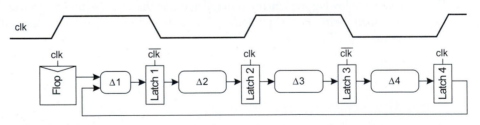

FIG 7.97 Example path

7.10   Determine the minimum clock period at which the circuit in Figure 7.98 will operate correctly for each of the following logic delays. Assume there is zero clock skew and that the latch delays are accounted for in the propagation delay Δ's.

a) $\Delta 1 = 300$ ps; $\Delta 2 = 400$ ps; $\Delta 3 = 200$ ps; $\Delta 4 = 350$ ps

b) $\Delta 1 = 300$ ps; $\Delta 2 = 400$ ps; $\Delta 3 = 400$ ps; $\Delta 4 = 550$ ps

c) $\Delta 1 = 300$ ps; $\Delta 2 = 900$ ps; $\Delta 3 = 200$ ps; $\Delta 4 = 350$ ps

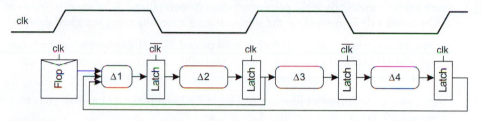

FIG 7.98 Another example path

7.11 Repeat Exercise 7.10 if the clock skew is 100 ps.

7.12 Label the timing types of each signal in the circuit from Figure 7.97. The flip-flop is constructed with back-to-back transparent latches—the first controlled by *clk_b* and the second by *clk*.

7.13 Using a simulator, compare the *D*-to-*Q* propagation delays of a conventional dynamic latch from Figure 7.17(d) and a TSPC latch from Figure 7.30(a). Assume each latch is loaded with a fanout of 4. Use 4 λ-wide clocked transistors and tune the other transistor sizes for least propagation delay.

7.14 Using a simulator, find the setup and hold times of a TSPC latch under the assumptions of Exercise 7.13.

7.15 Determine the maximum logic propagation delay available in a cycle for a traditional domino pipeline using a 500 ps clock cycle. Assume there is zero clock skew.

7.16 Repeat Exercise 7.15 if the clock skew between any two elements can be up to 50 ps.

7.17 Determine the maximum logic propagation delay available in a cycle for a four-phase skew-tolerant domino pipeline using a 500 ps clock cycle. Assume there is zero clock skew.

7.18 Repeat Exercise 7.17 if the clock skew between any two elements can be up to 50 ps.

7.19 How much time can one phase borrow into the next in Exercise 7.18 if the clocks each have a 50% duty cycle? Assume $t_{hold} = 0$.

7.20 Repeat Exercise 7.18 if the clocks have a 65% duty cycle.

7.21 Design a fast pulsed latch. Make the gate capacitance on the clock and data inputs equal. Let the latch drive an output load of four identical latches. Simulate your latch and find the setup and hold times and clock-to-*Q* propagation and contamination delays. Express your results in FO4 inverter delays.

7.22 Simulate the worst-case propagation delay of an 8-input dynamic NOR gate driving a fanout of 4. Report the delay in all 16 design corners (voltage, temperature, nMOS, pMOS). Also determine the delay of a fanout-of-4 inverter in each of these corners. By what percentage does the absolute propagation delay of the NOR gate vary across corners? By what percentage does its normalized delay vary (in terms of FO4 inverters)? Comment on the implications for circuits using matched delays.

7.23 A synchronizer uses a flip-flop with τ_s = 54 ps and T_0 = 21 ps. Assuming the input toggles at 10 MHz and the setup time is negligible, what is the minimum clock period for which the mean time between failures exceeds 100 years?

7.24 Simulate the synchronizer flip-flop of Figure 7.82 and make a plot analogous to Figure 7.80. From your plot, find Δ_{DQ}, h, τ, and T_0.

7.25 InferiorCircuits, Inc., wants to sell you a perfect synchronizer that they claim never produces a metastable output. The synchronizer consists of a regular flip-flop followed by a high-gain comparator that produces a high output for inputs above $0.25 \cdot V_{DD}$ and a low output for inputs below that point. The VP of marketing argues that even if the flip-flop enters metastability, its output will hover near $V_{DD}/2$ so the synchronizer will produce a good high output after the comparator. Why wouldn't you buy this synchronizer?

Design Methodology and Tools

8.1 Introduction

The manner in which you go about designing a particular system, chip, or circuit can have a profound impact on both the effort expended and the outcome of the design. IC designers have developed and adapted strategies from allied disciplines such as software engineering to form a cohesive set of principles to increase the likelihood of timely, successful designs. We will explore these principles in this chapter. While the broad principles of design have not changed in decades, the details of design styles and tools have evolved along with advances in technology and increasing levels of productivity. This chapter represents current CMOS design methods and provides an overview of a complex subject that could fill many books on its own. We encourage you to actively monitor the companies discussed and literature cited in the chapter to track the latest developments in this rapidly changing field.

As introduced in Section 1.6, an integrated circuit can be described in terms of three *domains*: (1) the *behavioral* domain, (2) the *structural* domain, and (3) the *physical* domain. The behavioral domain specifies what we wish to accomplish with a system. For instance, at the highest level, we might want to build an ultra-low-power radio for a distributed sensor network. The structural domain specifies the interconnection of components required to achieve the behavior we desire. Again, by way of example, our sensor radio might require a sensor, a radio transceiver, a processor and memory (with software), and a power source connected in a particular manner. Finally, the physical domain specifies how to arrange the components in order to connect them, which in turn allows the required behavior. Our example might start with the specification for an enclosure to hold the device, followed by a succession of physical drawings or specifications that may culminate in descriptions of geometry to be used to define a chip. Design flows from behavior to structure and ultimately to a physical implementation via a set of manual or automated transformations. At each transformation, the correctness of the transformation is tested by comparing the pre- and post-transformation design. For instance, if a power level is specified in the original behavioral description of the sensor radio, a test is run on the design in the structural domain with feedback from the physical domain to ensure this design goal is met.

In each of these domains there are a number of design options that can be selected to solve a particular problem. For instance, at the behavioral level, we can choose the wireless standard and the format in which data is transmitted by the sensor radio. In the structural domain, we can select which particular circuit style, logic family, or clocking strategy to use. At the physical level, we have many options about how the circuit is implemented in terms of chips, boards, and enclosures. These domains can further be hierarchically divided into different levels of *design abstraction*. Classically, these have included the following for digital chips:

- Architectural or functional level
- Logic or Register Transfer level (RTL)
- Circuit level

For analog and RF circuits, the block diagram level replaces the logic level.

The relationship between description domains and levels of abstraction is elegantly shown by the *Gajski–Kuhn Y chart* in Figure 8.1. In this diagram, the three radial lines represent the behavioral, structural, and physical domains. Along each line are enumerated types of objects in that domain. In the behavioral domain, we have represented conventional software and hardware description language categories. As we move out along any of the radial axes, the increasing level of design abstraction is able to represent greater

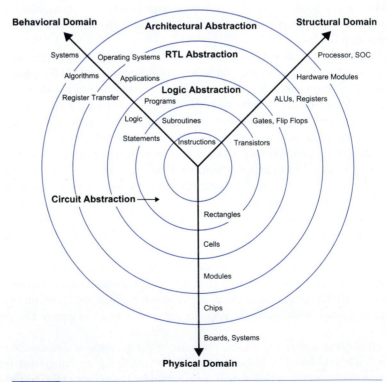

FIG 8.1 Gajski-Kuhn Y chart

complexity. Thus, in the behavioral domain, the lowest level of abstraction is an instruction or a statement in software or HDL descriptions, respectively. Circles represent levels of similar design abstraction: the architectural, RTL, logic, and circuit levels. The particular abstraction levels and design objects may differ slightly depending on the design method.

In this chapter, we will examine how to transform a description from one domain into another while maintaining the integrity of the design. It is only in this way that we can start with a behavior and successfully build a product.

We begin by discussing some of the guiding principles that apply to most engineering projects. Then we survey the various design strategies available to the CMOS IC designer; these range from rapid prototyping or small-volume approaches to those suitable for high-volume digital, analog, or RF design. We then examine the economics of design, which can guide us to the right selection of an implementation strategy, and review documentation requirements. In conclusion, we look at the reasons for the gap in performance between custom and ASIC design methods.

8.2   Structured Design Strategies

The viability of an IC is in large part affected by the productivity that can be brought to bear on the design. This in turn depends on the efficiency with which the design can be converted from concept to architecture, to logic and memory, to circuit, and ultimately to physical layout. A good VLSI design system should provide for consistent descriptions in all three description domains (behavioral, structural, and physical) and at all relevant levels of abstraction (e.g., architecture, RTL/block, logic, circuit). The means by which this is accomplished can be measured in various terms that differ in importance based on the application. These parameters can be summarized in terms of:

- Performance—speed, power, function, flexibility

- Size of die (hence, cost of die)

- Time to design (hence, cost of engineering and schedule)

- Ease of verification, test generation, and testability (hence, cost of engineering and schedule)

Design is a continuous tradeoff to achieve adequate results for all of the above parameters. As such, the tools and methodologies used for a particular chip will be a function of these parameters. Certain end results have to be met (i.e., the chip must conform to certain performance specifications), but other constraints may depend on economics (i.e., size of die affecting yield) or even subjectivity (i.e., what one designer finds easy, another might find incomprehensible).

Given that the process of designing a system on silicon is complicated, the role of good VLSI-design aids is to reduce this complexity, increase productivity, and assure the designer of a working product. A good method of simplifying the approach to a design is by the use of constraints and abstractions. By using constraints, the tool designer has some hope of automating procedures and taking a lot of the "legwork" (effort) out of a design. By using abstractions, the designer can collapse details and arrive at a simpler object to handle.

In this chapter, we will examine design methodologies that allow a variation in the freedom available in the design strategy. The choice, assuming all styles are equally available, should be entirely economic. According to function, suitable design methods are selected. Following these steps, the required chip cost is estimated and the quickest means of achieving that chip should be chosen. We will focus on structured approaches to design since they offer the most appropriate method of dealing with design complexity.

The successful implementation of almost any integrated circuit requires attention to the details of the engineering design process. Over the years, a number of structured design techniques have been developed to deal with complex hardware and software projects. Not surprisingly, the techniques have a great deal of commonality. Rigorous application of these techniques can drastically alter the amount of effort that has to be expended on a given project and also, in all likelihood, the chances of successful conclusion.

8.2.1 A Software Radio—A System Example

To guide you through the process of structured design, we will use as an example a hypothetical "software radio," as illustrated in Figure 8.2. This device is used to transmit and receive radio frequency (RF) signals. Information is modulated onto an RF *carrier* to transmit data, voice, or video. The RF carrier is demodulated to receive information. An ideal software radio could receive any frequency and decode or encode any type of information at any data rate. Some day, this might be possible, but given the limitations of current processes there are some bounds. To understand the impact of design methods on system solutions, we will examine the software radio in more detail. This system will then form the basis for discussion about structured approaches to design.

Figure 8.3 illustrates a typical transmit path for a generic radio transmitter, which is called an *IQ modulator*. An input data stream is encoded into *inphase* (I) and *quadrature* (Q) signals. The I and Q represent signal amplitudes of a (voltage) vector that vary instantaneously in time as shown in the bottom of Figure 8.3. For appropriate I and Q values, any form of modulated carrier can be synthesized. I is multiplied by an oscillator (sine)

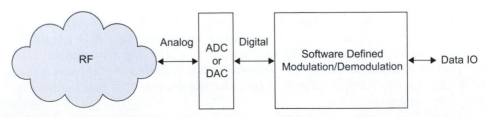

FIG 8.2 Software radio block diagram

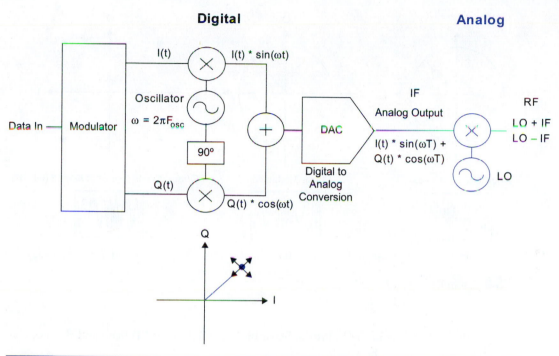

FIG 8.3 Software radio transmit path

operating at a frequency of F_{osc}. The quadrature (Q) signal is multiplied by the cosine of this frequency. The resultant signals are summed and passed to a digital-to-analog converter (DAC). In the design shown, this generates what we term an *Intermediate Frequency* or IF.

Typical IQ constellations are shown in Figure 8.4. *Amplitude Modulation* (AM), depicted in Figure 8.4(a), varies only in the magnitude of the carrier that varies in accordance with the amplitude of the modulation waveform. This is shown as a signal with an arbitrary phase angle (which we don't care about) and a vector that travels from the origin to a point on a circle that represents the maximum value of the carrier. In the case of an AM radio, the carrier frequency might be 800 KHz (in the AM band) and the modulation frequencies range from roughly 300 Hz to 6 KHz (voice and music frequencies). *Phase Modulation* is shown in Figure 8.4(b). Here, the vector travels around the maximum carrier amplitude circle varying the phase angle (δ) as the modulation changes. This is a constant amplitude modulation, which might be used with a carrier frequency of 100 MHz (in the FM broadcast band—we are loosely associating phase modulation with *frequency modulation* (FM) as they are closely related) and could have modulation frequencies of 200 Hz to 20 KHz (hi-fi audio). Finally, Figure 8.4(c) shows *Quadrature Phase Shift Keying* (QPSK) modulation, which is typical of data transmission systems. Two bits of data are encoded onto four phase points as shown in the diagram. A typical carrier frequency

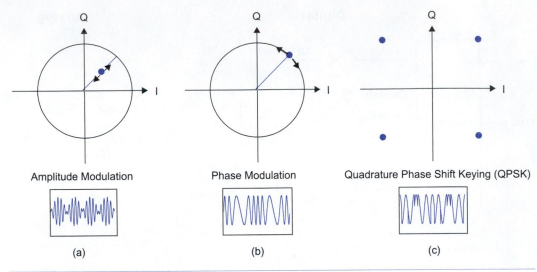

FIG 8.4 Examples of IQ modulation

might be 2.4 GHz in the Industrial Scientific and Medical (ISM) band and the modulation data rate might be 10 Megabits/second.

Clearly, the ranges of carrier and modulation frequencies vary considerably. Generally, for high carrier frequencies, the modulation can be performed at a moderate frequency and then "mixed" up to a higher frequency by analog multiplication. This is completed in the analog domain and is illustrated by the blue components on the right side of Figure 8.3. An analog multiplier (called a *mixer* in RF terminology) takes an analog Local Oscillator (LO) and the Intermediate Frequency (IF) signal that we have generated and produces sum and difference frequencies. (It is also possible to generate the desired RF frequency directly, but in this design we will use an intermediate frequency approach.) Analog bandpass filtering or a slightly more sophisticated mixer can be used to select the mixing component (LO+IF or LO-IF) that we desire. For instance, if we generate a data signal on a 20 MHz IF and mix it with a 2.4 GHz LO, we can generate a 2.402 or 2.398 GHz data signal. This is called *upconversion*.

To complete the software radio, the receive path is shown in Figure 8.5. It is roughly the reverse of the transmit path. As in the transmit case, higher frequencies can be *down-converted* to lower IF frequencies that are suitable for processing by practical ADCs. The RF signal is mixed with the LO and low pass filtered to produce the difference frequency. For example, if a 2.4 GHz LO is mixed with the 2.402 GHz RF signal, the 20 MHz IF signal is restored. An analog-to-digital converter (ADC) converts the modulated IF carrier into a digital stream of data. This data is mixed (multiplied) in the digital domain by an oscillator operating at the IF frequency. After digital low pass filtering (LPF), the original *I* and *Q* signals can be reconstructed and passed to a demodulator. For further details on digital radio, consult a communications theory text such as [Haykin00].

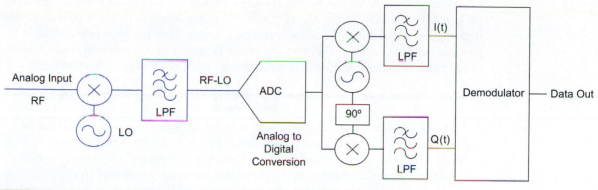

FIG 8.5 Software radio receive path

In summary, we see that multiplication, sine wave generation, and filtering are important for a software radio. While the modulation and demodulation have not been described in detail, operations can include equalization (multiplication), time to frequency conversion (fast Fourier transform), correlation, and other specialized coding operations. In the subsequent sections we will explore the design principles of hierarchy, regularity, modularity, and locality with concrete examples applied to the software radio.

8.2.2 Hierarchy

The use of hierarchy, or "divide and conquer," involves dividing a system into modules, then repeating this process on each module until the complexity of the submodules is at an appropriately comprehensible level of detail. This may entail stopping at a level where a prebuilt component is available for the particular function. The process parallels the software strategy in which large programs are split into smaller and smaller sections until simple subroutines with well-defined behavior and interfaces can be written. In the case of predefined modules, the design task involves using library code intended for the required function. The notion of "parallel hierarchy" can be used to aggregate descriptions in each of the behavioral, structural, and physical domains that represent a design (parallel hierarchy means a hierarchy—not necessarily identical—is used in each domain). Furthermore, equivalency tools can ensure the consistency of each domain. Because these tools can be applied hierarchically, you can progress in verification from the bottom to the top of a design, checking each level of hierarchy where domains are intended to correspond. For instance, a RISC processor core can have an HDL model that describes the behavior of the processor; a gate netlist that describes the type and interconnection of gates required to produce the processor; and a placement and routing description that describes how to physically build the processor in a given process. Later in the chapter, we will see how domain-to-domain comparisons are used to ensure consistency between domains.

Hierarchy allows the use of *virtual components*, soft versions of the more conventional packaged IC. Virtual components are placed into a chip design as pieces of code and come with support documentation such as verification scripts. They can be supplied by an independent *intellectual property* (IP) provider or can be reused from a previous product developed in your organization. Virtual components are discussed further in Section 8.5.7.

Example

The digital operations in the transmit path of the software radio (Figure 8.3) can be performed in software. Hence, a microprocessor can form the basis for the design. In this case, the design might have the hierarchy of a typical microprocessor, as shown in Figure 8.6. At the top level, the microprocessor contains an arithmetic logic unit (ALU), program counter (PC), register file, instruction decoder, and memory. The ALU can be further decomposed into an adder, a Boolean logic unit, and a shifter. The shifter and adder can together perform multiplication. The diagram illustrates how a relatively complex component can be rapidly decomposed into simple components within a few levels of hierarchy. Each level only has a few modules, which aids in the understanding of that level of the hierarchy.

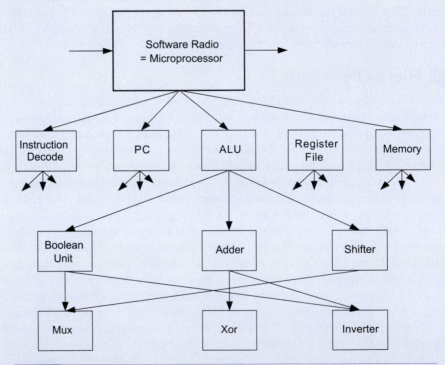

FIG 8.6   Possible hierarchy of software radio using a single microprocessor

Example

We can roughly estimate the performance required in the transmit path by noting that we require at least two multiplications, an addition, and two table lookups (sine and cosine). Another addition would be required to maintain a loop counter. An iterative multiply takes N cycles for an N-bit word, so for a 16-bit word width, the total number of cycles for the steps described would be approximately 16 + 16 + 1 + 2 + 2 + 1 (if table lookups take two clock cycles). This yields a total of roughly 40 clock cycles. For a 100 MHz processor, the fastest we could perform the IQ conversion would be approximately 400 ns, which, according to Nyquist's criteria ($F_{analog\_max} = F_{sample}/2$), would be capable of generating a 1.25 MHz IF signal. This is, of course, without any extra processing for modulating the carrier. While we could add another processor, this may be wasteful of area and power, given the operation that has to be performed.

A better approach is to use dedicated hardware for the computationally intensive fixed-function blocks. The trick is to notice that the IQ modulator portion of the software radio transmit and receive path for a given DAC and ADC resolution has a relatively fixed architecture. For the transmit path, the hierarchy shown in Figure 8.7 can be used where the blue sections have been converted to fixed function blocks. This is a relatively safe bet because the IQ upconversion is a generic communications building block. In addition to the multipliers, a device called a *Numerically Controlled Oscillator* (NCO) has been introduced [Lu93, Lu93b, Hwang02]. The NCO, described in detail in the next section, generates sine or cosine waveforms at a speed determined by the delay through an N-bit adder where N is in the range of 16 to 32 for typical NCOs. The

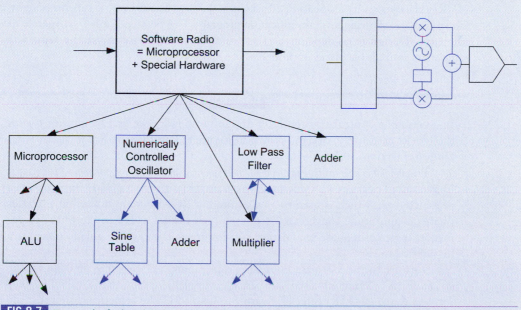

FIG 8.7 Transmit chain with dedicated IQ upconverter

move to dedicated hardware for the IQ upconversion allows the circuit to produce a new value once every clock cycle. If we conservatively say that the arithmetic blocks operate at the same speed that the microprocessor ALU does, then the circuit will now operate at 100 MHz. Taking into account sampling theory, this means that we can generate analog frequencies up to almost 50 MHz with a suitable DAC. The microprocessor now only has to respond at the modulation data rate, providing IQ values to the IQ upconverter.

8.2.3 Regularity

Hierarchy involves dividing a system into a set of submodules. However, hierarchy alone does not solve the complexity problem. For instance, we could repeatedly divide the hierarchy of a design into different submodules but still end up with a large number of different submodules. With regularity as a guide, the designer attempts to divide the hierarchy into a set of similar building blocks. Regularity can exist at all levels of the design hierarchy. At the circuit level, uniformly sized transistors can be used, while at the gate level, a finite library of fixed-height, variable-length logic gates can be used (see later in the chapter for a description of standard cells). At the logic level, parameterized RAMs and ROMs could be used in multiple places. At the architectural level, multiple identical processors can be used to boost performance.

Regularity aids in verification efforts by reducing the number of subcomponents to validate and by allowing formal verification programs (see Section 8.4.1.4) to operate more efficiently. Design reuse depends on the principle of regularity to use the same virtual component in multiple places or products.

Example

In an example of regularity applied to the software radio, we first look inside two of the blocks used in the designs shown in Figure 8.3 and Figure 8.5 to assess what kinds of functions are required.

The NCO is shown in Figure 8.8(a). It is composed of a registered adder that is incremented every clock cycle by a phase increment register. This implements a phase counter, which is used to step through a ROM lookup table that provides phase-to-amplitude conversion. A phase offset can be added to the phase incrementer to perform phase modulation. With this structure we are able to generate a digital sine wave.

Turning to the low-pass filter shown in Figure 8.5, Figure 8.8(b) shows the structure for a commonly used low-pass filter implementation that is called a *Finite Impulse Response* (FIR) filter [Edwards93, Choi97]. The structure computes the function:

$$Y[n] = \sum X[n-k]h[k] \qquad (8.1)$$

where $X[n]$ is the sampled input, $h[k]$ are the filter coefficients that characterize the particular filter, and $Y[n]$ is the output. As the structure indicates, the filter is composed of registers, multipliers, and an adder. Filters are characterized by the number of *taps* (coefficients). More taps yield better filters approaching an ideal "brick wall" filter with steeper cutoff and low ripple. This in turn requires more registers and more multipliers.

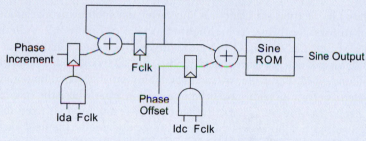

(a) Numerically Controlled Oscillator (NCO) Structure

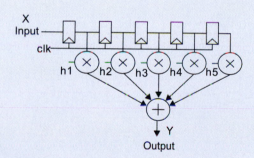

(b) Finite Impulse Response (FIR) Filter Structure

FIG 8.8 Structure of numerically controlled oscillator and low-pass filter (implemented as a finite impulse response (FIR) filter)

Having examined the detail of these blocks, we notice that the common functions are registers, adders, and multipliers with precisions as yet undefined. Parallel N-bit adders can be composed of N single-bit full adders. Multipliers are also built from full adders. N-bit registers are built from 1-bit flip-flops. Thus, one form of regularity might be to use the same full adder for all parallel adders and multipliers. Similarly, the same flip-flop would be used in all locations.

Typically, the phase counter adder in the NCO would be of the order of 16–32 bits wide. The phase increment adder might be 8–16 bits wide. The sizes of the multipliers and adders in the FIR filter vary widely, but depend on the input data width. This typically varies from 1–12 bits.

Example

As illustrated in the previous section, IQ upconversion and downconversion can be converted to fixed hardware, as highlighted in blue in Figure 8.9. Whether the hardware is shared (i.e., the NCO and the multipliers) is a determination that can be made at the time of design. Once this is decided, the IQ modulation and demodulation is still undefined. These blocks tend to be highly variable depending on the particular system. Software radios have been proposed in areas where the standards are likely to evolve as time progresses. Rather than have any product fixed to an old standard, a software radio allows the product to be updated in the field via a firmware update. Thus, in our quest for a software radio architecture, we still want programmability.

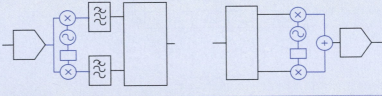

FIG 8.9   Common IQ blocks

A solution to maintaining programmability while increasing processing power might be to use a multiprocessor as shown in Figure 8.10. Here, the IQ up-and-down conversion has been retained and the IQ modulation/demodulation is performed by the four processors. The number of processors is arbitrary and would be ascertained by a detailed analysis of the required computational power.

Imagine that the computational power required slightly exceeds that provided by the four processors shown in Figure 8.10. Because multiplication is a frequently required operation in signal processing operations, it makes sense to build a multiplier into each microprocessor, as shown in Figure 8.11. Hence, we maintain regularity and improve processing power.

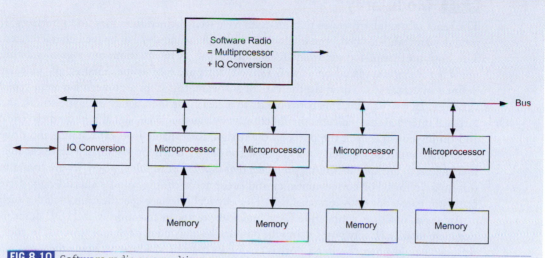

FIG 8.10 Software radio as a multiprocessor

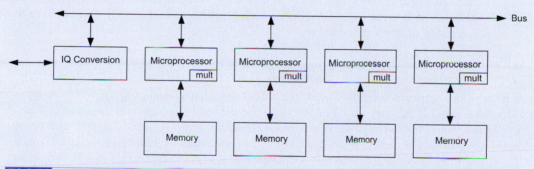

FIG 8.11 Enhanced multiprocessor for software radio

If the multiplication is a one-cycle operation, the throughput for multiplication-intensive operations can improve by a factor of up to M as compared to an M-bit processor with an iterative multiplication operation. This style of acceleration can be repeated for any operation that is computationally intensive. The application code is profiled, timing bottlenecks are identified, and custom hardware is added with appropriate instructions to access the hardware. In this manner the overall solution remains programmable while the speed of processing increases markedly. Tensilica sells extensible processors using such an approach. However, adding functional units increases die size and power dissipation, so tradeoffs are necessary.

8.2.4 Modularity

The tenet of modularity states that modules have well-defined functions and interfaces. If modules are "well-formed," the interaction with other modules can be well characterized. The notion of "well-formed" may differ from situation to situation, but a good starting point is the criteria placed on a "well-formed" software subroutine. First of all, a clearly defined interface is required. In the case of software, this is an argument list with typed variables. In the IC case, this corresponds to a clearly defined behavioral, structural, and physical interface that indicates the function as well as the name, signal type, and electrical and timing constraints of the ports on the design. Reasonable load capacitance and drive capability should be required for I/O ports. Too large a fan-in or too small a drive capability can lead to unexpected timing problems that take effort to solve, where we are trying to minimize effort. For noise immunity and predictable timing, inputs should only drive transistor gates, not diffusion terminals. The physical interface specification includes such attributes as position, connection layer, and wire width. In common with HDL descriptions, we usually classify ports as inputs, outputs, bidirectional, power, or ground. In addition, we would note whether a port is analog or digital. Modularity helps the designer clarify and document an approach to a problem, and also allows a design system to more easily check the attributes of a module as it is constructed (i.e., that outputs are not shorted to each other). The ability to divide the task into a set of well-defined modules also aids in System-On-Chip (SOC) designs where a number of IP sources have to be interfaced to complete a design.

Example

The low-pass filter in our software radio needs a large number of registers. Assuming that we wanted to implement a long filter (say 128 taps) with a 12-bit word, 3072 flip-flops are required for the I and Q filters. These can contribute significantly to the power of a portable communications IC, so it is tempting to prune transistors from the flip-flops. Figure 8.12 illustrates static and dynamic latches. The dynamic latch is much smaller but is sensitive to noise on the input and V_{DD}, suffers from the effects of leakage, and must be continuously clocked to preserve its state. The choice between the two options would depend to some extent on the design style that was being used. If the filter was being designed as a custom layout with known placement and hence wiring loads, the dynamic latch could be used. However, if the design was being automatically placed and routed, the static latch is the only prudent choice. For reasons discussed in Section 6.3, module inputs should only drive the gate terminals of MOS transistors used in a standard-cell place and route scenario. Usually, "discretion is the better part of valor" in these situations.

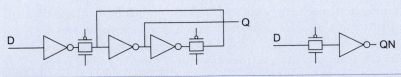

FIG 8.12 Cell level modularity example

Example

Another form of modularity, which aids in debugging and testability, is to ensure that all registers in a design are writable and readable. Figure 8.13 shows this principle applied to the NCO circuit. All registers have been made loadable via enables (lda, ldb, etc.) rather than the gated clocks shown in the original figure. Tristate buffers have been added so that the registers can also be read. Input and output registers have been added to the sine ROM to provide a temporally modular design. All modules should be registered on both their inputs and outputs. In this manner we know the input arrival time is a *D*-to-*clk* setup before the clock edge and the output timing is a *clk*-to-*Q* delay after the clock edge. The test task is identical for each register save for the number of bits that have to be written or read and the address of the register. In fact, by taking this approach, a table of register name, address, and bit width can be compiled and the HDL code to control the registers, documentation, and test programs automatically generated. With a more *ad hoc* approach, these productivity aids are not available.

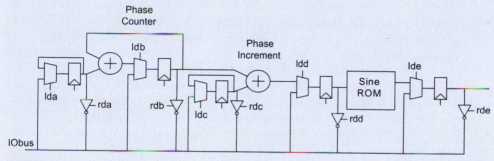

FIG 8.13 Modular test added to NCO

Example

In Figure 8.13, the phase-to-amplitude conversion is achieved using a sine ROM. The number of phase points determines the number of words in the ROM, and the desired precision determines the width of each word. A 7-bit ROM with an 8 bit-phase address will be 256 words long. Improving the phase precision involves increasing the length of the ROM. One trick to reduce the ROM length by a factor of four is to use the scheme shown in Figure 8.14. Only the first quadrant of the sine wave needs to be stored in the ROM. The sign bit (MSB) is used to conditionally invert the sine wave and the next bit in significance (*Phase[n-1]*) is used to count up or down through the ROM. Thus, a 256-word phase-to-amplitude table only needs 64 entries in the ROM.

While the sine lookup table is adequate for small ROMs, large ROMs can be slow and sometimes difficult to work into a standard cell design flow. An alternate structure that can

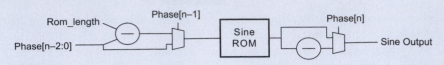

FIG 8.14 Sine ROM optimization

be used to generate a sine is called a *CORDIC processor*. The structure of the CORDIC processor is shown in Figure 8.15 [Volder59, Timmermann94, Gielis91, Grayver98, Sharma00]. It can be used to iteratively compute the functions:

$$x = x_0 \cos z_0 - y_0 \sin z_0$$
$$y = y_0 \cos z_0 + x_0 \sin z_0$$

(8.2)

Figure 8.15(a) shows a single CORDIC stage, which is composed of a parallel *m*-bit adder and two parallel *m*-bit adder/subtracters (an *m*-bit adder/subtracter is made up of an *m*-bit adder and *m* XOR gates, as described in Section 10.2.3.1). In addition, each stage contains two fixed shifters and a constant. The output of each stage is registered with three *m*-bit registers.

   k stages are cascaded to perform the computation, as shown in Figure 8.15(b). If $x_0 = 1$, $y_0 = 0$, and $z_0 = \phi$, the structure will iteratively calculate $x_k \approx \cos \phi$, $y_k \approx \sin \phi$.[1] The right shift amount is equal to the iteration number (*n*), as shown in the figure. The constant is an approximation to the arctangent of an angle that decreases as the iteration index increases. The sign of the constant and the choice of addition or subtraction at stage *n* depends on the sign of z_{n-1}. See [Timmermann94] for more details. The amplitude precision is determined by the size of the adders and the phase precision is determined by the number of stages. The CORDIC processor can also compute square root, sinh, and cosh, and can perform IQ upconversion.

   The objective of showing the CORDIC structure is to illustrate a possible improvement in regularity and modularity over the ROM implementation. The CORDIC only uses adders/subtracters and registers composed of regular logic gates and flip-flops, as opposed to a ROM structure. Moreover, the CORDIC is modular in three ways. First, it scales by simply extending the pipeline length or the bit width of the adders and registers. Second, it can provide a general class of polar functions, which might be of more utility in a situation where the maximum flexibility is required (i.e., as a coprocessor to a general-purpose microprocessor [Timmermann94]). If the cost is deemed satisfactory, the CORDIC provides a degree of future-proofing. Finally, it may be possible to eliminate the multipliers in the IQ

---

[1]Strictly speaking, CORDIC introduces a gain of approximately 1.647. To compensate, we can use $x_0 = 1/1.647$.

downconversion block by using an appropriately dimensioned CORDIC. Possible disadvantages include the pipeline delay through the CORDIC and perhaps power dissipation.

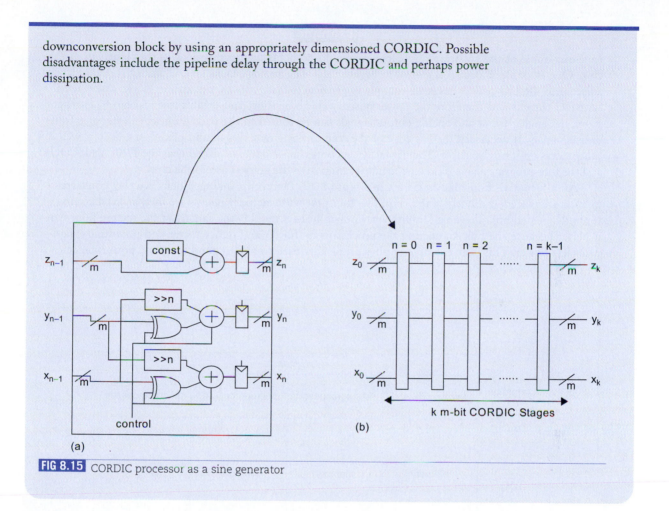

FIG 8.15 CORDIC processor as a sine generator

8.2.5 Locality

By defining well-characterized interfaces for a module, we are effectively stating that other than the specified external interfaces, the internals of the module are unimportant to other modules. In this way we are performing a form of "information hiding" that reduces the apparent complexity of the module. In the software and HDL world, this is paralleled by a reduction of global variables to a minimum (hopefully to zero). Increasingly, locality often means temporal locality or adherence to a clock or timing protocol. This is addressed in Chapter 7, where different clocking strategies are examined. One of the central themes of temporal locality is to reference all signals to a clock. Thus, input signals are specified with required setup and hold times relative to the clock, and outputs have delays related to the edges of the clock.

Example

In the example of the software radio, locality would probably be most evident in the floorplan of the chip. One example floorplan is shown in Figure 8.16. The analog blocks (ADC and DAC) are placed adjacent to the I/O pads. This is an example of physical locality because the analog blocks draw significant DC current and therefore the power busses have to be short and exhibit low resistance. Furthermore, the analog input and analog output signals can be routed to the pads without interference from digital signals. If necessary, the left edge of the chip can be guard-ringed and placed in a deep n-well if this process option is available. The digital IQ upconversion module is placed near the DAC and ADC, and the four programmable processor/memory composites are arrayed across the chip.

An alternative floorplan is shown in Figure 8.17. Here, the analog blocks and IQ conversion module are placed at the top of the chip. The four processor/memory blocks are then arrayed around a centrally located bus. The area for both array possibilities is roughly the same, but the second floorplan is better because the bus connecting the processors is shorter and hence faster and potentially dissipates less power. This is an example of physical locality used to obtain good temporal performance.

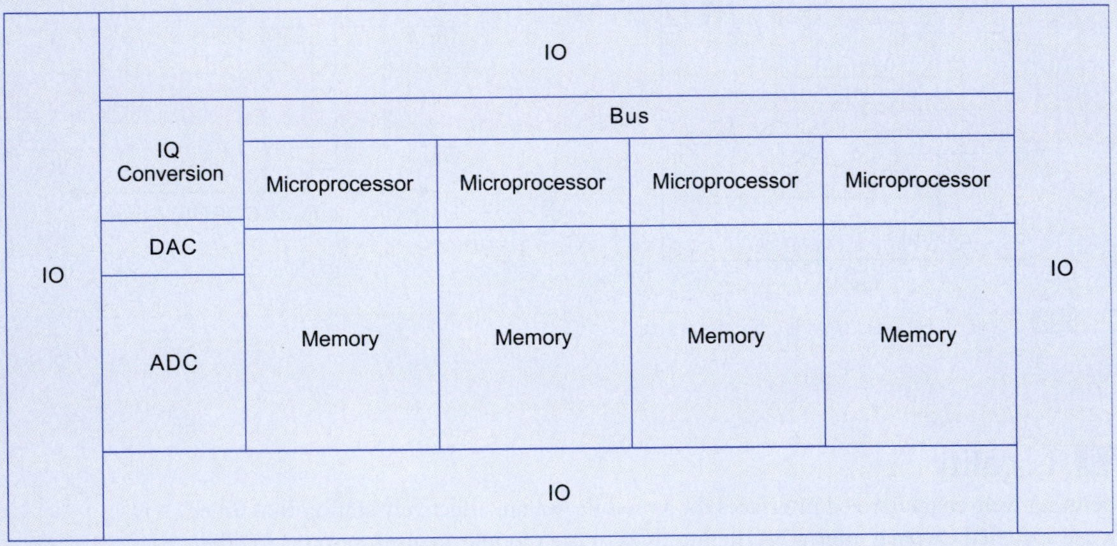

IO						
IO	IQ Conversion	Bus				IO
		Microprocessor	Microprocessor	Microprocessor	Microprocessor	
	DAC	Memory	Memory	Memory	Memory	
	ADC					
IO						

FIG 8.16 One possible floorplan for the software radio

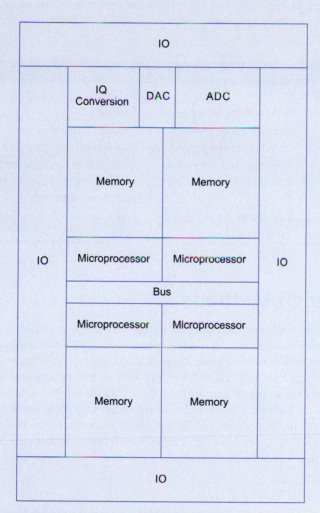

FIG 8.17 Alternate floorplan for software radio

8.2.6 Summary

There are strong parallels between the methods of design for software and hardware systems. Table 8.1 summarizes some of these parallels for the principles outlined above.

Table 8.1	Structured software and VLSI hardware design	
Design Principle	**Software**	**Hardware**
Hierarchy	Subroutines, libraries	Modules
Regularity	Iteration, code sharing, object-oriented procedures	Datapaths, module reuse, regular arrays, gate arrays, standard cells
Modularity	Well-defined subroutine interfaces	Well-defined module interfaces, timing and loading data for modules, registered inputs and outputs
Locality	Local scoping, no global variables	Local connections through floorplanning

8.3 Design Methods

In this section, we will examine a range of design methods that can be used to implement a CMOS system. This section will concentrate on the target of the design method, in contrast to the design flow used to build a chip. Design flows, which deal with how a design progresses through a set of tools, will be dealt with in the subsequent section. The base design methods are arranged roughly in order of "increased investment," which loosely relates to the time and cost it takes to design and implement the system. It is important to understand the costs, capabilities, and limitations of a given implementation technology to select the right solution. For instance, it is futile to design a custom chip when an off-the-shelf solution that meets the system criteria is available for the same or lower cost.

8.3.1 Microprocessor/DSP

Many times, the most practical method to solve a system design problem is to use a standard microprocessor or digital signal processor (DSP). There are many single-chip microprocessors with built-in RAM and EEROM/EPROM available in the market. For example, the PIC family of processors from Microchip offers a wide range of clock speeds, memory sizes, and analog I/O capability (ADCs) in a small package. For more signal-intensive problems, classical DSPs from vendors such as Analog Devices and Texas Instruments can be employed. Microprocessors provide great flexibility because systems can be upgraded in the field through software patches. Do not underestimate the cost of software development for microprocessor-based systems.

Even when you decide to build a system with an off-the-shelf microprocessor, you should consider the possibility of eventual integration. For example, if your product becomes very successful and you want to reduce costs by integrating it into a single system-on-chip rather than building it as a board with a microprocessor and various support chips, you will need a microprocessor that is available in embedded form so that you can keep your software. Examples of embedded commercial processor cores include ARM, MIPS, and IBM's PowerPC. The (Sun Microsystems) SPARC V8 architecture is also available in VHDL for research or commercial use from Gaisler Research.

8.3.2 Programmable Logic

Often, the cost, speed, or power dissipation of a microprocessor may not meet system goals and an alternative solution is required. A variety of programmable chips are available that can be more efficient than general purpose microprocessors yet faster to develop than dedicated chips:

- Chips with programmable logic arrays
- Chips with programmable interconnect
- Chips with reprogrammable logic and interconnect

The system designer should be familiar with these options for two reasons:

- First, it allows the designer to competently assess a particular system requirement for an IC and recommend a solution, given the system complexity, the speed of operation, cost goals, time-to-market goals, and any other top-level concerns.

- Second, it familiarizes the IC designer with methods of making any chip reprogrammable at the hardware level and hence both more useful and of wider spread use.

8.3.2.1 Programmable Logic Devices The devices covered in this section are descended from chips that implement two-level sum-of-product programmable logic arrays (PLAs) discussed in Section 11.7. They differ from the field-programmable gate arrays described in the next section in that they have limited routing capability. Historically, process densities did not allow the transistor count and routing resources found in modern field-programmable gate arrays. Programmable logic devices based on PLAs allowed a useful product to be fielded and well-established techniques allowed logic optimization to target PLA structures, so the associated CAD tools were relatively simple. They are still occasionally used because the regular array and interconnect make timing very predictable.

A PLA consists of an AND plane and an OR plane to compute any function expressed as a sum of products. Each transistor in the AND and OR plane must be capable of being programmed to be present or not. This can be achieved by fully populating the AND and OR plane with a NOR structure at each PLA location. Each node is programmed with a floating-gate transistor (see Section 3.4.3), a fusible link, or a RAM-controlled transistor, as illustrated in Figure 8.18. The first two versions were the way these types of devices were programmed when device densities were low. These devices are generally used for low-complexity problems (a few hundred gates), which require fairly high speed (100's MHz).

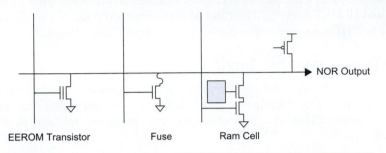

FIG 8.18 PLA NOR structure (one plane shown)

8.3.2.2 Field-programmable Gate Arrays (FPGAs)

Field-Programmable Gate Arrays (FPGAs) use the high circuit densities in modern processes to construct ICs that, as their name suggests, are completely programmable even after a product is shipped or "in the field." Two basic versions exist. The first uses a special process option such as a fuse or antifuse to permanently program interconnect and personalize logic. These are one-time programmable. The second type uses small static RAM cells to customize routing and logic functions. In general, an FPGA chip consists of an array of logic cells surrounded by programmable routing resources.

As an example of the first type of FPGA, devices manufactured by Actel embed an array of logic modules within an interconnect matrix that is formed on the top metal layers. Successive routing channels run vertically or horizontally, as shown in Figure 8.19. At the intersection of routing traces is placed a special one-time programmable contact called an *antifuse*. These normally have high resistance (effectively an open circuit). Upon application of a special programming voltage across the contact, the resistance permanently drops to a few ohms. CMOS switches allow the programming voltage to be directed to any antifuse in the chip. The advantage of this type of routing is that the size of the programmable interconnect is tiny—the intersection area of two metal traces. Moreover, the on-resistance is low compared to a CMOS switch, so the circuit speed is not compromised. The disadvantage is that the interconnect is not reprogrammable, so once a chip is programmed, its function is fixed to the extent that the interconnect has been personalized.

The floorplan of a typical Actel chip is shown in Figure 8.20. An array of logic modules (LM) is placed on the chip with the routing passing adjacent to the cells. Routes extend into and out of the logic modules and intersect vertical or horizontal routing channels. The cell is personalized in terms of functionality and connectivity (to other cells) by

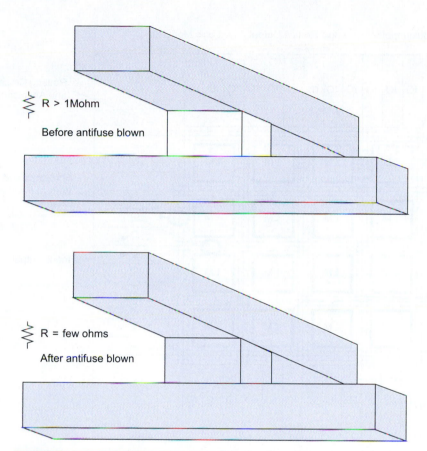

R > 1Mohm

Before antifuse blown

R = few ohms

After antifuse blown

FIG 8.19 Actel programmable interconnect

programming these crossing interconnects. The logic module consists of a combinational block, which can compute more than 4000 combinational functions of five primary inputs. A flip-flop cell provides for asynchronous set and reset, active low enable, and programmable clock polarity. With typical utilization, effective gate count (equivalent two input gates) can range from 82K to over 1 million gates.

This type of array would be suitable for reasonably complex, high-speed (100's MHz) applications where cost and power constraints were moderate. Antifuses are only available in specialized CMOS processes.

Figure 8.21 shows the floorplan of a simplified RAM-based FPGA. The chip is composed of an array of *configurable logic blocks* (CLBs). Similar to the one-time programmable devices, metal routing tracks run vertically and horizontally between the array of CLBs. These terminate at the gray blocks, which are routing switches that can be implemented using CMOS transmission gates or tristate buffers. The routing resources can also be con-

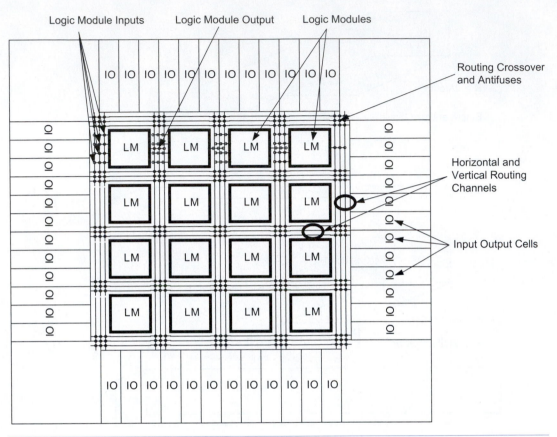

FIG 8.20 Representative Actel FPGA floorplan

nected to the inputs and outputs of the adjacent CLBs. CLBs use lookup tables to compute any function of several variables. The contents of the lookup table are stored in static RAM so the FPGA is programmed by reading the SRAM at startup from an external data source such as a serial EEPROM. Similarly, static RAM cells program the routing switches. Configurable I/O cells that can be used as input, output, or bidirectional pads surround the core array of CLBs.

A simple FPGA logic cell is shown in Figure 8.22. It is composed of a 16×1 static RAM as the logic element. This provides for any logic function of four variables merely by loading the RAM with the appropriate contents. Table 8.2 illustrates how the table should be loaded to perform various logic functions. A full adder can be implemented in two CLBs (one for carry and one for sum). The CLB shown also provides an optional output

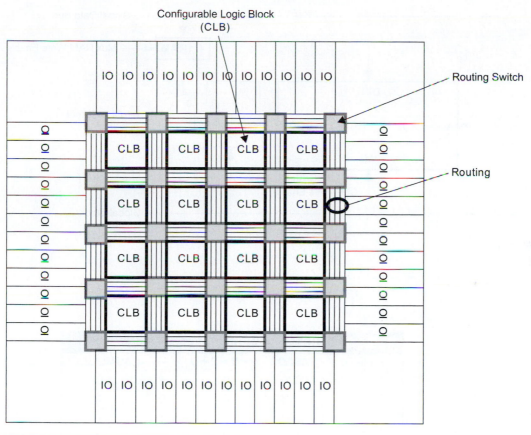

FIG 8.21 Simplified FPGA floorplan

register. While it may seem inefficient or slow to use a RAM to perform logic, specially designed single-data line RAMs are small and fast in current processes, and resources such as the routing tend to dominate modern designs from a density and speed viewpoint. The cell shown in Figure 8.22 is representative of first-generation FPGAs, and many optimizations have been made to improve performance. These include multiple dedicated clock lines, CLBs optimized for fast arithmetic, dedicated blocks of RAM, high-speed I/O interfaces, and even embedded RISC microprocessors. The Xilinx XCVP125 FPGA, state of the art in 2003, combines over 125,000 logic cells (approximately seven million two input gates), four RISC processors, multi-gigabit I/O and over 1000 user-available I/O pads (i.e., XILINX XC2VP125).

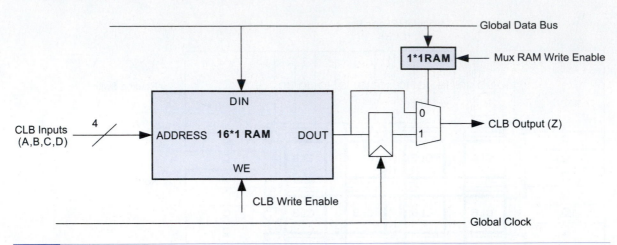

FIG 8.22 Simple FPGA logic cell

Table 8.2	RAM CLB functions			
Address	ABCD	A • B • C • D	~A	SUM(A,B,C)
0	0000	0	1	0
1	1000	0	0	1
2	0100	0	1	1
3	1100	0	0	0
4	0010	0	1	1
5	1010	0	0	0
6	0110	0	1	0
7	1110	0	0	1
8	0001	0	1	0
9	1001	0	0	1
10	0101	0	1	1
11	1101	0	0	0
12	0011	0	1	1
13	1011	0	0	0
14	0111	0	1	0
15	1111	1	0	1

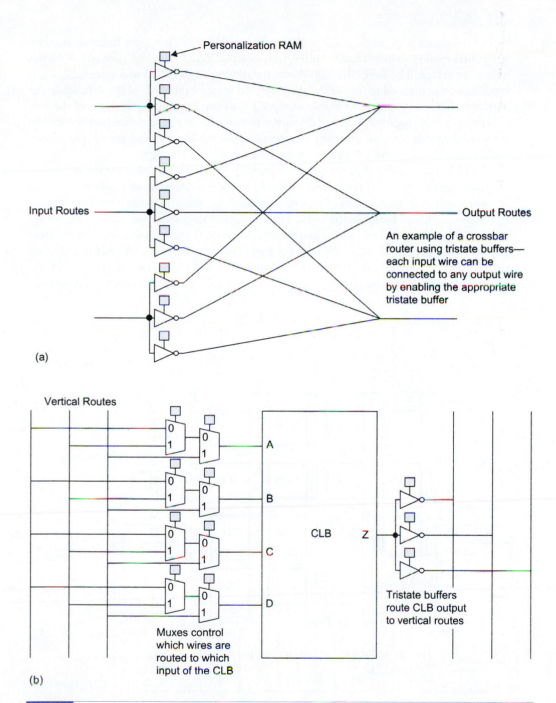

Personalization RAM

Input Routes

Output Routes

An example of a crossbar router using tristate buffers—each input wire can be connected to any output wire by enabling the appropriate tristate buffer

(a)

Vertical Routes

A

B

CLB   Z

C

D

Tristate buffers route CLB output to vertical routes

Muxes control which wires are routed to which input of the CLB

(b)

FIG 8.23 Simplified FPGA routing cell

One approach to routing cells is shown in Figure 8.23. The upper diagram shows a horizontal routing switch. Tristate buffers controlled by RAM bits determine which output routes are driven. The lower diagram shows the routing to the CLB of Figure 8.22. Each input can be connected to any of the three vertical routes to the left of the cell, while the tristate buffers can connect the output of the CLB to any route along the right of the cell.

Putting it all together, Figure 8.24 shows an example of a personalized array with the associated routing. Three inverting input pads enter on the left and are routed to two XOR gates, the output of which is routed out of the chip on the right.

FPGAs have matured to the point where they are the best choice for many low- to medium-volume custom logic applications. The economy of scale of modern processes allows extremely high densities of reprogrammable logic gates to be delivered at low cost to the designer, and hence to the consumer. Compared to conventional ICs, FPGAs are slower and are less efficient in area and power. However, the cost to manufacture is virtually zero (programming cost). The cost per part is moderate to high. As initial manufacturing costs for custom integrated circuits increase and FPGA costs decrease, FPGAs become attractive for more and more low- to mid-volume, moderate power applications.

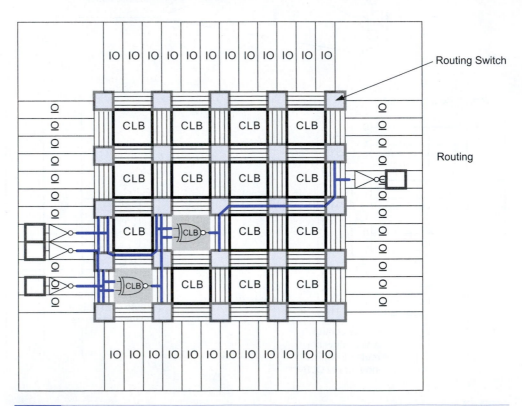

FIG 8.24 Personalized and routed FPGA

Note that (after sorting out the intellectual property rights with the appropriate patent holders) it is possible to implement FPGA blocks on any CMOS chip to provide some degree of programmability at the gate level.

8.3.3   Gate Array and Sea of Gates Design

The chips described in the previous section do not require a fabrication run. Designers typically strive to keep the nonrecurring engineering cost (NRE, see Section 8.5) as low as possible. One method of doing this is to construct a common base array of transistors and personalize the chip by altering the metallization (metal and via masks) that is placed on top of the transistors. This style of chip is called a Gate Array (GA). A particular subclass of a gate array is known as a Sea-of-Gates (SOG) chip. A vendor stocks *master* or *base* wafers that have been processed up to the polysilicon gate layer (i.e., the transistors have been formed). Contacts and metallization are then specified on a per-design basis to complete the chip. The cost can be kept low due to the following factors:

- The wafer cost can be reduced by producing many base wafers for a variety of different chips.

- Only the metallization masks are required to personalize the design; cost and processing time can be kept low for a small number of metal layers.

- Packaging cost is reduced by using standard packages and pinouts.

- Test and production costs are reduced by reusing common test fixtures.

Companies such as LSI Logic have pushed gate array technology to an advanced point by combining gate array structures with dedicated high-speed I/O circuitry. Their RapidChip technology boasts integrated RISC cores combined with nearly 20 million gates, 10 Mbits of RAM memory, and I/O speeds up to 4.25 Gbps. Speeds in the range of 300 MHz are possible with specially designed logic blocks.

Aside from their use as a System on Chip technology, it is worthwhile reviewing GA/SOG techniques because they can also be used on custom chips to provide an area of reprogrammable logic on an otherwise fixed function chip. The basic GA/SOG approach is shown in Figure 8.25. Rows of nMOS and pMOS transistors are arrayed across a chip area. Each logic row consists of an n row and p row. Gate arrays differ from SOG in that the array of transistors is not continuous and transistors can be grouped and perhaps individually sized for economic implementation of specific applications such as memories. Figure 8.26(a) shows an SOG structure, which features continuous rows of transistors. Figure 8.26(b) shows a gate array structure that uses groups of three transistor pairs. Because the SOG structure is a continuous array of transistors, it has to allow for the isolation of a group of transistors. Grounding the gate of the nMOS transistor or connecting the gate of the pMOS transistor to the V_{DD} rail provides this isolation.

Figure 8.27 shows a portion of an SOG structure programmed to be a 3-input NAND gate. Note that the nMOS and pMOS transistors at each end isolate the gate, as described previously. Personalization of this SOG structure commences at contact and

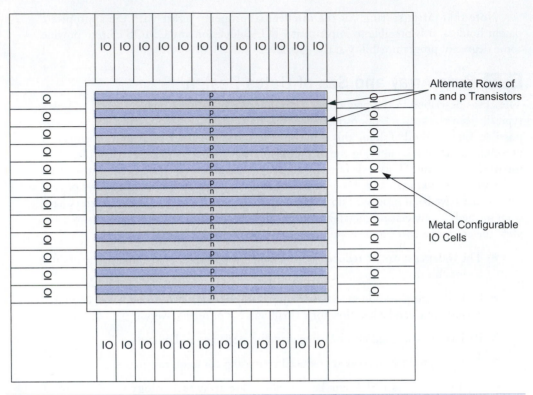

FIG 8.25 Sea of Gates floorplan and transistor layout

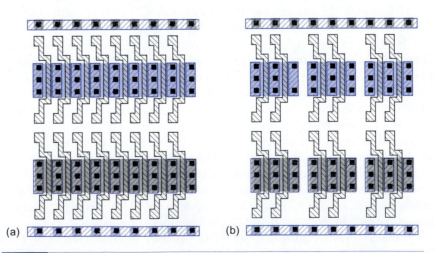

FIG 8.26 SOG and Gate Array cell layouts

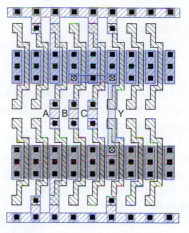

FIG 8.27 SOG programming example for 3-input NAND gate

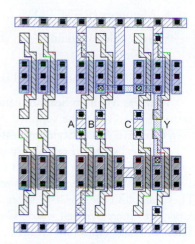

FIG 8.28 Gate Array programming example for 3-input NAND gate

metal1 masks, and can continue up for all metal layers available in the process. Figure 8.28 shows the 3-input NAND on a gate array.

Gate and SOG arrays have lower up-front costs because they require fewer masks. They are also faster to manufacture because only the metallization steps are required. Some vendors offer automatic conversions of field-programmable gate arrays into fixed gate arrays to reduce costs as a product moves from low- to mid-volume manufacturing.

Compared to an FPGA given the same process, a gate array dissipates less power and costs less in volume production. The NRE cost (i.e., cost to fabricate) can be a fraction of the cost of a custom mask set.

8.3.4 Cell-based Design

Cell-based design uses a standard cell library as the basic building blocks of a chip. The cells are placed in appropriate positions, then their interconnections are routed. Cell-based design can deliver smaller, faster, and lower-power chips than gate arrays or programmable logic but has high NRE costs to produce the custom mask set. Therefore, it is only economical for high volume parts or when the performance commands a lucrative sales price. As compared to full-custom design, cell-based design offers much higher productivity because it uses predesigned cells with layouts. Foundries and library vendors supply cells with a wide range of functionality. These include:

- Small-scale integration (SSI) logic (NAND, NOR, XOR, AOI, OAI, inverters, buffers, registers)

- Memories (RAM, ROM, CAM, register files)

- System level modules such as processors, protocol processors, serial interfaces, and bus interfaces

- Possibility of mixed-signal and RF modules

Whereas Medium Scale Integration (MSI) functions such as adders, multipliers, and parity blocks used to be supplied as cells, synthesis engines commonly construct these from base-level Small Scale Integration (SSI) gates in current design systems.

A typical standard cell library is shown in Table 8.3. A 1X (normal power) cell commonly is defined to use the widest transistors that fit within the vertical pitch of the standard cell. 2X and larger (high power) cells use wider transistors to deliver more current. They must fold the transistors (see Section 8.8.3) to fit within the cell; this comes at the expense of increased cell width. Gates are often available in low power versions as well. These cells use minimum-width transistors to reduce capacitance. Low-power cells tend to be slow because of the wire capacitance they must drive. Although they do not save area, they do reduce power consumption on noncritical paths.

Sophisticated libraries also generate memories of assorted sizes from a graphical user interface. The generators yield not only the physical layout but also a complete data sheet indicating access times, cycle times, and power dissipation.

In the event that a standard cell library may not be available for a process, it is worthwhile to review some of the approaches to standard cell design. Usually, standard cells are a fixed height with power and ground routed respectively at the top and bottom of the cells, as was shown in Figure 1.62. This allows the cells to be abutted end to end and to have the supply rails connect. A single row of nMOS transistors adjacent to GND (ground) and a single row of pMOS transistors adjacent to V_{DD} (power) are normally used. The polysilicon gate is connected from nMOS transistor to pMOS transistor and, in the case of multiplexers and registers, the polysilicon connection has to be crossed between vertically coincident nMOS and pMOS transistors. Decisions about the sizes of transistors have to be made. Following this decision, the cells are almost completely defined by the process design rules. Figure 8.29 illustrates this point. The height of the cell is defined by the sum of the nMOS and pMOS transistor widths, the separation on n and p regions, the spacing to V_{DD} and GND busses, and the width of these busses. The horizontal pitch is defined by the poly-to-metal2 contacted pitch, as shown in the figure. It is relatively easy to construct a software program to automatically generate cells like those shown in Figure 8.29. Cell delay is characterized through simulation to good agreement with silicon, as was discussed in Section 5.5.3. Fabrication of such cells to prove performance is rarely required. Options to standard cells include routing the clock with the power and ground busses and routing multiple supply voltages to each cell. The latter technique is sometimes used to reduce power by connecting gates that are not in the critical path to a lower than normal supply voltage. Recall that the power drops with the square of the supply voltage.

Table 8.3	Typical standard cell library	
Gate Type	**Variations**	**Options**
Inverter / buffer / tristate buffers		Wide range of power options, 1X, 2X, 4X, 8X, 16X, 32X, 64X minimum size inverter
NAND / AND	2–8 inputs	High, normal, low power
NOR / OR	2–8 inputs	High, normal, low power
XOR / XNOR		High, normal, low power
AOI / OAI		High, normal, low power
Multiplexers	Inverting/noninverting	High, normal, low power
Schmitt trigger		High, normal, low power
Adder / half adder		High, normal, low power
Latches		High, normal, low power
Flip-flops	D, with and without synch/asych set and reset, scan	High, normal, low power
I/O pads	Input, output, tristate, bidirectional, boundary scan, slew rate limited, crystal oscillator	Various drive levels (1–16 mA) and logic levels

8.3.5   Full Custom Design

A number of techniques can be used to design standard cells or larger circuit blocks at the mask level. The oldest and most traditional technique is termed *custom mask layout,* in which a designer sits in front of a graphics display running an interactive editor and pieces designs together at the geometry level one rectangle at a time. This work is sometimes called polygon pushing. Historically, when there were no interactive graphics displays or editors (yes, there was a time when this was true!!), designers manually cut masks from a material called Rubylith. Sheets of the material (a red strippable layer on an otherwise clear Mylar plastic background) were used directly to produce photomasks for chips. Mask design under such conditions was laborious and prone to error because there were no tools to check the designs other than human eyes.

A variation of custom mask design is called *symbolic layout*. Rather than dealing with rectangles and polygons on various mask levels, the primitives are transistors, contacts, wires, and ports (points of connection). These primitives can also be manipulated by a graphics editor. Some systems allow for a "design rule free" placement of symbolic entities. The actual placement occurs after a spacing process that compacts each primitive as close to its neighbor as possible according to the design rules of the process in use. By using a symbolic layout system, layout topologies can be transported from process to process without a huge amount of effort.

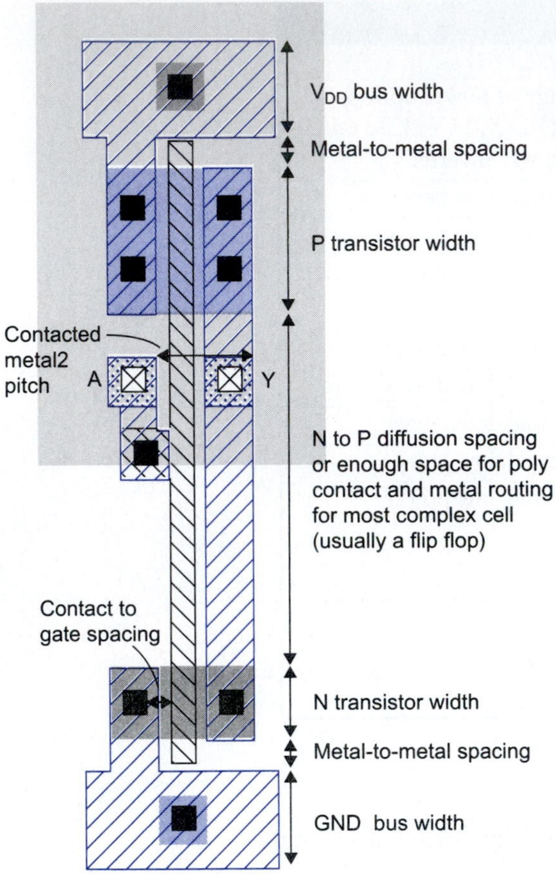

FIG 8.29 Typical standard cell layout with some of the constraints

In these times of cell-based design, digital CMOS ICs use custom mask design only for the highest of volume parts such as microprocessor datapaths. However, analog and RF designs, cell libraries, memories, and I/O cells still frequently use custom design. There are a variety of custom MOS layout hints in Section 8.8. Custom design is also worthwhile pedagogically because it completes the link from transistors to systems.

From time to time, we have mentioned software generators as a method of generating physical layout. This kind of idea has been around for a long time and was often referred to as *silicon compilation*. Complete microprocessors were typical of layouts that were generated. A "correct by construction" method was used to build the layouts hierarchically. In other words, only the mask description was generated, with perhaps a high-level instruction level simulator being the behavioral model. Generators are the most common method used today for library generation.

With modern design flows, many different "views" of a design are required to integrate with the regular path through the design system. For instance, in addition to the behavioral model, a timing view would be needed for timing verification, a logic view might be required for simulation, and a circuit view for layout versus schematic or netlist comparisons would be needed. Software generators can be used to provide all of these views automatically.

Modern versions of the venerable "silicon compiler" can be built in a structured hierarchical manner to generate memories, register files, and other special-purpose structures that can benefit from a customized layout. One of the most straightforward approaches is to write custom placement routines that in essence "hand place" certain standard cells within the row structure of a standard cell design. For instance, you may prefer a certain adder design and have a datapath layout for the adder. An algorithm can be written to place the cells on the standard cell grid. In addition, a linked algorithm can be written to generate a gate netlist in an HDL. In this way, both the physical and structural design are captured. The behavior can be represented by an HDL function or module call. Such custom placement can shorten wire lengths and thus improve speed and power.

To illustrate a memory generator, consider the sine ROM required for the NCO in the software radio. The ROM contents are available as a file with a list of numbers and we want to generate a layout and the Verilog. Generating the Verilog is straightforward and would proceed according to the following pseudo-code (loosely based on the C language):

```
function generate_ROM_verilog(ROM_length, ROM_width, ROM_filename);
    verilog_file_pointer = open("sinerom.v");
    generate_verilog_head(verilog_file_pointer, ROM_length, ROM_width);
    ROM_file_pointer = open(ROM_filename);
    index = 0;
    loop for index below 2<<ROM_length /* rom_length=6 -> 64 */
        ROM_entry = get_line_from_file(file_pointer);
        fprintf(verilog_file_pointer, " %d: q = %d'h%h",
                Index, ROM_width, ROM_entry);
        index++;
        file_pointer++;
    end_loop
    generate_verilog_tail(verilog_file_pointer);
end_function;

function generate_verilog_head(file_pointer, length, width);
    fprintf(file_pointer, "module quarter_wave(");
    fprintf(file_pointer, " input [%d:0] addr,", length-1);
    fprintf(file_pointer, " output reg [%d:0] q;)", width-1);
    fprintf(file_pointer, " always @(addr)");
    fprintf(file_pointer, " begin");
    fprintf(file_pointer, " case(addr)");
end_function;

function generate_verilog_tail(file_pointer);
    fprintf(file_pointer, " endcase");
    fprintf(file_pointer, " end");
    fprintf(file_pointer, "endmodule");
end_function;
```

The function is quite easy to follow. It first writes out the Verilog file header structure. This includes the module name, input/output ports, and the bit widths of the ports. These are specified to the function from the calling function. The program then sequences through the ROM file contents and formats each line of the case statement. Finally, the tail of the Verilog file is generated. The following code segment illustrates the output.

```
module quarter_wave(input      [5:0] addr,
                    output reg [6:0] q);
    always @(addr)
    begin
        case(addr)
            0: q = 7'h00;
            1: q = 7'h03;
            2: q = 7'h06;
            ...
            62: q = 7'h7f;
            63: q = 7'h7f;
        endcase
    end
endmodule
```

While this function is basic, the sophistication can be built up over time. There are a number of academic and commercial projects that have been aimed at HDL generation such as Verilog++ and Verilog2C++.

For the generation of the physical layout, we will concentrate on the generation of the ROM array. Details on ROM design may be found in Section 11.4. The main program is shown below. The program assumes that layout cells or tiles connect by abutment, i.e., the routing is implicit in the tile.

```
function generate_rom_layout(ROM_length, ROM_width, ROM_filename);
   lfp = open("sinerom.lay"); /*layout file pointer*/
   generate_rom(lfp, 0, 0, ROM_length, ROM_width, ROM_filename);
   x = length_of_rom_cell * ROM_length;
   generate_rom_right(lfp, x, ROM_length, ROM_width, ROM_filename);
endfunction
```

The notion of *tiling* is introduced in this code. The function `generate_rom` tiles two rectangular blocks horizontally. The first is the ROM array and row decoder. Then the `rom_right` is placed to the right of the array. This contains the ROM sense amplifier, which in this case is shown as a pMOS pull-up and buffer inverter for the pseudo-nMOS NOR structure used in the ROM. The bottom right of the `rom_right` structure can contain a column decoder, but it is not necessary in this design. The address buffers for the row decoder are placed in this corner. This physical structure is shown in Figure 8.30.

```
function generate_rom(lfp, xorigin, yorigin, length, width);
   generate_row_decoder(lfp, xorigin, yorigin, length, width);
   y = yorigin + row_decoder_height;
   generate_rom_array(lfp, xorigin, y, length, width);
end_function
```

The function `generate_rom` tiles two layout blocks vertically. The lower block is the ROM row decoder and the upper block is the ROM array proper. The row decoder gate is shown to the right of the floorplan. In this design, a pseudo-nMOS NOR gate is used for compactness and simplicity. This is shown in Figure 8.31.

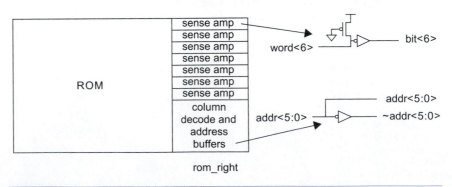

FIG 8.30 ROM layout—phase 1

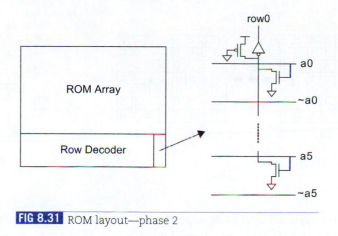

FIG 8.31 ROM layout—phase 2

Finally, the function `generate_rom_array` is invoked to build the ROM array itself.

```
function generate_rom_array(lfp, xorigin, yorigin, ROM_length, ROM_width, ROM_filename);
    ROM_file_pointer = open(ROM_filename);
    index = 0;
    x = xorigin;
    loop for index below 2<<ROM_length /* rom_length=6 -> 64 */
        ROM_entry = get_line_from_file(file_pointer);
        bit_index = 0;
        y = yorigin;
        loop for bit_index below ROM_width
            bit = get_bit_from_entry(bit_index, ROM_entry);
            if (bit) ROM_one(lfp, x, y);
            else ROM_zero(lfp, x, y);
            y = y + yorigin;
        end_loop
        x = x + rom_bit_width;
        index++;
        file_pointer++;
    end_loop
end_function;
```

This function tiles in both *x* and *y* directions and builds a ROM array as shown in Figure 8.32.

This builds a long, skinny ROM. For the sinc ROM example, the array would be 64 cells long by 7 cells high. The loop variables can be changed in the function to generate, for example, four ROM words vertically, which would result in a structure that was 16 cells long by 28 cells high. A column decoder would be added to the ROM sense amplifier in the `rom_right` block to select one bit out of each group of four.

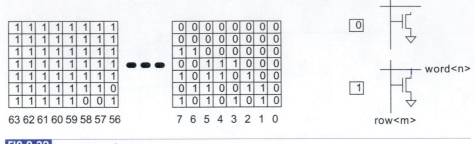

FIG 8.32 ROM array layout

Datapaths are another class of structure that can benefit from a software generator. The following code is representative of what might be used to place the cells in a CORDIC processor. Routing may be able to be incorporated into the tiled cells, or specific routing cells may have to be generated in addition to the pseudo-code shown below.

```
function generate_cordic_layout(bit_width, length);
    lfp = open("cordic.lay"); /*layout file pointer*/
    index = 0;
    x = 0;
    loop for index below length;
        generate_cordic_slice(lfp, x, width, index);
        index++;
        x = x + cordic_slice_width;
    end_loop
end_function
```

The function above arrays the CORDIC slices horizontally. This is shown in Figure 8.33.

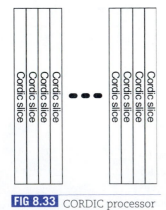

FIG 8.33 CORDIC processor

```
function generate_cordic_slice(lfp, x, width, index);
    y = 0;
    xleft = x;
    generate_cordic_shift(lfp, x, y, width, index);
    generate_cordic_constant(lfp, x, y + 2 * width * adder_height,
        width,index);
    x = x + shift_width;
    generate_cordic_adder(lfp, x, y, width,0); /* x slice */
    y = y + adder_height * width;
    generate_cordic_adder(lfp, x, y, width,0); /* y slice */
    y = y + adder_height * width;
    generate_cordic_adder(lfp, x, y, width,1); /* z slice */
    y = y + adder_height * width;
    generate_cordic_control(lfp, xleft, y); /* control on top */
end_function
```

The `generate_cordic_slice` function adds a shifter for the X and Y paths (`generate_cordic_shift`) and a constant for the Z path (`generate_cordic_constant`) and then builds the X and Y adder/subtracter and the Z adder vertically (`generate_cordic_adder`). Each is `width` bits wide. The control block resides on top of the three datapaths. This is shown in Figure 8.34. The shifter block would consist of wiring that would perform an n-bit shift and XY swizzle according to the horizontal index. Similarly, the constant block would generate a constant dependent on the horizontal index.

The next function generates the stage adders. This is shown in Figure 8.35.

```
function generate_cordic_adder(lfp, x, y, width, flag);
    index = 0;
    y = 0;
    loop for yindex below width
        generate_xyz_bit(lfp, x, y, flag);
        yindex++;
        y = y + adder_bit_height;
    end_loop
end_function
```

The X and Y adder bits have a multiplexer, XOR, ripple carry adder, and flip-flop, while the Z bits have a blank (null) cell, an adder, and a flip-flop. These bits are shown in Figure 8.36(a) and 8.36(b) respectively.

```
function generate_xyz_bit(lfp, x, y, flag);
    if (flag) place_null_cell(lfp, x, y);
    else place_xor(lfp, x, y);
    x = x + xor_width;
    place_adder(lfp, x, y);
    x = x + adder_width;
    place_flipflop(lfp, x, y);
end_function
```

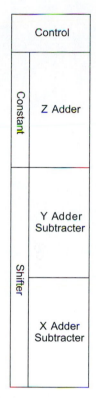

FIG 8.34 CORDIC processor slice

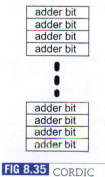

FIG 8.35 CORDIC XYZ adders

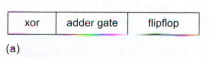

(a)

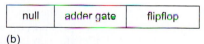

(b)

FIG 8.36 CORDIC XYZ adder bits

In the previous examples, the x and y locations for cell placement have been calculated using "magic" global dimensions (i.e., `adder_width` in the last code example). It is quite possible to build an automated tiling engine that just takes alternating ordered lists of cells and abuts them in the x or y direction, incrementing the origin of the cell to be placed by the dimensions of the previous cell. In this way, no global numbers need to be used in the code generation and the code is generic for a wide range of processes. The other detail omitted from the preceding examples is the generation of signal pins. These are usually needed eventually for layout routers to connect to the generated modules and for comparison with structural views.

8.3.6 Platform-based Design—System on a Chip

As systems have become more complex, the use of predefined intellectual property (IP) blocks has become commonplace. Designs frequently use a number of common blocks such as RISC processors, memory, and I/O functions attached to common busses. A *platform* can be used to implement a design by using common structures such as busses and common high-level languages (such as C) to program the processors. To a large extent, the RISC processor and memories can be interchanged and the number and type of peripherals can be changed while maintaining good design and verification times because the modules have been predesigned and the test and verification scripts come with the IP blocks. The design task is to put the blocks together, design any application-specific blocks, and place and route a correctly operational chip. Note that the last step, while automated, still takes considerable engineering effort.

As many current chips feature one or more embedded microprocessors, the task of writing software is added to the task of designing logic. Moreover, platform-based design poses the problem of partitioning the complete solution between hardware (HDL, gates) and software (programmed on the processor/s). This tends to remain a somewhat manual task, but is increasingly automated by CAD tools.

Platform-based systems typically consist of a basic RISC processor, which can be extended with multipliers, floating point units, or specialized DSP units. In addition (e.g., in Tensilica's Xtensa system), by profiling the executable code, special hardware can be added that corresponds to hardware-assisted instructions, which are introduced into the instruction set. In theory, additional hardware or extra processors can deal with a wide range of computational loads.

Manual techniques for hardware-software codesign mirror this approach. That is, the design begins with a software simulation (ideally on the embedded processor). Timing estimates are gathered, and manual decisions about what to commit to hardware are made. Special simulators to deal with embedded processors and logic have been developed.

With platform-based design, we have in essence come full circle from the first design method suggested: programming a microprocessor. This is the reason processor selection is important when starting out on a product design that may eventually be integrated. As the software effort will often exceed the hardware effort, you don't want to repeat that effort.

8.3.7 Summary

In this section, we have summarized a wide range of CMOS design options ranging from a software-based microprocessor to full custom design. The following table summarizes these options in terms of a variety of criteria. Each category is ranked in relation to each design method from low to high.

Table 8.4	Comparison of CMOS design methods						
Design Method	Non-recurring Engineering	Unit Cost	Power Dissipation	Complexity of Imple-mentation	Time to Market	Perfor-mance	Flexibility
Microprocessor/ DSP	low	medium	high	low	low	low	high
PLA	low	medium	medium	low	low	medium	low
FPGA	low	high	medium	medium	low	medium	high
Gate Array/SOG	medium	medium	low	medium	medium	medium	medium
Cell Based	high	low	low	high	high	high	low
Custom Design	high	low	low	high	high	very high	low
Platform Based	high	low	low	high	high	high	medium

In the next section, we review the design flows necessary to build a complete chip. While decisions can be difficult, the following diagram (Figure 8.37) should offer some assistance.

The diagram (Figure 8.37) indicates that the most cost-effective approach should be taken to hardware (or software) design given speed, power, and cost targets (occasionally, size will count as well). You should always use an off-the-shelf solution if system constraints are met, because the non-recurring engineering (NRE) costs are amortized over many units. The next most likely prospect is an FPGA design, especially for low-volume (100,000's) applications. Power and cost are the most likely attributes to be challenged in medium- to high-volume applications, and this is where standard cell designs will be used. Mixed-signal, RF, and high-speed digital designs require a cell-based or custom approach.

Increasingly, the NRE cost (predominantly mask cost) is approaching a level where even industry prototypes must be done using multi-project chips, amortizing the mask cost over multiple designs on the same reticle. Designs must be as re-programmable or adaptable as possible.

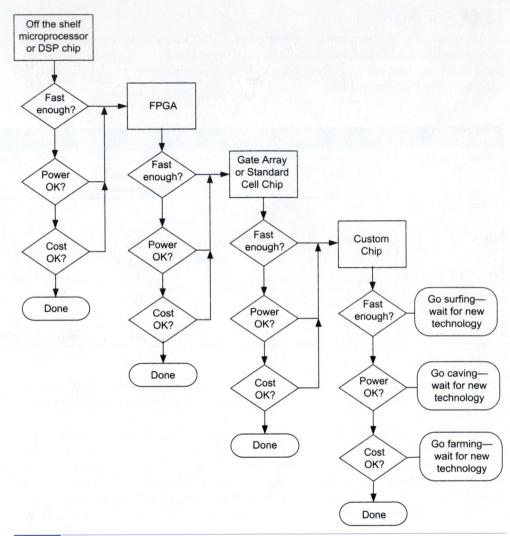

FIG 8.37 Design decision tradeoffs

8.4 Design Flows

A design flow is a set of procedures that allows designers to progress from a specification for a chip to the final chip implementation in an error-free way. In the previous section, we discussed the basic CMOS design methods without mentioning how we actually design an FPGA, gate array, or cell-based system. In this section, we will summarize the main design flows in use today.

A general design flow is shown in Figure 8.38. Design starts at the behavioral level and then proceeds to the structural level (gates and registers). This step is called behavioral or *Register Transfer Level* (RTL) synthesis because the designs are captured at the RTL (memory elements and logic) level in an HDL. The description is then transformed to a physical description suitable for chip fabrication. This step is called *physical synthesis* (or *layout generation*). Normally, the synthesis steps are automated, albeit guided by human judgment. The verification steps are also shown.

In Figure 8.38, the design has been partitioned into the *front end* stage at the behavioral level and the *back end* at the structural and physical levels. This is important because it illustrates a partitioning that is used to build *Application Specific Integrated Circuits* (ASICs). In an ASIC, the design can be developed at the HDL level and then passed to a company that completes the transition to an actual chip. In this way, the original design company does not have to invest the personnel or tools required to translate an HDL specification into a physical chip. Theoretically, in an ASIC flow, only a behavioral HDL needs to be designed and simulated (at the behavioral level). All subsequent operations can be completed by a third-party design service with only the final timing having to be verified by the back-end process. This is sometimes referred to as a "throw it over the wall" approach. While it works for moderately complex designs, the interaction between logic and layout is so important in more demanding circuits

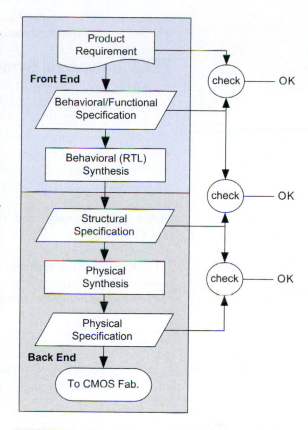

FIG 8.38 Generalized design flow

that such a flow becomes a schedule risk. Primarily, this occurs because the iteration time between logic design and physical placement takes too long when spread over two organizations. Multiple iterations are necessary because the prelayout timing estimates available to the HDL designer correlate poorly with the true postlayout timing because wire lengths are unpredictable before layout. Consider the case where the design cycle from logic to layout takes two hours when completed as an integrated task or one week if split into front-end and back-end tasks, as shown in the figure. If there are 100 iterations for the design, the integrated approach takes roughly 25 working days or five weeks, while the split approach takes two years (without vacations!). Having said this, companies are in business to make this approach work. If there are only ten iterations, the times are much more reasonable.

The next two sections summarize each of the tools required to perform the automatic transformation. We also will examine the verification tools required to guarantee the correctness of the transformation and look at specific design flows. Then, we will describe a manual flow that is typical of a mixed-signal or RF design. Finally, we will outline a method of transforming directly from the behavioral to the physical level.

8.4.1 Behavioral Synthesis Design Flow (ASIC Design Flow)

At the behavioral level, the operation of the system is captured without having to specify the implementation. This level provides the most independence from implementation details and is the most dependent on the tool flow for a good design.

The most popular style of tools for behavioral synthesis are those that directly transform a behavioral RTL description to a structural gate-level netlist. A typical behavioral flow for an ASIC is shown in Figure 8.39. Tool suppliers include Synopsys, Cadence Design Systems, Mentor Graphics, and Synplicity.

8.4.1.1 Logic Design and Verification The design starts with a specification, which might be a text description or a description in a system specification language. The designer(s) convert this to an RTL behavioral description in an HDL such as Verilog, or VHDL. A set of test benches are then constructed and the HDL is simulated to verify the correct behavior as defined by the specification and product requirements. Typical interactive design environments and simulators include NC-Verilog/SystemC/VHDL or Desk-

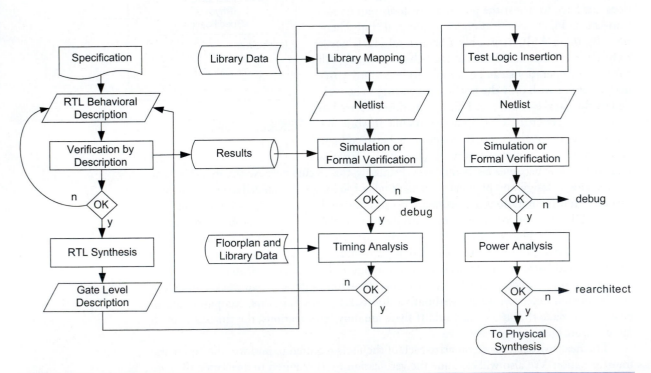

FIG 8.39 RTL synthesis flow

top Verilog/VHDL from Cadence Design Systems, VCS from Synopsys, ModelSim from Mentor Graphics and ActiveHDL from Aldec. Bear in mind that functional verification via simulation is usually carried out hierarchically. That is, after the overall architecture is defined, modules are successively built from the bottom up, verifying at each step. The design is iterated at this level until the correct behavior is evident. Test benches are covered further in Section 9.3.

Behavioral Verilog for an 8-bit implementation of the NCO previously introduced is presented below.

```verilog
module nco #(parameter size = 8,
            counter_size = 16,
            table_size = 64)
           (input fclock, reset,
            input [counter_size-1:0] initial_phase, phase_increment,
            output [size-1:0] q);

    reg   [counter_size-1:0] phase;
    wire  [size-3:0]         phase_part, inverted_adr, ROM_adr;
    wire  [size-2:0]         ROM_data;
    wire  [size-1:0]         wave_out;

    // numerically controlled oscillator
    // note that some constants are hardwired in the code below

    // phase counter
    always @(posedge fclock)
        if (reset) phase <= initial_phase;
        else phase <= phase + phase_increment;

    // add offset and determine ROM address
    assign phase_part = phase[counter_size-3:counter_size-8];
    assign inverted_adr = 7'3f - phase_part;
    assign ROM_adr = phase[counter_size-2] ? inverted_adr : phase_part;

    // look up data in ROM and negate if appropriate
    quarter_wave sine_table(ROM_adr, ROM_data);
    assign wave_out = phase[counter_size-1] ? ~ROM_data : ROM_data;
    assign q = wave_out + 8'h80 + phase[counter_size-1];
endmodule
```

8.4.1.2 RTL Synthesis The next step is to *synthesize* the behavioral description. This involves converting the RTL to generic gates and registers, then optimizing the logic to improve speed and area. Other steps involved at this stage are state machine decomposition, datapath optimization, and power optimization. Typical products include Design Compiler from Synopsys, BuildGates from Cadence, and Synplify Pro from Synplicity.

The behavioral code is synthesized into the following structural code using a generic gate library.

```verilog
module nco_struct(input              fclock, reset,
                  input  [15:0]  initial_phase, phase_increment,
                  output [7:0] q);

   wire [6:0] ROM_data;
   wire [15:0] phase_0;
   wire [5:0] nbus_5;
   wire [7:0] wave_out;
   wire [5:0] nbus_2;
   wire [15:0] nbus_1;
   wire [15:0] phase;

   xor i_7(wave_out[7], phase[15], 1'b0);
   xor i_6(wave_out[6], phase[15], ROM_data[6]);
   .
   .
   .
   xor i_2(wave_out[2], phase[15], ROM_data[2]);
   .
   .
   not i_10(nbus_5[2], phase[10]);
   not i_9(nbus_5[1], phase[9]);
   not i_8(nbus_5[0], phase[8]);
   generic_flip_flop phase_reg_15(.q(phase[15]), .d(phase_0[15]),
      .clk(fclock));
   .
   .
   .
   .
   quarter_wave sine_table(.addr(the_address), .q(ROM_Table));
   .
endmodule
```

The mapping to simple Verilog logic statements can be seen at the start of the program. Flip-flops are mapped to generic flip-flops that would implement a variety of flip-flop styles. The hierarchy is maintained, as illustrated by the call to the sine ROM.

8.4.1.3 Library Mapping *Library mapping* takes a generic HDL gate-level description and translates it to a netlist that specifies particular gates in the target library. This stage also maps predefined blocks such as memories to their appropriate descriptions. The following description is a portion of the mapped generic Verilog for the NCO shown above.

```verilog
module nco_struct_mapped(input              fclock, reset,
                        input  [15:0] initial_phase, phase_increment,
                        output [7:0]  q);
   .
   .
   BUFX4 i_506(.A(n_355), .Y(q[7]));
```

```
    .
MX2X1 i_00(.S0(reset), .B(initial_phase[15]), .A(nbus_1[15]),
   .Y(phase_0[15]));
NAND2BX1 i_8(.AN(n_102), .B(n_101), .Y(n_104));
XOR2X1 i_6(.A(phase[15]), .B(ROM_Table[6] ), .Y(n_103));
    .
    .
    .
DFFHQX1 phase_reg_0(.D(phase_0[15]), .CK(fclock), .Q(phase[15]));
    .
    .
    .

endmodule
```

Now the specific library elements are called out. XOR2X1 is a 2-input normal-power XOR gate and DFFHQX1 is a high-speed normal-power D flip-flop in the Artisan library used in the mapping.

8.4.1.4 Functional or Formal Verification

We must now prove that the structural netlist performs the same function as the original behavioral HDL. Ideally, the netlist would be correct-by-construction, but ambiguities in HDLs sometimes cause the synthesizer to produce incorrect netlists from poorly written behavioral code. One verification strategy is to rerun the logic test benches and check that they produce exactly the same output for the behavioral and structural descriptions.

Another strategy is to use a *formal verification* program that compares the logical equivalence of the two descriptions. Formal verification tools are still maturing, but offer the advantage that they mathematically prove both descriptions have exactly the same Boolean functions [Anastasakis02]. In contrast, simulation only is as good as the choice of test vectors. Formality from Synopsys and Incisive Conformal from Cadence are examples of formal verifiers.

Other types of verification that can be run are semantic and structural checks on the HDL. An example of a semantic check would be ensuring that all bus assignments match in bit width, while an example of a structural check would include making sure all outputs are connected.

8.4.1.5 Static Timing Analysis

At this point, the functional equivalence of the gate-level description and the original behavioral description has been established. Now the temporal requirements of the design have to be checked. For example, the adder may add, but does it add fast enough? At the behavioral level, clock cycle time is an abstract notion, but at the structural level, an actual cycle time has to be met by a particular set of gates. A *timing analyzer* is used to verify the timing.

The timing analyzer is a critical analytical tool in the arsenal of the modern CMOS digital designer. Timing can be verified in a cursory manner using a timing simulator, i.e., a simulator in which the actual gate timings are used rather than a cycle-based or unit delay simulator. While useful, this approach is usually neither complete nor rigorous and can take an extraordinary amount of time to run.

Static timing analysis, on the other hand, runs quickly and exhaustively evaluates *all* timing paths. The inputs to the timing analyzer at this point are derived from the basic timing of the library gates due to intrinsic gate delays and routing loads that can be either estimated statistically or derived from floorplanning data. (See Section 8.4.2.2 for a description of floorplanning.) Timing analyzers check for both *max-delay* (will all flip-flops meet their setup time at the required cycle time?) and *min-delay* (will any flip-flop violate its hold time?).

Static timing analysis can suffer from *false path* problems. Typical of this problem might be a reset line in a circuit that has many clock cycles to operate. The timing analyzer might report that it cannot complete in one cycle. The designer must manually flag such multicycle paths for the timing analyzer.

Typical timing analyzers include Pearl from Cadence and Pathmill and PrimeTime from Synopsys. The listing on the following page is a portion of a typical report for the NCO from a timing analyzer (this listing is from the Cadence BuildGates tool suite). The initial information specifies the required cycle time (11 ns) and then displays the calculated arrival time (5.89 ns). This results in what is termed a *slack time* of 5.11 ns. A table is then produced that lists the mapped Verilog instance, the signals that are changing and which direction the signals are toggling (by ^ and v characters for rising and falling), the cell type, the delay through that instance, and the arrival time. The required arrival time is also listed. Each line is examined for excessive timing and then normally the original Verilog is changed to improve delays through better pipelining or logic design. The synthesizer can also be directed to improve the timing. The sine table has been synthesized onto gates in this implementation.

8.4.1.6 Test Insertion   Logic and registers are then inserted/modified to aid in manufacturing tests (see Section 9.6). Two basic techniques are used. One involves inserting scannable registers so that the state of a circuit can be set and monitored. Accompanying this option is a technique called *Automatic Test Pattern Generation* (ATPG), which is used to generate tests for a scannable design. The other technique, called *Built-In Self-Test* (BIST), modifies registers to allow *in situ* testing within the chip. Figure 8.40 shows the NCO after a test insertion program has run. The difference between the original design shown in Figure 8.13 and this design is the fact that blue *scan registers* have been inserted.

Typical commercially available test programs include DFT Compiler from Synopsys for scan insertion and Tetramax for ATPG. LogicVision markets Logic BIST and Memory BIST for built-in self-test.

8.4.1.7 Power Analysis   The power consumption of the circuit is then estimated. Power consumption depends on the activity factors of the gates, which in turn depends on the inputs the chip receives. *Power analysis* can be performed for a particular set of test vectors by running a simulator and evaluating the total capacitance switched at each clock transition at each node. At this stage, if the power is too high, the design must return to the architectural level to rethink the solution. Commercial power analysis tools include PrimePower and Powermill from Synopsys.

```
Beginpoint: phase_reg_14/Q (^) triggered by leading edge of 'clock'
Required Time 11.00
- Arrival Time 5.89
= Slack Time 5.11
    Clock Rise Edge 0.00
    + Source Insertion Delay 0.20
    = Beginpoint Arrival Time 0.20
```

Instance	Arc	Cell	Delay	Arrival Time	Required Time
	fclock ^			0.20	5.31
phase_reg_14	CK ^ -> Q ^	DFFHQX1	0.58	0.78	5.89
i_34	A ^ -> Y ^	XOR2X1	0.78	1.57	6.68
sine_table	addr[3] ^	quarter_wave		1.57	6.68
sine_table/i_371	A ^ -> Y v	INVX1	0.45	2.02	7.13
sine_table/i_42	B v -> Y ^	NOR2X1	0.19	2.21	7.32
sine_table/i_44	AN ^ -> Y ^	NAND2BX1	0.16	2.37	7.48
sine_table/i_145	A0 ^ -> Y v	AOI211X1	0.10	2.47	7.58
sine_table/i_6	C0 v -> Y ^	OAI211X1	0.13	2.61	7.72
sine_table/i_484	A ^ -> Y ^	BUFX4	0.96	3.57	8.68
sine_table	q[0] ^	quarter_wave		3.57	8.68
i_120	B ^ -> Y v	NOR2BX1	0.22	3.79	8.90
i_7	AN v -> Y v	NOR2BX1	0.20	3.99	9.10
i_9	AN v -> Y v	NOR2BX1	0.21	4.20	9.31
i_20	B v -> Y ^	NAND2BX1	0.12	4.32	9.43
i_3	AN ^ -> Y ^	NAND2BX1	0.17	4.49	9.60
i_8	AN ^ -> Y ^	NAND2BX1	0.19	4.67	9.78
i_31	A ^ -> Y ^	XNOR2X1	0.28	4.96	10.07
i_504	A ^ -> Y ^	BUFX4	0.93	5.89	11.00
	q[6] ^		0.00	5.89	11.00

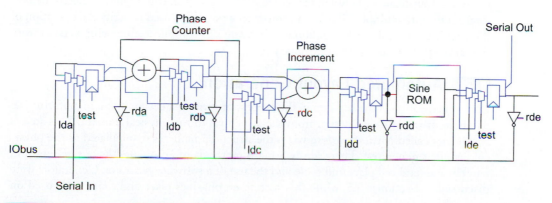

FIG 8.40 Scan register insertion for testing

Example

While automated tools are the solution to estimating power consumption in large circuits, sometimes you need a rapid, "back of the envelope" method of estimating power. How can we achieve this?

Solution: The following technique will give you a rough idea of the power dissipated in a circuit.

Step 1: Estimate the power in switched flip-flops and registers. Sum the clock input capacitance of all registers and, if possible, the internal switched capacitance. The latter is often hard to ascertain unless you have access to the internals of the cell library, but as a rough approximation, double the specified clock input capacitance. To this, add the capacitance of the clock routing. Again, conservatively, double the total clock load that you have previously calculated. The clock has an activity factor of 1. Now, given V_{DD} and the clock frequency, you can estimate the clock power using EQ (4.30). Thus, the power will be $2 \cdot 2 \cdot C_{clk} \cdot V_{DD}^2 \cdot f = 4C_{clk}V_{DD}^2 f$.

Step 2: Estimate the power in the combinational circuits. Sum all input and output capacitances for all gates. This approximates the internal capacitance of the logic gates. To this we must add the routing capacitance. This is best done after some experience with the process and design tools you are using, but as a guide, use a number between 2 and 10. The activity factor has to be estimated. Again experience helps here, but 0.5 is very conservative. Thus, the total logic power will be roughly $5 \cdot 0.5 \cdot C_{logic} \cdot V_{DD}^2 \cdot f = 2.5 C_{logic}V_{DD}^2 f$.

These two steps are very approximate. After a bit of experience with a technology and design flow, you will be able to put your own numbers on the approximations.

8.4.1.8 Summary Apart from increasing design productivity, logic synthesis systems are useful for transforming between technologies. For instance, you might synthesize behavioral HDL onto multiple FPGAs and construct a prototype used to verify the operation of the circuit under real-world conditions. Then you can compile a single-chip version from the same HDL using a gate-array library.

8.4.2 Automated Layout Generation

Layout generation is the last step in the process of turning a design into a manufacturable database. It transforms a design from the structural to the physical domain. This step is sometimes called *physical synthesis* when the structural netlist is manipulated as the physical layout is generated. There are two main methods of automated layout generation, namely, standard cell place and route and the use of a software generator. Generators were discussed in Section 8.3.5, while this section emphasizes place and route tools used on most ASICs.

Figure 8.41 shows a standard place and route layout generation design flow. It begins with the structural netlist describing gates, flip-flops, and their interconnections. The netlist might be provided in the *Design Exchange Format* (DEF) described in Section 8.9.4 or as a Verilog netlist like the one in Section 8.4.1.3. The placement tool also takes a standard cell library definition describing cell dimensions and port locations. Typically, this is in the *Library Exchange Format* (LEF), described in Section 8.11.3.

8.4.2.1 Placement The first step in Figure 8.41 is to place the standard cells. The key to automation of standard cell layouts is the use of constant-height, variable-width standard cells that are arrayed in rows across a chip, as shown in Figure 8.42. In contrast to SOG and gate array chips, standard cell chips can add application-specific custom blocks such as memories and analog blocks by allowing the standard cell rows to "flow" around the fixed-shape custom blocks. No separation has been shown between standard cell rows because routing takes place over the cells using multiple layers of metal. In older processes with two or three metal layers, a space between rows would be needed to allow routing. LEF summarizes the salient physical details of cells.

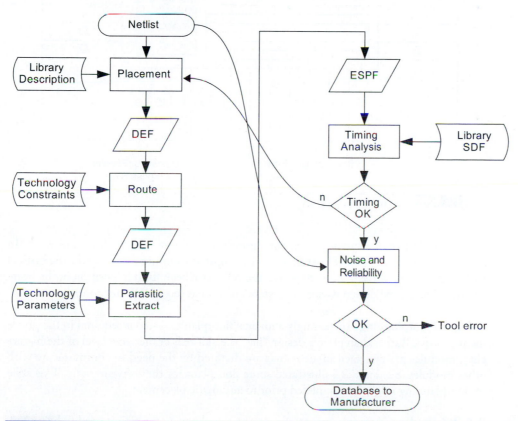

FIG 8.41 Standard cell place and route design flow

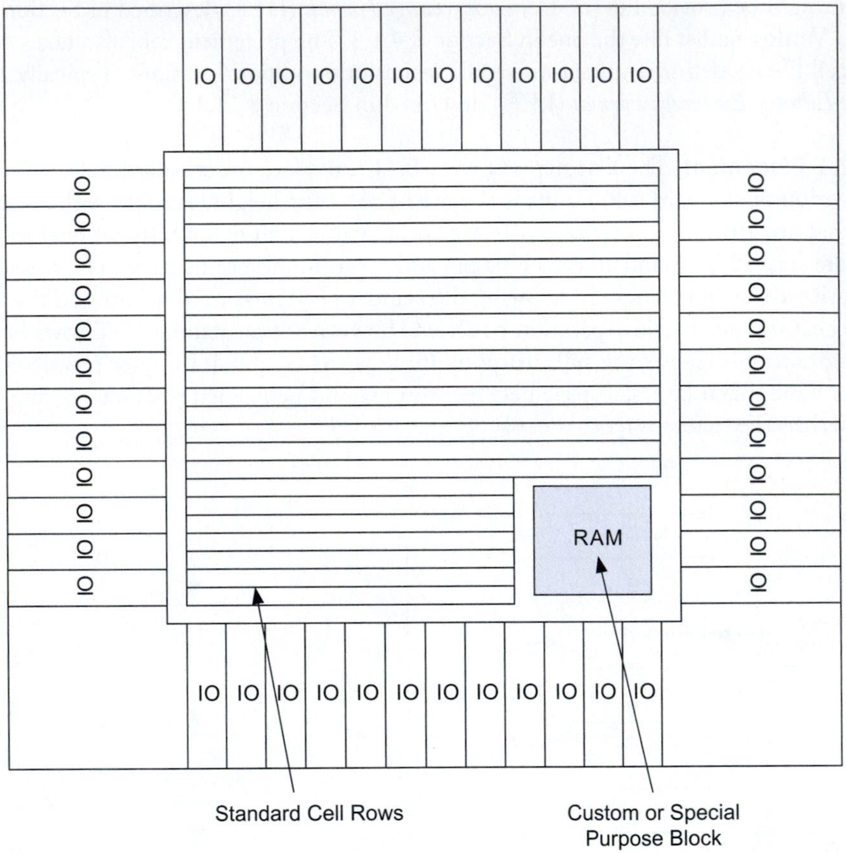

IO IO IO IO IO IO IO IO IO IO IO IO

IO IO IO IO IO IO IO IO IO IO IO IO

RAM

Standard Cell Rows

Custom or Special
Purpose Block

FIG 8.42 Standard cell chip layout

The objective of a simple placement algorithm is to minimize the length of wires. In *timing-driven placement*, the cost of wires is weighted to minimize delay on the critical paths. At the end of the placement phase, the cells have been fixed in position in the overall array. The placed design is saved in a standard format (e.g., DEF) for routing.

8.4.2.2 Floorplanning Increasingly a manual floorplanning step is required in the placement process. Rather than place a design "flat" (i.e., all cells at the same level of the hierarchy), modules are clustered in areas that are dictated by the need to communicate with other modules. Section 8.2.5 illustrated some floorplans for the software radio. This style of floorplanning might be completed prior to automatic placement.

8.4.2.3 Routing After placement of cells, the signal nets in the circuit need to be routed. Routing is normally divided into two steps: *global* routing and *detailed* routing.

A global router abstracts the routing problem to a notional set of abutting channels that cover the chip surface through which wires are routed. Routes are added to channels according to a cost function. Wires can be changed from channel to channel if the density of wires in a channel becomes too high. The detailed router places the actual geometry required to complete signal connections. Over time, a selection of detailed routers have been developed to automatically route signals. Older routers constrained signals to a grid of tracks, but newer *gridless* routers are more flexible for variable pitch wires. Moreover, they allow easy interface to foreign cells that may have I/O pin locations that are not on any specific routing grid. Routers also can route over the top of cells. LEF definitions are used to indicate obstructions on various layers in cell definitions. Advanced routers take into account manufacturability concerns such as redundant vias (more that one via inserted when space is available) and adjustable spacing (to separate wires and reduce coupling when there is room).

In the example of the flow shown in Figure 8.41, the router uses a technology file to specify routing layers and pitches for the process technology. It writes the results to another DEF file.

8.4.2.4 Parasitic Extraction

The placed and routed design is then passed to the circuit *parasitic extractor*. In the example shown in Figure 8.42, the placed and routed design is provided to the extractor in DEF format and the output is an *Extended Standard Parasitic Format* (ESPF), *Reduced Standard Parasitic Format* (RSPF), or *Standard Parasitic Exchange Format* (SPEF) that describes the R's and C's associated with all nets in the layout. The extractor uses another technology file defining the interlayer capacitances and layer resistances. The formats are covered in more detail in Section 8.11.6.

The capacitance extractor can be a 2D, 2.5D, or 3D extractor. Two-dimensional (2D) extractors look at a cross-section assuming wires extend uniformly outside the section. A 2.5D extractor uses lookup tables to more accurately estimate capacitance near nonuniformities. A 3D extractor solves Maxwell's equations in three dimensions to precisely determine capacitance of complex geometries. 3D extraction used to be prohibitively time-consuming, but new statistical algorithms, such as those in QuickCap from Magma Design, deliver good accuracy with faster runtimes.

8.4.2.5 Timing Analysis

Static timing analysis is now rerun with the actual routing loads placed on the gates. This is usually the bottleneck in the design process as the full reality of a physical realization is apparent. Multiple iterations of synthesis and placement & routing are usually necessary to converge on timing requirements.

Additionally, if possible (especially where dynamic circuits are used), a transistor-level timing simulation should be run. While this cannot usually be achieved using a SPICE-based simulator, a variety of transistor level simulators with "almost SPICE accuracy" have been in use since the late 1970s. These currently have the capacity to do whole-chip simulations at the transistor level, but at somewhat reduced transistor modeling accuracy. Pathmill and Nanosim from Synopsys and UltraSim from Cadence are examples of current simulators of this type.

8.4.2.6 Noise, V_{DD} Drop, and Electromigration Analysis

Analyses are now run to check noise, IR drop in supply lines, and electromigration limits. Noise analysis is run to evaluate crosstalk due to interlayer routing capacitance. For static circuits, aggressor nodes can alter the timing of victim nodes but not their function, as described in Section 4.5.4. For poorly designed dynamic circuits, crosstalk can cause irrecoverable errors. Signal-Storm, ElectronStorm, and VoltageStorm from Cadence are examples of such tools.

8.4.2.7 Timing-driven Placement

The trouble with a place-then-route strategy is that after the layout is completed, the parasitic routing capacitance is extracted and the timing analysis is done to estimate timing. The timing is not known until the physical layout is complete. If timing problems are found, the cycle has to be repeated with some kind of constraint placed on the problematic paths. With complex designs this quickly gets out of control, to the point where changing something on one iteration could undo something fixed on a previous iteration. There are stories of designs that never were completed because of this problem.

The solution is to use a technique called *timing-driven placement*, which takes into account the timing (speed) of the circuit as cells are placed. Cells on critical paths are given priority to minimize wire delay. This approach, illustrated in Figure 8.43, has been successful and often results in a one-pass approach for many designs.

8.4.2.8 Clock-tree Routing

Central to modern high-speed designs is the clock distribution strategy. In Section 12.5.4, a number of these approaches are explained. To minimize skew, it is often best to route the clock and its buffers before the main logic placement and routing is completed. This task is performed with a *clock tree router*. Any of the clocking approaches described in Chapter 7 can be implemented in this manner.

8.4.2.9 Power Analysis

Power estimation can be repeated for the extracted design now that real wire capacitances are available. Similar techniques to those used during RTL synthesis are used.

8.4.3 Mixed-signal or Custom-design Flow

In the previous section, we described a flow that would be used for a purely digital chip in which the procedure for converting from HDL to layout is highly automated. This flow offers high productivity for most large digital chips with moderate performance requirements. But what of smaller analog, RF, and high-speed digital sections of a chip? For these sections we use a custom-design flow, which is shown in Figure 8.44.

The designer begins by drawing a schematic (or possibly writing a netlist). An electrical rule check (ERC) verifies port connectivity and checks for unconnected inputs or outputs—the kind of simple connectivity errors that can occur easily in a manually drawn schematic. When the schematic is deemed correct, circuit simulation is then carried out using a SPICE-type simulator to verify DC, AC, transient, noise, and/or RF performance.

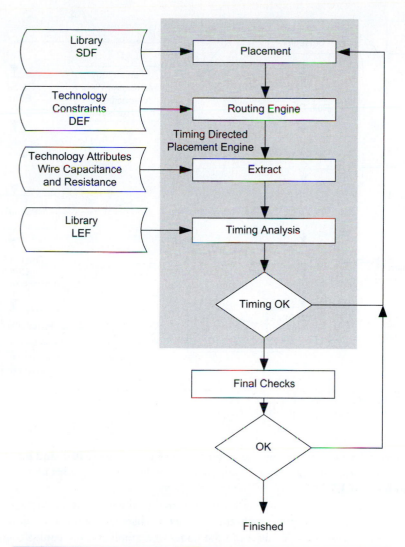

FIG 8.43 Timing directed placement design flow

Once the circuit behavior of the module has been verified, the layout can commence, starting with the floorplan. Floorplanning can be an iterative process that is refined as actual module sizes and critical paths become known. Custom layout is a very time-consuming task; for example, a large microprocessor can keep a hundred mask design technicians busy for two years. Automating noncritical parts of the layout is essential for productivity. When the layout for the module is complete, a layout circuit

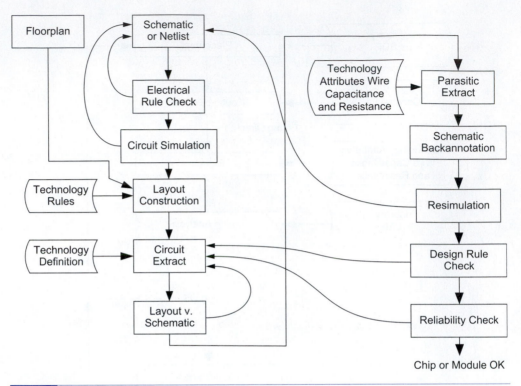

FIG 8.44 Mixed-signal or custom-design flow

extractor is invoked to determine the connectivity of primitives (MOS and bipolar transistors, diodes, resistors, capacitors, inductors) in the layout using rules like those illustrated in Section 3.5.2.

In the next step, the extracted netlist is compared to the schematic using a graph isomorphism program to determine whether the two netlists are identical in connectivity. This proceeds by assigning primitives to the nodes of a graph and the connections to the arcs in the graph. Graph coloring based on the connectivity and circuit parameters (i.e., transistor type, width, and length) determines the extent of the match. Once connectivity equivalence has been determined, each primitive attribute is checked for equivalence (i.e., capacitor or resistor value, transistor W/L). Discrepancies are reported to the user. Graphical feedback may be provided to help the designer find the source of any mismatch. This step is commonly called *layout versus schematic* (LVS).

Once the structural-to-physical equivalence has been established, the parasitic extract is completed. This adds the parasitic routing capacitance and resistance to the original primitive elements. In general, inductors are not extracted, but are dealt with by cookie cutting the inductor out of the layout and substituting a previously generated physical model. This is sometimes called *macro substitution*. The parasitic capacitance and resistance can be *back annotated* onto the schematic and the complete circuit resimulated. It must be pointed out that this step is extremely important. Matching simulated behavior to real device behavior is of critical importance in being able to accurately predict performance. It is too late when the circuit has been built!

The module layout can then be design-rule checked (DRC). Alternatively, this step can be completed just as the layout is completed. Normally, the AC performance is more important than tweaking the last design-rule error because running DRC on a circuit that does not meet performance goals is a waste of time.

Following this, a set of manufacturability verification steps needs to be completed. These can be manual or automated. In common with the standard cell design flow, power bus widths should be checked to ensure that they comply with metal migration and IR drop constraints. Power consumption can be found directly from circuit simulation. Adequate substrate and well contacts should be present in a bulk CMOS design, and all external I/O must be guard-ringed. At this stage, a check can also be made for substrate noise injection from digital to analog circuits. SubstrateStorm from Cadence performs this task.

This process can be completed hierarchically to build up large modules. Usually, the ultimate limitation comes from trying to simulate vast numbers of transistors accurately in SPICE. A variety of fast transistor-level simulators have been developed to deal with this problem, although there is always some upper limit to what can be simulated at the desired accuracy.

8.4.4 Programmed Behavioral Synthesis

An alternate class of behavioral synthesizer directly synthesizes blocks in a fashion reminiscent of silicon compilers. This type of synthesizer takes a fairly targeted application space (such as digital signal processing) and constructs a logic, circuit, and physical description based on a high-level input description. An example of such a system is the design system used at the Berkeley Wireless Research Center, which takes high-level descriptions defined in Simulink, and automatically translates them through a number of steps to a mask layout. The basis of this system is a set of predefined modules available in the Simulink library that have targets in the physical domain. This type of flow is illustrated in Figure 8.45. The translation between behavioral and physical layers of abstraction occurs directly. RTL and structural descriptions can also be generated as a side product to aid in verification. For example, a multiplier for complex numbers could be specified as a Simulink block. An instance of this multiplier would trigger the automatic construction of the appropriate RTL, gate, and mask-level description following a preprogrammed algorithm.

8.5 Design Economics

It is important for the IC designer to be able to predict the cost and the time to design a particular IC or sets of ICs. This can guide the choice of an implementation strategy. This section will summarize a simplified approach to estimate these values.

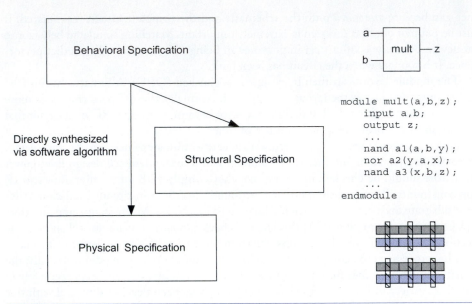

FIG 8.45 Direct behavioral-to-physical translation

In this section, we will concentrate on the cost of a single IC, although you should consider the overall system when making such decisions. System-level issues such as packaging and power dissipation can affect the cost of an IC.

The selling price S_{total} of an integrated circuit may be given by

$$S_{total} = C_{total} / (1-m) \qquad\qquad (8.3)$$

where

C_{total} is the manufacturing cost of a single IC to the vendor
m is the desired profit margin

The margin has to be selected to ensure a profit after overhead (G&A) and the cost of sales (marketing and sales costs) have been considered.

The costs to produce an integrated circuit are generally divided into the following elements:

- Non-recurring engineering costs (NREs)

- Recurring costs

- Fixed costs

8.5.1 Non-recurring Engineering Costs (NREs)

Non-recurring engineering costs are those that are spent once during the design of an integrated circuit. They include

- engineering design cost E_{total}
- prototype manufacturing cost P_{total}

These costs are amortized over the total number of ICs sold. F_{total}, the total non-recurring cost, is given by

$$F_{total} = E_{total} + P_{total} \tag{8.4}$$

The NRE costs can be amortized over the lifetime volume of the chips. Alternatively, the non-recurring costs can be viewed as an investment for which there is a required rate of return. For instance, if \$1M is invested in NRE for a chip, then \$10M has to be generated for a rate of return of 10.

8.5.1.1 Engineering Costs

The cost of designing the IC E_{total} hopefully will happen once during the chip design process. The costs include:

- personnel cost
- support costs

The personnel costs might include the labor for

- architectural design
- logic capture
- simulation for functionality
- layout of modules and chip
- timing verification
- DRC and tapeout procedures
- test generation

The support costs amortized over the life of the equipment and the length of the design project include

- computer costs
- CAD software costs
- education or re-education costs

Costs can be drastically reduced by reusing modules or acquiring fully completed modules from an intellectual property vendor. As a guide the per annum costs might break down as follows (these figures are in US dollars for engineers in the USA circa 2004):

Salary	$50–$100K
Overhead	$10–$30K
Computer	$10K
CAD Tools (digital front end)	$10K
CAD Tools (analog)	$100K
CAD Tools (digital back end)	$1M

The cost of the back-end tools clearly must be shared over the group designing the chips.

8.5.1.2 Prototype Manufacturing Costs

These costs (P_{total}) are the fixed costs to get the first ICs from the vendor. They include

- the mask cost
- test fixture costs
- package tooling

The photo-mask cost is proportional to the number of steps used in the process. Mask costs increase as the process dimensions are reduced, so although newer, smaller processes generally have increased mask costs, masks on the metallization layers can be less expensive than on the lower layers. A mask currently costs between $500 and $30,000 or more. A complete mask set for 130 nm CMOS costs in the vicinity of $500K to $1M. These prices will increase as process linewidths are reduced.

A test fixture consists of a printed wiring board probe assembly to probe individual die at the wafer level and interface to a tester. Costs range from $1000 to $50,000, depending on the complexity of the interface electronics.

If a custom package is required, it may have to be designed and manufactured (tooled). The time and expense of tooling a package depends on the sophistication of the package. Where possible, standard packages should be used.

An economical way of prototyping chips is to use a multi-project reticle that combines a number of different chip designs onto one mask set. Thus, if there were 200 sites available on a mask set and 20 projects were implemented, each project would get 10 die per wafer and the mask cost per project would be 1/20th of the cost of a complete mask set. This kind of service is provided by many of the silicon vendors and also MOSIS. For modest technology this can be quite cheap (~ $1000 per mm^2 for 0.6 μm). Some commercial users worry about protection of intellectual property when they share a mask set.

Example

You are starting a company to commercialize your brilliant research idea. Estimate the cost to prototype a mixed-signal chip. Assume you have seven digital designers, three analog designers, and five support personnel and that the prototype takes two fabrication runs and two years.

Solution: The seven digital designers will cost 7 • ($70K + $30K + $10K + $10K) = $840K. The three analog designers will cost 3 • ($100K + $30K + $10K + $100K) = $720K. The five support personnel cost 5 • ($40K + $20K + $10K) = $350K. One fabrication run with the back-end tools will cost $2M. Thus, the cost is $3.91M per year with one fab run. The total predicted cost here is nearly $8M.

You may see ways to improve this. Clearly, you can reduce the number of people and the labor cost. You might reduce the CAD tool cost and the fabrication cost by doing multiproject chips. However, the latter approach will not get you to a pre-production version, because issues such as yield and behavior across process variations will not be proved. Figure 8.46 shows the breakdown of the overall cost.

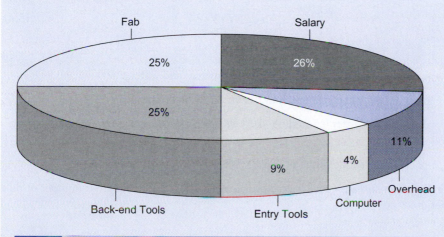

FIG 8.46 Pie chart showing prototyping costs for a mixed-signal IC

8.5.2 Recurring Costs

Once the development cost of an IC has been determined, the IC manufacturer will arrive at a price for the specific IC. A few large companies such as Intel, TI, STMicroelectronics, Toshiba, and IBM have in-house manufacturing divisions. Many fabless semiconductor companies outsource their manufacturing to a silicon foundry such as TSMC, Hitachi/

UMC, IBM, LSI Logic, or ST. This is a recurring cost; that is, it recurs every time an IC is sold. Another component of the recurring cost is the continuing cost to support the part from a technical viewpoint. Finally, there is what is called "the cost of sales," which is the marketing, sales force, and overhead costs associated with selling each IC. In a captive situation such as the IBM microelectronics division selling CPUs to the mainframe division, this might be zero.

The IC manufacturer will determine a part price for an IC based on the cost to produce that IC and a profit margin. The margin generally falls as the volume increases. An expression for the cost to fabricate an IC is as follows:

$$R_{\text{total}} = R_{\text{process}} + R_{\text{package}} + R_{\text{test}} \tag{8.5}$$

where

R_{package} = package cost

R_{test} = test cost—the cost to test an IC is usually proportional to the number of vectors and the time to test.

$$R_{\text{process}} = W / (N \cdot Y_w \cdot Y_{pa}) \tag{8.6}$$

where

W = wafer cost ($500–$3000 depending on process and wafer size)

N = gross die per wafer (the number of complete die on a wafer)

Y_w = die yield per wafer (should be ~70%–90+% for moderate sized dice in a mature process)

Y_{pa} = packaging yield (should be ~95%–99%)

If a die has area A and is fabricated on a wafer with radius r, the gross number of dice per wafer is

$$N = \pi \left[\frac{r^2}{A} - \frac{2r}{\sqrt{2A}} \right] \tag{8.7}$$

where the second term accounts for wasted area around the edges of a circular wafer.

The packaging yield is the number of chips that pass testing after the wafer has been diced and the parts packaged. The die yield is affected by defects randomly distributed around the wafer. The probability of a random defect causing a particular die to fail depends on the size of the die A and average number of defects per unit area D. If defects obey a Poisson distribution, Y_w may be given by [Seeds67]

$$Y_w = e^{-AD} \tag{8.8}$$

For small dice ($AD \ll 1$), Y_w is nearly 1 and R_{process} grows linearly with A. For large dice ($AD \gg 1$), Y_w drops off rapidly because most chips will have defects and R_{process} grows exponentially with A.

8.5.3   Fixed Costs

Once a chip has been designed and put into manufacture, the cost to support that chip from an engineering viewpoint may have a few sources. *Data sheets* describing the characteristics of the IC have to be written, even for application-specific ICs that are not sold outside the company that developed them. From time to time, *application notes* describing how to use the IC may be needed. In addition, specific application support may have to be provided to help particular users. This is especially true for ASICs, where the designer usually becomes the walking, talking, data sheet and application note. Another ongoing task may be failure or yield analysis if the part is in high volume and you want to increase the yield.

As a side comment, every chip or test chip designed should have accompanying documentation that explains what it is and how to use it. This even applies to chips designed in the academic environment because the time between design submission and fabricated chip can be quite large and can tax even the best memory.

Example

Suppose your startup seeks a return on investment of 5. The wafers cost $2000 and hold 400 gross die with a yield of 70%. If packaging, test, and fixed costs are negligible, how much do you need to charge per chip to have a 60% profit margin? How many chips do you need to sell to obtain a 5-fold return on your $8M investment?

Solution: $R_{\text{total}} = R_{\text{process}} = \$2000/(400 \cdot 0.7) = \$7.14$. For a 60% margin, the chips are sold at $\$7.14 / (1 - 0.6) = \$17.86$ with a profit of $10.72 per unit. The desired ROI implies a profit of $8M $\cdot$ 5 = $40M. Thus, $40M / $10.72 = 3.73M chips must be sold. Clearly, a large market is necessary to justify the investment in custom chip design.

8.5.4   Schedule

At the outset of a system design project involving newly designed ICs, it is important to estimate the design cost and design time for that system. Estimating the cost can help you determine the method by which the ICs will be designed. Estimating the schedule is essential to be able to select a strategy by which the ICs will be available in the right time and at the right price. This second task is usually the least well specified and requires some experience.

If we assume that fixed costs are kept reasonable and that for a given IC size, $R_{process}$ is constant, the variables left in determining the cost of an IC are E_{total}, the engineering design cost, and P_{total}, the prototype manufacturing cost. P_{total} depends on the way in which the IC is implemented. We examined a variety of strategies for the design of CMOS systems earlier in the chapter. The fixed costs of prototyping P_{total} are relatively constant, given an implementation technology. The engineering costs depend on the complexity of the chip, the design strategy, and the amount of sustaining engineering needed. Usually, the design and verification engineering costs dominate. For this reason, it is important to be able to estimate a schedule for the design of an IC and then manage the available resources to bring the project to a successful conclusion.

Increased engineering effort can reduce the size of the die, which reduces $R_{process}$. Hence, it is important to be able to trade off the reduction in die cost with the increase in engineering effort. Opinions vary, but it is usually best to get a product first to market and then shrink the die when the product becomes successful. Optimizing without market feedback is usually a recipe for loss of market share or even failure to gain any market share at all.

[Paraskevopoulos87] suggests a number of fairly obvious methods for increasing productivity, thereby improving schedules:

- using a high-productivity design method
- improving the productivity of a given technique
- decreasing the complexity of the design task by partitioning

A final caution: Adding people to a project that is already late tends to make it later [Brooks95].

8.5.5 Personpower

To estimate the schedule, you must have some idea of the amount of effort required to complete the design. As we have seen, typical IC projects will involve the following tasks

1. architectural design
2. HDL capture
3. functional verification
4. place and route
5. timing verification, signal integrity, reliability verification
6. DRC and tapeout procedures (ERC, LVS, mask generation)
7. test generation

While some researchers have attempted to derive analytical formulae for productivity, the best predictor of design schedule for a team is previous performance. Design time for a

Example

While it is hard to predict the design and test time for a chip, we can at least identify the main tasks and corresponding fixed periods in a chip design project. A representative Gantt chart is shown in Figure 8.47 for a project running over one year. The logic design time is shown as 12 weeks, which would be appropriate for an extremely simple chip. Double this time would be representative of moderately complex digital chips. The fixed times tend to be the fabrication time and packaging time, which are shown to be 10 weeks in the example. The design, debug, and test times will expand or contract to fit the complexity of the chip. And, if you are meticulous and lucky, you will not have to respin the chip.

ID	Task Name	Start	Finish	Duration	Q1 04	Q2 04	Q3 04	Q4 04	Q1 05
1	Specification	1/1/2004	1/28/2004	4w					
2	Digital Design	1/29/2004	4/21/2004	12w					
3	Place and Route	4/22/2004	6/16/2004	8w					
4	Fabrication	6/17/2004	8/11/2004	8w					
5	Packaging	8/12/2004	8/25/2004	2w					
6	Lab Test	8/26/2004	10/20/2004	8w					
7	Respin	10/21/2004	11/17/2004	4w					
8	Lab Test	11/18/2004	12/29/2004	6w					

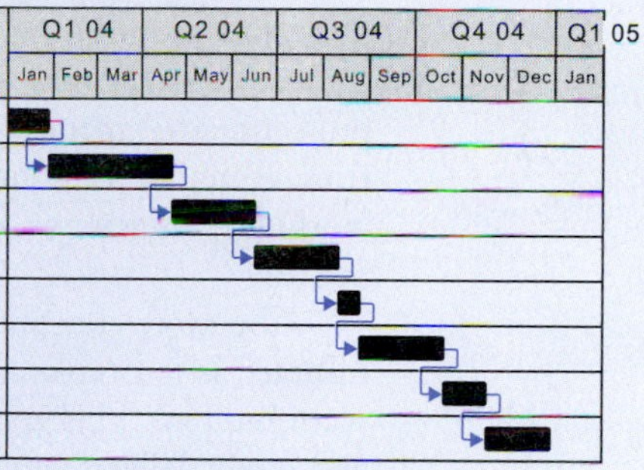

FIG 8.47 Gantt chart for simple chip

given team can be improved by design reuse or component-based design. It would seem that the time to design is proportional to the number of "modules" that are in the design raised to some power. That is, a four-module design is more than four times as complex as a single-module design. A module in this instance refers to a significant section of a chip such as a microprocessor, serial interface, or special functional unit.

Normally, projects are schedule-driven. In this case, it is important to make maximal use of design aids to meet the required schedule. Of importance is the cycle time of the so-called "edit-compile-debug" loop: i.e., the time it takes to make a change to the HDL; synthesize, place, and route it; and have a timing-verified final design. This can depend strongly on the efficiency of the design tools used, but if it is more than a day, design productivity can suffer. Ideally, the cycle is a few hours so that multiple bugs can be fixed each day.

Broadly speaking, schedules on the order of 18–24 months for a completely new chip seem to fit current average-complexity chips and state-of-the-art tools. For respins to slightly differentiate products, this can be reduced to six months or less, but there are certain fixed times such as IC fabrication and packaging that set hard limits on the complete design cycle time. Of course, for technologies such as FPGAs, design turnaround can be minutes (which is why FPGA verification is so important to ASIC or custom IC designs).

New microprocessors seem to take three to five years, and most experience one or more schedule slips.

8.5.6 Project Management

Project management is the overall supervision of the project. Tasks include making certain sufficient resources are available at the appropriate time, ensuring communication between different groups assigned to the project, and summarizing progress and risks to management. The development of processes for the conception, design, and ultimate manufacture of products is also the purview of the project manager.

There are two main ways to manage a chip design. The first is what might be called the rapid prototyping approach that is typical of startup companies, where a full-time project manager may be a luxury (and probably is more aptly named "seat-of-the-pants project management"). In this approach, a time goal is set and the workload is set to fit the time available. It is vital to rapidly get to the point where a prototype of the design is working—in essence, the skeleton—and the meat (detail) is gradually added. This can be risky.

The more conventional approach, which is appropriate for large companies and the military, is to preplan everything, estimating task times and putting these into a project planning tool. This approach, while necessary for large groups, tends to be feature-driven and rarely delivers products in shorter time scales than the rapid prototyping approaches. It is suitable when the tasks are well-defined and have been done before (then you know what the task times should be). The approach is stable and, depending on the team, often delivers products within budget and on time.

8.5.7 Design Reuse

Rarely is an IC designed as a single event. Rather, companies wish to amortize the development effort of a particular IC over several generations of products. This normally means that the design has to be transferred between several different processes. When design was mainly manual and at the mask level, a great deal of effort was expended on techniques to allow porting of designs between processes with the minimum of human intervention. Techniques used here include the use of symbolic layout methods and mask resizing software.

With the emergence of cell-based design, design migration falls into two steps:

- Acquiring or building a standard cell library in the new technology
- Retargeting the HDL description to the new cell library

The design and test generation does not have to be redone, although timing analysis and regression test bench simulation should definitely be completed.

In design flows where these steps cannot be followed, strict use of structured design techniques and software generator technologies can markedly improve porting times. Maintaining accurate and clear documentation will alleviate many problems downstream.

With the maturation of cell-based design, especially standard cell libraries and the use of hardware description languages, the notion of virtual components has become important as a method of transferring and reusing designs. Virtual components on an IC are notionally the same as discrete ICs used on a printed wiring board design. Each component has precisely defined behavior and a well-defined interface represented by a set of I/O pins and corresponding specifications for loading, setup and hold times, and delays. Components can be relatively simple or as complex as a RISC processor, MPEG decoder, or Wireless LAN modem. Virtual components can be classified as *hard*, *firm*, or *soft*. A hard module is normally defined at the mask level in a particular process. Thus, it will have a fixed floorplan, size, and a well-known set of timing parameters. A firm block will normally have a specific or generic netlist that describes each gate or register that must be used in the design (i.e., a 3-input NAND gate of normal power). This allows the design to be ported to multiple processes purely by netlist translation. The timing is dictated by the process and the final physical placement, however. A soft block is normally defined at the RTL level in the HDL. This captures the function of the block, but the detailed implementation is left to automated tools. Again, timing is dependent on the specific implementation. The Virtual Socket Interface Alliance monitors and encourages standards governing the implementation and use of virtual components.

8.6   Data Sheets and Documentation

A data sheet for an IC describes what it does and outlines the specifications for making the IC work in a system, such as power supply voltages, currents, input setup times, output delay times, and clock cycle times. The data sheet also includes package and pinout details.

A good habit to acquire is that of compiling a data sheet for any chip you might design. Not only is it the interface between the chip designer and the board-level designer, but also it is the interface to other members of the design team. In particular, it is good practice and is mandatory in industry to compile the data sheet for the chip and give it to the ultimate customer before the chip is fabricated. This prevents many undesirable scenarios that can arise when a perfectly designed chip meets a perfectly designed system. In this section, an outline of a typical data sheet will be reviewed by way of example.

8.6.1   The Summary

A summary of the chip includes the following details to orient the user:

- the designation and descriptive name of the chip
- a concise description of what the chip does
- a features list (optional for an internal product—but good for your ego!)
- a high-level block diagram of the chip function

8.6.2 Pinout

The pinout section should contain a description of the following pin attributes to document the external interface of the chip:

- name of the pin
- type of pin (i.e., whether input, output, tristate, digital, analog, etc.)
- a brief description of the pin function
- the package pin number

8.6.3 Description of Operation

This section should outline the operation of the chip as far as the user of the chip is interested. Programming options, data formats, and control options should be summarized.

8.6.4 DC Specifications

This section communicates the power dissipation and required voltages for the chip to correctly operate. The absolute maximum ratings should be stated for:

- supply voltage
- pin voltages
- junction temperature

The style of each I/O (i.e., TTL, CMOS, LVDS, ECL) should be summarized and the following DC specifications should be given over the operating range (temperature and voltage—i.e., mins and maxes):

- V_{IL} and V_{IH} for each input
- V_{OL} and V_{IH} for each output (at a given maximum drive current level)
- the input loading for each input
- quiescent current
- leakage current
- power-down current (if applicable)
- any other relevant voltages and currents

8.6.5 AC Specifications

The following timing specifications should be presented:

- setup and hold times on all inputs
- clock (and all other relevant inputs) to output delay times
- other critical timing such as minimum pulse widths

This data should be tabulated in table form and supported by a timing diagram where necessary. This is probably the most important section and an area where data provided ahead of the chip fabrication will aid the board designer. Designs are frequently snagged—for instance, when chip designers assume infinitely fast external memories and do not allow enough time between outputs changing and the next rising edge of the clock.

8.6.6 Package Diagram

A diagram of the package with the pin names attached should be supplied.

8.6.7 Principles of Operation Manual

Although the data sheet provides enough data to familiarize a user of a particular chip with the device, it is good practice to provide a Principles of Operation manual for internal users that have to test the chip or build support systems.

8.6.8 User Manual

A User Manual should also be provided. This is designed for use outside the group that designed the chip and can be a "cut down" version of the Principles of Operation manual.

8.7   Closing the Gap between ASIC and Custom

High-performance microprocessors and digital signal processing chips are typically designed in a custom methodology in which designers use a wide variety of circuit techniques and carefully tune all of the critical parts of the chip to achieve high performance at the expense of great design effort. Many other logic chips are built in a high-productivity ASIC methodology in which HDL code is synthesized onto a standard cell library and automatically placed and routed, as was described in Sections 8.4.1-8.4.2.

In a 180 nm process, custom microprocessors commonly operate at 1–2 GHz. In contrast, high-performance ASICs in comparable processes tend to operate at 200–350 MHz. Overall, custom design has consistently offered a 3–8x frequency advantage over ASIC designs. This is a remarkable difference, comparable to nearly 4–5 process generations of transistor improvement. [Chinnery02] offers a fascinating comparison of the two methodologies, explaining why custom flows offer so much higher performance and how ASIC designs can approach custom while maintaining high productivity. The remainder of this section summarizes the study's conclusions.

Measuring delays in terms of an FO4 inverter eliminates the process dependency and helps predict performance in the future. Overall, ASICs typically have cycle times of 50–100 FO4 inverter delays. High-performance ASICs such as the 520 MHz iCORE from STMicroelectronics can approach 25 FO4 delays per cycle. In comparison, the Pentium II and III operate at 20–24 FO4 delays, Alpha microprocessors at 13–16 FO4 delays, and the Pentium 4 at about 10 FO4 delays per cycle.

Chinnery identifies seven primary factors contributing to the custom performance advantage. Table 8.5 compares the effect of each of these factors. Custom design offers a large advantage over poor ASIC design in which performance is a secondary concern. Best practice ASIC methodologies can close much of the gap while still delivering far higher productivity than custom design. Poor is a relative term; often, economics demands fastest time to market rather than highest clock rate. Note that some factors affect only parts of the design so they cannot be simply multiplied to give the performance difference. The remainder of this section explores each of the factors.

Table 8.5	Relative performance of custom vs. ASIC design methodologies	
Factor	vs. Poor ASIC	vs. Best Practice ASIC
Microarchitecture (e.g., pipelining)	1.8x	1.3x
Sequencing overhead: elements, skew, time borrowing	1.45x	1.1x
Circuit families (e.g., domino)	1.4x	1.2x
Logic design	1.3x	1.0x
Cell design, cell sizing, and wire sizing	1.45x	1.1x
Layout: floorplanning, placement, wire management	1.4x	1.0x
Exploiting process variation and accessibility	2x	1.2x

8.7.1 Microarchitecture

Many problems have inherent parallelism. Pipelining or performing multiple operations simultaneously can speed them up. The benefits depend on the degree of parallelism in the problem and in the sequencing overhead of the pipeline registers. Increasing pipelining or issuing more operations at once also increases the area, power consumption, and design effort. In the next section, we will see that ASICs have greater sequencing overhead and thus cannot benefit from pipelining as much as custom designs. However, many ASICs still have plenty of room to benefit from aggressive microarchitecture.

8.7.2 Sequencing Overhead

ASIC designers have little control of layout so they need sequencing elements that avoid races and are tolerant of noise. Most ASICs use very conservative flip-flops with a sequencing overhead (setup time + clock-to-Q delay) of 4–6 FO4 inverter delays.

In comparison, custom designs tend to use more aggressive sequential circuits. Some use pulsed latches or transparent latches. Some integrate logic into the latches to reduce sequencing overhead or use faster, but less robust elements when noise on the input or output can be better controlled. These improvements drive the sequencing overhead down to 2–3 FO4 delays.

Moreover, most systems are designed with a clock skew budget of about 10% of the cycle time. This corresponds to about 5 FO4 delays for a typical ASIC as compared to 1-2 for a custom design. The ASICs use clock distribution networks that are automatically generated by CAD tools. These networks are not as well balanced as in a good custom design and are much more susceptible to process variation. Moreover, they are not analyzed as carefully as in a custom design, so the clock skew budget must be conservative to account for uncertainties.

Finally, flip-flops impose hard edges, so clock skew directly impacts performance. They also are incapable of borrowing time to balance logic between pipeline stages or to opportunistically adjust for unpredicted delay variations. Custom latches have some skew tolerance. Custom designs also sometimes intentionally delay clocks to balance logic.

In summary, ASICs tend to have a sequencing overhead of about 10 FO4 delays, while custom designs have overhead as small as 3 FO4 delays. ASICs can reduce sequencing overhead by substituting pulsed latches or transparent latches for flip-flops on critical paths whenever tools determine that no races will occur. Better clock distribution networks and filtered power supplies on the PLL reduce skew.

8.7.3 Circuit Families

Most custom designs use skew-tolerant domino circuits on critical paths. Domino gates are 1.5–2x faster than complementary CMOS. Skew-tolerant domino pipelines use overlapping clocks to eliminate latches and their sequencing overhead. Custom designs occasionally use other circuit families such as CPL or pseudo-nMOS for specialized applications.

Domino is sensitive to noise and must be shielded from crosstalk. The monotonicity requirement tends to force the use of dual-rail gates. Domino has a high activity factor and thus high power consumption. It is not well understood by ASIC timing analyzers and test tools. While some groups are developing domino synthesis methodologies, many observers doubt that domino will gain widespread use in ASIC flows.

8.7.4 Logic Design

Custom designs tend to carefully optimize the logic design. Many functions have clever implementations that can be significantly faster. For example, lookahead adders are much faster than ripple carry adders, and Booth-encoded multipliers can be faster than those using simple arrays.

ASICs are synthesized from HDL code. Older synthesis tools took the code literally, so a Booth-encoded multiplier had to be specified manually. Modern tools are better at choosing an appropriate implementation of common blocks like adders or multipliers (as long as you've purchased the vendor's intellectual property libraries!), but still need careful coding to produce good results for more exotic functions.

8.7.5 Cell and Wire Design

ASICs are synthesized onto a standard cell library. Some libraries have a small selection of gates. A library with only two drive strengths for NANDs, NORs, and inverters may be 25% slower than a library with a rich selection of drive strengths and buffer sizes and a full selection of compound and noninverting gates. The richer library can also reduce circuit area. Good libraries also offer asymmetric and skewed gates. While custom designs may use an unlimited set of cell sizes, this only offers a small percentage improvement over a good ASIC library with discrete sizes.

Wire capacitance and RC delay also are important. Custom designers choose the metal layer, wire width, and wire spacing and may even shield critical wires. ASIC designers have little control over the wires. Better ASIC tools are needed to automatically optimize wires on critical nets.

8.7.6 Layout

Global wires can account for a large fraction of total path delay and are getting worse with each process generation. Custom designers plan for these wires in advance and pipeline where necessary to tolerate the latency. They also add buffers and repeaters as needed. ASIC designers often do not think about the physical implementation and thus are surprised when long wires make a path much longer than simply counting gates would predict. Poor wire models in synthesis may mean these slow paths are not detected until after place and route has completed.

The key to managing wire delay is to carefully floorplan the chip. Units should be partitioned so that most communication happens inside the unit with short wires. Units that must communicate rapidly should be placed near each other. To reduce congestion, units communicating over wide busses should also be near each other. ASIC floorplanning tools are still maturing, but should eventually achieve parity with custom tools.

8.7.7 Process Variation

Custom designs are often fabricated on aggressive processes. For example, the Intel 130 nm process uses 60 nm effective channel lengths, while the TSMC process uses 80 nm effective lengths. Thus, an identical design will deliver a higher frequency on the Intel process. Intel also sometimes shrinks the effective channel length by another 5% as the process matures, improving performance even further. Foundries with higher performance processes tend to charge correspondingly higher prices.

Design corners also impact performance. Synchronous chips must be designed to operate at the worst temperature and voltage for which they are rated. Many ASICs are designed for the SS process corner so all of the chips fabricated without defects will meet timing. Thus, they can be tested only for functionality, not performance. Unfortunately, the distribution of chips may have a long tail at the slow end, as shown in Figure 8.48.

Custom chips are often designed at the TT process corner to give a 17%–28% speedup at the expense of rejecting a few chips that fall in the tail. Moreover, the fastest 10% of the chips (the FF corner) may be up to 30% faster than TT. Microprocessors are commonly sorted by speed, and the fastest ones are sold at a premium. This is sometimes called *binning*. Over time, process improvements may yield even faster chips.

ASIC performance is being crippled by the lack of testing at speed. High-performance ASICs can be designed for the TT corner of a fast process rather than the SS corner of a slower process. In some markets, they may be binned to deliver some very fast chips.

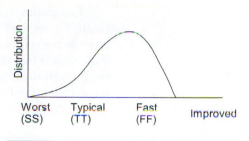

FIG 8.48 Distribution of chip behavior

8.7.8 Summary

Floorplanning and circuit families are commonly cited as the major advantages of custom over ASIC methodologies. Overall, [Chinnery02] concludes that these factors, while significant, are relatively overstated in their significance. Pipelining and process variation are of equal or greater significance. With attention to all of these details, ASICs may reduce their gap to within 2–3x the performance of custom designs.

8.8 CMOS Physical Design Styles

Basic gate layout was introduced in Section 1.5.4. In this section, we will examine the physical layout of CMOS gates in a general sense to understand the impact of the physical structure on the behavior and performance of circuits. For more extensive treatment by one of IBM's mask design instructors, see [Saint02].

8.8.1 Static CMOS Gate Layout

Complementary static CMOS gates can be designed using a single row of nMOS transistors below (or above) a single row of pMOS transistors, aligned at common gate connections. Most "simple" gates can be designed using an unbroken row of transistors in which abutting source/drain connections are made. This is sometimes called the "*line of diffusion*" rule, referring to the fact that the transistors form a line of diffusion intersected by polysilicon gate connections.

If we adopt this layout style, we can use automated techniques for designing such gates [Uehara81]. The CMOS circuit is converted to a graph where

1. the vertices in the graph are the source/drain connections

2. the edges in the graph are transistors gates that connect particular source/drain vertices

Two graphs, one for the pull-down network (n), and one for the pull-up network (p), result. Figure 8.49 shows an example of the graph transformation. The connection of edges in the graphs mirrors the series-parallel connection of the transistors in the circuits. Each edge is named with the gate signal name for that particular transistor. For example, the p-graph (light lines and circles) has four vertices: Y, I1, I2, and V_{DD}. It has four edges, representing the four transistors in the pull-up structure. Transistor A (A connected to gate) is an edge from the vertex Y to I2. The other transistors are similarly arranged in Figure 8.49(b). Note that the graphs are duals of each other because the pull-up and pull-down networks are the dual of each other. The n-graph (dark lines and crosses) overlays the p-graph in Figure 8.49(b) to illustrate this point. If two edges are adjacent in the p- or n-graph, then they can share a common source/drain connection and can be connected by abutment. Furthermore, if there exists a sequence of edges (containing all edges) in both graphs that have identical labeling, then the gate can be designed with no breaks in the line of diffusion. This path is known as a *Euler path*. The main points of the algorithm are:

1. Find all Euler paths that cover the graph.

2. Find a p- and n-Euler path that have identical labeling (a labeling is an ordering of the gate labels on each vertex).

3. If the paths in step 2 are not found, then break the gate in the minimum number of places to achieve step 2 by separate Euler paths.

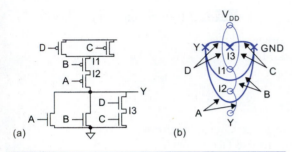

(a)

(b)

FIG 8.49 Circuit graphs

The original graph with a possible Euler path is shown in Figure 8.50(a). The sequence of gate signal labels in the Euler path is (A, B, C, D). To complete a layout, the transistors are arranged in the order of the labeling in parallel rows, as shown in stick diagram form in Figure 8.50(b). Vertical polysilicon lines form the gate connections. Metal routing wires complete the layout. This procedure can be followed when manually designing a gate, although good layouts usually become possible by inspection with a bit of practice.

A variation of the "line of diffusion" style occurs in circuits where a signal is applied to the gates of multiple transistors. In this case, transistors can be stacked on the appropriate gate signal using multiple rows of diffusion in a style called *gate matrix layout* [Wing82, Hu90]. This also

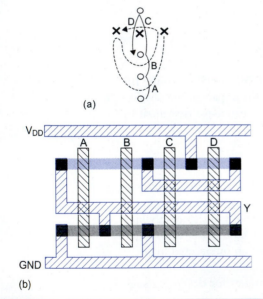

FIG 8.50 Stick diagram derived from Euler path

occurs in cascaded gates that cannot be constructed from a single row of transistors. A good example of this is the complementary XNOR gate. A schematic for this gate is shown in Figure 8.51(a). According to the style of layout that we have used to date, two possible layouts are shown in Figure 8.51(b) and Figure 8.51(c). The layout in Figure 8.51(b) uses the single row of n- and p-diffusion with a break, while that of Figure 8.51(c) uses a gate matrix layout. The selection of styles would depend on the overall layout—whether a short fat or long thin cell were needed. Note that the gate segments that are maximally connected to the power and ground rails should be placed adjacent to these signals.

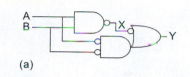

(a)

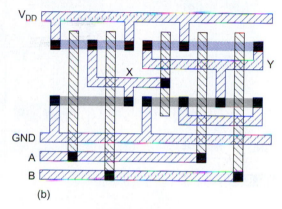

(b)

8.8.2 General CMOS Layout Guidelines

Layout can consume an unlimited amount of time because there are so many degrees of freedom and there is so much opportunity to squeeze a lambda here or there. In general, time to market is much more important than reducing chip area by a few percent, so it is important to settle on a simple and consistent layout design methodology. The following general layout guidelines can be stated:

1. Complete the electrical gate design and verification before layout. Circuit changes after layout is started become schedule busters.

2. Run V_{DD} and GND horizontally in metal at the top and bottom of the cell. Often these wires are wider than minimum to carry large DC currents without electromigration problems.

3. Run a vertical polysilicon line for each gate input.

4. Order the polysilicon gate signals to allow the maximal connection between transistors via abutting source/drain connections. These form gate segments.

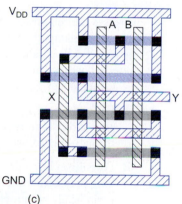

(c)

FIG 8.51 Broken line of diffusion and gate matrix cell layout styles

5. Place n-diffusion segments close to GND and p-diffusion segments close to V_{DD}, as dictated by connectivity requirements.

6. Make connections to complete the logic gate in polysilicon (for short connections between gates) or metal. Squeeze transistors together to minimize diffusion between transistors.

7. Place well and substrate contacts under the supply lines in each cell.

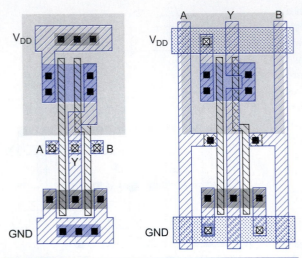

FIG 8.52 Standard cell metal usage

In general, metal layers should run perpendicular to each other to avoid "routing oneself into a corner." Exceptions are sometimes made to allow limited use of metal1 in the "wrong direction" to shorten connections or avoid the need for metal2 within a cell. Figure 8.52 shows two styles of standard cell layout for 2-input NOR gates. The first uses metal1 horizontally. The second uses metal1 vertically. Observe that the polysilicon gates are bent to minimize the diffusion between series transistors. The layouts assume that metal1–metal2 vias can be stacked on top of poly-metal1 contacts, as is common in modern planarized processes. If this is not allowed, the contacts must be placed adjacent to each other, sometimes increasing cell area.

For standard cells, inputs and outputs must usually be routed to contacts near the center in current processes or the top or bottom of the cell where they can connect to the routing channels for older processes with few metal layers. In the vertical metal1 style, this often increases the cell area because the metal1 cannot run over the top of other contacts within the cell. In datapath cells, however, inputs and outputs can contact bitlines running over the top of cells parallel with V_{DD} and GND. In this case, the vertical metal1 style may be preferred because metal2 bitlines are free to run horizontally over the cells.

Other layout guidelines include:

1. Diffusion has high resistance and capacitance. Never wire in diffusion. Minimize the area of diffusion regions. Fully contact large transistors to avoid series resistance through the diffusion between the contact and the edge of the transistor.

2. Polysilicon has high resistance, so use it only for short connections within cells. When long polysilicon lines are required (e.g., in the word line of a memory), strap the poly periodically with metal.

3. Lower levels of metal are thin and on a tight pitch. They are best for shorter connections (e.g., within a functional block) where density is important.

4. Upper levels of metal are thicker and on a wider pitch. They are faster and well-suited to global interconnections, the clock, and the global power/ground network. However, they are a scarce resource and must be carefully allocated.

5. Probe points should be placed on the top metal layer where they will be accessible during test (see Section 9.4).

6. Consider adding an assortment of unused gate array "happy gates" scattered through random logic. This facilitates making metal-only changes to fix logic bugs during silicon debug.

Note that the style of layout discussed involves optimizing the interconnection at the transistor level rather than the gate level. As a rule, smaller and perhaps faster layouts result by taking logic blocks with 10- to 100-transistor complexities rather than designing individual gates and trying to piece them together. For example, Figure 8.53(a) shows a transparent latch schematic. Figure 8.53(b) shows the latch layout built from simple standard cells, while Figure 8.53(c) shows an optimized layout with 2/3 the area. This improvement in density is due to a number of factors, including:

1. Better use of routing layers—routes can occur over cells

2. More "merged" source/drain connections

3. More use of "white space" (blank areas with no devices or connections) in sparse gates

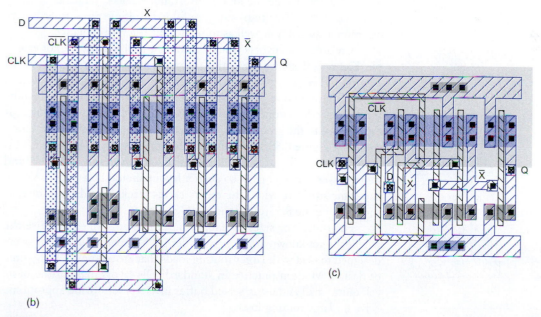

FIG 8.53 Transparent latch layouts

Improvements gained by optimizing at this level over a poorly implemented standard-cell approach can be up to 100% or more in area. However, such an approach is quite labor-intensive. These days, it is only worth investing manual effort in highly repetitive and reused structures like datapaths and widely used standard cells. Implementing random control logic manually in this manner is clearly a mistake because this type of logic often changes and the manual effort has to be continually spent to keep up with the changes. With modern multi-level metallization processes and optimized standard cell libraries, the density difference between custom-designed cells and hand or algorithmically placed standard cells is minimal if the same circuits are used because the transistor area fits under any routing. Density differences for custom circuits occur where the circuit is optimized to reduce the number of transistors (i.e., taking out buffer inverters in a latch). The point is that "in the old days" there was a much greater difference between custom and even a well-implemented standard cell design than there is today (given the same circuits).

8.8.3 Layout Optimization for Performance

Layout choices strongly impact the parasitic delay of logic gates. Good layout minimizes diffusion capacitance. Compact cells also lead to shorter wires between cells and lower wire parasitics. Figure 8.54 shows two implementations of a 24/12 λ inverter. Figure 8.54(a) is a conventional design. Figure 8.54(b) is a *folded* design in which each transistor is constructed from two parallel devices of half the width. Table 8.6 shows that the area and perimeter of the diffusion on the output Y has been reduced by nearly a factor of two. This does not change the logical effort, but reduces the parasitic delay by almost 50%. Wide transistors are folded multiple times to fit in layout more reasonably as well as to reduce diffusion capacitance.

A related case involves the design of parallel transistors. For example, Figure 8.55 shows three stick diagrams for a 2-input NOR gate. In Figure 8.55(a), the two nMOS drains are shared on the output as a single diffusion node, resulting in low parasitic capacitance on the output. The parallel connection of the two sources to the ground rail adds capacitance to the ground rail, but does not affect the output switching speed. In Figure 8.55(b), the two drains are separately contacted. This contributes more parasitic capacitance and results in a slower (and more power-hungry) gate, so such a layout should be avoided. When the transistors are wide, the entire series stack should be folded, as shown in Figure 8.55(c) for the pMOS stack.

The transistor sizes in datapaths can be optimized because the wire loads are known at the time of design. Standard cells can be over-designed to deal with larger routing loads that are unknown at the time of design. Modern practice in standard cells favors making the basic cell quite small and using scaled buffer inverters to provide appropriate drive for large routing loads.

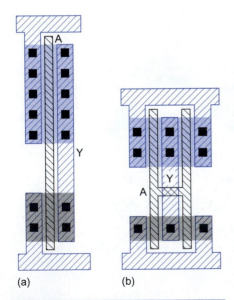

(a) (b)

FIG 8.54 Folding transistors

Table 8.6	Diffusion dimensions for unfolded and folded transistors					
Transistor	**Unfolded**			**Folded**		
	Width (λ)	Drain Area (λ^2)	Drain Perimeter (λ)	Width (λ)	Drain Area (λ^2)	Drain Perimeter (λ)
N1	12	60	22	6 + 6	36	12
P1	24	120	34	12 + 12	72	12

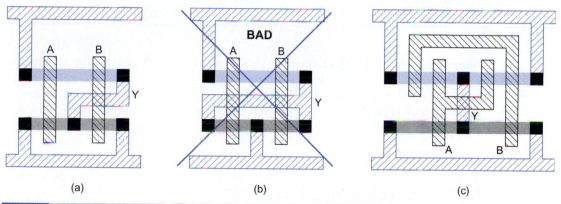

FIG 8.55   Parallel transistor layout

In long series stacks of transistors, the parasitic delay grows quadratically with the number of transistors, as shown in EQ (4.6). For gates driving small loads, the delay may be dominated by this parasitic component. For example, Figure 8.56(a) shows a 4-input footless dynamic NAND gate. It may be possible to reduce the parasitic delay by *tapering* transistors, as shown in Figure 8.56(b) [Shoji85, Wurtz93]. Tapering reduces the size of the inner transistors to reduce their diffusion capacitance at the expense of greater resistance, as shown in Table 8.7. Because the capacitance is discharged through all the series transistors while the resistance is only increased for the inner transistors, tapering reduces the parasitic delay. The greater resistance results in higher logical effort, so the technique is only appropriate for lightly loaded gates. Unfortunately, under most layout design rules, the spacing

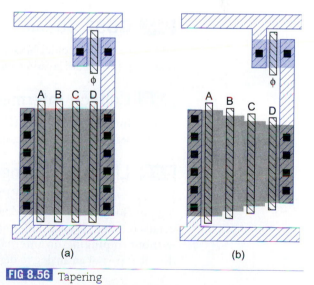

FIG 8.56   Tapering

Table 8.7	Diffusion dimensions in untapered and tapered stacks					
Transistor	Untapered			Tapered		
	Width (λ)	Drain Area (λ^2)	Drain Perimeter (λ)	Width (λ)	Drain Area (λ^2)	Drain Perimeter (λ)
N1	30	90	6	30	117	11
N2	30	90	6	27	105	11
N3	30	90	6	24	94	10
N4	30	150	40	22	110	32
P1	8	40	18	8	40	18

between uncontacted series transistors must increase when the transistors are tapered. Notice that the internal diffusion areas were larger in the tapered stack, as shown in Figure 8.56. As a result, tapering series stacks in practice may result in comparable or even greater parasitic delay [Hoppe90]. Tapering compound gates at contacted diffusion nodes does not increase the diffusion area and may still be a reasonable practice.

8.9 Interchange Formats

Throughout this chapter, reference has been made to a variety of (mostly) text interchange formats. These are summarized in the following sections.

8.9.1 GDS2 Stream

GDS2 stream is an older binary format that is the *de facto* standard for describing mask geometry. It includes layering and geometric primitives such as rectangles and polygons.

8.9.2 Caltech Intermediate Format (CIF)

Caltech Intermediate Format (CIF) is an alternative text-based mask description language largely used by the academic community [Mead80].

8.9.3 Library Exchange Format (LEF)

Library Exchange Format (LEF) describes the physical attributes of library cells, including port locations, layers, and via definitions. LEF abstracts the lower-level geometric details of a cell. Enough information is specified to allow a router to connect to the cell without impinging on internal cell constraints. The following LEF code segment describes part of an analog to digital converter.

```
LAYER metal1
    TYPE ROUTING ;
    WIDTH 0.2 ;
    SPACING 0.4 ;
    PITCH 1.0 ;
    DIRECTION HORIZONTAL ;
    CAPACITANCE CPERSQDIST 0.00003 ;
END metal 1
...
MACRO adc
    ORIGIN 0 0 ;
    SIZE 100 BY 200 ;
    PIN in
    DIRECTION INPUT ;
    PORT
    LAYER METAL4 ;
    RECT 10 10 10.5 10.5
    END in
    ...
    OBS
    LAYER METAL1 ;
    RECT 0 100 20 150 ;
    END
END adc
```

The description is somewhat human-readable. First, some technology specifications are evident. These specify which layer does what on the process. For instance, metal1 is a routing layer with the width, spacing, and pitch shown. The direction for routing is specified as horizontal. Alternate layers usually route in orthogonal directions. Finally, the capacitance of the layer is specified, which allows tools to make rough estimates of delays if necessary.

The description then proceeds with a declaration of the cell name (adc). The origin and size are then specified. A typical pin (in) is specified with the direction, layer, size (and position) designated. The size is specified by the RECT 10 10 10.5 10.5 command.

Finally, the OBS statement describes an obstruction—the areas in a given layer that a router must avoid within a cell.

8.9.4 Design Exchange Format (DEF)

Design Exchange Format (DEF) describes an actual design by listing the library elements and their placement and connectivity. DEF is used to pass designs between different design systems while maintaining the design intent. This is opposed to just passing geometric information such as GDS2 stream format. A segment of DEF is shown below.

```
VERSION 5.2 ;
NAMESCASESENSITIVE ON ;
DIVIDERCHAR "/" ;
BUSBITCHARS "[]" ;
DESIGN chip ;
UNITS DISTANCE MICRONS 1000 ;

COMPONENTS 1 ;
-adc_inst adc + FIXED ( -1000 -1000) N ;
END COMPONENTS

END DESIGN
```

Again, after some housekeeping, this is declared to be part of the design "chip." There is one placement of the adc block (`adc_inst` would be the Verilog role name). This block has been "fixed," i.e., the placement cannot be changed. Blocks can be unplaced, in which case, a placement program may move the blocks to improve timing or chip size.

8.9.5 Standard Delay Format (SDF)

SDF is an IEEE standard (P1497) to describe timing information in designs. It specifies pin-to-pin delays of modules, clock-to-data delays, and interconnect delays. A segment of SDF resulting from a timing analysis run is shown below.

```
(DELAYFILE
(SDFVERSION    "2.1")
(DESIGN        "chip")
(DATE          "March 24, 2003 11:57:3")
(VENDOR        "")
(PROGRAM       "PEARL")
(VERSION       "PEARL 5.1-s072 (64 bit)")
(DIVIDER       /)
(VOLTAGE       1.080:1.20:1.320)
(PROCESS       "slow=1.5:nom=1.0:fast=0.75")
(TEMPERATURE   80.000:25.000:-40.000)
(TIMESCALE     1ns)
(CELL
(CELLTYPE      "chip")
(INSTANCE)
(DELAY
   (ABSOLUTE
      (INTERCONNECT clk_mdi u_pad_clk_mdi/PAD (0.1:0.2:0.3) (0.4:0.5:0.6))
      (INTERCONNECT u_pad_clk_mdi/PAD clk_mdi (0.4:0.5:0.6) (0.7:0.8:0.9))
      )
      )
   (CELL
      (CELLTYPE "DFlipFlop")
      (INSTANCE foo/bar/reg)
      (DELAY
```

```
            (ABSOLUTE
                (IOPATH CK Q (.2:.25:.3) (.3:.35:.4))
            )
        )
    (TIMINGCHECK
        (WIDTH (posedge CK) (.06:.06:.06))
        (WIDTH (negedge CK) (.12:.12:.12))
        (SETUP (posedge D) (posedge CK) (0.1:0.1:0.1))
        )
    )
)
```

The segment begins with the design name and a log of what program ran to produce the SDF. Some interconnect delays are then given. The statement

```
(INTERCONNECT clk_mdi u_pad_clk_mdi/PAD (0.1:0.2:0.3) (0.4:0.5:0.6))
```

specifies an interconnect delay path between `clk_mdi` and `u_pad_clk_mdi/PAD` with rising and falling delays respectively of (`0.1:0.2:0.3`) and (`0.4:0.5:0.6`) ns, where the (`a:b:c`) format relates to (`slow:nominal:fast`) process corners. The specification of delay values can be extended from that shown here to include transitions between all states (0, 1, Z, X).

Next, the delay within a DFlipFlop cell is specified. This starts with an absolute delay from the clk-to-Q specified by

```
(IOPATH CK Q (.2:.25:.3) (.3:.35:.4))
```

Again the delay numbers noted here are for rise and fall times and slow, nominal, and fast process corners. The clock constraints on the DFlipFlop are then listed. The minimum clock widths are specified for each process corner. Finally, D-to-*clk* setup time is specified.

8.9.6 DSPF and SPEF

Detailed Standard Parasitic Format (DSPF) and Standard Parasitic Exchange Format (SPEF) are formats are used to pass parasitic RC values between extraction tools and timing verification tools. They are modeled after SPICE decks, as shown below.

```
*|DSPF 1.5
*|DESIGN "chip"
*|DATE "Mon Mar 24 07:26:34 2003"
*|VENDOR "Cadence Design Systems — HLD"
*|PROGRAM "hyperExtract 4.5.0"
*|VERSION "3.4E"
*|DIVIDER |
*|DELIMITER .
*|BUSBIT []

.SUBCKT chip in out
+ Vdd gnd
```

```
*Net Section
*
*|GROUND_NET gnd
*
*|NET name1 0.003PF
*I ???
*S ????
C1 node1 node2 1.2
C2 node3 node4 0.85
R1 node4 node5 0.5
*
.ENDS
```

8.9.7 Advanced Library Format (ALF)

Advanced Library Format (ALF) is an alternate format used to describe primitive library elements. It combines electrical performance characteristics such as capacitive pin loading and slew rates (e.g., SDF), physical modeling characteristics (such as pin layers and other geometric data, e.g., LEF) and behavioral specifications (such as the logic implemented by a cell, e.g. Verilog and VHDL).

8.9.8 WAVES Waveform and Vector Exchange Specification

WAVES is IEEE Std. 1029.1 and a subset of the VHDL standard. It provides for the graphical definition of stimulus and response patterns for digital systems. The following code is an example generated by the Aldec Active HDL suite.

```
% ****************************************************************
% * This file is automatically generated WAVES vector file,   *
% * and can be used in Test Bench generator.                  *
% * ACTIVE-HDL Testbench Generator ver. 3.5.                  *
% * Copyright (C) ALDEC Inc.                                  *
% *                                                           *
% * This file was generated on: 10:29:49 AM 12/9/2003         *
% ****************************************************************
%
% Begin Comment
%
% reset : sig reg
% clk : sig reg
%
% 4.00 us : END_SIMULATION_TIME
% End Comment
%
% Begin of Vectors
%
%@ 0 0
```

```
x 0 : 15 ns ; % 0 fs
x 1 : 15 ns ; % 15 ns
1 0 : 15 ns ; % 30 ns
1 1 : 15 ns ; % 45 ns
.
1 0 : 15 ns ; % 330 ns
1 1 : 15 ns ; % 345 ns
1 0 : 15 ns ; % 360 ns
1 1 : 15 ns ; % 375 ns
0 0 : 15 ns ; % 390 ns
0 1 : 15 ns ; % 405 ns
0 0 : 15 ns ; % 420 ns
0 1 : 15 ns ; % 435 ns
0 0 : 15 ns ; % 450 ns
0 1 : 15 ns ; % 465 ns
```

The preceding code describes two signals: *reset* and *clk*, which are shown in Figure 8.57.

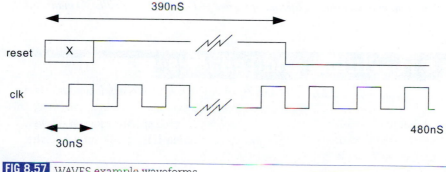

FIG 8.57 WAVES example waveforms

8.9.9 Physical Design Exchange Format (PDEF)

Physical Design Exchange Format (PDEF) is a Synopsys format to pass information between front-end and back-end tools. It describes hierarchical groupings of cells and the relevant interconnect information such as RC delays and cell loadings.

8.9.10 OpenAccess

Standard text file formats are used to exchange information between tools, designers and companies wishing to cooperate. A more effective manner of sharing data between tools running on the same machine is to provide an Application Programming Interface (API) or a set of software routines that communicate with a central database and extract data and insert data into that database. The OpenAccess API is one example of such an interface that has been applied to LEF and DEF descriptions.

8.10   Historical Perspective

When the second edition of this text was published, a few designers (including the author) were still using tools that were "home brewed," that is, cobbled together from publicly available or custom software. The designs were at a level of complexity where the ability to customize a tool to perform a function was quicker than waiting for a vendor to have a commercially available tool. The cost of entry into even custom IC design could be kept quite low. Now it is virtually impossible not to use a commercially available design flow without seriously impinging on schedules and design quality and integrity. Most large US, APAC, and EMEA universities now have commercially produced tools that are used for teaching. So, while the availability of advanced tools is an expensive startup cost for a small company, the basic tools and flows are available at a selection of tertiary institutes.

Small, well-defined designs can still be completed with a skeleton of tools; it just takes longer and is harder to prove correct. A layout editor, 2D extractor, LVS, and SPICE simulator can design a lot of circuits, especially in the analog mixed-signal area. But remember to engineer things properly and abide by the advice of this quote:

"A moment of convenience, a lifetime of regret." – *David Moon on TECO madness.*

8.11   Pitfalls and Fallacies

Inadequate design flow

In the previous section, we stated that it is possible to design small circuits without the benefit of a commercial design flow. On the other hand, using an inadequate design flow for a large chip is a recipe for failure. While commercial ASIC design flows are fairly well exercised, mixed-signal and advanced processes can tend to break tools (even commercial tools). Consider the case of moving from a 0.25 µm process to a 130 nm process without a signal integrity checker. In the latter technology case, this tool is almost mandatory to prevent timing anomalies due to unwanted wire-to-wire coupling.

Insufficient verification

Synopsys found that 82% of design spins for chips with functional flaws were due to lack of verification [Schutten03]. Another 47% of re-spins had incorrect specifications. And 14% had errors in imported IP. This outlines the need for good specifications and a well-thought-out verification plan. Verification is further covered in Chapter 9.

Inaccurate parasitic extraction

Parasitic extractions programs output reams of data relating to C and R values in a design. Unless these are guaranteed by your vendor, it is prudent to do a small design and compare the values with hand-calculated values. You can never be too careful when it comes to designing a chip. When the chip comes back, compare a known path with what was predicted by the tool set.

Exercises

8.1 The gate mix for a family of chips has found that 30% of the gates are scannable Dflip-flops, 20% are 3- or 4-input gates, and 40% are 2-input gates (NAND or NOR) and 10% are buffers of various sizes. Propose a gate array cell that best fits this gate mix for density. Is an SOG array a better choice from an area point of view?

8.2 An FIR filter for a GSM receiver with sigma-delta converter as shown in Figure 8.8(b) has a single-bit input. To what structure do the multipliers degenerate? If the coefficients are a single bit and a 288-tap filter has to operate at 13 MHz, what architecture would you use for the overall design?

8.3 What kind of RAM cell would you use to control a configurable logic block in an FPGA? Design the cell and outline the reasons for your choice.

8.4 Explain the tradeoffs between using a transmission gate or a tristate buffer to implement an FPGA routing block.

8.5 Using the MOSIS scalable rules, design a standard cell library where the nMOS transistor is 4λ wide and the pMOS is 6λ wide. What is the minimum vertical and horizontal pitch that you can achieve?

8.6 Write a software generator in a familiar high-level language to implement a programmable inverter (1x to 64x basic inverter size) for the standard cell designed in Exercise 8.5. The output should be terms of rectangles with the following syntax:

```
begin inv_Mx cell_name

  .

  .

    rect layer lower-x lower-y upper-x upper-y

  .

  .

end cell_name
```

layers are n-diff, p-diff, contact, poly, metal1, via12, metal2.

M is the number of paralleled unit inverters.

Research the syntax of LEF and write a LEF generator for the inverters.

8.7 Complete the 64-entry 7-bit sine ROM layout generator introduced in Section 8.3.5.

8.8 Write a software generator that generates a Verilog description for the sine ROM. Parameterize the ROM in amplitude and phase resolution.

8.9 If the ROM generator had to be implemented with static CMOS standard cells (as opposed to transistors), what cells would you use for each block in the example given in Section 8.3.5?

8.10 Show how power can be measured with SPICE for a digital circuit.

8.11 Research the detail of a CORDIC processor and design a CORDIC capable of generating an 8-bit sine wave with 8-bit phase accuracy. Code in an HDL. Compare this in terms of size, speed, and power dissipation with a ROM-based NCO.

8.12 Construct a data sheet for the CORDIC designed in the previous question.

8.13 Estimate the die cost of a 4×4 mm die, with Y_w = 80% and Y_{pa} = 98% for an 8″ wafer costing $2,200 each. The die may be shrunk to 3.3×3.3 mm in a more advanced process that costs $3000 per wafer. Is it worth moving to the new process if the volume is large enough?

8.14 Sketch a stick diagram for a large inverter with an 80 λ pMOS transistor and 40 λ nMOS transistor. Fold the transistors so that no single transistor is wider than 20 λ.

8.15 Using the RC delay model, estimate the worst-case falling delay of the untapered and tapered 4-input dynamic NAND gates in Figure 8.56. Let the load be a 24/6 λ inverter. Assume the diffusion capacitance between two series unit transistors is $C/2$ if the transistors are the same width, $3C/4$ if the transistors are of different widths, and C if the node requires a contact. Does tapering produce a faster gate in this example?

8.16 Using the same assumptions as in Exercise 8.15, calculate the falling parasitic delay and logical effort from each input of the tapered and untapered 4-input dynamic NAND gates.

Testing and Verification

9

9.1 Introduction

While in real estate the refrain is "Location! Location! Location!" the comparable advice in IC design should be "Testing! Testing! Testing!" For many chips, testing accounts for more effort than design.

Tests fall into three main categories. The first set of tests verifies that the chip performs its intended function. These tests are run before tapeout to verify the functionality of the circuit and are called *functionality tests* or *logic verification*. The second set of tests are run on the first batch of chips that return from fabrication. These tests confirm that the chip operates as it was intended and help debug any discrepancies. They can be much more extensive than the logic verification tests because the chip can be tested at full speed in a system. For example, a new microprocessor can be placed in a prototype motherboard to try to boot the operating system. This *silicon debug* requires creative detective work to locate the cause of failures because the designer has much less visibility into the fabricated chip compared to during design verification. The third set of tests verify that every transistor, gate, and storage element in the chip functions correctly. These tests are conducted on each manufactured chip before shipping to the customer to verify that the silicon is completely intact. These will be called *manufacturing tests*. In some cases, the same tests can be used for all three steps, but often it is easier to use one set of tests to chase down logic bugs and another, separate set optimized to catch manufacturing defects.

In Section 8.5.2, we noted that the yield of a particular IC was the number of good die divided by the total number of die per wafer. Because of the complexity of the manufacturing process, not all die on a wafer function correctly. Dust particles and small imperfections in starting material or photomasking can result in bridged connections or missing features. These imperfections result in what is termed a *fault*. Later in the chapter we will examine a number of fault mechanisms. The goal of a manufacturing test procedure is to determine which die are good and should be shipped to customers.

Testing a die (chip) can occur at the

- wafer level
- packaged chip level
- board level
- system level
- field level

By detecting a malfunctioning chip early, the manufacturing cost can be kept low. For instance, the approximate cost to a company of detecting a fault at the various levels [Williams86] is:

- wafer $0.01–$0.10
- packaged chip $0.10–$1
- board $1–$10
- system $10–$100
- field $100–$1000

Obviously, if faults can be detected at the wafer level, the cost of manufacturing is lower. In an extreme example, Intel failed to correct a logic bug in the Pentium floating-point divider until more than 4 million units had shipped in 1994. IBM halted sales of Pentium-based computers and Intel was forced to recall the flawed chips. The mistake and lack of prompt response cost the company an estimated $450 million.

It is interesting to note that most failures of first-time silicon result from problems with the functionality of the design; that is, the chip does exactly what the simulator said it would do, but for some reason (almost always human error) this functionality is not what the rest of the system expects.

The remainder of this section will provide an overview of the processes involved in logic verification, chip debug, and manufacturing test. Section 9.2 discusses the mechanics of testing and test programs. Sections 9.3–9.5 address the principles behind each phase of testing. If testing is not considered in advance, the manufacturing test can be extremely time consuming and hence expensive. Some chips have even proved impossible to debug because designers have so little visibility into the internal operation. Sections 9.6–9.9 focus on how to design chips to facilitate debug and manufacturing test at the chip and board level. Even if a chip works when it is shipped to the customer, it may wear out over time. Section 9.10 examines reliability and reliability testing.

9.1.1 Logic Verification

Verification tests are usually the first ones a designer might construct as part of the design process. Does this adder add? Does this counter count? Does this state-machine yield the right outputs each cycle? Does this modem decode data correctly?

In Chapter 8, we noted that verification tests were required to prove that a synthesized gate description was functionally equivalent to the source RTL. Figure 9.1 shows that we may want to prove that the RTL is equivalent to the design specification at a higher behavioral or specification level of abstraction. The behavioral specification might be a verbal description; a plain language textual specification; a description in some high-level computer language such as C, FORTRAN, Pascal, or LISP; a program in a system-modeling language such as SystemC; or a hardware description language such as VHDL or Verilog; or simply a table of inputs and required outputs. Often designers produce a *golden model* in one of the previously mentioned formats and it becomes the reference

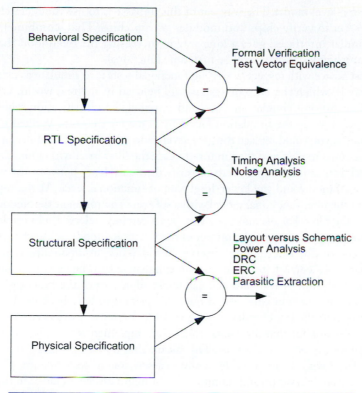

FIG 9.1 Functional equivalence at various levels of abstraction

against which all other representations are checked. Functional equivalence involves running a simulator at some level on the two descriptions of the chip (e.g., one at the gate level and one at a functional level) and ensuring that the outputs are equivalent at some convenient check points in time for all inputs applied. This is most conveniently done in an HDL by employing a *test bench*, i.e., a wrapper that surrounds a module and provides for stimulus and automated checking. The most detailed check might be on a cycle-by-cycle basis. Increasingly, verification involves real-time or near real-time emulation in an FPGA-based system to confirm system-level performance *in situ*, i.e., in the actual system that will use the end chip. This is recommended because of the increasing level of complexity of chips and the systems they implement. As an example, in the area of wireless local area network chips, without a real-time emulation system, it is virtually impossible to simulate the unseen effects of an unreliable channel with out-of-band interferers.

You can check functional equivalence through simulation at various levels of the design hierarchy. If the description is at the RTL level, the behavior at a system level may be able to be fully verified. For instance, in the case of a microprocessor, you can boot the operating system and run key programs for the behavioral description. However, this might be impractical (due to long simulation times) for a gate-level model and even harder for a transistor-level model. The way out of this impasse is to use the hierarchy inherent within a system to verify chips and modules within chips. That, combined with well-defined modular interfaces, goes a long way in increasing the likelihood that a system composed of many VLSI chips will be first-time functional.

The best advice with respect to writing functional tests is to simulate as closely as possible the way in which the chip or system will be used in the real world. Often this is impractical due to slow simulation times and extremely long verification sequences. One approach is to move up the simulation hierarchy as modules become verified at lower levels. For instance, you could replace the gate-level adder and register modules in a video filter with functional models and then in turn replace the filter itself with a functional model. At each level, you can write small tests to verify the equivalence between the new higher-level functional model and the lower-level gate or functional level. At the top level, you can surround the filter functional model with a software environment that models the real-world use of the filter. For instance, you can feed a carefully selected subsample of a video frame to the filter and compare the output of the functional model with what the designer expected theoretically. You can also observe the video output on a video frame buffer to check that it looks correct (by no means an exhaustive test, but a confidence builder). Finally, if enough time is available, you can apply all or part of the functional test to the gate level and even the transistor level if transistor primitives have been used.

Verification at the top chip level using an FPGA emulator offers several advantages over simulation and, for that matter, the final chip implementation. Most noticeably, the emulation times can be near real time. This means that the actual analog signals (if used) can be interfaced with the chip. Additionally, to assess system performance, you can introduce fine levels of observation and monitoring that might not be included in the final chip. For instance, you could include a bit-error rate circuit in a communication modem to aid performance optimization.

In most projects the amount of verification effort greatly exceeds the design effort. Remember the following statement, culled from many years of IC design experience, whenever you are tempted to minimize verification effort to meet tight schedules:

"If you don't test it, it won't work! (guaranteed)"

9.1.2 Basic Digital Debugging Hints

Many times, when a chip returns from fabrication, the first set of tests are run in a lab environment, so you need to prepare for this event. You can begin by constructing a circuit board that provides the following attributes:

- Power for the IC with ability to vary V_{DD} and measure power dissipation

- Real-world signal connections (i.e., analog and digital inputs and outputs as required)

- Clock inputs as required (it is helpful to have a stable variable-frequency clock generator)

- A digital interface to a PC (either serial or parallel ports for slow data or PCI bus for fast data interchanges)

You can write software routines to interface with the chip through the serial or parallel port or the bus interface. The chip should have a serial UART port or some other interface that can be used independently of the normal operation of the chip. The lowest level of the software should provide for peeking (reading) and poking (writing) registers in the chip. An alternate or complementary approach is to provide interfaces for a logic analyzer. These are easily added to a PCB design in the form of multi-pinned headers. Figure 9.2 shows a typical test board, illustrating the *zero insertion force* (ZIF) socket for the chip (in the center of the board), an area for analog circuitry interface (on the left), a set of headers for logic analyzer connection (at the top and bottom) and a set of programmable power supplies (on the right). In addition, an interface is provided for control by a serial port of a PC (at the bottom left).

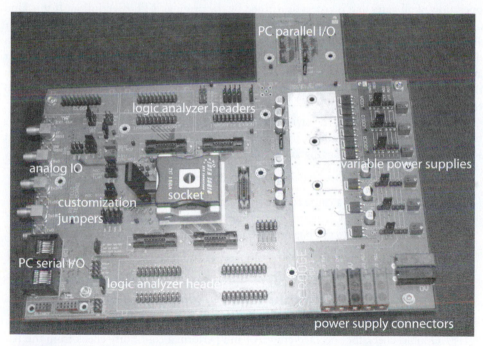

FIG 9.2 Typical test board

You should start with a "smoke test." This involves ramping the supply voltages from zero to V_{DD} while monitoring the current without any clocks running. For a fully static circuit, the current should remain at zero. Analog circuits will draw their quiescent current.

Following this, you can enable the clock(s); some dynamic current should be evident. Beware that many CMOS chips appear to operate when the clock is connected but the power supply is turned off because the clock may partially power the chip through the input protection diodes on the input pads. If possible, you should initially run the clock at reduced speed so that setup time failures are not the initial culprit in any debug operation.

In the case of a digital circuit, you should examine various registers for health using PC-based peek and poke software. This checks the integrity of the signal path from the PC to the chip. Often, designers place an ID in the register at address zero. Peeking at this register proves the read path from the chip. If the chip registers are reset to a known state, the registers can be read sequentially and compared with the design values. In the case of the logic analyzer, you can download the equivalent test pattern to exercise the chip. Frequently, these patterns can be automatically generated from the verification test bench. Up to this point, no functionality of the chip has been exercised apart from register reads and writes.

Where the chip has built-in self-test (see Sections 9.6–9.7), you can run the commercial software that provides for this functionality over a boundary scan interface. This type of system automatically runs a set of tests on the chip that completely verify the correct operation of all gates and registers as defined by the original RTL description. If this kind of a test interface was not used, you should pursue a manual effort in which the functionality of the chip is checked from the bottom-up. Of course, if you are a gambler, you can do a top-level test like running a piece of code or trying to boot the operating system right away. Experience shows that this often does not work, usually because of problems with the test fixture, and so you must revert to the bottom-up method to prove that one piece of the design works at a time.

If you detect anomalous behavior, you must go about debugging. The basic method is to postulate a method of failure, then test the hypothesis. Debug is an art in itself, but some pointers for sane debugging are as follows:

- Keep an annotated and dated logbook for all tests done.

- When postulating a cause for the bug and a test, do one change at a time and observe the result: Changing many things and then seeing if they work will not logically lead you to the bug and is commonly called the "shotgun approach."

- Check everything two or three times; never assume anything unless it is measured and logged in a notebook.

- Check signals and supply voltages at the pins of the IC; frequently, new test boards have errors.

- Double-check the specified chip I/O and perform a continuity check from the IC pins to expected places (i.e., test pins, supplies) on the board.

- Never count out a possible reason for a bug, however crazy, unless you can prove it isn't the cause.

- Use freeze spray or a heat gun to cool down or heat up a circuit to check for temperature problems.

- Check the state of any internal registers against that noted in the documentation.

- Evaluate the timing of any inputs and outputs with respect to the clock; often setup or hold times can be violated in a new test setup.

- When a bug is discovered and corrected, hunt for other portions of the design that might have a similar bug that hasn't been detected yet. Where there is one rat, there are many rats!

- Never assume anything—question everything—a slight touch of paranoia helps!!

When the chip is demonstrated to be operational, you can measure more subtle aspects of the design such as performance (power, speed, analog characteristics). This involves normal lab techniques of configure, measure, and record. Where possible, store all results as computer readable results (i.e., stored images from digital oscilloscope, screen dumps from logic analyzer) for communication with colleagues.

For the most part, if a digital chip simulates at the gate level and passes timing analysis checks during design, it will do exactly the same in silicon. Possible deviations from the simulated circuit occur in the following cases:

- Circuit is slower than predicted—fix—slow clock or raise V_{DD}

- Circuit has a race condition—fix—heat with heat gun if a logic gate caused race

- Circuit has dynamic logic problems—fix—don't do it again…

- Gnarly crosstalk problems—fix—get better tools

- Wrong functionality—fix—do a better job of verification

With analog circuitry, a wide range of issues can affect performance over and above what was simulated. These include power and ground noise, substrate noise, and temperature and process effects. However, you can employ the same basic debug approaches.

9.1.3 Manufacturing Tests

Whereas verification or functionality tests seek to confirm the function of a chip as a whole, manufacturing tests are used to verify that every gate operates as expected. The need to do this arises from a number of manufacturing defects that might occur during either chip fabrication or accelerated life testing (where the chip is stressed by over-voltage and over-temperature operation). Typical defects include:

- layer-to-layer shorts (e.g., metal-to-metal)

- discontinuous wires (e.g., metal thins when crossing vertical topology jumps)

- missing or damaged vias

- shorts through the thin gate oxide to the substrate or well

These in turn lead to particular circuit maladies, including:

- nodes shorted to power or ground

- nodes shorted to each other

- inputs floating/outputs disconnected

Tests are required to verify that each gate and register is operational and has not been compromised by a manufacturing defect. Tests can be carried out at the wafer level to cull out bad die, or can be left until the parts are packaged. This decision would normally be determined by the yield and package cost. If the yield is high and the package cost low (i.e., a plastic package), then the part can be tested only once after packaging. However, if the wafer yield was lower and the package cost high (i.e., an expensive ceramic package), it is more economical to first screen bad dice at the wafer level. The length of the tests at the wafer level can be shortened to reduce test time based on experience with the test sequence.

Apart from the verification of internal gates, I/O integrity is also tested, with the following tests being completed:

- I/O levels (i.e., checking noise margin for TTL, ECL, or CMOS I/O pads)

- speed test

With the use of on-chip test structures described in Section 9.6, full-speed wafer testing can be completed with a minimum of connected pins. This can be important in reducing the cost of the wafer test fixture.

In general, manufacturing test generation assumes the function of the circuit/chip is correct. It requires ways of exercising all gate inputs and monitoring all gate outputs.

Example

Consider testing the MIPS microprocessor from Chapter 1. Explain the difference between the tests you would use for logic verification or silicon debug and the tests you would use for manufacturing.

Solution: Logic verification should test that each operation can be performed. For example, a test program might exercise all of the instructions to demonstrate that each one behaves as intended. Logic verification will not necessarily prove that the instruction works for all possible addresses and data values. In contrast, manufacturing tests must prove that every gate operates correctly. They ideally stimulate each gate to produce both a 0 and a 1 to ensure the gate is not damaged. The manufacturing tests may be the only tests applied to a microprocessor prior to it being placed in a system and used. Clearly, it is a challenge to devise a set of tests that is both complete enough that customers receive very few defective chips and short enough to keep testing economical.

9.2   Testers, Test Fixtures, and Test Programs

To test a chip after it is fabricated, you need a tester, a test fixture, and a test program.

9.2.1   Testers and Test Fixtures

A tester is a device that can apply a sequence of stimuli to a chip or system under test and monitor and/or record the results of those operations. Testers come in various shapes and sizes.

To test a chip, one or more of four general types of test fixtures may be required. These are as follows:

- A *probe card* to test at the wafer level or unpackaged die level with a chip tester

- A *load board* to test a packaged part with a chip tester

- A printed circuit board (PCB) for bench-level testing (with or without a tester)

- A PCB with the chip *in situ,* demonstrating the application for which the chip is used

We will concentrate first on the cases where a general-purpose production tester is to be used. Production testers are usually expensive pieces of equipment with configurable I/O ports (drive current, output levels, input levels) and huge amounts of RAM behind each test pin. The tester drives input pins from this memory on a cycle-by-cycle basis and samples and stores the levels on output pins. Figure 9.3 shows a typical production tester. In the background, you can see the four-bay cabinet holding the drive electronics. To the right in the background is the controlling workstation. The test head is shown on the front center. This is where the chip is placed in the load board to be tested.

The probe card or load board for the *device under test* (DUT) is connected to the tester, as shown in Figure 9.4. The test program is compiled and downloaded into the tester and the tests are applied to the bare die or packaged chip. The tester samples the chip outputs and compares the values with those provided by the test program. If there are any differences, the chip is marked as faulty (with an ink dot) and the failing tests may be displayed for reference and stored for later analysis. In the case of a probe card, the card is raised, moved to the next die on the wafer, lowered, and the test procedure repeated. In the case of a load board with automatic part handling, the tested part is removed from the board and sorted into a good or bad bin. A new part is fed to the load board and the test is repeated. In most cases, these procedures take a few seconds for each part tested.

The ability to vary the voltage and timing on a per-pin basis with a tester allows a process known as "Shmooing" to be carried out. For instance, you could sweep V_{DD} from 3 V to 6 V on a 5 V part while varying the tester cycle time. This yields a graph called a *shmoo plot* that shows the speed sensitivity of the part with respect to voltage. Another

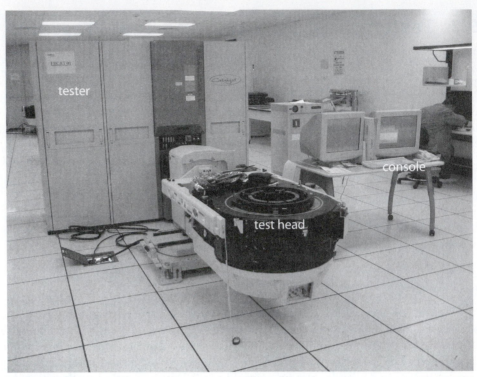

FIG 9.3   The Teradyne Catalyst: A typical production tester
(Photo: John Haddy, Cisco Systems)

shmoo that is frequently performed is to skew the timing on inputs with respect to the chip clock to look for setup and hold variations. Examples of shmoo plots and their interpretations are given in Section 9.4.

Testers can be very expensive, especially for high-frequency and/or analog/RF chips. The cost of testing each chip depends on the amount of time it must be in the tester. Applying tests to check every node on the chip may be prohibitively costly, so some designs face a tradeoff between test cost and the fraction of defective chips that slip through testing. Tester usage is charged by time, so the shorter a test runs, the cheaper a part is to test.

> ## Example
>
> **S**uppose a $2 million tester has an expected useful life of two years before it becomes inadequate to test faster next-generation parts. How much does the tester cost per second?
>
> **Solution:** Dividing the tester cost by the number of seconds in two years gives 3.2 cents/second.

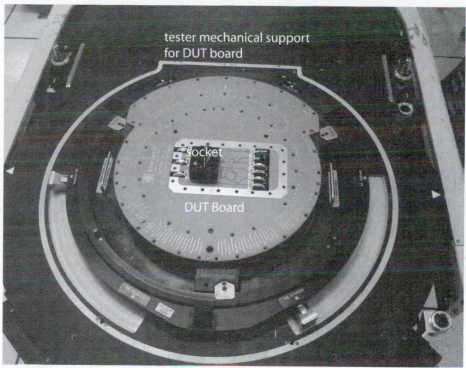

FIG 9.4   Tester load board in test head
(Photo: John Haddy, Cisco Systems)

Testers are available that can be used to test an IC in a laboratory environment. They mirror large production testers, but generally have less functionality (e.g., slower, less memory per pin, less expandability) and are markedly less expensive. A probe card that allows wafer probing or a socketed load board is required for each design. A good logic analyzer with a pattern generator and a socketed test board can also be used to test a chip. Some groups effectively design their own logic analyzers by surrounding a chip with FPGAs and using the logic and RAM within the FPGA to apply and observe test patterns.

9.2.2   Test Programs

The tester requires a *test program* (in verification and test, this is an overloaded term). This program is normally written in a high-level language (for instance, the IMAGE language used by Teradyne is based on C) that supports a library of primitives for a particular tester. The test program specifies a set of input patterns and a set of output *assertions*. If an output does not match the asserted value at the corresponding time, the tester will report an error. Before the patterns and assertions are applied, the test program has to set up the various attributes of a tester such as:

- Set the supply voltages

- Assign mapping between stimulus file signal names and physical tester pins

- Set the pins on the tester to be inputs or outputs and their V_{OH}/V_{IH} levels

- Set the clock on the tester

- Set the input pattern and output assertion timing

And then on a per chip basis:

- Apply supply voltages

- Apply digital stimulus and record responses

- Check responses against assertions

- Report and log errors

A stimulus or pattern file can be derived from running a simulation on the design. Special *vector change descriptions* (VCDs) are used to compact simulation results. An example of a simple stimulus/pattern file for the case of a full adder is shown below:

```
              III            OO
                             SC
                             UA
                             MR
                              R
              ABC             Y
0             000            00
1             001            10
2             010            10
3             011            01
4             100            10
5             101            01
6             110            01
7             111            11
```

The first line designates the signal directions and shows three inputs (I) and two outputs (O). Reading downwards, the next five lines designate the signal names (A, B, C, SUM, CARRY). Thereafter, each line designates a new *test vector*. The first column is the test vector number. The next three columns are the binary value of the inputs and the following two columns are the expected output values. Each line represents a certain length clock cycle that is asserted by the tester. Signals change after a specified period in relation to an internal clock running at the required test period. Clock generation can be carried out in two different ways. First, the clock can be treated like any other signal, in which case, it takes two tester cycles to complete a single clock cycle: one for the clock low and one for the clock high. Alternatively, a timing generator can be used, which allows the clock rising edge (for instance) to be placed anywhere in the tester cycle. So for instance, if the inputs are changed at the start of the tester cycle, the clock might be programmed to rise at the middle of the cycle.

Each pin on the tester is connected to a function memory, which is used to either drive an input or check an output at a DUT pin. Multiple bits may be required per pin to control tristate input pins or mask outputs when they should be ignored. These memories have finite length, so sometimes with older testers, more than one vector load has to be used to test a part. This normally slows testing as the reload procedure may be slow. Modern testers seldom suffer from this problem.

The clock speed, T_c, is specified, as are supply voltage levels. The time at which pins are driven and sampled is also specified on a pin-by-pin basis (T_s). The format of the test data is usually chosen from Non Return to Zero (NRZ), Return To Zero (RTZ), or other formats such as Surround By Zero (SBZ).

9.2.3 Handlers

An IC handler is responsible for feeding ICs to a test fixture attached to a tester. Chutes or trays containing packaged chips can be used to gravity-feed the devices to the handler, which uses a variety of mechanical means to pick the chips up and place them in the test socket on the load board. The tester stimulus is then applied and chips are binned depending on whether or not they passed the test. It is possible to heat and cool a chuck to test the chip at temperature. However, package-level testing is not normally carried out at temperature because of the time it takes to temperature-cycle the chuck.

An example of a handler is shown in Figure 9.5. This is the NS-6040 from Seiko-Epson. The body of the machine holds the mechanical positioning equipment, while the upper central section supports the test fixture. The light on top indicates a functioning or stopped machine and is designed to be visible across a production floor where many machines might be operating. (The light of another is visible top center.) A screen at the top right provides status information to the operator. The unit has wheels for easy movement, but also has firm footings, which are lowered when the machine is in use.

Handlers add a constant time to the test process, typically around 1 second. Thus, load boards and handlers are often constructed to deal with two or four chips at once to reduce the cost of testing. Because a load board must be designed to fit to a given handler, select the handler before starting design of the load board.

9.3   Logic Verification Principles

9.3.1 Test Benches and Harnesses

A verification *test bench* or *harness* is a piece of HDL code that is placed as a wrapper around a core piece of HDL. In the simplest test bench, inputs are applied to the module under test and at each cycle, the outputs are examined to determine whether they comply with a predefined expected data set. The data set can be derived from another model and available as a file or the value can be computed on the fly.

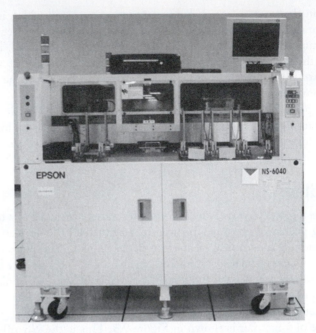

FIG 9.5   Photograph of an Epson NS-6040 IC handler
(Source: John Haddy—Cisco Systems)

Example

For the NCO introduced in Sections 8.2.3 and 8.4.1.1, the following is the core of a test bench that writes the sine values to a file. These may be subsequently used with a program such as MATLAB to calculate distortion or other attributes of interest.

```verilog
// Test Bench for NCO test structure
`timescale 1ns/10ps

module test_nco;
   parameter delay = 50;
   parameter MAX_COUNT = 512;

   reg         reset, clk;
   reg [15:0] init_phase, phase_delta;
   wire [7:0]  sine;
   integer     plot_file;
   integer     i;

   initial begin
      init_phase = 0; phase_delta = 0; clk = 0;
   end
```

```
nco dut(.reset(reset), .fclk(clk), .initial_phase(init_phase),
        .phase_increment(phase_delta), .q(sine));

    always #(delay/2) clk = ~clk; // define a clock
    always @(posedge clk)
        $fwrite(plot_file, "%d %d \n", $time, sine);

    // main test loop just cycles for MAX_COUNT cycles
    initial begin
        plot_file = $fopen("plot.out");
        for (i = 0; i < MAX_COUNT; i = i + 1)
            @(posedge clk);
        $finish;
    end
endmodule
```

Alternatively, the desired sine and cosine values can be externally calculated and stored in a file and compared "on the fly." The following code segment (similar to that shown in Appendix A.8) demonstrates that method of testing.

```
...
reg [7:0] expected_sine[511:0];
...
// read in the expected sine values to a file
initial
    begin
        $readmemh("hexsine.dat", expected_sine);
    end
always #(delay/2) clk = ~clk; // define a clock
    ...
// get the expected value and compare
for (i = 0; i < MAX_COUNT; i = i + 1)
always @(posedge clk)
    begin
        e_sine = expected_sine[i];
    if (sine !== expected_sine)
        begin
            $display("Error");
            $stop;
        end
    end
...
```

The sine file could be generated in MATLAB using the following code:

```
% MATLAB code generates one cycle of an 8 bit sine wave
t = [1:256];
t = t * 2 * pi / 256;
```

```
% approximates 8 bit resolution using fix
s = fix(sin(t) * 127) + 128;

f = fopen("sine.dat", "w");
f1 = fopen("hexsine.dat", "w");

for i1 = 1:256
  fprintf(f, "%d\n", s(i1));
  fprintf(f1, "%#.2x\n", s(i1));
end

fclose(f);
fclose(f1);
```

This produces the following `hexsine.dat` file:

```
0x80
0x83
0x86
0x89
...
0x71
0x74
0x77
0x7a
0x7d
```

The file comparison method is applicable to a wide range of simulation scenarios as files form a common basis for I/O between different design systems. The notion of a *golden model* is frequently used as the reference for establishing functional equivalence. A golden model might be a model for the system being designed in a high-level language such as C or in a design tool such as MATLAB. The golden model writes expected output files that are used as the basis for comparison.

Simulators usually provide settable break points and single or multiple stepping abilities to allow the designer to step through a test sequence while debugging discrepancies.

9.3.2 Regression Testing

High-level language scripts are frequently used when running large test benches, especially for *regression testing*. Regression testing involves performing a suite of simulations to automatically verify that no functionality has inadvertently changed in a module or set of modules. During a design, it is common practice to run a regression script every night after design activities have concluded to check that bug fixes or feature enhancements have not broken completed modules.

Example

In the software radio example, Figure 8.11 showed a possible architecture that used a combination of an IQ conversion block and a multiplier-based multiprocessor. The following regression testing might be done.

```
Test IQ Conversion
    Test Upconverter
        Test NCO
            Test Read and Write of All Registers
            Test Phase Incrementer
            Test Phase Adder
            Test Sine ROM (Read Contents)
            Test Overall NCO at a set of frequencies
        Test Multiplier
    Test Downconverter
        Test NCO
        ...
        Test Mulitplier
        ...
        Test Low Pass Filter
        ...
Test Microprocessor Memory Core
    Test Microprocessor
        Test ALU
        Test Instruction Decode
        Test Program Counter
        Test Register File Read/Write
        Exhaustive Instruction Test
    Test Memory Read/Write
Test Interprocessor Bus IO
Test IQ Conversion to Processor pathways
Test Overall Software Radio Functionality
```

Note the way in which the correctness of modules is slowly built up by verifying lower-level models first. The low-level tests are gradually built up in complexity until the complete functionality can be verified. At low levels, it is easier to exhaustively verify that logic is correct. For instance, we can verify that the sine ROM is in fact generating a sine wave for one frequency. We then use this knowledge to postulate that it generates correct sine waves for all input frequencies when we verify at the levels above the NCO. At the chip level, we assume that IQ conversion is correct for all combinations of signal frequency and local oscillator frequency even though we may only check a small subset. If we started at the top level and ran a simulation for a few frequencies, we could never have confidence that the lower levels were correct. In addition, if there is a problem, trying to locate the problem by debugging at the top level is futile. Running regression tests from the bottom up is designed to overcome this verification nightmare.

9.3.3 Version Control

Combined with regression testing is the use of versioning, that is, the orderly management of different design iterations. Unix/Linux tools such as CVS are useful for this.

> **Example**
>
> In the software radio example, the regression testing halts at the ALU test in the example given above. Working late, the design leader, Vanessa Eagleeye, examines the CVS history and discovers that Fred Codechanger has made an edit to the ALU design to try a new adder during the day. She is able to revert the code to what was previously working and then rerun the regression test and have a peaceful night's sleep. Fred corrects his mistake the next day and is advised to remember to run the regression verification step before making such hurried edits.

9.3.4 Bug Tracking

Another important tool to use during verification (and in fact the whole design cycle) is a bug-tracking system. Bug-tracking systems such as the Unix/Linux based GNATS allow the management of a wide variety of bugs. In these systems, each bug is entered and the location, nature, and severity of the bug noted. The bug discoverer is noted, along with the perceived person responsible for fixing the bug.

> **Example**
>
> From the example described in the previous section, Vanessa enters a bug report describing the bug. She cites Fred as the person responsible and the level as severe. The next day, Fred fixes the problem and changes the bug status to fixed. The bug report is kept in the system, but does not appear in any listing of outstanding bugs. It is kept to track the re-introduction of bugs, as this might give managers an idea of a problem area in the design management.
>
> Tracking the number of bugs can give you an idea of the rate at which a design is converging toward a finished state. If the trend is downward, the design is converging. On the other hand, an upward trend tends to indicate a design early in its verification cycle.

9.4 Silicon Debug Principles

The area of basic digital debugging was introduced in Section 9.1.2. A major challenge in silicon debugging is when the chip operates incorrectly, but you cannot ascertain the cause by making measurements at the chip pins or scan chain outputs (see Section 9.6.2).

There are a number of techniques for directly accessing the silicon. First, specific signals can be brought to the top of the chip as *probe points*. These are small squares (5–10 μm on a side) of top-level metal that connect to key points in the circuit that the designer has had the foresight to include before debug. The overglass cut mask should specify a hole in the passivation over the probe pads so the metal can be reliably contacted. Typical of these kinds of test points might be internal bias points in linear circuits or perhaps key points in a high-speed signal chain (be careful not to excessively load the circuit to be probed). The exposed squares can be probed with a picoprobe (fine-tipped probe) in a fixture under a microscope. During design, the load of the picoprobe has to be taken into account by providing buffers if necessary. The Model 35 probe from GGB Industries has a capacitance of .05 pF, input resistance of 1.25 MΩ, and frequency response from DC to 26 GHz. It can probe down to a 10 μm by 10 μm window.

The die can also be probed electrically or optically if mechanical contact is not feasible. An *electron beam* (ebeam) probe uses a scanning electron microscope to produce a tightly focused beam of electrons to measure on-chip voltages. Similarly, *Laser Voltage Probing* (LVP) [Lasserre99] involves shining a laser at a circuit and observing the reflected light. The reflections are modulated by the electric fields so switching waveforms can be deduced. However, the probing can be invasive; the stream of photons may disturb sensitive dynamic nodes. *Picosecond Imaging Circuit Analysis* (PICA) [Knebel98] captures faint light emission naturally produced by switching transistors and hence is noninvasive. Silicon is partially transparent to infrared light, so both LVP and PICA can be performed through the substrate from the backside of a chip in a flip-chip package.

On a more coarse scale, infrared (IR) imaging can be used to examine "hot spots" in a chip, which may be the source of problems (for instance, a resistive short between power rails). There are also liquid crystal materials, which can be "painted on" to a die to indicate temperature problems at a coarse resolution.

If the location of the fault is known, a *Focused Ion Beam* (FIB) can be used to cut wires or lay new conductors down. Even with plastic-packaged parts, the plastic can be carefully ground off and these repairs completed. The reason for this kind of tool is that normally in any chip project, time is of the essence and FIB runs are quicker (and cheaper for a few parts) than frequent mask changes. Laser cutting is also possible. Commercial providers such as MEFAS offer these services.

Example

A short between V_{DD} and GND has rendered a chip just back from tapeout non-functional. The position of the fault is known and it can be corrected by a cut to the top level metal. Several packaged parts are sent to the FIB house with a location from a given fiducial mark and an accompanying plot of the position of the metal to be cut. The FIB house exposes the die (i.e., by grinding a plastic package). The operator then locates the cut position manually using a microscope and runs the FIB machine. The modified packages are then returned to the designers, where hopefully they celebrate an otherwise useless chip.

Debugging logic circuits will often involve extremely fast or novel circuits that are largely analog in nature. In this case, it is advisable to have a model of the circuit in question available in SPICE. Debugging analog circuits, as with purely digital circuits, involves making an assertion and then trying to prove the assertion is correct. This can begin with a SPICE simulation and then progress to silicon measurement.

Failures causes may be *manufacturing*, *functional*, or *electrical*. Manufacturing failures occur when a chip has a defect or is outside of the parametric specifications. Debug can reject chips with manufacturing problems, although circuits sensitive to weaknesses in the manufacturing process can be changed to improve yield, as will be discussed in Section 9.6.5. Functional failures are logic bugs or physical design errors that cause the chip to fail under all conditions. They arise from inadequate logic verification and are usually the easiest to fix. Electrical failures occur when the chip is logically correct, but malfunctions under certain conditions such as voltage, temperature, or frequency. Section 6.3 addressed many causes of electrical failures. Some electrical failures can be so severe that they appear as functional failures, while others occur rarely and are extremely difficult to reproduce and diagnose.

So-called Shmoo plots can help to debug electrical failures in silicon [Baker97]. A Shmoo plot is often made with voltage on the *X* axis and speed as the *Y* axis. The test vectors are applied at each combination of voltage and clock speed, and the success of the test is recorded. Often, only a set of vectors applicable to a particular module is applied to diagnose a problem in that module. The Shmoo plots on the facing page show a variety of conditions [Josephson02].

A healthy normal chip should operate at increasing frequency as the voltage increases. The brick wall pattern suggests that the chip may be randomly initialized in one of two states, only one of which is correct. For example, a register without a reset signal may randomly have an initial state of 0 or 1. The wall pattern in which the chip fails to operate at any frequency above or below a particular voltage can indicate charge sharing, coupling noise, or a race condition. The reverse speedpath behavior indicates a leakage problem in which a weakly held node leaks to an invalid level before the end of the cycle. At higher voltage, the leakage is exacerbated and appears at shorter clock periods. The floor is a variant on the leakage problem where the part fails at low frequency independent of the voltage. A finger indicates coupling problems dependent on the alignment of the aggressor and victim, where at certain frequencies the alignment always causes a failure.

A shmoo can also plot operating speed against temperature. At cold temperature, FETs are faster, have lower effective resistance, and have higher threshold voltages. A normal shmoo should show speed increasing as temperature decreases. Failures at low temperature could indicate coupling or charge sharing noise exacerbated by faster edge rates. Failures at high temperature could indicate excessive leakage or noise problems exacerbated by the lower threshold voltages. Walls at either temperature could indicate race conditions where the path that wins the race varies with temperature.

Figure 9.6 shows an actual shmoo from the 433 MHz Alpha 21164 [Gronowski96]. At 1.8 V, the chip works for clock periods of 2.3 ns and greater. At higher voltage, the chip can operate at shorter periods, as one would expect.

Clock period in ns on the left, frequency increases going up
Voltage on the bottom, increase left to right

* indicates a failure

```
1.0   *   *   *   *   *   *            1.0   *       *       *
1.1   *   *   *   *   *                1.1       *       *       *
1.2   *   *   *   *                    1.2   *       *       *
1.3   *   *   *                        1.3       *       *       *
1.4   *   *                            1.4   *       *       *
1.5   *                                1.5       *       *       *
     1.0 1.1 1.2 1.3 1.4 1.5                1.0 1.1 1.2 1.3 1.4 1.5
```

Normal   "Brick Wall"
Well-behaved shmoo   Bistable
Typical Speedpath   Initialization

```
1.0           *   *   *                1.0   *   *   *   *   *
1.1           *   *   *                1.1       *   *   *   *
1.2           *   *   *                1.2           *   *   *
1.3           *   *   *                1.3               *   *
1.4           *   *   *                1.4                   *
1.5           *   *   *                1.5                   *
     1.0 1.1 1.2 1.3 1.4 1.5                1.0 1.1 1.2 1.3 1.4 1.5
```

"Wall"   "Reverse speedpath"
Fails at a certain voltage   Increase in voltage reduces frequency
Coupling, charge share, races   Speedpath, leakage

```
1.0                                    1.0
1.1                                    1.1
1.2                                    1.2   *   *   *   *
1.3                                    1.3           *   *
1.4   *   *   *   *   *   *            1.4
1.5   *   *   *   *   *   *            1.5
     1.0 1.1 1.2 1.3 1.4 1.5                1.0 1.1 1.2 1.3 1.4 1.5
```

"Floor"   "Finger"
Works at high but not low frequency   Fails at a specific point in the shmoo
Leakage   Coupling

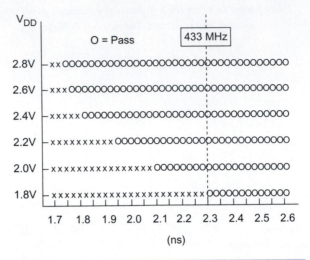

FIG 9.6 **FIG 9.6**   Alpha 21164 shmoo

9.5 Manufacturing Test Principles

A critical factor in all VLSI design is the need to incorporate methods of testing circuits. This task should proceed concurrently with any architectural considerations and not be left until fabricated parts are available (as is a recurring temptation to designers).

Figure 9.7(a) shows a combinational circuit with N inputs. To test this circuit exhaustively, a sequence of 2^N inputs (or test vectors) must be applied and observed to fully exercise the circuit. This combinational circuit is converted to a sequential circuit with addition of M registers, as shown in Figure 9.7(b). The state of the circuit is determined by the inputs and the previous state. A minimum of 2^{N+M} test vectors must be applied to exhaustively test the circuit. As observed by [Williams83] more than two decades ago,

> *With LSI, this may be a network with N = 25 and M = 50, or 2^{75} patterns, which is approximately 3.8 x 10^{22}. Assuming one had the patterns and applied them at an application rate of 1 μs per pattern, the best time would be over a billion years (10^9).*

Clearly, exhaustive testing is infeasible for most systems, even in the age of multi-GHz processors. Fortunately, the number of potentially nonfunctional nodes on a chip is much smaller than the number of states. Manufacturing test engineers must cleverly devise test vectors that detect any (or nearly any) defective node without requiring so many patterns.

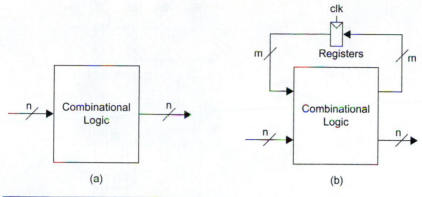

FIG 9.7   The combinational explosion in test vectors

9.5.1 Fault Models

To deal with the existence of good and bad parts, it is necessary to propose a *fault model*, i.e., a model for how faults occur and their impact on circuits. The most popular model is called the *Stuck-At* model. The *Short Circuit/Open Circuit* model can be a closer fit to reality, but is harder to incorporate into logic simulation tools.

9.5.1.1 Stuck-at Faults
In the Stuck-At model, a faulty gate input is modeled as a *stuck at zero* (Stuck-At-0, S-A-0) or *stuck at one* (Stuck-At-1, S-A-1). This model dates from board-level designs, where it was determined to be adequate for modeling faults. Figure 9.8 illustrates how an S-A-0 or S-A-1 fault might occur. These faults most frequently occur due to gate oxide shorts (the nMOS gate to GND or the pMOS gate to V_{DD}) or metal-to-metal shorts.

9.5.1.2 Short-circuit and Open-circuit Faults
Other models include *stuck–open* or *shorted* models [Jayasumana91]. Two bridging or shorted faults are shown in Figure 9.9. The short *S1* results in an S-A-0 fault at input A, while short *S2* modifies the function of the gate. It is evident that to ensure the most accurate modeling, faults should be modeled at the transistor level because it is only at this level that the complete circuit structure is known. For instance, in the case of a simple NAND gate, the intermediate node between the series nMOS transistors is hidden by the schematic. This implies that test generation should ideally take account of possible shorts and open circuits at the switch level [Galiay80]. Expediency dictates that most existing systems rely on Boolean logic representations of circuits and stuck-at fault modeling.

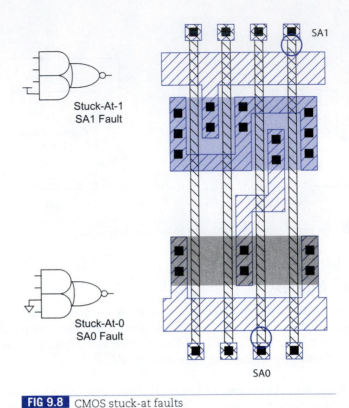

FIG 9.8 CMOS stuck-at faults

A particular problem that arises with CMOS is that it is possible for a fault to convert a combinational circuit into a sequential circuit. This is illustrated in Figure 9.10 for the case of a 2-input NOR gate in which one of the transistors is rendered ineffective. If nMOS transistor A is stuck open, then the function displayed by the gate will be

$$Z = \overline{A + B} + \overline{B}Z' \qquad (9.1)$$

where Z' is the previous state of the gate. As another example, if either pMOS transistor is missing, the node would be arbitrarily charged (i.e., it might be high due to some weird charging sequence) until one of the nMOS transistors discharged the node. Thereafter, it would remain at zero, barring charge leakage effects.

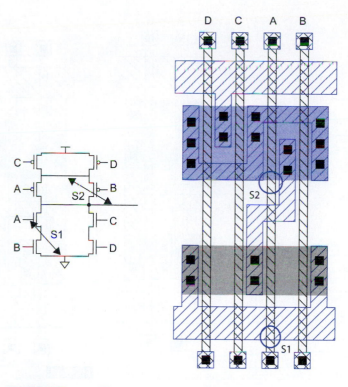

FIG 9.9 CMOS bridging faults

It is also possible for transistors to exhibit a stuck-open or stuck-closed state. Stuck-closed states can be detected by observing the static V_{DD} current (I_{DD}) while applying test vectors. Consider the fault shown in Figure 9.11, where the drain connection on a pMOS transistor in a 2-input NOR gate is shorted to V_{DD}. This could physically occur if stray metal (caused by a speck of dust at the photolithography stage) overlapped the V_{DD} line and drain connection as shown. If we apply the test vector 01 or 10 to the A and B inputs and measure the static I_{DD} current, we will notice that it rises to some value determined by size of the nMOS transistors.

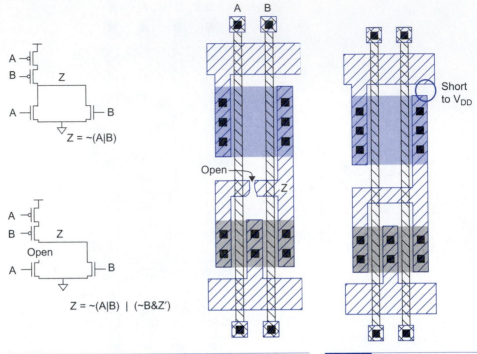

FIG 9.10 A CMOS open fault that causes sequential faults   **FIG 9.11** A defect that causes static I_{DD} current

9.5.2 Observability

The *observability* of a particular circuit node is the degree to which you can observe that node at the outputs of an integrated circuit (i.e., the pins). This metric is relevant when you want to measure the output of a gate within a larger circuit to check that it operates correctly. Given the limited number of nodes that can be directly observed, it is the aim of good chip designers to have easily observed gate outputs. Adoption of some basic design for test techniques can aid tremendously in this respect. Ideally, you should be able to observe directly or with moderate indirection (i.e., you may have to wait a few cycles) every gate output within an integrated circuit. While at one time this aim was hindered by the expense of extra test circuitry and a lack of design methodology, current processes and design practices allow you to approach this ideal. Section 9.6 examines a range of methods for increasing observability.

9.5.3 Controllability

The *controllability* of an internal circuit node within a chip is a measure of the ease of setting the node to a 1 or 0 state. This metric is of importance when assessing the degree of difficulty of testing a particular signal within a circuit. An easily controllable node would be directly settable via an input pad. A node with little controllability might require many hundreds or thousands of cycles to get it to the right state. Often you will find it impossible to generate a test sequence to set a number of poorly controllable nodes into the right state. It should be the aim of good chip designers to make all nodes easily controllable. In common with observability, the adoption of some simple design for test techniques can aid in this respect tremendously. Making all flip-flops resettable via a global reset signal is one step toward good controllability.

9.5.4 Fault Coverage

A measure of goodness of a set of test vectors is the amount of *fault coverage* it achieves. That is, for the vectors applied, what percentage of the chip's internal nodes were checked? Conceptually, the way in which the fault coverage is calculated is as follows. Each circuit node is taken in sequence and held to 0 (S-A-0), and the circuit is simulated with the test vectors comparing the chip outputs with a *known good machine*—a circuit with no nodes artificially set to 0 (or 1). When a discrepancy is detected between the *faulty machine* and the good machine, the fault is marked as detected and the simulation is stopped. This is repeated for setting the node to 1 (S-A-1). In turn, every node is stuck (artificially) at 1 and 0 sequentially. The fault coverage of a set of test vectors is the percentage of the total nodes that can be detected as faulty when the vectors are applied. To achieve world-class quality levels, circuits are required to have in excess of 98.5% fault coverage.

9.5.5 Automatic Test Pattern Generation (ATPG)

Historically, in the IC industry, logic and circuit designers implemented the functions at the RTL or schematic level, mask designers completed the layout, and test engineers wrote the tests. In many ways, the test engineers were the Sherlock Holmes of the industry, reverse engineering circuits and devising tests that would test the circuits in an adequate manner. For the longest time, test engineers implored circuit designers to include extra circuitry to ease the burden of test generation. Happily, as processes have increased in density and chips have increased in complexity, the inclusion of test circuitry has become less of an overhead for both the designer and the manager worried about the cost of the die. In addition, as tools have improved, more of the burden for generating tests has fallen on the designer. To deal with this burden, *Automatic Test Pattern Generation* (ATPG) methods have been invented. The use of some form of ATPG is standard for most digital designs.

Commercial ATPG tools can achieve excellent fault coverage. However, they are computation-intensive and often must be run on servers or compute farms with many parallel processors. Some tools use statistical algorithms to predict the fault coverage of a set of vectors without performing as much simulation. Adding scan and built-in self-test, as described in Section 9.6, improves the observability of a system and can reduce the number of test vectors required to achieve a desired fault coverage.

9.5.6 Delay Fault Testing

The fault models dealt with until this point have neglected timing. Failures that occur in CMOS could leave the functionality of the circuit untouched, but affect the timing. For instance, consider the layout shown in Figure 9.12 for an inverter gate composed of paralleled nMOS and pMOS transistors. If an open circuit occurs in one of the nMOS transistor source connections to GND, then the gate would still function but with increased t_{pdf}. In addition, the fault now becomes sequential as the detection of the fault depends on the previous state of the gate.

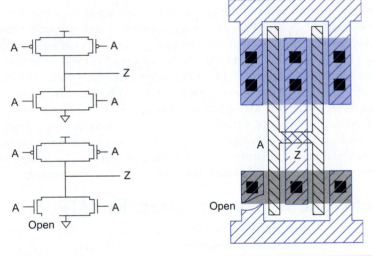

FIG 9.12 An example of a delay fault

Delay faults may be caused by crosstalk [Paul02]. Delay faults can also occur more often in SOI logic through the history effect. Software has been developed to model the effect of delay faults and is becoming more important as a failure mode as processes scale.

9.6 Design for Testability

The keys to designing circuits that are testable are controllability and observability. Restated, controllability is the ability to set (to 1) and reset (to 0) every node internal to the circuit. Observability is the ability to observe, either directly or indirectly, the state of any node in the circuit. Good observability and controllability reduce the cost of manufacturing test because they allow high fault coverage with relatively few test vectors. Moreover, they can be essential to silicon debug because physically probing most internal signals has become so difficult.

We will first cover three main approaches to what is commonly called *Design for Testability* (DFT). These may be categorized as:

- *Ad hoc* testing
- Scan-based approaches
- Built-in self-test (BIST)

Following this, we will look at the application of these techniques to particular types of circuits. In this treatment we will look at:

- Random logic (multilevel standard cell, two-level PLA)
- Regular logic arrays (datapaths)
- Memories (RAM, ROM, CAM)

9.6.1 *Ad hoc* testing

Ad hoc test techniques, as their name suggests, are collections of ideas aimed at reducing the combinational explosion of testing. They are summarized here for historical reasons. They are only useful for small designs where scan, ATPG, and BIST are not available. A complete scan-based testing methodology is recommended for all digital circuits. Having said that, common techniques for ad hoc testing involve:

- Partitioning large sequential circuits
- Adding test points
- Adding multiplexers
- Providing for easy state reset

A technique classified in this category is the use of the bus in a bus-oriented system for test purposes. Each register has been made loadable from the bus and capable of being driven onto the bus. Here, the internal logic values that exist on a data bus are enabled onto the bus for testing purposes.

Frequently, multiplexers can be used to provide alternative signal paths during testing. In CMOS, transmission gate multiplexers provide low area and delay overhead.

Any design should always have a method of resetting the internal state of the chip within a single cycle or at most a few cycles. Apart from making testing easier, this also makes simulation faster as a few cycles are required to initialize the chip.

In general, *ad hoc* testing techniques represent a bag of tricks developed over the years by designers to avoid the overhead of a systematic approach to testing, as will be described in the next section. While these general approaches are still quite valid, process densities and chip complexities necessitate a structured approach to testing.

9.6.2 Scan Design

The *scan-design* strategy for testing has evolved to provide observability and controllability at each register. In designs with scan, the registers operate in one of two modes. In *normal mode*, they behave as expected. In *scan mode*, they are connected to form a giant shift register called a *scan chain* spanning the whole chip. By applying *N* clock pulses in scan mode, all *N* bits of state in the system can be shifted out and new *N* bits of state can be shifted in. Therefore, scan mode gives easy observability and controllability of every register in the system.

Modern scan is based on the use of scan registers, as shown in Figure 9.13. The scan register is a *D* flip-flop preceded by a multiplexer. When the *SCAN* signal is deasserted, the register behaves as a conventional register, storing data on the *D* input. When *SCAN* is asserted, the data is loaded from the *SI* pin, which is connected in shift register fashion to the previous register *Q* output in the scan chain.

For the circuit shown, to load the scan chain, *SCAN* is asserted and *CLK* is pulsed eight times to load the first two ranks of 4-bit registers with data. *SCAN* is deasserted and *CLK* is asserted for one cycle to operate the circuit normally with predefined inputs. *SCAN* is then reasserted and *CLK* asserted eight times to read the stored data out. At the same time, the new register contents can be shifted in for the next test. Testing proceeds in this manner of serially clocking the data through the scan register to the right point in the circuit, running a single system clock cycle and serially clocking the data out for observation. In this scheme, every input to the combinational block can be controlled and every output can be observed. In addition, running a random pattern of 1's and 0's through the scan chain can test the chain itself.

Test generation for this type of test architecture can be highly automated. ATPG techniques can be used for the combinational blocks and, as mentioned, the scan chain is easily tested. The prime disadvantage is the area and delay impact of the extra multiplexer in the scan register. Designers (and managers alike) are in widespread agreement that this cost is more than offset by the savings in debug time and production test cost.

9.6.2.1 Parallel Scan

You can imagine that serial scan chains can become quite long, and the loading and unloading can dominate testing time. A fairly simple idea is to split the chains into smaller segments. This can be done on a module-by-module basis or completed automatically to some specified scan length. Extending this to the limit yields an extension to serial scan called *random access scan* [Ando80]. To some extent, this is similar to that used inside FPGAs to load and read the control RAM.

The basic idea is shown in Figure 9.14. The figure shows a two-by-two register section. Each register receives a column (column<m>) and row (row<n>) access signal along with a row data line (data<n>). A global write signal (write) is connected to all registers. By asserting the row and column access signals in conjunction with the write signal, any register can be read or written in exactly the same method as a conventional RAM. The notional logic is shown to the right of the four registers. Implementing the logic required at the transistor level can reduce the overhead for each register.

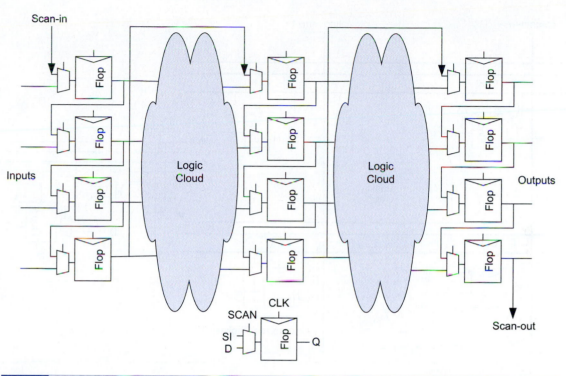

FIG 9.13 Scan-based testing

9.6.2.2 Partial Scan Sometimes, making every register scannable is too expensive, so only a partial set is scanned. In the CORDIC structure introduced in Section 8.2.4, each CORDIC slice has three m-bit registers. Converting all of these to scan registers may not be desirable. As the structure is a data pipeline, the test registers can be placed on the input and output of the pipeline as shown below in Figure 9.15. Partial scan is a throwback to *ad hoc* testing and should be avoided except in unusual circumstances.

9.6.2.3 Circuit Design of Scannable Elements As we have seen, an ordinary flip-flop can be made scannable by adding a multiplexer on the data input, as shown in Figure 9.16(a). Figure 9.16(b) shows a circuit design for such a scan register using a transmission-gate multiplexer. The setup time increases by the delay of the extra transmission gate in series with the D input as compared to the ordinary static flip-flop in Figure 7.19(b). Figure 9.16(c) shows a circuit using clock gating to obtain nearly the same setup time as the ordinary flip-flop. In either design, if a clock enable is used to stop the clock to unused portions of the chip, care must be taken that ϕ always toggles during scan mode.

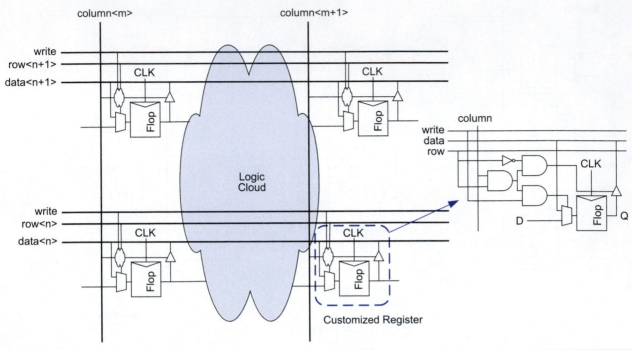

FIG 9.14 Parallel scan—basic structure

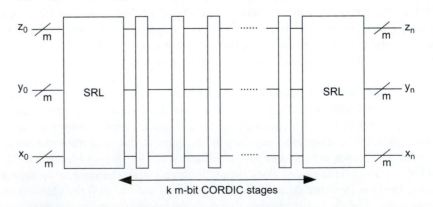

FIG 9.15 Example of partial scan as applied to the CORDIC processor

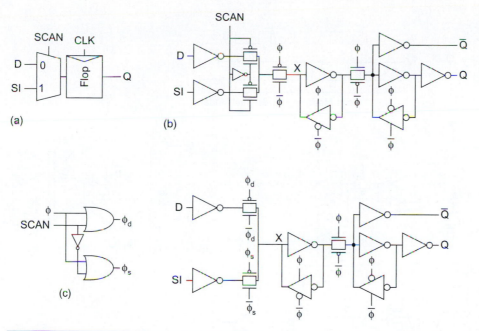

FIG 9.16 Scannable flip-flops

During scan mode, the flip-flops are connected back-to-back. Clock skew can lead to hold time problems in the scan chain. These problems can be overcome by adding delay buffers on the *SI* input to flip-flops that might see large clock skews. Another approach is to use nonoverlapping clocks to ensure hold times. For example, the *Level Sensitive Scan Design* (LSSD) methodology developed at IBM uses flip-flops with two-phase nonover-lapping clocks like those of Figure 7.21. During scan mode, a scan clock ϕ_s is toggled in place of ϕ_2, as shown in Figure 9.17. The nonoverlapping clocks also prevent hold time problems in normal operation, but increase the sequencing overhead of the flip-flop. Alternatively, ϕ_1 and ϕ_2 can be complementary clocks, but ϕ_s can be nonoverlapping to prevent races. Figure 9.17(c) shows a conventional design using a weak feedback inverter on the master latch that can be overpowered when either the ϕ_2 or ϕ_s transmission gates are on. Figure 9.17(d) shows a design from the PowerPC 603 microprocessor using a general-ized tristate feedback [Gerosa94]. Figure 9.17(e) shows another gate-level LSSD flip-flop design [Eichelberger78]. Such a design is substantially larger and slower than a conven-tional pass-transistor circuit, so it is primarily of historical interest. In the IBM LSSD methodology, ϕ_s, ϕ_1, ϕ_2, and *SI* are often called *A*, *B*, *C*, and *I*, respectively.

Systems using latches can also be modified for scan. Typically, a scan input and an extra *slave scan latch* are added to convert the latch into a scannable flip-flop. Figure 9.18

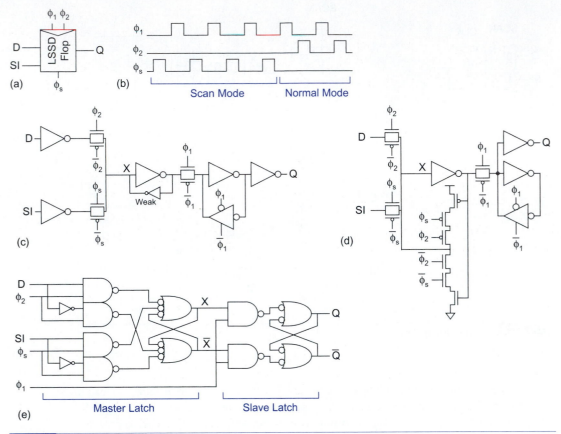

FIG 9.17 LSSD flip-flops

shows a scannable transparent latch. During scan, the global clock is stopped low, so ϕ_1 is low and the latch is opaque. Then a two-phase nonoverlapping scan clock ϕ_{1s} and ϕ_{2s} is toggled to march the data through the scan chain. The *SO* scan-out terminal of each latch connects to the *SI* scan-in terminal of the next latch. Figure 9.18(c) shows a faster and more compact, but less robust version of the scannable latch suitable for custom datapaths [Harris01a]. Scanning one latch in each cycle is adequate to provide good observability and controllability in a system; there is no need to scan the ϕ_2 latch.

The same principle applies to pulsed latches. Figure 9.19 shows the scannable Naffziger pulsed latch used on the Itanium 2 [Naffziger02] (see also Section 7.3.3). It uses a single-phase scan clock. The global clock is stopped during scan so the pulsed latches remain opaque. The scan input overpowers the feedback node Y to avoid loading the critical path from D to $\overline{Q}$. The transmission gate latch driving *SO* has a dynamic node Z, so ϕ_s has a limit on how long it can be high to properly retain data during scan. This is handled

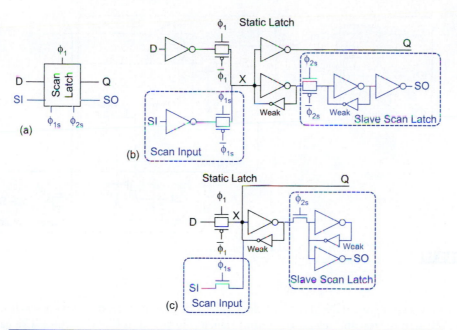

FIG 9.18 Scannable transparent latches

on-chip with a clock chopper that converts the external low-frequency scan clock into an on-chip ϕ_s with short pulses. The scan chain must also be checked for hold time races. Note that the *SO* transmission gate is ON during normal operation, loading the $\overline{Q}$ output and increasing power consumption through spurious transitions on *Z* and *SO*. Many designers would elect to use a second scan clock wire to avoid these problems.

   Domino pipelines also can be scanned. Traditional domino pipelines incorporate scan into the two-phase transparent latches on the half-cycle boundaries. Skew-tolerant domino eliminates the latches and must include scan directly in the domino gates. One natural point to scan is the last gate of each cycle.

   Figure 9.20(a) shows how to make the last ϕ_4 gate of each cycle in a skew-tolerant domino pipeline scannable [Harris01a]. The last dynamic gate has a full keeper and thus will retain its state when either high or low. The scan technique resembles that of a transparent latch from Figure 9.18(c). The key is to turn off both the precharge and the evaluation transistors so the output node floats and behaves like a master latch. Then a two-phase scan clock is toggled to shift data first onto the master node and then into a slave scan latch. These scan clocks are again called ϕ_{1s} and ϕ_{2s} and bear no relationship to the domino clocks ϕ_1 and ϕ_2. *gclk* is stopped low, so ϕ_4 is high and the precharge transistor is off. A special clock gater forces ϕ_{4s} low during scan to turn the evaluation transistor off. When scan is complete, *gclk* rises so the next ϕ_1 domino gate resumes normal operation. This scan approach adds a small amount of loading on the critical path through the

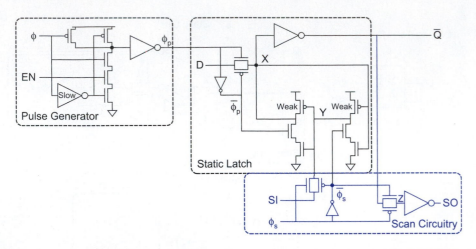

FIG 9.19 Scannable Naffziger pulsed latch

dynamic gate. Figure 9.20(b) shows a clock gater that produces the domino phases. It uses an SR latch to stop and release ϕ_{4s} during scan, as illustrated in Figure 9.20(c). The gater also accepts an enable to stop the domino clocks when the pipeline is idle.

The Itanium 2 provides domino scan in a similar fashion, but with a single-phase scan clock that is compatible with scan of the Naffziger pulsed latches [Naffziger02]. The last domino gate in each half-cycle uses a dynamic latch converter, as was discussed in Section 7.5.5.2. Scan circuitry can be added to the DLC in much the same way as it is added to a latch.

Robust scan circuitry obeys a number of rules to avoid electrical failures. *SI* is locally buffered to prevent problems with directly driving diffusion inputs and overdriving feedback inside the latch. The output is also buffered so noise cannot back drive the state node. Two-phase nonoverlapping scan clocks prevent hold-time problems, and static feedback on the state node allows low-frequency operation. All internal nodes should swing rail-to-rail. These rules can be bent to save area at the expense of greater electrical verification on the scan chain, as was done for the Itanium 2.

9.6.3 Built-in Self-Test (BIST)

Self-test and built-in test techniques, as their names suggest, rely on augmenting circuits to allow them to perform operations upon themselves that prove correct operation. These techniques add area to the chip for the test logic, but reduce the test time required and thus can lower the overall system cost. [Stroud02] offers extensive coverage of the subject from the implementer's perspective.

One method of testing a module is to use *signature analysis* [Frowerk77, Nadig77] or *cyclic redundancy checking*. This involves using a *pseudo-random sequence generator* (PRSG)

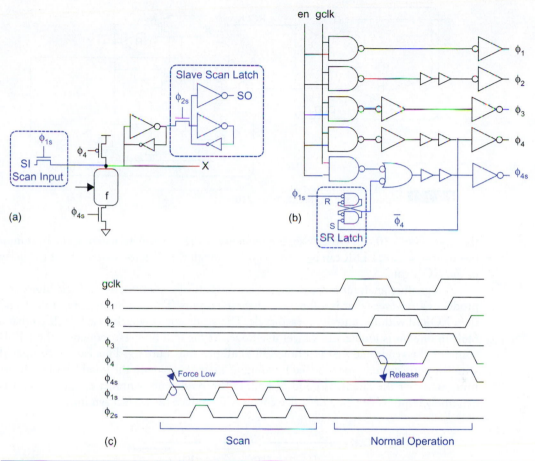

FIG 9.20 Scannable dynamic gate for four-phase skew-tolerant domino

to produce the input signals for a section of combinational circuitry and a *signature ana-lyzer* to observe the output signals.

A PRSG is defined by a polynomial of some length *n*. It is constructed from a *linear feedback shift register* (LFSR), which in turn is made of *n* flip-flops connected in a serial fashion, as shown in Figure 9.22(a). The XOR of particular outputs are fed back to the input of the LFSR. An *n*-bit LFSR will cycle through 2^n-1 states before repeating the sequence. LFSRs are discussed further in Section 10.5.2. They are described by a *characteristic polynomial* indicating which bits are fed back. A *complete feedback shift register* (CFSR), shown in Figure 9.22(b), includes the zero state that may be required in some test situations [Wang86]. An *n*-bit LFSR is converted to an *n*-bit CFSR by adding an $n-1$ input NOR gate. When in state 0...01, the next state is 0...00. When in state 0...00,

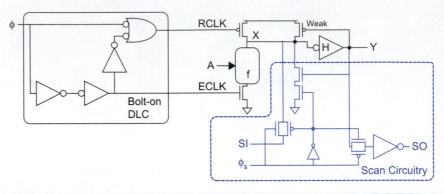

FIG 9.21   Itanium 2 scannable domino gate

the next state is 10...0. Otherwise, the sequence is the same. Alternatively, the bottom n bits of an $n + 1$-bit LFSR can be used to cycle through the all zeros state without the delay of the NOR gate.

A signature analyzer receives successive outputs of a combinational logic block and produces a *syndrome* that is a function of these outputs. The syndrome is reset to 0, and then XORed with the output on each cycle. The syndrome is swizzled each cycle so that a fault in one bit is unlikely to cancel itself out. At the end of a test sequence, the LFSR contains the syndrome that is a function of all previous outputs. This can be compared with the correct syndrome (derived by running a test program on the good logic) to determine whether the circuit is good or bad. If the syndrome contains enough bits, it is extremely improbable that a defective circuit will produce the correct syndrome.

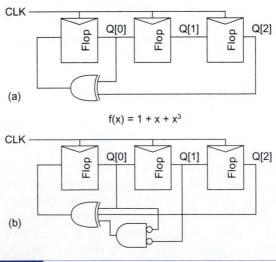

FIG 9.22   Pseudo-random sequence generator

9.6.3.1 BILBO The combination of signature analysis and the scan technique creates a structure known as *BILBO*—for *Built-In Logic Block Observation* [Koenemann79]—or *BIST*—for *Built-In Self-Test*. The 3-bit BILBO register shown in Figure 9.23 is a scannable, resettable register that also can serve as a pattern generator and signature analyzer. $C[1:0]$ specifies the mode of operation. In the reset mode (10), all the flip-flops are synchronously initialized to 0. In normal mode (11), the flip-flops behave normally with their D input and Q output. In scan mode (00), the flip-flops are configured as a 3-bit shift register between *SI* and *SO*. Note that there is an inversion between each stage. In test mode (01), the register behaves as a pseudo-random sequence generator or signature analyzer. If all the D inputs are held low, the Q outputs loop through a pseudo-random bit sequence, which can serve as the input to the combinational logic. If the D inputs are taken from the combinational logic output, they are swizzled with the existing state to produce the syndrome. In summary, BIST is performed by first resetting the syndrome in the output register. Then both registers are placed in the test mode to produce the pseudo-random inputs and calculate the syndrome. Finally, the syndrome is shifted out through the scan chain.

Various companies have commercial design aid packages that support BIST. LogicVision has a package called Logic BIST, which takes a synthesized netlist and adds the scan registers and PRSG circuits automatically. It then checks the fault coverage and provides

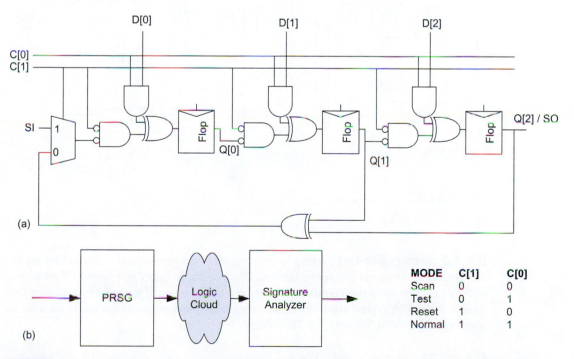

FIG 9.23 Built-In Logic Block Observation (BILBO/BIST): (a) individual register (b) use in a system

generated scripts to use boundary scan to run tests on the final chip. As an example, on a WLAN modem chip comprising roughly 1 million gates, a full at-speed test takes under a second with BIST. This comes with roughly a 7.3% overhead in the core area (but actually zero because the design was pad limited) and a 99.7% fault coverage level. The WLAN modem parts designed in this way were fully tested in less than ten minutes on receipt of first silicon. This kind of test method is incredibly valuable for productivity in manufacturing test generation.

Example

To test a 16-bit CORDIC using BIST, three identical 16-bit LFSRs are designed using the characteristic polynomial to be defined in Table 10.7. These LFSRs are placed at the start of the pipeline as shown in Figure 9.24. Similarly, three signature analyzers are placed at the end of the pipeline. When in BIST mode, the CORDIC can test at full operational speed. For instance, if we run 16384 cycles at 100 MHz, the total test takes 163.84 μs.

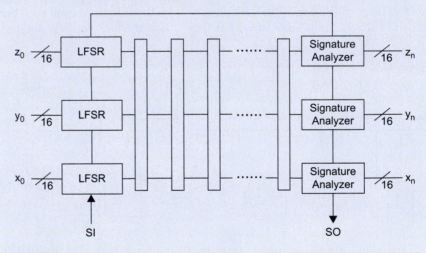

FIG 9.24   CORDIC with BIST

9.6.3.2 Memory Self-test Testing large memories on a production tester can be expensive because they contain so many bits and thus require so many test vectors. Embedding self-test circuits with the memories can reduce the number of external test vectors that have to be run. A typical read/write memory (RAM) test program for an M-bit address memory might be as follows [Nair78, Dekker90]:

```
FOR i=0 to M-1 write(~data)
FOR i=0 to M-1 read(~data) then write(data)
```

```
FOR i=0 to M-1 read(data) then write(~data)
FOR i=M-1 to 0 read(~data) then write(xFF)
FOR i=M-1 to 0 read(data) then write(~data)
```

where `data` is 1 and `~data` is 0 for a single-bit memory or a selected set of patterns for an *n*-bit word. For an 8-bit memory, `data` might be `x00`, `x55`, `xAA`, and `xFF`. These patterns test writing all zeroes, all ones, and alternating ones and zeroes. An address counter, some multiplexers, and a simple state machine result in a low-overhead self-test structure for read/write memories. [Oshawa87] describes a 4-Mbit RAM with self-test. The self-test consists of 256 K cycles that input a checkerboard pattern of alternating 1's and 0's to test for cell-to-cell interference. This is followed by 256 K cycles in which the data is read out. Then a complemented checkerboard is written and read. A total of 1 million cycles provide a test sufficient for system maintenance.

ROM memories can be tested by placing a signature analyzer at the output of the ROM and incorporating a test mode that cycles through the contents of the ROM. A significant advantage of all self-test methods is that testing can be performed when the part is in the field. With care, self-test can even be done during normal system operation.

Example

For the ROM used in the NCO, BIST circuitry can be added as shown in Figure 9.25. Here, a signature analyzer has been added to the output of the ROM as part of the output register. The phase counter is used to cycle through the ROM. The advantage of this style of test is that it is self-contained and can be embedded in the circuitry and so can form part of a self-test capability. The circuit is initialized and run for 256 cycles, and then the resulting signature can be read out and compared with the required value. Note that the signature analyzer is part of the "active" circuitry and so can be run at extremely high speeds—much higher than might be possible over a test bus that extends across the entirety of a chip. Again, this emphasizes the efficacy of a self-test strategy.

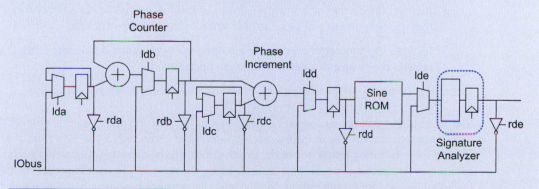

FIG 9.25 NCO ROM testing

9.6.4 IDDQ Testing

Bridging faults were introduced in Section 9.5.1.2. A method of testing for bridging faults is called IDDQ test (V_{DD} supply current Quiescent) or supply current monitoring [Acken83, Lee92]. This relies on the fact that when a complimentary CMOS logic gate is not switching, it draws no DC current (except for leakage). When a bridging fault occurs, then for some combination of input conditions, a measurable DC I_{DD} will flow. Testing consists of applying the normal vectors, allowing the signals to settle, and then measuring I_{DD}. As potentially only one gate is affected, the IDDQ test has to be very sensitive. In addition, to be effective, any circuits that draw DC power such as pseudo-nMOS gates or analog circuits have to be disabled. Dynamic gates can also cause problems. As current measuring is slow, the tests must be run slower (of the order of 1 ms per vector) than normal, which increases the test time.

IDDQ testing can be completed externally to the chip by measuring the current drawn on the V_{DD} line or internally using specially constructed test circuits. This technique gives a form of indirect massive observability at little circuit overhead. However, as sub-threshold leakage current increases, IDDQ testing ceases to be effective because variations in subthreshold leakage exceed currents caused by the faults.

9.6.5 Design for Manufacturability

Circuits can be optimized for manufacturability to increase their yield. This can be done in a number of different ways.

9.6.5.1 Physical At the physical level (i.e., mask level), the yield and hence manufacturability can be improved by reducing the effect of process defects. The design rules for particular processes will frequently have guidelines for improving yield. The following list is representative.

- Increase the spacing between wires where possible—this reduces the chance of a defect causing a short circuit.

- Increase the overlap of layers around contacts and vias—this reduces the chance that a misalignment will cause an aberration in the contact structure.

- Increase the number of vias at wire intersections beyond one if possible—this reduces the chance of a defect causing an open circuit.

Increasingly, design tools are dealing with these kinds of optimizations automatically.

9.6.5.2 Redundancy Redundant structures can be used to compensate for defective components on a chip. For example, memory arrays are commonly built with extra rows. During manufacturing test, if one of the words is found to be defective, the memory can be reconfigured to access the spare row instead. Laser-cut wires or electrically programmable fuses can be used for configuration. Similarly, if the memory has many banks and one or more are found to be defective, they can be disabled, possibly even under software control.

9.6.5.3 Power Elevated power can cause failure due to excess current in wires, which in turn can cause metal migration failures. In addition, high-power devices raise the die temperature, degrading device performance and, over time, causing device parameter shifts. The method of dealing with this component of manufacturability is to minimize power through design techniques described elsewhere in this text. In addition, a suitable package and heat sink should be chosen to remove excess heat.

9.6.5.4 Process Spread We have seen that process simulations can be carried out at different process corners. Monte Carlo analysis, which was introduced in Section 5.5.6, can provide better modeling for process spread and can help with centering a design within the process variations.

9.6.5.5 Yield Analysis When a chip has poor yield or will be manufactured in high volume, dice that fail manufacturing test can be taken to a laboratory for yield analysis to locate the root cause of the failure. If particular structures are determined to have caused many of the failures, the layout of the structures can be redesigned. For example, during volume production ramp-up for a major microprocessor, the silicide over long thin polysilicon lines was found to often crack and raise the wire resistance. This in turn led to slower-than-expected operation for the cracked chips. The layout was modified to widen polysilicon wires or strap them with metal wherever possible, boosting the yield at higher frequencies.

9.7 Boundary Scan

Up to this point we have concentrated on the methods of testing individual chips. Many system defects occur at the board level, including open or shorted printed circuit board traces and incomplete solder joints. At the board level, "bed-of-nails" testers historically were used to test boards. In this type of a tester, the board-under-test is lowered onto a set of test points (nails) that probe points of interest on the board. These can be sensed (the observable points) and driven (the controllable points) to test the complete board. At the chassis level, software programs are frequently used to test a complete board set. For instance, when a computer boots, it might run a memory test on the installed memory to detect possible faults.

The increasing complexity of boards and the movement to technologies such as multichip modules (MCMs) and surface mount technologies (with an absence of through-board vias) resulted in system designers agreeing on a unified scan-based methodology called *boundary scan* for testing chips at the board (and system) level. Boundary scan was originally developed by the Joint Test Access Group and hence is commonly referred to as JTAG. Boundary scan has become a popular standard interface for controlling BIST features as well.

The IEEE 1149 boundary scan architecture [IEEE1149.1-01, Parker03] is shown in Figure 9.26. All of the I/O pins of each IC on the board are connected serially in a standardized scan chain accessed through the *Test Access Port* (TAP) so that every pin can be observed and controlled remotely through the scan chain. At the board level, ICs obeying the standard can be connected in series to form a scan chain spanning the entire board. Connections between ICs are tested by scanning values into the outputs of each chip and checking that those values are received at the inputs of the chips they drive. Moreover, chips with internal scan chains and BIST can access those features through boundary scan to provide a unified testing framework.

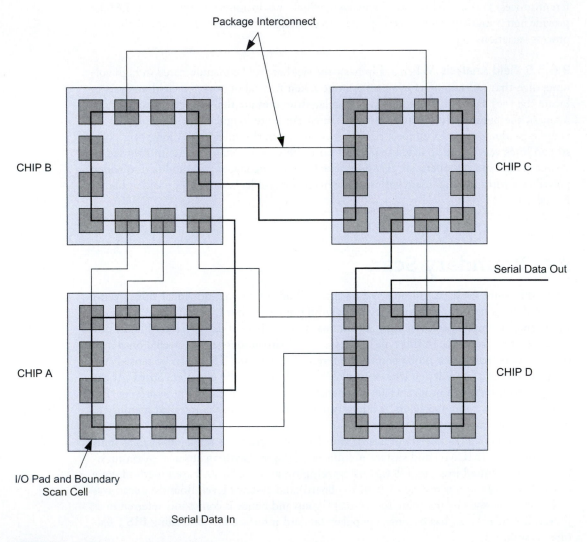

FIG 9.26 Boundary scan architecture

9.7.1   The Test Access Port (TAP)

The *Test Access Port* has four or five single-bit connections:

- *TCK*   Test Clock   Input   clocks tests into and out of the chip
- *TMS*   Test Mode Select   Input   controls test operations
- *TDI*   Test Data In   Input   test data into the chip
- *TDO*   Test Data Out   Output   test data out of the chip; driven only when TAP controller is shifting out test data
- *TRST** Test Reset Signal   Input   optional active low signal to asynchronously reset the TAP controller if no power-up reset signal is automatically generated by the chip

When the chip is in normal mode, *TRST** and *TCK* are held low and *TMS* is held high to disable boundary scan. To prevent race conditions, inputs are sampled on the rising edge of *TCK* and outputs toggle on the falling edge.

9.7.2   The Test Logic Architecture and Test Access Port

The basic test architecture is shown in Figure 9.27. It consists of

- The TAP interface pins

- A set of two or more test-data registers (DR) to collect data from the chip

- An instruction register (IR) specifying the type of test to perform

- A TAP controller, which controls the scan of bits through the instruction and test-data registers

The TAP controller is a small finite-state machine that configures the system. In one mode, it scans an instruction into the instruction register specifying what boundary scan should do. In another mode, it scans data in and out of the test-data registers. The specification requires at least two test-data registers: the boundary scan register and the bypass register. The boundary scan register is associated with all the inputs and outputs on the chip so that boundary scan can observe and control the chip I/Os. The bypass register is a single flip-flop used to accelerate testing by avoiding shifting data into the boundary scan registers of idle chips when only a single chip on the board is being tested. Internal scan chain, BIST, or configuration registers can be treated as optional additional data registers controlled by boundary scan.

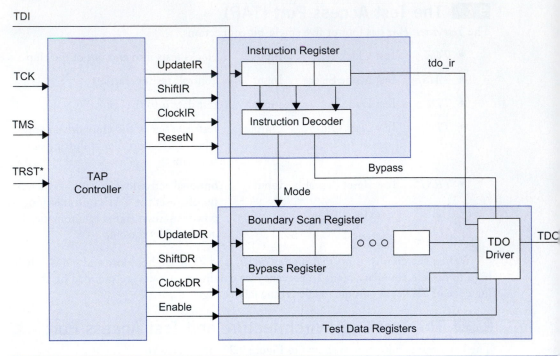

FIG 9.27 TAP architecture

9.7.3 The TAP Controller

The TAP controller is a 16-state FSM that proceeds from state to state based on the *TCK* and *TMS* signals. It provides signals that control the test-data registers and the instruction register. These include serial shift clocks and update clocks.

The state transition diagram is shown in Figure 9.28. The TAP controller is initialized to Test-Logic-Reset on power-up by *TRST** or an internal power-up detection circuit. It moves from one state to the next on the rising edge of *TCK* based on the value of *TMS*.

A typical test sequence will involve clocking *TCK* at some rate and setting *TRST** to 0 for a few cycles and then returning this signal to 1 to reset the TAP controller state machine. *TMS* is then toggled to traverse the state machine for whatever operation is required. These operations include serially loading an instruction register or serially loading or reading data registers that are used to test the chip. A variety of these operations will be described as this section unfolds.

The following Verilog code implements the TAP controller. The *TRST** is named `trstn`. Note that the controller produces gate clocks to control the data and instruction registers at the appropriate times.

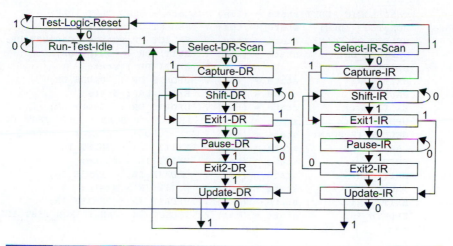

FIG 9.28 TAP controller state diagram

```
// TAP Controller States
`define TEST_LOGIC_RESET  4'b1111
`define RUN_TEST_IDLE     4'b1100
`define SELECT_DR_SCAN    4'b0111
`define CAPTURE_DR        4'b0110
`define SHIFT_DR          4'b0010
`define EXIT1_DR          4'b0001
`define PAUSE_DR          4'b0011
`define EXIT2_DR          4'b0000
`define UPDATE_DR         4'b0101
`define SELECT_IR_SCAN    4'b0100
`define CAPTURE_IR        4'b1110
`define SHIFT_IR          4'b1010
`define EXIT1_IR          4'b1001
`define PAUSE_IR          4'b1011
`define EXIT2_IR          4'b1000
`define UPDATE_IR         4'b1101

module tapcontroller(input       tms, tck, trstn,
                     output reg ShiftIR, ShiftDR,
                     output     ClockIR, ClockDR,
                     output     UpdateIR, UpdateDR,
                     output reg Resetn, Enable);

   reg [3:0] state;

   // next state logic
   always @(posedge tck, negedge trstn)
      if (~trstn) state = `TEST_LOGIC_RESET;
      else case (state)
```

```verilog
    `TEST_LOGIC_RESET:state = (tms) ? state : `RUN_TEST_IDLE;
    `RUN_TEST_IDLE:    state = (tms) ? `SELECT_DR_SCAN : state;
    `SELECT_DR_SCAN:   state = (tms) ? `SELECT_IR_SCAN : `CAPTURE_DR;
    `CAPTURE_DR:       state = (tms) ? `EXIT1_DR : `SHIFT_DR;
    `SHIFT_DR:         state = (tms) ? `EXIT1_DR : state;
    `EXIT1_DR:         state = (tms) ? `UPDATE_DR : `PAUSE_DR;
    `PAUSE_DR:         state = (tms) ? `EXIT2_DR : state;
    `EXIT2_DR:         state = (tms) ? `UPDATE_DR : `SHIFT_DR;
    `UPDATE_DR:        state = (tms) ? `SELECT_DR_SCAN : `RUN_TEST_IDLE;
    `SELECT_IR_SCAN:   state = (tms) ? `TEST_LOGIC_RESET : `CAPTURE_IR;
    `CAPTURE_IR:       state = (tms) ? `EXIT1_IR : `SHIFT_IR;
    `SHIFT_IR:         state = (tms) ? `EXIT1_IR : state;
    `EXIT1_IR:         state = (tms) ? `UPDATE_IR : `PAUSE_IR;
    `PAUSE_IR:         state = (tms) ? `EXIT2_IR : state;
    `EXIT2_IR:         state = (tms) ? `UPDATE_IR : `SHIFT_IR;
    `UPDATE_IR:        state = (tms) ? `SELECT_DR_SCAN : `RUN_TEST_IDLE;
  endcase

// Clock registers on rising edge of tck at end of state
// otherwise idle clock high
assign ClockIR  = tck | ~((state == `CAPTURE_IR) | (state == `SHIFT_IR));
assign ClockDR  = tck | ~((state == `CAPTURE_DR) | (state == `SHIFT_DR));

// Update registers on falling edge of tck
assign UpdateIR = ~tck & (state == `UPDATE_IR);
assign UpdateDR = ~tck & (state == `UPDATE_DR);

// Change control signals on falling edge of tck
always @(negedge tck, negedge trstn)
  if (~trstn) begin
     ShiftIR <= 0;
     ShiftDR <= 0;
     Resetn  <= 0;
     Enable  <= 0;
  end else begin
     ShiftIR <= (state == `SHIFT_IR);
     ShiftDR <= (state == `SHIFT_DR);
     Resetn  <= ~(state == `TEST_LOGIC_RESET);
     Enable  <= (state == `SHIFT_IR) | (state == `SHIFT_DR);
  end
endmodule
```

9.7.4 The Instruction Register

The instruction register has to be at least two bits long. Recall that boundary scan requires at least two data registers. The instruction register specifies which data register will be placed in the scan chain when the DR is selected. It also determines where the DR will load its value from in the Capture-DR state and whether the values will be driven to output pads or core logic. Three instructions are required to be supported:

- BYPASS—This instruction places the bypass register in the DR chain so that the path from *TDI* to *TDO* involves only a single flip-flop. This allows specific chips to be tested in a serial scan chain without having to shift through the lengthy shift register stages in all the chips. This instruction is represented with all 1's in the IR.

- SAMPLE/PRELOAD—This instruction places the boundary scan registers (i.e., at the chip's I/O pins) in the DR chain. In the Capture-DR state, it copies the chip's I/O values into the DRs. They can then be scanned out in successive Shift-DR states. New values are shifted into the DRs, but not driven onto the I/O pins yet.

- EXTEST—This instruction allows for the testing of off-chip circuitry. It is similar to SAMPLE/PRELOAD, but also drives the values from the DRs onto the output pads. By driving a known pattern onto the outputs of some chips and checking for that pattern at the input of other chips, the integrity of connections between chips can be verified.

In addition to these instructions, the following are also recommended (others can be defined as needed):

- INTEST—This instruction allows for single-step testing of internal circuitry via the boundary scan registers. It is similar to EXTEST, but also drives the chip core with signals from the DRs rather than from the input pads.

- RUNBIST—This instruction is used to activate internal self-testing procedures within a chip.

Note that the instruction encodings are not part of the specification (except that BYPASS is all 1's). The component designer must document what encodings were selected for each instruction.

A typical IR bit is shown in Figure 9.29. Observe that it contains two flip-flops. The ClockIR flip-flops of each bit are connected to form a shift register. They are loaded with a constant value from the Data input in the Capture-IR state, and then are shifted out in the Shift-IR state while new values are shifted in. The constant value is user-defined, but must have a 01 pattern in the least significant two bits so that the integrity of the scan chain can be verified. In the Update-IR state, the contents of the shift register are copied in parallel to the IR output to load the entire instruction at once. This prevents the IR from momentarily having illegal values while new instructions are shifted in. On reset, the IR should be asynchronously loaded with an innocuous instruction such as BYPASS that does not interfere with the normal behavior of the core logic.

A minimal implementation of a 3-bit control register is shown below. Notice the instruction encoding definitions. This implements the six registers required for a 3-bit instruction. The instruction is decoded to produce mode_in, mode_out, and bypass signals to control the data registers, as will be discussed in the next sections.

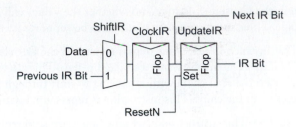

FIG 9.29 Instruction bit implementation

```
// Instructions
`define BYPASS          3'b111
`define SAMPLE_PRELOAD  3'b101
`define EXTEST          3'b110
`define NOP             3'b001
`define INTEST          3'b100

module inst_reg(input   tdi,
                input   Resetn, ClockIR, UpdateIR, ShiftIR,
                output  tdo_ir, mode_in, mode_out, bypass);

   reg [2:0] shiftreg, instreg;

   always @(posedge ClockIR)
      shiftreg <= ShiftIR ? {tdi, shiftreg[2:1]} : `NOP;
   always @(posedge UpdateIR, negedge Resetn)
      if (~Resetn) instreg <= `BYPASS;
      else instreg <= shiftreg;

   assign tdo_ir = shiftreg[0];
   assign bypass = (instreg == `BYPASS);
   assign mode_in = (instreg == `INTEST);
   assign mode_out = (instreg == `INTEST) || (instreg == `EXTEST);
endmodule
```

9.7.5 Test Data Registers

The test data registers are used to set the inputs of modules to be tested and collect the results of running tests. The simplest data register configuration consists of a boundary scan register (passing through all I/O pads) and a bypass register (1 bit long). Figure 9.30 shows a generalized view of the data registers in which an internal data register has been added. This register might represent the scan chain within the chip or a BILBO signature register. Thus, boundary scan elegantly incorporates other built-in test structures. A multiplexer under the control of the TAP controller selects which data register is routed to the *TDO* pin. When internal data registers are added, the IR decoder must produce extra control signals to select which one is in the DR chain for a particular instruction.

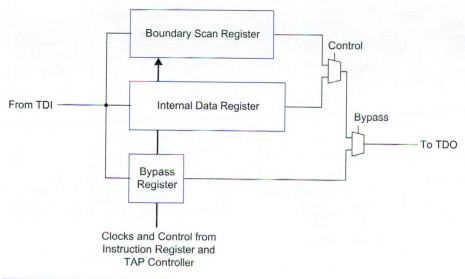

FIG 9.30 Test-data registers

9.7.5.1 Boundary Scan Register The boundary scan register connects to all of the I/O circuitry. Like the instruction register, it internally consists of a shift register for the scan chain and an additional bank of flip-flops to update the outputs in parallel. An extra multiplexer on the output allows the boundary scan register to override the normal path through the I/O pad so it can observe and control inputs and outputs. The schematic and symbol for a single bit of the boundary scan register are shown in Figure 9.31.

The boundary scan register can be configured as an input pad or output pad, as shown in Figure 9.32(a and b). As an input, the register receives DataIn from the pad and sends Qout to the core logic in the chip. As an output, the register receives DataIn from the core logic and drives Qout to a pad. Tristate and bidirectional pads use two or three boundary scan register cells, as shown in Figure 9.32(c and d).

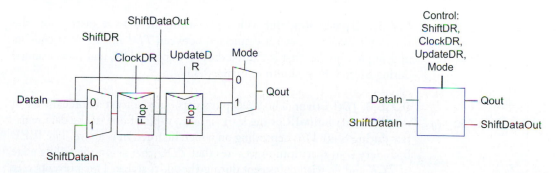

FIG 9.31 Boundary scan register bit

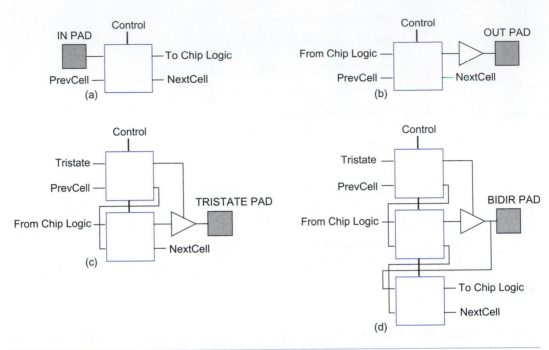

FIG 9.32 Boundary scan pad configuration

The Mode signal determines whether Qout should be taken from DataIn or the boundary scan register. Separate mode_in and mode_out signals are used for input and output pads so they can be controlled separately. In normal chip operation, both mode signals are 0, so the boundary scan registers are ignored. For the EXTEST instruction, mode_out = 1, so the outputs can be controlled by the boundary scan registers. For INTEST or RUNBIST instructions, mode_in and mode_out are both 1, so the core logic receives its inputs from the boundary scan registers and the outputs are also driven to known safe values by the boundary scan registers.

9.7.5.2 Bypass Register When executing the BYPASS instruction, the single-bit Bypass register is connected between *TDI* and *TDO*. It consists of a single flip-flop that is cleared during Capture-DR, and then scanned during Shift-DR, as shown in Figure 9.33.

9.7.5.3 *TDO* Driver The *TDO* pin shifts out the least significant bit of the IR during Shift-IR or the least significant bit of one of the data registers during Shift-DR, depending on which instruction is active. The IEEE boundary scan specification requires that *TDO* change on the falling edge of *TCK* and be tristated except during the Shift states. This prevents race conditions when the value is clocked into the next chip in the rising edge of

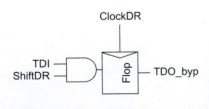

FIG 9.33 Bypass register

TCK and allows multiple chips to be connected in parallel with their *TDO* pins tied together to reduce the length of the boundary scan chain.

Figure 9.34 shows a possible implementation of the *TDO* driver. The multiplexers choose among the possible shift registers including the instruction register, boundary scan register, and bypass register. Additional multiplexers would be used if more data registers were included. A flip-flop or latch delays the *TDO* signal until the falling edge of *TCK*. The tristate drives *TDO* during Shift-IR or Shift-DR.

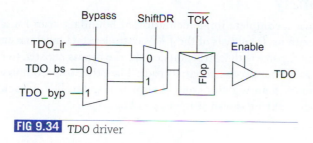

FIG 9.34 *TDO* driver

9.7.5.4 Complete Test Data Register Logic   The Verilog code below describes the complete Test Data Register for a chip with four inputs a[3:0] and four outputs y[3:0]. The four input and four output boundary scan register bits are collected into a single 8-bit shift register. mode_in serves the four most significant bits connected to the inputs, while mode_out serves the four least significant bits connected to the outputs.

```
module data_reg(input   [3:0] a, fromlogic,
                input         tck, tdi, tdo_ir,
                input         ClockDR, UpdateDR, ShiftDR, Enable,
                input         mode_in, mode_out, bypass,
                output  [3:0] y, tologic,
                output        tdo);

   reg   [7:0] shiftreg, datareg;
   wire        tdo_selected;
   reg         tdo_byp, tdo_delayed;

   // Boundary scan registers
   // four input registers and four output registers connected in 8-bit chain
   always @(posedge ClockDR)
      shiftreg <= ShiftDR ? {tdi, shiftreg[7:1]} : {a, fromlogic};
   always @(posedge UpdateDR)
      datareg <= shiftreg;
   assign tologic = mode_in ? datareg[7:4] : a;
   assign y = mode_out ? datareg[3:0] : fromlogic;

   // Bypass register
   always @(posedge ClockDR)
      tdo_byp <= tdi & ShiftDR;
```

```
    // tdo output driver
    // select appropriate register to shift out, delay to negative edge of tck
    assign tdo_selected = ShiftDR ? (bypass ? tdo_byp : shiftreg[0]) : tdo_ir;
    always @(negedge tck)
       tdo_delayed <= tdo_selected;
    assign tdo = Enable ? tdo_delayed : 1'bz;
endmodule
```

9.7.6 Summary

Figure 9.35 shows a complete implementation of boundary scan for a chip with four inputs and four outputs. It consists of the TAP controller state machine and state decoder, a 3-bit instruction register with instruction decode, the bypass register, four boundary scan input pads, and four boundary scan output pads. The other pads comprise the test access port. The boundary scan register control signals (UpdateDR, ClockDR, ShiftDR, mode_in, and mode_out) are shown as the Control bus.

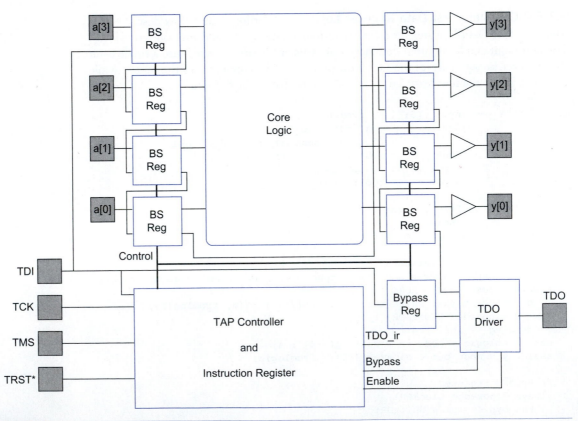

FIG 9.35 Complete boundary scan implementation

The Verilog for this design is shown below.

```
module core(input   [3:0] tologic,
            output [3:0] fromlogic);

   // a silly chip logic function
   assign fromlogic = {&tologic, |tologic, ^tologic, ~tologic[0]};
endmodule

module top(input       tck, tms, tdi, trstn,
           input   [3:0] a,
           output      tdo,
           output [3:0] y);

   wire [3:0] tologic, fromlogic;
   wire       UpdateIR, ShiftIR, ClockIR;
   wire       UpdateDR, ShiftDR, ClockDR;
   wire       Resetn, Enable;
   wire       mode_in, mode_out, bypass;
   wire       tdo_ir;

   // Core Logic
   core core(tologic, fromlogic);

   // TAP Controller
   tapcontroller tc(tms, tck, trstn, ShiftIR, ShiftDR, ClockIR, ClockDR,
                    UpdateIR, UpdateDR, Resetn, Enable);

   // Instruction register
   inst_reg ir(tdi, Resetn, ClockIR, UpdateIR, ShiftIR,
               tdo_ir, mode_in, mode_out, bypass);

   // Test data registers
   data_reg dr(a, fromlogic, tck, tdi, tdo_ir,
               ClockDR, UpdateDR, ShiftDR, Enable, mode_in, mode_out, bypass,
               y, tologic, tdo);
endmodule
```

Boundary scan testing typically begins with the SAMPLE/PRELOAD instruction. Then a data value is preloaded into the boundary scan registers. Next, the EXTEST or INTEST instruction is applied to activate the loaded value. Subsequent data values are shifted into the boundary scan registers and the results of the tests are shifted out.

Figure 9.36 shows waveforms for this operation. The TAP controller is initially reset. At this point, the core logic operates normally with an input pattern of 0000 and an output pattern of 0001. Then the IR is loaded with 101 (SAMPLE/PRELOAD). The data pattern 0111 is shifted in. The IR is loaded with 1000 (INTEST). This sends the 0111 pattern to the core logic, producing an output pattern of 0110. Finally, the data pattern 1111 is shifted in and the old output 0110 is shifted out. Because the INTEST is still active, the 1111 is applied to the core, producing a new output of 1100.

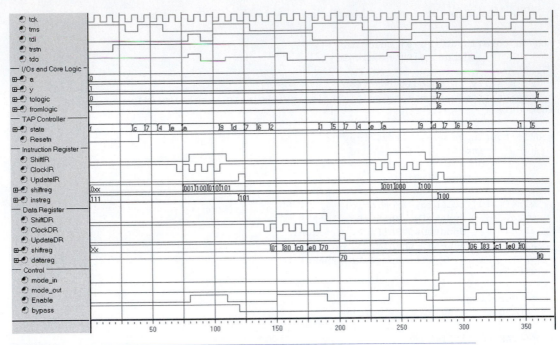

FIG 9.36 Boundary scan example waveforms

Boundary scan is in widespread use in chips today. It provides a uniform interface to single- and multiple-chip testing and circuit-board testing.

9.8 System-on-chip (SOC) Testing

System-on-Chip testing uses the techniques described so far to test the disparate system blocks that might exist on a typical complex chip. The IEEE P1500 Standard for Embedded Core Test (SECT) is being developed to standardize methods for testing SOCs.

Figure 9.37 shows the basic approach that P1500 takes. Each large chip module (i.e., a processor, memory, accelerator) is surrounded with a *wrapper* that resembles a boundary scan chain. The Wrapper Serial Port (*WSP*) has a serial input (*WSI*—Wrapper Serial Input), a serial output (*WSO*—Wrapper Serial Output), and a control port (*WSC*—Wrapper Serial Control). Because serial control and testing can be slow, an optional set of parallel ports is also specified. The Wrapper Parallel Port (*WPP*) consists of the Wrapper Parallel Input (*WPI*), Wrapper Parallel Output (*WPO*), and the Wrapper Parallel Control (*WPC*). The interface to the serial ports is standardized to enable "plug and play" operation. The parallel ports are user-defined to allow test flexibility.

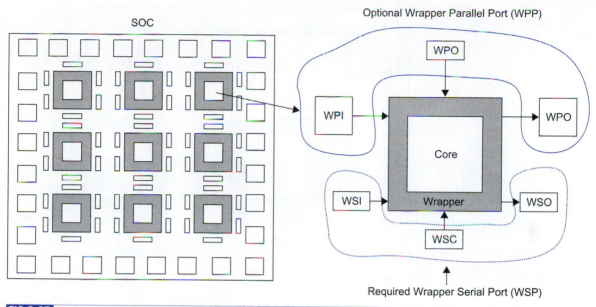

FIG 9.37 IEEE P1500 embedded core test strategy

Embedded cores are interconnected, as shown in Figure 9.38. The *WSP* blocks are daisy-chained in a serial fashion, as shown in the diagram. Global *WSI* and *WSC* signals are provided to control the embedded cores with test data. In addition, a global *WSO* signal is available for observation or daisy-chaining chips together. The serial wrapper ports can optionally control wrapper parallel ports via an enable (*ENA*) signal. The *test access method* (TAM) of the *WPP* can follow one of several architectures.

Some of these parallel TAM methods are shown in Figure 9.39. Figure 9.39(a) shows a daisy-chained method where blocks are interconnected in the same way as the required serial port wrappers. Figure 9.39(b) shows a bus-based TAM. This would allow faster and more direct access to modules. Figure 9.39(c) shows a parallel (direct access) method. This may have the disadvantage that the number of signals or control pins may increase beyond that desired. Finally, Figure 9.39(d) shows autonomous local TAM controllers that are designed to access and test each module. For example, these could be BIST modules that self-test each module and report success or failure over the serial port.

Together, serial scan, BIST, boundary scan, and P1500 form an ordered set of techniques to facilitate testing and verification of large systems on one or more chips.

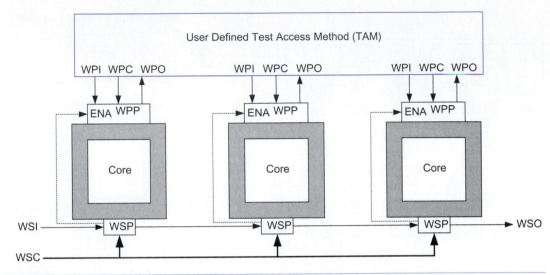

FIG 9.38 P1500 user-defined test access method (TAM)

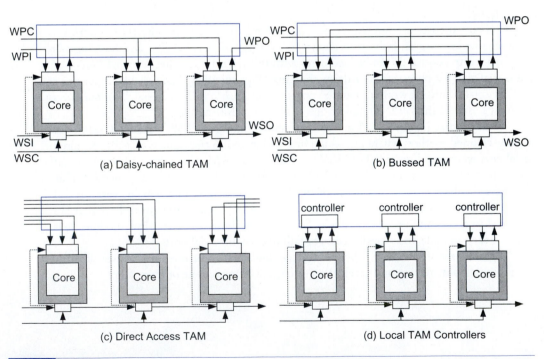

FIG 9.39 Various methods of TAM

9.9 Mixed-signal Testing

OPTIONAL

Mixed-signal chips contain both analog and digital circuitry. A full exposé on mixed-signal testing is beyond the scope of this book, but we will address a few points. When designing mixed-signal systems, thought should be put into choosing suitable methods of self-test or at least making the analog test engineer's job easier. Many times, analog and RF blocks can be used to help in the testing and verification of digital blocks and vice-versa. Often, the best way to view a dynamically changing situation is via an oscilloscope. On-chip sampling circuits can function as a high-speed equivalent-time oscilloscope to transfer waveforms out over a lower-bandwidth interface [Ho98]. If high-speed digital/analog converters (DACs) are available in the signal path of the chip, then they can be used to aid in system test and debug.

Example

Consider the IQ upconversion path in the software radio example. If multiplexers are placed in the signal path from the NCO to the DAC (Figure 9.40), a variety of signals can be routed to the DACs. The Phase Counter can be used as a ramp generator and fed to the DAC to aid in characterizing of the DAC (a linear ramp is required to measure DAC parameters—see Chapter 12). In addition, any other signal of interest can be routed to the DAC for observation in real time on an oscilloscope. For instance, the receive IQ signals at any point in the receive path can be routed to the DAC and observed on the DAC. This method of test and debug is extremely powerful. If possible, remembering the principle of regularity (and modularity)—making it possible to route any signal register to the DACs—can be invaluable. Circuits that don't even have a DAC can benefit from this technique with the inclusion of a DAC (they can be small and only take one extra pad).

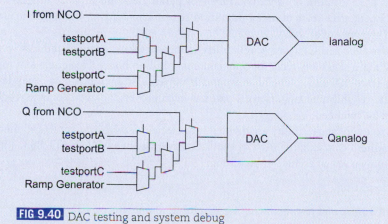

FIG 9.40 DAC testing and system debug

Analog/digital converter testing requires real-time access to the digital output of the ADC. Providing parallel digital test ports by reassigning pins on the chip I/O can facilitate this testing. If this is impossible, a "capture RAM" on chip can be used to capture results in real-time and then the contents can be transferred off-chip at a slower rate for analysis.

If both ADCs and DACs are present, a loopback strategy can be employed, as shown in Figure 9.41. Both analog and digital signals can loop back. Communication and graphics systems frequently have I/O systems that can be configured as shown. It is often worthwhile to add either a DAC or a small ADC to a system to allow a level of analog self-test.

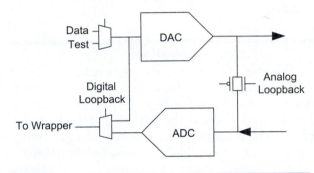

FIG 9.41 Analog and digital loopback

9.10 Reliability Testing

ICs do not have an infinite lifetime. Transistor parameters can drift and metal lines can eventually break after years of metal migration stress. This leads to the desire to assess the reliability (lifetime) of an IC.

As introduced in Chapter 4, reliability is measured in terms of failures in time or FITs. A FIT is a reliability term for one failure in 1 billion hours. Designers aim for failure rates from 10 to 100 FIT. 100 FIT is around 1000 years, so methods have been developed to accelerate reliability testing. Once a part has been proven to have adequate reliability, it is regarded as "qualified."

Tests to establish reliability are divided into electrical, environmental, and mechanical tests and adhere to standards such as the U.S. government's MIL-STD-883. This standard covers a variety of scenarios, including:

- Life testing

- Environmental testing, including vibration, temperature cycling, and shock and moisture testing

- Radiation testing

- Electrical tests, including noise margin, power consumption, and analog tests

- Package tests, including mechanical tests on wire bond strengths and ultrasonic die inspection

The reader is referred to the standard for a complete listing of all tests.

A part has to be *qualified* prior to shipping in production. Qualification can occur at various levels. A process has to be qualified before companies will use it for product. The qualification process guarantees that devices will not fail if they are exposed to normal voltages over a reasonable lifetime. Possible failure modes can include threshold voltage shifts and open circuits from electromigration. A particular standard cell logic and I/O library can be qualified. In particular, this means that the IP can stand up to prescribed conditions for the lifetime of the IC. Finally, the individual design has to be qualified. This involves assessing the reliability of the part.

One of the methods of achieving a reliability metric is to use accelerated life testing (ALT). This process runs a set of chips at elevated temperature and voltage for an extended period. At periodic intervals in this period the chips are re-tested (or tested *in situ*) and any failures logged. The time of any failures leads to a reliability number at that voltage and temperature. Lifetime typically decreases exponentially with voltage and temperature, so measurements made at a few elevated levels over modest periods of time are extrapolated back to predict failure rates at normal levels. ALT can be dynamic or static. Static ALT just supplies the chip with supply voltages and perhaps a clock. Dynamic ALT operates the chip as it would in a system. ALT uses what is termed a *burn-in board*. This has a number (usually around 40) of chips running at temperatures of 150°–200° C. The chips are periodically checked over a period of 4000 hours. Failures can be mapped to an MTBF number.

A burn-in test can be applied to each chip to ensure higher than normal reliability (i.e., for space-borne applications). This operates the chip at elevated temperatures for a particular period. The process is designed to cull out infant mortality failures (see Section 4.8.1). Normally, high-volume chips do not undergo burn-in because of time and cost pressures.

9.11   Testing in a University Environment

Industry environments are usually well-funded, and the appropriate testability tools are available to ensure a product-grade test effort. But what do you do in a university environment when the infrastructure might not be quite as affluent as in the industry setting? Not only may test tools be unavailable, but also the very act of building a test board can be a daunting extra amount of work on top of the chip design. The following are some tips that might help in this situation.

Taking the time to include circuitry to aid in testing on the chip is usually much easier than adding it at the board level. For a start, the integrated environment available for most IC design flows allows the designer to simulate the test circuitry. So, while it might seem superfluous to the task at hand, including test circuitry can save a huge amount of effort after the chip returns. Moreover, on-chip circuitry can often test at speeds that are impossible off-chip without extremely expensive production test machines. The main point is to think ahead.

BIST is straightforward to integrate into a library of registers or I/O pads. In the case of data pipelines such as the NCO and CORDIC used as examples in this chapter and the previous one, we have shown how to add testing capability to the registers. The University of Tennessee has developed a "SmartFrame" pad frame compatible with the MOSIS AMI 0.6 mm process that incorporates a PRSG and signature analyzer into the pads [Bouldin03]. Alternatively, if a boundary scan interface that supports INTEST is incorporated onto the chip, the chip can be tested from a PC using a commercial boundary scan controller. For example, the Corelis NetUSB-1149.1/E can drive the scan chains at up to 80 MHz.

In the absence of BIST, there are several ways to test a chip. One is to breadboard or wirewrap a test board with switches for inputs and LEDs for outputs. This is tedious for all but the simplest chips. A custom-printed circuit board test fixture is even more labor-intensive, but often necessary for high-performance research chips. Another strategy is to use a logic analyzer with pattern generator. This approach requires a specialized test fixture to hold the chip and often has a steep learning curve for students, but it can perform tests at 10's to 100's of MHz. Yet another approach is to use a low-cost *functional chip tester* that reads the test vectors from pretapeout simulation, applies the digital patterns, and checks the results at low speed. Figure 9.42 shows the TestosterICs functional chip tester from One Hot Logic [Harris03] used by the author to reduce the time students spend testing their chips. The DUT boards contain zero insertion force (ZIF) sockets for 40-pin DIP and 16x16 PGA packages commonly used in class projects.

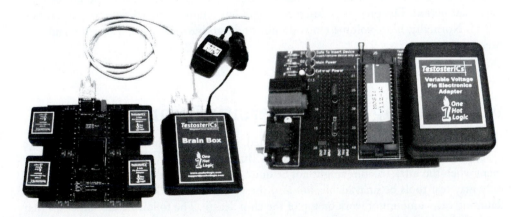

FIG 9.42 TestosterICs functional chip tester

The following "war stories" are collected from real products at a wide variety of companies and published with permission, often under the condition of anonymity. They are presented to illustrate some of the pitfalls that can happen to smart people who are dealing with complex systems on a tight schedule. The skilled engineer learns from these mistakes; in most cases, the company extended their verification flow to ensure that similar problems would be caught before wreaking havoc on future products. Could one of these happen to you?

A Product in the Field Hangs Unpredictably

A microprocessor had been in the field for several years when reports began arriving from major customers that certain programs would cause the system to hang at unpredictable times with intervals of hours to days. The manufacturer appointed a tiger team to resolve the error. The hang rate proved to be insensitive to power supply voltage, operating temperature, and clock rate. It was observed on all versions of the chip regardless of foundry, manufacturing technology, or motherboard. The programs that failed all involved a mix of floating point and integer operations; none were purely integer codes.

After several months of work, the problem was isolated to a particular unit in the processor. By this point, 30 engineers were involved in chasing the problem. Picoprobing showed that when the hang occurred, an instruction was left stuck in the pipeline waiting to issue. A logic simulation of the RTL is orders of magnitude slower than running the actual code, but an engineer developed a simple test case that could trigger the hang on real hardware in a matter of seconds, and thus it could trigger the failure in simulation in a practical amount of time. Simulations showed that the RTL ran flawlessly, suggesting the error involved a circuit that did not match the RTL.

On this processor, the circuits had been verified against the RTL using a technique called "shadow-mode simulation." A "circuit understanding" tool parsed the

transistor-level netlist into gates and identified the logic function of each gate. Circuits were verified to match the RTL by replacing a module of the RTL with the corresponding extracted circuit and simulating to check that the system produced identical results as the original RTL. The simulation is time-consuming, so each module is typically checked over tens of thousands of cycles, rather than the billions of cycles used in primary RTL verification.

A shadow-mode simulation using circuits from the failing unit still ran flawlessly. However, an engineer observed that a long wire crossing a large schematic was driven from both ends to reduce the RC delay. The signals X1 and X2 driving each end were intended to be identical (Figure 9.43). The engineer experimented with splitting the wire and checking that both drivers produced identical results, and on certain test cases they did not. This led to the wire experiencing contention and being driven to an indeterminate logic value. The invalid result propagated through other logic and hung the processor. Unfortunately, the circuit-understanding tool had incorrectly determined that the logic for the two ends was identical and had never detected the error. Even if the tool had been correct, the original test cases never would have exercised the patterns that caused the drivers to produce different results. A simple modification to the driver fixed the problem, but many units were already in the field. Fortunately, a software patch was developed to prevent the operations that caused the hang from ever being issued.

Hanging is a serious problem, but not as severe as unknowingly calculating the wrong answer. After the problem was corrected, engineers spent several more weeks proving to customers that the failure mode

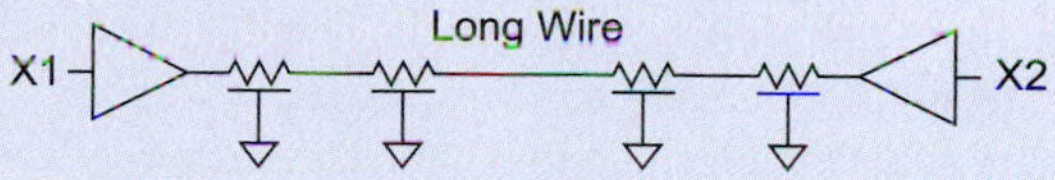

FIG 9.43   Long wire driven from both ends

would hang the machine but could never result in an incorrect calculation.

To avoid repeating this problem in the future, engineers have turned to formal verification tools that prove that RTL and schematics are equivalent in their Boolean function. Such tools are not susceptible to incomplete test patterns. However, the tools are often expensive, proprietary, and difficult to use.

A Product Fails after the Manufacturing Process Matures

A team designing a data communications product was comfortable with a particular microprocessor that was at the end of its production run. The team negotiated to order several thousand units of the discontinued microprocessor before production was shut down. The data communications product became successful and was shipped in large quantity. After it had been in the field for some time, major customers reported that the product would crash in large networks. These customers included large financial, government, and Internet service provider organizations who were adversely affected by the crashes. It took the data communications company weeks to isolate the problem to hanging of the microprocessor, and then a team of engineers at the microprocessor company began investigating the issue.

The microprocessor team investigated potential signal and power supply integrity issues. Although no signal integrity problems were apparent, a shmoo plot showed unusual sensitivity of minimum clock period to supply voltage. An engineer had recently read the application note for the power regulator on the system board and had learned that it had a propensity for oscillation if not properly bypassed. The system board lacked the bypass capacitors recommended in the application note, so the engineer wrote a memo to the product manager suggesting a change to the board. The memo was misinterpreted as a solution to the problem and customers were informed that a fix was on its way. Unfortunately, further testing showed that bypassing the regulator did not fix the crashes.

When the system crashed, it wrote its state to a core file. An engineer began reading a hexadecimal dump of the file and noticed a pattern that led to solving the crash. The pattern was associated with simultaneous access to many banks in an 8-way associative instruction cache. The cache had fuses associated with each bank, so banks containing bad blocks could be disabled during manufacturing test. During original product debug, the manufacturing process was relatively immature and most processors only had five operational cache banks. However, the processors manufactured at the end of the production run were built on a more mature process and often had all eight banks functional. Simultaneous access to all the banks tickled a signal integrity problem, resulting in power supply droop from excessive IR drops caused by poor contacts to the V_{DD} plane. The solution was a software change to disable three of the banks at system startup.

Better power supply analysis is performed to avoid repeating this problem.

A Wasted Spin

A microprocessor was taped out and came back nearly fully operational. Minor changes were made to the layout and documentation was developed; then a second revision (colloquially called a second *spin* of the chip) was taped out. The second revision came back completely nonfunctional, with a short between power and ground. Optical inspection while manufacturing the polysilicon layer showed that there was no field oxide on the chip.

Inspection of the masks showed that the active area mask specified active area (i.e., diffusion) for the entire chip rather than just where transistors belonged. The layout tool assigned each layer—such as active area or metal1—a unique number. However, although the layout for active area layer was correct, the mask did not appear to match the active layer.

Layout documentation had been annotated on an unused layer by drawing rectangles and text to indicate functional blocks. A larger rectangle defined the entire chip area. Careful tracing of the mask-generation software found that the "unused" layer had been used for active area many years ago and that the documentation rectangles were merged with the true active area to form a blob of active covering the entire chip.

Another microprocessor from a different vendor also failed when it was first built. Visual inspection of the die showed that the entire cache was missing. The cache had been removed from the design database to speed up final verification because it had already been checked separately. An engineer neglected to put it back in before tapeout.

Both of these wasted fabrication runs could have been avoided by using more rigorous verification methods at both the design and mask fabrication facilities. Validation of dataset size by the designer would have caught the missing geometries. Use of the industry standard mask database inspection tools would have caught the error after mask build. Although in the past, fabrication of a modest number of parts for testing was a small part of the design cost, with the escalation of mask and wafer fabrication costs, these mistakes can be a multi-million-dollar error. The extra time to market has a large opportunity cost as well.

At High Voltage, a Chip Only Operates at Low Frequency

While booting the operating system during silicon debug, a microprocessor operated as expected at low voltage. At high voltage, the part only functioned at low

frequency. The high-voltage roof is an indication of a potential coupling problem in which the coupling is exacerbated by the fast edge rates associated with high-voltage operation. Test cases revealed that the problem resulted from incorrect operation of the register file when certain instructions executed. When the designers inspected the scan latches, they found that the correct 0 value was sent to the register file to write, but that an incorrect 1 was read. This indicated that either read or write operation was failing at high voltage. Trying one operation at high voltage and the other at low voltage proved the problem was in the write path.

A schematic of the register file write circuitry is shown in Figure 9.44. The register file uses predischarged write bitlines that are conditionally pulled high, depending on the data. The appropriate cell is written by turning on the corresponding write access transistor. The register cell is intentionally unstable so that the value on the bitline can overpower the cell and write the appropriate value. A weak keeper holds the metal2 bitline low when writing a 0. However, the register file is large and the keeper is at the opposite end from the data transistor. The resistance of the long thin wire further reduces the effectiveness of the keeper against noise on the bitline.

When the neighboring bitlines switch high, they couple onto the victim line and tend to pull it high. The circuit fails if the aggressors introduce too much coupling noise. At high voltage, the aggressor drivers are stronger

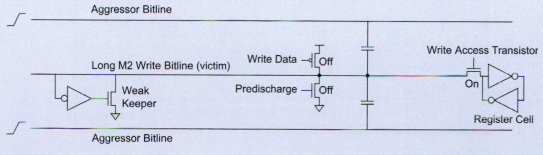

FIG 9.44 Register file write circuitry

and cause a momentary glitch on the victim. At low frequency, the keeper is sufficient to restore the victim to a low level.

The coupling problem had been flagged during design by an automated noise-checking tool. However, the tool is conservative and the area of the register file would have increased significantly if the bitlines were spaced far enough apart to satisfy the tool. Therefore, the designer checked for excessive coupling with a SPICE simulation. The simulation apparently did not properly model the combination of circumstances that caused the failure. A second engineer cross-checked all circuits that waived the noise-checker warning, but also did not discover the excessive coupling. The problem was solved by placing a second keeper near the write data transistor to fight against the coupling.

Another Funny Shmoo

During silicon debug, a microprocessor cache only functioned correctly over the peculiar range of voltages and frequencies shown in the shmoo in Figure 9.45[1]. Test code exercising the cache revealed that failures were

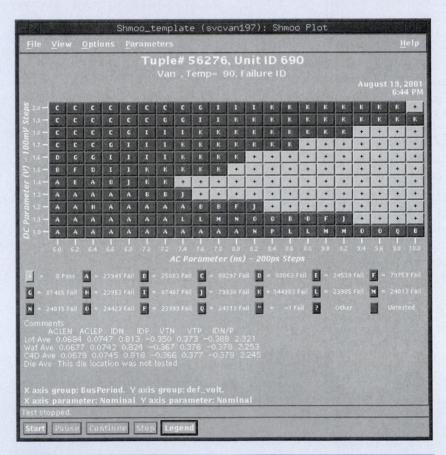

FIG 9.45 Flying saucer shmoo

---

[1]A shmoo of this type is sometimes called a *flying saucer*.

caused by bad data being read from the cache. Scan isolated the problem to a dynamic multiplexer choosing one of the global bitlines, as shown in Figure 9.46.

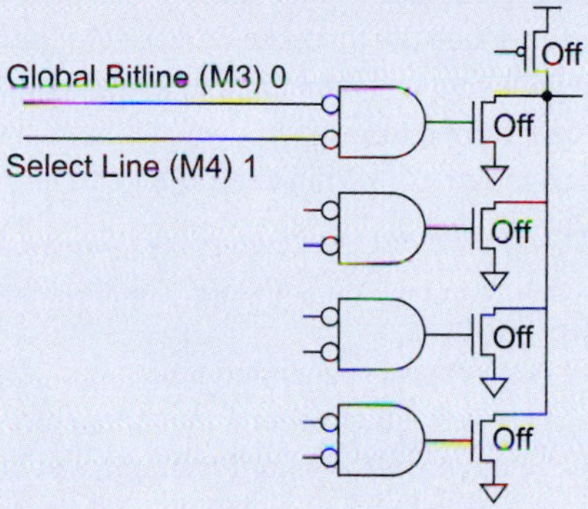

FIG 9.46 Dynamic bitline multiplexer

The multiplexer inputs were the NORs of dynamic metal3 global bitlines and corresponding select signals. The metal4 select lines were early and did not need to be dynamic, but were implemented as dynamic nodes anyway. All of the transistors in the dynamic multiplexer were supposed to remain OFF in this particular test case, leaving the multiplexer output high.

One input of the multiplexer had a low value on the global bitline, but was not selected, as shown. Therefore, the transistor should have been OFF. Nevertheless, the output of the multiplexer incorrectly discharged. One neighbor of the select line was ground; the other fell low. Coupling from a single neighbor is generally not enough to cause noise failure. However, many global bitlines ran over the top of the select line and also fell low. Laser voltage probing showed that the select line was incorrectly pulled low, apparently from coupling caused by these falling bitlines as well as the neighbor line. The odd shape of the shmoo happened because the failures only occurred when the neighbor and overhead lines both fell at about the same time; otherwise, the keeper on the select line was strong enough to recover from one noise event before the other arrived. Because the bitline and

control paths were different, the noise events only happened simultaneously for certain voltages.

Noise analysis tools usually check only neighbors, and the single switching neighbor was not sufficient to trigger an error. In this circumstance, so many global bitlines ran over the top of the select wire that their coupling could not be neglected. The problem was fixed by converting the control line into a static signal more resistant to coupling noise. A better noise analyzer could have considered coupling from neighbors above and below, especially on dynamic nets. However, it is difficult to extract information about such orthogonal neighbors because they are often drawn at different levels of the layout hierarchy. Moreover, assuming all neighbors switch in the worst possible direction is usually pessimistic for long wires. Nevertheless, such a data-dependent failure mechanism is a source of nightmares for designers.

Incorrect Operation at Low Temperature

A floating-point coprocessor was tested by running the LINPACK benchmark. The benchmark performs a series of floating-point operations and generates a checksum to verify the result. The chip would occasionally produce the wrong checksum. One of the engineers heated the coprocessor by removing the heatsink and found that the coprocessor became reliable at higher temperature.

This suggested that the problem might be caused by coupling, which is generally more serious at lower temperature where the edge rates are faster. The error was tracked to a long on-chip bus with many wires laid out on a tight pitch. Although the wires were subject to coupling noise, they were not on a critical path and should have had plenty of time to settle to the correct value. Unfortunately, they drove the diffusion input of a latch. When crosstalk drove an input below $-V_t$, it would turn on the pass transistor and incorrectly discharge the latch (see Section 6.3.9).

The floating-point unit bug was holding up lucrative product shipments. While a corrected coprocessor was being fabricated, the old unit was shipped in products with a bolt-on thermostat/heater unit used to guarantee a minimum operating temperature.

An obvious lesson of this experience is to avoid driving diffusion inputs with potentially noisy signals. More fundamentally, however, this bug demonstrated a marginal design of the cell library that should have been caught in the library review. Moreover, humans are inherently prone to errors. Electrical rules like no noisy diffusion inputs aren't worth the paper they are printed on unless computer code exists to enforce them.

Slower Than Expected Performance

An application-specific integrated circuit (ASIC) was fabricated on a gate array by a third-party gate array manufacturer. Although static timing analysis predicted that the chip would function fast enough, the manufacturer found that most of the chips would not operate at the desired frequency and instead had to be derated by about 20%.

The designer examined a die plot, looking for the source of the unexpectedly slow performance. The plot showed that the horizontal power and ground lines were only strapped along the edges of the chip, as shown in Figure 9.47(a). Some rows of gates consumed large amounts of power, causing large IR drops along their power lines. Measurements showed that the power supply sometimes drooped below 2 V, despite the nominal 3.3 V power supply. When the wide vertical

power supply straps were added, as shown in Figure 9.47(b), most chips met target speed.

Modern chips require low-resistance on-chip power distribution networks and often use power and ground pads distributed across the die rather than just at the periphery to reduce the distance and resistance between the pads and the gates. Power integrity analysis should be performed to verify that the static or dynamic voltage droops remain within their budget everywhere on the chip.

Class Chip Failures

One of the authors has supervised a number of class project chips. Some of the reasons that chips have come back partially or completely nonfunctional include:

- *Insufficient simulation*
  A ring oscillator was placed on the chip as a test structure to verify that the hardware was at least partially functional even if the rest of the chip might not work. It didn't oscillate. It had not been simulated because it was "too simple." Inspection during debug found that the oscillator had an even number of inverters!

  Another chip was designed with a new CAD tool that had a buggy simulator. Most of the chip operated correctly, but the chip as a whole would not simulate. The problem was attrib-

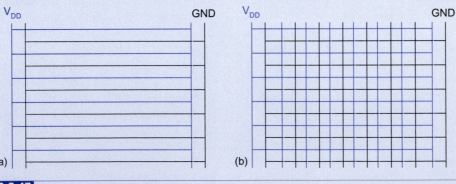

FIG 9.47 Power supply network

uted to a bug in the simulator and was taped out anyway. The chip came back nonfunctional.

● *Incomplete top-level verification*
One year, a pad frame was used that was incompatible with the normal verification flow. The chip cores were verified, placed in the pad frame, and then routed to the pads. DRC and simulation were not performed on the connections to the pads, so students carefully scrutinized their routing by hand. Upon testing, three of the four different designs were found to have errors in the routing to the pads. No errors were found in the cores that had been verified. "If you don't test it, it won't work! (guaranteed)"

— A neural network chip seemed to have a defective scan chain because the scan data out line never budged from 0 as configuration data was scanned into the chip. Testing found that the chip was correctly configured except in the last bit of the scan chain. Inspection of the layout revealed that the scan data out line (which came from the last bit of the scan chain) had been shorted to ground while being routed to the pads.

— A carry-lookahead adder produced incorrect results on certain input patterns. The least significant bits were always correct. Inspection of the layout revealed that the A[4] input was routed from the pad most of the way to the core but part of the wire was missing, probably because the designer accidentally hit UNDO after finishing the route.

— A GPS searcher chip had an inverter connected to a pair of pins to verify that the chip showed basic functionality. The output was stuck low. Inspection of the layout revealed that the input was attached to an output pad and the output to an input pad. The GPS searcher itself was fully operational.

While some of these may represent class situations, the same type of reasons for partial failure also plague industry chips. In particular, when time scales are stressed, the boundary conditions are often overlooked, which leads to problems when the chips are fabricated. Once a good verification methodology is put in place that includes a known-good pad frame, top-level DRC, and full-chip simulation, students have had a 100% success rate on class chips.

Summary

This chapter has summarized the important issues in CMOS chip testing and has provided some methods for incorporating test considerations into chips from the start of the design. Scan is now an indispensable technique to observe and control registers because probing signals directly has become extremely difficult. The importance of writing adequate tests for both the functional verification and manufacturing verification cannot be understated. It is probably the single most important activity in any CMOS chip design cycle and usually takes the longest time no matter what design methodology is used. If one message is left in your mind after this chapter, it should be that you should be absolutely rigorous about the testing activity surrounding a chip project and it should rank first in any design trade-offs.

Exercises

9.1   A circuit does not operate at the desired frequency. Cooling the circuit with freeze spray fixes the problem. A shmoo shows the circuit operates correctly at higher than nominal V_{DD}. What is the general nature of the likely problem and why?

9.2   You have to test a large die (1 cm × 1 cm) that is housed in a package that costs $5. Would you do wafer testing? Why?

9.3   A verification script detects a single discrepancy between the golden model and your design out of 400,000 vectors. Would you proceed to fabrication? Explain your decision.

9.4   Explain what is meant by a Stuck-at-1 fault and a Stuck-at-0 fault.

9.5   How are sequential faults caused in CMOS? Give an example.

9.6   Explain the different kinds of physical faults that can occur on a CMOS chip and relate them to typical circuit failures.

9.7   Explain the terms controllability, observability, and fault coverage.

9.8   Why is it important to have a high fault coverage for a set of test vectors?

9.9   Explain how serial-scan testing is implemented.

9.10 Explain the principles of Built-In Self-Test (BIST). What are the advantages and disadvantages of BIST?

9.11 You have to design an extremely fast divide by eight frequency divider that taxes the capabilities of the process you are using. What test strategy would you employ to test the divider? Explain the reasons for your choice.

9.12 Design a register that minimizes transistor count, but allows parallel scan to be implemented, as outlined in Figure 9.14.

9.13 Explain how a Pseudo-Random Sequence Generator (PRSG) can be used to test a 16-bit data path. How would the outputs be collected and checked?

9.14 Design a block diagram of a test generator for a 4K × 32 static RAM.

9.15 Using P1500, describe in block diagram form how you would perform SOC testing of the software radio shown in Figure 8.11. What architecture of TAM would you use and why?

Datapath Subsystems 10

10.1 Introduction

Most chip functions can be divided into the following categories:

- Datapath operators
- Memory elements
- Control structures
- Special-purpose cells
  - I/O
  - Power distribution
  - Clock generation and distribution
  - Analog

CMOS system design consists of partitioning the system into subsystems of the types listed above. Many options exist that make tradeoffs between speed, density, programmability, ease of design, and other variables. This chapter addresses design options for common datapath operators. The next chapter addresses arrays, especially those used for memory. Control structures are most commonly coded in a hardware description language and synthesized. Special-purpose subsystems are considered in Chapter 12.

As introduced in Chapter 1, datapath operators benefit from the structured design principles of hierarchy, regularity, modularity, and locality. They may use N identical circuits to process N-bit data. Related data operators are placed physically adjacent to each other to reduce wire length and delay. Generally, data is arranged to flow in one direction, while control signals are introduced in a direction orthogonal to the dataflow.

Common datapath operators considered in this chapter include adders, one/zero detectors, comparators, counters, Boolean logic units, error-correcting code blocks, shifters, and multipliers.

10.2 Addition/Subtraction

"Multitudes of contrivances were designed, and almost endless drawings made, for the purpose of economizing the time and simplifying the mechanism of carriage."

—Charles Babbage, on Difference Engine No. 1, 1864 [Morrison61]

Addition forms the basis for many processing operations, from counting to multiplication to filtering. As a result, adder circuits that add two binary numbers are of great interest to digital system designers. An extensive, almost endless, assortment of adder architectures serve different speed/area requirements. This section begins with half adders and full adders for single-bit addition. It then considers a plethora of carry-propagate adders (CPAs) for the addition of multi-bit words. Finally, related structures such as subtracters and multiple-input adders are discussed.

10.2.1 Single-bit Addition

The *half adder* of Figure 10.1(a) adds two inputs, A and B. The result is 0, 1, or 2, so two bits are required to represent the value; they are called the sum S and carry-out C_{out}. The carry-out is equivalent to a carry-in to the next more significant column of a multi-bit adder, so it can be described as having double the *weight* of the other bits. If multiple adders are to be cascaded, each must be able to receive the carry-in. Such a *full adder* shown in Figure 10.1(b) has a third input called C or C_{in}.

The truth tables for the half adder and full adder are given in Table 10.1 and Table 10.2. For a full adder, it is sometimes useful to define *Generate* (G), *Propagate* (P) and perhaps *Kill* (K) signals. The adder generates a carry when C_{out} is true independent of C_{in}, so $G = A \cdot B$. The adder kills a carry when C_{out} is false independent of C_{in}, so $K = \overline{A} \cdot \overline{B} = \overline{A + B}$. The adder propagates a carry, i.e., produces a carry-out if and only if it receives a carry-in, when exactly one input is true: $P = A \oplus B$.

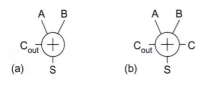

FIG 10.1 Half and full adders

Table 10.1	Truth table for half adder		
A	*B*	*C*_{out}	*S*
0	0	0	0
0	1	0	1
1	0	0	1
1	1	1	0

Table 10.2 Truth table for full adder

A	*B*	*C*	*G*	*P*	*K*	C_{out}	*S*
0	0	0	0	0	1	0	0
		1				0	1
0	1	0	0	1	0	0	1
		1				1	0
1	0	0	0	1	0	0	1
		1				1	0
1	1	0	1	0	0	1	0
		1				1	1

From the truth table, the half adder logic is:

$$\begin{aligned} S &= A \oplus B \\ C_{out} &= A \cdot B \end{aligned} \qquad \textbf{(10.1)}$$

and the full adder logic is:

$$\begin{aligned} S &= A\overline{B}\,\overline{C} + \overline{A}B\overline{C} + \overline{A}\,\overline{B}C + ABC \\ &= (A \oplus B) \oplus C = P \oplus C \\ C_{out} &= AB + AC + BC \\ &= AB + C(A + B) \\ &= \overline{\overline{A}\,\overline{B} + \overline{C}(\overline{A} + \overline{B})} \\ &= \text{MAJ}(A,\ B,\ C) \end{aligned} \qquad \textbf{(10.2)}$$

The most straightforward approach to designing an adder is with logic gates. Figure 10.2 shows a half adder. Figure 10.3 shows a full adder at the gate (a) and transistor (b) levels. The carry gate uses the factored definition and is also called a *majority* gate. Full adders are used most often, so they will receive the attention of the remainder of this section.

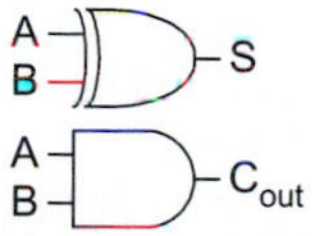

FIG 10.2 Half adder design

The full adder of Figure 10.3(b) employs 32 transistors (6 for the inverters, 10 for the majority gate, and 16 for the 3-input XOR). A more compact design is based on the observation that S can be factored to reuse the C_{out} term.

$$S = ABC + (A + B + C)\overline{C}_{out} \qquad \textbf{(10.3)}$$

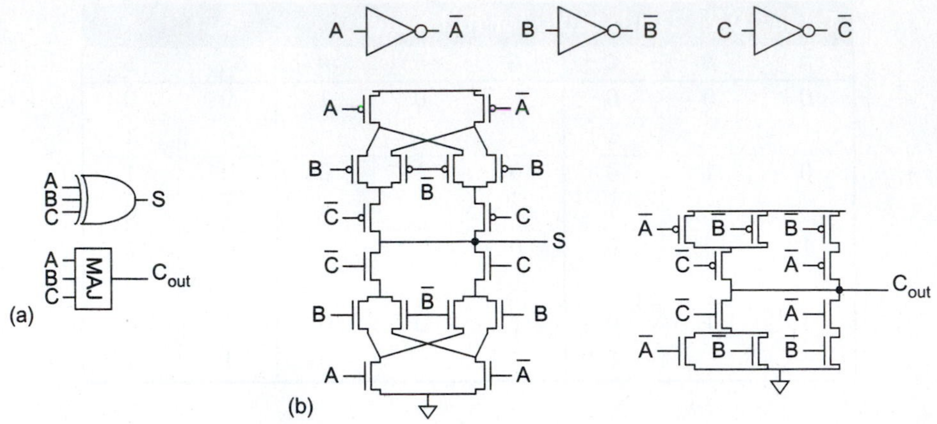

FIG 10.3 Full adder design

Such a design is shown at the gate (a) and transistor (b) levels in Figure 10.4 and uses only 28 transistors. Note that the pMOS network is identical to the nMOS network rather than being the conduction complement. This simplification reduces the number of series transistors and makes the layout more uniform. It is possible because the addition function is *symmetric*, i.e., the function of complemented inputs is the complement of the function.

This design has a greater delay to compute S than C_{out}. In carry-ripple adders (Section 10.2.2.1), the critical path goes from C (C_{in}) to C_{out} through many full adders, so the extra delay computing S is unimportant. Figure 10.4(c) shows the adder with transistor sizes optimized to favor the critical path using a number of techniques:

- Feed the carry-in signal (C) to the inner inputs so the internal capacitance is already discharged.

- Make all transistors in the sum logic whose gate signals are connected to the carry-in and carry logic minimum size (1 unit, e.g., $4\lambda / 2\lambda$). This minimizes the branching effort on the critical path. Keep routing on this signal as short as possible to reduce interconnect capacitance.

- Determine widths of series transistors by logical effort and simulation. Build an asymmetric gate that reduces the logical effort from C to $\overline{C}_{out}$ at the expense of effort to S.

- Use relatively large transistors on the critical path so that stray wiring capacitance is a small fraction of the overall capacitance.

- Remove the output inverters and alternate positive and negative logic to reduce delay and transistor count to 24 (see Section 10.2.2.1).

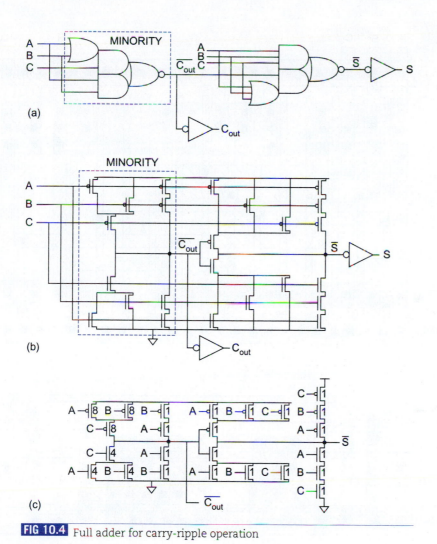

FIG 10.4 Full adder for carry-ripple operation

Figure 10.5 shows two layouts of the adder (see also the inside front cover). The choice of the aspect ratio depends on the application. In a standard-cell environment, the layout of Figure 10.5(a) might be appropriate when a single row of nMOS and pMOS transistors is used. The routing for the *A*, *B*, and *C* inputs is shown inside the cell, although it could be placed outside the cell because external routing tracks have to be assigned to these signals anyway. Figure 10.5(b) shows a layout of Figure 10.4(c) that might be appropriate for a datapath. Here, the transistors are rotated and all of the wiring is completed in polysilicon and metal1. This allows metal2 bus lines to pass over the cell horizontally. Moreover, the widths of the transistors can increase without impacting the bit-pitch (height) of the datapath. In this case, the widths are selected to reduce the C_{in} to C_{out} delay that is on the critical path of a carry-ripple adder.

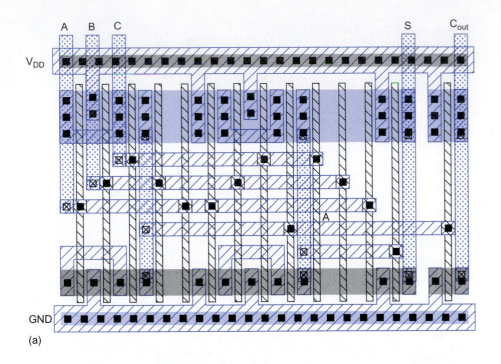

(a)

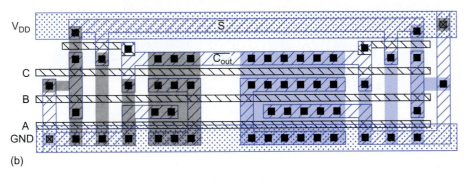

(b)

FIG 10.5 Full adder layouts. Color version on inside front cover.

A rather different full adder design uses transmission gates to form multiplexers and XORs. Figure 10.6(a) shows the transistor-level schematic using 24 transistors and providing buffered outputs of the proper polarity with equal delay. The design can be understood by parsing the transmission gate structures into multiplexers and an "invertible inverter" XOR structure (see Section 10.7.4), as drawn in Figure 10.6(b)[1]. Note that the multiplexer choosing S is configured to compute $P \oplus C$, as given in EQ (10.2).

---

[1]Some switch-level simulators, notably IRSIM, are confused by this XOR structure and may not simulate it correctly.

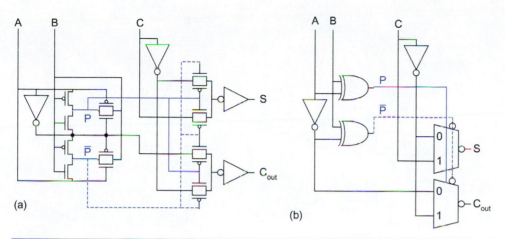

FIG 10.6 Transmission gate full adder

The transistor count can be reduced to 22 at the expense of an extra inverter delay by computing $\overline{P}$ from P rather than using a separate XOR. [Zhuang92] proposes such a design, shown in Figure 10.7. Note that in the Zhuang full adder, the multiplexer for S is replaced with an XOR, but the effect is the same.

Another transmission-gate approach uses complementary pass-transistor logic (CPL) [Yano90], shown in Figure 10.8. In comparison to a poorly optimized 40-transistor static CMOS full adder, the author finds CPL is twice as fast, 30% lower in power, and slightly smaller. On the other hand, in comparison to a careful implementation of the static CMOS full adder in Figure 10.4(b), [Zimmermann97a] finds the CPL delay slightly better, the power comparable, and the area much larger.

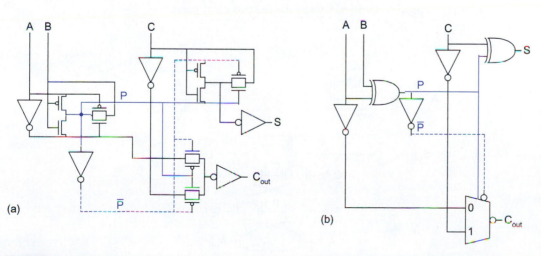

FIG 10.7 Zhuang full adder

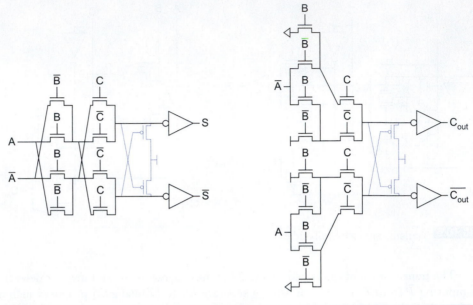

FIG 10.8 CPL full adder

Dynamic full adders are widely used in fast multipliers. As the sum logic inherently requires true and complementary versions of the inputs, dual-rail domino is necessary. Figure 10.9 shows such an adder using footless dual-rail domino XOR/XNOR and MAJORITY/MINORTY gates [Heikes94]. The delays to the two outputs are reasonably well balanced, which is important for multipliers where both paths are critical. It shares transistors in the sum gate to reduce transistor count and takes advantage of the symmetric property to provide identical layouts for the two carry gates.

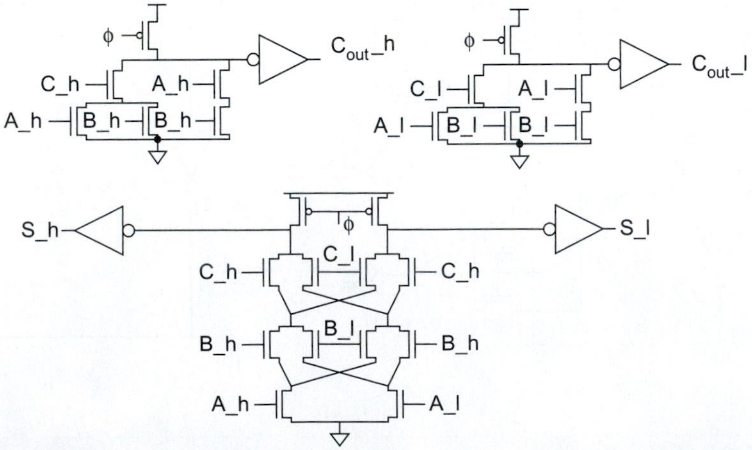

FIG 10.9 Dual-rail domino full adder

Static CMOS full adders typically have a delay of 2–3 FO4 inverters, while domino adders have a delay of about 1.5.

10.2.2 Carry-propagate Addition

N-bit adders take inputs $\{A_N, \ldots, A_1\}$, $\{B_N, \ldots, B_1\}$, and carry-in C_{in}, and compute the sum $\{S_N, \ldots, S_1\}$ and the carry-out of the most significant bit C_{out}, as shown in Figure 10.10. They are called *carry-propagate adders* (CPAs) because the carry into each bit can influence the carry into all subsequent bits. For example, Figure 10.11 shows the addition $1111_2 + 0000_2 + 0/1$, in which each of the sum and carry bits is influenced by C_{in}. The simplest design is the carry-ripple adder in which the carry-out of one bit is simply connected as the carry-in to the next. Faster adders look ahead to predict the carry-out of a multi-bit group. This is usually done by computing group PG signals to indicate whether the multi-bit group will propagate a carry-in or will generate a carry-out. Long adders use multiple levels of lookahead structures for even more speed.

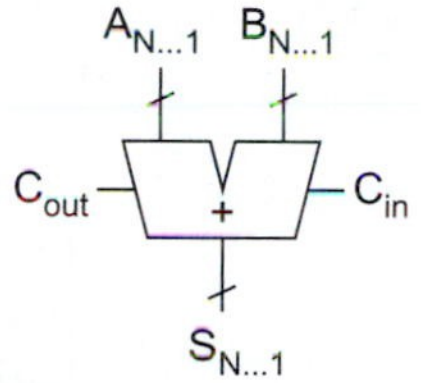

FIG 10.10 Carry-propagate adder

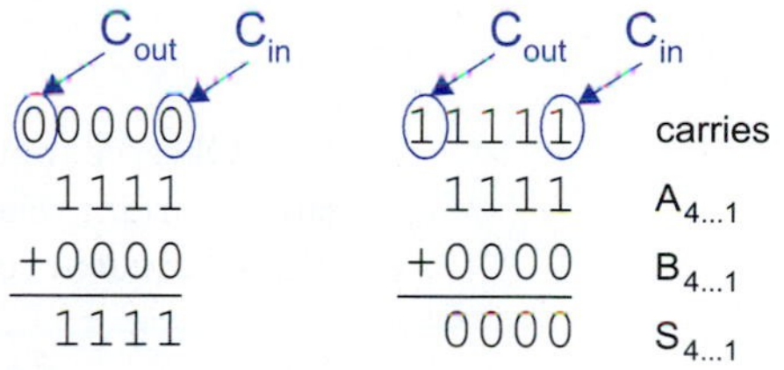

10.2.2.1 Carry-ripple Adder
An N-bit adder can be constructed by cascading N full adders, as shown in Figure 10.12(a) for $N = 4$. This is called a *carry-ripple adder* (or *ripple-carry adder*). The carry-out of bit i, C_i, is the carry-in to bit $i +1$. This carry is said to have twice the *weight* of the sum S_i. The delay of the adder is set by the time for the carries to ripple through the N stages, so the $t_{C \rightarrow Cout}$ delay should be minimized.

This delay can be reduced by omitting the inverters on the outputs, as was done in Figure 10.4(c). Because addition is a self-dual function, an inverting full adder receiving complementary inputs produces true outputs. Figure 10.12(b) shows a carry-ripple adder built from inverting full adders. Every other stage operates on complementary data. The delay in inverting the adder inputs or sum outputs is finessed out of the critical ripple-carry path.

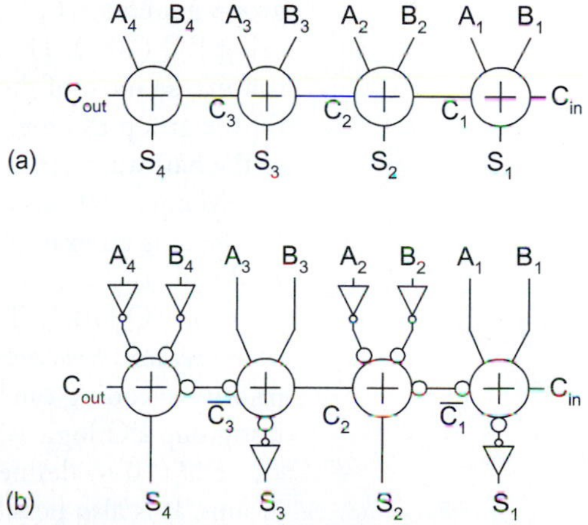

FIG 10.11 Example of carry propagation

10.2.2.2 Carry Generation and Propagation
This section introduces notation commonly used in describing faster adders. Recall that the P and G *propagate*[2] and *generate* signals were defined in Section 10.2.1. We can generalize these signals to describe whether a group spanning bits $i \ldots j$, inclusive, generate a carry or propagate a carry. A group of bits generates a carry if its carry-out is true independent of the carry-in; it propagates a carry if its carry-out is true when there is a carry-in. These signals can be defined recursively for $i \geq k > j$ as

(a)

(b)

FIG 10.12 4-bit carry-ripple adder

---

[2]Do not confuse the *propagate* term P with the *carry-propagate adder* (CPA).

$$G_{i:j} = G_{i:k} + P_{i:k} \bullet G_{k-1:j}$$
$$P_{i:j} = P_{i:k} \bullet P_{k-1:j}$$

(10.4)

with the base case

$$G_{i:i} \equiv G_i = A_i \bullet B_i$$
$$P_{i:i} \equiv P_i = A_i \oplus B_i$$

(10.5)

In other words, a group generates a carry if the upper (more significant) or the lower portion generates and the upper portion propagates that carry. The group propagates a carry if both the upper and lower portions propagate the carry.[3]

The carry-in must be treated specially. Let us define $C_0 = C_{in}$ and $C_N = C_{out}$. Then we can define generate and propagate signals for bit 0 as

$$G_{0:0} = C_{in}$$
$$P_{0:0} = 0$$

(10.6)

Observe that the carry into bit i is the carry-out of bit i-1 and is $C_{i-1} = G_{i-1:0}$. This is an important relationship; *group generate* signals and *carries* will be used synonymously in the subsequent sections. We can thus compute the sum for bit i using EQ (10.2) as

$$S_i = P_i \oplus G_{i-1:0}$$

(10.7)

Hence, addition can be reduced to a three-step process of computing bitwise generate and propagate signals using EQs (10.5) and (10.6), combining these signals to determine group generates $G_{i-1:0}$ for all $N \geq i \geq 1$ using EQ (10.4), and then calculating the sums using EQ (10.7). These steps are illustrated in Figure 10.13. The first and third steps are routine, so most of the attention in the remainder of this section is devoted to alternatives for the group PG logic with different tradeoffs between speed, area, and complexity. Some of the hardware can be shared in the bitwise PG logic, as shown in Figure 10.14.

Many notations are used in the literature to describe the group PG logic. In general, PG logic is an example of a *prefix* computation [Leighton92]. It accepts inputs $\{P_{N:N}, \ldots, P_{0:0}\}$ and $\{G_{N:N}, \ldots, G_{0:0}\}$ and computes the *prefixes* $\{G_{N:0}, \ldots, G_{0:0}\}$ using the relationship given in EQ (10.4). This relationship is given many names in the literature including the *delta operator, fundamental carry operator,* and *prefix operator.* Many other problems such as priority encoding can be posed as prefix computations and all the techniques used to build fast group PG logic will apply, as we will explore in Section 10.10.

EQ (10.4) defines *valency-2* group PG logic because it combines pairs of smaller groups. It is also possible to define higher-valency group logic to use fewer stages of more complex gates [Beaumont-Smith99], as shown in EQ (10.8). For example, in valency-4

---

[3]Alternatively, some adders use $\overline{K}_i = A_i + B_i$ in place of P_i because OR is faster than XOR. The group logic becomes $G_{i:j} = G_{i:k} + \overline{K}_{i:k} \bullet G_{k-1:j}$ and $K_{i:j} = K_{i:k} \bullet K_{k-1:j}$. However, P_i is still required in EQ (10.7) to compute the final sum.

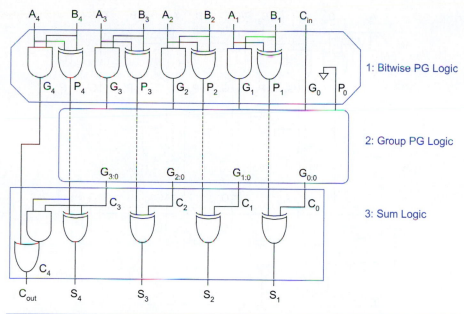

FIG 10.13 Addition with generate and propagate logic

group logic, a group propagates the carry if all four portions propagate. A group generates a carry if the upper portion generates, the second portion generates and the upper propagates, the third generates and the upper two propagate, or the lower generates and the upper three propagate. Logical effort teaches us that the best stage effort is about 4. Therefore, it is not necessarily better to build fewer stages of higher-valency gates; simulations or calculations should be done to compare the alternatives for a given technology.

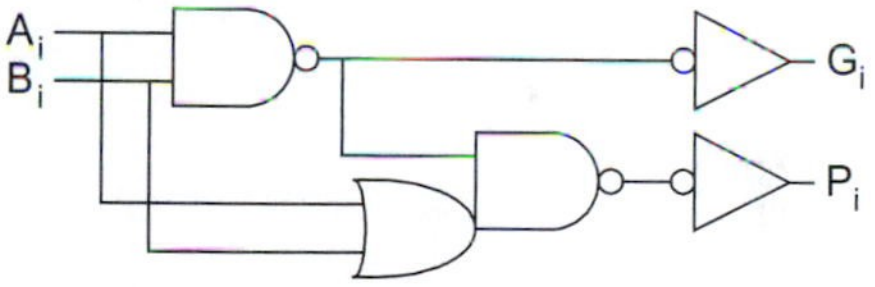

FIG 10.14 Shared bitwise PG logic

$$
\begin{aligned}
G_{i:j} &= G_{i:k} + P_{i:k} \cdot G_{k-1:l} + P_{i:k} \cdot P_{k-1:l} \cdot G_{l-1:m} + P_{i:k} \cdot P_{k-1:l} \cdot P_{l-1:m} \cdot G_{m-1:j} \\
&= G_{i:k} + P_{i:k}\big(G_{k-1:l} + P_{k-1:l}\big(G_{l-1:m} + P_{l-1:m}G_{m-1:j}\big)\big) \\
P_{i:j} &= P_{i:k} \cdot P_{k-1:l} \cdot P_{l-1:m} \cdot P_{m-1:j}
\end{aligned}
\right\} \quad (i \geq k > l > m > j) \quad (10.8)
$$

10.2.2.3 PG Carry-ripple Addition The critical path of the carry-ripple adder passes from carry-in to carry-out along the carry chain majority gates. As the P and G signals will have already stabilized by the time the carry arrives, we can use them to simplify the majority function into an AND-OR gate[4]:

---

[4]Whenever positive logic such as AND-OR is described, it is implicit that you can also use an AOI gate and alternate positive and negative polarity stages as was done in Figure 10.12(b) to save area and delay.

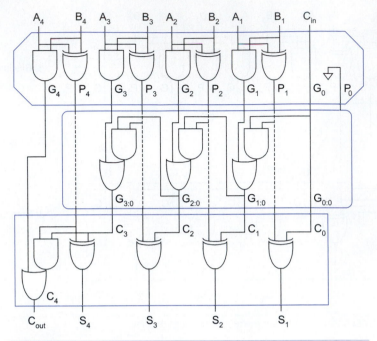

FIG 10.15 4-bit carry-ripple adder using PG logic

$$C_i = A_i B_i + \left(A_i + B_i \right) C_{i-1}$$
$$= A_i B_i + \left(A_i \oplus B_i \right) C_{i-1}$$
$$= G_i + P_i C_{i-1}$$

(10.9)

Because $C_i = G_{i:0}$, carry-ripple addition can now be viewed as the extreme case of group PG logic in which a 1-bit group is combined with an i-bit group to form an $(i+1)$-bit group

$$G_{i:0} = G_i + P_i \bullet G_{i-1:0}$$

(10.10)

In this extreme, the group propagate signals are never used and need not be computed. Figure 10.15 shows a 4-bit carry-ripple adder. The critical carry path now proceeds through a chain of AND-OR gates rather than a chain of majority gates. Figure 10.16 illustrates the group PG logic for a 16-bit carry-ripple adder, where the AND-OR gates in the group PG network are represented with gray cells.

Diagrams like these will be used to compare a variety of adder architectures in subsequent sections. The diagrams use black cells, gray cells, and white buffers defined in

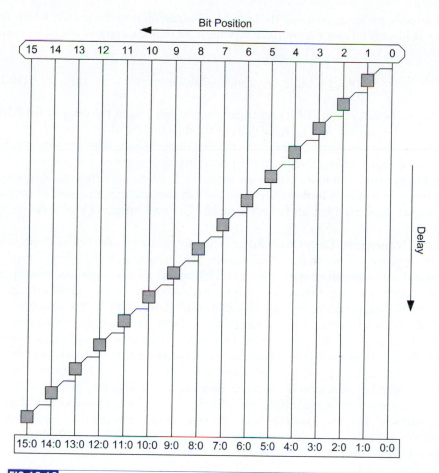

FIG 10.16 Carry-ripple adder group PG network

Figure 10.17(a) for valency-2 cells. Black cells contain the group generate and propagate logic (an AND-OR gate and an AND gate) defined in EQ (10.4). Gray cells containing only the group generate logic are used at the final cell position in each column because only the group generate signal is required to compute the sums. Buffers can be used to minimize the load on critical paths. Each line represents a *bundle* of the group generate and propagate signals (propagate signals are omitted after gray cells). The bitwise PG and sum XORs are abstracted away in the top and bottom boxes and it is assumed that an AND-OR gate operates in parallel with the sum XORs to compute the carry-out:

$$C_{\text{out}} = G_{N:0} = G_N + P_N G_{N-1:0} \tag{10.11}$$

The cells are arranged along the vertical axis according to the time at which they operate [Guyot97]. From Figure 10.16 it is apparent that the carry-ripple adder critical path delay is:

$$t_{\text{ripple}} = t_{pg} + (N-1)t_{AO} + t_{\text{xor}} \tag{10.12}$$

where t_{pg} is the delay of the 1-bit propagate/generate gates, t_{AO} is the delay of the AND-OR gate in the gray cell, and t_{xor} is the delay of the final sum XOR.

Often, using noninverting gates leads to more stages of logic than are necessary. Figure 10.17(b) shows how to alternate two types of inverting stages on alternate rows of the group PG network to remove extraneous inverters. For best performance, $G_{k-1:j}$ should drive the inner transistor in the series stack. You can also reduce the number of stages by using higher-valency cells, as shown in Figure 10.17(c) for a valency-4 black cell.

10.2.2.4 Manchester Carry Chain Adder

The carry chain can also be built from switch logic using propagate, generate, and kill signals. The majority gate of Figure 10.18(a) can be replaced with a switch network. Figure 10.18(b) shows a static implementation operating on a complementary carry. The complementary carry can be propagated through the transmission gate, generated with the nMOS transistor, or killed with the pMOS transistor. Figure 10.18(c) shows a dynamic version that is faster and requires less hardware.

Multiple stages are directly connected to build a *Manchester carry chain*, as shown in Figure 10.19(a) [Kilburn59]. The resistance and capacitance of the carry chain grow with the length, so the delay grows with the square of length. This is clearly not viable for long adders. As with long wires, the delay can be made linear with length by periodically breaking the chain and inserting an inverter to buffer the signals. The best chain length depends on the parasitic capacitance and can be determined through simulation or calculations for a particular technology (see Exercise 10.4), but is typically 3 or 4; Figure 10.19(b) shows a valency-4 carry chain. The widths of the transistors along the chain can be tapered to reduce parasitic delay.

Observe that the Manchester carry chain computes the functions

$$
\begin{aligned}
C_0 &= G_{0:0} = C_0 \\
C_1 &= G_{1:0} = G_1 + P_1 C_0 \\
C_2 &= G_{2:0} = G_2 + P_2\big(G_1 + P_1 C_0\big) \\
C_3 &= G_{3:0} = G_3 + P_3\big(G_2 + P_2\big(G_1 + P_1 C_0\big)\big)
\end{aligned}
\tag{10.13}
$$

$G_{3:0}$ is like the valency-4 group generate circuit of EQ (10.8), while the other outputs are the generate signals for smaller groups (including a simple buffer of the input). In other words, the carry chain can be viewed as a buffer and three gray cells of increasing valency, as shown in Figure 10.19(c). If the carry chain of Figure 10.19(b) is redrawn in a more conventional form (Figure 10.20), it can also be seen to be another representation of a footless multiple-output domino gate, as was discussed in Section 6.2.4.6.

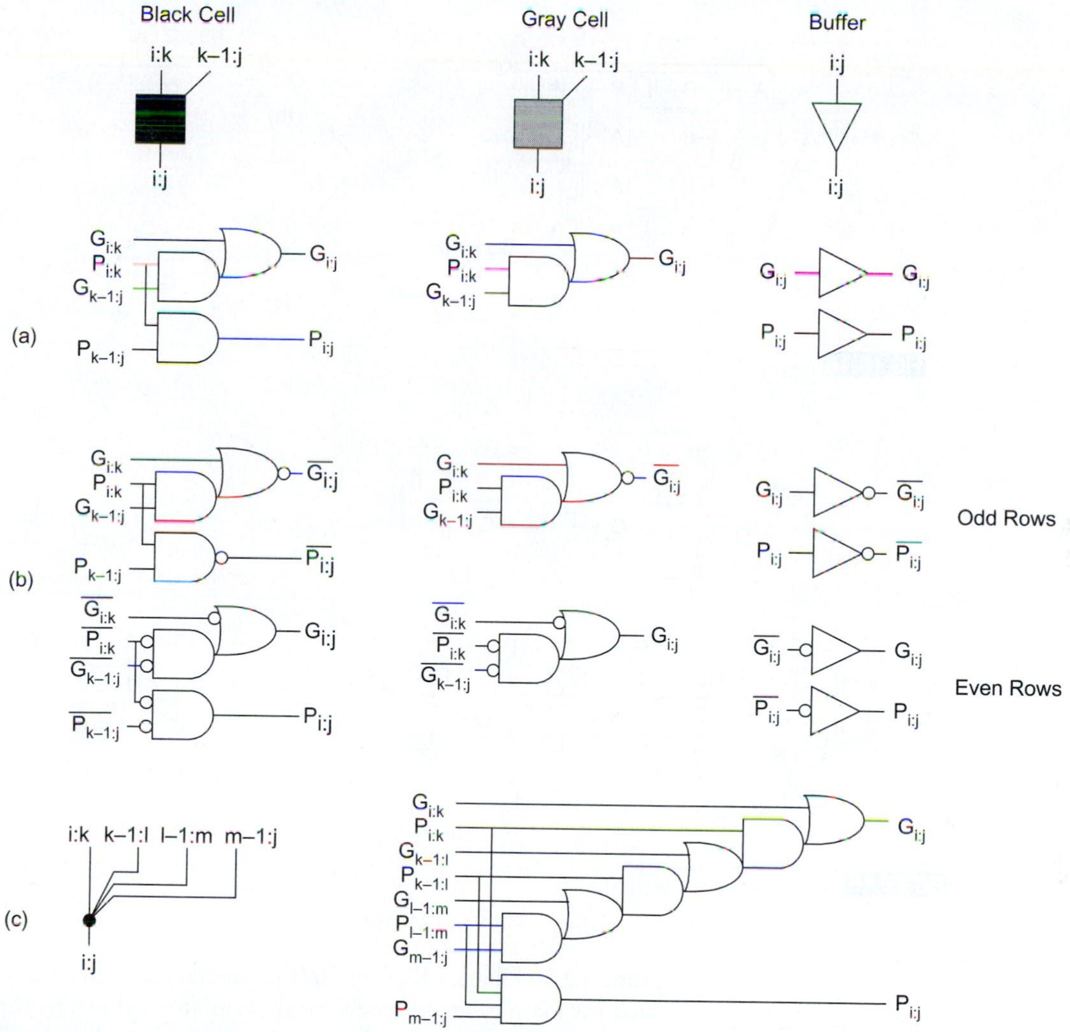

FIG 10.17 Group PG cells

Figure 10.21 shows a Manchester carry chain adder using valency-4 stages. It is similar to the carry-ripple adder, but uses $N/3$ stages.

10.2.2.5 Carry-skip Adder The critical path of a Manchester carry chain adder involves a series propagate transistor for each bit of the adder. This is a significant improvement over the carry-ripple adder, which uses a majority gate or AND-OR gate for each bit, but still may be too slow for large adders. The *carry-skip* (also called *carry-bypass*) adder, first

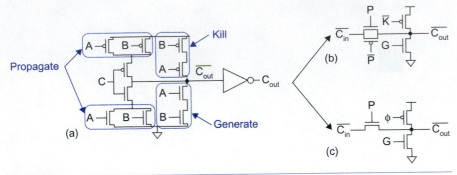

FIG 10.18 Carry chain designs

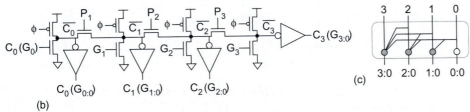

FIG 10.19 Manchester carry chains

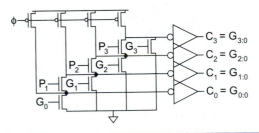

FIG 10.20 Equivalence of Manchester carry chain and multiple-output domino gate

proposed by Charles Babbage in the nineteenth century and used for many years in mechanical calculators, shortens the critical path by computing the group propagate signals for each carry chain and using this to skip over long carry ripples [Morgan59, Lehman61]. Figure 10.22 shows a carry skip adder built from 4-bit groups. The rectangles compute the bit-wise propagate and generate signals (as in Figure 10.15), and also contain a 4-input AND gate for the propagate signal of the 4-bit group. The skip multiplexer selects the group carry-in if the group propagate is true or the ripple adder carry-out otherwise.

The critical path through Figure 10.22 begins with generating a carry from bit 1, and then propagating it through the remainder of the adder. The carry must ripple through the

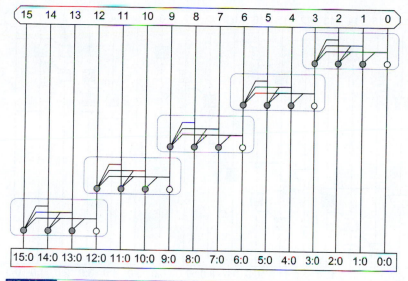

FIG 10.21 Manchester carry chain adder group PG network

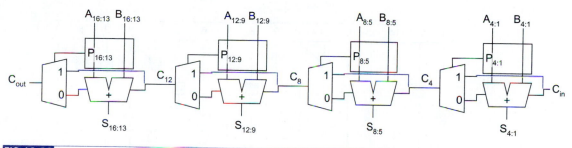

FIG 10.22 Carry-skip adder

next three bits, but then may skip across the next two 4-bit blocks. Finally, it must ripple through the final 4-bit block to produce the sums. This is illustrated in Figure 10.23. The 4-bit ripple chains at the top of the diagram determine if each group generates a carry. The carry skip chain in the middle of the diagram skips across 4-bit blocks. Finally, the 4-bit ripple chains with the blue lines represent *the same adders* that can produce a carry-out when a carry-in is bypassed to them. Note that the final AND-OR and column 16 are not strictly necessary because C_{out} can be computed in parallel with the sum XORs using EQ (10.11).

In general, an N-bit adder using k n-bit groups ($N = n \cdot k$) has a delay of

$$t_{skip} = t_{pg} + \left[2(n-1) + (k-1)\right]t_{AO} + t_{xor} \tag{10.14}$$

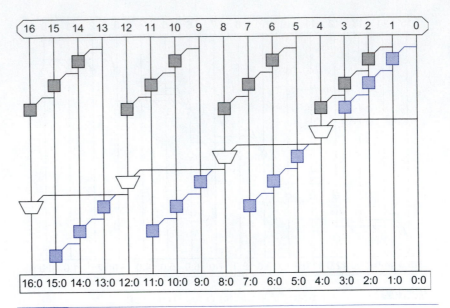

FIG 10.23 Carry-skip adder PG network

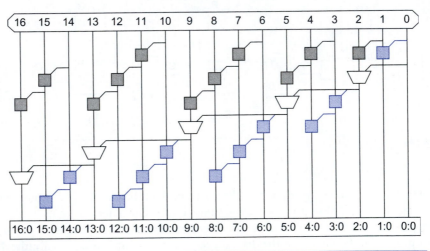

FIG 10.24 Variable group size carry-skip adder PG network

This critical path depends on the length of the first and last group and the number of groups. In the more significant bits of the network, the ripple results are available early. Thus, the critical path could be shortened by using shorter groups at the beginning and end and longer groups in the middle. Figure 10.24 shows such a PG network using groups of length [2, 3, 4, 4, 3], as opposed to [4, 4, 4, 4], which saves two levels of logic in a 16-bit adder.

The hardware cost of a carry-skip adder is equal to that of a simple carry-ripple adder plus k multiplexers and k n-input AND gates. It is attractive when ripple-carry adders are too slow, but the hardware cost must still be kept low. For long adders, you could use a multi-level skip approach to skip across the skips. A great deal of research has gone into choosing the best group size and number of levels [Majerski67, Oklobdžija85, Guyot87, Chan90, Kantabutra91], although now, parallel prefix adders are generally used for long adders instead.

It might be tempting to replace each skip multiplexer in Figure 10.22 and Figure 10.23 with an AND-OR gate combining the carry-out of the n-bit adder or the group carry-in and group propagate. Indeed, this works for domino-carry skip adders in which the carry out is precharged each cycle; it also works for carry-lookahead adders and carry-select adders covered in the subsequent section. However, it introduces a sneaky long critical path into an ordinary carry-skip adder. Imagine summing 111...111 + 000...000 + C_{in}. All of the group propagate signals are true. If $C_{in} = 1$, every 4-bit block will produce a carry-out. When C_{in} falls, the falling carry signal must ripple through all N bits because of the path through the carry out of each n-bit adder. Domino-carry skip adders avoid this path because all of the carries are forced low during precharge, so they can use AND-OR gates.

Figure 10.25 shows how a Manchester carry chain can be modified to perform carry skip [Chan90]. A valency-5 chain is used to skip across groups of 4 bits at a time.

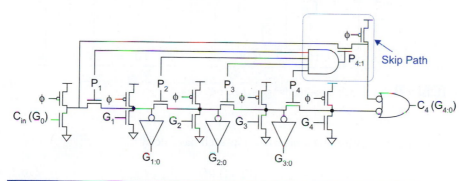

FIG 10.25 Carry-skip adder Manchester stage

10.2.2.6 Carry-lookahead Adder

The *carry-lookahead adder* (CLA) [Weinberger58] is similar to the carry-skip adder, but computes group generate signals as well as group propagate signals to avoid waiting for a ripple to determine if the first group generates a carry. Such an adder is shown in Figure 10.26 and its PG network is shown in Figure 10.27 using valency-4 black cells to compute 4-bit group PG signals.

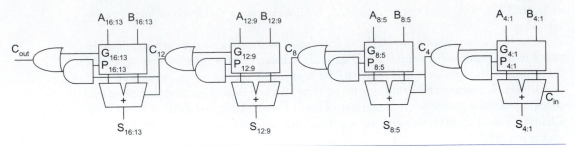

FIG 10.26 Carry-lookahead adder

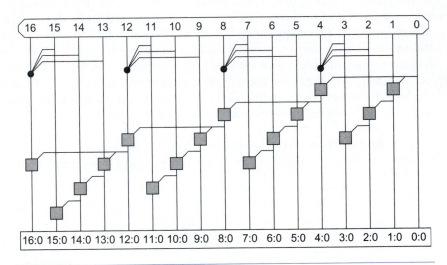

FIG 10.27 Carry-lookahead adder group PG network

In general, a CLA using k groups of n bits each has a delay of

$$t_{cla} = t_{pg} + t_{pg(n)} + \left[(n-1) + (k-1)\right]t_{AO} + t_{xor} \qquad (10.15)$$

where $t_{pg(n)}$ is the delay of the AND-OR-AND-OR-...-AND-OR gate computing the valency-n generate signal. This is no better than the variable-length carry-skip adder in Figure 10.24 and requires the extra n-bit generate gate, so the simple CLA is seldom a good design choice. However, it forms the basis for understanding faster adders presented in the subsequent sections.

CLAs often use higher-valency cells to reduce the delay of the n-bit additions by computing the carries in parallel. Figure 10.28 shows such a CLA in which the 4-bit adders are built using Manchester carry chains or multiple static gates operating in parallel.

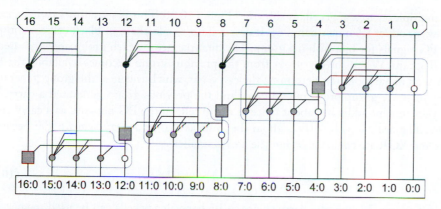

FIG 10.28 Improved CLA group PG network

10.2.2.7 Carry-select, Carry-increment, and Conditional-sum Adders The critical path of the carry-skip and carry-lookahead adders involves calculating the carry into each n-bit group, and then calculating the sums for each bit within the group based on the carry-in. A standard logic design technique to accelerate the critical path is to precompute the outputs for both possible inputs, and then use a multiplexer to select between the two output choices. The *carry-select adder* [Bedrij62] shown in Figure 10.29 does this with a pair of n-bit adders in each group. One adder calculates the sums assuming a carry-in of 0 while the other calculates the sums assuming a carry-in of 1. The actual carry triggers a multiplexer that chooses the appropriate sum. The critical path delay is

$$t_{\text{select}} = t_{pg} + \left[n + (k-2)\right]t_{AO} + t_{\text{mux}} \tag{10.16}$$

The two n-bit adders are redundant in that both contain the initial PG logic and final sum XOR. [Tyagi93] reduces the size by factoring out the common logic and simplifying

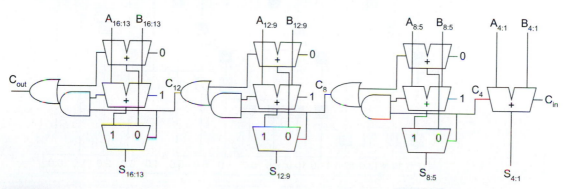

FIG 10.29 Carry-select adder

the multiplexer to a gray cell, as shown in Figure 10.30. This is sometimes called a *carry-increment* adder [Zimmermann96]. It uses a short ripple chain of black cells to compute the PG signals for bits within a group. The bits spanned by each group are annotated on the diagram. When the carry-out from the previous group becomes available, the final gray cells in each column determine the carry-out, which is true if the group generates a carry or if the group propagates a carry and the previous group generated a carry. The carry-increment adder has about twice as many cells in the PG network as a carry-ripple adder. The critical path delay is about the same as that of a carry-select adder because a mux and XOR are comparable, but the area is smaller.

$$t_{\text{increment}} = t_{pg} + \left[(n-1) + (k-1) \right] t_{AO} + t_{\text{xor}} \tag{10.17}$$

Of course, Manchester carry chains or higher-valency cells can be used to speed the ripple operation to produce the first group generate signal. In that case, the ripple delay is replaced by a group PG gate delay and the critical path becomes:

$$t_{\text{increment}} = t_{pg} + t_{pg(n)} + \left[(k-1) \right] t_{AO} + t_{\text{xor}} \tag{10.18}$$

As with the carry-skip adder, the carry chains for the more significant bits complete early. Again, we can use variable-length groups to take advantage of the extra time, as shown in Figure 10.31(a). With such a variable group size, the delay reduces to:

$$t_{\text{increment}} \approx t_{pg} + \sqrt{2N} t_{AO} + t_{\text{xor}} \tag{10.19}$$

The delay equations do not account for the fanout that each stage must drive. The fanouts in a variable-length group can become large enough to require buffering between stages. Figure 10.31(b) shows how buffers can be inserted to reduce the branching effort

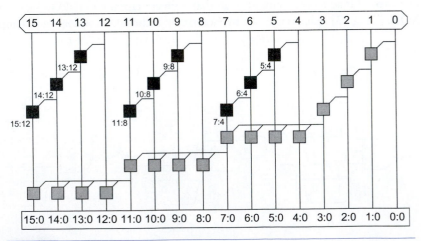

FIG 10.30 Carry-increment adder PG network

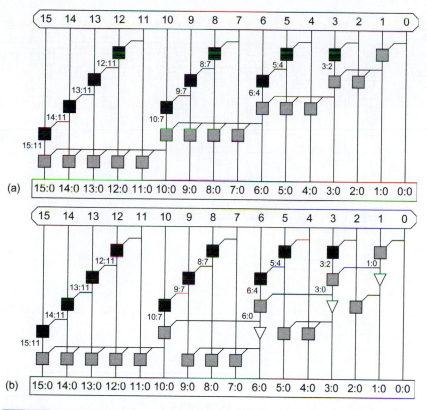

FIG 10.31 Variable-length carry-increment adder

while not impeding the critical lookahead path; this is a useful technique in many other applications.

In wide adders, we can recursively apply multiple levels of carry-select or carry-increment. For example, a 64-bit carry-select adder can be built from four 16-bit carry-select adders, each of which selects the carry-in to the next 16-bit group. Taking this to the limit, we obtain the *conditional-sum* adder [Sklansky60] that performs carry-select starting with groups of 1 bit and recursively doubling to $N/2$ bits. Figure 10.32 shows such an adder. In the first two rows, full adders compute the sum and carry-out for each bit assuming carries-in of 0 and 1, respectively. In the next two rows, multiplexer pairs select the sum and carry-out of the upper bit of each block of two, again assuming carries-in of 0 and 1. In the next two rows, multiplexers select the sum and carry-out of the upper two bits of each block of four, and so forth.

Figure 10.33 shows the operation of a conditional-sum adder in action for $N = 16$ with $C_{in} = 0$. In the block width 1 row, a pair of full adders compute the sum and carry-out for each column. One adder operates assuming the carry-in to that column is 0, while the other assumes it is 1. In the block width 2 row, the adder selects the sum for the upper half

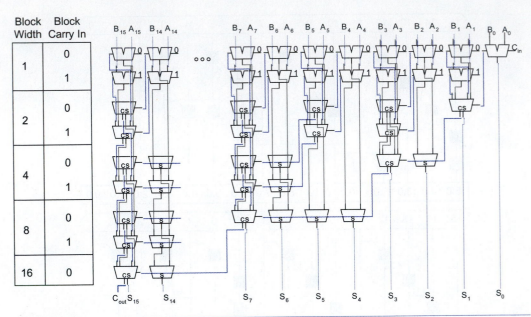

FIG 10.32 Conditional-sum adder

Block Width	Block Carry In		16	15	14	13	12	11	10	9	8	7	6	5	4	3	2	1	
		a	1	0	1	1	1	0	1	1	0	1	1	0	1	1	0	1	0
		b	0	0	0	1	1	0	0	1	1	0	1	1	0	1	1	0	
			\>\> Block Sum and Carry Out \<\<																C_{in}
1	0	s	1	0	1	0	0	0	1	0	1	1	0	1	1	0	1	1	0
		c	0	0	0	1	1	0	0	1	0	0	1	0	0	1	0	0	
	1	s	0	1	0	1	1	1	0	1	0	0	1	0	1	0	1	0	
		c	1	0	1	1	1	0	1	1	1	1	0	1	1	1	1	1	
2	0	s	1	0	0	0	0	0	0	0	1	1	0	1	0	0	1	1	1
		c	0		1		1		1		0		1		1		0		
	1	s	1	1	0	1	0	1	0	1	0	0	1	0	0	1			
		c	0		1		1		1		1		1		1				
4	0	s	1	1	0	0	0	1	0	0	0	0	0	1	0	0	1	1	
		c	0				1				1				1				
	1	s	1	1	0	1	0	1	0	1	0	0	1	0					
		c	0				1				1								
8	0	s	1	1	0	1	0	1	0	0	0	0	1	0	0	0	1	1	
		c	0								1								
	1	s	1	1	0	1	0	1	0	1									
		c	0																
16	0	s	1	1	0	1	0	1	0	1	0	0	1	0	0	0	1	1	Sum
		c	0																C_{out}
	1	s																	
		c																	

FIG 10.33 Conditional-sum addition example

of each block (the even-numbered columns) based on the carry-out of the lower half. It also computes the carry-out of the pair of bits. Again, this is done twice, for both possibilities of carry-in to the block. In the block width 4 row, the adder again selects the sum for the upper half based on the carry-out of the lower half and finds the carry-out of the entire block. This process is repeated in subsequent rows until the 16-bit sum and the final carry-out are selected.

The conditional-sum adder involves nearly $2N$ full adders and $2N\log_2 N$ multiplexers. As with carry-select, the conditional-sum adder can be improved by factoring out the sum XORs and using AND-OR gates in place of multiplexers. This leads us to the Sklansky tree adder discussed in the next section.

10.2.2.8 Tree Adders For wide adders ($N > {\sim}16$ bits), the delay of carry-lookahead (or carry-skip or carry-select) adders becomes dominated by the delay of passing the carry through the lookahead stages. This delay can be reduced by looking ahead across the lookahead blocks [Weinberger58]. In general, you can construct a multilevel tree of look-ahead structures to achieve delay that grows with $\log N$. Such adders are variously referred to as *tree* adders, *logarithmic* adders, *multilevel-lookahead* adders, *parallel-prefix* adders, or simply *lookahead* adders. The last name appears occasionally in the literature, but is not recommended because it does not distinguish whether multiple levels of lookahead are used.

There are many ways to build the lookahead tree that offer tradeoffs among the number of stages of logic, the number of logic gates, the maximum fanout on each gate, and the amount of wiring between stages. Three fundamental trees are the Brent-Kung, Sklansky, and Kogge-Stone architectures. We begin by examining each in the valency-2 case that combines pairs of groups at each stage.

The *Brent-Kung* tree [Brent82] (Figure 10.34(a)) computes prefixes for 2-bit groups. These are used to find prefixes for 4-bit groups, which in turn are used to find prefixes for 8-bit groups, and so forth. The prefixes then fan back down to compute the carries-in to each bit. The tree requires $2(\log_2 N) - 1$ stages. The fanout is limited to 2 at each stage. The diagram shows buffers used to minimize the fanout and loading on the gates, but in practice, the buffers are generally omitted.

The *Sklansky* or *divide-and-conquer* tree [Sklansky60] (Figure 10.34(b)) reduces the delay to $(\log_2 N)$ stages by computing intermediate prefixes along with the large group prefixes. This comes at the expense of fanouts that double at each level: The gates fanout to [8, 4, 2, 1] other columns. These high fanouts cause poor performance on wide adders unless the gates are appropriately sized or the critical signals are buffered before being used for the intermediate prefixes. Transistor sizing can cut into the regularity of the layout because multiple sizes of each cell are required, although the larger gates can spread into adjacent columns. Note that the recursive doubling in the Sklansky tree is analogous to the conditional-sum adder of Figure 10.32. With appropriate buffering, the fanouts can be reduced to [8, 1, 1, 1], as explored in Exercise 10.7.

The *Kogge-Stone* tree [Kogge73] (Figure 10.34(c)) achieves both $(\log_2 N)$ stages and fanout of 2 at each stage. This comes at the cost of many long wires that must be routed between stages. The tree also contains more PG cells; while this may not impact the area if the adder layout is on a regular grid, it will increase power consumption. Despite these

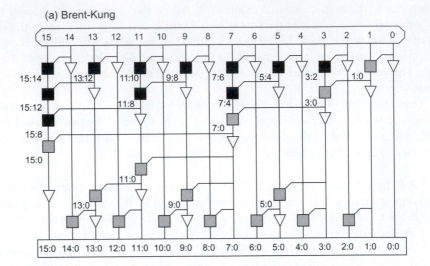

(a) Brent-Kung

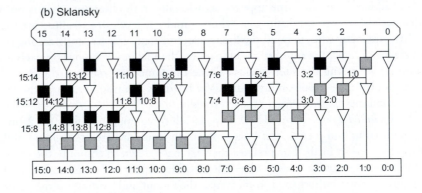

(b) Sklansky

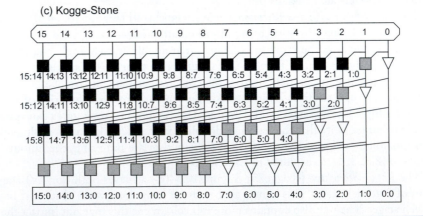

(c) Kogge-Stone

FIG 10.34 Tree adder PG networks

(d) Han-Carlson

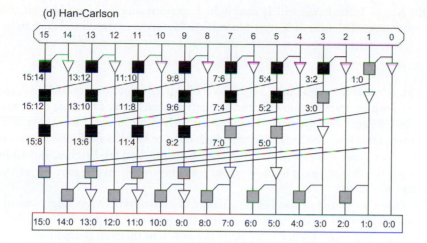

(e) Knowles [2,1,1,1]

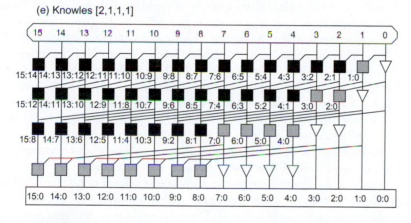

(f) Ladner-Fischer

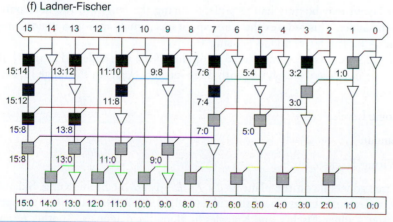

costs, the Kogge-Stone tree is widely used in high-performance 32-bit and 64-bit adders.

In summary, a Sklansky or Kogge-Stone tree adder reduces the critical path to:

$$t_{\text{tree}} \approx t_{pg} + \lceil \log_2 N \rceil t_{AO} + t_{\text{xor}} \tag{10.20}$$

An ideal tree adder would have $\log_2 N$ levels of logic, fanout never exceeding 2, and no more than 1 wiring track ($G_{i:j}$ and $P_{i:j}$ bundle) between each row. The basic tree architectures represent cases that approach the ideal, but each differ in one respect. Brent-Kung has too many logic levels. Sklansky has too much fanout. And Kogge-Stone has too many wires. Between these three extremes, the Han-Carlson, Ladner-Fischer, and Knowles trees fill out the design space with different compromises between number of stages, fanout, and wire count.

The *Han-Carlson* trees [Han87] are a family of networks between Kogge-Stone and Brent-Kung. Figure 10.34(d) shows such a tree that performs Kogge-Stone on the odd-numbered bits, and then uses one more stage to ripple into the even positions.

The *Knowles* trees [Knowles01] are a family of networks between Kogge-Stone and Sklansky. All of these trees have $\log_2 N$ stages, but differ in the fanout and number of wires. If we say that 16-bit Kogge-Stone and Sklansky adders drive fanouts of [1, 1, 1, 1] and [8, 4, 2, 1] other columns, respectively, the Knowles networks lie between these extremes. For example, Figure 10.34(e) shows a [2, 1, 1, 1] Knowles tree that halves the number of wires in the final track at the expense of doubling the load on those wires.

The *Ladner-Fischer* trees [Ladner80] are a family of networks between Sklansky and Brent-Kung. Figure 10.34(f) is similar to Sklansky, but computes prefixes for the odd-numbered bits and again uses one more stage to ripple into the even positions. Cells at high-fanout nodes must still be sized or ganged appropriately to achieve good speed. Note that some authors use Ladner-Fischer synonymously with Sklansky.

An advantage of the Brent-Kung network and those related to it (Han-Carlson and the Ladner-Fischer network with the extra row) is that for any given row, there is never more than one cell in each pair of columns. These networks have low gate count. Moreover, their layout may be only half as wide, reducing the length of the horizontal wires spanning the adder. This reduces the wire capacitance, which may be a major component of delay in 64-bit and larger adders [Huang00].

Figure 10.35 shows a 3-dimensional taxonomy of the tree adders [Harris03]. If we let $L = \log_2 N$, we can describe each tree with three integers (l, f, t) in the range $[0, L-1]$. The integers specify:

- Logic Levels: $L + l$
- Fanout: $2^f + 1$
- Wiring Tracks: 2^t

The tree adders lie on the plane $l + f + t = L - 1$. 16-bit Brent-Kung, Sklansky, and Kogge-Stone represent vertices of the cube (3, 0, 0), (0, 3, 0) and (0, 0, 3), respectively. Han-Carlson, Ladner-Fischer, and Knowles lie along the diagonals. Another recently discovered family of adders lies on the plane inside the cube.

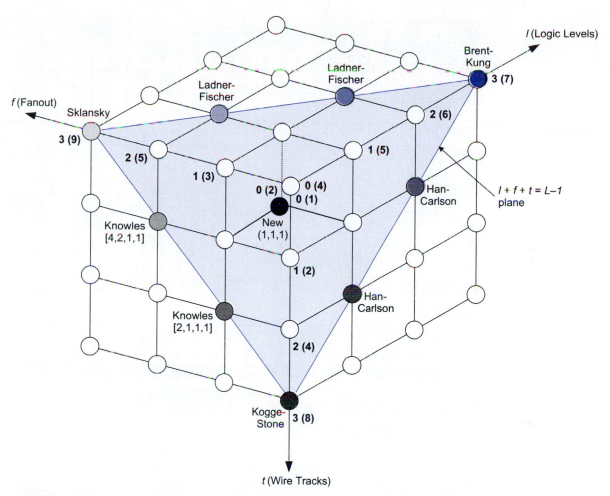

FIG 10.35 Taxonomy of prefix networks

10.2.2.9 Higher-valency Tree Adders Any of the trees described so far can combine more than two groups at each stage [Beaumont-Smith01]. The number of groups combined in each gate is called the *valency* or *radix* of the cell. For example, Figure 10.36 shows 27-bit valency-3 Brent-Kung, Sklansky, Kogge-Stone, and Han-Carlson trees. The rounded boxes mark valency-3 carry chains (that could be constructed using a Manchester carry chain, multiple-output domino gate, or several discrete gates). The trapezoids mark carry-increment operations. The higher-valency designs use fewer stages of logic, but each stage has greater delay. This tends to be a poor tradeoff in static CMOS circuits because the stage efforts become much larger than 4, but is good in domino because the logical efforts are much smaller so fewer stages are necessary.

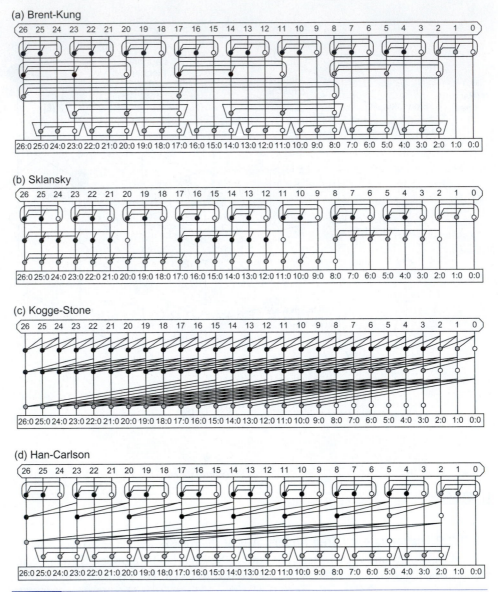

FIG 10.36 Higher-valency tree adders

Nodes with large fanouts or long wires can use buffers. The prefix trees can also be internally pipelined for extremely high-throughput operation. Some higher-valency designs combine the initial PG stage with the first level of PG merge. For example, the Ling adder described in Section 10.2.2.12 computes generate and propagate for 4-bit groups from the primary inputs in a single stage.

10.2.2.10 Hybrid Tree/Select Adders

Tree adders can be merged with carry-select adders to form a hybrid with lower delay or fewer gates. The carry-select adder precomputes the sum for both carries-in of 0 and 1 for short groups (e.g., 4, 8, or 16 bits). Meanwhile, the tree adder computes the carry-in to each group. Finally, a multiplexer selects the correct sum for each group when the carry-in becomes available. The group length should be balanced such that the carry-in and precomputed sums become available at about the same time.

The *spanning-tree adder* [Lynch92] is one such hybrid based on a higher-valency Brent-Kung tree of Figure 10.36(a). Figure 10.37 shows a simple valency-3 version that precomputes sums for 3-bit groups and saves one logic level by selecting the output based on the carries into each group. The carry-out (C_{out}) is explicitly shown. Note that the least significant group requires a valency-4 gray cell to compute $G_{3:0}$, the carry-in to the second select block.

[Lynch92] describes a 56-bit spanning-tree design from the AMD AM29050 floating-point unit using valency-4 stages and 8-bit carry select groups. [Kantabutra93] and [Blackburn96] describe optimizing the spanning-tree adder by using variable-length carry-select stages and appropriately selecting transistor sizes.

A carry-select box spanning bits $i...j$ is shown in Figure 10.38(a). It uses short carry-ripple adders to precompute the sums assuming carry-in of 0 and 1 to the group, and then selects between them with a multiplexer, as shown in Figure 10.38(b). The adders can be simplified somewhat because the carry-ins are constant, as shown in Figure 10.38(c) for a 4-bit group.

The Brent-Kung adder involves more than the minimum number of logic levels. Kogge-Stone and Sklansky use fewer levels, but require many gates and either many wires or high fanout. [Mathew03] describes a 32-bit *sparse-tree adder* using a valency-2 tree similar to Sklansky to compute only the carries into each 4-bit group, as shown in Figure 10.39. This reduces the gate count and power consumption in the tree. The tree can be viewed as a (2, 2, 0) Ladner-Fischer tree with the final two tree levels and XOR replaced by the select multiplexer. The adder assumes the carry-in is 0 and does not produce a carry-out, saving one input to the least-significant gray box and eliminating the prefix logic in the four most significant columns.

These hybrid approaches are widely used in high-performance 32–64-bit higher-valency domino adders because they offer the small number of logic levels of higher-

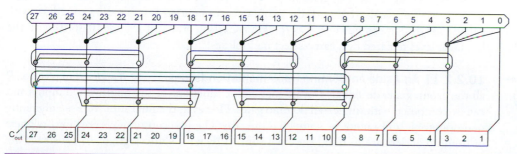

FIG 10.37 Spanning-tree adder

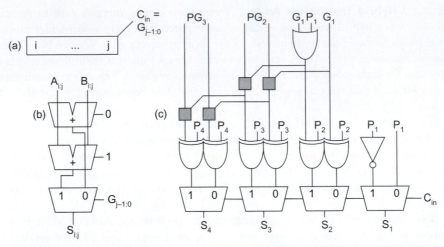

FIG 10.38 Carry-select implementation

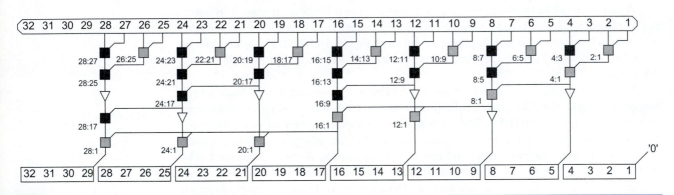

FIG 10.39 Sparse-tree adder

valency trees while reducing the gate count and power consumption in the tree. Figure 10.40 shows a 27-bit valency-3 Kogge-Stone design with carry-select on 3-bit groups. Observe how the number of gates in the tree is reduced threefold. Moreover, because the number of wires is also reduced, the extra area can be used for shielding to reduce path delay. This design can be viewed as the Han-Carlson adder of Figure 10.36(d) with the last logic level replaced by a carry-select multiplexer.

10.2.2.11 An Aside on Domino Implementation Issues Using $\overline{K} = A + B$ in place of P, all the group generate signals $G_{i:0}$ are monotonic functions of the noninverted inputs and can be computed with single-rail domino gates. However, the final sum XOR is inherently nonmonotonic and cannot be computed this way. The two common choices for designers of domino adders are to build the final sum XOR with static logic or to construct the entire adder out of dual-rail domino.

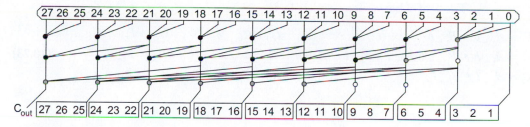

FIG 10.40 Hybrid Kogge-Stone/select adder

Domino adders with a static sum XOR produce nonmonotonic outputs that must be stabilized on a clock edge before driving subsequent domino gates. As adders are often used in self-bypass loops where the output of the adder serves as one of the inputs on the next cycle, this introduces a hard edge and the associated costs of setup time and clock skew into the critical path.

The alternative is to build a dual-rail domino sum XOR accepting monotonic true and complementary (_h and _l) versions of the carries. Producing these carries in turn requires extra hardware all the way back to the adder inputs, which also must be provided in dual-rail form. If the sum is also computed in dual-rail form, the outputs can be directly bypassed to the inputs in a skew-tolerant fashion. The drawback of such adders is the extra hardware involved in the group PG network. Again, there are two common cell designs. One is to build dual-rail group propagate and generate signals, i.e., four signals per bundle. Another is to use monotonic one-of-three hot propagate-generate-kill (PGK) signals.

The first approach uses the following logic:

$$
\begin{aligned}
G_i\_h &= A_i\_h \cdot B_i\_h & G_0\_h &= C_{in}\_h & G_{i:j}\_h &= G_{i:k}\_h + K_{i:k}\_l \cdot G_{k-1:j}\_h \\
K_i\_h &= A_i\_l \cdot B_i\_l & K_0\_h &= C_{in}\_l & K_{i:j}\_h &= K_{i:k}\_h + G_{i:k}\_l \cdot K_{k-1:j}\_h & \text{(10.21)} \\
G_i\_l &= A_i\_l + B_i\_l & G_0\_l &= 0 & G_{i:j}\_l &= G_{i:k}\_l \cdot G_{k-1:j}\_l \\
K_i\_l &= A_i\_h + B_i\_h & K_0\_l &= 0 & K_{i:j}\_l &= K_{i:k}\_l \cdot K_{k-1:j}\_l
\end{aligned}
$$

$$
\begin{aligned}
P_i\_h &= A_i \oplus B_i = A_i\_h \cdot B_i\_l + A_i\_l \cdot B_i\_h = G_i\_l \cdot K_i\_l \\
P_i\_l &= \overline{A_i \oplus B_i} = A_i\_h \cdot B_i\_h + A_i\_l \cdot B_i\_l = G_i\_h + K_i\_h \\
S_i\_h &= G_{i-1:0} \oplus P_i = G_{i-1:0}\_h \cdot P_i\_l + K_{i-1:0}\_h \cdot P_i\_h & \text{(10.22)} \\
S_i\_l &= \overline{G_{i-1:0} \oplus P_i} = G_{i-1:0}\_h \cdot P_i\_h + K_{i-1:0}\_h \cdot P_i\_l
\end{aligned}
$$

Observe that the group generate _h and _l and kill _h and _l signals are not truly complementary; they are sometimes called *pseudo-complements* [Wang93]. They take advantage of the symmetry of the addition function so that the same type of gate can be reused. It is left to the reader to recursively verify that $G_{i-1:0}\_h$ and $K_{i-1:0}\_h$ are the true and complementary versions of the carries into bit i. Schematics of each gate are shown in Figure 10.41(a).

The 1-of-3 hot version uses less logic:

$$G_i = A_i\_h \cdot B_i\_h \qquad\qquad G_0 = C_{\text{in}}\_h \quad G_{i:j} = G_{i:k} + P_{i:k} \cdot G_{k-1:j}$$
$$P_i = A_i\_h \cdot B_i\_l + A_i\_l \cdot B_i\_h = A \oplus B \qquad P_0 = 0 \qquad\quad P_{i:j} = P_{i:k} \cdot P_{k-1:j} \qquad\qquad \textbf{(10.23)}$$
$$K_i = A_i\_l \cdot B_i\_l \qquad\qquad K_0 = C_{\text{in}}\_l \quad K_{i:j} = K_{i:k} + P_{i:k} \cdot K_{k-1:j}$$

$$P_i' = G_i + K_i$$
$$S_i\_h = G_{i-1:0} \cdot P_i' + K_{i-1:0} \cdot P_i \qquad\qquad \textbf{(10.24)}$$
$$S_i\_l = K_{i-1:0} \cdot P_i' + G_{i-1:0}\_h \cdot P_i$$

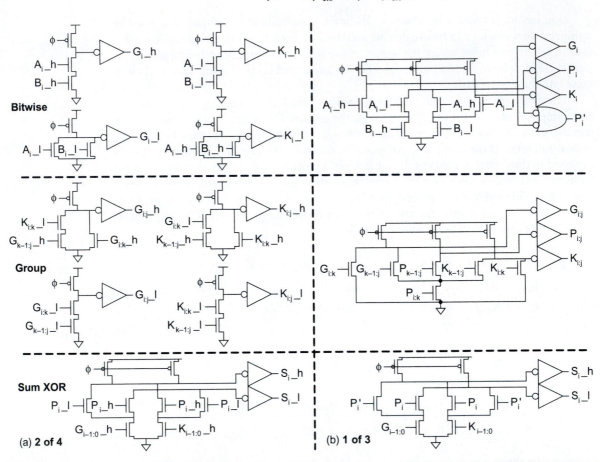

FIG 10.41 Domino adder circuit components

The approach gets its name because exactly one of the three signals P, G, or K is true for any group. The complementary propagate signal $P' = G + K$ is required for the final sum XOR, but nowhere earlier. Again notice that the G and K functions are identical, simplifying design and layout. The group kill prefixes are the complements of the group generates ($G_{i-1:0} = \overline{K_{i-1:0}}$); this is used in EQ (10.24) to reduce the loading on each signal in the sum XOR. The 1-of-n hot technique can be useful for other domino applications, such as multiplexer select signals and shifter control signals; it also reduces switching activity and power consumption. Figure 10.41(b) shows how transistors can be shared between gates.

Manchester carry chains can also generate both polarities of carries. Figure 10.42 shows how the same type of carry chain can be used for both _h and _l carries as well as to find the group propagate signal using 1-of-3 hot encoding.

The domino gates are fast, but power-hungry. Figure 10.43 shows an interesting static version of the same circuit using triple-rail "push-pull" CPL to obtain nearly the same speed as domino at much lower power. The UltraSparc VI uses this cell (without $C1$ and $C2$ carry inverters) for a low-power ALU [Chillarige03].

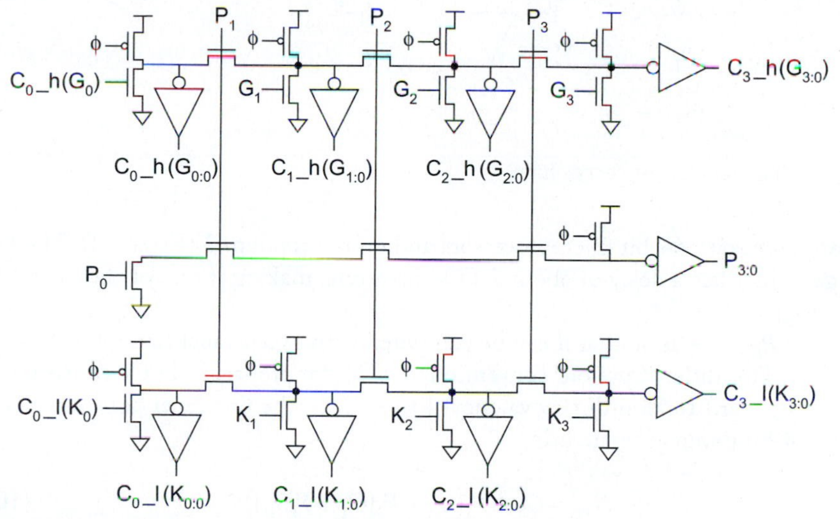

FIG 10.42 1-of-3 hot domino Manchester carry chain

10.2.2.12 Ling and Naffziger Adders A 64-bit tree adder using valency-2 cells requires at least six levels of group PG logic plus the initial bitwise PG logic and final XOR for a total of eight levels of logic. Using valency-4 cells reduces this to three levels of group PG logic and a total of five levels of logic. The bitwise PG logic can be merged into the first stage of group PG logic, but this leads to an excessively complex gate with tall transistor stacks for valency-4. The Ling adder [Ling81] defines a *pseudo-generate* (sometimes called *pseudo-carry*) signal that is simple enough that the bitwise PG logic can be merged into the initial valency-4 stage, resulting in only four levels of logic on the critical path. The actual carries are produced later in the path with an OR operation. The Ling adder was originally developed for bipolar transistors where a wired-OR contributes negligible delay. Hewlett-Packard adapted it to domino logic in which the OR operation is just another transistor in parallel; this Naffziger adder [Naffziger96, Naffziger98] has been widely used

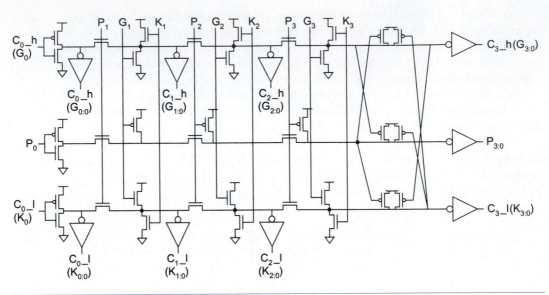

FIG 10.43 1-of-3 hot CPL Manchester carry chain

in Hewlett-Packard 64-bit microprocessors and on the Itanium 2 [Fetzer02]. The 64-bit Naffziger adder has a delay of about 7 FO4 inverters, making it one of the fastest published designs.

Define $P_i = A_i + B_i$ so that it can be computed with two parallel transistors in a domino circuit. (This differs from our convention of XOR for propagate, but is consistent with Naffziger's paper.) In an ordinary valency-4 tree adder, the first level of group PG cells computes 4-bit group generate using

$$G_{3:0} = G_3 + P_3\left(G_2 + P_2\left(G_1 + P_1 G_0\right)\right) \tag{10.25}$$

If we were to compute $G_{3:0}$ directly from $A_3 \dots A_0$ and $B_3 \dots B_0$, the gate would be complicated and quite slow. Instead, let us define a group *pseudo-generate* signal $H_{3:0}$ in which the P_3 term is removed.

$$H_{3:0} = G_3 + \left(G_2 + P_2\left(G_1 + P_1 G_0\right)\right) = G_3 + G_{2:0} = G_{3:0} + G_{2:0} \tag{10.26}$$

We will also use a group *pseudo-propagate* signal I that is a shifted copy of the actual group propagate signal. In general, we can define pseudo-generate and pseudo-propagate signals H and I as

$$\begin{aligned} H_{i:j} &= G_{i:j} + G_{i-1:j} \\ I_{i:j} &= P_{i-1:j-1} \end{aligned} \tag{10.27}$$

These signals are recursively combined into bigger groups (for $i \geq k > j$) in the same fashion as actual group generate and propagate signals (EQ (10.4)).

$$H_{i:j} = H_{i:k} + I_{i:k}H_{k-1:j}$$
$$I_{i:j} = I_{i:k}I_{k-1:j}$$

(10.28)

The actual group generate signal is formed from pseudo-generate using

$$P_{i:i}H_{i:j} = P_{i:i}\left(G_{i:j} + G_{i-1:j}\right)$$
$$= G_{i:j} + P_{i:i}G_{i-1:j}$$
$$= G_{i:j}$$

(10.29)

It is left to the reader (Exercise 10.11) to prove EQ (10.28) from EQ (10.27) and EQ (10.29). Now that we can compute the $G_{i-1:0}$ prefixes, all that remains is to compute the final sums with the usual XOR.

The 64-bit Naffziger adder combines these Ling equations with a hybrid tree/select adder (from Section 10.2.2.10), as shown in Figure 10.44. It uses the Ling equations to find the pseudo-generate signal into each 16-bit block in three levels of compound valency-4 dual-rail domino gates. Meanwhile, it computes the generate prefixes for each bit in a 16-bit block assuming pseudo-carry in signals to the block of both 0 and 1. The prefix logic consists of a short carry-ripple and Manchester carry chain. Finally, the adder selects the appropriate generate given the pseudo-carries and performs an XOR to find the sum. By doing the select based on the pseudo-carry instead of the actual carry, it saves one stage of logic.

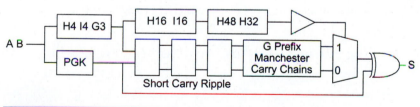

FIG 10.44 Naffziger adder

The Naffziger adder uses dual-rail domino throughout. The equations below describe only the true (_h) logic. The complementary (_l) logic uses identical gates on complementary inputs, as in EQ (10.21). As usual, let $A_0 = B_0 = C_{in}$.

The (H4 I4 G3) block computes the H and I signals for each 4-bit group using EQ (10.30) ($k = 0...15$). As a byproduct, we find G for 3-bit groups, which we will use later when computing the generate prefixes. Figure 10.45 shows a domino gate that calculates G and H. It uses a clever combined static NOR/INV gate to produce both outputs. We use an nMOS secondary precharge transistor to prevent charge sharing on the heavily loaded internal node.

$$G_{4k+3:4k+1} = G_{4k+3} + P_{4k+3}\left(G_{4k+2} + P_{4k+2}G_{4k+1}\right)$$

$$= A_{4k+3}B_{4k+3} + \left(A_{4k+3} + B_{4k+3}\right)\left(A_{4k+2}B_{4k+2} + \left(A_{4k+2} + B_{4k+2}\right)A_{4k+1}B_{4k+1}\right)$$

$$H_{4k+4:4k+1} = G_{4k+4} + \left(G_{4k+3} + P_{4k+3}\left(G_{4k+2} + P_{4k+2}G_{4k+1}\right)\right) \tag{10.30}$$

$$= A_{4k+4}B_{4k+4} + G_{4k+3:4k+1}$$

$$I_{4k+4:4k+1} = \left(A_{4k+3} + B_{4k+3}\right)\left(A_{4k+2} + B_{4k+2}\right)\left(A_{4k+1} + B_{4k+1}\right)\left(A_{4k} + B_{4k}\right)$$

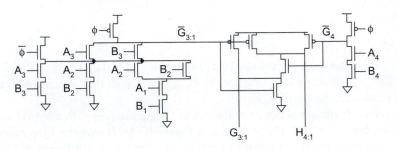

FIG 10.45 GH gate

At the same time, the 1-of-3 hot P_i, G_i, and K_i signals are computed in the PGK block for each bit ($i = 1\dots64$) using EQ (10.23). They will also be used later.

The next step is to compute the H and I signals for each 16-bit group (H16 I16 block). For the least significant group, the carry-in is incorporated into the pseudo-generate using $I_{4:1}$ so no pseudo-propagate is needed (EQ (10.31)). For the other three groups, both H and I are found using valency-4 black cells (EQ (10.32), $k = 1\dots3$).

$$H_{16:1} = H_{16:13} + I_{16:13}\left(H_{12:9} + I_{12:9}\left(H_{8:5} + I_{8:5}\left(H_{4:1} + I_{4:1}\right)\right)\right) \tag{10.31}$$

$$H_{16k+16:16k+1} = H_{16k+16:16k+13} +$$

$$I_{16k+16:16k+13}\left(H_{16k+12:16k+9} + I_{16k+12:16k+9}\left(H_{16k+8:16k+5} + I_{16k+8:16k+5}H_{16k+4:16k+1}\right)\right) \tag{10.32}$$

$$I_{16k+16:16k+1} = I_{16k+16:16k+13}I_{16k+12:16k+9}I_{16k+8:16k+5}I_{16k+4:16k+1}$$

Now we can find the pseudo-carry signals into each 16-bit block using EQ (10.33) (H48 H32 block).

$$H_{0:1} = C_{\text{in}}$$

$$H_{16:1} = H_{16:1}$$

$$H_{32:1} = H_{32:17} + I_{32:17}H_{16:1} \tag{10.33}$$

$$H_{48:1} = H_{48:33} + I_{48:33}\left(H_{32:17} + I_{32:17}H_{16:1}\right)$$

Meanwhile, we need to compute the generate signals for each bit within the 16-bit block assuming pseudo-carries of both 0 and 1 into the 16-bit block. This is done in two steps. First, we use a series of fast ripple gates to find the generate out of the third, seventh, eleventh, and fifteenth bits within the block using EQ (10.34) ($k = 0...3$). The critical path goes through the 11th bit. The signal $G^0_{16k+i:0}$ indicates whether the group spanning bits $16k + 1:0$ will generate a carry if the pseudo-carry $H_{16k:1}$ into the kth 16-bit block is 0. Observe how the 1-bit generate signals $G_{16k+4,8,12}$ can be folded in without adding to the number of series transistors in the domino gates.

$$
\begin{aligned}
G^0_{16k-1:0} &= 0 \\
G^1_{16k-1:0} &= 1 \\
G^0_{16k+3:0} &= G_{16k+3:16k+1} \\
G^1_{16k+3:0} &= G_{16k+3:16k+1} + I_{16k+4:16k+1} \\
G^0_{16k+7:0} &= G_{16k+7:16k+5} + I_{16k+8:16k+5}\left(G^0_{16k+3:0} + G_{16k+4}\right) \\
G^1_{16k+7:0} &= G_{16k+7:16k+5} + I_{16k+8:16k+5}\left(G^1_{16k+3:0} + G_{16k+4}\right) \\
G^0_{16k+11:0} &= G_{16k+11:16k+9} + I_{16k+12:16k+9}\left(G^0_{16k+7:0} + G_{16k+8}\right) \\
G^1_{16k+11:0} &= G_{16k+11:16k+9} + I_{16k+12:16k+9}\left(G^1_{16k+7:0} + G_{16k+8}\right) \\
G^0_{16k+15:0} &= G_{16k+15:16k+13} + I_{16k+16:16k+13}\left(G^0_{16k+11:0} + G_{16k+12}\right) \\
G^1_{16k+15:0} &= G_{16k+15:16k+13} + I_{16k+16:16k+13}\left(G^1_{16k+11:0} + G_{16k+12}\right)
\end{aligned}
\tag{10.34}
$$

Then pairs of Manchester carry chains compute the generate signal into each bit for both possibilities of pseudo-carries into the kth 16-bit block ($k = 0...3$). Four pairs ($n = 0...3$) are required for each block. The outputs of the carry chains are given in EQ (10.35). For the least significant chains ($n = 0$), the $G^{0/1}_{16k+1:0}$ signals are constants and can be optimized away.

$$
\begin{aligned}
G^0_{16k+4n:0} &= G_{16k+4n} + P_{16k+4n}G^0_{16k+4n-1:0} \\
G^1_{16k+4n:0} &= G_{16k+4n} + P_{16k+4n}G^1_{16k+4n-1:0} \\
G^0_{16k+4n+1:0} &= G_{16k+4n+1} + P_{16k+4n+1}\left(G_{16k+4n} + P_{16k+4n}G^0_{16k+4n-1:0}\right) \\
G^1_{16k+4n+1:0} &= G_{16k+4n+1} + P_{16k+4n+1}\left(G_{16k+4n} + P_{16k+4n}G^1_{16k+4n-1:0}\right) \\
G^0_{16k+4n+2:0} &= G_{16k+4n+2} + P_{16k+4n+2}\left(G_{16k+4n+1} + P_{16k+4n+1}\left(G_{16k+4n} + P_{16k+4n}G^0_{16k+4n-1:0}\right)\right) \\
G^1_{16k+4n+2:0} &= G_{16k+4n+2} + P_{16k+4n+2}\left(G_{16k+4n+1} + P_{16k+4n+1}\left(G_{16k+4n} + P_{16k+4n}G^1_{16k+4n-1:0}\right)\right)
\end{aligned}
\tag{10.35}
$$

Finally, we use a carry select multiplexer to choose the appropriate carry into each bit and an XOR to find the sum. These two functions can be merged into another compound domino gate using EQ (10.36) for blocks $k = 0...3$ and bits $i = 1...16$ within each block. Figure 10.46 shows a dual-rail domino implementation of the sum function taking advantage of the fact that $\overline{P}_i = G_i + K_i$. The H signals drive a large fanout and thus are first buffered with a pair of inverters.

$$S_{16k+i} = P_{16k+i} \oplus \left(\overline{H}_{16k:1} G^0_{16k+i-1:0} + H_{16k:1} G^1_{16k+i-1:0} \right) \tag{10.36}$$

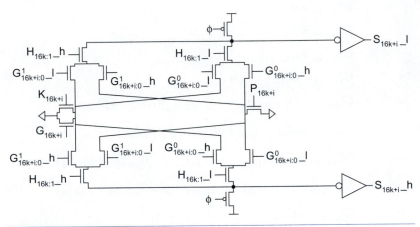

FIG 10.46 Dual-rail sum gate

10.2.2.13 Summary Having examined so many adders, you probably want to know which adder should be used in which application. Table 10.3 compares the various adder architectures that have been illustrated with valency-2 prefix networks. The category "logic levels" gives the number of AND-OR gates in the critical path, excluding the initial PG logic and final XOR. Of course, the delay depends on the fanout and wire loads as well as the number of logic levels. The category "cells" refers to the approximate number of gray and black cells in the network. Carry-lookahead is not shown because it uses higher-valency cells. Carry-select is also not shown because it is larger than carry-increment for the same performance.

In general, carry-ripple adders should be used when they meet timing constraints because they are compact and easy to build. When faster adders are required, carry-increment and carry-skip architectures work well, particularly for 8–16 bit lengths. Hybrids combining these techniques are also popular. At word lengths of 32 and especially 64 bits, tree adders are distinctly faster.

Table 10.3 Comparison of adder architectures

Architecture	Classification	Logic Levels	Max Fanout	Tracks	Cells
Carry-Ripple		$N-1$	1	1	N
Carry-Skip ($n=4$)		$N/4+5$	2	1	$1.25N$
Carry-Increment ($n=4$)		$N/4+2$	4	1	$2N$
Carry-Increment (variable group)		$\sqrt{2N}$	$\sqrt{2N}$	1	$2N$
Brent-Kung	$(L-1,0,0)$	$2\log_2 N-1$	2	1	$2N$
Sklansky	$(0,L-1,0)$	$\log_2 N$	$N/2+1$	1	$0.5\,N\log_2 N$
Kogge-Stone	$(0,0,L-1)$	$\log_2 N$	2	$N/2$	$N\log_2 N$
Han-Carlson	$(1,0,L-2)$	$\log_2 N+1$	2	$N/4$	$0.5\,N\log_2 N$
Ladner Fischer ($l=1$)	$(1,L-2,0)$	$\log_2 N+1$	$N/4+1$	1	$0.25\,N\log_2 N$
Knowles [2,1,...,1]	$(0,1,L-2)$	$\log_2 N$	3	$N/4$	$N\log_2 N$

Good logic synthesis tools automatically map the "+" operator onto an appropriate adder to meet timing constraints while minimizing area. For example, the Synopsys DesignWare libraries contain carry-ripple adders, carry-select adders, carry-lookahead adders, and a variety of prefix adders. Figure 10.47 shows the results of synthesizing 32-bit and 64-bit adders under different timing constraints. As the latency decreases, synthesis selects more elaborate adders with greater area. The results are for a 0.18 µm commercial cell library with an FO4 inverter delay of (a rather slow) 89 ps in the TTTT corner and the area includes estimated interconnect as well as gates. The fastest designs use tree adders and achieve remarkably fast (prelayout) delays of 7.0 and 8.5 FO4 for 32-bit and 64-bit adders, respectively, by creating nonuniform designs with side loads carefully buffered off the critical path. The carry-select adders achieve an interesting area/delay tradeoff by using carry-ripple for the lower 3/4 of the bits and carry-select only on the upper 1/4. The results will be somewhat slower when true layout parasitics are included. In comparison, a good handcrafted 64-bit domino adder has a delay of about 7 FO4 and an area of under 3 Mλ^2 [Naffziger96].

Other surveys of addition include [Ercegovac04, Parhami00, Koren02, and Zimmermann97b].

10.2.3 Adder Variants

We conclude the study of addition by examining subtraction and addition of multiple inputs. Incrementers are covered in Section 10.10.

10.2.3.1 Subtraction An N-bit subtracter uses the 2's complement relationship

$$A-B = A+\overline{B}+1 \tag{10.37}$$

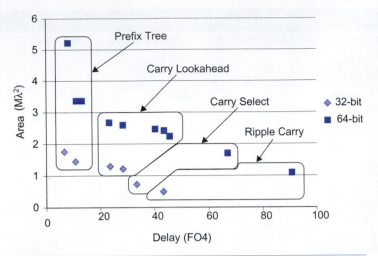

FIG 10.47 Area vs. delay of synthesized adders

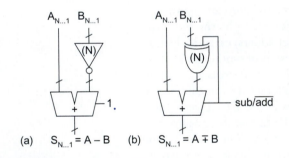

(a) $S_{N...1} = A - B$ **(b)** $S_{N...1} = A \mp B$

FIG 10.48 Subtracters

This involves inverting one operand to an N-bit CPA and adding 1 via the carry input, as shown in Figure 10.48(a). An adder/subtracter uses XOR gates to conditionally invert B, as shown in Figure 10.48(b). In prefix adders, the XOR gates on the B inputs are sometimes merged into the bitwise PG circuitry.

10.2.3.2 Multiple-input Addition The most obvious method of adding k N-bit words is with $k-1$ cascaded CPAs as illustrated in Figure 10.49(a) for 0001 + 0111 + 1101 + 0010, but this consumes a large amount of hardware and is slow. A better technique is to note that a full adder sums three inputs of unit weight and produces a sum output of unit weight and a carry output of double weight. If N full adders are used in parallel, they can accept three N-bit input words $X_{N...1}$, $Y_{N...1}$, and $Z_{N...1}$, and produce two N-bit output words $S_{N...1}$ and $C_{N...1}$, as shown in Figure 10.49(b). The results correspond to the sums and carries-out of each adder. This is called *carry-save redundant format* because the carry outputs are preserved rather than propagated along the adder. The full adders in this application are sometimes called *(3,2) counters* or *carry-save adders* (CSA) because they accept three inputs and produce two outputs in carry-save form. When the carry word C is shifted left by one position (because it has double weight) and added to the sum word S with an ordinary CPA, the result is $X + Y + Z$. Alternatively, a fourth input word can be added to the carry-save redundant result with another row of CSAs, again resulting in a carry-save redundant result. Such carry-save addition of four numbers is illustrated in Figure 10.49(c), where the underscores in the carry outputs serve as reminders that the carries must be shifted left

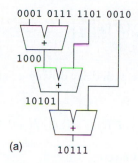

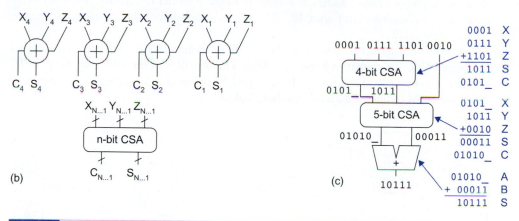

(b)

FIG 10.49 Multiple-input adders

one column on account of their greater weight. In general, k numbers can be summed with $k–2$ CSAs and only one CPA. This approach will be exploited in Section 10.9 to rapidly add many partial products in a multiplier. The technique dates back to von Neumann's early computer [Burks46].

10.3   One/Zero Detectors

Detecting all 1's or 0's on wide N-bit words requires large fan-in AND or NOR gates. Recall that by DeMorgan's law, AND, OR, NAND, and NOR are fundamentally the same operation except for possible inversions of the inputs and/or outputs. You can build a tree of AND gates, as shown in Figure 10.50(a). Here, alternate NAND and NOR gates have been used. The path has log N stages. In general, the minimum logical effort is achieved with a tree alternating NAND gates and inverters and the path logical effort is:

$$G_{\text{and}}(N) = \left(\frac{4}{3}\right)^{\log_2 N} = N^{\log_2 \frac{4}{3}} = N^{0.415} \qquad (10.38)$$

A rough estimate of the path delay driving a path electrical effort of H using static CMOS gates is:

$$D \approx \left(\log_4 F\right) t_{FO4} = \left(\log_4 H + 0.415 \log_4 N\right) t_{FO4} \qquad (10.39)$$

where t_{FO4} is the fanout-of-4 inverter delay.

If the word being checked has a natural skew in the arrival time of the outputs (such as at the output of a ripple adder), the designer might consider mimicking the adder delay in the detector, as shown in Figure 10.50(b). Here the delay from the last changing output A_7 is a single delay.

Another fast detector uses a pseudo-nMOS or dynamic NOR structure to perform the "wired-OR," as shown in Figure 10.50(c). This works well for words up to about 16 bits; for larger words, the gates can be split into 8–16-bit chunks to reduce the parasitic delay and avoid problems with subthreshold leakage.

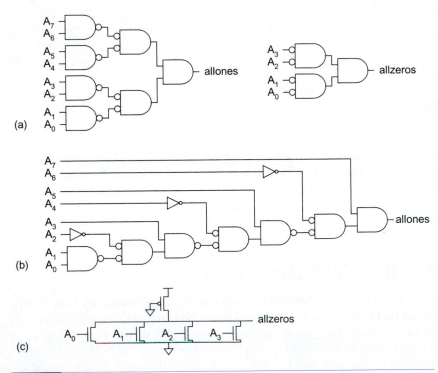

FIG 10.50 One/zero detectors

10.4   Comparators

10.4.1   Magnitude Comparator

A *magnitude comparator* determines the larger of two binary numbers. To compare two unsigned numbers A and B, compute $B - A = B + \overline{A} + 1$. If there is a carry-out, $A \leq B$. A zero detector indicates that the numbers are equal. Figure 10.51 shows a 4-bit unsigned comparator built from a carry-ripple adder and 2's complementer. The relative magnitude is determined from the carry-out (C) and zero (Z) signals according to Table 10.4. For wider inputs, any of the faster adder architectures can be used.

Comparing signed 2's complement numbers is slightly more complicated because of the possibility of overflow when subtracting two numbers with different signs. Instead of simply examining the carry-out, we must determine if the result is negative (N, indicated by the most significant bit of the result) and if it overflows the range of possible signed numbers. The overflow signal V is true if the inputs had different signs (most significant bits) and the output sign is different than the sign of B. The actual sign of the difference $B - A$ is $(N \oplus V)$ because overflow flips the sign. If this sign is negative, we know $A > B$. Again, the other relations can be derived from this corrected sign and the Z signal.

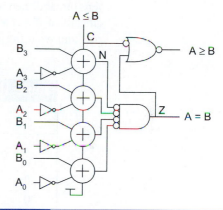

FIG 10.51 Unsigned magnitude comparator

Table 10.4	Magnitude comparison	
Relation	**Unsigned Comparison**	**Signed Comparison**
$A = B$	Z	Z
$A \neq B$	$\overline{Z}$	$\overline{Z}$
$A < B$	$\overline{C} + Z$	$(N \oplus V) + Z$
$A > B$	$\overline{C}$	$(N \oplus V)$
$A \leq B$	C	$(\overline{N \oplus V})$
$A \geq B$	$\overline{C} + Z$	$(N \oplus V) + Z$

10.4.2   Equality Comparator

An *equality comparator* determines if ($A = B$). This can be done more simply and rapidly with XNOR gates and a 1's detector, shown in Figure 10.52. Again, this could be done with a pseudo-nMOS or dynamic gate in which the XOR functions were combined with a "wired-OR."

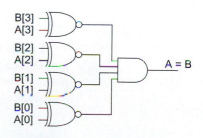

FIG 10.52 Equality comparator

10.4.3 $K = A + B$ Comparator

Sometimes it is necessary to determine if $(A + B = K)$. For example, the sum-addressed memory [Heald98] described in Section 11.2.2.3 contains a decoder that must match against the sum of two numbers, such as a register base address and an immediate offset. Remarkably, this comparison can be done faster than computing $A + B$ because no carry propagation is necessary. The key is that if you know A and B, you also know what the carry into each bit must be if $K = A + B$ [Cortadella92]. Therefore, you only need to check adjacent pairs of bits to verify that the previous bit produces the carry required by the current bit, and then use a 1's detector to check that the condition is true for all N pairs. Specifically, if $K = A + B$, Table 10.5 lists what the carry-in c_{i-1} must have been for this to be true and what the carry-out c_i will be for each bit position i.

Table 10.5			Required and generated carries if $K = A + B$	
A_i	B_i	K_i	c_{i-1} (required)	c_i (produced)
0	0	0	0	0
0	0	1	1	0
0	1	0	1	1
0	1	1	0	0
1	0	0	1	1
1	0	1	0	0
1	1	0	0	1
1	1	1	1	1

From this table, you can see that the required c_{i-1} for bit i is:

$$c_{i-1} = A_i \oplus B_i \oplus K_i \tag{10.40}$$

and the c_{i-1} produced by bit $i-1$ is:

$$c_{i-1} = \left(A_{i-1} \oplus B_{i-1}\right)\overline{K}_{i-1} + A_{i-1} \cdot B_{i-1} \tag{10.41}$$

Figure 10.53 shows one bitslice of a circuit to perform this operation. The XNOR gate is used to make sure that the required carry matches the produced carry at each bit position; then the AND gate checks that the condition is satisfied for all bits.

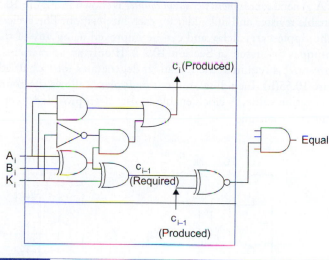

FIG 10.53 $A + B = K$ comparator

10.5 Counters

Two commonly used types of counters are *binary counters* and *linear-feedback shift registers*. An N-bit binary counter sequences through 2^N outputs in binary order. It has a minimum cycle time that increases with N. An N-bit linear-feedback shift register sequences through up to 2^N-1 outputs in pseudo-random order. It has a short minimum cycle time independent of N, so it is useful for extremely fast counters as well as pseudo-random number generation.

In general, divide-by-M counters ($M < 2^N$) can be built using an ordinary N-bit counter and circuitry to reset the counter upon reaching M. M can be a programmable input if an equality comparator is used.

10.5.1 Binary Counters

The simplest binary counter is the *asynchronous ripple-carry counter*, shown in Figure 10.54. It is composed of N registers connected in toggle configuration, where the falling transition of each register clocks the subsequent register. Therefore, the delay can be quite long. It has no reset signal, making it extremely difficult to test. In general, asynchronous circuits introduce a whole assortment of problems, so the ripple-carry counter is shown mainly for historical interest and would not be recommended for commercial designs.

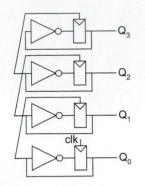

FIG 10.54 Asynchronous ripple-carry counter

A general *synchronous up/down counter* is shown in Figure 10.55(a). It uses a resettable register and full adder for each bit position. The cycle time is limited by the ripple-carry delay and can be improved using any of the faster adder techniques discussed in Section 10.2.2. If only an *up counter* (also called an *incrementer*) is required, the full adder degenerates into a half adder, shown in Figure 10.55(b). Including an input multiplexer allows the counter to load an initialization value. A clock enable is also often provided to each register for conditional counting.

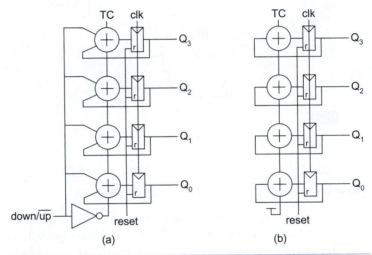

FIG 10.55 Synchronous counters

10.5.2 Linear-feedback Shift Registers

A linear-feedback shift register (LFSR) consists of N registers connected together as a shift register. The input to the shift register comes from the XOR of particular bits of the register, as shown in Figure 10.56 for a 3-bit LFSR. On reset, the registers must be initialized to a nonzero value (e.g., all 1's). The pattern of outputs for the LFSR is shown in Table 10.6.

This LFSR is an example of a *maximal-length* shift register because its output sequences through all 2^n-1 combinations (excluding all 0's). The inputs fed to the XOR are called the *tap sequence* and are often specified with a *characteristic polynomial*. For example, this 3-bit LFSR has the characteristic polynomial $1 + x^2 + x^3$ because the taps come after the second and third registers.

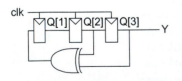

FIG 10.56 3-bit LFSR

Table 10.6	LFSR sequence		
Cycle	$Q[1]$	$Q[2]$	$Q[3]/Y$
0	1	1	1
1	0	1	1
2	0	0	1
3	1	0	0
4	0	1	0
5	1	0	1
6	1	1	0
7	1	1	1
	repeats forever		

The output Y follows the 7-bit sequence [1110010]. This is an example of a *pseudo-random bit sequence* (PRBS) because it is spectrally random. LFSRs are used for high-speed counters and pseudo-random number generators. The pseudo-random sequences are handy for built-in self-test and bit-error-rate testing in communications links. They are also used in many spread-spectrum communications systems such as GPS and CDMA where their correlation properties make other users look like uncorrelated noise.

Table 10.7 lists characteristic polynomials for some commonly used maximal-length LFSRs. For certain lengths, N, more than two taps may be required. For many values of N, there are multiple polynomials resulting in different maximal-length LFSRs. Observe that the cycle time is set by the register and XOR delays, independent of N. [Golomb81] offers the definitive treatment on linear-feedback shift registers.

Table 10.7	Characteristic polynomials
N	**Polynomial**
3	$1 + x^2 + x^3$
4	$1 + x^3 + x^4$
5	$1 + x^3 + x^5$
6	$1 + x^5 + x^6$
7	$1 + x^6 + x^7$
8	$1 + x^1 + x^6 + x^7 + x^8$
9	$1 + x^5 + x^9$
15	$1 + x^{14} + x^{15}$
16	$1 + x^4 + x^{13} + x^{15} + x^{16}$
23	$1 + x^{18} + x^{23}$
24	$1 + x^{17} + x^{22} + x^{23} + x^{24}$
31	$1 + x^{28} + x^{31}$
32	$1 + x^{10} + x^{30} + x^{31} + x^{32}$

Example

Sketch an 8-bit linear-feedback shift register. How long is the pseudo-random bit sequence that it produces?

Solution: Figure 10.57 shows an 8-bit LFSR using the four taps after the first, sixth, seventh, and eighth bits, as given in Table 10.7. It produces a sequence of $2^8-1 = 255$ bits before repeating.

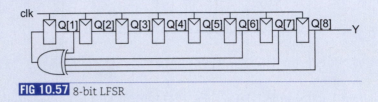

FIG 10.57 8-bit LFSR

10.6   Boolean Logical Operations

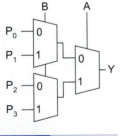

FIG 10.58 Boolean logical unit

Boolean logical operations are easily accomplished using a multiplexer-based circuit shown in Figure 10.58. Table 10.8 shows how the inputs are assigned to perform different logical functions. By providing different P values, the unit can perform other operations such as XNOR(A, B) or NOT(A). An *Arithmetic Logic Unit* (ALU) requires both arithmetic (add, subtract) and Boolean logical operations. You can either multiplex between an adder and Boolean unit or merge the Boolean unit into the adder as in the classic TTL 181 ALU [Weste93].

Table 10.8	Functions implemented by Boolean unit			
Operation	P_0	P_1	P_2	P_3
AND(A, B)	0	0	0	1
OR(A, B)	0	1	1	1
XOR(A, B)	0	1	1	0
NAND(A, B)	1	1	1	0
NOR(A, B)	1	0	0	0

10.7   Coding

Error-detecting and error-correcting codes are commonly used to increase system reliability. Memory arrays are particularly susceptible to soft errors caused by alpha particles or cosmic rays flipping a bit. Such errors can be detected or even corrected by adding a few extra *check bits* to each word in the array. Codes are also used to reduce the bit error rate in communication links.

The simplest form of error-detecting code is *parity*, which detects single-bit errors. More elaborate *error-correcting codes* (ECC) are capable of single-error correcting and double-error detecting (SECDED). Gray codes are another useful alternative to the standard binary codes. All of the codes are heavily based on the XOR function, so we will examine a variety of CMOS XOR designs.

10.7.1 Parity

A parity bit can be added to an N-bit word to indicate whether the number of 1's in the word is even or odd. In *even parity*, the extra bit is the XOR of the other N bits, which ensures the (N+1)-bit coded word has an even number of 1's:

$$A_n = \text{PARITY} = A_0 \oplus A_1 \oplus A_2 \oplus ... \oplus A_{n-1} \qquad \textbf{(10.42)}$$

Figure 10.59 shows a conventional implementation. Multi-input XOR gates can also be used. In a datapath, the XORs can be placed in a linear column with a tree-routing channel for connections.

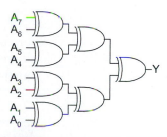

10.7.2 Error-correcting Codes

The *Hamming distance* [Hamming50] between a pair of binary numbers is the number of bits that differ between the two numbers. A single-bit error transforms a data word into another word separated by a Hamming distance of 1. Error-correcting codes add check bits to the data word so that the minimum

FIG 10.59 8-bit parity generator

Hamming distance between valid words increases. Parity is an example of a code with a single check bit and a Hamming distance of 2 between valid words, so that single-bit errors lead to invalid words and hence are detectable. If more check bits are added so that the minimum distance between valid words is 3, a single-bit error can be corrected because there will be only one valid word within a distance of 1. If the minimum distance between valid words is 4, a single-bit error can be corrected and an error corrupting two bits can be detected (but not corrected). If the probability of bit errors is low and uncorrelated from one bit to another, such SECDED codes greatly reduce the overall error rate of the system. Larger Hamming distances improve the error rate further at the expense of more check bits.

In general, you can construct a distance-3 Hamming code of length 2^c-1 with c check bits and $N = 2^c-c-1$ data bits [Wakerly00] using a simple procedure. If the bits are numbered from 1 to 2^c-1, each bit in a position that is a power of 2 serves as a check bit. The value of the check bit is chosen to obtain even parity for all bits with a 1 in the same position as the check bit, as illustrated in Figure 10.60(a) for a 7-bit code with 4 data bits and 3 check bits. The bits are traditionally reorganized into contiguous data and check bits, as shown in Figure 10.60(b). The structure is called a *parity-check matrix* and each check bit can be computed as the XOR of the highlighted data bits:

$$C_0 = D_3 \oplus D_1 \oplus D_0$$
$$C_1 = D_3 \oplus D_2 \oplus D_0 \tag{10.43}$$
$$C_2 = D_3 \oplus D_2 \oplus D_1$$

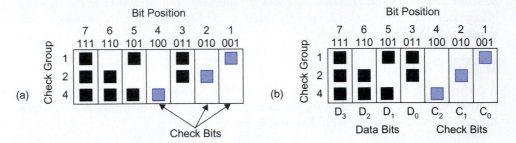

FIG 10.60 Parity-check matrix

The error-correcting decoder examines the check bits. If they all have even parity, the word is considered to be correct. If one or more groups have odd parity, an error has occurred. The pattern of check bits that have the wrong parity is called the *syndrome* and corresponds to the bit position that is incorrect. The decoder must flip this bit to recover the correct result.

Example

Suppose the data value 1001 were to be transmitted using a distance-3 Hamming code. What are the check bits? If the data bits were garbled into 1101 during transmission, explain what the syndrome would be and how the data would be corrected.

Solution: According to EQ (10.43), the check bits should be 100, corresponding to a transmitted word of 1001100. The received word is 1101100. The syndrome is 110, i.e., odd parity on check bits C_2 and C_1, which indicates an error in bit position 110 = 6. This position is flipped to produce a corrected word of 1001100 and the check bits are discarded, leaving the proper data value of 1001.

Many texts provide extensive information on a variety of error-correcting codes. For example, see [Lin83, Sweeney02].

10.7.3 Gray Codes

The *Gray codes*, named for Frank Gray, who patented their use on shaft encoders [Gray53], has a useful property that consecutive numbers differ in only one bit position. While there are many possible Gray codes, one of the simplest is the *binary-reflected Gray code* that is generated by starting with all bits zero and successively flipping the right-most bit that produces a new string. Table 10.9 compares 3-bit binary and binary-reflected Gray codes. Finite state machines that typically move through consecutive states can save power by Gray-coding the states to reduce the number of transitions. When a counter value must be synchronized across clock domains, it can be Gray-coded so that the synchronizer is certain to receive either the current or previous value because only one bit changes each cycle.

Table 10.9	3-bit Gray code	
Number	**Binary**	**Gray Code**
0	000	000
1	001	001
2	010	011
3	011	010
4	100	110
5	101	111
6	110	101
7	111	100

Converting between N-bit binary B and binary-reflected Gray code G is remarkably simple.

$$\begin{array}{ll} \text{Binary-> Gray} & \text{Gray-> Binary} \\ G_{N-1} = B_{N-1} & B_{N-1} = G_{N-1} \\ G_i = B_{i+1} \oplus B_i & B_i = B_{i+1} \oplus G_i \quad N-1 > i \geq 0 \end{array} \tag{10.44}$$

10.7.4 XOR/XNOR Circuit Forms

One of the chronic difficulties in CMOS circuit design is to construct a fast, compact, low-power XOR or XNOR gate. Figure 10.61 shows a number of common static single-rail 2-input XOR designs; XNOR designs are similar. Figure 10.61(a) and Figure 10.61(b) show gate-level implementations; the first is cute, but the second is slightly more efficient. Figure 10.61(c) shows a complementary CMOS gate. Figure 10.61(d) slightly improves the gate by optimizing out two contacts and is a commonly used standard cell design. Figure 10.61(e) shows a transmission gate design. Figure 10.61(f) is the 6-transistor "invertible inverter" design. When A is '0,' the transmission gate turns on and B is passed to the output. When A is '1,' the A input powers a pair of transistors that invert B. It is compact, but nonrestoring. Some switch-level simulators such as IRSIM cannot handle this unconventional design. Figure 10.61(g) [Wang94] is a compact and fast 4-transistor pass-gate design, but does not swing rail to rail.

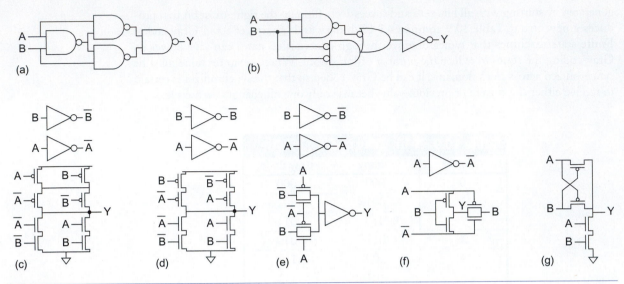

FIG 10.61 Static 2-input XOR designs

XOR gates with 3 or 4 inputs can be more compact, although not necessarily faster than a cascade of 2-input gates. Figure 10.62(a) is a 4-input static CMOS XOR [Griffin83] and Figure 10.62(b) is a 4-input CPL XOR/XNOR, while Figure 6.20(c) showed a 4-input CVSL XOR/XNOR. Observe that the true and complementary trees share most of the transistors. As mentioned in Chapter 6, CPL does not perform well at low voltage.

Dynamic XORs pose a problem because both true and complementary inputs are required, violating the monotonicity rule. The common solutions mentioned in Section 10.2.2.11 are to either push the XOR to the end of a chain of domino logic and build it with static CMOS or to construct a dual-rail domino structure. The dual-rail domino 2-input XOR was shown in Figure 6.30(c).

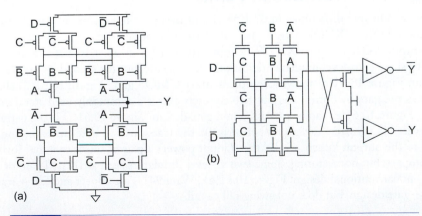

FIG 10.62 4-input XOR designs

10.8   Shifters

There are several commonly used shifters:

- *Logical shifter*: Shifts the number to the left or right and fills empty spots with 0's. Specified by << or >> in Verilog.

    ○ Example: 1011 LSR 1 = 0101; 1011 LSL 1 = 0110

- *Arithmetic shifter*: Same as logical shifter, but on right shifts fills the most significant bits with copies of the sign bit (to properly sign, extend 2's complement numbers when using right shift by k for division by 2^k). Specified by <<< or >>> in Verilog.

    ○ Example: 1011 ASR 1 = 1101; 1011 ASL 1 = 0110

- *Barrel shifter* (rotator): Rotates numbers in a circle such that empty spots are filled with bits shifted off the other end

    ○ Example: 1011 ROR 1 = 1101; 1011 ROL 1 = 0111

The *funnel shifter* can perform all three such operations. It concatenates two N-bit inputs and selects an N-bit subfield Y, as shown in Figure 10.63. Table 10.10 lists what inputs should be applied to perform a left or right shift of an N-bit word A by k bits. Generating the offset for left shifts requires a 2's complement subtraction. If N is a power of 2, the funnel shifter can be further simplified by accepting a $(2N-1)$-bit input Z and simply taking the 1's complement (i.e., bitwise inversion) of the shift amount for left shifts, as shown in Table 10.11. The funnel shifter can support all types of shifts and rotates by choosing suitable inputs Z, or it can be hardwired for only particular shifts.

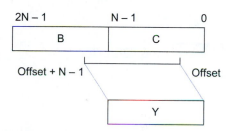

FIG 10.63 Funnel shifter function

Table 10.10	Funnel shifter operation		
Shift Type	**B**	**C**	**Offset**
Logical Right	0...0	$A_{N-1}...A_0$	k
Logical Left	$A_{N-1}...A_0$	0...0	$N-k$
Arithmetic Right	$A_{N-1}...A_{N-1}$ (sign extension)	$A_{N-1}...A_0$	k
Arithmetic Left	$A_{N-1}...A_0$	0	$N-k$
Rotate Right	$A_{N-1}...A_0$	$A_{N-1}...A_0$	k
Rotate Left	$A_{N-1}...A_0$	$A_{N-1}...A_0$	$N-k$

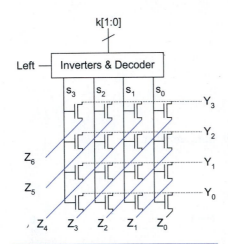

FIG 10.64 Array funnel shifter

Table 10.11	Simplified funnel shifter	
Shift Type	**Z**	**Offset**
Logical Right	$0..0, A_{N-1}...A_0$	k
Logical Left	$A_{N-1}...A_0, 0..0$	$\overline{k}$
Arithmetic Right	$A_{N-1}...A_{N-1}, A_{N-1}...A_0$	k
Arithmetic Left	$A_{N-1}...A_0, 0..0$	$\overline{k}$
Rotate Right	$A_{N-2}...A_0, A_{N-1}...A_0$	k
Rotate Left	$A_{N-1}...A_0, A_{N-1}..A_1$	$\overline{k}$

The simplest funnel shifter design consists of an array of N N-input multiplexers accepting 1-of-N-hot select signals (one multiplexer for each output bit). Such an array shifter is shown in Figure 10.64 using nMOS pass transistors for a 4-bit shifter. The shift amount is conditionally inverted and decoded into select signals that are fed vertically across the array. The outputs are taken horizontally. Each row of transistors attached to an output forms one of the multiplexers. The $2N-1$ inputs run diagonally to the appropriate mux inputs. Figure 10.65 shows a stick diagram for one of the N^2 transistors in the array. nMOS pass transistors suffer a threshold drop, but the problem can be solved by precharging the outputs (done in the Alpha 21164 [Gronowski96]) or by using full CMOS transmission gates.

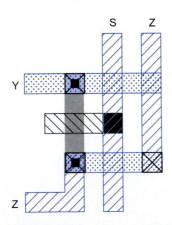

FIG 10.65 Array funnel shifter cell stick diagram

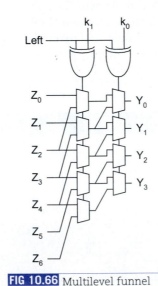

FIG 10.66 Multilevel funnel shifter

The array shifter works reasonably well for medium-sized shifters in transistor-level designs, but has high parasitic capacitance in larger shifters and is not amenable to standard cell designs. Figure 10.66 shows a 4-bit shifter based on multiple levels of smaller multiplexers (which, of course, can be transmission gates) [Lim72]. In general, the first level shifts by $N/2$, the second by $N/4$, and so forth until the final level shifts by 1. With such a structure, no decoder is necessary. The XOR gates on the control inputs conditionally invert the shift amount for left shifts. [Tharakan92] describes a domino implementation using 3:1 multiplexers to reduce the number of logic levels.

Other shift options, including shuffles, bit-reversals, and interchanges, are sometimes required. These are also built from appropriate combinations of multiplexers. The speed of an N-bit shifter is proportional to log N, so shifting can be a fast operation. 4:1 multiplexers work well in larger shifters to reduce the number of logic levels.

10.9   Multiplication

Multiplication is a less common operation than addition, but is still essential for microprocessors, digital signal processors, and graphics engines. Multiplication algorithms will be used to illustrate methods of designing different cells so that they fit into a larger structure. The most basic form of multiplication consists of forming the product of two unsigned (positive) binary numbers. This can be accomplished through the traditional technique taught in primary school, simplified to base 2. For example, the multiplication of two positive 4-bit binary integers, 12_{10} and 5_{10}, proceeds as shown in Figure 10.67.

```
   1100  :  12₁₀        Multiplicand
   0101  :   5₁₀        Multiplier
   ─────
   1100
   0000                  Partial
   1100                  Products
   0000
   ─────
00111100  :  60₁₀        Product
```

FIG 10.67 Multiplication example

$M \times N$-bit multiplication can be viewed as forming N partial products of M bits each, and then summing the appropriately shifted partial products to produce an $M+N$-bit result P. Binary multiplication is equivalent to a logical AND operation. Therefore, generating partial products consists of the logical ANDing of the appropriate bits of the multiplier and multiplicand. Each column of partial products must then be added and, if necessary, any carry values passed to the next column. We denote the multiplicand as $Y = (y_{M-1}, y_{M-2}, \ldots, y_1, y_0)$ and the multiplier as $X = (x_{N-1}, x_{N-2}, \ldots, x_1, x_0)$. For unsigned multiplication, the product is given in EQ (10.45). Figure 10.68 illustrates the generation, shifting, and summing of partial products in a 6 × 6-bit multiplier.

$$P = \left(\sum_{j=0}^{M-1} y_j 2^j \right) \left(\sum_{i=0}^{N-1} x_i 2^i \right) = \sum_{i=0}^{N-1} \sum_{j=0}^{M-1} x_i y_j 2^{i+j} \qquad (10.45)$$

				y_5	y_4	y_3	y_2	y_1	y_0			Multiplicand
				x_5	x_4	x_3	x_2	x_1	x_0			Multiplier
				x_0y_5	x_0y_4	x_0y_3	x_0y_2	x_0y_1	x_0y_0			
			x_1y_5	x_1y_4	x_1y_3	x_1y_2	x_1y_1	x_1y_0				
		x_2y_5	x_2y_4	x_2y_3	x_2y_2	x_2y_1	x_2y_0					Partial
	x_3y_5	x_3y_4	x_3y_3	x_3y_2	x_3y_1	x_3y_0						Products
x_4y_5	x_4y_4	x_4y_3	x_4y_2	x_4y_1	x_4y_0							
x_5y_5	x_5y_4	x_5y_3	x_5y_2	x_5y_1	x_5y_0							
p_{11}	p_{10}	p_9	p_8	p_7	p_6	p_5	p_4	p_3	p_2	p_1	p_0	Product

FIG 10.68 Partial products

Large multiplications can be more conveniently illustrated using *dot diagrams*. Figure 10.69 shows a dot diagram for a simple 16×16 multiplier. Each dot represents a placeholder for a single bit that can be a 0 or 1. The partial products are represented by a horizontal boxed row of dots, shifted according to their weight. The multiplier bits used to generate the partial products are shown on the right.

There are a number of techniques that can be used to perform multiplication. In general, the choice is based upon factors such as latency, throughput, area, and design complexity. An obvious approach is to use an $M+1$-bit carry-propagate adder (CPA) to add the first two partial products, then another CPA to add the third partial product to the running sum, and so forth. Such an approach requires $N-1$ CPAs and is slow, even if a fast CPA is employed. More efficient parallel approaches use some sort of array or tree of full adders to sum the partial products. We begin with a simple array for unsigned multipliers, and then modify the array to handle signed 2's complement numbers using the Baugh-Wooley algorithm. The number of partial products to sum can be reduced using Booth encoding and the number of logic levels required to perform the summation can be reduced with Wallace trees. Unfortunately, Wallace trees are complex to lay out and have long, irregular wires, so hybrid array/tree structures may be more attractive. For completeness, we consider a serial multiplier architecture. This was once popular when gates were relatively expensive, but is now less often necessary. Multipliers are discussed in more detail in many computer arithmetic texts [Ercegovac04, Parhami00, Koren02, Flynn01].

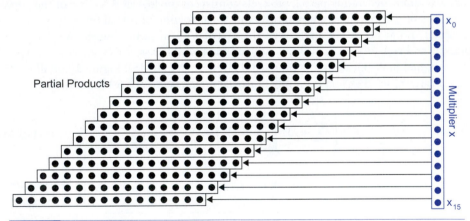

FIG 10.69 Dot diagram

10.9.1 Unsigned Array Multiplication

Fast multipliers use carry-save adders (CSAs, see Section 10.2.3.2) to sum the partial products. A CSA typically has a delay of 1.5–2 FO4 inverters independent of the width of the partial product, while a carry-propagate adder (CPA) tends to have a delay of 4–15+ FO4 inverters depending on the width, architecture, and circuit family. Figure 10.70 shows a 4×4 array multiplier for unsigned numbers using an array of CSAs. Each cell contains a 2-input AND gate that forms a partial product and a full adder (CSA) to add the partial product into the running sum. The first row converts the first partial product into carry-save redundant form. Each later row uses the CSA to add the corresponding partial product to the carry-save redundant result of the previous row and generate a carry-save redundant result. The least significant N output bits are available as sum outputs directly from CSAs. The most significant output bits arrive in carry-save redundant form and require an M-bit carry-propagate adder to convert into regular binary form. In the figure, the CPA is implemented as a carry-ripple adder. The array is regular in structure and uses a single type of cell, so it is easy to design and lay out. Assuming the carry output is faster than the sum output in a CSA, the critical path through the array is marked on the figure with a dashed line. The adder can easily be pipelined with the placement of registers between rows. In practice, circuits are assigned rectangular blocks in the floorplan so the parallelogram shape wastes space. Figure 10.71 shows the same adder squashed to fit a rectangular block.

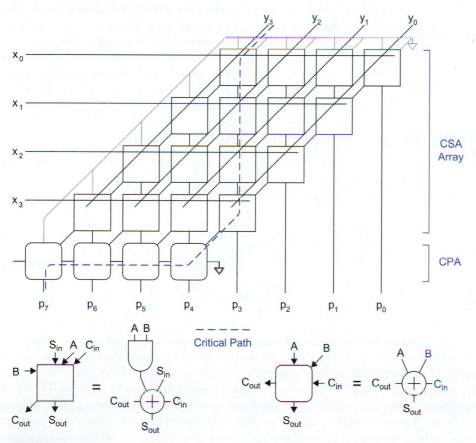

FIG 10.70 Array multiplier

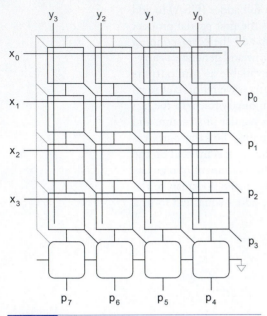

FIG 10.71 Rectangular array multiplier

A key element of the design is a compact CSA. This not only benefits area but also helps performance because it leads to short wires with low wire capacitance. An ideal CSA design has approximately equal sum and carry delays because the greater of these two delays limits performance. Note that the first row of CSAs adds the first partial product to a pair of 0's. This leads to a regular structure, but is inefficient. At a slight cost to regularity, the first row of CSAs can be used to add the first three partial products together. This reduces the number of rows by two and correspondingly reduces the adder propagation delay. It is also possible to use the first row of CSAs to add one or two other inputs with no extra delay. For example, digital signal processing chips frequently offer multiply-accumulate units (MACs) that compute $Y = A \cdot B + C$, a key operation for filters and transforms. Yet another way to improve the multiplier array performance is to replace the bottom row with a faster CPA such as a lookahead or tree adder. In summary, the critical path of an array multiplier involves N-2 CSAs and a CPA.

The CSA is an asymmetric circuit: certain inputs have more logical effort than others. The multiplier can be designed so that the sum output (with greater logical effort) drives the input to the next stage with lower capacitance [Sutherland99].

10.9.2 2's Complement Array Multiplication

Multiplication of 2's complement numbers at first might seem more difficult because some partial products are negative and must be subtracted. Recall that the most significant bit of a 2's complement number has a negative weight. Hence, the product is:

$$P = \left(-y_{M-1}2^{M-1} + \sum_{j=0}^{M-2} y_j 2^j\right)\left(-x_{n-1}2^{N-1} + \sum_{i=0}^{N-2} x_i 2^i\right)$$

$$= \sum_{i=0}^{N-2}\sum_{j=0}^{M-2} x_i y_j 2^{i+j} + x_{N-1}y_{M-1}2^{M+N-2} - \left(\sum_{i=0}^{N-2} x_i y_{M-1} 2^{i+M-1} + \sum_{j=0}^{M-2} x_{N-1}y_j 2^{j+N-1}\right)$$

$$\tag{10.46}$$

In EQ 10.46, two of the partial products have negative weight and thus should be subtracted rather than added. The *Baugh-Wooley* [Baugh73] multiplier algorithm handles subtraction by taking the 2's complement of the terms to be subtracted (i.e., inverting the bits and adding one). Figure 10.72 shows the partial products that must be summed. The upper parallelogram represents the unsigned multiplication of all but the most significant bits of the inputs. The next row is a single bit corresponding to the product of the most significant bits. The next two pairs of rows are the inversions of the terms to be subtracted. Each term has implicit leading and trailing 0's, which are inverted to leading and trailing 1's. Extra 1's must be added in the least significant column when taking the 2's complement.

The partial products for the 2's complement multiplier are summed according to:

$$\sum_{i=0}^{N-2}\sum_{j=0}^{M-2} x_i y_j 2^{i+j}$$

$$x_{N-1} y_{M-1} 2^{M+N-2}$$

$$-\sum_{i=0}^{N-2} x_i y_{M-1} 2^{i+M-1}$$

$$-\sum_{j=0}^{M-2} x_{N-1} y_j 2^{j+N-1}$$

					y_5	y_4	y_3	y_2	y_1	y_0	
					x_5	x_4	x_3	x_2	x_1	x_0	
							x_0y_4	x_0y_3	x_0y_2	x_0y_1	x_0y_0
						x_1y_4	x_1y_3	x_1y_2	x_1y_1	x_1y_0	
					x_2y_4	x_2y_3	x_2y_2	x_2y_1	x_2y_0		
				x_3y_4	x_3y_3	x_3y_2	x_3y_1	x_3y_0			
			x_4y_4	x_4y_3	x_4y_2	x_4y_1	x_4y_0				
	x_5y_5										
1	1	$\overline{x_4y_5}$	$\overline{x_3y_5}$	$\overline{x_2y_5}$	$\overline{x_1y_5}$	$\overline{x_0y_5}$	1	1	1	1	1
											1
1	1	$\overline{x_5y_4}$	$\overline{x_5y_3}$	$\overline{x_5y_2}$	$\overline{x_5y_1}$	$\overline{x_5y_0}$	1	1	1	1	1
											1
p_{11}	p_{10}	p_9	p_8	p_7	p_6	p_5	p_4	p_3	p_2	p_1	p_0

FIG 10.72 Partial products for 2's complement multiplier

The multiplier delay depends on the number of partial product rows to be summed. The *modified Baugh–Wooley multiplier* [Hatamian86] reduces this number of partial products by precomputing the sums of the constant 1's and pushing some of the terms upward into extra columns. Figure 10.73 shows such an arrangement. The parallelogram-shaped array can again be squashed into a rectangle as shown in Figure 10.74, giving a design almost identical to the unsigned multiplier of Figure 10.71. The AND gates are replaced by NAND gates in the hatched cells and 1's are added in place of 0's at a few of the unused inputs. The signed and unsigned arrays are so similar that a single array can be used for both purposes if XOR gates are used to conditionally invert some of the terms depending on the mode.

					y_5	y_4	y_3	y_2	y_1	y_0	
					x_5	x_4	x_3	x_2	x_1	x_0	
					1	$\overline{x_5y_0}$	x_0y_4	x_0y_3	x_0y_2	x_0y_1	x_0y_0
					$\overline{x_5y_1}$	x_1y_4	x_1y_3	x_1y_2	x_1y_1	x_1y_0	
				$\overline{x_5y_2}$	x_2y_4	x_2y_3	x_2y_2	x_2y_1	x_2y_0		
			$\overline{x_5y_3}$	x_3y_4	x_3y_3	x_3y_2	x_3y_1	x_3y_0			
		$\overline{x_5y_4}$	x_4y_4	x_4y_3	x_4y_2	x_4y_1	x_4y_0				
1	x_5y_5	$\overline{x_4y_5}$	$\overline{x_3y_5}$	$\overline{x_2y_5}$	$\overline{x_1y_5}$	$\overline{x_0y_5}$					
p_{11}	p_{10}	p_9	p_8	p_7	p_6	p_5	p_4	p_3	p_2	p_1	p_0

FIG 10.73 Simplified partial products for 2's complement multiplier

10.9.3 Booth Encoding

The array multipliers in the previous sections compute the partial products in a radix-2 manner, i.e., by observing one bit of the multiplier at a time. Radix 2^r multipliers produce N/r partial products, each of which depend on r bits of the multiplier. Fewer partial products leads to a smaller and faster CSA array. For example, a radix-4 multiplier produces $N/2$ partial products. Each partial product is 0, Y, $2Y$, or $3Y$, depending on a pair of bits of X. Computing $2Y$ is a simple shift, but $3Y$ is a *hard* multiple requiring a slow carry-propagate addition of $Y + 2Y$ before partial product generation begins.

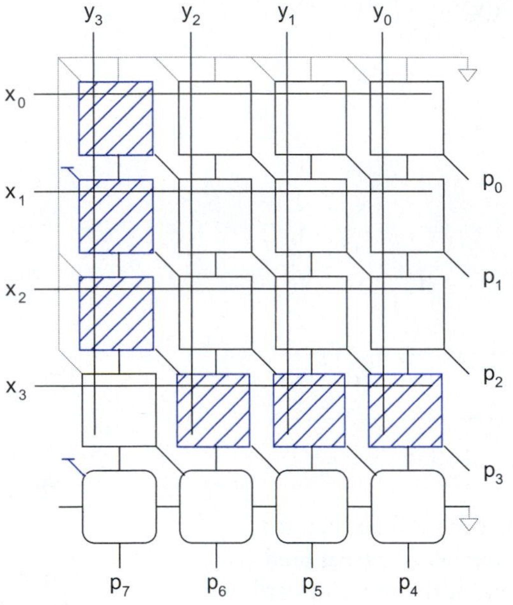

Booth encoding was originally proposed to accelerate serial multiplication [Booth51]. *Modified Booth encoding* [MacSorley61] allows higher radix parallel operation without generating the hard $3Y$ multiple by instead using negative partial products. Observe that $3Y = 4Y - Y$ and $2Y = 4Y - 2Y$. However, $4Y$ in a radix-4 multiplier array is equivalent to Y in the next row of the array that carries four times the weight. Hence, partial products are chosen by considering a pair of bits along with the most significant bit from the previous pair. If the most significant bit from the previous pair is true, Y must be added to the current partial product. If the most significant bit of the current pair is true, the current partial product is selected to be negative and the next partial product is incremented. Table 10.12 shows how the partial products are selected, based on bits of the multiplier. Negative partial products are generated by taking the 2's complement of the multiplicand (possibly left-shifted by one column).

FIG 10.74 Modified Baugh-Wooley 2's complement multiplier

Table 10.12	Radix-4 modified Booth encoding values				
Inputs			**Partial Product**	**Booth Selects**	
x_{2i+1}	x_{2i}	x_{2i-1}	PP_i	X_i	$2X_i$	M_i
0	0	0	0	0	0	0
0	0	1	Y	1	0	0
0	1	0	Y	1	0	0
0	1	1	$2Y$	0	1	0
1	0	0	$-2Y$	0	1	1
1	0	1	$-Y$	1	0	1
1	1	0	$-Y$	1	0	1
1	1	1	$-0 (= 0)$	0	0	1

In a radix-4 Booth-encoded multiplier, each group of three bits (a pair, along with the most significant bit of the previous pair) is decoded into several select lines (X_i, $2X_i$, and M_i, given in the rightmost columns of Table 10.12) and driven across the partial product row as shown in Figure 10.75. The multiplier Y is distributed to all the rows. The select lines control Booth selectors that choose the appropriate multiple of Y for each partial product. The Booth selectors substitute for the AND gates of a simple array multiplier. Figure 10.75 shows a conventional Booth selector design that computes the jth partial product bit of the ith partial product [Chandrakasan01]. If the partial product has a magnitude of Y, y_i is selected. If it has a magnitude of $2Y$, y_{i-1} is selected. If it is negative, the multiple is inverted (and a 1 is added to the least significant column elsewhere in the array to form the 2's complement). Exercise 10.18 explores an alternative Booth encoding that simplifies the encoder logic.

Even in an unsigned multiplier, negative partial products must be sign-extended to be summed correctly. Figure 10.76 shows a 16-bit radix-4 Booth partial product array for an unsigned multiplier using the dot diagram notation. Each dot in the Booth-encoded multiplier is produced by a Booth selector rather than a simple AND gate. For each partial product i, the sign bit $s_i = M_i = x_{2i+1}$ is 1 for negative multiples (those in the bottom half of Table 10.12) or 0 for positive multiples. Observe how an extra 1 is added to the least significant bit in the next row to form the 2's complement of negative multiples. Inverting the implicit leading 0's generates leading 1's on negative multiples. The extra terms increase the size of the multiplier. PP_8 is required in case PP_7 is negative; this partial product is always 0 or Y because x_{16} and x_{17} are 0.

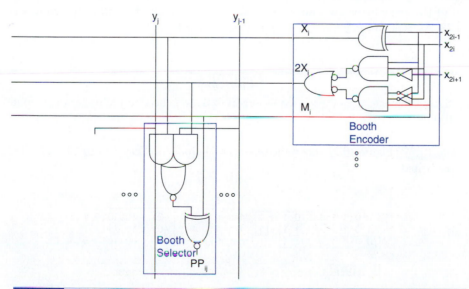

FIG 10.75 Radix-4 Booth encoder and selector

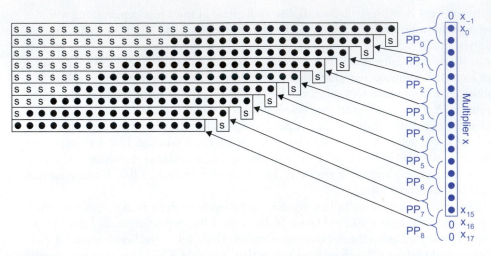

FIG 10.76 Radix-4 Booth-encoded partial products with sign extension

Observe that the sign extension bits are all either 1's or 0's. If a single 1 is added to the least significant position in a string of 1's, the result is a string of 0's plus a carry-out the top bit that may be discarded. Therefore, the large number of s bits in each partial product can be replaced by an equal number of constant 1's plus the inverse of s added to the least significant position, as shown in Figure 10.78. These constants mostly can be optimized out of the array by precomputing their sum. The simplified result is shown in Figure 10.78. As usual, it can be squashed to fit a rectangular floorplan.

Example

Sketch the partial products used by a radix-4 Booth-encoded multiplier to compute $01110_2 \times 01101_2$.

Solution: Figure 10.77 shows the three partial products. They sum to $14 \times 13 = 182$ as desired.

	X_{2i+1}	X_{2i}	X_{2i-1}	P_i
0 1 1 1 0 0 1 0 1	1	0	n/a	$-2Y = 100101 + 1$
1 0 1 1 1 1 1 1 $\mid$ 1	1	1	1	$-0Y = 111111 + 1$
0 0 1 1 0 1 $\mid$ 1	n/a	0	1	$Y\ \ = 001101 + 0$

FIG 10.77 Example of Booth-encoded partial products

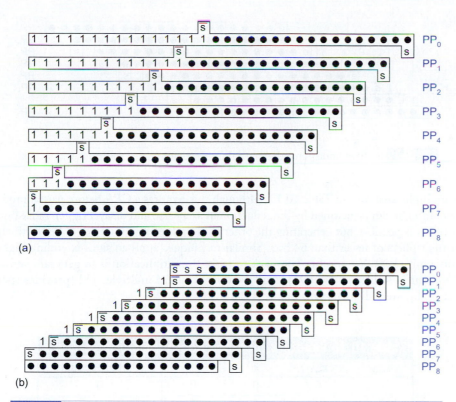

FIG 10.78 Radix-4 Booth-encoded partial products with simplified sign extension

The critical path of the multiplier involves the Booth decoder, the select line drivers, the Booth selector, approximately $N/2$ CSAs, and a final CPA. Each partial product fills about $M + 5$ columns. 53×53-bit radix-4 Booth multipliers for IEEE double-precision floating-point units are typically 20%–50% smaller (and arguably up to 20% faster) than nonencoded counterparts, so the technique is widely used. The multiplier requires $M \cdot N/2$ Booth selectors. Because they account for a substantial portion of the area and only a small fraction of the critical path, they should be optimized for size over speed.

Signed 2's complement multiplication is similar, but the sign extension is based on the sign of the partial product (i.e., the most significant bit) rather than simply M_i because the multiplicand might have been negative [Bewick94]. Figure 10.79 shows such an array, where the sign extension bit is $e_i = M_i \oplus y_{15}$. Also notice that PP_8, which was either Y or 0 for unsigned multiplication, is always 0 and can be omitted for signed multiplication because the multiplier x is sign-extended such that $x_{17} = x_{16} = x_{15}$.

Large multipliers can use Booth encoding of higher radix. For example, ordinary radix-8 multiplication reduces the number of partial products by a factor of three, but requires hard multiples of $3Y$, $5Y$, and $7Y$. Radix-8 Booth-encoding only requires the hard

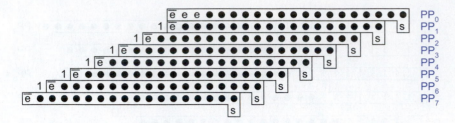

FIG 10.79 Radix-4 Booth-encoded partial products for signed multiplication

$3Y$ multiple, as shown in Table 10.13. Although this requires a CPA before partial product generation, it can be justified by the reduction in array size and delay. Higher-radix Booth encoding is possible, but generating the other hard multiples appears not to be worthwhile for multipliers of fewer than 64 bits. Similar techniques apply to sign-extending higher-radix multipliers. Yet another approach to radix-8 multiplication is to generate *partially redundant* multiples [Bewick94], thus avoiding the hard multiple. This provides speed, area, and power advantages over radix-4 Booth.

Table 10.13	Radix-8 modified Booth encoding values			
x_{i+2}	x_{i+1}	x_i	x_{i-1}	Partial Product
0	0	0	0	0
0	0	0	1	Y
0	0	1	0	Y
0	0	1	1	$2Y$
0	1	0	0	$2Y$
0	1	0	1	$3Y$
0	1	1	0	$3Y$
0	1	1	1	$4Y$
1	0	0	0	$-4Y$
1	0	0	1	$-3Y$
1	0	1	0	$-3Y$
1	0	1	1	$-2Y$
1	1	0	0	$-2Y$
1	1	0	1	$-Y$
1	1	1	0	$-Y$
1	1	1	1	-0

10.9.4 Wallace Tree Multiplication

Observe that a CSA is effectively a "1's counter" that adds the number of 1's on the A, B, and C inputs and encodes them on the sum and carry outputs, as summarized in Table 10.14. A CSA is therefore also known as a *(3,2) counter* [Dadda65] because it converts three inputs into a count encoded in two outputs. The carry-out is passed to the next more significant column, while a corresponding carry-in is received from the previous column. Therefore, for simplicity, a carry is represented as being passed directly down the column. Figure 10.80 shows a dot diagram of an array multiplier column that sums N partial products sequentially using N–2 CSAs. The output is produced in carry-save redundant form suitable for the final CPA.

Table 10.14	An adder as a 1's counter				
A	**B**	**C**	**Carry**	**Sum**	**Number of 1's**
0	0	0	0	0	0
0	0	1	0	1	1
0	1	0	0	1	1
0	1	1	1	0	2
1	0	0	0	1	1
1	0	1	1	0	2
1	1	0	1	0	2
1	1	1	1	1	3

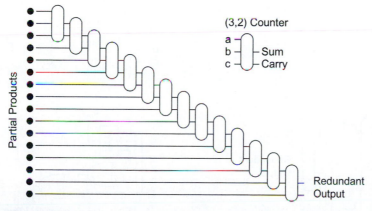

FIG 10.80 Dot diagram for array multiplier

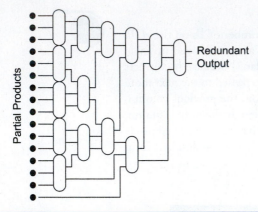

FIG 10.81 Dot diagram for Wallace tree multiplier

The column addition is slow because only one CSA is active at a time. Another way to speed the column addition is to sum partial products in parallel rather than sequentially. Figure 10.81 shows a *Wallace tree* using this approach [Wallace64]. The Wallace tree requires

$$\left\lceil \log_{3/2}\left(N/2\right) \right\rceil$$

levels of (3,2) counters to reduce N inputs down to 2 carry-save redundant form outputs. Unfortunately, the routing between levels becomes much more complicated. The longer wires have greater wire capacitance and the irregular tree is difficult to lay out.

[4:2] compressors can be used in a binary tree to produce a much more regular layout, shown in Figure 10.82 [Weinberger81, Santoro89]. A [4:2] compressor takes four inputs of equal weight and produces two outputs. It can be constructed from two (3,2) counters as shown in Figure 10.83(a). Along the way, it generates an intermediate carry into the next column and accepts a carry from the previous column, so it may more aptly be called a *(5,3) counter*. Only

$$\left\lceil \log_{2}\left(N/2\right) \right\rceil$$

levels of [4:2] compressors are required, although each has greater delay than a CSA. The regular layout and routing also make the binary tree attractive. Figure 10.83(b) shows a [4:2] compressor design using fewer levels of logic [Ohkubo95]. [Itoh01n] describes layout issues in a tree multiplier.

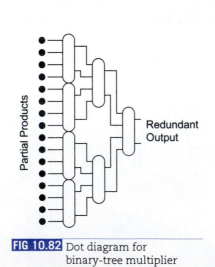

FIG 10.82 Dot diagram for binary-tree multiplier

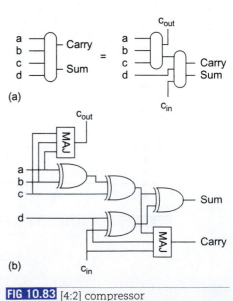

FIG 10.83 [4:2] compressor

10.9.5 Hybrid Multiplication

Arrays offer regular layout, but many levels of CSAs. Trees offer fewer levels of CSAs, but less regular layout and some long wires. A number of hybrids have been proposed that offer tradeoffs between these two extremes. These include *odd/even arrays* [Hennessy90], *arrays of arrays* [Dhanesha95], *balanced delay trees* [Zuras86], and *overturned-staircase trees* [Mou90]. They can achieve nearly as few levels of logic as the Wallace tree while offering more regular (and faster) wiring.

10.9.6 Fused Multiply-Add

Many algorithms, particularly in digital signal processing, require computing $P = X \cdot Y + Z$. While this can be done with a multiplier and adder, it is much faster to use a *fused multiply-add* unit, which is simply an ordinary multiplier modified to take another partial product Z. The extra partial product increases the delay of an array multiplier by just one extra CSA.

10.9.7 Serial Multiplication

Large parallel multipliers consume huge numbers of transistors. While transistor budgets have expanded to the point that this is often acceptable, designers of low-cost systems still may find serial multiplication attractive. Serial multiplication uses far less hardware, but requires multiple clock cycles to operate. Multiplication can be performed in a word-serial or bit-serial fashion.

Figure 10.84(a) shows a word-serial multiplication unit that only requires an M-bit adder and an $(M+N)$-bit shift register [Patterson04]. On each step, it conditionally adds the multiplicand Y to the running product if the appropriate bit of the multiplier X is 1. It is based on the observation that on the kth step, the running product has a length of $M+k$ bits and that bits $0\ldots k-1$ of X have already been considered and are no longer necessary. The unit is initialized by loading all of X into the lower portion of the shift register and a running product of 0's into the upper portion. On step k, Y is added to the running product if $x_k = 1$. The shift register then shifts right, discarding x_k and doubling the weight at which the next partial product will be added to the running product. After N steps, the

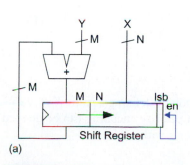

(a)

Step	Shift Reg	Notes
	0000|0101	initialize
0a	1100|0101	add 1*Y
0b	01100|010	shift right
1a	01100|010	add 0*Y
1b	001100|01	shift right
2a	111100|01	add 1*Y
2b	0111100|0	shift right
3a	0111100|0	add 0*Y
3b	00111100|	shift right

(b)

FIG 10.84 Word-serial multiplier

shift register will contain the final product. Figure 10.84(b) demonstrates multiplying $1100 \times 0101 = 00111100$. The vertical bar separates the running product from the remaining bits of X.

The cycle time of word-serial multiplication is set by the M-bit carry-propagate addition on each step. This CPA delay can be shortened to a CSA delay by maintaining the partial product in carry-save redundant form. The cost is doubling the number of registers to hold the redundant partial product and a final CPA to convert the redundant result into a 2's complement number at the end of the multiplication.

10.10 Parallel-prefix Computations

Many datapath operations involve calculating a set of outputs from a set of inputs in which each output bit depends on all the previous input bits. Addition of two N-bit inputs $A_N...A_1$ and $B_N...B_1$ to produce a sum output $Y_N...Y_1$ is a classic example; each output Y_i depends on a carry-in c_{i-1} from the previous bit, which in turn depends on a carry-in c_{i-2} from the bit before that, and so forth. At first, this dependency chain might seem to suggest that the delay must involve about N stages of logic, as in a carry-ripple adder. However, we have seen that by looking ahead across progressively larger blocks, we can construct adders that involve only $\log N$ stages. Section 10.2.2.2 introduced the notion of addition as a prefix computation that involves a bitwise precomputation, a tree of group logic to form the prefixes, and a final output stage, shown in Figure 10.13. In this section, we will extend the same techniques to other prefix computations with associative group logic functions.

Let us begin with the *priority encoder* shown in Figure 10.85. A common application of a priority encoder circuit is to arbitrate among N units that are all requesting access to a shared resource. Each unit i sends a bit A_i indicating a request and receives a bit Y_i indicating that it was granted access; access should only be granted to a single unit with highest priority. If the least significant bit of the input corresponds to the highest priority, the logic can be expressed as:

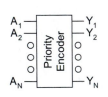

FIG 10.85 Priority encoder

$$Y_1 = A_1$$
$$Y_2 = A_2 \cdot \overline{A_1}$$
$$Y_3 = A_3 \cdot \overline{A_2} \cdot \overline{A_1} \tag{10.47}$$
$$...$$
$$Y_N = A_N \cdot \overline{A_{N-1}} \cdot ... \cdot \overline{A_1}$$

We can express priority encoding as a prefix operation by defining a prefix $X_{i:j}$ indicating that none of the inputs $A_i...A_j$ are asserted. Then priority encoding can be defined with bitwise precomputation, group logic, and output logic with $i \geq k > j$:

$$X_{i:i} = \overline{A_i} \qquad \text{bitwise precomputation}$$
$$X_{i:j} = X_{i:k} \cdot X_{k-1:j} \qquad \text{group logic} \qquad\qquad \textbf{(10.48)}$$
$$Y_i = A_i \cdot X_{i-1:1} \qquad \text{output logic}$$

Any of the group networks (e.g., ripple, skip, lookahead, select, increment, tree) discussed in the addition section can be used to build the group logic to calculate the $X_{i:0}$ prefixes. Short priority encoders use the ripple structure. Medium-length encoders may use a skip, lookahead, select, or increment structure. Long encoders use prefix trees to obtain log N delay. Figure 10.86 shows four 8-bit priority encoders illustrating the different group

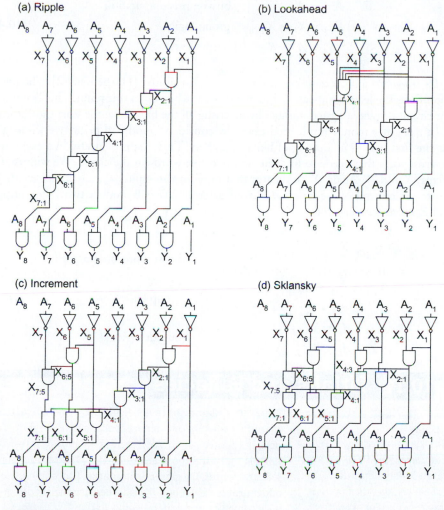

FIG 10.86 Priority encoder trees

logic. Each design uses an initial row of inverters for the $X_{i:i}$ precomputation and a final row of AND gates for the Y_i output logic. In between, ripple, lookahead, increment, and Sklansky networks form the prefixes with various tradeoffs between gate count and delay. Compare these trees to Figure 10.16, Figure 10.27, Figure 10.30, and Figure 10.34(b), respectively. [Wang00j, Delgado-Frias00, Huang02] describe a variety of priority encoder implementations.

An *incrementer* can be constructed in a similar way. Adding 1 to an input word consists of finding the least significant 0 in the word and inverting all the bits up to this point. The X prefix plays the role of the propagate signal in an adder. Again, any of the prefix networks can be used with varying area-speed tradeoffs.

$$
\begin{aligned}
X_{i:i} &= A_i & \text{bitwise precomputation} \\
X_{i:j} &= X_{i:k} \bullet X_{k-1:j} & \text{group logic} \\
Y_i &= A_i \oplus X_{i-1:1} & \text{output logic}
\end{aligned}
\tag{10.49}
$$

Decrementers and *2's complement* circuits are also similar [Hashemian92]. The decrementer finds the least significant 1 and inverts all the bits up to this point. The 2's complement circuit negates a signed number by inverting all the bits above the least significant 1.

In a slightly more complicated example, consider a modified priority encoder that finds the first two 1's in a string of binary numbers. This might be useful in a cache with two write ports that needs to find the first two free words in the cache. We will use two prefixes: X and W. Again $X_{i:j}$ indicates that none of the inputs $A_i...A_j$ are asserted. $W_{i:j}$ indicates exactly one of the inputs $A_i...A_j$ are asserted. We will produce two 1-hot outputs, Y and Z, corresponding to th

$$
\begin{aligned}
X_{i:i} &= \overline{A_i} & \text{bitwise precomputation} \\
W_{i:i} &= A_i \\
X_{i:j} &= X_{i:k} \bullet X_{k-1:j} \\
W_{i:j} &= W_{i:k} \bullet X_{k-1:j} + X_{i:k} \bullet W_{k-1:j} & \text{group logic} \\
Y_i &= A_i \bullet X_{i-1:1} \\
Z_i &= A_i \bullet W_{i-1:1} & \text{output logic}
\end{aligned}
\tag{10.50}
$$

10.11　Pitfalls and Fallacies

Equating logic levels and delay

Comparing a novel design with the best existing design is difficult. Some engineers cut corners by merely comparing logic levels. Unfortunately, delay depends strongly on the logical effort of each stage, the fanout it must drive, and the wiring capacitance. For example,

[Srinivas92] claims that a novel adder is 20%–28% faster than the fastest known binary lookahead adder, but does not present simulation results. Moreover, it reports some of the speed advantages to three or four significant figures. On closer examination [Dobson95], the adder proves to just be a hybrid tree/carry-select design with some unnecessary precomputation.

Designing circuits with threshold drops

In modern processes, single-pass transistors that pull an output to $V_{DD} - V_t$ are generally unacceptable because the threshold drop (amplified by the body effect) results in an output with too little noise margin. Moreover, when they drive the gate terminals of a subsequent stage, the stage turns partially ON and consumes static power. Many 10-transistor full-adder cells have been proposed that suffer from such a threshold drop problem.

Reinventing adders

There is an enormous body of literature on adders with various tradeoffs among speed, area, and power consumption. The design space has been explored fairly well and many designers (one of the authors included) have spent quite a bit of time developing a "new" adder, only to find that it is only a minor variation on an existing theme. Similarly, a number of recent publications on priority encoders reinvent prefix network techniques that have already been explored in the context of addition.

10.12 Historical Perspective

Given the wide variety of adders in the literature, which ones are used in practice? The carry-ripple adder is compact and fast enough for many noncritical paths. For high-performance microprocessors, however, the adder is often in the critical path in the ALU or the address generation units. In these applications, a tree adder of logarithmic depth is desirable. Ideally, an N-bit tree adder would have $\log N$ logic levels, fanout never exceeding 2, and a single wiring track between each logic level. Brent-Kung has extra logic levels. Sklansky has large fanout. Kogge-Stone has more wiring tracks. The best choice depends on the particular implementation technology. For example, in older processes when wires were cheaper than gates, Kogge-Stone was most attractive. In modern processes, the ratio of wire capacitance to gate capacitance has become larger, so designs with fewer tracks are preferred.

The Kogge-Stone and Han-Carlson adders have been widely used on 32-bit and 64-bit microprocessors including the UltraSparc III and several high-speed demonstration units [Silberman98, Heald00, Vangal02]. The Naffziger adder is used throughout the Itanium 2 and Hewlett Packard PA-RISC lines of 64-bit microprocessors [Fetzer02]. The AMD29050 microprocessor used the 64-bit "Redundant Cell" adder [Lynch91], which can be viewed as a hybrid valency-4 tree adder driving 8-bit carry-select multiplexers.

Summary

This chapter has presented a range of datapath subsystems. How one goes about designing and implementing a given CMOS chip is largely affected by the availability of tools, the schedule, the complexity of the system, and the final cost goals of the chip. In general, the simplest and least expensive (in terms of time and money) approach that meets the target goals should be chosen. For many systems, this means that synthesis and place-and-route is good enough. Modern synthesis tools draw on a good library of adders and multipliers with various area/speed tradeoffs that are sufficient to cover a wide range of applications. For systems with the most stringent requirements on performance or density, custom design at the schematic level still provides an advantage. Domino parallel-prefix trees provide the fastest adders when the high power consumption can be tolerated. Domino CSAs are also used in fast multipliers. However, in multiplier design, the wiring capacitance is paramount and a multiplier with compact cells and short wires can be fast as well as small and low in power.

Exercises

10.1 Design a fast 8-bit adder. The inputs may drive no more than 30 λ of transistor width each and the output must drive a 20/10 inverter. Simulate the adder and determine its delay.

10.2 When adding two unsigned numbers, a carry-out of the final stage indicates an overflow. When adding two signed numbers in 2's complement format, overflow detection is slightly more complex. Develop a Boolean equation for overflow as a function of the most significant bits of the two inputs and the output.

10.3 Repeat Exercise 10.2 for a signed add/subtract unit like that shown in Figure 10.48(b). Your overflow output should be a function of the subsignal and the most significant bits of the two inputs and the output.

10.4 Develop equations for the logical effort and parasitic delay with respect to the C_0 input of an n-stage Manchester carry chain computing $C_1 \ldots C_n$. Consider all of the internal diffusion capacitances when deriving the parasitic delay. Use the transistor widths shown in Figure 10.87 and assume the P_i and G_i transistors of each stage share a single diffusion contact.

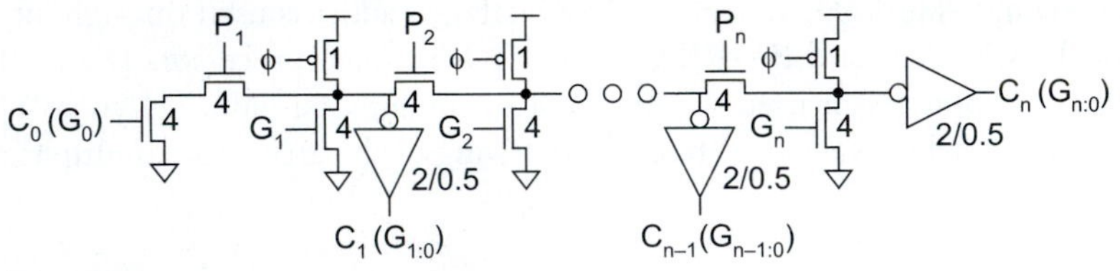

FIG 10.87 Manchester carry chain

10.5 Using the results of Exercise 10.4, what Manchester carry chain length gives the least delay for a long adder?

10.6 The carry increment adder in Figure 10.31(b) with variable block size requires five stages of valency-2 group PG cells for 16-bit addition. How many stages are required for 32-bit addition? For 64-bit addition?

10.7   Sketch the PG network for a modified 16-bit Sklansky adder with fanout of [8, 1, 1, 1] rather than [8, 4, 2, 1]. Use buffers to prevent the less-significant bits from loading the critical path.

10.8   Figure 10.34 shows PG networks for various 16-bit adders and Figure 10.35 illustrates how these networks can be classified as the intersection of the $l + f + t = 3$ plane with the face of a cube. The plane also intersects one point inside the cube at $(l, f, t) = (1, 1, 1)$ [Harris03]. Sketch the PG network for this 16-bit adder.

10.9   Sketch a diagram of the group PG tree for a 32-bit Ladner-Fischer adder.

10.10 Write a Boolean expression for C_{out} in the circuit shown in Figure 10.6(b). Simplify the equation to prove that the pass-transistor circuits do indeed compute the majority function.

10.11 Prove EQ (10.28) from EQ (10.27) and EQ (10.29).

10.12 Sketch a design for a comparator computing $A - B = k$.

10.13 Show how the layout of the parity generator of Figure 10.59 can be designed as a linear column of XOR gates with a tree-routing channel.

10.14 Design an ECC decoder for distance-3 Hamming codes with $c = 3$. Your circuit should accept a 7-bit received word and produce a 4-bit corrected data word. Sketch a gate-level implementation.

10.15 How many check bits are required for a distance-3 Hamming code for 8-bit data words? Sketch a parity-check matrix and write the equations to compute each of the check bits.

10.16 Find the 4-bit binary-reflected Gray code values for the numbers 0–15.

10.17 Design a Gray-coded counter in which only one bit changes on each cycle.

10.18 Table 10.12 and Figure 10.75 illustrated radix-4 Booth encoding using X_i, $2X_i$, and M_i. An alternative encoding is to use POS, NEG, and DOUBLE. POS is true for the multiples Y and $2Y$. NEG is true for the multiples $-Y$ and $-2Y$. DOUBLE is true for the multiples $2Y$ and $-2Y$. Design a Booth encoder and selector using this encoding.

10.19 Adapt the priority encoder logic of EQ (10.50) to produce three 1-hot outputs corresponding to the first three 1's in an input string.

10.20 Sketch a 16-bit priority encoder using a Kogge-Stone prefix network.

10.21 Use logical effort to estimate the delay of the priority encoder from Exercise 10.20. Assume the path electrical effort is 1.

10.22 Write equations for a prefix computation that determines the second location in which the pattern 10 appears in an N-bit input string. For example, 010010 should return 010000.

Array Subsystems

11

11.1 Introduction

Memory arrays often account for the majority of transistors in a CMOS system-on-chip. Arrays may be divided into categories as shown in Figure 11.1. *Programmable Logic Arrays* (PLAs) perform logic rather than storage functions, but are also discussed in this chapter.

Random access memory is accessed with an *address* and has a latency independent of the address. In contrast, *serial access memories* are accessed sequentially so no address is necessary. *Content addressable memories* determine which address(es) contain data that matches a specified *key*.

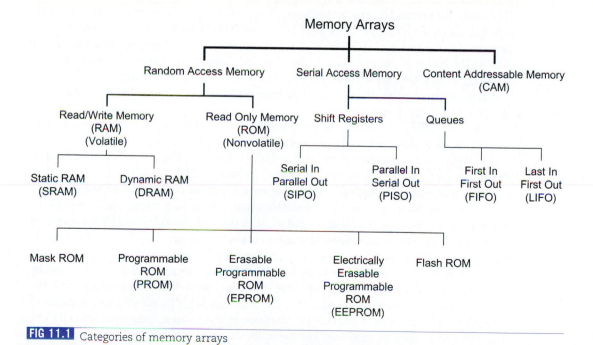

FIG 11.1 Categories of memory arrays

Random access memory is commonly classified as *read-only memory* (ROM) or *read/write memory* (confusingly called RAM). Even the term ROM is misleading because many ROMs can be written as well. A more useful classification is *volatile* vs. *nonvolatile* memory. Volatile memory retains its data as long as power is applied, while nonvolatile memory will hold data indefinitely. RAM is synonymous with volatile memory, while ROM is synonymous with nonvolatile memory.

Like sequencing elements, the memory cells used in volatile memories can further be divided into *static* structures and *dynamic* structures. Static cells use some form of feedback to maintain their state, while dynamic cells use charge stored on a floating capacitor through an access transistor. Charge will leak away through the access transistor even while the transistor is OFF, so dynamic cells must be periodically read and rewritten to refresh their state. Static RAMs (SRAMs) are faster and less troublesome, but require more area per bit than their dynamic counterparts (DRAMs).

Some nonvolatile memories are indeed read-only. The contents of a mask ROM are hardwired during fabrication and cannot be changed. But many nonvolatile memories can be written, albeit more slowly than their volatile counterparts. A *programmable* ROM (PROM) can be programmed once after fabrication by blowing on-chip fuses with a special high programming voltage. An *erasable programmable* ROM (EPROM) is programmed by storing charge on a floating gate. It can be erased by exposure to ultraviolet (UV) light for several minutes to knock the charge off the gate. Then the EPROM can be reprogrammed. *Electrically erasable programmable* ROMs (EEPROMs) are similar, but can be erased in microseconds with on-chip circuitry. *Flash* memories are a variant of EEPROM that erases entire blocks rather than individual bits. Sharing the erase circuitry across larger blocks reduces the area per bit. Because of their good density and easy in-system reprogrammability, Flash memories have replaced other nonvolatile memories in most modern CMOS systems.

Memory cells can have one or more *ports* for access. On a read/write memory, each port can be read-only, write-only, or capable of both read and write.

A typical small memory array architecture is shown in Figure 11.2. Central to the design is a memory array consisting of 2^n *words* of storage of 2^m bits each. In the simplest design, the array is organized with one row per word and one column per bit in each word. Often there are far more words in the memory than bits in each word, which would lead to a very tall, skinny memory that is hard to fit in the chip floorplan and slow because of the long vertical wires. Therefore, the array is often folded into fewer rows of more columns. After folding, each row of the memory contains 2^k words, so the array is physically organized as 2^{n-k} *rows* of 2^{m+k} *columns* or bits. The row decoder activates one of the rows by asserting one of the *wordlines*. During a read operation, the cells on this wordline drive the *bitlines*, which may have been conditioned to a known value in advance of the memory access. The column decoder controls a multiplexer in the column circuitry to select 2^m bits from the row as the data to access. The figure illustrates an 8-word by 4-bit 2-way multiplexed memory folded into a 4-row by 8-column array with $n = 3$, $m = 2$, $k = 1$. Larger

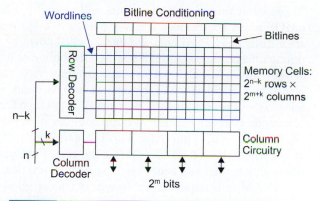

FIG 11.2 General memory array architecture

memories are generally built from multiple smaller subarrays so that the wordlines and bitlines remain reasonably short, fast, and low in power dissipation.

We begin in Section 11.2 with SRAM, the most widely used form of on-chip memory. SRAM also illustrates all the issues of cell design, decoding, and column circuitry design. Subsequent sections address DRAMs, ROMs, serial access memories, CAMs, and PLAs.

11.2   SRAM

The fundamental building block of a static RAM is the SRAM memory cell. The cell is activated by raising the wordline and is read or written through the bitline. Figure 11.3(a) shows a 12-transistor SRAM cell built from a simple static latch and tristate inverter. The cell has a single bitline. True and complementary read and write signals are used in place of a single wordline. A representative layout in Figure 11.3(b) has an area of $46 \times 75 \lambda$. The power and ground lines can be shared between mirrored adjacent cells, but the area is still limited by the wires and is undesirably large. However, the cell is easy to design because all nodes swing rail-to-rail and it is fast when used in small RAMs and register files.

Figure 11.4 shows a 6-transistor (6T) SRAM commonly used in practice. Such a cell uses a single wordline and both true and complementary bitlines. The complementary bitline is often called bit_b or $\overline{bit}$. The cell contains a pair of cross-coupled inverters and an *access transistor* for each bitline. True and complementary versions of the data are stored on the cross-coupled inverters. If the data is disturbed slightly, positive feedback around the loop will restore it to V_{DD} or GND. The wordline is asserted to read or write the cell.

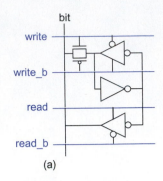

(a)

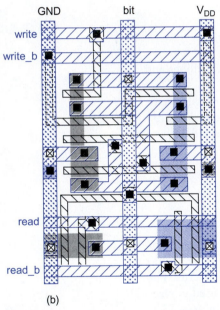

(b)

FIG 11.3   12-transistor SRAM cell

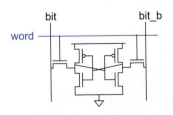

FIG 11.4   6-transistor SRAM cell

Throughout this chapter, wordlines will be highlighted in blue. The nMOS access transistors are best at passing '0's. For reads, the bitlines are initially precharged high and one is pulled down by the SRAM cell through the access transistor. For writes, the bitline or its complement is actively driven low and this low value overpowers the cell to write the new value. Careful choice of transistor sizes is necessary for correct operation, as will be examined in Section 11.2.1. The 6T cell achieves its compactness at the expense of more complex peripheral circuitry for reading and writing the cells. This is a good tradeoff in large RAM arrays where the cell size dominates the area. The small cell size also offers shorter wires and hence lower power consumption.

SRAM cells require clever layout to achieve good density. Figure 11.5(a) shows a stick diagram of a typical design. The cell is designed to be mirrored and overlapped to share V_{DD} and GND lines between adjacent cells along the cell boundary, as shown in Figure 11.5(b). Note how a single diffusion contact to the bitline is shared between a pair of cells. This halves the diffusion capacitance, and hence reduces the delay discharging the bitline during a read access. The wordline is run in both metal1 and polysilicon; the two layers must occasionally be strapped (e.g., every four or eight cells). Sample layouts derived from the stick diagram are shown in Figure 11.6. Figure 11.6(a) shows a conservative cell of $26 \times 45 \lambda$, obeying the MOSIS submicron design rules (see also inside front cover). In this layout, the metal1 and polysilicon wordlines are contacted in each cell. The substrate and well are also contacted in each cell. Figure 11.6(b) shows a scanning electron micrograph of the polysilicon and diffusion layers in a more aggressive ($20 \times 22 \lambda$) cell in the LSI Logic 130 nm process [Kong01w]. SRAM is so important that design rules are very carefully studied and bent where possible to minimize cell area in commercial processes. Notice that diagonal lines are also used. Moreover, each substrate contact can be shared among multiple cells to save area at the expense of regularity.

SRAM operation is divided into two phases. As described in Section 7.4.3, the phases will be called ϕ_1 and ϕ_2, but may actually be generated from *clk* and its complement *clkb*. Assume that in phase 2, the SRAM is precharged. In phase 1, the SRAM is written or read by raising the appropriate wordline and either driving the bitlines to the value that should be written or leaving the bitlines floating and observing which one is pulled down. Reading a large SRAM can be slow because the capacitance of all the cells sharing the bitline is large. Sense amplifiers accelerate reads by detecting small differences between the bitline and its complement. The following sections discuss the role of each block from Figure 11.2 in the SRAM operation.

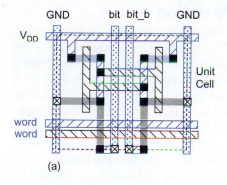
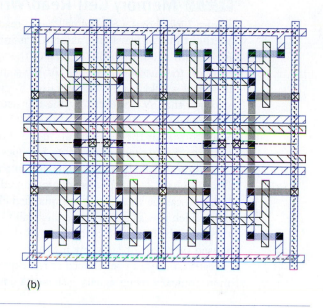

FIG 11.5 Stick diagram of 6T SRAM cell

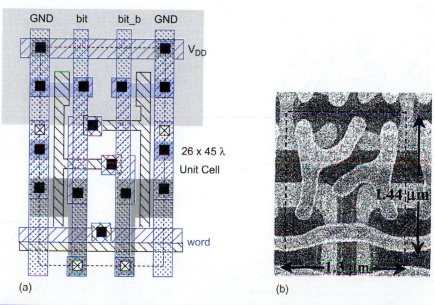

FIG 11.6 Layout of 6T SRAM cell. © IEEE 2001. Color version on inside front cover.

11.2.1 Memory Cell Read/Write Operation

Figure 11.7 shows a SRAM cell being read. The bitlines are both initially floating high. Without loss of generality, assume *A* is initially '0' and thus *A_b* is initially '1.' *A_b* and *bit_b* both should remain '1'. When the wordline is raised, *bit* should be pulled down through transistors *N1* and *N2*. At the same time *bit* is being pulled down, node *A* tends to rise. *A* is held low by *N1*, but raised by current flowing in from *N2*. Hence, *N1* must be stronger than *N2*. Specifically, the transistors must be ratioed such that node *A* remains below the switching threshold of the *P2/N3* inverter. This constraint is called *read stability*. Waveforms for the read operation are shown in Figure 11.7(b) as a 0 is read onto *bit*. Observe that *A* momentarily rises, but does not glitch badly enough to flip the cell.

Figure 11.8 shows the same cell in the context of a full column from the SRAM. During phase 2, the bitlines are precharged high. The wordline only rises during phase 1; hence, it can be viewed as a _q1 qualified clock. Many SRAM cells share the same bitline pair, which acts as a distributed dual-rail footless dynamic multiplexer. The capacitance of the entire bitline must be discharged through the access transistor. The output can be sensed by a pair of HI-skew inverters. By raising the switching threshold of the sense inverters, delay can be reduced at the expense of noise margin. The outputs are dual-rail monotonically rising signals, just as in a domino gate.

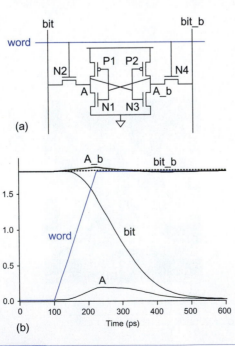

(a)

(b)

FIG 11.7 Read operation for 6T SRAM cell

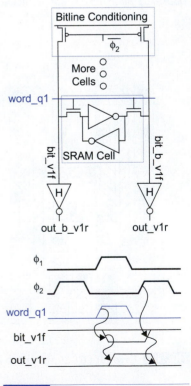

FIG 11.8 SRAM column read

The waveforms of Figure 11.9 show the SRAM cell being written. Again, assume *A* is initially '0' and that we wish to write a '1' into the cell. *bit* is precharged high and left floating. *bit_b* is pulled low by a write driver. We know on account of the read stability constraint that *bit* will be unable to force *A* high through *N2*. Hence, the cell must be written by forcing *A_b* low through *N4*. *P2* opposes this operation; thus, *P2* must be weaker than *N4* so that *A_b* can be pulled low enough. This constraint is called *writeability*. Once *A_b* falls low, *N1* turns OFF and *P1* turns ON, pulling *A* high as desired.

Figure 11.10(a) again shows the cell in the context of a full column from the SRAM. During phase 2, the bitlines are precharged high. Write drivers pull the bitline or its complement low during phase 1 to write the cell. The write drivers can consist of a pair of transistors on each bitline for the data and the write enable, or a single transistor driven by the appropriate combination of signals (Figure 11.10(b)). In either case, the series resistance of the write driver, bitline wire, and access transistor must be low enough to overpower the pMOS transistor in the SRAM cell.

In summary, to ensure both read stability and writeability, the nMOS pull-down transistor in the cross-coupled inverters must be strongest. The access transistors are of intermediate strength, and the pMOS pull-up transistors must be weak. To achieve good layout density, all of the transistors must be relatively small. In Figure 11.6(a), the pull-downs are 8/2 λ, the access transistors 4/2, and the pull-ups 3/3. The SRAM cells must operate correctly in all process corners at all voltages and temperatures. This requires thorough simulation.

It is no longer common for designers to develop their own SRAM cells. Usually, the fabrication vendor will supply cells that are carefully tuned to the particular manufacturing process. In high-performance processes, two or more cells may be provided with different speed/density tradeoffs. [Glasser85] describes SRAM cells with four transistors and two large resistors, but the 6T cell is almost universally used in contemporary standard CMOS processes.

11.2.2 Decoders

The simplest decoder is a collection of AND gates using true and complementary versions of the address bits. Figure 11.11 shows several straightforward implementations. The first implementation in Figure 11.11(a) is a static NAND gate followed by an inverter. This structure is useful for up to 5–6 inputs or more if speed is not critical. The NAND transistors are usually made minimum size to reduce the load on the buffered address lines because there are 2^{n-k} transistors on each true and complementary address line in the row

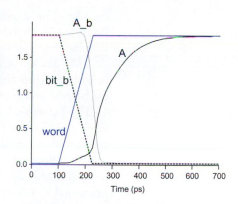

FIG 11.9 Write operation for 6T SRAM cell

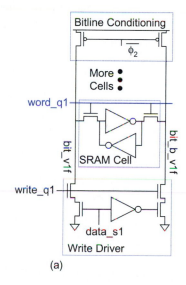

(a)

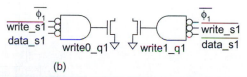

(b)

FIG 11.10 SRAM column write

decoder. The second implementation shown in Figure 11.11(b) uses a pseudo-nMOS NOR gate buffered with two inverters. The NOR gate transistors can be made minimum size and the inverters can be scaled appropriately to drive the wordline.

The layout of the decoder must be pitch-matched to the memory array, i.e., the height of each decoder gate must match the height of the row it drives. This can be tricky for SRAM and even harder for ROMs and other arrays with small memory cells. Figure 11.12(a) shows a layout of a conventional standard-cell style approach. The minimum-sized transistors in the NAND gate drive a larger buffer inverter. The decoder height grows with the number of inputs. The AND gates are easily programmed by connecting the polysilicon inputs to the appropriate address inputs. Figure 11.12(b) shows a layout on a pitch that is tighter and independent of the number of inputs. The decoder is programmed by placement of transistors and metal straps; this is best done with scripting software that generates layout, as discussed in Section 8.3.5. The polysilicon address lines should be strapped with metal2 to reduce their resistance, but the metal2 is left out of the figure for readability. The decoder pitch is 5 tracks or 40 λ. If every other row is mirrored to share V_{DD} and GND, the pitch can be reduced to 4 tracks or 32 λ.

11.2.2.1 Predecoding Decoders with many inputs can be formed from a cascade of smaller gates. For example, Figure 11.13(a) shows a 16-word decoder in which the 4-input AND function is built from a pair of 2-input NANDs followed by a 2-input NOR.

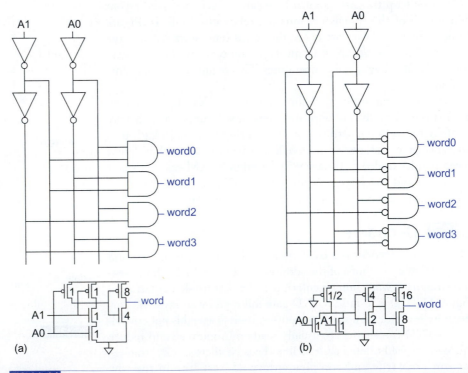

FIG 11.11 Decoders

Many NAND gates share exactly the same inputs and are thus redundant. The decoder area can be improved by factoring these common NANDs out, as shown in Figure 11.13(b). This technique is called *predecoding*. It does not change the path effort of the decoder, but does improve area. In general, blocks of p address bits can be predecoded into 1-of-2^p-hot predecoded lines that serve as inputs to the final stage decoder. For example, Figure 11.13(b) shows a $p = 2$-bit design that decodes each pair of address bits into a 1-of-4-hot code.

The wordline generally must be qualified with the clock for proper bitline timing. This is often performed with another AND gate after the decoder or with an extra clk input to the final stage of decoding.

11.2.2.2 Faster Decoders The logical effort of a decoder can be reduced by observing that only one of the outputs will be high so the pMOS transistors can be shared among many outputs [Lyon87]. A NOR gate pulls low efficiently through parallel transistors, but

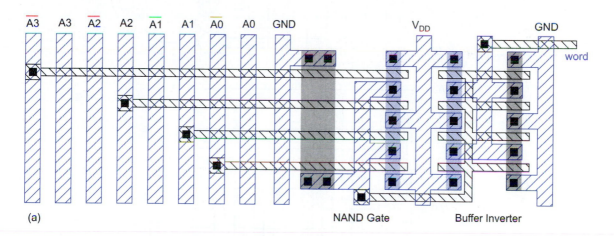

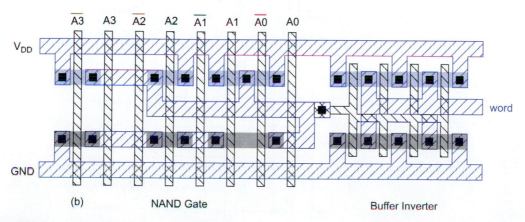

FIG 11.12 Stick diagrams of two decoder layouts

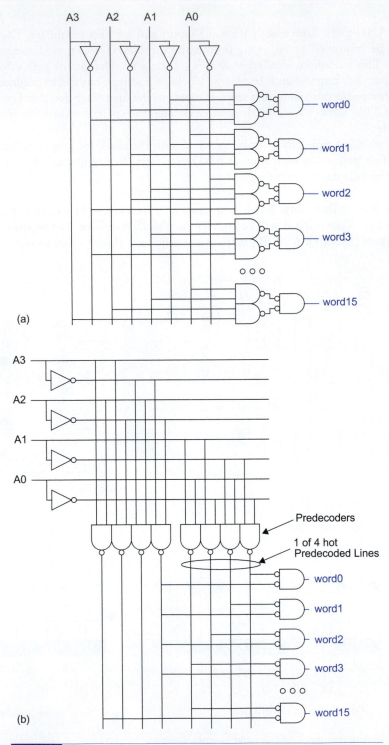

FIG 11.13 Ordinary and predecoding circuits

has a poor logical effort because the output is pulled high through wide series transistors. The Lyon-Schediwy decoder can be viewed as 2^n n-input NOR gates sharing pMOS pull-ups, as shown in Figure 11.14 for a 3:8 decoder. The cost of the wide transistors is amortized across many outputs. Relative transistor widths are chosen to present the same capacitance to each input while providing current drive equal to a unit inverter. The logical effort of each input is only $(1 + 3.5)/3 = 1.5$ per wordline output, as compared to 7/3 for an ordinary NOR3 (see Exercise 11.4).

Decoders typically have high electrical and branching effort. Therefore, they need many stages, so the fastest design is the one that minimizes the logical effort. A tree of 2- and 3-input NAND gates and inverters offers the lowest logical effort to build high fan-in gates in static CMOS [Sutherland99]. Dynamic gates are attractive for fast decoders because they have lower logical effort.

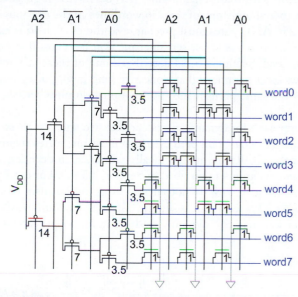

FIG 11.14 Lyon-Schediwy decoder

Example

Estimate the delays of 8:256 decoders using static CMOS and footed domino gates. Assume the decoder has an electrical effort of $H = 10$ and that both true and complementary inputs are available.

Solution: The decoder consists of 256 8-input AND gates. It has a branching effort of $B = 256/2 = 128$ because each of the true inputs and each of the complementary inputs are used by half the gates. Assuming the logical effort of the path G is close to 1, the path effort is $F = GBH = 1280$ and the best number of stages is

$\log_4 F = 5.16$. Let us consider a six-stage design using three levels of 2-input AND gates, each constructed from a 2-input NAND and an inverter.

The static CMOS design has a logical effort of $G = [(4/3) \cdot (1)]^3 = 64/27$. Therefore, the stage effort is $F = 3034$. The parasitic delay is $P = 3 \cdot (2 + 1) = 9$. The total delay is $D = NF^{1/N} + P = 31.8\ \tau$ or 6.4 FO4 inverter delays.

The footed domino design using HI-skew inverters has a logical effort of $[(1) \cdot (5/6)]^3 = 125/256$ and a stage effort of 625. The parasitic delay is $P = 3 \cdot (4/3 + 5/6) = 6.5$. The total delay is 4.8 FO4 inverter delays. In general, domino decoders are about 33% faster than static CMOS.

A major problem with traditional domino decoders is the high power consumption. For example, even though only one of the 256 wordlines in the previous example will rise on each cycle, all 256 AND gates must precharge so the clock load is extremely large. A much lower-power approach is to use self-resetting domino gates that only precharge the wordline that evaluated. Section 7.5.2.4 describes some of these self-resetting gates and [Amrutur01] shows some variations that work with long input pulses. Self-resetting domino has essentially the same performance as traditional domino because it uses the same basic gates.

Yet another approach for dynamic decoders is to use wide NOR structures in which N–1 of the N outputs discharge on each cycle. As most memories require monotonically rising outputs but the NORs are monotonically falling, such decoders require the race-based nonmonotonic techniques described in Section 7.5.4.3. For example, Figure 11.15 shows a 4-input AND gate with monotonically rising output using a race-based NOR structure [Nambu98]. This technique is slightly faster than a domino AND tree, but dissipates more power because the dynamic node X must be precharged on each cycle [Amrutur01].

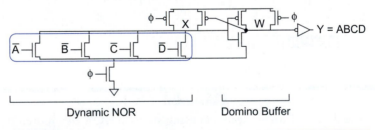

FIG 11.15 4-input AND using race-based NOR

11.2.2.3 Sum-addressed Decoders Many microprocessor instruction sets include addressing modes in which the effective address is the sum of two values, such as a base address and an offset. In conventional SRAMs used as caches, the two values must first be added, and then the result decoded to determine the cache wordline. If access latency needs to be minimized, these two steps can be combined into one in a *sum-addressed memory* [Heald98].

Recall from Section 10.4.3 that checking if $A + B = K$ is faster than actually computing $A + B$ because no carry propagation need occur. A *sum-addressed decoder* for an N-word memory accepts two inputs, A and B. In a simple form, it contains N comparators driving the N wordlines. The first checks if $A + B = 0$. The second checks if $A + B = 1$, and so forth. The comparators contain redundant logic repeated across wordlines. [Heald98] shows how to reduce the area by factoring out common terms in a predecoder.

11.2.3 Bitline Conditioning and Column Circuitry

The bitline conditioning circuitry is used to precharge the bitlines high before operation. A simple conditioner consists of a pair of pMOS transistors, as shown in Figure 11.16(a). It is also possible to construct pseudo-nMOS SRAMs with weak pull-up transistors in place of the precharge transistors (Figure 11.16(b)) where no clock is available. Another technique is to precharge through nMOS transistors to $V_{DD} - V_t$ (Figure 11.16(c)). This results in faster single-ended bitline sensing because the bitlines do not swing as much, but reduces noise margins and may require more precharge time.

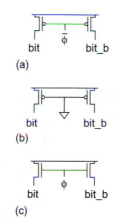

FIG 11.16 Bitline conditioning circuits

Each column must also contain write drivers and read sensing circuits. Figure 11.10 showed two examples of write drivers and Figure 11.8 showed the use of HI-skew inverters to sense reads. Many *sense amplifiers* have been invented to provide faster sensing by responding to a small voltage swing. The differential sense amplifier in Figure 11.17(a) is based on an analog differential pair and requires no clock. The differential gain of the amplifier is $g_{mN1} (r_{oN1} \| r_{oP1})$, as will be discussed in Section 12.6.5. However, the circuit consumes a significant amount of DC power. The clocked sense amplifier in Figure 11.17(b) only consumes power while activated, but requires a timing chain to activate at the proper time. When the sense clock is low, the amplifier is inactive. When the sense amplifier rises, it effectively turns on the cross-coupled inverter pair, which pulls one output low and the other high through *regenerative feedback*. The *isolation transistors* speed up the response by disconnecting the outputs from the highly capacitive bitlines during sensing. See Section 6.4.2 for more discussion of sense amplifier circuits. Power dissipation can be reduced for read operations by turning off the wordlines once sufficient differential voltage has been achieved on the bitlines. This reduces the bitline swing and hence the charge required to restore the bitlines to V_{DD} after sensing. *Current sense amplifiers* that detect a differential current rather than a small voltage swing are another promising technique [Wicht01].

Sense amplifiers are very susceptible to differential noise on the bitlines because they detect small voltage differences. If bitlines are not precharged long enough, residual voltages on the lines from the previ-

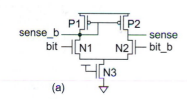

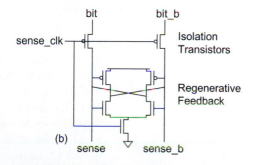

FIG 11.17 Sense amplifiers

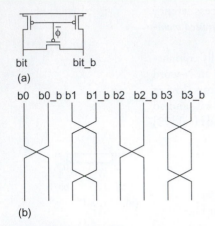

FIG 11.18 Bitline noise reduction through equalizers and twisting

ous read may cause pattern-dependent failure. An equalizer transistor (Figure 11.18(a)) can be added to the bitline conditioning circuits to reduce the required precharge time by ensuring that *bit* and *bit_b* are at nearly equal voltage levels even if they have not precharged quite all the way to V_{DD}. Coupling from transitioning bitlines in neighboring cells may also introduce noise. The bitlines can be *twisted* or *transposed* to cause equal coupling onto both the bitline and its complement, as shown in Figure 11.18(b). For example, careful inspection shows that *b1* couples to *b0_b* for the first quarter of its length, *b2* for the next quarter, *b2_b* for the third quarter, and *b0* for the final quarter. *b1_b* also couples to each of these four aggressors for a quarter of its length, so the coupling will be the same onto both lines.

The sense amplifier offset voltage is the differential input voltage (*bit* − *bit_b*) necessary to produce zero differential output voltage (*sense* − *sense_b*). If *N1* is identical to *N2* and *P1* to *P2*, the sense amplifier will ideally have zero offset voltage. In practice, the offset voltage is nonzero because of statistical dopant fluctuations that affect V_t. The differential input must substantially exceed the offset voltage to be sensed reliably. A typical budget for offset voltage is 50 mV [Amrutur00]. Unfortunately, the threshold variations and offset voltage are not changing very much with technology scaling, so the offset voltage is becoming a larger fraction of the supply voltage, making sense amplifiers less effective [Mizuno94].

Clocked sense amplifiers must be activated at just the right time. If they fire too early, the bitlines may not have developed enough voltage difference to operate reliably. If they fire too late, the SRAM is unnecessarily slow. The clock is generated by circuitry that must match the delay of the decoder, wordlines, and bitlines. This leads to all of the delay matching challenges discussed in Section 7.5.4.1. Many arrays use a chain of inverters, but inverters do not track the delay of the access path very well across process and environmental corners: A margin of more than 30% is often necessary in the typical corner for reliable operation in all corners. Alternatively, the array may use dummy or *replica* cells and bitlines to more closely track the access path. For example, [Amrutur98] describes a 16-kbit SRAM array that eliminates 3 FO4 inverter delays of margin by replacing the inverter chain with an extra replica row and column configured to operate no faster than the slowest real access.

In general, 2^k:1 column multiplexers may be required to extract 2^m bits from the 2^{m+k} bits of each row. The column multiplexers can either act as their own tree decoder or require a separate column decoder to generate select signals. Figure 11.19 shows an 8:1 column multiplexer using a tree decoder. The data is routed through pass transistors enabled by the column address lines. The address decoding is in essence distributed. Decoders for both polarities of the bitline are shown, although one of these can be omitted for single-ended read operations. The read (and usually of lesser importance, write) operations are somewhat delayed by the series pass transistors. Figure 11.20 shows a single-ended 4:1 column mux using a separate column decoder. The multiplexer is faster because data from the bitline must propagate through only one series transistor. The column decoding takes place in parallel with row decoding so it does not impact delay. In both

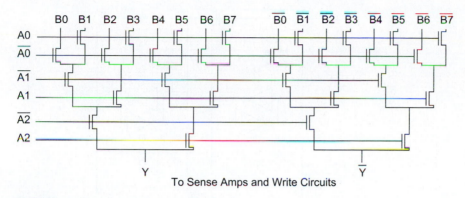

FIG 11.19 Tree decoder column multiplexer

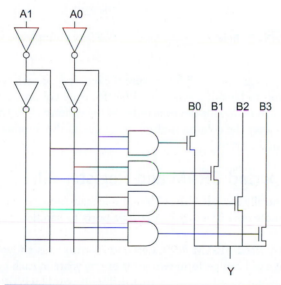

FIG 11.20 Column multiplexer with separate decoder

cases, the outputs may need to be precharged because only nMOS pass transistors are used. Column multiplexers can also use full transmission gates. pMOS pass transistors can be used if a sense amplifier responds to voltages near V_{DD}.

Figure 11.21 shows a complete pair of bits and associated column circuitry for a 2-way multiplexed SRAM. The output of the nMOS-only multiplexer is precharged high. Both the write drivers and the read sensing inverter are connected to the multiplexer outputs.

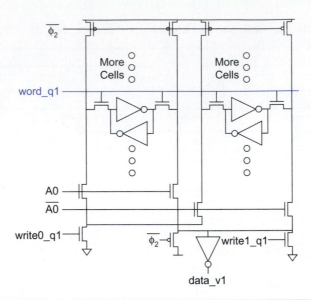

FIG 11.21 Complete pair of columns for 2-way multiplexed SRAM

Column multiplexing is also helpful because the bit pitch of each column is so narrow that it can be difficult to lay out a sense amplifier for each column. After multiplexing, multiple columns are available for the remainder of the column circuitry. Moreover, placing sense amplifiers after the column multiplexers reduces the number of power-hungry amplifiers required in the array.

11.2.4 Multi-ported SRAM and Register Files

Register files are generally fast SRAMs with multiple read and write ports. Data caches in superscalar microprocessors often require multiple ports to handle multiple simultaneous loads and stores.

A simple dual-ported SRAM adds a second wordline, as shown in Figure 11.22 [Horowitz87]. Such a cell can perform two reads or one write in each cycle. The reads are performed by independently selecting different words with the two wordlines. Read becomes a single-ended operation; one read appears on *bit*, while the other appears in complementary form on *bit_b*. For example, asserting *wordA*[7] and *wordB*[3] reads the third word onto *bit* and the complement of the seventh onto *bit_b*. Write still requires both *bit* and *bit_b*, so only a single write can occur. With careful timing, accesses can be performed each half-cycle, permitting two reads in the first phase and a write in the second phase, as commonly required for pipelined processors.

A more general multi-ported SRAM with three write ports and four read ports is shown in Figure 11.23. Merely adding more access transistors causes problems for read stability when multiple ports attempt to read the same cell because all

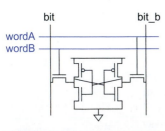

FIG 11.22 Simple dual-ported SRAM

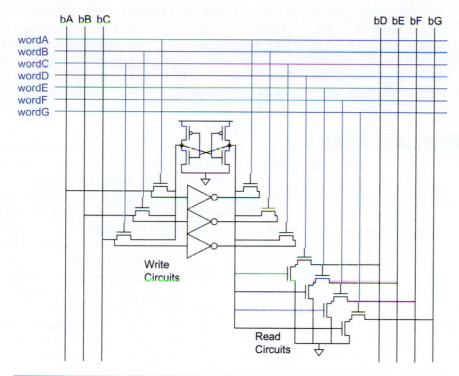

FIG 11.23 General multi-ported SRAM

the access transistors will be driving the cell high. Instead, the bitlines are isolated from the cell during reads. The multi-port SRAM cell still is built around the cross-coupled inverters in the center. Ports *A*, *B*, and *C* perform writes by driving a value onto one side of the inverter pair and its complement onto the other side. Ports *D*, *E*, *F*, and *G* perform reads by discharging the bitline if the cell stores a 0.

Register files for superscalar processors often require an enormous number of ports. For example, the Itanium 2 processor issues up to six integer instructions in a cycle, each of which requires two source registers and a destination. The register file requires two more write ports for late cache data returns, leading to a total of 12 read ports and 8 write ports [Fetzer02]. The area of the large register file is dominated by the mesh of wordlines and bitlines. A rough rule for estimating multiport SRAM cell area is to count the number of tracks for the wordlines and bitlines and then add three in each dimension for internal wiring. The number of bitlines can be reduced by performing single-ended read operations. Similarly, writes can use a local inverter placed in empty space under the wires so only one, rather than two bitlines are required for write ports. The Itanium 2 further reduces the area to only 12 wordlines by time-multiplexing, reading in one half-cycle, and then writing on the other using pulses on the wordlines. [Golden99] and [Heald00] show other designs for the large register files of the AMD Athlon and Sun UltraSparc III, respectively.

A single-ended register file with 16 read ports and 4 write ports has a cell size of 23×23 tracks, or about $184 \times 184 \, \lambda = 33856 \, \lambda^2$. The area can be improved by partitioning the register file into two parts, each with 8 read ports and 4 write ports. Write operations update both register files so the data remains consistent, but half the reads use one register file and half use the other. The cell size is now 15×15 tracks with an area of $14400 \, \lambda^2$ per file, or $28800 \, \lambda^2$ all together. The partitioned register file is not only smaller but also faster because of the shorter bitlines and wordlines.

11.2.5 Large SRAMs

The critical path in a static RAM read cycle includes the clock to address delay time, the row address driver time, row decode time, bitline sense time, and the setup time to any data register. The write operation is usually faster than the read cycle because the bitlines are actively driven by large transistors. However, the bitlines may have to be allowed to recover to their quiescent values before any more access cycles take place.

If the memory array becomes large, the wordlines and bitlines become rather long. The long lines have high capacitance, leading to long delay and high power consumption. Thus, large memories are partitioned into multiple smaller memory arrays, sometimes called *banks* or *subarrays*. Each subarray presents some area overhead for its periphery circuitry, so the size of the subarrays represents a tradeoff between area and speed.

Reading data onto a bitline is much like activating a wide footless dynamic multiplexer because the bitline is pulled down through two series transistors. The wordline transistor acts as the select and the nMOS in the cross-coupled inverter acts as the data signal. Recall that a dynamic multiplexer has a constant logical effort but a parasitic delay proportional to the number of inputs (i.e., words). Without sense amplifiers, the bitlines become quite slow for more than 32 words. Sense amplifiers permit smaller bitline swings, reducing the parasitic delay. With sense amplifiers, bitlines reasonably accommodate up to about 128 words.

The wordline presents an RC delay from the resistance of the wire and gate capacitance of the transistors it drives. This increases with the square of the number of bits on a wordline. Typical SRAMs use up to about 256 bits on each wordline.

Figure 11.24 shows a typical large SRAM partitioned into S 4-kbyte (128-word × 256-bit) subarrays [Amrutur00]. The *divided wordline* decoder operates in three stages: predecoding, global wordline decoding, and local wordline decoding [Yoshimoto83]. The 128 global wordlines are long, but lightly loaded and can use wide, upper-level metal wires with low resistance. The local wordline decoder gates the global wordline with a bank select line to activate the appropriate subarray. During a read, the subarray sense amplifiers detect the swing of the bitline and drive the results onto global datalines. Global sense amplifiers detect small swings on these datalines and drive the results out of the SRAM. During a write, the datalines and bitlines are driven in reverse. Subarrays with many words can also use *divided bitlines* to reduce the diffusion capacitance and improve the delay and power of the subarray.

Large memories with multiple subarrays can simulate more than one access port even if each subarray is single-ported. For example, in a system with two subarrays, even-num-

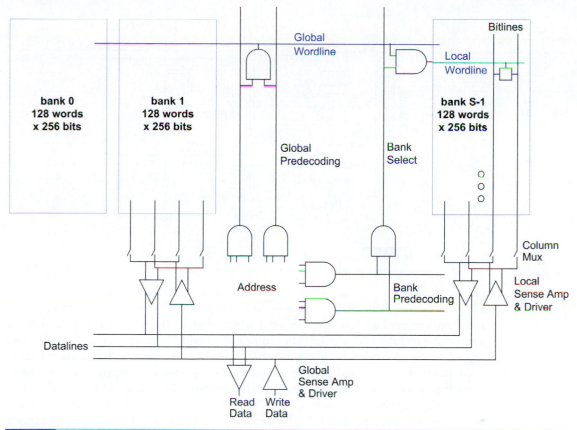

FIG 11.24 Large memory array architecture with subarrays

bered words could be stored in one subarray while odd-numbered words are stored in the other. Two accesses could occur simultaneously if one addresses an even word and another an odd word. If both address an even word, we encounter a *bank conflict* and one access must wait. Increasing the number of banks offers more parallelism and lower probability of bank conflicts.

11.2.6 Logical Effort of RAMs and Register Files

The method of logical effort is helpful to estimate the delay of a static RAM or register file. The critical read path for a small single-ported RAM with no column multiplexing involves the decoder to drive the wordline and the SRAM cell that pulls down the bitline. Figure 11.25 highlights this path for a 2^n word by 2^m-bit memory with total storage of 2^N bits ($N = m + n$).

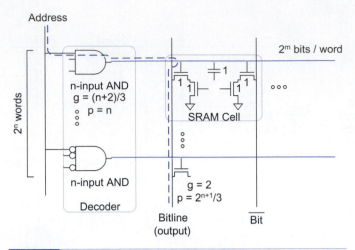

FIG 11.25 Critical path for read of small SRAM

The decoder is modeled as an n-input AND gate taking some combination of true and complemented address inputs. It has a logical effort of $(n + 2)/3$ and parasitic delay of n according to Tables 4.2 and 4.3. The bitline is discharged in the SRAM cell through two series transistors that behave like a dynamic multiplexer. Suppose each cell has two unit-sized access transistors and stray wire capacitance approximately equal to another unit-sized transistor, for a total capacitance of $3C$ presented by each cell to the wordline. Because there are two transistors in series, the cell delivers about half the current of a unit inverter with input capacitance $3C$. Hence, the logical effort is 2 because the cell delivers half the current of an inverter with the same input capacitance. Suppose each cell presents $1C$ of diffusion capacitance on the bitline, so the total bitline capacitance is 2^nC. The cell has an effective resistance of $2R$ discharging the bitline through two series unit transistors. Hence, the bitline has a parasitic delay of $2^{n+1}RC$. Normalized by $\tau = 3RC$, this gives $p = 2^{n+1}/3$.

Putting these two stages together, the path logical effort is $G = (n + 2)/3 \cdot 2$. If the true and complementary bitline outputs each drive capacitance equal to half that seen by the address inputs, the path electrical effort is $H = 1/2$. Within the path are a 2^n-way branch as each address bit is needed by each wordline decoder and another 2^m-way branch as each wordline drives all the bits on that word. Hence, the branching effort is $B = 2^N$. The path effort delay is $F = GBH = 2^N(n + 2)/3$. The parasitic delay is $P = n + 2^{n+1}/3$. The best number of stages is approximately $\log_4 F = N/2 + \log_4 [(n + 2)/3]$. These stages would include buffers in the address driver, multiple levels of gates in the decoder, buffers to drive the wordline, and an inverter on the bitline output. The path delay is $D = 4\log_4 F + P = 2N + 4\log_4 [(n + 2)/3] + n + 2^{n+1}/3$. For a 32-word × 32-bit register file, $n = 5$, $N = 10$, and $D = 48.8 \tau = 9.8$ FO4 inverter delays.

This model is clearly an oversimplification. The n-input AND gate is usually constructed out of a chain of low fan-in gates, but this only slightly improves its logical effort. We also neglect the effort of the clock gating to drive the wordlines on the clock edge. We assume the RAM is small enough that sense amplifiers are not used and that the wire resistance is negligible. The pull-down transistor inside the SRAM cell may be larger than the access transistor. Nevertheless, the model offers insights into the number of stages that the memory should use and its approximate delay. For example, it shows that, without sense amplifiers, putting too many words on a bitline causes excessive parasitic delay.

[Amrutur00] models the delay of large SRAMs using logical effort in substantially more detail than can be repeated here. The overall delay includes components contributed by both the gates and the wire RC. In a well-designed 2^N-bit SRAM ($N \geq 16$) using static CMOS decoders, the gate delay component is approximately $1.2 \cdot N - 4$ FO4 inverter delays. More aggressive decoders using domino or race-based NOR techniques from Section 11.2.2 can reduce this delay by about 15% [Amrutur01]. Wire delay becomes important for RAMs beyond the 1-Mbit ($N = 20$) capacity. A lower bound for wire delay is set by the speed of light at about 1.75 FO4 for 4-Mbit memories. This delay doubles for each quadrupling in memory size. In practice, the wire delay depends on the wire width and thickness and repeater strategy, but can be several times this lower bound. In processes beyond the 100 nm generation, sense amplifiers will need relatively larger bitline swings because their offset voltages are not scaling with the supply voltage. This will add several FO4 inverter delays to the bitline-sensing time.

11.2.7 Case Study: Itanium 2 Cache

Figure 7.92 showed a die photograph of the 1 GHz Itanium 2 processor [Weiss02] illustrating a large (24 Mbit) embedded SRAM used as the Level 3 cache in a 180 nm process. The cache is partitioned into 135 subarrays: 128 are used for data, while 5 are used for ECC and 2 are spares available to replace bad data. All of the subarrays are accessed in parallel through 8-bit paths, so the cache line size is 1024 bits.

Each subarray in turn is divided into eight 24-kbit banks organized as 96 words by 256 bits. The subarray decoder produces 96 global wordlines and 8 bank select lines. The local decoder activates the local wordline in the selected bank and the column multiplexer selects 8 bits from the 256-bit bank. One more level of 4:1 multiplexers returns the data in 2-bit chunks over four cycles to reduce the amount of global wiring. One load and one store can take place every four cycles at 1.2 GHz. Of this time, roughly half is used for the subarray access and half is used for distributing the address and data.

The individual 6T SRAM cells have an area of approximately 5.9 μm^2. The cache has a flexible floorplan to flow around the processor core and accommodate floorplan changes late in the design cycle. Overall, the cache has an 85% area efficiency, i.e., the 6T SRAM cells account for 85% of the area, while the decoders and other periphery circuitry account for only 15%.

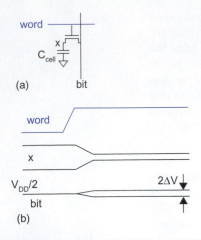

(a)

(b)

FIG 11.26 DRAM cell read operation

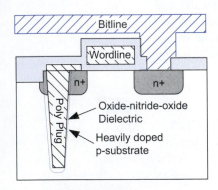

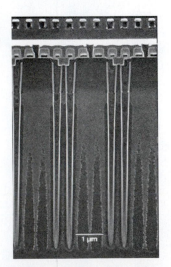

FIG 11.27 Trench capacitor

11.3 DRAM

Dynamic RAMs (DRAMs) store their contents as charge on a capacitor rather than in a feedback loop. Thus, the basic cell is substantially smaller than SRAM, but the cell must be periodically read and refreshed so that its contents do not leak away. Commercial DRAMs are built in specialized processes optimized for dense capacitor structures. They offer an order of magnitude greater density (bits/cm²) than high-performance SRAM built in a standard logic process, but they also have much higher latency. DRAM circuit design is a very specialized art and is the topic of excellent books such as [Keeth01]. This section provides an overview of the general technique.

A 1-transistor (1T) dynamic RAM cell consists of a transistor and a capacitor, as shown in Figure 11.26(a). Like SRAM, the cell is accessed by asserting the wordline to connect the capacitor to the bitline. On a read, the bitline is first precharged to $V_{DD}/2$. When the wordline rises, the capacitor shares its charge with the bitline, causing a voltage change ΔV that can be sensed, as shown in Figure 11.26(b). The read disturbs the cell contents at x, so the cell must be rewritten after each read. On a write, the bitline is driven high or low and the voltage is forced onto the capacitor. Some DRAMs drive the wordline to $V_{DDP} = V_{DD} + V_t$ to avoid a degraded level when writing a '1.'

The DRAM capacitor C_{cell} must be as physically small as possible to achieve good density. However, the bitline is contacted to many DRAM cells and has a relatively large capacitance C_{bit}. Therefore, the cell capacitance is typically much smaller than the bitline capacitance. According to the charge-sharing equation, the voltage swing on the bitline during readout is

$$\Delta V = \frac{V_{DD}}{2} \frac{C_{cell}}{C_{cell} + C_{bit}} \tag{11.1}$$

We see that a large cell capacitance is important to provide a reasonable voltage swing. It also is necessary to retain the contents of the cell for an acceptably long time and to minimize soft errors, as was mentioned in Section 4.8.7. For example, 30 fF is a typical target. The most compact way to build such a high capacitance is to extend into the third dimension. For example, Figure 11.27 shows a cross-section and scanning electron microscope (SEM) image of *trench capacitors* etched under the source of the transistor. The walls of the trench are lined with an oxide-nitride-oxide dielectric. The trench is then filled with a polysilicon conductor that serves as one terminal of the capacitor attached to the transistor drain, while the heavily doped substrate serves as the other terminal. A variety of three-dimensional capacitor structures have been used in specialized DRAM processes that are not available in conventional CMOS processes.

11.3.1 Subarray Architectures

Like SRAMs described in Section 11.2.5, large DRAMs are divided into multiple subarrays. The subarray size represents a tradeoff between density and performance. Larger subarrays amortize the decoders and sense amplifiers across more cells and thus achieve better density. But they also are slow and have small bitline swings because of the high wordline and bitline capacitance. A typical subarray size is 256 words by 512 bits, as shown in Figure 11.28.

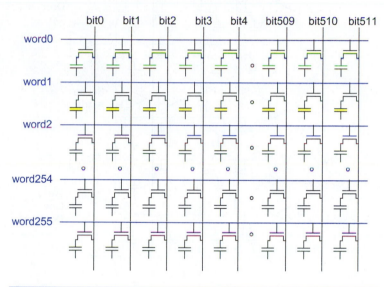

FIG 11.28 DRAM subarray

A subarray of this size has an order of magnitude higher capacitance on the bitline than in the cell, so the bitline voltage swing ΔV during a read is very small. The array uses a sense amplifier to compare the bitline voltage to that of an idle bitline (precharged to $V_{DD}/2$). The sense amplifier must also be very compact to fit the tight pitch of the array. The low-swing bitlines are very sensitive to noise. Three bitline architectures, *open*, *folded*, and *twisted*, offer different compromises between noise and area.

Until the 64 kbit generation, DRAMs used the open bitline architecture shown in Figure 11.29. In this architecture, the sense amplifier receives one bitline from each of two subarrays. The wordline is only asserted in one array, leaving the bitlines in the other array floating at the reference voltage. The arrays are very dense. However, any noise that affects one array more than the other will appear as differential noise at the sense amplifier. Thus, open bitlines have unacceptably low signal-to-noise ratios for high-density DRAM.

The folded bitline architecture is shown in Figure 11.30. In this architecture, each bitline connects to only half as many cells. Adjacent bitlines are organized in pairs as inputs to the sense amplifiers. When a wordline is asserted, one bitline will switch while

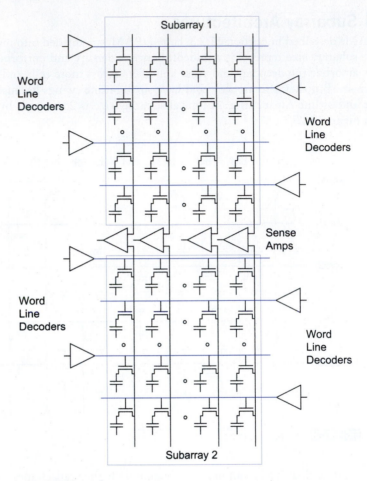

FIG 11.29 Open bitlines

its neighbor serves as the quiet reference. Many noise sources will couple equally onto the two adjacent bitlines so they tend to appear as common mode noise that is rejected by the sense amplifier. This noise advantage comes at the expense of greater layout area. Figure 11.31 shows a clever layout for a 6 × 8 folded bitline subarray that is only 33% larger than an open bitline layout. Observe how DRAM processes push the design rules and use diagonal polysilicon to reduce area. Notice how pairs of cells in the layout share a single bitline contact to minimize the bitline capacitance.

Unfortunately, the folded bitline architecture is still susceptible to noise from a neighboring switching bitline that capacitively couples more strongly onto one of the bitlines in the pair. Capacitive coupling is very significant in modern processes. The twisted bitline architecture [Hidaka89] solves this problem by swapping the positions of the folded bitlines part way along the array in much the same way as SRAM bitlines were twisted in Figure 11.18(b). The twists cost a small amount of extra area within the array.

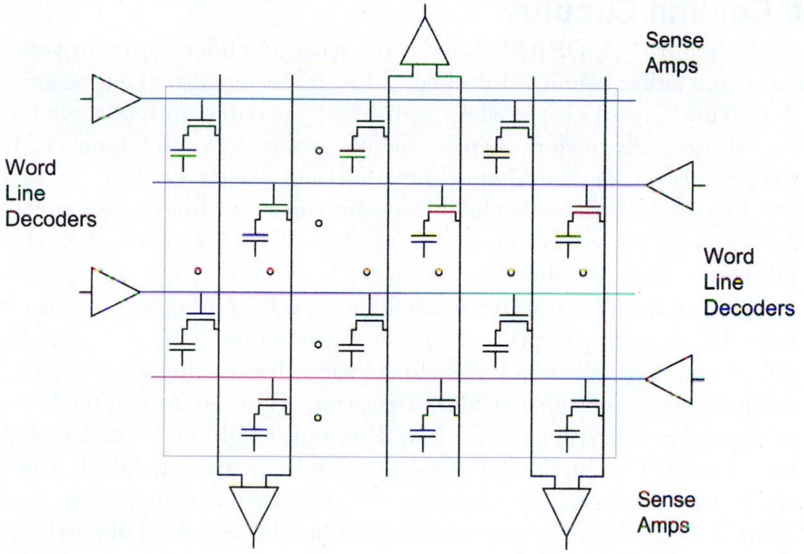

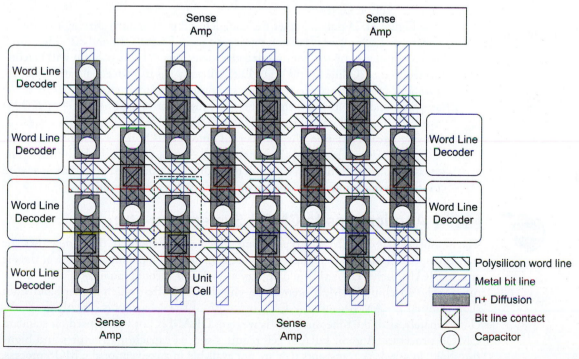

FIG 11.30 Folded bitlines

FIG 11.31 Layout of folded bitline subarray

Sense Amps

Word Line Decoders

Word Line Decoders

Sense Amps

Sense Amp

Sense Amp

Word Line Decoder

Word Line Decoder

Word Line Decoder

Word Line Decoder

Word Line Decoder

Word Line Decoder

Sense Amp

Sense Amp

Unit Cell

Polysilicon word line

Metal bit line

n+ Diffusion

Bit line contact

Capacitor

11.3.2 Column Circuitry

The column circuitry in a DRAM includes the sense amplifiers, write drivers, column multiplexing, and bitline conditioning circuits. In a folded or twisted bitline architecture, the column circuitry is placed on both sides of the array so that it can be laid out on four times the pitch of a single column, as was shown in Figure 11.31. Part of the circuitry can be shared between two adjacent subarrays.

Figure 11.32(a) shows a basic sense amplifier built from cross-coupled inverters with supplies tied to control voltages. Initially, the two bitlines *bit* and *bit** are precharged to $V_{DD}/2$, the bottom voltage V_n is at $V_{DD}/2$, and the top voltage V_p is at 0 so all of the transistors in the amplifier are OFF. During a read, one of the bitlines will change by a small amount while the other floats at $V_{DD}/2$. V_n is then pulled low. As it falls to a threshold voltage below the higher of the two bitline voltages, the cross-coupled nMOS transistors will begin to pull the lower bitline voltage down to 0. After a small delay, V_p is pulled high. The cross-coupled pMOS transistors pull the higher bitline voltage up to V_{DD}. For example, Figure 11.32(b) shows the waveforms while reading a '0' on *bit* while using *bit** as a reference. Driving the active bitline to one of the rails has the side effect of rewriting the cell with the value that was just read.

Figure 11.33 shows a bitline conditioning circuit that precharges and equalizes a pair of bitlines to $V_{DD}/2$ when *EQ* is asserted. This consumes very little power because the voltage is reached by sharing charge between one bitline at V_{DD} and the other at GND.

Figure 11.34 puts together the complete column circuitry serving two folded subarrays. Each subarray column produces a pair of signals, *bit* and *bit**. The *CSEL* signal, produced by the column decoder, determines if this column will be connected to the I/O line for the array. Each subarray has its own equalization transistors and pMOS portion of the sense amplifier. However, the nMOS sense amplifier and I/O lines are shared between the subarrays. Either *ISO1* or *ISO2* is asserted to connect one subarray to the I/O lines while leaving the other isolated. During a read operation, the data is read onto the I/O lines. During a write, one I/O line is driven high and the other low to force a value onto the bitlines. The cross-coupled pMOS transistors pull the bitlines to a full logic level during a write to compensate for the threshold drop through the isolation transistor.

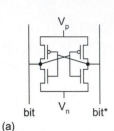

(a)

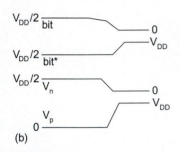

(b)

FIG 11.32 Sense amplifier

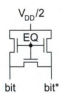

FIG 11.33 Bitline conditioning

11.3.3 Applications to CMOS Systems-on-chip

The number of bits of RAM on a typical chip is increasing faster than the number of logic transistors. Recall that the Itanium II in Figure 7.92 devotes approximately 50% of the die area to SRAM cache. DRAM would be very attractive if it could more compactly substitute for the on-chip SRAM. Moreover, on-chip DRAM offers much higher bandwidth than external DRAM because the full bandwidth of the subarrays is directly available. Unfortunately, as of the time this book is written, DRAM is not widely used on standard CMOS processes. A good DRAM cell requires special capacitor structures and high-threshold, low-leakage transistors that are not available in a conventional CMOS process.

CMOS logic requires low-threshold transistors and many metal layers not available in a conventional DRAM process. Adding these process steps increases the manufacturing cost. If cost-effective solutions can be found, DRAM may become an important technology for the CMOS designer [Chandrakasan01].

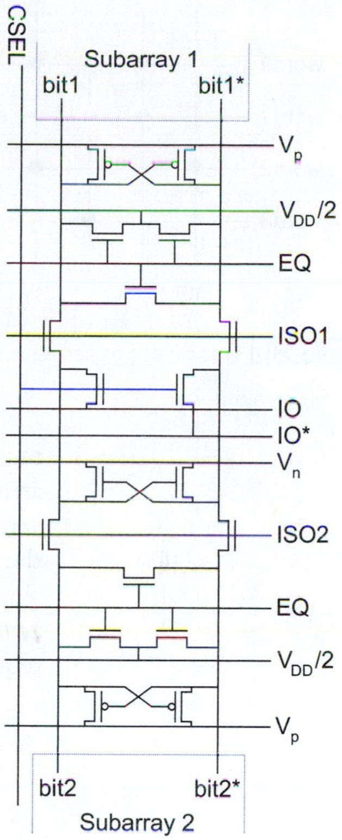

11.4   Read-only Memory

Read-only Memory (ROM) cells can be built with only one transistor per bit of storage. A ROM is a nonvolatile memory structure in that the state is retained indefinitely — even without power. A ROM array is commonly implemented as a single-ended NOR array using any of the NOR gate structures studied so far, including the pseudo-nMOS and the footless dynamic NOR gate. As in SRAM cells and other footless dynamic gates, the wordline input must be low during precharge on dynamic NOR gates. In situations where DC power dissipation is acceptable and the speed is sufficient, the pseudo-nMOS ROM is the easiest to design, requiring no timing. The DC power dissipation can be significantly reduced in multiplexed ROMs by placing the pull-up transistors after the column multiplexer.

Figure 11.35 shows a 4-word by 6-bit ROM using pseudo-nMOS pull-ups with the following contents:

```
word0: 010101
word1: 011001
word2: 100101
word3: 101010
```

FIG 11.34 Column circuitry

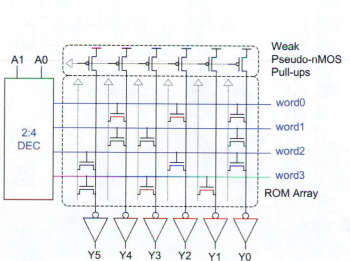

FIG 11.35 Pseudo-nMOS ROM

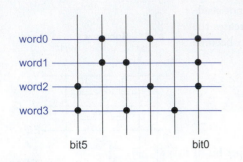

FIG 11.36 Dot diagram representation of ROM

The contents of the ROM can be symbolically represented with a dot diagram in which dots indicate the presence of 1's, as shown in Figure 11.36. The dots actually correspond to nMOS pull-down transistors connected to the bitlines, but the outputs are inverted.

Mask-programmed ROMs can be configured by the presence or absence of a transistor or contact, or by a threshold implant that turns a transistor permanently OFF where it is not needed. Omitting transistors has the advantage of reducing capacitance on the wordlines and power consumption. Programming with metal contacts was once popular because such ROMs could be completely manufactured except for the metal layer, and then programmed according to customer requirements through a metallization step. The advent of EEPROM and Flash memory chips has reduced demand for such mask-programmed ROMs. Figure 11.37 shows a layout for the 4-word by 6-bit ROM array. The wordlines run horizontally in polysilicon, while the bitlines and grounds run vertically in metal1. Notice how each ground is shared between a pair of cells. Each bit of the ROM occupies a $12 \times 8 \ \lambda$ cell[1]. Polysilicon wordlines are only appropriate for small or slow ROMs. A larger ROM can run metal2 straps over the polysilicon and contact the two periodically (e.g., every eight columns). Occasional substrate contacts are also required.

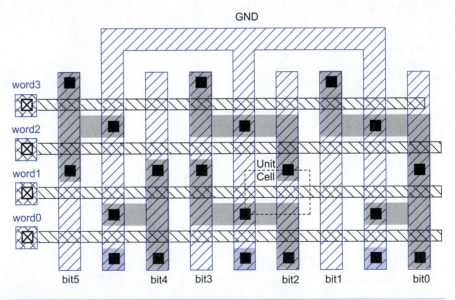

FIG 11.37 ROM array layout

---

[1]The cell can be reduced to 11 x 7 λ by running the ground line in diffusion and by reducing the width and spacing to 3 λ.

Row decoders for ROMS are similar to those for RAMs except that they are usually very constrained by the ROM wordline pitch. Figure 11.38 shows how each output of a 2:4 decoder can be shoehorned into a single horizontal track using vertical polysilicon true and complementary address lines and metal supply lines. Column decoders for ROMs are usually simpler than those for RAMs because single-ended sensing is commonly employed.

Figure 11.39 shows a complete pseudo-nMOS ROM including row decoder, cell array, pMOS pull-ups, and output inverters.

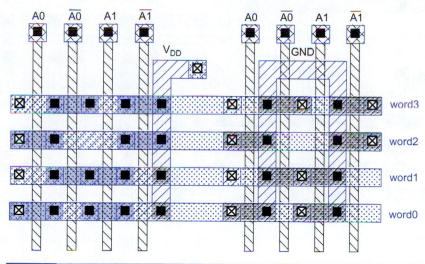

FIG 11.38 Row decoder layout on tight pitch

11.4.1 Programmable ROMs

It is often desirable for the user to be able to program or reprogram a ROM after it is manufactured. As discussed in Section 11.1, this contradicts the "Read Only" nature of the device; ROM has in practice become synonymous with *nonvolatile*, not read-only memory. Programming/writing speeds are generally slower than read speeds for ROMs. Four types of nonvolatile memories include *Programmable* ROMs (PROMs), *Erasable Programmable* ROMs (EPROMs), *Electrically Erasable Programmable* ROMs (EEPROMs), and *Flash* memories. All of these memories require some enhancements to a standard CMOS process: PROMs use fuses while EPROMs, EEPROMs, and Flash use charge stored on a floating gate.

Programmable ROMs can be fabricated as ordinary ROMs fully populated with pull-down transistors in every position. Each transistor is placed in series with a fuse made of polysilicon, nichrome, or some other conductor that can be burnt out by applying a high current. The user typically configures the ROM in a specialized PROM programmer before putting it in the system. As there is no way to repair a blown fuse, PROMs are also referred to as *one-time programmable* memories.

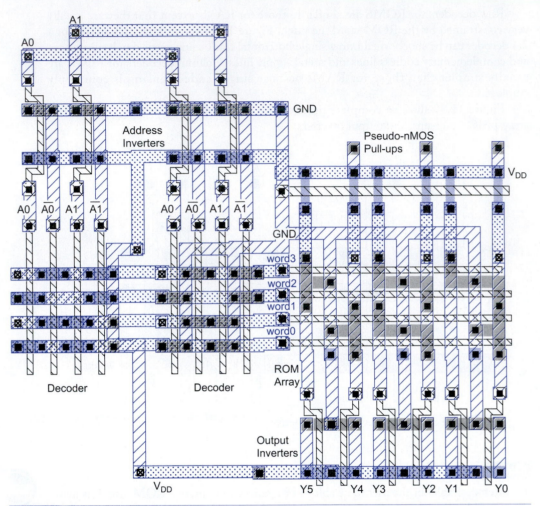

FIG 11.39 Complete ROM layout

As technology has improved, reprogrammable nonvolatile memory has largely displaced PROMs. These memories, including EPROM, EEPROM, and Flash, use a second layer of polysilicon to form a floating gate between the primary gate and the channel, as shown in Figure 11.40. The floating gate is a good conductor, but it is not attached to anything. Applying a high voltage to the upper gate causes electrons to jump through the thin oxide onto the floating gate through the processes called *avalanche injection* or *Fowler–Nordheim tunneling*. Injecting the electrons induces a negative voltage on the floating gate, effectively increasing the threshold voltage of the transistor to the point that it is always OFF.

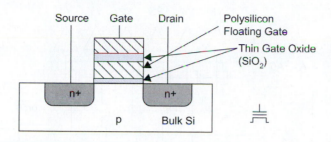

FIG 11.40 Cross-section of floating gate nMOS transistor

[Itoh01k] and [Rabaey03] offer a good overview of these memories. In brief, EPROM is programmed electrically, but it is erased through exposure to ultraviolet light that knocks the electrons off the floating gate. It offers a dense cell, but it is inconvenient to erase and reprogram. EEPROM and Flash can be erased electrically without being removed from the system. EEPROM offers fine-grained control over which bits are erased, while Flash is erased in bulk. EEPROM cells are larger to enable them to provide this versatility, so Flash has become the most economical form of convenient nonvolatile storage. For example, Flash memory cards are widely used in digital cameras to store pictures even after the camera is turned off. Flash is also useful for firmware or configuration data because it can be rewritten to upgrade a system in the field without opening the case or removing parts.

11.4.2 NAND ROMs

The ROM from Figure 11.35 is called a NOR ROM because each of the bitlines is just a pseudo-nMOS NOR gate. The bitline pulls down when a wordline attached to any of the transistors is asserted high. The size of the cell is limited by the ground line. Figure 11.41 shows a NAND ROM that uses active-low wordlines. Transistors are placed in series and the transistors on the nonselected rows are ON. If no transistor is associated with the selected word, the bitline will pull down. If a transistor is present, the bitline will remain high.

Figure 11.42(a) shows a layout of the NAND ROM. The cell size is only $7 \times 8 \lambda$. The contents are specified by using either a transistor or a metal jumper in each bit position. The contacts limit the cell size. Figure 11.42(b) shows an even smaller layout in which transistors are located at every position. In this design, an extra implantation step can be used to create a negative threshold voltage, turning certain transistors permanently ON where they are not needed. In such a process, the cell size reduces to only $6 \times 5 \lambda$, assuming that the decoder and bitline circuitry can be built on such a tight pitch.

A disadvantage of the NAND ROM is that the delay grows quadratically with the number of series transistors discharging the bitline. NAND structures with more than 8–16 series transistors become extremely slow, so NAND ROMs are often broken into multiple small banks with a limited number of series transistors. Nevertheless, these NAND

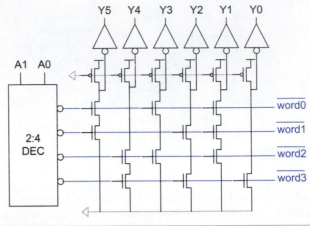

FIG 11.41 Pseudo-nMOS NAND ROM

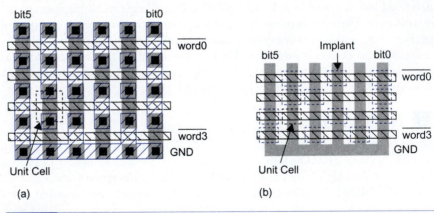

FIG 11.42 NAND ROM array layouts

structures are attractive for Flash memories in which density and cost are more important than access time.

11.5 Serial Access Memories

Using the basic SRAM cell and/or registers, we can construct a variety of serial access memories including shift registers and queues. These memories avoid the need for external logic to track addresses for reading or writing.

11.5.1 Shift Registers

A *shift register* is commonly used in signal-processing applications to store and delay data. Figure 11.43(a) shows a simple 4-stage 8-bit shift register constructed from 32 flip-flops. As there is no logic between the registers, particular care must be taken that hold times are satisfied. Flip-flops are rather big, so large, dense shift registers use dual-port RAMs instead. The RAM is configured as a circular buffer with a pair of counters specifying where the data is read and written. The read counter is initialized to the first entry and the write counter to the last entry on reset, as shown in Figure 11.43(b). Alternately, the counters in an *N*-stage shift register can use two 1-of-*N* hot registers to track which entries should be read and written. Again, one is initialized to point to the first entry and the other to the last entry. These registers can drive the wordlines directly without the need for a separate decoder, as shown in Figure 11.43(c).

One variant of a shift register is a *tapped delay line* that offers a variable number of stages of delay. Figure 11.44 shows a 64-stage tapped delay line that could be used in a video processing system. Delay blocks are built from 32-, 16-, 8-, 4-, 2-, and 1-stage shift registers. Multiplexers control pass-around of the delay blocks to provide the appropriate total delay.

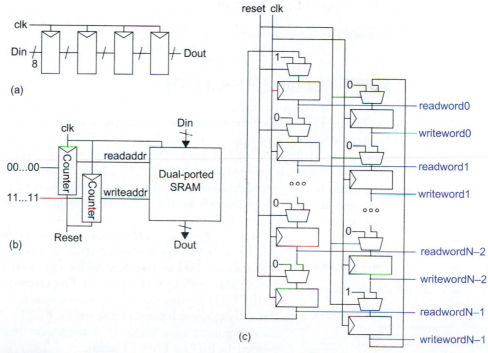

FIG 11.43 Shift registers

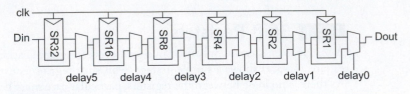

FIG 11.44 Tapped delay line

Another variant is a serial/parallel memory. Figure 11.45(a) shows a 4-stage Serial In Parallel Out (SIPO) memory and Figure 11.45(b) shows a 4-stage Parallel In Serial Out (PISO) memory. These are also often useful in signal processing and communications systems.

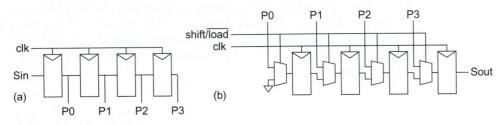

FIG 11.45 Serial/parallel memories

11.5.2 Queues (FIFO, LIFO)

Queues allow data to be read and written at different rates. Figure 11.46 shows an interface to a queue. The read and write operations each are controlled by their own clocks that may be asynchronous. The queue asserts the *FULL* flag when there is no room remaining to write data and the *EMPTY* flag when there is no data to read. Because of other system delays, some queues also provide ALMOST-FULL and ALMOST-EMPTY flags to communicate the impending state and halt write or read requests. The queue internally maintains read and write pointers indicating which data should be accessed next. As with a shift register, the pointers can be counters or 1-of-N hot registers.

First In First Out (*FIFO*) queues are commonly used to buffer data between two asynchronous streams. Like a shift register, the FIFO is organized as a circular buffer. On reset, the read and write pointers are both initialized to the first element and the FIFO is EMPTY. On a write, the write pointer advances to the next element. If it is about to catch the read pointer, the FIFO is FULL. On a read, the read pointer advances to the next element. If it catches the write pointer, the FIFO is EMPTY again.

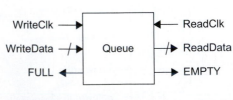

FIG 11.46 Queue

Last In First Out (*LIFO*) queues, also known as *stacks*, are used in applications such as subroutine or interrupt stacks in microcontrollers. The LIFO uses a single pointer for both read and write. On reset, the pointer is initialized to the first element and the LIFO is EMPTY. On a write, the pointer is incremented. If it reaches the last element, the LIFO is FULL. On a read, the pointer is decremented. If it reaches the first element, the LIFO is EMPTY again.

11.6 Content-addressable Memory

Figure 11.47 shows the symbol for a content-addressable memory (CAM) [Grosspietsch92, Schultz95, Miyatake01]. The CAM acts as an ordinary SRAM that can be read or written given adr and data, but also performs *matching* operations. Matching asserts a *matchline* output for each word of the CAM that contains a specified *key*.

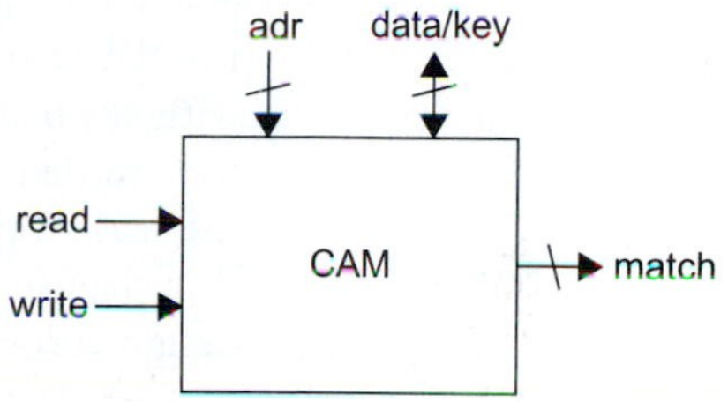

A common application of CAMs is translation lookaside buffers (TLBs) in microprocessors supporting virtual memory. The virtual address is given as the key to the TLB CAM. If this address is in the CAM, the corresponding matchline is asserted. This matchline can serve as the wordline to access a RAM containing the associated physical address, as shown in Figure 11.48. A NOR gate processing all of the matchlines generates a *Miss* signal for the CAM. Note that the *read*, *write*, and *adr* lines for updating the TLB entries are not drawn.

10T and 9T implementations of the CAM cell are shown in Figure 11.49. The cells consist of a normal SRAM cell with additional transistors to perform the match. Multiple CAM cells in the same word are tied to the same matchline. The matchline is either precharged or pulled high as a distributed pseudo-nMOS gate. The key is placed on the bitlines. In Figure 11.49(a), if the key and the value stored in the cell differ, the matchline will be pulled down. Only if all of the key bits match all of the bits stored in the word of memory will the matchline for that word remain high. The key can contain a "don't care"

FIG 11.47 Content-addressable memory

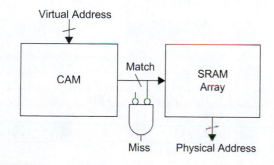

FIG 11.48 Translation Lookaside Buffer (TLB) using CAM

by setting both *bit* and *bit_b* low. Figure 11.50 shows a layout of this cell in a 56 × 43 λ area; CAMs generally have about twice the area of SRAM cells. Figure 11.49(b) shows another CAM cell design with one fewer transistor. *N1* and *N2* perform an XOR of the key and cell data. If the values disagree, *N3* is turned on to pull down the wordline. However, the gate of *N3* sees a degraded high logic level.

Figure 11.51 shows a complete 4 × 4 CAM array. Like an SRAM, it consists of an array of cells, a decoder, and column circuitry. However, each row also produces a dynamic matchline. The matchlines are precharged with the clocked pMOS transistors. The *miss* signal is produced with a distributed pseudo-nMOS NOR.

When the matchlines are used to access a RAM, the monotonicity problem must be considered. Initially, all the matchlines are high. During CAM operation, the lines pull down, leaving at most one line asserted to indicate which row contains the key. However, the RAM requires a monotonically rising wordline. Figure 11.52 refines Figure 11.48 with strobed AND gates driving the wordlines as early as possible after the matchlines have settled. The strobe can be timed with an inverter chain or replica delay line in much the same way that the sense amplifier clock for a SRAM was generated in Section 11.2.3. As usual, self-timing margin must be provided so the circuit operates correctly across all design corners.

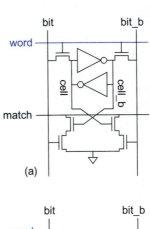

(a)

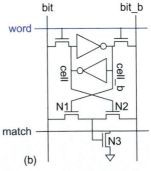

(b)

FIG 11.49 CAM cell implementations

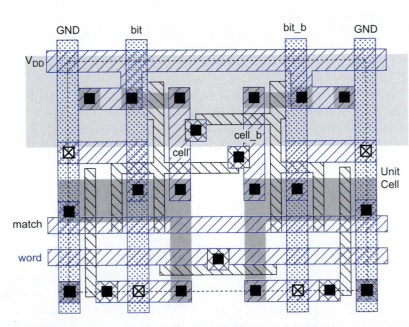

FIG 11.50 CAM cell layout. Color version on inside front cover.

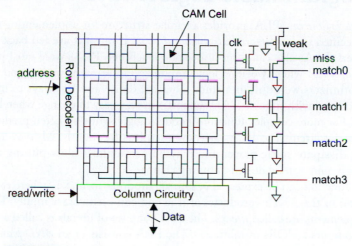

FIG 11.51 4 × 4 CAM array

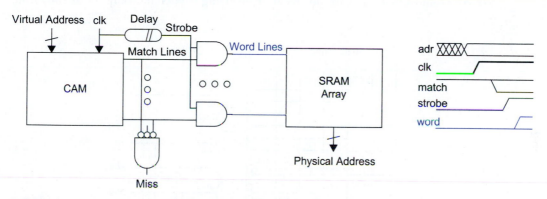

FIG 11.52 Refined TLB path with monotonic wordlines

Large CAMs can use many of the same techniques as large RAMs, including sense amplifiers and multiple subarrays. They tend to consume relatively large amounts of power because the bitlines and matchlines may all transition during a parallel search. [Miyatake01] describes a design of a mid-sized CAM using a pMOS match-line driver to reduce the swing of the matchlines and save power.

11.7   Programmable Logic Arrays

A *programmable logic array* (PLA) provides a regular structure for implementing combinational logic specified in *sum-of-products canonical form*. If outputs are fed back to inputs through registers, PLAs also can form finite state machines. PLAs were most popular in the early days of VLSI when two-level logic minimization was well understood, but multilevel logic optimizers were still immature. They are dense and fast ways to implement simple functions, and with suitable CAD support, are very easy to change when logic bugs are discovered. For more complex functions, modern logic synthesis often produces faster and more compact circuits, so PLAs are now used less frequently. Moreover, pseudo-nMOS PLAs dissipate static power and may be particularly slow pulling up, while dynamic PLAs require careful design of timing chains.

Any logic function can be expressed in sum-of-products form, i.e., where each output is the OR (sum) of the ANDs (products) of true and complementary inputs. The inputs and their complements are called *literals*. The AND of a set of literals is called a *product* or *minterm*. The outputs are ORs of minterms. The PLA consists of an *AND plane* to compute the minterms and an *OR plane* to compute the outputs.

NOR gates are particularly efficient in pseudo-nMOS and dynamic logic because they use only parallel, never series, transistors. Hence, we use DeMorgan's law to replace the AND and OR gates with NORs after inverting inputs and outputs, as shown in

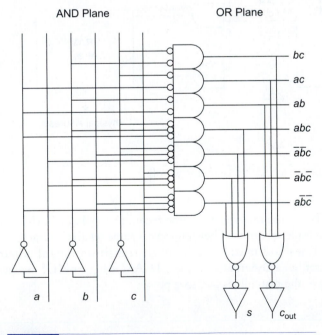

FIG 11.53 NOR/NOR representation of PLA

Example

Write the equations for a full adder in sum-of-products form. Sketch a 3-input, 2-output PLA implementing this logic.

Solution: Figure 11.54 shows the PLA. The logic equations are

$$s = a\bar{b}\bar{c} + \bar{a}b\bar{c} + \bar{a}\bar{b}c + abc$$

$$c_{out} = ab + bc + ac$$

(11.2)

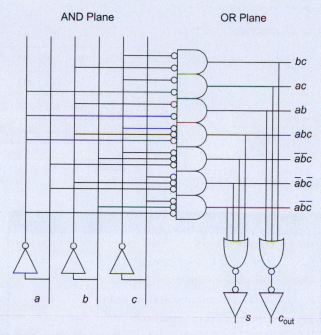

FIG 11.54 AND/OR representation of PLA

Figure 11.53. For brevity, we often represent the PLA with a dot diagram, shown in Figure 11.55. Observe that a ROM and a PLA are very similar in form. The ROM decoder is equivalent to an AND plane generating all 2^n minterms. The ROM array corresponds to an OR plane producing the outputs.

A generic floorplan for a simple PLA is shown in Figure 11.56. This has been designed as a set of tiles, designated by letters given in Table 11.1. Experienced designers often add a few unused rows and columns to their PLAs to accommodate last-minute design changes without changing the overall footprint of the PLA.

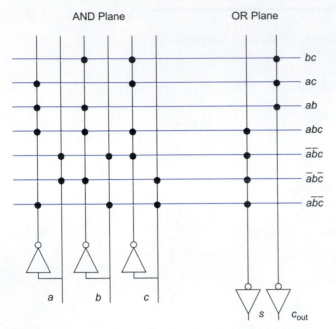

AND Plane   OR Plane

bc
ac
ab
abc
$\overline{ab}c$
$\overline{a}b\overline{c}$
$a\overline{b}\overline{c}$

a   b   c

s   c_{out}

FIG 11.55 Dot diagram representation of PLA

TL	TA	TA	TM	TO	TO	TR
LA	AN	AN	AO	OR	OR	RO
LA	AN	AN	AO	OR	OR	RO
BL	BA	BA	BM	BO	BO	BR

FIG 11.56 Generic PLA floorplan

Table 11.1	PLA cell tiles
Tile	**Function**
AN	AND plane programming cell
OR	OR plane programming cell
AO	AND-OR communication cell
TA	Top AND plane input cell
BA	Bottom AND plane input cell
TO	Top OR plane output cell
BO	Bottom OR plane output cell
LA	Left AND plane cell
RO	Right OR plane cell
BL	Bottom left cell
BM	Bottom middle cell
BR	Bottom right cell
TL	Top left cell
TM	Top middle cell
TR	Top right cell

The most straightforward PLA design uses a pseudo-nMOS NOR gate. Figure 11.57 shows the circuit diagram for the full adder PLA. Design of the pseudo-nMOS NOR gates should follow the guidelines given in Section 6.2.2. Advantages of this PLA include simplicity and small size. Disadvantages include the static power dissipation of the NOR gates and slow pull-up response. A clocked register can be added if desired. Figure 11.58 shows a layout for the pseudo-nMOS PLA. The transistor gates are run in polysilicon and could be strapped with metal2. Observe how ground lines can be shared between pairs of minterms and outputs so that each minterm and output can be placed on a 1.5 track pitch. The inverters require careful layout to fit the tight pitch. You can see how the layout could be assembled by tiling the various cells listed in Table 11.1.

Dynamic PLAs eliminate the contention current and are slightly faster than their pseudo-nMOS counterparts. Figure 11.59(a) shows a PLA using footed dynamic NORs for both the AND and OR planes. Unfortunately, the AND plane must drive the OR plane directly, violating monotonicity. The OR plane must take a clock phase that is delayed until the minterms adequately discharge (to below V_t). This clock is often generated with a replica delay line that is guaranteed to be no faster than the slowest minterm in the AND plane. Moreover, the OR plane outputs must be captured before the AND plane precharges so that the results are not corrupted. To accomplish this, the PLA may be supplied by clocks similar to those shown in Figure 11.59(b).

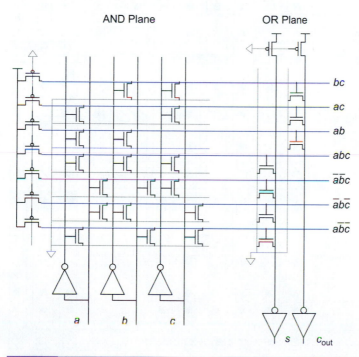

FIG 11.57 Pseudo-nMOS PLA schematic

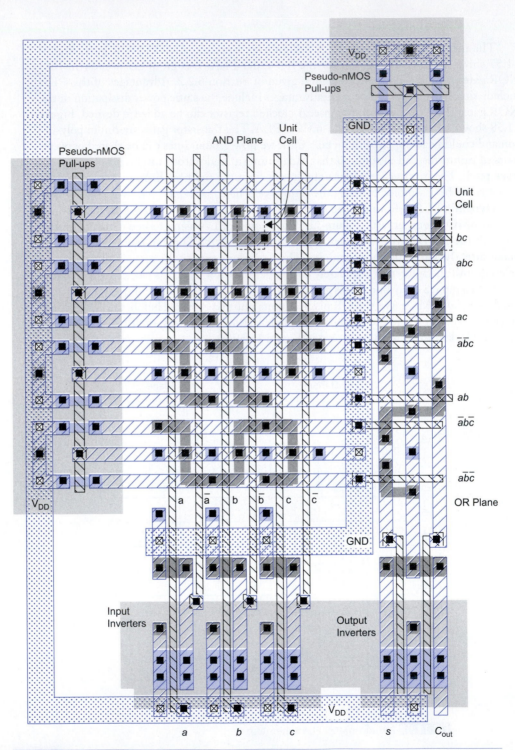

FIG 11.58 Pseudo-nMOS PLA layout

Figure 11.59(c) shows a self-timed dynamic PLA using two dummy rows as replica delay lines. Assume that the inputs arrive from flip-flops and settle shortly after the rising edge of the clock. The clocked circuitry acts as a pulse generator, producing a low-going precharge pulse on ϕ_{AND} shortly after the clock edge. The width of the pulse is equal to the delay of dummy AND row 1 plus two inverters and should be great enough to fully precharge all of the real AND rows. Thus, the loading on the dummy AND row is chosen to equal or exceed the worst loading of any real row. This worst loading consists of one nMOS drain for each input and one gate for each output. In this figure, the size of the inverter loading the AND line can be selected to contribute the desired gate load. Once the AND plane enters evaluation, the second dummy AND row starts to discharge through a single transistor. Again, this row is loaded to equal or exceed the delay of the worst real AND row. The three inverters provide some self-timing margin to ensure that ϕ_{OR} will not rise until the AND plane has fully evaluated. The output of the OR plane can be sampled into flip-flops on the next rising edge of the clock. [Wang01] surveys a variety of other PLA designs.

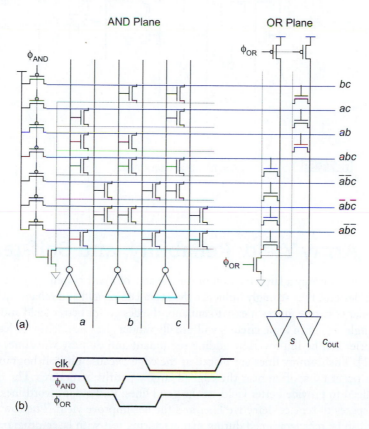

FIG 11.59 Dynamic PLA schematic (continues)

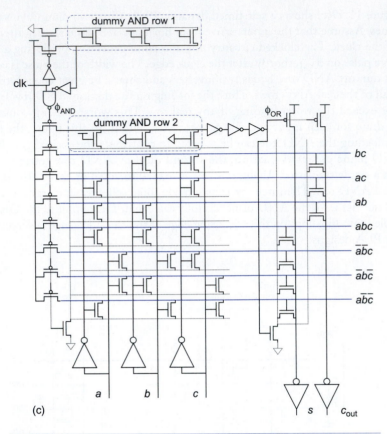

FIG 11.59 Dynamic PLA schematic (continued)

11.8 Array Yield, Reliability, and Self-test

Because arrays occupy a large fraction of the die area of many system-on-chip and microprocessor designs, they strongly influence the overall chip yield and reliability. Fortunately, their regular structure makes it easy to enhance the design for better yield and reliability.

A single defect in logic circuits will usually render the entire chip useless. The yield of memories can be improved by adding redundant and dummy wordlines and bitlines [Keeth01]. The dummy lines are placed at the edge because photolithography and etch problems occur most often near the edge of large repetitive structures. The dummy lines are sacrificed to provide better yield on the good lines. Redundant wordlines and bitlines serve as spares to replace defective lines and further improve yield. The row and column decoders can be reprogrammed during manufacturing test with laser-programmable fuses to eliminate the bad lines and substitute the redundant lines, or can be configured at start-up using built-in self-test. Large memories add entire redundant subarrays to replace sub-arrays with defective decoders or excessive numbers of defects. For example, some microprocessors have many banks in a set-associative cache. A processor can be manufactured with 8 banks and sold with the guarantee that at least 6 are operational. The bad ones are located during manufacturing test and disabled with fuses.

Memories are also prone to soft errors that spontaneously flip a bit stored in one of the cells, as discussed in Section 4.8.7. Error-detecting and correcting codes (ECC) are commonly used to recover from such errors, as discussed in Section 10.7.2. For example, adding 8 check bits to a 64-bit word in a memory is sufficient to correct any error in the word and detect any pair of errors. ECC supplements redundant bitlines to dramatically improve yield as well.

Memories are generally tested by writing known patterns to each of the bits, reading the patterns, and then rewriting the opposite pattern and reading it back to ensure each bit can be set to 0 and to 1. Performing these tests on an external tester is time-consuming and expensive, particularly for embedded memories that are not directly accessible from the chip I/Os. Many memories now contain built-in self-test (BIST) that places multi-plexers in the address and data paths to take over the memory during test mode, as discussed in Section 9.6.6.7. A small BIST controller containing an address counter and data pattern generator drives the memory at full speed during test. Self-test can be performed every time the chip is reset.

11.9 Historical Perspective

MOS memory made a splash in 1970 when Intel announced sales of the first 1103 1-kb DRAM chip and IBM replaced magnetic core memories with semiconductor memories in its 370-series mainframe computers. Since then, DRAM has become a commodity business characterized by ferocious price competition among a rather small number of manufacturers. Indeed, in 1986, Intel left what was then its core business when the market was flooded by cheap chips from Japan. DRAM capacity per chip has increased by 60% per year and cost per bit has decreased by 27% per year. Feature size improvement accounts for part but not all of the capacity gains. The area per bit has shrunk faster than feature size because of clever cell designs such as the 1T DRAM cell, innovative layout, and three-dimensional capacitor structures. Larger dice have become economical because of manufacturing yield improvements. Growing DRAM capacity has benefited system designers as much as the advances in processor performance.

DRAM density has quadrupled approximately every three years. Table 11.2 lists some of the innovations at each DRAM generation [Itoh01k]. The first generations of DRAMs used 3T or 4T memory cells shown in Figure 11.60. IBM patented the 1T cell in 1968 [Dennard68] and it soon became ubiquitous because the small cell size justified the complexity of the capacitor. Early DRAMs were built in nMOS processes requiring high supply voltages. V_{DD} standardized at 5 V through the 1980s and 1990s and CMOS peripheral circuitry was eventually adopted to save power. Other improvements addressed the signal-to-noise ratio, bandwidth and latency, power consumption, and test time.

Table 11.2	DRAM generations			
Capacity	Years of Volume Shipment	Power Supply (V)	Memory Cell	Circuit Innovations
1 kb	1970s	> 12	3T or 4T	MOS technology Differential sensing
4 kb	1970s	> 12	3T or 4T	Multiplexed addresses
16 kb	–1984	12	1T	Dynamic amplifier Dynamic driver
64 kb	1981–1987	5	1T	Folded bitline Word bootstrapping Substrate bias generator
256 kb	1984–1992	5	1T	Shared amplifier Metal-strapped wordline Redundancy
1 Mb	1987–1997	5	1T	CMOS peripheral circuits Half-V_{DD} precharge Multidivided data line BIST
4 Mb	1991–2000	5	1T	3-D capacitor structure
16 Mb	1994–2003	5	1T	On-chip voltage converter Twisted bitlines
64 Mb	1997–	3.3	1T	Synchronous small-signal I/O Multidivided wordlines
256 Mb	2001–	1.8–3.3	1T	Double data rate interface

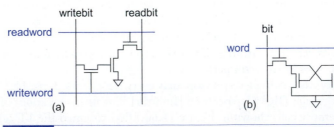

FIG 11.60 3T and 4T DRAM cells

Summary

Arrays repeat a basic cell in two dimensions. The cell is carefully optimized to provide very high density. For performance or density reasons, the nodes within the array do not always swing from rail to rail. Periphery circuitry restores the output swings to full digital logic levels.

The static RAM is very widely used in CMOS systems. The ubiquitous 6T cell consists of a cross-coupled inverter pair to hold the state and two access transistors for differential reads and writes. The bitlines are first preconditioned to a known value. A decoder asserts one of the wordlines. That word is read onto the bitlines and sensed. A column multiplexer may select only a subset of the bits as outputs. SRAMs are used in caches and other embedded memories. Multi-ported SRAMs are used in register files.

Content-addressable memories are similar to SRAMs. However, they also provide a lookup mode in which a key is placed on the bitlines and each word that contains that key asserts its matchline. CAMs are important for looking up addresses in translation look-aside buffers and network routers.

Dynamic RAMs store information on a capacitor using a single access transistor. With specialized process steps to build compact capacitors, they offer an order of magnitude higher density of data storage than SRAM. However, the data gradually leaks off the capacitors, so DRAMs must be periodically refreshed to maintain their state. DRAMs are usually built in specialized processes on dedicated chips, but potentially may be useful for high-capacity embedded memories on digital CMOS processes.

Read-only memories also use a single access transistor, but their contents are wired to a constant value. They are commonly used to store code and are convenient because they can be easily changed late in the design process to correct bugs or add features. EEPROM and Flash memories are even more convenient because they can be changed after fabrication or even in the field. Flash is also widely used for nonvolatile data storage.

A ROM can also be viewed as a lookup table. In general, a ROM of 2^x words by y bits can serve as a lookup table to perform any function of x inputs and y outputs. If a function is written in sum-of-products form, the ROM decoder performs the AND operation while the ROM array performs the OR. Many functions are relatively sparse. A programmable logic array optimizes out the unnecessary entries by replacing the decoder with an AND plane. In some cases, PLAs are smaller than ROMs, yet provide the same flexibility of easy changes late in the design cycle. PLAs were commonly used for microcoded finite state machines in the 1980s. They are still occasionally used, but good logic synthesis tools now deliver the same ease of change for random logic while avoiding the complicated circuit design needed for an efficient PLA.

A good design flow should provide automatic generators for simple SRAMs and ROMs. The designer should be comfortable with using these arrays where they are appropriate. High-performance designs need more elaborate multi-ported SRAM, large memory arrays, and CAMs. Most of these arrays demand skilled circuit design and careful simulation.

For more advanced reading on memory design, see [Keeth01, Prince99, or Itoh01k].

Exercises

11.1 An embedded SRAM contains 2048 8-bit words. If it is physically arranged in a square fashion, how many inputs does each column multiplexer require?

11.2 Estimate the dimensions of the SRAM array in Exercise 11.1 using the SRAM cell from Figure 11.6b, assuming periphery circuitry adds 10% to each dimension of the core.

11.3 Sketch designs for a 6:64 decoder with and without predecoding. Comment on the pros and cons of predecoding.

11.4 Sketch a 4:16 Lyon-Schediwy Decoder. Label transistor sizes so each input sees the same capacitance. Compute the logical effort and branching effort seen by the decoder.

11.5 Estimate the minimum delay of a 10:1024 decoder driving an electrical effort of $H = 20$ using

(a) static CMOS gates

(b) footless domino gates

11.6 Design the footless domino decoder from Exercise 11.5(b) using self-resetting domino gates. Assume the inputs are available in true and complementary form as pulses with a duration of 3 FO4 inverters and can each drive 48 λ of gate width. Indicate transistor sizes and estimate the delay of the decoder.

11.7 Develop a model of wordline decoder delay for a RAM with 2^n rows and 2^m columns. Assume true and complementary inputs are available and that the input capacitance equals the capacitance of one of the columns so $H = 2^m$. Use static CMOS gates and express your result in terms of n and m.

11.8 Explain the tradeoffs between open, closed, and twisted bitlines in a dynamic RAM array.

11.9 Sketch a dot diagram for a 2-input XOR using a ROM.

11.10 Sketch a dot diagram for a 2-input XOR using a PLA.

11.11 Sketch a schematic for an 8-word × 2-bit NAND ROM that serves as a lookup table to implement a full adder.

11.12 Explain the advantages and disadvantages of NAND ROMs as compared to NOR ROMs.

11.13 Develop a model for the read time of a ROM with 2^n rows and 2^m columns analogous to that of the SRAM from Section 11.2.6. Assume the wire capacitance in the ROM array is negligible compared to the gate and diffusion capacitance. Assume the ROM cells are laid out such that two cells share a single diffusion contact and hence each contributes only $C/2$ of diffusion capacitance.

Special-purpose Subsystems | 12

12.1 Introduction

This chapter describes a variety of special-purpose subsystems that a digital designer may encounter. These subsystems are usually designed by a specialist or obtained from a third-party vendor, and each is the subject of entire books. However, the skilled digital designer should be conversant in each area in order to understand the impact of the other subsystems on a core digital design.

The chapter begins with packaging. The package strongly impacts the power distribution and I/O subsystems, which are discussed next. Then we look at clock generation and distribution, which are vital to achieve low clock skew and communicate synchronously with the external world. Finally, the chapter addresses rudimentary analog circuit design.

12.2 Packaging

The chip package provides a mechanical and electrical connection between the chip and a circuit board. It is no longer possible to separate the design of a high-performance integrated circuit from the design of its package. An ideal package has the following properties:

- Connects signals and power between the chip and board with little delay or distortion

- Removes heat produced by the chip

- Protects the chip from mechanical damage and thermal expansion stress

- Is inexpensive to manufacture and test

To provide good signal and power connections, the package must offer short wires with low resistance and inductance. The impacts of the package on the power supply and I/O are discussed further in Sections 12.3 and 12.4, respectively. The remainder of this section describes some of the types of packages commonly available and how they remove heat from the chip.

12.2.1 Package Options

Table 12.1 lists a variety of common integrated circuit packages. Figure 12.1 shows photographs of these packages. The I/O count includes connections for both signals and power. I/O spacing is typically specified in the archaic unit of mils (1 mil = 0.001 inch = 25.4 μm). Packages come in both ceramic and plastic varieties; plastic is cheaper, but cannot remove as much heat. Older packages tended to use *through-hole* pins, which pass through holes in a printed circuit board and are soldered from below. The pins contribute inductance, and the size of the holes limits the density of the pins. *Surface mount* (SMT) packages are soldered to the surface of a printed circuit board to alleviate these problems. Dual Inline Packages (DIP), Pin Grid Array (PGA), and Plastic Leadless Chip Carrier (PLCC) packages are easy to insert into low-cost sockets, so they are convenient for components that might be removed for reprogramming or replacement. Ball Grid Array (BGA) packages have recently become the preferred approach for parts that require a large number of high-bandwidth signals in a compact form factor. Package design is a rapidly advancing field and new packages are being adopted each year.

Table 12.1 Package options

Package	# I/Os	Description
Dual Inline Package (DIP)	8, 14, 16, 20, 28, 40, 64	Two rows of through-hole pins on 100 mil centers. Low cost. Long wires between chip and corner pins.
Pin Grid Array (PGA)	65–391+	Array of through-hole pins on 100 mil centers. Low thermal resistance and high pin counts.
Small Outline IC (SOIC)	8,10, 14, 16, 20, 24, 28	Two rows of SMT pins on 50 mil centers. Low cost, good for low-power parts with small pin counts.
Thin Small Outline Package (TSOP)	28–86+	Two rows of SMT pins on 0.5 or 0.8 mm centers in a thin package. Commonly used for DRAMs.
Plastic Leadless Chip Carrier (PLCC)	20, 28, 44, 68, 84	J-shaped SMT pins on all four sides on 50 mil centers. Sturdy leads are convenient for socketing.
Quad Flat Pack (QFP)	44–240	SMT pins on all four sides on 15.7–50 mil centers. High density of I/Os. Available in thin (TQFP) and very thin (VQFP) forms as thin as 1.6 mm.
Ball Grid Array (BGA)	49–2000+	Array of SMT solder balls on underside of package on 15.7–50 mil centers. Extremely high density of I/Os with low parasitics. Requires specialized assembly and inspection equipment to blindly attach to array of pads on printed circuit board.
Flip-Chip	Many	Direct connection of chip to printed circuit board through solder balls on top metal layer of chip. Even higher I/O density and lower parasitics than BGA.

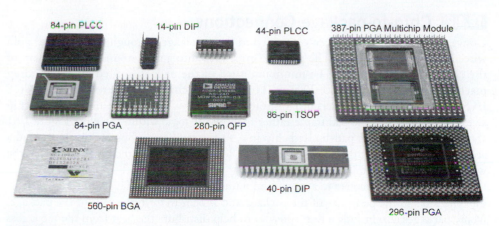

FIG 12.1 Integrated circuit packages. © 2003 Harvey Mudd College. Reprinted with permission.

Some high-performance systems are packaged in multi-chip modules (MCMs) that contain more than one die within the package. The MCM substrate connects the dice directly without the need for a printed circuit board, offering potentially higher speed interconnections. For example, the Pentium Pro shown in Figure 12.2 used an MCM to connect the microprocessor and one or two external cache dice. In another extreme example, the IBM z900 mainframe uses a multichip module containing 20 CPUs, 8 cache chips, and a kilometer of interconnect [Harrer02]. The module measures 127 mm on a side and dissipates 1.3 kW. MCMs are expensive to build and depend on the availability of tested but unpackaged dice, so they have not achieved the popularity anticipated in the mid-1990s.

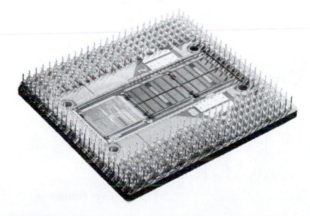

FIG 12.2 Pentium Pro MCM. Reprinted with permission of Intel Corporation.

12.2.2 Chip-to-package Connections

Conventionally, chips have been connected to their packages through thin (25 μm) gold wires bonded to metal pads. The pads are organized into a ring around the periphery of the chip called a *pad frame*. The minimum pitch of the pads is limited by the bonding machine to approximately 100–200 μm. Thus, a 1-cm² chip is limited to several hundred I/Os. Chips with large numbers of I/Os sometimes are *pad-limited,* meaning that the chip size is determined by the pad frame rather than by the logic within the chip. Figure 1.61 showed an example of a pad-limited chip in a 40-pin pad frame. Some chips have used a second ring of pads, but this approach results in longer bond wires and greater risk that the wires will accidentally touch.

The bond wires connect to a metal lead frame in the package. This lead frame distributes the I/Os to the periphery of the package and is bent to form the pins of the package. Many packages also include a heat spreader to help distribute the heat from the die across the package and ultimately out to the heat sink. Figure 12.3 shows a cutaway of a dual-in-line package showing a corner of the chip with bond wires connecting to the lead frame [Mahalingam85]. The metal leads contribute parasitic inductance and coupling capacitance to their neighbors. More advanced packages internally resemble printed circuit boards, using multiple layers of signals and power/ground planes to distribute the I/Os on controlled-impedance transmission lines.

Since the late 1990s, many manufacturers have adopted *flip-chip* connections. This technology, also called *Controlled Collapse Chip Connection* (C4), was developed by IBM in the 1960s and has been used on their mainframes for decades. In a flip-chip design, the surface of the chip is covered with an array of pads on the top level of metal. Lead solder balls are bonded to these pads in a final process step called *wafer bumping*. The chip is flipped upside down and connected to the package by heating the balls until they melt. The bonding requires careful alignment, but surface tension from the solder helps pull the

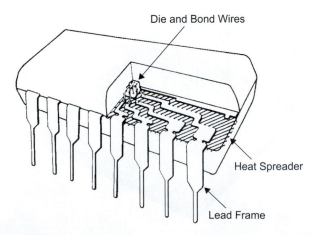

FIG 12.3 Cutaway view of dual-in-line package. © IEEE 1985. [Mahalingam85]

chip into place. The chip is in nearly direct contact with the package, eliminating the inductance associated with the bond wires. The bumps can be placed on a pitch of 200 µm or less, offering thousands of connections between the die and package. Flip-chip technology introduces new testing problems because the top-level metal wires are no longer accessible for probing during debug.

12.2.3 Package Parasitics

Figure 12.4 shows a model of an integrated circuit package. The bond wires and lead frame contribute parasitic inductance to the signal traces. They also have some mutual inductive and capacitive coupling to nearby signal traces, potentially causing crosstalk when multiple signals switch. The V_{DD} and GND wires also have inductance from both bond wires and the lead frame. Moreover, they have nonzero resistance, which becomes important for chips drawing large supply current. High-performance packages often include bypass capacitors between V_{DD} and GND. As we will see in Section 12.3.5, the bypass capacitors have their own parasitic resistance and inductance that limit their effectiveness at high frequencies.

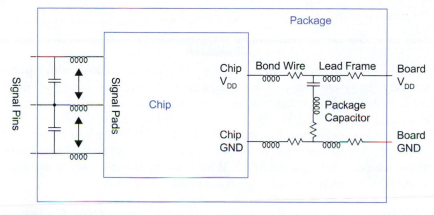

FIG 12.4 Package parasitics

12.2.4 Heat Dissipation

A 60-watt light bulb has a surface area of about 120 cm^2 and is too hot to touch. In comparison, the Itanium 2 dissipates 130 W on a 4 cm^2 die, resulting in a power density 60 times as great! Clearly, removing heat from chips is a major challenge for the package.

The heat generated by a chip flows from the transistor junctions where it is generated through the substrate and package. It can be spread across a heat sink, and then carried away through the air by means of convection. Just as current flow is determined by voltage difference and electrical resistance, the heat flow is determined by temperature difference and thermal resistance. Thus, the temperature difference ΔT between the transistor junctions and the ambient air is:

$$\Delta T = \Theta_{ja}P \qquad (12.1)$$

where Θ_{ja} is the thermal resistance (in °C/W) between the junction and ambient and P is the power consumption of the chip. The thermal resistance in turn can be modeled as the series resistance from the die to the package Θ_{jp} and from the package to the air Θ_{pa}.

$$\Theta_{ja} = \Theta_{jp} + \Theta_{pa} \qquad (12.2)$$

For most low-cost packages, Θ_{pa} dominates the resistance. Still air can transfer about 0.001 W/(cm²°C) [Glasser85]. Thus, a package with a surface area of 10 cm² has a thermal resistance of about Θ_{pa} = 100° C/W. Such a package cannot handle chips dissipating more than about 1 watt. Forced air transfers 0.01–0.03 W/(cm²° C). High-power chips add a large heat sink and a fan to the package to reduce the thermal resistance. For example, a 72-pin ceramic PGA package has a thermal resistance Θ_{pa} of 34° C/W in still air, 18° C/W in 400 ft/minute airflow, and 10° C/W in 400 ft/minute airflow with a good heat sink.

Example

You are planning to package an ASIC in a ball grid array package with a passive heat sink. The system box contains a large fan providing 250 linear feet/minute (LFM) of airflow. The package vendor specs the thermal resistance from the junction to package at 0.9° C/W. The heat sink vendor specs the thermal resistance from the package to ambient for this airflow at 4.0° C/W for the heat sink plus 0.1° C/W for the heat sink adhesive between the package and heat sink. The system box ambient temperature may reach 55° C. What is the maximum power dissipation of your ASIC if its junction temperature is not to exceed 100° C?

Solution: The thermal resistance is Θ_{ja} = 0.9 + 0.1 + 4.0 = 5° C/W. The temperature difference between the junction and ambient must not exceed ΔT = 100 − 55 = 45° C. Therefore, the maximum power dissipation is $P = \Delta T / \Theta_{ja}$ = 9 W.

Advances in heat sinks, fans, and packages have raised the practical limit for heat removal from about 8 W in 1985 to nearly 100 W today for affordable packaging. Forced-air cooling appears to be reaching its limits, setting a cap on the power consumption of chips. Liquid cooling is used in some mainframes and supercomputers to remove even greater amounts of heat. Perhaps mass production of liquid cooling systems will eventually drive the cost down to acceptable levels for mainstream systems.

12.3   Power Distribution

The power distribution subsystem of a chip consists of metal wires or planes on the chip, in the package, and on the printed circuit board. It also includes bypass capacitors to supply the instantaneous current requirements of the system. An ideal power distribution network has the following properties:

- Maintains a stable voltage with little noise

- Provides average and peak power demands

- Provides current return paths for signals

- Avoids wearout from electromigration and self-heating

- Consumes little chip area and wiring

- Is easy to lay out

Real networks must balance these competing demands, meeting targets of noise and reliability as inexpensively as possible. The noise goal is typically ±5% or ±10%; for example, a system with nominal V_{DD} = 1.8 V may guarantee the actual supply remains within 1.62–1.98 V. Reliability goals demand enough vias and metal cross-sectional area to carry the supply current, as was discussed in Section 4.8. The two fundamental sources of power supply noise are IR drops and L di/dt noise.

Figure 12.5 plots the power consumption versus time for a microprocessor [Gauthier02]. The power varies on a number of time scales. While the processor is active, the power depends on the operations and data. It also spikes near the clock edges when

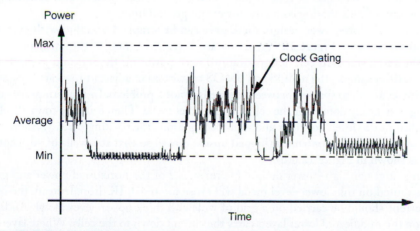

FIG 12.5 Time-dependent power consumption of microprocessor. Reprinted with permission of Sun Microsystems.

the large clock loads switch. When the processor becomes idle, clock gating turns off the clock to unused units, driving the power down significantly. As the supply voltage is nearly constant, the supply current I (also called I_{DD}) is proportional to the instantaneous power demand. As this current flows through the resistance R of the power distribution network, it causes a voltage droop proportional to IR. Moreover, as the changing current flows through the inductance of the printed circuit board and package, it also causes a voltage drop proportional to the rate of change: L di/dt.

This section begins by examining the physical design of a power distribution network. It then discusses IR drops and L di/dt noise. The key to controlling noise from current spikes is to provide adequate bypass capacitance on and off the chip to provide low supply impedance at all frequencies. The power network is complicated enough that manual analysis is inadequate; instead, it typically must be modeled in a finite element simulation. The power network also provides return paths for current flowing in signal wires. The geometry of the network affects the inductance of on-chip signals. Some critical circuits such as phase-locked loops and analog blocks require a quiet supply for good performance. RC filters can reduce much of the supply noise. In sensitive circuits, noise carried through the substrate is also important. For other introductions to power distribution subsystems, see [Bakoglu90, Dally98, Chandrakasan01, Mezhiba03].

12.3.1 On-chip Power Distribution Network

The on-chip power distribution network consists of power and ground wires within the cells and more wires connecting the cells together. Most cells contain internal power and ground busses routed on metal1 or metal2. These wires are typically wider than minimum to provide lower resistance and better electromigration immunity. For example, Figure 1.62 showed a cell library with 8 λ metal1 power/ground busses. These wires are normally connected between adjacent cells by abutment. Standard cell designs and datapaths both can use rows of cells sharing common power and ground lines.

In a small, low-power design, these rows can be strapped together with even wider vertical metal wires. Figure 12.6(a) shows an abstract diagram of this strapping. Figure 1.63 showed a standard cell design strapped with power on the left and ground on the right. In this example, the nMOS and pMOS transistors in adjacent rows are separated by a routing channel, so spacing between the wells is not a problem. In modern processes, the routing is typically done over the cell in upper-level metal. Therefore, the rows of cells can be packed more closely together and well spacing limits the packing density. Alternatively, every other row can be *mirrored* (flipped upside down) so that the wells of adjacent rows abut, as shown in Figure 12.6(b).

In a larger or high-power design, the resistance of the horizontal power and ground busses routed on thin lower-level metal will cause too much IR drop. Instead, the bulk of the current should be carried on a grid of wide and thick upper-level metal. Additional grids on the middle and lower layers carry the current down to the cells. Where layers connect, plenty of vias should be used to carry the high currents. Figure 12.7 shows such a power distribution grid in a six-level metal process with mirrored cells [Ayers03] and metal2 power busses within the cell. Notice that the metal3 lines share vertical tracks to

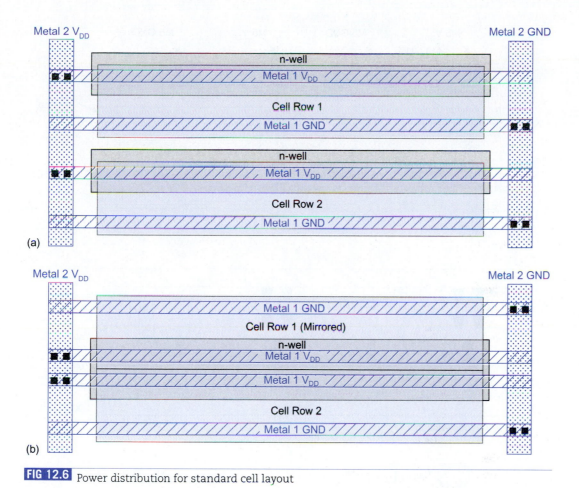

FIG 12.6 Power distribution for standard cell layout

conserve metal resources. Metal1 does not form a significant part of the power network and is left out of the figure.

The power grid extends across the entire chip. Ultimately, it must connect to the package through the I/O pads. When a pad ring is used, the connections are all near the periphery of the chip. Thus, the biggest IR drops occur near the center of the chip where the current flows through the longest wires and greatest resistance. Indeed, the Alpha 21264 devoted solid planes of metal to power and ground distribution to achieve acceptable IR drops at the center. C4 solder bumps distributed across the die are much better for power distribution because they can deliver the current from the low-resistance power plane in the package directly to the area of the chip where the current is needed. Thus, less on-chip metal resources are needed for power distribution.

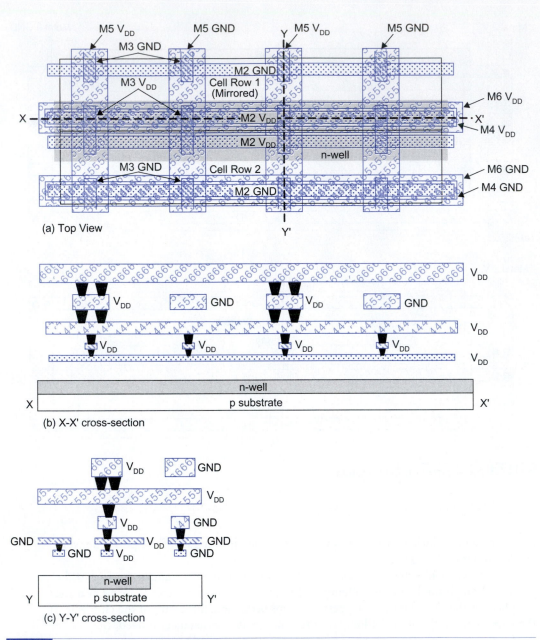

FIG 12.7 Power grid layout

12.3.2 IR Drops

The resistance of the power supply network includes the resistance of the on-chip wires and vias, the resistance of the bond wires or solder bumps to the package, the resistance of the package planes or traces, and the resistance of the printed circuit board places. Because the package and printed circuit board typically use copper that is much thicker and wider than on-chip wires, the on-chip network dominates the resistive drop.

IR drops arise from both average and instantaneous current requirements. The instantaneous current may be much larger than the average drop because current draw tends to locally spike near the clock edge when many registers and gates switch simultaneously. Bypass capacitance near the switching gates can supply much of this instantaneous current, so a well-bypassed power supply network only needs low enough resistance to deliver the average current demand, not necessarily the peak.

Example

Suppose a row of 64 repeaters share a common metal2 power bus like that shown in Figure 12.6(a). The bus is 320 μm long and 1 μm wide. The metal2 has a sheet resistance of 0.05 Ω/□. If the repeaters drive 0.4 pF wire loads with 200 ps transition times, estimate the power supply droop seen by the repeater for a 1.8 V nominal supply.

Solution: Each repeater draws a current of approximately

$$I = C\frac{\Delta V}{\Delta t} = \left(0.4 \text{ pF}\right)\frac{1.8 \text{ V}}{200 \text{ ps}} = 3.6 \text{ mA}$$

The power and ground busses each have a length of 320 squares and thus a resistance of $R = 16\ \Omega$. The supply droop at the end of the wire caused by the 64 repeaters is 64 IR/2 = 1.85 V, or more than V_{DD}, which is obviously impossible. Instead, as the power supply begins to droop, the repeaters deliver less current, reducing the droop, but increasing the transition time and delay. One way to alleviate this problem is to use a power grid so that each repeater obtains its current from its own vertical wire rather than sharing the single horizontal wire with all of the simultaneously switching neighbors. Figure 12.8 shows a simulation of one of the repeaters. It compares the two power bus layouts. When all the repeaters share a single power wire, the power supply droops by nearly 30% and the propagation delay is more than doubled. When each repeater has its own power wire so the supply noise is negligible, the output is much crisper.

(continues)

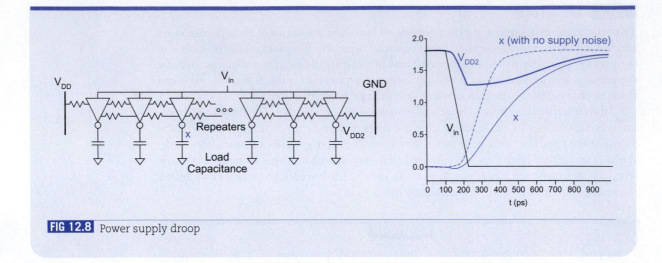

FIG 12.8 Power supply droop

12.3.3 L di/dt Noise

The inductance of the power supply is typically dominated by the inductance of the bond wires or C4 bumps connecting the die to the package. A typical bond wire has an inductance of about 1 nH/mm, while a C4 ball is on the order of 100 pH. Recall that the inductance of multiple inductors in parallel is reduced. Modern packages devote many (often 50% or more) of their pins to power and ground to minimize supply inductance. The two largest sources of current transients are switching I/O signals and changes between idle and active mode in the chip core.

Example

A 1 GHz chip transitions from idle to full power operation in a single cycle. The idle mode draws 20 A and the full power mode draws 60 A. If the power supply has 20 pH of series inductance, estimate the power supply noise caused by this transition if the chip has no internal bypass capacitance.

Solution: The current transient is:

$$\frac{\Delta I}{\Delta t} = \frac{\left(60 \text{ A} - 20 \text{ A}\right)}{1 \text{ ns}} = 40 \text{ GA / s}$$

The inductive noise is $L \, \Delta I/\Delta t = 0.8$ V. This is clearly unacceptable in a low-voltage process. Once again, the chip needs internal bypass capacitance to supply the instantaneous current, reducing the transient seen by the I/O pins.

L di/dt noise is becoming enough of a problem that some high-power systems must resort to microarchitectural solutions that prevent the chip from transitioning between minimum and maximum power in a single cycle. For example, a pipeline may enter or exit idle mode one stage at a time rather than all at once to spread the current change over many cycles.

12.3.4 On-chip Bypass Capacitance

As we have seen, chips need a substantial amount of capacitance between power and ground to provide the instantaneous current demands of the chip. This is called *bypass* or *decoupling* capacitance. The bypass capacitance is distributed across the chip so that a local spike in current can be supplied from nearby bypass capacitance rather than through the resistance of the overall power grid. It also greatly reduces the di/dt drawn from the package.

Example

How much bypass capacitance is needed to supply a sudden current spike of 40 A for 1 ns with no more than a 200 mV supply droop?

Solution: We solve

$$I = C \frac{\Delta V}{\Delta t} \Rightarrow C = \frac{(40 \text{ A})(1 \text{ ns})}{0.2 \text{ V}} = 200 \text{ nF}$$

Fortunately, the inherent gate capacitance of quiescent transistors provides a significant amount of *symbiotic* bypass capacitance [Dally98]. For example, Figure 12.9 shows one inverter driving another. The gate-to-source capacitances of the load inverter are shown explicitly. When $A = 1$ and $B = 0$, $M1$ is ON, charging up C_{gs4}. Similarly, when $A = 0$ and $B = 1$, $M2$ is ON, charging up C_{gs3}. The charged capacitor stores energy that can be released to supply sudden current demands. At any given time, approximately half of the gate capacitance of any quiescent circuit behaves as symbiotic bypass capacitance. Moreover, because only a small fraction of the gates are likely to be switching at any given time, nearly half of the entire gate capacitance on the chip will serve as bypass capacitance.

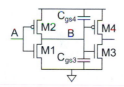

FIG 12.9 Symbiotic bypass capacitance

In most low- and medium-power chips, this symbiotic capacitance provides adequate bypassing to filter instantaneous IR drops and L di/dt noise. In high-power chips, additional explicit capacitance is sometimes necessary. The only dielectric available in a standard CMOS process to build compact high-capacitance structures is gate oxide, so the extra bypass capacitance is commonly built with an nMOS transistor with the gate tied to V_{DD} and the source and drain tied to GND, as shown in Figure 12.10(a).

Decoupling capacitor layout should maximize the capacitance per unit area. The transistor length is ideally as long as possible so that most of the area is devoted to gate oxide

Example

Estimate the symbiotic bypass capacitance per square millimeter for a chip with feature size f if gate capacitance is 2 fF/μm of transistor width and transistor gates occupy 9% of chip area.

Solution: The capacitance density of a 1 μm wide transistor of length f is $2/f$ fF/μm^2. At 9% utilization, this corresponds to $0.18/f$ nF/mm^2. Half of that, or $0.09/f$ nF/mm^2, serves as symbiotic bypass capacitance on average. In an f = 180 nm process, this means the symbiotic bypass capacitance is approximately 0.5 nF/mm^2.

capacitance rather than contacts. However, if the length is too great, the resistance of the ON transistor causes a slow response and ineffective bypass [Larsson97]. In practice, a length of 4–12 times minimum is acceptable. Figure 12.10(b) shows a "waffle" layout of a thin oxide decoupling capacitor. Most of the area consists of polysilicon over diffusion acting as capacitance. Occasional openings in the polysilicon are used to contact the diffusion to ground, while occasional field oxide bumps sit below polysilicon V_{DD} contacts (unless process design rules allow direct connection of metal to polysilicon over gate oxide). Small decoupling capacitors can be treated as unit cells. They are assembled into arrays by CAD tools. The high density of polysilicon and metal1 may increase process variation in nearby devices, so some tradeoffs in density or spacing to logic may be required. Decoupling capacitors must also be placed close (e.g., within about 200 μm in a 90 nm process) to the switching circuits they serve so they can deliver charge before much droop occurs. The direct V_{DD} connections to only gate oxide violate antenna rules in some processes. This can be avoided by coupling V_{DD} through a pMOS device, as shown in Figure 12.10(c). A typical cell using 0.35 μm I/O transistors achieves 320 fF/cell.

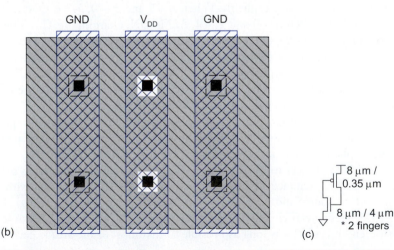

FIG 12.10 Waffle bypass capacitor layout

As gate oxide becomes thinner and tunneling increases, thin oxide capacitors may become undesirable because of their high leakage. This may force process engineers to offer an extra mask step to build specialized decoupling capacitors using either a medium-thickness oxide or a more exotic material with higher dielectric constant.

12.3.5 Power Network Modeling

Figure 12.11 shows a lumped model of the power distribution network for a system including the voltage regulator, the printed circuit board planes, the package, and the chip. The network also includes bypass capacitors near the voltage regulator, near the chip package, possibly inside the chip package, and definitely on chip. The external capacitors are modeled as an ideal capacitor with an effective series resistance (ESR) and effective series inductance (ESL) representing the parasitics of the capacitor package. Larger capacitors have bigger effective series inductances.

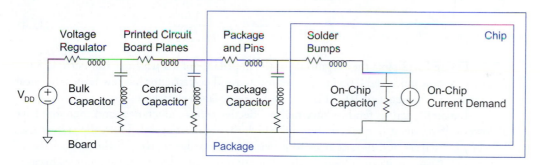

FIG 12.11 Power distribution system model

The voltage regulator seeks to produce a constant output voltage independent of the load current. It is modeled as an ideal voltage source in series with a small resistance and the inductance of its pins. Near the regulator is a large bulk capacitor (typically electrolytic or tantalum). Power and ground planes on the printed circuit board carry the supply current to the package, contributing some resistance and inductance. Typically, the board designer places several small ceramic capacitors near the package. The package and its pins again contribute resistance and inductance. High-frequency packages often contain small capacitors inside the package for further decoupling. Finally, the chip connects to the package through solder bumps or bond wires with additional resistance and inductance. The dynamic and static current demands of the chip are modeled as a variable current source with a waveform that might resemble Figure 12.5. The on-chip bypass capacitance consists of the symbiotic capacitance and possibly some explicit decoupling capacitance. It typically has negligible inductance because it is located so close to the switching loads.

12.3.5.1 Power Supply Impedance
A good power distribution network should offer a low impedance at all frequencies of interest so that the supply voltage remains steady independent of the changing chip current demands. If the system had no bypass capacitance,

the distribution network would consist of only the resistance and inductance, so it would have an impedance of $Z = R + j\omega L$. This impedance increases with frequency ω and becomes unacceptably high for most systems by about 1 MHz.

The bypass capacitors in parallel with the supply provide an alternative low-impedance path at higher frequencies. An ideal capacitor has impedance that decreases with frequency as $Z = 1/j\omega C$. Unfortunately, the effective series inductance of the capacitors limits the useful frequency range of the real capacitor. The impedance of a capacitor C with effective series resistance R and inductance L is:

$$Z = \frac{1}{j\omega C} + R + j\omega L \qquad (12.3)$$

This impedance has a minimum of $Z = R$ at the self-resonant frequency of

$$f_{\text{resonant}} = \frac{\omega_{\text{resonant}}}{2\pi} = \frac{1}{2\pi\sqrt{LC}} \qquad (12.4)$$

Figure 12.12 plots the magnitude of the impedance of a 1 µF capacitor with 0.25 nH of series inductance and 0.03 Ω of series resistance. The capacitor has low impedance near its resonant frequency of 10 MHz, but higher impedance elsewhere.

Larger capacitors tend to have higher effective series inductances and therefore have lower self-resonant frequencies beyond which they are not useful. Thus, the system uses many capacitors of different sizes to provide low impedance over all the frequencies of interest. Also, capacitors closer to the chip are more useful at high frequencies because they have less inductance in the board and package between them and the chip. The bulk

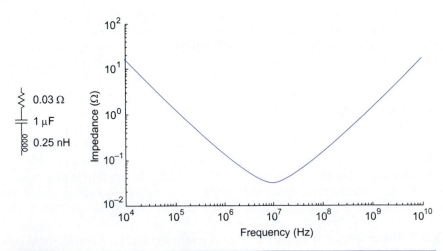

FIG 12.12 Impedance of bypass capacitor

and ceramic capacitors are most effective over the 1–10 MHz range. Capacitors in the package tend to be useful in the 10–200 MHz range. Above a few hundred MHz, the inductance of the solder bumps or bond wires renders all but the on-chip decoupling capacitors ineffective. Figure 12.13 plots an idealized supply impedance vs. frequency on a log-log scale for the complete power network model showing multiple resonances.

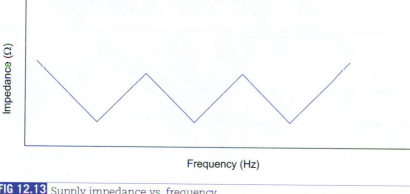

FIG 12.13 Supply impedance vs. frequency

The bypass capacitors also resonate with the series inductance of the package and solder bumps. This makes achieving low impedance across all frequencies difficult. Some packages have a high impedance somewhere in the 10–100 MHz range that is caused by such resonances. This is below the clock frequency of many chips. Power supply noise could be quite severe if the chip were to alternate between high- and low-power operations at the resonant frequency. This may be unlikely to occur by accident, but raises the threat of viruses that crash a computer by executing code that alternates power demand at just the wrong rate.

Another way to think about the need for nearby bypass capacitance is to imagine a sudden spike in current on the chip. Some round-trip propagation delay must occur before the spike reaches the power supply, the supply adjusts the current it is delivering, and that current returns to the chip. A lower bound on this delay is the speed of light. Therefore, when a gate switches, the voltage regulator will not know about the event until sometime after the transition has completed. The current to charge the load must come from the nearby bypass capacitors rather than the voltage regulator. This will cause a small droop on the power supply. Somewhat later, this droop will propagate to the regulator, which will boost its current and recharge the bypass capacitors.

12.3.5.2 Distributed Power Supply Models

The model presented so far is a lumped approximation that is convenient for analysis and facilitates gaining intuition about chip behavior. Chip designers also are concerned about the variation in supply voltage across the chip. This requires a distributed model, which we can approximate with a mesh of small elements as shown in Figure 12.14. The mesh represents the resistance and induc-

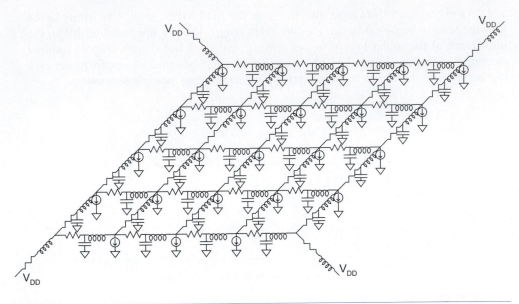

FIG 12.14 Distributed power supply model

tance of the on-chip power supply grid. Symbiotic or explicit decoupling capacitors are distributed across the chip. At each node, a current source represents the local current demand of the circuitry. The solder bumps or bond wires to the package are modeled with additional resistance and inductance. In this model, the package is treated as a perfect V_{DD} connected to the corners of the grid. In a more complex model, you also could add the distributed resistance, inductance, and bypass capacitance of the package itself.

For high-power chips, the designer can extract a mesh model as a SPICE netlist based on the power grid wiring and the amount of local decoupling capacitance. Different current waveforms can be applied at different nodes; for example, the current signatures of synthesized logic, SRAM, repeater banks, and domino logic are all quite different. The full-chip power grid simulation often takes many days to run and results in a map of voltage vs. time for the current pattern applied. Figure 6.56 shows a snapshot of the voltage droop on the Itanium 2 microprocessor. The droop was greatest in the integer execution unit, where several power-hungry domino adders all contribute to the *IR* drop.

12.3.6 Signal Return Paths

All current flows in loops. Whenever a signal is sent down a wire, the current eventually must return to the driver through the power or ground wires. Therefore, the power supply network also serves as a *signal return path*.

As we have seen, the impedance between V_{DD} and GND should be low at high frequency because of the decoupling capacitors. Hence, the V_{DD} wires are sometimes called an *AC ground* because high-frequency alternating current can flow readily between power and ground through the capacitors. Therefore, both V_{DD} and GND grids serve as possible signal return paths regardless of whether the signal was 0 or 1.

As discussed in Section 4.5.5, the inductance of a signal wire is proportional to the area of its current loop. This inductance could increase delay or cause noise if it is too large, so a good power network provides return paths near all of the signal paths to keep the area of the current loops small. Signals sharing overlapping loop areas also couple onto each other through their mutual inductance.

Figure 12.15 shows a circuit with a pathologically large current loop covering a large fraction of the chip area. The power and ground networks use a fingered layout [Johnson93]. Suppose the output of gate *A* rises. Some current *I* is drawn from the V_{DD} supply to charge up node *x* and gate *B*. The current flows through C_{gs} of the nMOS transistor in gate *B* to the GND wire and back along the GND fingers. It eventually returns to V_{DD} through the on-chip decoupling capacitance. This large loop indicated by the dashed line results in high inductance. As the speed of light in a wire is set by its $\sqrt{LC}$ product, the signal becomes unusually slow. To avoid this problem, a power grid should occupy at least two orthogonal layers of metal rather than using fingers on one layer.

A common strategy is to place one V_{DD} or GND line for each *N* signal lines on each level of metal. A *signal-to-return* (SR) ratio such as *N* = 10 on at least two orthogonal upper levels of metal ensures that no path has outrageously long return paths that might cause "slow mode" wave propagation from the high inductance. Some engineers argue that an SR ratio of 2:1 is good practice at high frequencies [Morton02]. This ratio has the added benefit that every signal has either V_{DD} or GND as one of its neighbors, reducing capacitive crosstalk by a factor of two. However, it is quite expensive in terms of wiring resources.

FIG 12.15 Fingered power network with long current return path

12.3.7 Power Supply Filtering

Certain structures such as the phase-locked loop (PLL), clock buffers, and analog circuits are particularly sensitive to power supply noise. For example, supply noise on the clock buffers can directly increase clock jitter. Figure 12.16 shows an RC power supply filter circuit that eliminates the high-frequency noise on the local supply. The local filtered power supply is typically connected to the power grid through a single wire or solder bump. The resistance of this wire must be low enough to carry the current demand of the local circuitry without excessive *IR* drop, yet low enough to produce an RC time constant that will filter noise at frequencies of interest. Typically, this requires a huge filter capacitor as well, making power supply filtering expensive in terms of chip area.

For example, the Pentium 4 uses a power supply filter on the clock buffers to reduce clock jitter [Kurd01]. The filter attenuates typical supply noise from 10% to 2% of V_{DD} using a pMOS transistor as the resistor. It has an RC time constant of 2.5 ns with an *IR* drop of 70 mV.

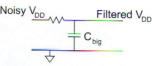

FIG 12.16 Power supply filter

12.3.8 Substrate Noise

The body terminal of a bulk CMOS transistor is connected to the substrate or well. The p-type substrate for an nMOS transistor is normally connected to GND and the n-well for a pMOS transistor is normally connected to V_{DD}. The connection is made through a relatively high-resistance substrate or well contact. Current flow in the substrate causes noise on the body terminal. This current may come from capacitive coupling through the reverse-biased source/drain to substrate diodes or from impact ionization as current flows through an ON transistor. The substrate noise modulates threshold voltages by means of the body effect.

Substrate noise is also a problem for *mixed-signal* designs where separate power supplies are used for noisy digital circuits and quiet analog circuits. The large number of rapidly switching digital circuits creates noise on the digital ground that propagates to the sensitive analog circuitry via the common substrate.

The substrate and well should use plenty of contacts to guarantee a low-resistance path to the power network. Guard rings, described in Section 4.8.5, provide some protection against noise caused by currents in the nearby substrate. Analog circuits should be physically separated from digital circuits and protected by guard rings connected to a quiet analog supply. Twin-tub or triple-well processes and SOI also experience much less substrate coupling because transistors are isolated in their own wells.

Modeling and analyzing substrate noise is beyond the scope of this book. See [Donnay03] for extensive coverage of the subject.

12.4 I/O

The input/output (I/O) subsystem is responsible for communicating data between the chip and the external world. A good I/O subsystem has the following properties:

- Drives large capacitances typical of off-chip signals
- Operates at voltage levels compatible with other chips
- Provides adequate bandwidth
- Limits slew rates to control high-frequency noise
- Protects chip against damage from electrostatic discharge (ESD)
- Protects against over-voltage damage
- Has a small number of pins (low cost)

I/O pad design requires specialized analog expertise and knowledge of process-specific ESD structures. [Dabral98] describes many of these issues. Thus, the system designer should obtain a well-characterized pad library from the processor or library vendor that is suited to the manufacturing process. This section summarizes some of the basic design options in I/O subsystems.

A pad consists of a square of top-level metal of approximately $100\,\mu m$ on a side that is either soldered to a bond wire connecting to the package or coated with a lead solder ball. The term *pad* sometimes refers to just the metal square and other times to the complete cell containing the metal, ESD protection circuitry, and I/O transistors. Input and output pads usually contain built-in receiver and driver circuits to perform level conversion and amplification. Section 12.4.1 addresses some of these basic circuits, along with their ESD and overvoltage protection circuits. Section 12.4.2 examines the MOSIS $1.6\,\mu m$ I/O pads to illustrate basic pad design.

12.4.1   Basic I/O Pad Circuits

Basic I/O pads include V_{DD} and GND, digital input, output, and bidirectional pads, and analog pads. This section is primarily concerned with simple digital pads that operate on the same CMOS voltage levels as the core.

12.4.1.1 V_{DD} and GND Pads

Power and ground pads are simply squares of metal connected to the package and the on-chip power grid. Most high-performance chips devote about half of their pins to power and ground. This large number of pins is required to carry the high current and to provide low supply inductance.

One of the largest sources of noise in many chips is the ground bounce caused when output pads switch. The pads must rapidly charge the large external capacitive loads, causing a big current spike and high L di/dt noise. The problem is especially bad when many pins switch simultaneously, as could be the case in a 64-bit off-chip data bus. Such busses should be interdigitated with many power and ground pins to supply the output current through a low-inductance path. In many designs, the *dirty* power and ground lines serving the output pads are separated from the main power grid to reduce the coupling of I/O-related noise into the core.

12.4.1.2   Output Pads

First and foremost, an output pad must have sufficient drive capability to deliver adequate rise and fall times into a given capacitive load. If the pad drives resistive loads, it must also deliver enough current to meet the required DC transfer characteristics. Given a load capacitance (typically 2–50 pF) and a rise/fall time specification, the output transistor widths can be calculated or determined through simulation. Typically, these transistors must be very wide and are folded into many legs.

Output pads generally contain additional buffering to reduce the load seen by the on-chip circuitry driving the pad. The method of logical effort tells us that the fastest buffers are built from strings of inverters with fanouts of about 4. In practice, a higher fanout (e.g., 6–8) gives nearly as good delay while reducing the area and power consumption of the buffer. The final stage may have an especially high fanout because the edge rates in the external world are normally an order of magnitude longer than those on chip. However, the final stage must be large enough to source or sink reasonable amounts of current with a small voltage drop.

Latchup is a particular problem near output pads, especially when the pads experience voltage transients above V_{DD} or below GND. These transients are likely to occur because of ringing from the bond wire inductance and/or from driving improperly terminated

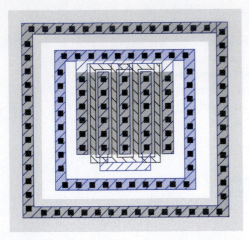

FIG 12.17 Double guard rings around folded nMOS output transistor

transmission lines. These transients cause the drain-to-body diodes to become forward-biased, forcing current to flow into the substrate or well and potentially causing latchup.

To avoid latchup, the nMOS and pMOS transistors should be separated by substantial distances and surrounded by guard rings. If possible, the output transistors (i.e., those whose drains connect directly to external circuitry) should be doubly guard-ringed, as shown in Figure 12.17. This means that an n-transistor should be encircled with p+ substrate contacts connected to GND, and then further encircled with n+ well contacts in an n-well connected to V_{DD}. The rings should be continuous in diffusion with frequent contacts to metal. Furthermore, dummy collectors consisting of p+ connections to GND and n+ in n-well connections to V_{DD} should be placed between the output transistors and any internal circuitry. These dummy collectors and guard rings serve to capture most of the stray carriers injected into the substrate when the diodes are forward-biased.

The output transistors also often have gates longer than normal to prevent avalanche breakdown damage when overvoltage is applied to the drains. Non-silicided gates are also preferable because the polysilicon gate resistance better distributes overvoltage across the legs of the output transistor, preventing damage.

12.4.1.3 Input Pads

Input pads also contain an inverter or buffer as a level of protection between the pad and core circuitry. The buffer can perform some level conversion or noise filtering as well. For example, an input pad might receive a TTL signal with V_{OL} = 0.4 V and V_{OH} = 2.4 V. If the pad operates at 5 V, it should use a LO-skewed inverter with a trip point well below 2.4 V to properly receive the signal. Pads can include pull-up or pull-down resistors to place an unconnected pad in a certain state. Remember that even these structures must be guard-ringed if implemented in diffusion.

Some input pads also contain *Schmitt triggers*, as shown in Figure 12.18 [Schmitt38]. A Schmitt trigger has hysteresis that raises the switching point when the input is low and lowers the switching point when the input is high. This helps filter out glitches that might occur if the input rises slowly or is rather noisy.

On a dry day, you have probably experienced a shock when you touch a metal object because you have built up so much charge on your body while walking on carpets. Such shocks can destroy integrated circuits. Input pads have transistor gates connected directly to the external world. These gates are subject to damage from electrostatic discharge that can puncture and break down the oxide. The breakdown voltage was 40–100 V for older processes with thick (> 100 Å) oxides but now is 5 V or less for modern thin oxides.

Usually a combination of resistance and diode clamps is used to protect the gate oxides, as shown in Figure 12.19. The

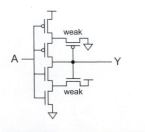
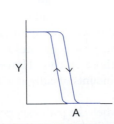

FIG 12.18 Schmitt trigger

diodes turn on if the pad voltage becomes greater than about V_{DD} + 0.7 V or less than –0.7 V. The resistor limits the peak current that flows into these diodes, preventing burnout. Resistor values anywhere from 100 Ω to 3 kΩ are used. This resistance, in conjunction with any input capacitance C, will lead to an RC time constant that can be important for high-speed circuits. The resistors are sometimes made from several squares of unsilicided p+ diffusion in an n-well. Clamping diodes are formed using n+ diffusion to the substrate and p+ diffusion to n-wells. As with output transistors, these diodes should be double guard-ringed so that they do not inject charge into the substrate and cause latchup. Some processes offer extra mask steps to build transistors with thicker oxides. These thick-oxide transistors are more robust against ESD and can operate at I/O voltages greater than the core voltage.

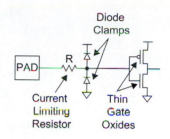

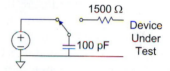

FIG 12.19 Input protection circuitry

ESD protection circuits are tested by zapping the pin with an external high voltage. Figure 12.20 shows a test circuit for the *Human Body Model* (HBM) of electrostatic discharge. This model simulates the discharge that takes place when a person who is improperly grounded touches a pin of a chip. The human is represented as a 100 pF capacitor in series with a 1500 Ω resistor. During test, the capacitor is charged to a high voltage, and then discharged through the resistor into the pad of the chip. The ESD robustness of the pad is measured as the maximum voltage that the pad can endure. For example, ±15 kV is good for parts such as serial port transceivers that might be exposed to ESD by an end user who handles a cable. Parts in an enclosed system are only subject to damage during assembly and can allow limits in the 2–4 kV range.

FIG 12.20 Human Body Model ESD test circuit

12.4.1.4 Bidirectional Pads
Figure 12.21 shows a bidirectional pad with an output driver that can be tristated and an input receiver. The output driver consists of independently controlled nMOS and pMOS transistors. When the enable is '1,' one of the two transistors turns ON. When the enable is '0,' both transistors are OFF so the pad is tristated. This design is preferable to the four-transistor "totem pole" tristate from Section 1.4.7 when driving large capacitances because it has only two rather than four huge transistors in the final stage and the transistors need only be half as wide. Figure 12.22 shows a clever variation on this design in which the NAND and NOR are merged together into a single 6-transistor network with two outputs. Such a tristate buffer is smaller and presents less input capacitance on the D_{out} terminal.

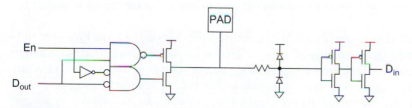

FIG 12.21 Bidirectional pad circuitry

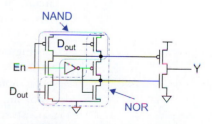

FIG 12.22 Improved tristate buffer

Many pad libraries provide only a bidirectional pad. By hardwiring the enable signal to '1' or '0', the pad can be used as an output or input.

12.4.1.5 Analog Pads   Analog inputs and outputs connect to simple metal pads and then directly to the on-chip analog circuitry without any digital buffer or driver. Analog pads still require protection circuitry. The protection circuitry must be carefully designed so it does not degrade the bandwidth or signal integrity of the analog components. This is achieved by minimizing the protection diode area. RF pads are extremely demanding because any extra load can compromise performance.

12.4.2   Example: MOSIS I/O Pads

Figure 12.23 shows a layout of a bidirectional pad from the MOSIS service for a 1.6 μm two-metal layer process (see also the inside front cover). The overall cell is about 200 μm on a side. The pad is the large (100 x 75 μm) rectangle consisting of a sandwich of metal1 and metal2 connected with many vias. The SiO_2 overglass covering the metal2 is etched away over the pad so the bond wire can be connected directly to the pad. Two large metal2 rectangles cover most of the pad. The upper one with the legs sticking up is GND, while the lower is V_{DD}.

The bidirectional pad schematic is shown in Figure 12.24. The input protection circuitry consists of some resistance, a thick oxide transistor, and the drain diffusion diodes of the wide output transistors. The resistors are n+ and p+ diffusion wires, each 3.5 squares long. They have nominal sheet resistances of 53 and 75 Ω/□, so the parallel combination of resistance is 150 Ω. To the left and right of the metal pad are thick oxide nMOS transistors consisting of interdigitated fingers. They consist of a source and drain separated by 3 λ, but have no gate. They help protect the pad from ESD because high voltages will punch through the channel and dissipate. The effectiveness of thick oxide transistors is process-dependent. The pad uses many substrate/well contacts and is surrounded by double guard rings to prevent latchup during ESD events. The tristate driver and receiver use extensively folded transistors to fit in the space available.

12.4.3   Level Converters

Many chips require a low core voltage for the logic transistors, yet must interface with other chips operating at higher voltages. The I/O pads thus can include level converter circuits to translate between different voltage standards. As introduced in Section 3.2.7, these circuits often use transistors with longer channels and thicker oxides to endure the higher voltages.

Figure 12.25 shows some simple level converters for chips using a low V_{DDL} core voltage and higher V_{DDH} I/O voltage. Figure 12.25(a) is an output driver that takes a low-swing input voltage and produces a higher-swing output voltage. It uses a CVSL structure consisting of four high-voltage transistors indicated in bold. The inverter uses low-voltage transistors and the low-voltage power supply. The output Y can be followed by a high-voltage inverter or buffer to deliver more uniform rise/fall times. Figure 12.25(b) is an input receiver that takes a high-swing input voltage and produces a lower-swing voltage

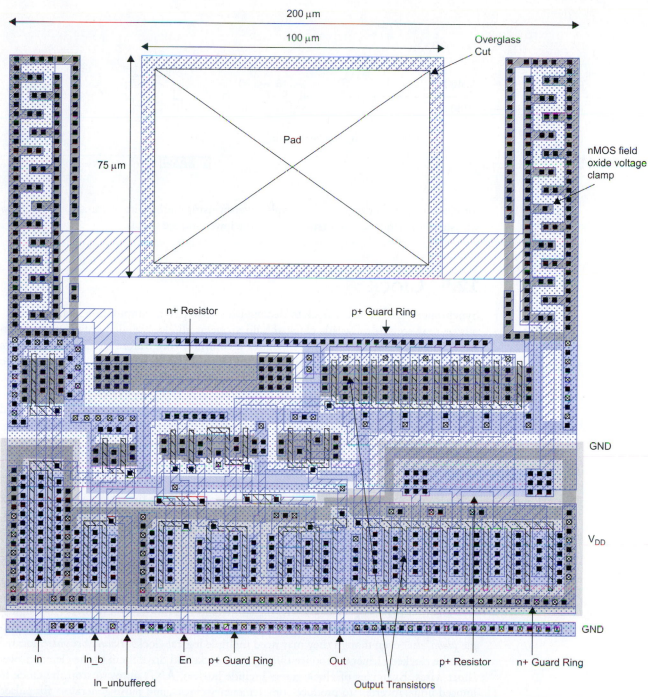

FIG 12.23 MOSIS 1.6 μm bidirectional pad. Color version on inside front cover.

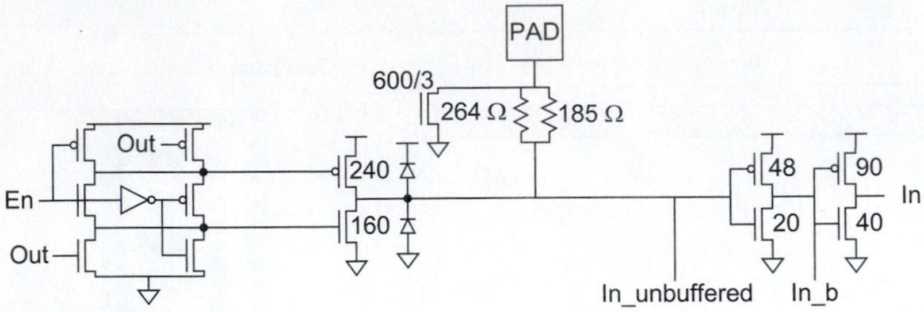

FIG 12.24 MOSIS bidirectional pad schematic

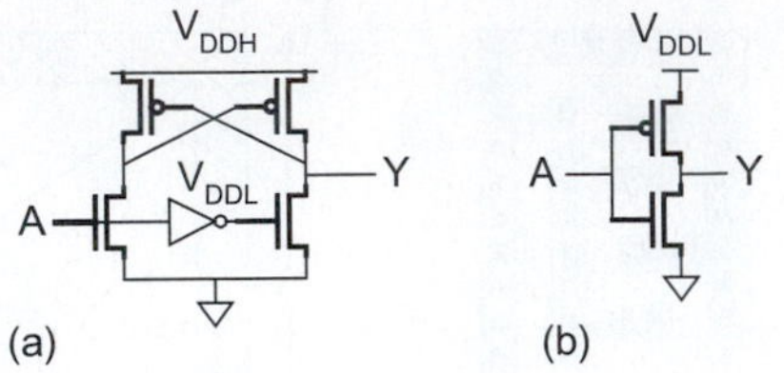

FIG 12.25 Level converters

for core circuits. It consists of a simple inverter using high-voltage transistors that can withstand the large gate voltages, but uses the lower voltage supply.

12.5 Clock

Synchronous systems use a clock to distinguish one step in a computation from the previous or next step. Ideally, this clock should arrive at all clocked elements in the system simultaneously so that the system shares a common time reference. These elements include latches and flip-flops, memories, and dynamic gates. In practice, the arrival time differs somewhat from one point to another; this difference is called *clock skew*. The central challenge in clock system design is to deliver the clock to all the clocked elements on the chip while finding an acceptable compromise among skew, power consumption, metal resource usage, and design effort.

12.5.1 Definitions

A system is designed to use one or more *logical clocks*. The logical clocks are idealized signals with no skew used by the logic designer when describing the system with a hardware description language. For example, a system with flip-flops requires a single logical clock, usually called *clk*. A system using two-phase transparent latches requires two logical clocks ϕ_1 and ϕ_2 (or ph1 and ph2 in a hardware description language). Unfortunately, mismatched clock network paths and processing and environmental variations make it impossible for all clocks to arrive at their ideal times, so the designer must settle for actually receiving a multitude of skewed *physical clocks*.

Distributing a single clock across the entire chip in a low-skew fashion is challenging. Distributing more than one is nearly impossible. Therefore, most systems distribute a single *global clock* even though they may need multiple logical clocks. *Local clock gaters* located near the clocked elements produce the physical clocks and drive them to the elements over short wires. Examples of clock gaters include buffers, AND gates to stop the clock to unused units, inverters to produce complementary clocks, and pulse generators for pulsed latches.

The term *clock skew* has been used informally in many ways. We will define skew precisely as the *difference between the nominal and actual interarrival time of a pair of physical clocks*. For example, Figure 12.26(a) shows a system with two flip-flops. Both should receive the logical clock *clk* with zero interarrival time, but they actually receive physical clocks clk_1 and clk_2. Because of differences in the delay of the clock distribution wires and the local clock buffers, clk_1 arrives 25 ps before clk_2. Therefore, we say the clock skew is 25 ps. Figure 12.26(b) shows a system with three transparent latches. The latches use complementary logical clocks ϕ_1 and ϕ_2 with a nominal interarrival time of $T_c/2$ between rising edges. They actually receive physical clocks ϕ_{1a}, ϕ_{1b}, and ϕ_{2a}. We see the clock skews are:

$$t_{skew}^{\phi_{1a},\phi_{1b}} = 5 \text{ ps}; \quad t_{skew}^{\phi_{1a},\phi_{2a}} = 15 \text{ ps}; \quad t_{skew}^{\phi_{1b},\phi_{2a}} = 10 \text{ ps}$$

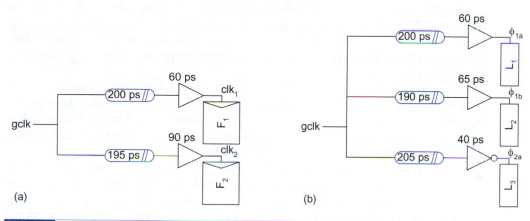

FIG 12.26 Clock skew example

Sometimes designers intentionally delay clocks to solve setup or hold time problems. For example, suppose that a critical path existed between F_1 and F_2 in Figure 12.26(a). The designer might intentionally delay the clock to F_2 by 30 ps to give the path more time by using the slower local clock buffer on clk_2. In this case, the nominal interarrival time of clk_1 and clk_2 is 30 ps. The actual interarrival time is 25 ps, so the clock skew is 5 ps. Some designers call this 30 ps delay *intentional skew*. We prefer to call it *intentional delay* and reserve the term *clock skew* to account for unintentional differences in clock arrival times.

Clock skew can also be measured between different edges of the clock or between different cycles. For example, Figure 12.27 shows two physical clock waveforms in which the edges differ from their nominal timing. The clock skews are defined based on the edge (*rising/falling*) and the number of intervening cycles as well as the physical clock:

$$t_{skew}^{clk_1,clk_1,(r,r,0)} = 0 \text{ ps}; \quad t_{skew}^{clk_1,clk_1,(r,f,0)} = 30 \text{ ps}; \quad t_{skew}^{clk_1,clk_1,(r,r,1)} = 70 \text{ ps}$$
$$t_{skew}^{clk_1,clk_2,(r,r,0)} = 0 \text{ ps}; \quad t_{skew}^{clk_1,clk_2,(r,f,0)} = 0 \text{ ps}; \quad t_{skew}^{clk_1,clk_2,(r,r,1)} = 40 \text{ ps}$$

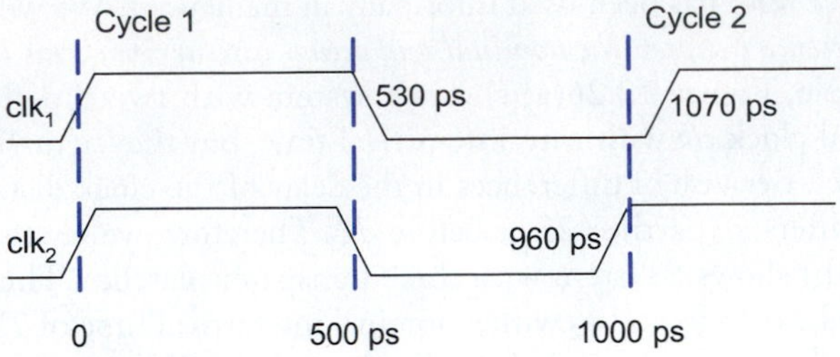

FIG 12.27 Skewed clock waveforms

For a path between two flip-flops, the hold time constraint depends on the skew between the same rising edges of both physical clocks. The setup time constraint depends on the skew between the rising edge of one physical clock and the subsequent rising edge of the other. We will see that clock distribution networks tend to introduce more skew from one cycle to the next so setup and hold time constraints can budget different amounts of skew.

The actual clock skew between two clocked elements varies with time and is different from one chip to another. Moreover, it is unknowable at design time. From the engineering perspective, a more useful parameter is the *clock skew budget*. The clock skew budget should be larger than the actual skew encountered on any long or short path on any working chip, yet no larger than necessary lest the chip be overdesigned.

While in principle designers could tabulate clock skew budgets between physical clocks at every pair of clocked elements on the chip, the table would be unreasonably large and unwieldy. Instead, they group physical clocks into *clock domains* and use a single skew budget to describe the entire domain. For example, you could define two latches to be in a local clock domain if their physical distance is no more than 500 μm. Then you could just define local and global skews, with the local skew being smaller than the global skew. If the clock period is long compared to the maximum skew, you can define only a single global skew budget and pessimistically assume all clocked elements might see this worst-case skew.

Clock skew sources can be classified as *systematic*, *random*, *drift*, and *jitter*. Figure 12.28(a) illustrates these sources in a simple clock distribution network. The global clock is distributed along wires to two gaters. One wire is 3 mm, while the other is 3.1 mm. The gaters are nominally identical, but one drives a lumped load of 1.3 pF while the other

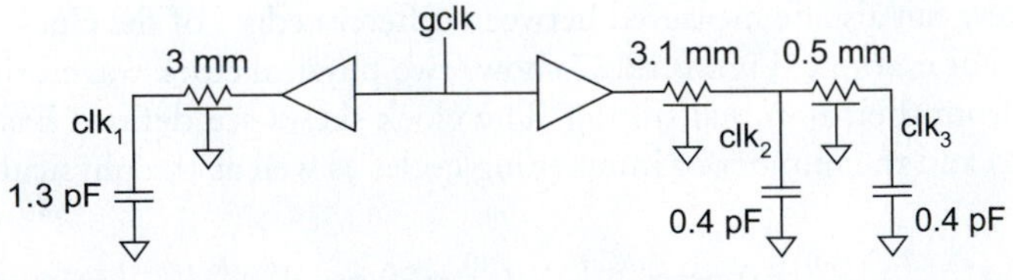

FIG 12.28 Simple clock distribution network

drives a load of 0.8 pF distributed along a 0.5 mm wire. The *systematic* clock skew is the portion that exists even under nominal conditions; this component can be predicted by simulation. By adjusting the size of one of the gaters, the systematic skew between clk_1 and clk_2 could be driven to zero. However, some systematic skew will always exist between clk_2 and clk_3 because of the flight time along the wire after the gater.

The *random* component of skew is caused by manufacturing variations that could affect the wire width, thickness, or spacing and the transistor channel length, threshold voltage, or oxide thickness. These cause unpredictable changes in resistance, capacitance, and transistor current, introducing additional skew. In principle, the actual random skew could be measured during chip test or on startup, and adjustable delay elements could be calibrated to compensate for the random skew.

Drift is caused by time-dependent environmental variations that occur relatively slowly. For example, after the chip turns on, it will heat up. The temperature affects gate and wire delay differently. Also, a temperature gradient across the chip leads to skew. Drift can also be nulled out with adjustable delay elements. Unlike random skew, compensating for drift must take place periodically rather than just once at startup. The frequency of calibration depends on the thermal time constant of the chip.

Jitter is caused by high-frequency environmental variation, particularly power supply noise. This noise leads to delay variation in the clock buffers and gaters in both time and space. Jitter is particularly insidious because it occurs too rapidly for compensation circuits to be able to counter it.

Some engineers do not report jitter as part of the skew. In such a case, they must include both jitter and skew in the setup and hold time budgets.

12.5.2 Clock System Architecture

Figure 12.29 shows an overview of a typical clock subsystem. The chip receives an external clock signal through the I/O pads. The clock generation unit may include a *phase-locked loop* (PLL) or *delay-locked loop* (DLL) to adjust the frequency or phase of the global clock, as shall be discussed in Section 12.5.3. This global clock is then distributed across the chip to points near all of the clocked elements. The clock distribution network must be carefully designed to minimize clock skew. Local clock gaters receive this global clock and drive the physical clock signals along short wires to small groups of clocked elements.

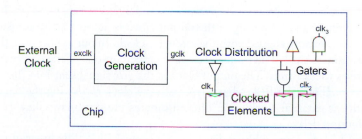

FIG 12.29 Clock subsystem

12.5.3   Global Clock Generation

The global clock generator receives an external clock signal and produces the global clock that will be distributed across the die. In this simplest case, the clock generator is simply a buffer to drive the large capacitance of the clock distribution network. However, the input pad, buffering, distribution network, and gaters have significant delay that leads to a large skew between the external clock and the physical clocks received at the clocked elements. Moreover, this delay varies with processing and environment. Because of this skew, clocked elements on the chip are no longer synchronized with the external I/O signals. Guaranteeing setup and hold times becomes problematic, particularly at high frequencies where the skew exceeds half of the clock period. More sophisticated clock generators use *phase-locked loops* (PLLs) to compensate for this delay. Moreover, phase-locked loops can perform frequency multiplication to provide an on-chip clock at a higher frequency than the external clock.

Figure 12.30 shows a typical synchronous chip interface using a phase-locked loop to compensate for on-chip clock delays [Chandrakasan01]. In this example, Chip 1 sends a clock and D_{out} to Chip 2 and receives D_{in} back from Chip 2. The data should be synchronized to the clock so each chip can sample it on the positive clock edge. However, the internal clock on Chip 2 is delayed through the clock distribution network. The PLL adjusts the internal clock in Chip 2 to correct for this delay and keep the clock synchronized with the data.

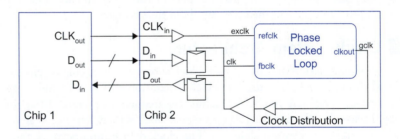

FIG 12.30 Synchronous chip interface with PLL

A phase-locked loop receives a *reference clock* and a *feedback clock* and produces an *output clock*. The phase and frequency of the output clock is automatically adjusted until the feedback clock is exactly aligned with the reference clock. In our application, the reference clock comes from Chip 1. The feedback clock is tapped off one of the physical clock wires that also drives the registers. The output clock is the *gclk* signal sent into the clock distribution network. Thus, the PLL adjusts *gclk* until the clock *clk* driving the data registers is aligned with the external clock *exclk*. This eliminates the systematic skew caused by the clock distribution delay.

Figure 12.31(a) shows a block diagram of a typical PLL. The heart of the PLL is a *voltage-controlled oscillator* (VCO). The control voltage is adjusted until the oscillator produces an output clock of the proper phase at the same frequency as the reference clock. This control is performed with a charge pump and RC filter. A phase detector determines whether the feedback clock leads or lags the reference clock. The charge pump consists of a pair of current sources enabled by the up and down signals to adjust the voltage on V_{ctrl} until the feedback clock becomes aligned with the reference.

Figure 12.31(b) shows a PLL that performs frequency multiplication. It uses a divide-by-N counter on the *fbclk* to generate a clock aligned in phase but at N times the frequency of the reference clock. For example, the Pentium 4 uses a 100 MHz external system clock that can be multiplied up to a 3 GHz core clock. With another divide-by-M counter on the *refclk* terminal, the PLL can produce any rational N/M multiple of the input frequency.

A PLL is a feedback system and requires careful analysis to ensure stability over all process and environmental corners. The RC loop filter behaves as a second-order system, although stray capacitance on V_{ctrl} often leads to third-order effects. Noise on V_{ctrl} causes the VCO to change frequency, which leads to a phase error that increases over time; this problem is called *phase-error accumulation* and appears as jitter in the clock skew budget. Therefore, the loop filter should have high enough bandwidth to rapidly correct for noise. Phase-error accumulation impacts the timing budget for one chip communicating with another. However, only cycle-cycle jitter is important for on-chip paths between flip-flops. RC loop filters are difficult to build because of uncertainties in the passive component values caused by process variation. Thus, many PLLs use an active filter.

Figure 12.32 shows a circuit-level implementation of a simple PLL. The phase detector is a *phase-frequency detector* consisting of a pair of flip-flops with asynchronous reset that pro-

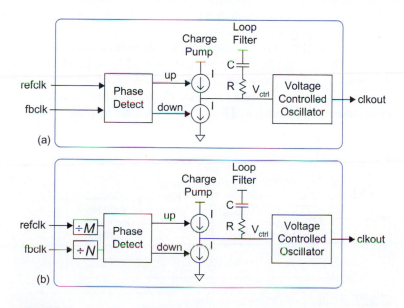

FIG 12.31 Phase-locked loop block diagram

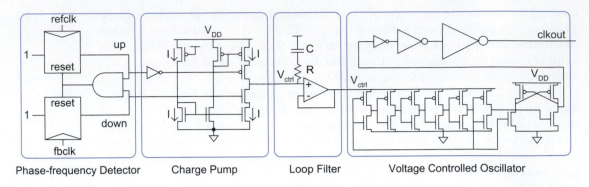

FIG 12.32 Simple PLL implementation

duces pulses to drive the frequency up or down depending on which clock arrived first. The charge pump uses a pair of current sources switched on by the *up* and *down* signals. Current sources are discussed further in Section 12.6.4. V_{ctrl} is buffered with an amplifier such as that from Figure 12.63 hooked up as a unity-gain follower. The VCO consists of an ordinary ring oscillator running off V_{ctrl} rather than V_{DD} so that its period increases as V_{ctrl} decreases. The number of stages determines the range over which the frequency can be adjusted. A level converter and some buffers amplify the signal to drive *clkout* across the chip.

Power supply and substrate noise are the primary sources of PLL jitter, so the PLL should use a filtered power supply and generous guard rings to limit jitter. [Maneatis96, Maneatis03] describe self-biased delay elements that minimize sensitivity to supply noise and process variation. [Ingino01] describes a PLL with a filtered power supply operating up to 4 GHz with a ±25 ps peak-to-peak jitter.

A delay-locked loop (DLL) is a variant of a PLL that uses a voltage-controlled delay line rather than a voltage-controlled oscillator, as shown in Figure 12.33. It can be viewed as a control system that adjusts phase rather than frequency on *clkout*. DLLs are not capable of frequency multiplication. However, noise on V_{ctrl} creates only a steady phase error, so DLLs avoid phase-error accumulation. DLLs use a first-order loop filter, so ensuring stability is also easier.

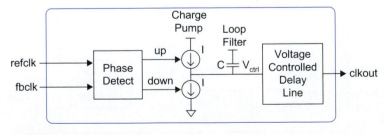

FIG 12.33 Delay-locked loop block diagram

PLLs and DLLs are notoriously difficult to build correctly. They require expertise in both feedback control systems and analog design. They must be carefully designed to acquire lock successfully and operate correctly with low jitter across process and environmental variations. A number of texts including [Chandrakasan01, Baker98, Dally98, Best03] offer introductions to loop design. For many applications, loops are best obtained as predesigned cells from a third party that specializes in loop design and offers a guarantee. True Circuits is a well-regarded PLL/DLL supplier.

12.5.4 Global Clock Distribution

The global clock must be distributed across the chip in a way that reaches all of the clocked elements at nearly the same time. In antiquated processes with slow transistors and fast wires, the clock wire had negligible delay and any convenient routing plan could be used to distribute the clock. In modern processes, the RC delay of the resistive clock wire driving its own capacitance and the clock load capacitance tends to be close to 1 ns for a well-designed distribution network covering a 15 mm square die. If the clock were routed randomly, this would lead to a clock skew of about 1 ns between physical clocks near and far from the clock generator. This could be several times the cycle time of the system. Thus, the clock distribution system must be carefully designed to equalize the *flight time* between the clock generator and the clocked receivers. Global clock distribution networks can be classified as *grids*, *H-trees*, *spines*, *ad-hoc*, or *hybrid* [Restle98].

Random skew, drift, and jitter from the clock distribution network are proportional to the delay through the network because they are caused by process or environmental variations in the distribution elements. Therefore, the designer should try to keep this distribution delay low. Unfortunately, as chips are getting larger, wires are getting slower, and clock loads are increasing, the distribution delay tends to go up even as cycle times are going down. In the past, systematic clock skew was the dominant component. Now, good clock distribution networks achieve low systematic skews, but the random, drift, and jitter components are becoming an increasing fraction of the cycle time.

12.5.4.1 Grids A clock grid is a mesh of horizontal and vertical wires driven from the middle or edges. The mesh is fine enough to deliver the clock to points nearby every clocked element. The resistance is low between any two nearby points in the mesh so the skew is also low between nearby clocked elements. This reduces the chance of hold-time problems because such problems tend to occur between nearby elements where the propagation delay between elements is also small. Grids also compensate for much of the random skew because shorting the clock together makes variations in delays irrelevant. They can be routed early in the design without detailed knowledge of latch placement. However, grids do have significant systematic skew between the points closest to the drivers and the points furthest away. They also consume a large amount of metal resources and hence have a high switching capacitance and power consumption. Section 12.8 traces the evolution of clock grids in the Alpha series of microprocessors.

12.5.4.2 H-Trees

An H-tree is a fractal structure built by drawing an H shape, then recursively drawing H shapes on each of the vertices, as shown in Figure 12.34. With enough recursions, the H-tree can distribute a clock from the center to within an arbitrarily short distance of every point on the chip while maintaining exactly equal wire lengths. Buffers are added as necessary to serve as repeaters. If the clock loads were uniformly distributed around the chip, the H-tree would have zero systematic skew. Moreover, the trees tend to use less wire and thus have lower capacitance than grids [Restle98].

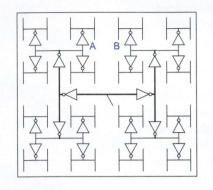

FIG 12.34 H-tree

In practice, the H-tree still shows some skew because the clock loads are not uniform, loading some leaves of the tree more than others. Moreover, the tree often must be routed around obstructions such as memory arrays. The leaves of the H do not reach every point on the chip, so some short physical clock wires are required after the local clock gater. Nevertheless, with careful tapering of the wires and sizing of the clock gaters, H-trees can deliver nearly zero systematic skew. A drawback of H-trees is that they may have high random skew, drift, and jitter between two nearby points that are leaves of different legs of the tree. For example, the points A and B in Figure 12.34 might experience large skews. As the points are close, this is a particular problem for hold times.

Figure 12.35 shows a modified H-tree used on the Itanium 2. The primary clock driver in the center of the chip sends a differential output to four differential repeaters on the leaves of the H. These repeaters drive a somewhat irregular pattern of wiring to second-level clock buffers (SLCBs) serving units all across the chip. The wiring and SLCB placement is determined by the nonuniform clock loads and obstructions on the chip. A custom clock router automatically generated the tree based on the actual clock loads so that the tree could be easily rerouted when loads change late in the design process. The SLCBs drive local clock gaters, producing the multitude of clock waveforms used on the microprocessor. Some of these waveforms were shown in Section 7.9.2.

Figure 12.36 shows the differential driver used as a primary clock buffer and repeater on the Itanium 2 [Anderson02]. The input stage is a differential amplifier sensitive to the point where the differential inputs cross over. The repeater pulses either p_1 or n_1 and p_2 or n_2 to switch the internal nodes y and $\bar{y}$. The small tristate keeper prevents these nodes from floating after the pulse terminates. The SLCB uses the same structure, but produces only a single-ended output. It also provides a current-starved adjustable delay line to compensate for systematic skew and to help locate critical paths during debug. The repeater provides a high drive capability with a low input capacitance. Thus, few stages of clock buffering are needed in the network. With so few repeaters, the area overhead of providing a filtered power supply is modest. Although the repeaters are relatively slow, their jitter is controlled with supply filtering.

12.5.4.3 Spines

Figure 12.37 shows a clock distribution scheme using a pair of spines. As with the grid, the clock buffers are located in a few rows across the chip. However, instead of driving a single clock grid across the entire die, the spines drive length-matched serpentine wires to each small group of clocked elements. If the loads are uniform, the

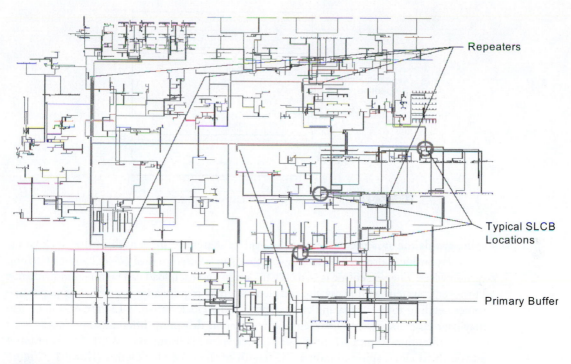

FIG 12.35 Itanium 2 modified H-tree

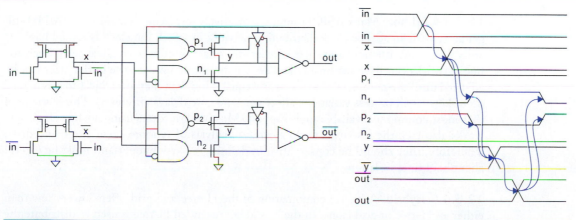

FIG 12.36 Itaniuim 2 repeater

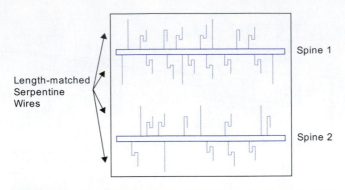

FIG 12.37 Clock spines with serpentine routing

spine avoids the systematic skew of the grid by matching the length of the clock wires. Each serpentine is driven individually so gaters can be used to save power by not switching certain wires. The serpentine is also easy to design and each load can be tuned individually. However, a system with many clocked elements may require a large number of serpentine routes, leading to high area and capacitance for the clock network. Like trees, spines also may have large local skews between nearby elements driven by different serpentines.

The Pentium II and III use a pair of clock spines [Geannopoulos98]. The Pentium 4 adds a third clock spine to reduce the length of the final clock wires [Kurd01]. Figure 12.38(a) shows the global clock buffers distributing the clock to the three spines on the Pentium 4 with zero systematic skew while Figure 12.38(b) shows a photograph of the chip annotated with the clock spine locations. The spines drive 47 independent clock domains, each of which can be gated individually. The clock domain gaters also contain adjustable delay buffers used to null out systematic and random skew and even to force deliberate clock delay to improve performance.

12.5.4.4 Ad-hoc Many ASICs running at relatively low frequencies (100's of MHz) still get away with an ad-hoc clock distribution network in which the clock is routed haphazardly with some attempt to equalize wire lengths or add buffers to equalize delay. Such ad-hoc networks can have reasonably low systematic skews because the buffer sizes can be adjusted until the nominal delays are nearly equal. However, they are subject to severe random skew when process variations affect wire and gate delays differently. This is the level that most commonly available tools support. Most design teams using ad-hoc clock networks also lack the resources to do a careful analysis of random skew, jitter, and drift. Therefore, they should be conservative in defining a skew budget and must be careful about hold time violations.

12.5.4.5 Hybrid A hybrid combination of the H-tree and grid offers lower skew than either an H-tree or grid alone. In the hybrid approach, an H-tree is used to distribute the clock to a large number of points across the die. A grid shorts these points together. Com-

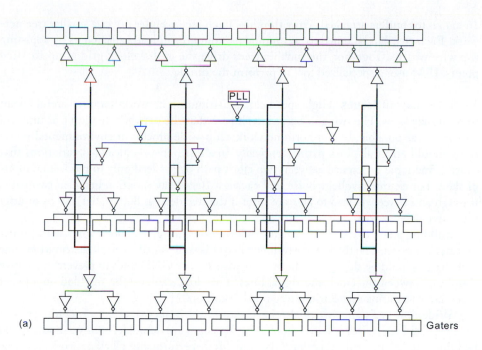

(a)   Gaters

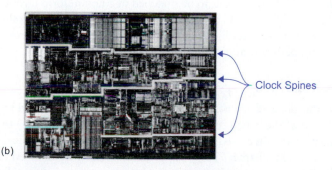

(b)

FIG 12.38 Pentium 4 clock spines. (b) © 2001 IEEE.

pared to a simple grid, the hybrid approach has lower systematic skew because the grid is driven from many points instead of just the middle or edge. Compared to an H-tree, the hybrid approach is less susceptible to skew from nonuniform load distributions. The grid also reduces local skew and brings the clock near every location where it is needed. Finally, the hybrid approach is regular, making layout of well-controlled transmission line structures easier.

IBM has used such a hybrid distribution network on a variety of microprocessors including the Power4, PowerPC, and S/390 [Restle01]. A primary buffered H-tree drives

16–64 sector buffers arranged across the chip. Each sector buffer drives a smaller tree network. Each tree can be tuned to accommodate nonuniform load capacitance by adjusting the wire widths. Together, the tunable trees drive the global clock grid at up to 1024 points. IBM uses a specialized tool to perform the tuning.

12.5.4.6 Layout Issues   High-speed clock distribution networks require careful layout to minimize skew. The two guiding principles are that the network should be as uniform and as fast as possible. In a uniform network, chip-wide process or environmental variations should affect all clock paths identically. In a fast network, localized variations that cause a fractional difference between two clock path delays lead only to modest amounts of skew. For example, voltage noise that causes a 10% delay variation between two paths through an H-tree will lead to 80 ps of jitter if the tree delay is 800 ps, but 160 ps of jitter if the tree delay is 1600 ps.

Building a fast clock network requires low-resistance global clock wires with proper repeater insertion. The thick, top-level metal layer is well-suited to clock distribution. The wide wires should be shielded on both sides with V_{DD} or GND lines to prevent capacitive coupling between the clock and signal lines. The clock can even be shielded on a lower metal layer to form a microstrip waveguide [Anderson02].

Wide, low-resistance wires also have significant inductive effects, including faster than expected edge rates and overshoot. The fast edges are desirable, but overshoot should be minimized to prevent overvoltage damage. High-performance clock networks must be extracted with a field solver and modeled as transmission lines [Huang03]. Uniformity is again important: Even if the RC delays appear to be matched in a nonuniform layout, the *RLC* delays can be significantly different. As discussed in Section 4.5.5, wide wires should be split into multiple narrower traces interdigitated with V_{DD}/GND wires that provide a low-inductance current return path and minimize skin effect.

12.5.5   Local Clock Gaters

Local clock gaters receive the global clock and produce the physical clocks required by the clocked elements. The output of the gaters typically run a short distance (< 1 mm) to the clocked elements. Clock gaters are often used to stop or *gate* the clock to unused blocks of logic to save power. As discussed in Chapter 7, they can produce a variety of modified clock waveforms including pulsed clocks, delayed clocks, stretched clocks, nonoverlapping clocks, and double-frequency pulsed clocks. When used to modify the clock edges, they are sometimes called *clock choppers* or *clock stretchers*. Figure 12.39 shows a variety of clock gaters.

Most systems require a large number of clock gaters, so it is impractical to filter the power supply at every one. Variations in clock gater delay caused by voltage noise, cross-die process variation, and nonuniform temperature distribution cause skew between clocks produced by different gaters. The best way to limit this skew is to make the gater delay as short as possible. Variations in the input threshold of the clocked elements also causes skew. The best way to limit this skew is to produce crisp rise/fall times at the clock gaters. The final stage should have a fanout of no more than about 4.

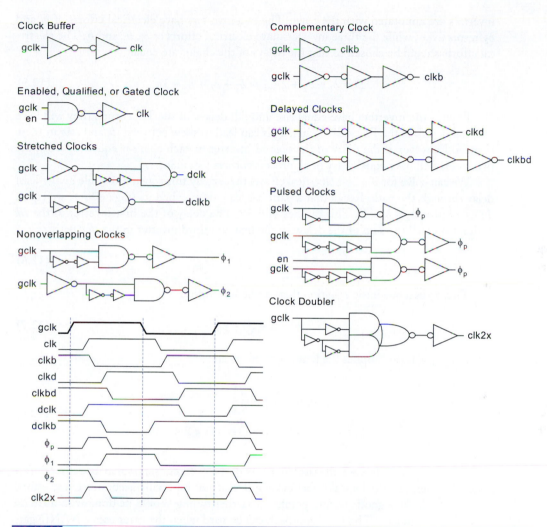

FIG 12.39 Clock gaters

Clock gaters may introduce some systematic delay between phases. For example, if *clkb* is produced with three inverters while *clk* is produced with only two, *clkb* may be delayed slightly from *clk*. The designer can either choose to carefully size the inverters such that the net delay is equal or accept that the delays are unequal and simply roll the systematic difference into timing analysis.

Figure 12.40 shows a circuit in which the delay of two inverters is matched against the delay of three when driving a fanout of *F*. The

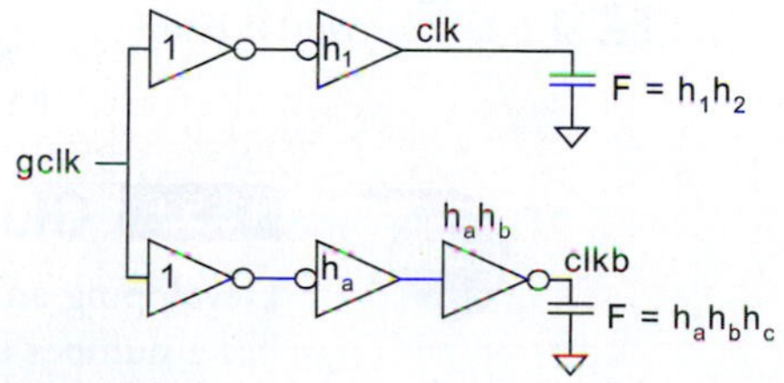

FIG 12.40 2- and 3-inverter paths

inverters are annotated with their size. The two inverters have electrical efforts of h_1 and h_2, respectively, while the three inverters have electrical efforts of h_a, h_b, and h_c. The electrical efforts should be chosen so that the delays of the chains are equal:

$$D = h_1 + h_2 + 2p_{\text{inv}} = h_a + h_b + h_c + 3p_{\text{inv}} \tag{12.5}$$

Even if the inverters have equal rise and fall delays in the TT corner, they will have unequal delays in the FS or SF corner. This can lead to skew between *clk* and *clkb* in these corners. However, if the delay of the second inverter in each chain is equal ($h_2 = h_b$), the two gaters will have equal delay in all process corners [Shoji86].

We can solve for the best electrical efforts that satisfy this constraint while giving least delay through the path. Recall that a path has least delay when its stage efforts are equal. Thus, choose $h_a = h_c = h^*$. This implies $h_1 = h^{*2}$. The delay of the first inverter in the *clk* path must equal the sum of the delays of the first and third inverter in the *clkb* path:

$$h^{*2} + p_{\text{inv}} = 2h^* + 2p_{\text{inv}} \tag{12.6}$$

This gives a quadratic equation that can be solved for h*:

$$h^* = 1 + \sqrt{1 + p_{\text{inv}}} \tag{12.7}$$

For $p_{\text{inv}} = 1$, this implies the best stage efforts are:

$$
\begin{aligned}
h_1 &= 5.8 \quad h_2 = \frac{F}{5.8} \\
h_a &= 2.4 \quad h_b = \frac{F}{5.8} \quad h_c = 2.4
\end{aligned}
\tag{12.8}
$$

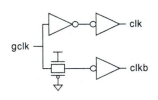

In this case, the rise/fall times of the different stages may be rather different, so the logical effort delay model is not especially accurate. These efforts make a good starting point, but further tuning should be done with a circuit simulator. The same approach can be used when the gater uses a NAND gate in place of one of the inverters.

Another approach is to try to match the delay of two inverters against one inverter and a transmission gate, as shown in Figure 12.41. This matching will not be perfect across all process corners. However, the gater may have less overall delay and hence produce less jitter from power supply noise.

FIG 12.41 2- and 1-inverter paths

12.5.6 Clock Skew Budgets

Developing an appropriate clock skew budget for design is a tricky process. The designer has a number of choices, including ignoring clock skew, budgeting worst-case clock skew everywhere, or budgeting different amounts of clock skew between different clock domains. Ultimately, the designer's objective is to build a system that achieves performance targets and has no hold time failures while consuming as little area, power, and design effort as possible. The performance target can be a fixed number set by a standard or can simply be "as fast as possible."

It is possible to ignore clock skew if you are conservative about hold times and simply want the system to run as fast as possible. You must take reasonable care in the clock distribution network so that the skew between back-to-back flip-flops is unlikely to be too large. Many ASIC and FPGA flip-flops are designed with long contamination delays so they can tolerate significant skew before violating hold times. Build the system to run as fast as possible. When it is manufactured, clock skew will cause it to run slower than expected. The advantage of this methodology is that designers can be more productive because they do not need to think about clock skew. A disadvantage is that it uses slow flip-flops. Another drawback is that some paths really will have more skew than others. If all paths are designed to have equal delay, the paths with more skew will limit performance, while the other paths will be overdesigned and will consume more area and power than necessary. Moreover, if skew-tolerant circuit techniques are used in some places but not others, the non-tolerant circuits will tend to form the critical paths.

A related approach is to estimate the worst-case clock skew and budget it everywhere. In systems using only flip-flops, this can be done by designing to a shorter clock period. For example, if an ASIC must meet a 4 ns clock period and is predicted to have 500 ps of skew, it can be designed to meet a 3.5 ns clock period with no skew. This method requires work on the part of the clock designer to predict the clock skew, but still protects most of the designers from worrying about skew.

As cycle times get shorter than about 25 FO4 inverter delays, budgeting worst-case skew everywhere makes design impossible. Instead, multiple skew budgets must be developed that reflect smaller amounts of skew between elements in a local clock domain. This method entails more thought on the part of designers to take advantage of locality and requires a static timing analyzer that applies the appropriate skew. A good timing analyzer also properly handles skew-tolerant techniques such as transparent latches and domino gates with overlapping clocks [Harris99]. Be careful—some commercial timing analyzers such as PrimeTime do not handle these circuits well.

12.5.6.1 Clock Skew Sources

As discussed earlier, clock skew comes from many sources. The output of the phase-locked loop has some jitter because of noise in the PLL and jitter in the external clock source. The clock distribution network introduces more skew from variations in the buffers and wire. The buffers may have different delays because of differences in V_{DD} and temperature, as well as random variations in their channel length and threshold voltages. The wire length and loading between buffers may not be perfectly matched. Each gater drives a physical clock along a wire, so clocked elements at different ends of the wire will see different RC delays. As mentioned in Section 2.3.2, the effective gate capacitance of the clocked loads depends on the switching activity of the source and drain. For some clocked elements, this causes significant data dependence in the clocked capacitance and the local wire delay.

For hold time checks, we are concerned with the skew between two consecutive clocked elements at a particular moment in time. For setup time checks, we are concerned with the skew between elements from one cycle to the next. Jitter in the clock distribution network can affect the instantaneous clock period, so setup time skew budgets must include the cycle-to-cycle jitter of the entire clock distribution system even for elements in the same local clock domain. Hence, we can define separate clock skew budgets for setup time and hold time analyses.

The sources can be categorized as systematic, random, drift, and jitter. Recall that systematic skews can be modeled as extra delay and taken out of the skew budget if you are willing to do the modeling. Good clock distribution networks have close to zero systematic skew. Systematic and random skews can also be eliminated by calibrating delay lines, as will be discussed in Section 12.5.7. Drift occurs slowly enough that it can be eliminated by periodic recalibration of the delay lines. Ultimately, jitter is the most serious source of skew because it changes too rapidly to predict and counteract.

12.5.6.2 Statistical Clock Skew Budgeting

The most conservative approach to estimating clock skew is to find the worst-case value of each skew source and sum these values. A real chip is unlikely to simultaneously see all of these worst cases, so such a sum is pessimistic and makes design of high-speed chips nearly impossible.

Most skew sources do not have Gaussian distributions, so taking the root sum square of the sources is inappropriate. A better approach is to perform a Monte Carlo simulation of the different skew sources to find the likely distribution of skews. The skew budget is selected at some point in this distribution. For hold times, the skew must be budgeted conservatively because the chip will not work if a hold time is violated. For example, the hold time skew budget can be selected so that 95–99% of chips will have no hold time violations.

If the goal is to build a chip that operates as fast as possible, any fixed amount of skew that affects all paths equally is irrelevant to the designer because there is nothing to do about it from the point of view of meeting setup times. However, if different paths experience different amounts of skew, a path that sees less skew can contain more logic than a path that sees a larger skew. Moreover, a path using skew-tolerant sequencing elements can contain more logic than a path between flip-flops. Hence, it is useful to predict the median skew seen in various clock domains for the purposes of setup time analysis.

As the systematic clock skew tends to be low, most clock skew sources occur from random process variations and noise. However, critical paths also experience random process variation and noise, so some will be slower than simulation predicts while others will be faster. If the chip is tuned until many critical paths have nearly the same cycle time in simulation, it is likely that a few paths will be slower than expected in the fabricated part and will limit the chip speed. It is improbable that the paths with worst-case variations in data delay are also those affected by the worst clock skew. Hence, a Monte Carlo simulation considering both variations in delay of the data paths and clock network will predict a smaller and more realistic clock skew budget [Harris01b].

Overall, choosing the appropriate clock skew budget is an ongoing source of research and debate among designers. In practice, many design teams seem to perform some calculations, and then fudge the numbers until the clock skew budget is about 10% of the cycle time. This strategy has historically led to functional chips most of the time, but becomes more risky as cycle times decrease. Measured clock skew numbers reported in publications are notoriously optimistic; for example, [Mule02] finds an average reported skew of 3.2% of the cycle time in recent microprocessors. Part of the reason is that measuring the worst case skew is difficult. Measurements tend to be made at only a few clocked elements for a small number of clock cycles, while the chip must be designed to operate correctly for the largest skew seen anywhere on the chip anytime during its ~10^{17} cycle life span.

12.5.6.3 Case Study: Itanium 2 Skew Budget

As illustrated in Section 12.5.4, the 1.0 GHz Itaniuim 2 uses an H-tree for clock distribution. The distribution network uses four levels of buffering between the PLL and the clocked elements. These buffers are called the primary driver (PD), repeater, second-level clock buffer (SLCB), and gater. The clocks are distributed on wide, shielded upper-level metal.

Table 12.2 lists the sources of variation impacting clock skew. Two adjacent pulsed latches sharing the same gater will see much less skew than two latches on opposite corners of the chip that have only the primary driver in common. Hence, we can define a hierarchy of four clock domains characterized by whether two elements share the same gater, SLCB, repeater, or only the PD. Setup times are concerned with cycle-to-cycle jitter, while hold times are not. In summary, we will define eight distinct skew budgets for these different scenarios [Harris01b]. The sources affecting each of these budgets are indicated in the eight columns.

The dominant components in the skew budget are the PLL jitter, the voltage, temperature, channel length, and threshold variations at each buffer, and the systematic mismatches in wire length and loading. Wire capacitance and resistance variations were considered negligible in the manufacturing process. Note that certain variations such as channel length of the primary driver have no effect on clock skew because they affect all physical clocks equally.

The buffer delays are 150 ps for the primary driver and repeater, 280 ps for the SLCB, and 180 ps for a simple gater, excluding wire RC flight times. The impact of the skew sources on each of these buffer delays includes:

- **Voltage**: The power network is designed to have less than ±100 mV noise on a 1.2 V supply. This full noise can be seen between any two points on the chip or from one cycle to the next at any given point; it exhibits little temporal or spatial locality. The voltage variation leads to a 13% delay change/100 mV.

- **Temperature**: The full-chip power simulation gives a temperature map shown in Figure 6.57 that predicts a variation of 20° C across the core (excluding the caches). This causes a 1.5% delay variation.

- **Channel length**: Transistors can experience systematic variations of up to ±12.5 nm from their nominal drawn L_e of 180 nm. This leads to a ±10% delay variation.

- **Threshold voltage:** The standard deviation in threshold voltages is 16.8 mV for small nMOS transistors (< 12.5 μm wide), 14.6 mV for small pMOS, 7.9 mV for wide nMOS, and 6.5 mV for wide pMOS. Monte Carlo simulations of the clock buffers show that this leads to a distribution of delays with a standard deviation of 2%.

Table 12.2   Clock skew components

	Component	Type	Same Clock Edge (hold)				Cycle-to-cycle (setup)			
			PD	Repeater	SLCB	Gater	PD	Repeater	SLCB	Gater
PLL	PLL Jitter	Jitter					x	x	x	x
PD	Voltage	Jitter					x	x	x	x
	Temperature	Drift								
	L_e	Rand								
	V_t	Rand								
	Wire	Syst	x				x			
Repeater	Voltage	Jitter	x				x	x	x	x
	Temperature	Drift	x				x			
	L_e	Rand	x				x			
	V_t	Rand	x				x			
	Loading	Syst	x				x			
	Wire	Syst	x	x			x	x		
SLCB	Voltage	Jitter	x	x			x	x	x	x
	Temperature	Drift	x	x			x	x		
	L_e	Rand	x	x			x	x		
	V_t	Rand	x	x			x	x		
	Loading	Syst	x	x			x	x		
	Wire	Syst	x	x	x		x	x	x	
Gater	Voltage	Jitter	x	x	x		x	x	x	x
	Temperature	Drift	x	x	x		x	x	x	
	L_e	Rand	x	x	x		x	x	x	
	V_t	Rand	x	x	x		x	x	x	
	Loading	Syst	x	x	x		x	x	x	
	Wire RC	Syst	x	x	x	x	x	x	x	x

● **Wire and loading**: The worst case differences in wire length and load capacitance were found through simulation of the routed H-tree.

Table 12.3 summarizes the magnitude of each of these sources of delay variation. Some sources are described by the half-range of a uniform distribution. Others are described by a standard deviation of a Gaussian distribution.

The skew budget was selected by performing a Monte Carlo simulation. $N = 400$ chips were simulated. For each chip, the systematic, random, and drift sources of skews were assigned a random value from their distribution. Jitter sources were assigned their worst case because they vary rapidly in time and the chip must work on all cycles. The

Table 12.3   Magnitude of skew sources (ps)

	Component	Half-range	Standard Deviation
PLL	PLL Jitter	7.5	
PD	Voltage	19.5	
	Temperature	n/a	
	L_e	n/a	
	V_t	n/a	
	Wire	1	
Repeater	Voltage	19.5	
	Temperature	1.1	
	L_e	15	
	V_t		5
	Loading	5	
	Wire	1.5	
SLCB	Voltage	36.4	
	Temperature	2.1	
	L_e	28	
	V_t		5.6
	Loading	10	
	Wire	4	
Gater	Voltage	23.4	
	Temperature	1.4	
	L_e	18	
	V_t		3.6
	Loading	7.5	
	Wire RC	10	

assignments were done for every clock buffer on the chip and the skew budget for that chip for each domain was defined as the maximum skew between any two elements in that domain. This gives another distribution of skews in each domain across the N chips. The setup time skew was taken from the median of the distribution to represent a "typical" chip. The hold time skew was selected at the 95th percentile in the distribution so that nearly all chips would have no hold time problems. Table 12.4 lists the skew budgets. Jitter represents more than 200 ps of the setup time skew budgets, indicating that power supply noise on the PLL and clock buffers is a major problem for the H-tree. A more careful investigation into the actual voltage noise seen at the buffers and a better effort to filter the supply noise at the buffers could offer a substantial improvement in this jitter. Hold time skew budgets are smaller because they suffer less from jitter.

Table 12.4	Clock skew budgets (ps)						
Same Clock Edge (hold)				Cycle-to-cycle (setup)			
PD	Repeater	SLCB	Gater	PD	Repeater	SLCB	Gater
280	229	106	20	312	302	267	232

A more realistic budget considers that variation takes place in the logic paths as well as the clock distribution network. Thus, some critical paths will be longer than predicted while others will be shorter. This variation is called *data skew*. The total skew impacting a path is the sum of its clock and data skew. The cycle-limiting path is the one that has the worst total skew, not necessarily the one with just the worst clock skew. Another Monte Carlo simulation incorporating data skew gives total skew budgets shown in Table 12.5. The total skew is slightly higher than the clock skew alone in most cases. However, the difference between the skew in the PD and gater delays is substantially smaller. When the design goal is to maximize frequency, the difference between global and local setup time skew is most important because it indicates how much less logic can go in a global path than in a local path if all paths should be equally critical. Also note that the hold time skew in the PD domain decreased slightly because there were relatively few hold time paths crossing this domain and it was uncommon that one with bad data skew also experienced bad clock skew.

Table 12.5	Total skew budgets (ps)						
Same Clock Edge (hold)				Cycle-to-cycle (setup)			
PD	Repeater	SLCB	Gater	PD	Repeater	SLCB	Gater
275	226	117	44	322	313	309	285

12.5.7 Adaptive Deskewing

Just as a PLL or a DLL can compensate for the overall clock distribution delay, additional adjustable delay buffers can compensate for mismatches in clock distribution delay along various paths. For example, the Pentium II and 4 use such buffers at the leaves of the clock spine to eliminate systematic and random variations in the clock distribution network. Figure 12.42 shows an example of a digitally adjustable delay line with eight levels of adjustment. The select signals use a *thermometer code*[1] to produce a monotonically decreasing propagation delay as more pass transistors are turned on.

In the Pentium II, a phase comparator checks the arrival times of the physical clocks and adjusts the digitally controlled delay lines to make all clocks arrive simultaneously. The loop bandwidth is low enough to ignore jitter, but high enough to compensate   for tem-

---

[1]In an N-bit thermometer code, a number $n \in [0,N]$ is represented with n 1's in the least significant positions. For example, the number 3 is represented in an 8-bit thermometer code as 00000111.

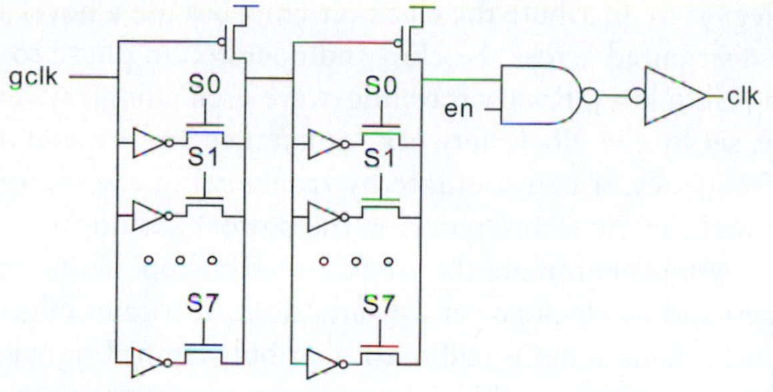

FIG 12.42 Digitally adjustable delay line

perature drift. This technique is sometimes called *adaptive deskewing* [Geannopoulos98]. In the Pentium 4, the delay line is adjusted using a scan chain through the boundary scan test access port. 46 phase comparators measure the phase of the clock gaters. Their results can be shifted out through the TAP. The delay lines can be adjusted to reduce systematic and random skew to ±8 ps, as compared to approximately 64 ps before adjustment. The delay lines can also deliberately delay certain clocks to improve performance or assist with debug [Kurd01]. The Itanium series of microprocessors uses similar deskew techniques [Tam00, Anderson02, Stinson03, Tam04]. In the 1.5 GHz Itanium 2, deskew takes place during manufacturing test; on-chip fuses are blown to eliminate the systematic and random skew without needing calibration upon reset or during normal operation.

A drawback of adaptive deskewing is that the buffers introduce extra delay. Voltage noise on the buffers appears as jitter. Unless all of the deskew buffers use well-filtered power supplies, the extra jitter from the deskew buffers can overwhelm the improvement in systematic and random skew.

12.5.8 Clocking Alternatives

A number of radical alternatives to standard clocking have been proposed. [Mule02] offers a survey of many of these methods.

Clock distribution consumes significant amounts of power and introduces much of the clock skew. On-chip metal wires have high resistance, contributing to the problems. Alternatively, the clock could be distributed on the printed circuit board or in the package using low-resistance transmission lines, and then brought onto the chip in many local regions. The clock could also be distributed optically to photodetectors around the chip or broadcast wirelessly using microwaves. Each of these methods suffers from the difficulty of testing the chip before it is packaged. If the package is expensive and the yield is not extremely high, testing chips after packaging adds significantly to the average part cost. Optical or wireless clocking also requires the development of on-chip photodetectors or antennae.

Another strategy is to distribute the clock on-chip, but use a novel architecture. Many oscillators can be distributed across the chip and operated in phase so clock distribution becomes localized [Gutnik00]. Rotary traveling-wave oscillator arrays exploit the inherent natural frequency, set by the clock network inductance and capacitance [Wood01]. If pumped at this frequency, it can oscillate by transferring the energy back and forth between magnetic and electric forms, reducing the power consumption.

Asynchronous systems eliminate the clock entirely. Proponents argue that no clocks means no clock skew and no clock power consumption. Advocates of synchronous systems point out that asynchronous systems still must distribute control signals to all the sequential elements and that variation in this delay appears as sequencing overhead in just the same way as clock skew. Moreover, the control signals dissipate power. [Sparsø01] provides a good tutorial introduction to some of the advantages of, challenges in, and techniques for asynchronous design. Synchronous designers have borrowed many techniques such as self-timed memories, self-resetting domino, and source-synchronous clocking that might once have been considered asynchronous. The debate has raged for decades, but nearly all commercial systems are still synchronous and will likely remain so indefinitely.

12.6 Analog Circuits

Although the emphasis of this book has been on digital circuits, system-on-chip designers commonly need to use some analog or radio-frequency (RF) circuitry to interface with the real world. This section offers a brief introduction to the subject of analog circuit design for digital designers. [Baker02, Gray01, Johns96] offer excellent coverage of CMOS analog circuit design while [Lee04, Razavi98] pioneered RF circuits in CMOS. The combination of analog and digital circuitry is naturally called *mixed-signal* design.

12.6.1 MOS Small-signal Model

Although the MOS transistor is a nonlinear device, it can be approximated as linear for small changes around a bias point. This is useful for understanding the behavior of amplifiers and other analog circuits. The currents and voltages can be written as

$$V_{gs} = V_{GS} + v_{gs}$$
$$V_{ds} = V_{DS} + v_{ds} \tag{12.9}$$
$$I_{ds} = I_{DS} + i_{ds}$$

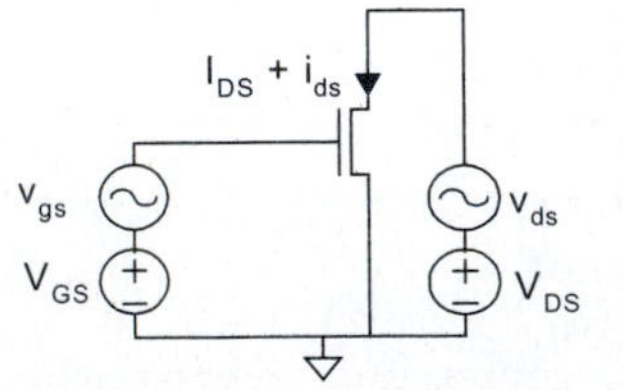

FIG 12.43 Small-signal variations around bias point

where the gate source voltage V_{gs} is expressed as a *bias-point* voltage V_{GS} plus a *small-signal* offset v_{gs}, as shown in Figure 12.43. The drain current also is expressed as a bias-point current I_{DS} plus a small-signal offset i_{ds} proportional to v_{gs} and v_{ds}.

MOS transistors are typically used in their saturation region in analog circuits. By expanding the saturation current model of EQ (2.8) around the operating point, we find the dependence of output current on small changes in input voltage.

$$
\begin{aligned}
I_{ds} &= \frac{\beta}{2}\left(V_{gs} - V_t\right)^2 \\
&= \frac{\beta}{2}\left(V_{GS} + v_{gs} - V_t\right)^2 \\
&= \frac{\beta}{2}\left[\left(V_{GS} - V_t\right)^2 + 2\left(V_{GS} - V_t\right)v_{gs} + v_{gs}^2\right] \\
&= \underbrace{\frac{\beta}{2}\left(V_{GS} - V_t\right)^2}_{I_{DS}} + \underbrace{\beta\left(V_{GS} - V_t\right)v_{gs} + O\left(v_{gs}^2\right)}_{i_{ds}}
\end{aligned}
\tag{12.10}
$$

If v_{gs} is small enough, the $O(v_{gs}^2)$ term is negligible.

In general, we can find the sensitivity of current to small changes in voltage by taking the first-order Taylor series around the operating point:

$$
I_{ds} \approx I_{DS} + \underbrace{\left.\frac{\partial I_{ds}}{\partial V_{gs}}\right|_{V_{gs}=V_{GS}} v_{gs} + \left.\frac{\partial I_{ds}}{\partial V_{ds}}\right|_{V_{ds}=V_{DS}} v_{ds}}_{i_{ds}}
\tag{12.11}
$$

We commonly write the sensitivities as

$$
i_{ds} = g_m v_{gs} + g_{ds} v_{ds}
\tag{12.12}
$$

where

$$
g_m = \left.\frac{\partial I_{ds}}{\partial V_{gs}}\right|_{V_{gs}=V_{GS}} = \beta\left(V_{GS} - V_t\right)
\tag{12.13}
$$

and because the saturation current is ideally independent of V_{ds},

$$
g_{ds} = \left.\frac{\partial I_{ds}}{\partial V_{ds}}\right|_{V_{ds}=V_{DS}} = 0
\tag{12.14}
$$

This Taylor series approach gives the same results as the direct expansion in EQ (12.10).

In a real MOSFET, the output current does increase with V_{ds} because of channel length modulation. Considering EQ (2.29),

$$
g_{ds} = \left.\frac{\partial I_{ds}}{\partial V_{ds}}\right|_{V_{ds}=V_{DS}} = \lambda\beta\frac{\left(V_{GS} - V_t\right)^2}{2} = \lambda I_{DS}
\tag{12.15}
$$

g_m is called the *transconductance* because it reflects the dependence of the drain current on the gate voltage. g_{ds} is the *output conductance*, reflecting the dependence of the drain current on the drain voltage. Here λ is the channel length modulation coefficient, not the unit of distance. Recall that λ is inversely dependent on channel length. Current sources and high-gain analog amplifiers require low output conductance and thus often use longer than minimum transistors. Often it is convenient to think about the reciprocal, *output resistance*.

$$r_o = r_{ds} = \frac{1}{g_{ds}} = \frac{1}{\lambda I_{DS}} \qquad (12.16)$$

Figure 12.44 shows the I-V characteristics of an nMOS transistor annotated with the bias-point and small-signal parameters. The transistor is biased in saturation at $V_{GS} = 0.9$, $V_{DS} = 1.0$. The transconductance and output conductance are the slope of the I_{ds} curve with respect to small changes in V_{gs} and V_{ds} around the operating point.

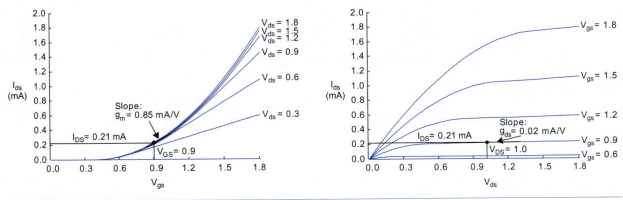

FIG 12.44 Bias-point and small-signal behavior

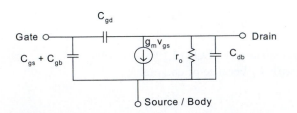

FIG 12.45 Small-signal model for an MOS transistor

Assuming the body is at the same potential as the source, we can model the transistor with the small-signal equivalent circuit of Figure 12.45 to relate small-signal voltages and currents. The current source reflects the dependence of i_{ds} on v_{gs} and the resistor reflects the dependence of i_{ds} on v_{ds}. The capacitors can be considered for high-frequency operation or ignored for low-frequency operation. The results of the small-signal model are added to the results of the bias point to find the overall behavior of the circuit.

12.6.2 Common Source Amplifier

Figure 12.46 shows a common source amplifier with a resistive load. The amplifier is biased at $V_{GS} = 0.9$ V, setting an output bias point V_{OUT}. If we increase V_{gs} by some small v_{in}, the transistor will turn on harder, pulling the output down by some small v_{out}. The gain of the amplifier is thus $A = v_{out} / v_{in}$.

Figure 12.47 shows the simulated DC transfer characteristics of the common source amplifier. The bias point is indicated on the graph. The slope of the transfer characteristic around the operating point is the gain. The transistor is a nonlinear device so its gain varies with the operating point. However, the slope and gain are relatively constant for modest values of v_{in}.

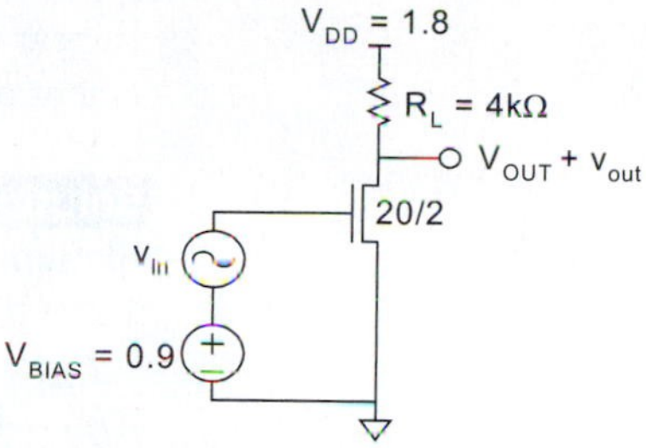

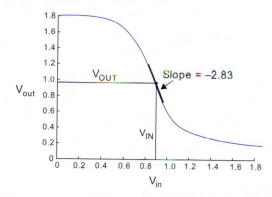

FIG 12.46 Common source amplifier

FIG 12.47 DC transfer characteristics of common source amplifier

Example

Calculate the bias point V_{OUT} and small-signal low-frequency gain if $V_t = 0.4$, $\beta = 1550$ μA/V^2 and the nMOS output impedance is infinite.

Solution: At the bias point of $V_{GS} = 0.9$, $I_{DS} = 1550 \cdot 10^{-6} \cdot (0.9 - 0.4)^2/2 = 193$ μA^2 and thus the output voltage is $V_{OUT} = 1.8 - I_{DS}R_L = 1.03$ V. Figure 12.48 shows a small-signal equivalent circuit around this bias point. The model omits the capacitors because they act as open circuits at low frequency and also leaves out r_o because it is infinite. The load resistor ties to a small-signal ground because V_{DD} is constant. According to EQ (12.13), $g_m = 1550 \cdot 10^{-6} \cdot (0.9 - 0.4) = 0.77$ mA/V. Using Kirchhoff's current law at the output node shows $v_{out} = -g_m R_L v_{in}$, or the gain $A = v_{out}/v_{in} = -g_m R_L = -3.1$.

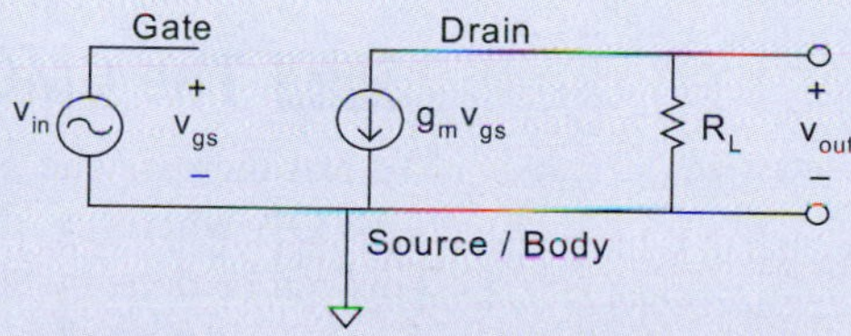

FIG 12.48 Small-signal model of common source amplifier

[2]Note that this differs slightly from Figure 12.44 because channel length modulation is neglected.

Example

Repeat the previous example if the transistor has a channel length modulation coefficient of $\lambda = 0.1$ V^{-1}.

Solution: The bias point changes only slightly on account of channel length modulation. Assuming from the previous example that $V_{DS} \sim 1.0$, we find $I_{DS} = 1550 \cdot 10^{-6} \cdot (0.9 - 0.4)^2 \cdot (1 + 1.0 \cdot 0.1)/2 = 212$ μA. The output voltage is now $V_{OUT} = 1.8 - I_{DS}R_L = 0.95$, justifying our assumption. The output resistance is $r_o = 1/(0.1 \cdot 212$ μA$) = 47$ kΩ. Figure 12.49 shows a small-signal equivalent circuit around this bias point including r_o. Now the gain depends on the parallel combination of the output and load resistances, $A = -g_m(R_L \,||\, r_o) = -2.83$, where the parallel combination of two resistors is

$$R_1 \,||\, R_2 \equiv \frac{R_1 R_2}{R_1 + R_2} \tag{12.17}$$

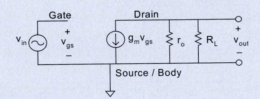

FIG 12.49 Small-signal model of common source amplifier including output resistance

The pull-up resistor can be built as a pMOS transistor of width P operating in its linear region, as shown in Figure 12.50. The gain now depends on the effective resistance of the pMOS transistor. This is simply a pseudo-nMOS inverter. Its DC transfer characteristics were discussed in Section 2.5.4.

12.6.3 The CMOS Inverter as an Amplifier

We can increase the gain of the common source amplifier by using an active load that turns ON when the nMOS transistor turns OFF and OFF when the nMOS turns ON. This can be done by building the load from a pMOS transistor connected to the input, as shown in Figure 12.51. Our high-gain amplifier is simply a CMOS inverter. Hence, the CMOS inverter is an analog amplifier operated under saturating conditions. It can also be viewed as an nMOS common-source amplifier driving a pMOS common-source amplifier. Near the input threshold voltage, the CMOS inverter acts as an inverting linear amplifier with a characteristic of

$$v_{out} = -A v_{in} \tag{12.18}$$

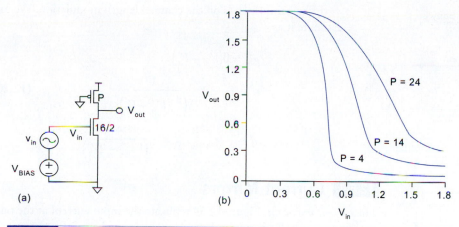

FIG 12.50 Pseudo-nMOS inverter viewed as common source amplifier with pMOS load

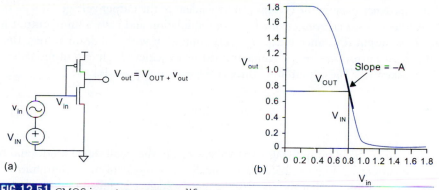

FIG 12.51 CMOS inverter as an amplifier

where A is the amplifier gain.

We can further examine this region with a circuit simulator by using the circuit shown in Figure 12.52 with a high-value resistor (10 MΩ) between input and output to *DC bias* the inverter at V_{inv}. The input is *AC coupled* using a capacitor. The gain of this amplifier is estimated using the small-signal transistor model from Figure 12.45 to construct an equivalent circuit (Figure 12.53) valid for small-signal swings around the linear operating point of the amplifier. The gain is approximately given by

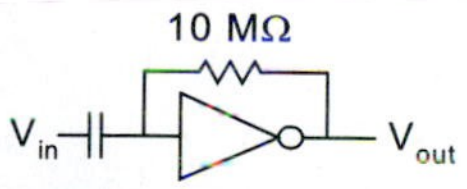

$$
\begin{aligned}
A &= g_{m\text{-total}} r_{o\text{-effective}} \\
&= \left(g_{mn} + g_{mp} \right) \left(r_{on} \mid r_{op} \mid 10\text{M}\Omega \right) \\
&\approx g_m r_o \ (\text{if } g_{mn} = g_{mp} \text{ and } r_{on} = r_{op} \ll 10\text{M}\Omega)
\end{aligned}
\tag{12.19}
$$

FIG 12.52 AC-coupled CMOS inverter

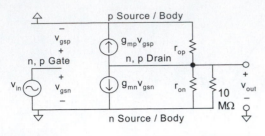

FIG 12.53 Small-signal model of amplifier

The gain depends on the channel length modulation. We can estimate it as

$$g_m = \beta\left(V_{gs} - V_t\right)$$

$$r_o = \frac{1}{\lambda I_{ds}} = \frac{1}{\lambda\left(\frac{\beta}{2}\right)\left(V_{gs} - V_t\right)^2} \qquad \textbf{(12.20)}$$

$$A \approx g_m r_o\Big|_{V_{gs} = \frac{V_{DD}}{2}} = \frac{4}{\lambda\left(V_{DD} - 2V_t\right)}$$

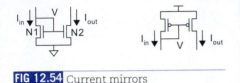

FIG 12.54 Current mirrors

12.6.4 Current Mirrors

The *current mirrors* in Figure 12.54 replicate the input current at the output. The voltage V adjusts itself to the correct level to sink or source I_{in} through $N1$. This voltage also controls the gate of $N2$. If both transistors operate in saturation where I_{ds} depends only on the gate voltage, not the drain voltage, then $I_{out} = I_{in}$. Such an ideal current source has infinite output impedance because the current is independent of the output voltage.

Real devices suffer from channel length modulation and have a finite output resistance. This output resistance makes I_{out} vary somewhat with the drain voltage on $N2$. Good current mirrors use long-channel transistors to achieve high output resistance and nearly constant I_{out}. The output impedance at the operating point is:

$$R_{out} = \frac{v_{out}}{i_{out}} \qquad \textbf{(12.21)}$$

This can be found by applying a test voltage v_{out} to the small signal model and computing the current i_{out}. Figure 12.55 shows a small-signal model of the current mirror with the test source. Observe that the small-signal gate voltage v is pulled to ground through the parallel combination of g_{m1} and r_{o1}, indicating that the gate voltage will not stray from the bias point. Applying Kirchhoff's current law (KCL) on the drain of $N2$ shows

$$i_{out} - \frac{v_{out}}{r_{o2}} = 0 \qquad \textbf{(12.22)}$$

so the output impedance is $R_{out} = r_{o2}$.

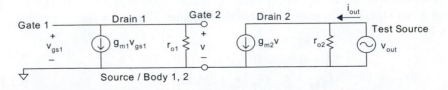

FIG 12.55 Small-signal model for current mirror output impedance

Example

Figure 12.56(a) shows a simple current source constructed from a resistor and a current mirror. It can be modeled as an ideal current source in parallel with a finite output resistance, as shown in Figure 12.56(b). Find I_{out}, the current source output impedance R_{out}, and the range of V_{out} over which the current source operates correctly. Again assume $V_t = 0.4$, $\beta = 1550\ \mu A/V^2$, and $\lambda = 0.1\ V^{-1}$.

Solution: The effect of channel length modulation on bias point is small, so we will neglect it in our analytical solution. We find the bias point by solving the nonlinear equation for input current $I_1 = 103\ \mu A$ at $V_1 = 0.765$.

$$I_1 = \frac{\beta}{2}\left(V_1 - V_t\right)^2 = \frac{\beta}{2}\left(1.8 - I_1 R - V_t\right)^2 \qquad \textbf{(12.23)}$$

At the bias point, $I_{out} = I_1$ and $R_{out} = r_o = 1/(0.1 \cdot I_1) = 97\ k\Omega$. The current source operates correctly as long as the output transistor remains in saturation. This is true as long as $V_{out} > V_1 - V_t = 0.365$ V. A good current source should have high—ideally infinite—output impedance.

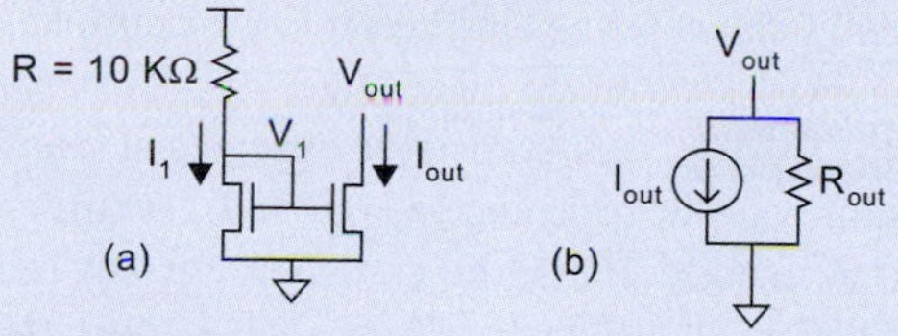

FIG 12.56 Simple current source

The output impedance can be raised by using a *cascoded* current mirror shown in Figure 12.57. In this design, V_1 and V_2 adjust to whatever they must be to sink I_{in} through N1 and N3. As $I_{out} \approx I_{in}$, the gate-to-source voltages for N3 and N4 must be nearly equal. Hence $V_3 \approx V_1$. Thus, N1 and N2 have nearly the same drain voltage as well as gate voltage, making their currents nearly equal even if the transistors have significant output conductance. By KCL, N4 must have the same current as N2, so I_{out} is very weakly sensitive to V_{out}. In other words, the cascoded current mirror has high output impedance.

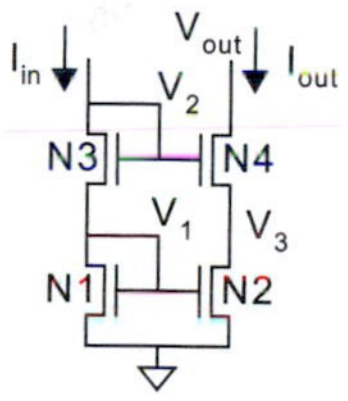

Figure 12.58 shows a small-signal model for the cascoded current mirror[3]. The output impedance increases to (see Exercise 12.5)

$$R_{out} = r_{o2} r_{o4}\left[\frac{1}{r_{o2}} + \frac{1}{r_{o4}} + g_{m4}\right] \approx \left(g_{m4} r_{o2}\right) r_{o4} \qquad \textbf{(12.24)}$$

FIG 12.57 Cascoded current mirror

---

[3]Note that we ignore the body effect in this and subsequent examples. The body effect leads to another transconductance element g_{mb} between drain and source dependent on V_{bs}.

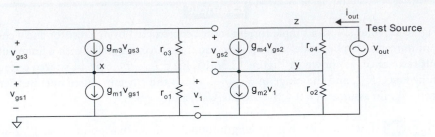

FIG 12.58 Small-signal model for cascode output impedance

Example

Find the output impedance of a cascoded current mirror using the same parameters and bias current as in the previous example.

Solution: Each transistor operates at the same current $I_{out} = 103\ \mu A$ and has the same output resistance $r_o = 97\ k\Omega$. At this current, $g_m = 1550 \cdot 10^{-6} \cdot (0.765 - 0.4) = 0.57\ mA/V$. Thus, the cascoded current mirror has an output impedance of $R_{out} = (0.57 \cdot 10^{-3})(97 \cdot 10^3)^2 = 5.3\ M\Omega$, more than 50 times that of an ordinary current mirror.

Current mirrors can use multiple output transistors to create multiple copies of an input current. The mirror can also multiply a current by N by using an output transistor N times as wide as the input, as shown in Figure 12.59. Better yet, the output can be driven by N identical transistors in parallel because identical transistors match more closely.

FIG 12.59 Current mirrors with multiple outputs and with current gain

12.6.5 Differential Pairs

A differential pair steers current to two outputs. In Figure 12.60, the current I_{ref} is divided between the two outputs depending on the difference between the two input voltages. If the input voltages are equal, the output currents are equal. If one input is substantially higher than the other, it draws all of the current. *Common mode* noise that affects both inputs equally causes no change in the output current. Hence, differential pairs are widely used because they are insensitive to many noise sources. For this reason, differential pairs are used in sense amplifiers on RAM bitline circuitry.

Differential pairs are easiest to analyze by finding the input voltage difference from the output current difference rather than vice versa. In analog circuits, the transistors normally operate in saturation. We define $V_{go} = V_{gs} - V_t$ as the *gate overdrive* of a transistor. An ideal transistor would deliver saturation current proportional to the square of the gate overdrive. According to the α-power law model, we can estimate the current of a real transistor with velocity saturation as

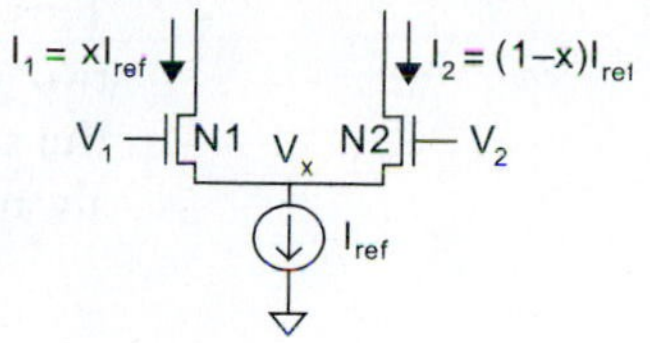

$$I_{ds} = kV_{go}^{\alpha} \tag{12.25}$$

for some $1 < \alpha < 2$.

When $V_1 = V_2$, the amplifier is in its balanced condition and $I_1 = I_2 = I_{ref}/2$. Both transistors have some gate overdrive V_{go}. We use this to express k in terms of I_{ref} and V_{go}, so we will be able to eliminate it from subsequent equations.

$$\frac{I_{ref}}{2} = kV_{go}^{\alpha} \Rightarrow k = \frac{I_{ref}}{2V_{go}^{\alpha}} \tag{12.26}$$

We would like to know how much differential voltage ΔV is required between V_1 and V_2 so that $N1$ draws xI_{ref} and $N2$ draws $(1 - x)I_{ref}$. If we increase the gate overdrive of $N1$ by some fraction δ_1, its current becomes

$$xI_{ref} = k\left[V_{go}\left(1 + \delta_1\right)\right]^{\alpha} \tag{12.27}$$

Substituting EQ (12.26), we can solve for the δ_1 required to obtain x

$$xI_{ref} = \frac{I_{ref}}{2V_{go}^{\alpha}}\left[V_{go}\left(1 + \delta_1\right)\right]^{\alpha}$$

$$\delta_1 = \left(2x\right)^{\frac{1}{\alpha}} - 1 \tag{12.28}$$

Similarly, the gate overdrive of $N2$ must go down by some fraction δ_2.

$$\delta_2 = 1 - \left(2(1 - x)\right)^{\frac{1}{\alpha}} \tag{12.29}$$

Combining EQ (12.28) and (12.29) shows that the fractional currents will be x and $1 - x$ when a differential voltage ΔV is applied.

$$\Delta V = \left[\delta_1 + \delta_2\right]V_{go} = \left[\left(2x\right)^{\frac{1}{\alpha}} - \left(2(1 - x)\right)^{\frac{1}{\alpha}}\right]V_{go} \tag{12.30}$$

FIG 12.60 Differential pair

This relationship is plotted in Figure 12.61 for several values of α. Values of α closer to two give higher gain. In any case, if one voltage is up and the other down by an overdrive, the second transistor will be completely OFF. Hence, a voltage differential of $2V_{go}$ is always enough for one transistor to hog all the current.

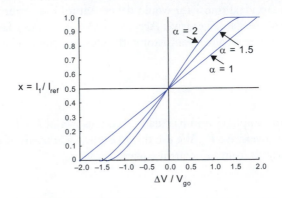

FIG 12.61 Differential pair transfer characteristics

Taking the derivative of EQ (12.30) around $x = 1/2$ and manipulating gives the small-signal transconductance of the differential pair (see Exercise 12.11).

$$\left.\frac{\partial I_1}{\partial \Delta V}\right|_{\Delta V=0} = \frac{\alpha}{4}\frac{I_{ref}}{V_{go}} = \frac{g_m}{2} \tag{12.31}$$

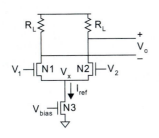

FIG 12.62 Differential amplifier

Figure 12.62 shows a differential amplifier. The amplifier consists of a differential pair ($N1$ and $N2$) driving a resistive load. The differential pair uses a transistor $N3$ with an adjustable bias voltage as the current source; this voltage could be set by a current mirror elsewhere. The high impedance load is often built from a pMOS current mirror because it is much more compact than large passive resistors.

The output impedance is $R_{out} = (R_L \| r_o)$. Because the output is taken differentially, the voltage gain is twice that of each half.

$$A = \frac{V_o}{\Delta V} = 2\frac{\partial I_1}{\partial \Delta V} R_{out} = g_m(R_L \| r_o) \tag{12.32}$$

An ideal differential amplifier is sensitive only to the voltage difference between the two inputs, not the average (*common mode*) input voltage. A real amplifier works well only while all the transistors remain in saturation. The gain drops off when one of the transistors enters the linear region. This means the headroom available is diminishing as V_{DD} shrinks.

12.6.6 Simple CMOS Operational Amplifier

Figure 12.63 shows a simple CMOS operational amplifier. It consists of a differential amplifier followed by a common source amplifier to achieve high gain. The differential amplifier uses a pMOS current mirror as a load to get high impedance r_{op2} in a compact area. The output is taken single-ended to a pMOS common source amplifier $P3$ loaded by nMOS current mirror $N5$. $N3$ and R set the bias voltage and current for the op-amp.

The small-signal gain of the differential stage is computed from EQ (12.32).

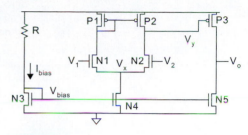

FIG 12.63 Two-stage CMOS operational amplifier

$$A_{\text{diff}} = \frac{v_y}{v_1 - v_2} = g_{mn2}\left(r_{on2} \, || \, r_{op2}\right) \tag{12.33}$$

The small-signal gain of the common source amplifier is:

$$A_{\text{commonsource}} = \frac{v_o}{v_y} = -g_{mp3}\left(r_{op3} \, || \, r_{on5}\right) \tag{12.34}$$

Hence, the overall gain of the amplifier is:

$$A = \frac{v_o}{v_2 - v_1} = g_{mn2}g_{mp3}\left(r_{on2} \, || \, r_{op2}\right)\left(r_{op3} \, || \, r_{on5}\right) \tag{12.35}$$

This operational amplifier works well as a comparator. It senses a small difference between the two inputs and drives V_o high or low depending on which input is higher. It is only suited to driving capacitive loads; it does not deliver enough current to drive resistive loads well. It also has a limited common-mode input range to keep all of the transistors in saturation. If the circuit is used in a feedback application, a compensation capacitor may need to be connected between V_y and V_o to ensure stability. [Baker98] and [Gray01] describe op-amp design and frequency response in much more detail.

12.6.7 Digital-to-analog and Analog-to-digital Converter Basics

Digital-to-analog converters (DACs) are relatively easy to design if the target resolution and speed are moderate. Speed, linearity, power dissipation, size, and ease of design are of importance when selecting a DAC architecture. Analog-to-digital converters (ADCs) are more of a challenge, but converters with low speed and precision can be implemented quite easily. The descriptions given in the following two sections can form the start of investigations if implementation is envisaged.

DACs and ADCs are sampled data systems. As such, some more circuits are normally required apart from the basic converter. Figure 12.64 illustrates a DAC in use in a system.

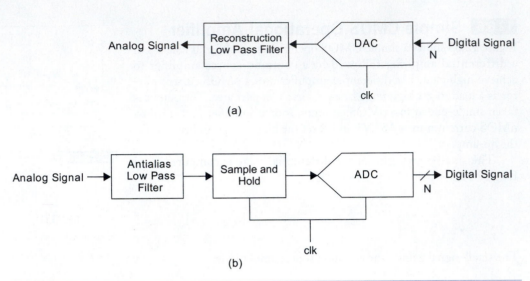

FIG 12.64 DAC and ADC in system

The DAC is followed by a reconstruction filter that converts the quantized DAC output to a smoothed analog value. The complete ADC circuit is similar, but converts from analog to digital. It also has a low-pass filter on the input and usually a sample and hold circuit to hold the input while the ADC does a conversion. The purpose of these blocks will become more apparent as the next two sections unfold.

Before discussing DACs and ADCs in detail, a few DAC metrics will be explained [Hoeschele94]. The parameters are also applicable to ADCs, except rather than a specification on a digital-to-analog transformation, the ADC parameters refer to the deviation from an ideal quantized analog-to-digital transformation.

12.6.7.1 Resolution and Full-scale Range   The first parameter of interest in a DAC (or ADC) is the *resolution*. This specifies how many individual quantized steps the DAC possesses. For instance, an N-bit DAC has 2^N individual steps. Thus, an 8-bit DAC/ADC has 256 steps. The *full-scale range* (FSR) is the maximum output voltage (or current) of the DAC/ADC. The resolution (R) of the DAC/ADC is given by R = FSR/2^N. So for a 1 V FSR 8-bit DAC/ADC, the resolution is roughly 4 mV while a 10-bit DAC/ADC would have a resolution of approximately 1 mV. The FSR is related to the voltage supply. In the case of a DAC or ADC, the FSR can not be greater than V_{DD} and common implementations have FSRs of the order of V_{DD} minus one or two V_t drops. As processes scale and V_{DD} decreases, analog performance becomes more difficult to achieve because the resolution for a given size converter decreases in relation to any noise that might be present.

12.6.7.2 Linearity   A primary concern with a DAC/ADC is the linearity or accuracy of transformation from a digital code to a quantized analog value. Linearity of a DAC/ADC is determined by component linearity, where the components in CMOS are transistors, resistors, and capacitors. It is also affected by the introduction of unwanted signals that are classified as noise. Static (DC) linearity is normally calculated using measures called the *Integral Nonlinearity* (INL) and *Differential Nonlinearity* (DNL).

- DNL: The departure from one LSB in transitioning from one digital code to the next. For instance, in Figure 12.65(a) the DNL highlighted is approximately 0.5 least significant bits (LSB) for the 3-bit DAC shown.

- INL: The maximum deviation from a straight line drawn between the endpoints of the DAC (ADC) output (input) characteristic. Again, the INL illustrated is around 0.5 LSB.

- Offset: The difference between (nominally) zero and the actual value when the digital code for zero is applied. This is a fraction of an LSB in the figure.

A good DAC is monotonic; i.e., for each increase (or decrease) in digital code input the DAC, the analog output increases (or decreases) in value. Correspondingly, for each increase in analog input, an ADC would expect to see an increase in the digital code. A non-monotonic DAC step is shown in Figure 12.65(b), where the analog output decreases to an increase in the applied digital code. Depending on the application, this may be undesirable.

Typical INL and DNL plots are shown in Figure 12.66. These are usually measured at DC. In the plots, the INL is +0.2/−0.25 LSBs and the DNL is ±0.11 LSBs.

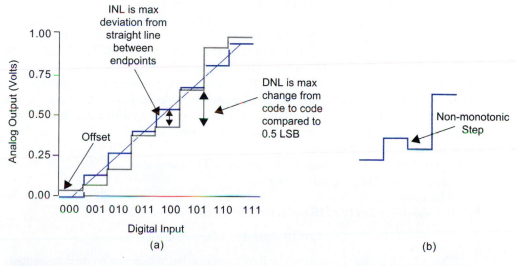

FIG 12.65 DAC linearity measures

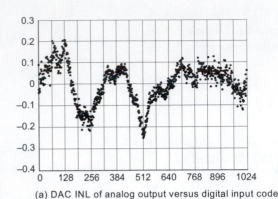

(a) DAC INL of analog output versus digital input code

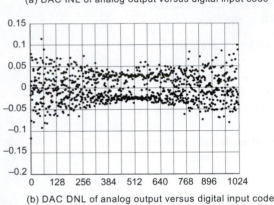

(b) DAC DNL of analog output versus digital input code

FIG 12.66 Typical DAC INL and DNL plots

12.6.7.3 Noise and Distortion Measures

With a (digital) sine wave applied to the DAC, the ratio of the desired signal energy to harmonics and noise is called the *Signal-to-Noise Ratio* (SNR). In practice, to measure this in a DAC, the DAC is fed a digital sine wave from a numerically controlled oscillator. The DAC is low-pass filtered, as shown in Figure 12.64, and then fed to a spectrum analyzer. The undesired signals are then apparent.

In comparison, an ADC measures this by applying a high-quality sine wave to the ADC and doing an FFT on the stored digital samples for a number of cycles of the input.

In addition, dynamic measures such as the *Total Harmonic Distortion* (THD) and *Spurious Free Dynamic Range* (SFDR) are also of interest. The THD measures all unwanted harmonic content and expresses this as a ratio of desired-to-undesired outputs. For instance, in a DAC, for a desired frequency of f_{signal}, there may be significant second harmonic ($2f_{signal}$) and third harmonic ($3f_{signal}$) noise caused by device non-linearity. These unwanted harmonics distort the required signal and degrade the THD.

SNR and THD are classical measures of analog linearity, which originally were applied to totally analog systems. A more popular measure with DACs and ADCs that takes into consideration digital artifacts is the *spurious free dynamic range* (SFDR). The SFDR is a measure over a specified frequency range of the desired output versus unwanted signals. These include noise from sources such as switches and noise injected from other sources on chip. SNR, THD, and SFDR are measured in dB (decibels).

The *Intermodulation Distortion* (IMD) is measured by outputting two simultaneous sine waves and measuring the amplitude of spurious products. This is a measure of the extent of unwanted multiplication products that may not show up with a simple single frequency (tone) test. ADCs measure this by applying two analog sine waves to the ADC and analyzing the digital output.

Figure 12.67 shows a typical DAC spectrum. The harmonics and SFDR are shown. The "fuzz" is typical of sampled data systems.

The noise and distortion values can be transformed into a single value called the *Effective Number of Bits* (ENOB), which is defined by:

$$ENOB = (SNR - 1.76) / 6.02$$

This rolls all performance numbers into a single value, which can be used to compare DAC implementations. The effective SNR can also be evaluated from an ENOB number.

$$SNR = ENOB \cdot 6.02 + 1.76 \text{ dB}$$

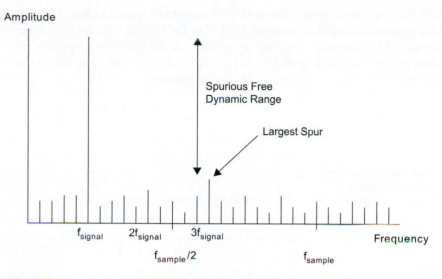

FIG 12.67 Typical DAC frequency plot showing harmonics and noise

Example

An 8-bit converter has a maximum SNR of 49.92 dB. If an implementation has an ENOB value of 7.1 bits, what is the SNR (in dB)?

Solution: SNR = 7.1 • 6.02 + 1.76 = 44.5 dB

Example

If the required SFDR is 47 dB, how many ENOBs are required?

Solution: ENOB = (47 − 1.76)/6.02 = 7.51 bits

As mentioned previously, DACs and ADCs are sampled data systems. Two major artifacts are present in DACs and ADCs. The first effect is due to sampling theory. This results in the replication of copies of the baseband signals at multiples of the clock frequency, as is shown in Figure 12.68(a). This requires that a low-pass filter follow a DAC to eliminate the unwanted signals. This is commonly called a *reconstruction filter*. In the figure, the Nyquist frequency ($f_{sample}/2$) is noted. This is the maximum signal frequency that can be generated by a DAC clocked at f_{sample}.

The second artifact results from the frequency response of a pulse, which is a $\sin(x)/x$ function as shown in Figure 12.68(b). This results in a roll-off with frequency of the desired (and undesired) signals. This is also illustrated in Figure 12.68(a). This may have to be compensated for by the digital circuits driving the DAC if a flat response is required. Similarly, an ADC responds to signals at multiples of the sampling frequency. Thus, an *antialiasing filter* has to be provided so that the ADC does not see these signals. In addition, compensation may be needed for the $\sin(x)/x$ response.

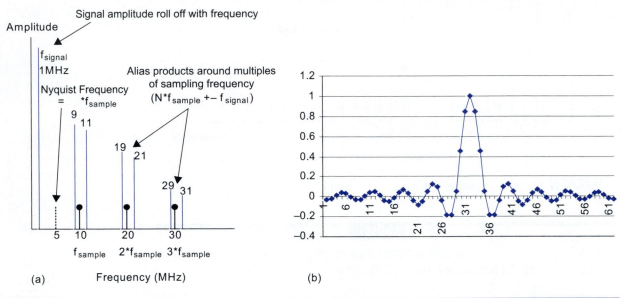

FIG 12.68 Frequency response of a DAC

Example

For a sample frequency of 10 Mhz, what are the alias products for a 1 MHz signal?

Solution: Products appear at the signal frequency offset from multiples of the sample frequency. So the first is at $f_{sample} - f_{signal} = 9$ MHz and the next at $f_{sample} + f_{signal} = 11$ MHz. The next two are offset from the second harmonic, i.e., 19 MHz and 21 Mhz, and so on. These are illustrated in Figure 12.68(a).

12.6.8 Digital-to-analog Converters

This section will present a selection of DACs roughly in order of increasing implementation complexity. All DACs presented here are within the capability of a careful CMOS circuit designer. DACs are simpler than ADCs.

A DAC can be implemented in software if a processor or DSP is available on the chip already, as shown in Figure 12.69. The processor provides a stream of 1's and 0's, which, when integrated (low-pass filtered), provides the required analog output. The RC time constant of the low-pass filter is designed to cut out the harmonics present in the output at multiples of the sampling frequency. Typically, a simple DAC such as this is highly *oversampled*; i.e., the clock frequency of the serial bit stream is many times the eventual analog output frequency. For the simple first-order filter shown, this oversampling ratio might be in the range of 256–1024.

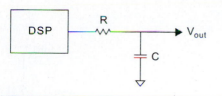

FIG 12.69 A DSP or processor-controlled DAC

If no processor is available, a converter called the *Pulse Width Modulated* (PWM) DAC, shown in Figure 12.70, can be implemented. This DAC employs a digital counter and a comparator. The counter cycles through a set of 2^N values and the comparator is set high when the count is less than the digital input B. This results in a waveform with a varying duty cycle. If B is small, the waveform will spend the majority of the time low. Conversely, if B is large, the output is high for most of the counter cycle. The linearity of this and the previous converter can be quite high, as it is limited by the linearity of the on-chip R and C in the filter, which can be reasonably good.

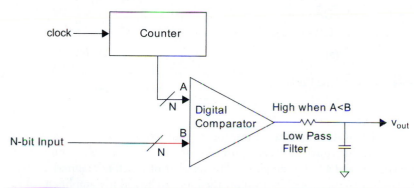

FIG 12.70 A pulse width modulated DAC

The *resistor string DAC* is perhaps the most straightforward Nyquist-rate DAC to design. A Nyquist-rate DAC is one in which the analog output values change at the DAC update rate and the usable analog frequencies extend to half the Nyquist (clock) rate. Two versions of this DAC are shown in Figure 12.71. A reference *voltage ladder*, consisting of 2^N resistors for an N-bit DAC, is connected from the supply to ground. A 2-bit converter using four resistors is shown in Figure 12.71(a). CMOS switches are used to switch the appropriate reference voltage to the output. A decoder switch can be used as shown, in which case the switch depth is N. Alternatively, each switch can be individually decoded as shown in Figure 12.71(b), in which case, the switch depth is one. While simple, this DAC is slower than other designs due to the RC time constant through the ladder and switches. In addition, the load resistance has to be high compared to the resistor string. Typically, it is useful as a reference DAC driving a CMOS buffer, op-amp, or comparator. It is suitable for moderate speed applications. Resistors can be constructed using polysilicon. The DAC is inherently monotonic due to the resistor string.

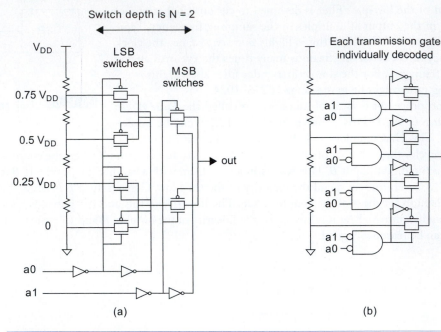

FIG 12.71 Resistor string DACs

The *R-2R DAC* uses an array of resistors of value R and $2R$, as the name suggests. A 4-bit version of this converter is shown in Figure 12.72. The DAC employs resistors, CMOS switches, and a buffer amplifier. The number of resistors required is $2N + 1$. The buffer has to be designed carefully to allow the input to be linearly amplified.

The fastest DACs in CMOS are those built from current sources. The basic principle of a 1-bit current DAC is shown in Figure 12.73(a). A digitally controlled current is switched into a resistor to create a voltage swing proportional to the product of the resistance and current. A simple current source ($P1$) is shown in Figure 12.73(b) along with a differential switch ($P2$, $P3$). The switch toggles the current from one leg to the other of a differential resistor network ($R1$, $R2$) under the control of complementary signals S and SN. The current source $P1$ is set with V_{bias}. The differential voltage ($V_{out} - V_{outn}$) is the DAC output. Figure 12.73(c) shows an improved cascode current source ($P1$, $P2$) with better linearity. Although pMOS devices have been shown to allow a GND-referenced system, nMOS devices can also be used for a V_{DD}-referenced system.

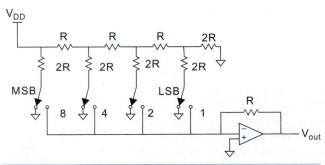

FIG 12.72 R-2R DAC

FIG 12.73 A current DAC

There are two basic methods of building a *current DAC*. The first is shown in Figure 12.74(a). Here, N weighted current sources are used to build an N-bit DAC. Each current source is built by sizing the current source in line with the bit weighting. For a 4-bit converter, a 1X, 2X, 4X, and 8X current source would be required. These could be built from

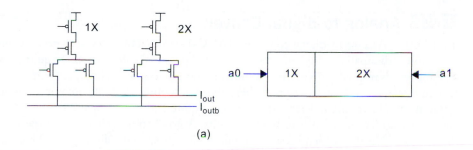

(a)

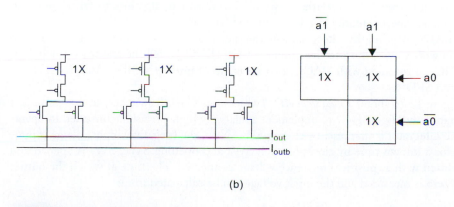

(b)

FIG 12.74 Current DAC architectures

1X, 2X, 4X, and 8X the basic current source using current mirroring and appropriately sizing the transistors. This style of current DAC is suitable for small DACs or DACs where speed is paramount. Problems can surface when designers try to match the scaled current sources to each other.

Figure 12.74(b) shows an alternative architecture which uses $2^N - 1$ identical current sources. This is called a *fully segmented DAC*. The linearity can be quite good if the matching of the current sources can be maintained. A 2-bit converter is shown in schematic form and a possible floorplan is shown.

Figure 12.75 shows a full implementation of a 4-bit current DAC. To the basic cascode current source and switch, a decode OAI gate is added in each current cell. An array of fifteen current cells is used. To the left of the current array, a row decoder, and at the bottom a column decoder, have been added. These decode the two LSBs and two MSBs respectively, and drive row and column lines into the array. Depending on which lines are activated, a number of current sources are turned on in proportion to the presented digital value. Each current source is biased by the common V_{bias1} and V_{bias2} voltages. These can be generated using a *replica bias* generator (using a replica of the unit current source) as shown (at the top left) and these in turn can be buffered by operational amplifiers. Some design hints have been included in the figure. Examples of this style of DAC can be found in [Bugeja00, Tiilikainen01].

12.6.9 Analog-to-digital Converters

Like DACs, Analog-to-digital converters (ADCs) are rated by precision, speed of conversion, power dissipation, chip size, and ease of design. ADCs also feature Nyquist converters and oversampled converters. Achieving state of the art ADC performance is an exacting science and the descriptions here about designing an ADC are not presented in great detail. However, moderate ADC design is quite within the capability of the CMOS designer who has an interest in analog circuits. A good strategy is to gain experience by "having a go" in the academic environment by designing and following an ADC design through fabrication and measurement. With the accuracy of simulation and layout extraction tools available for CMOS processes, many ADC (and DAC) architectures can be simulated in extreme detail, with the simulation results being close to fabricated chip results. Thus, experimentation can be completed "virtually."

An ADC is primarily categorized by the speed of conversion and number of bits. In common with DACs, the effective number of bits (ENOBs) tell the real story of a particular ADC. In common with DACs, other measures of interest are the offset, INL, DNL, SNR, and SFDR.

Figure 12.76 shows a *dual slope ADC,* which uses CMOS switches, an integrator and comparator, and digital logic to implement a relatively high-precision converter. In operation, the integrator is first reset by closing SWc. Next, the (negative) input voltage is integrated for a known time by closing switch SWa and opening SWc. SWa is opened and SWb closed with a (positive) reference voltage connected. The time at which the output crosses zero is measured and the input voltage can be calculated from:

$$V_{in}/|V_{ref}| = t_{measure}/t_{integrate}$$

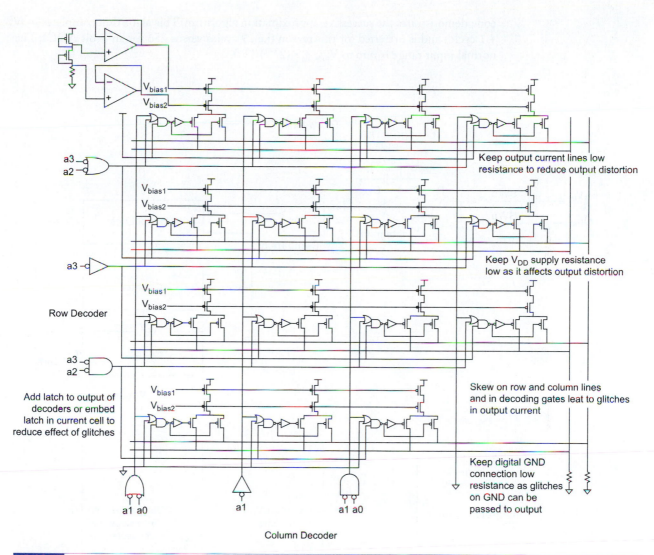

Keep output current lines low resistance to reduce output distortion

Keep V_{DD} supply resistance low as it affects output distortion

Skew on row and column lines and in decoding gates leat to glitches in output current

Keep digital GND connection low resistance as glitches on GND can be passed to output

Row Decoder

Add latch to output of decoders or embed latch in current cell to reduce effect of glitches

Column Decoder

FIG 12.75 A 4-bit current DAC

This converter is relatively slow due to the integration times, but can achieve good precision with simple analog components. CMOS inverters can be used for the amplifiers, as shown later in this section. An example of a dual slope ADC can be found in [Rodgers89].

Figure 12.77 shows an ADC that compares the input to a value set by a DAC. The easiest algorithm for conversion is to ramp the DAC from zero to full scale and observe when the comparator switches. This would take 2^N clock cycles for an N-bit converter. A quicker method is to use a successive approximation algorithm. The following pseudo-code demonstrates the successive approximation algorithm. This algorithm completes in $N + 1$ cycles and is preferred for this reason (i.e., 9 cycles versus 256 for an 8-bit ADC). The normal input range is zero to $V_{fullscale} \cdot (2^{N-1})/(2^N)$.

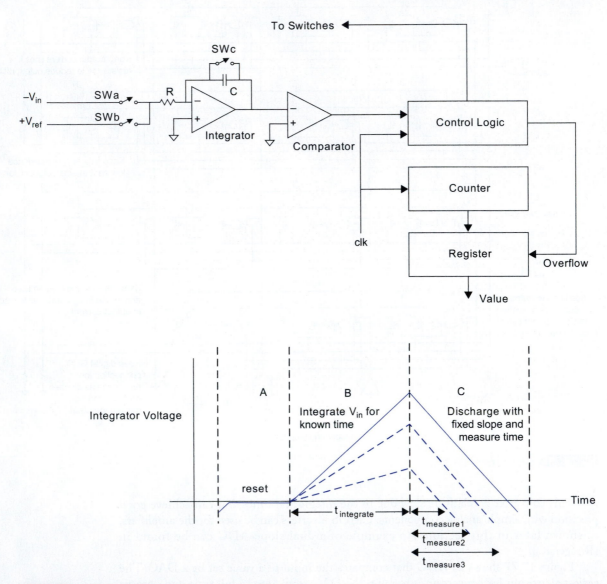

FIG 12.76 Dual slope ADC

```
Initialize:   Vdac = Vfullscale/2
              Vstep = Vfullscale/4
              Vout = 0

    loop repeat N
    write(Vdac) ;;output DAC value
    flag = read(comparator) ;; flag = Vin>Vdac
    if (flag)
       {
           Vdac = Vdac + Vstep ;; DAC value is high so reduce
           Vout = Vout + 1 ;;note contribution to output
       }
    else
       {
           Vdac = Vdac - Vstep ;; DAC is low so increase
       }
       Vout = Vout << 1 ;; multiply output by two
       Vstep = Vstep >> 1 ;; divide step by two
    repeat
```

Typical operation of the *successive approximation ADC* is shown in Figure 12.78 for a 3-bit converter with a 1 V full scale and 0.28 V applied to the input. In time slot 0, V_{dac} is set to half scale and the comparator sampled. As V_{dac} is greater than V_{in}, V_{step} of 0.25 V is subtracted from V_{dac}. In cycle 2, the process is repeated and 0.125 V is added to V_{dac}. Finally, 0.0625 V is subtracted for a final value 0.3125 V. The next cycle successively approximates 0.6875 V for a 0.72 V input. V_{out} lags the input by a conversion cycle of four clock cycles.

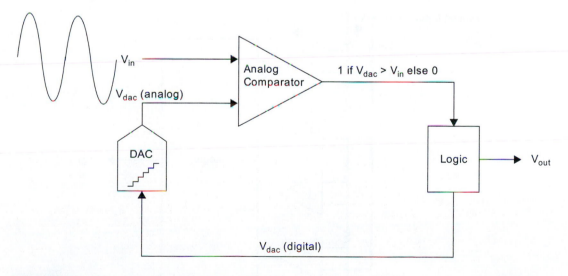

FIG 12.77 Successive approximation ADC

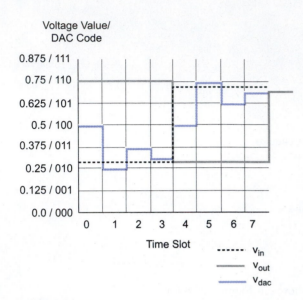

FIG 12.78 Successive approximation ADC operation

Representative successive approximation ADCs can be found in [Sauerbrey03, Promizter01, Mortezapour00]. One of the original commercial implementations can be found in [Timko80].

The *flash* converter is shown in Figure 12.79. This is the fastest converter architecture. The design consists of 2^N parallel comparators, each of which is fed with the input and a monotonically increasing reference voltage. This is most easily generated by using a string of resistors that is grounded at one end and fed at the other end with a reference voltage. When the analog input rises above the reference value fed to a particular comparator, the comparator turns on. Thus, a thermometer-like sequence of 1's will appear at the output of the comparators, rising and falling in sympathy with the analog input. The whole system is clocked at some frequency determined by the fastest time that the signal can be sampled and the comparator switched. The *thermometer code* is then fed to digital logic to extract a binary number corresponding to the analog input.

While much design effort can go into the comparator and associated sampling circuitry, a simple but still useful cir-

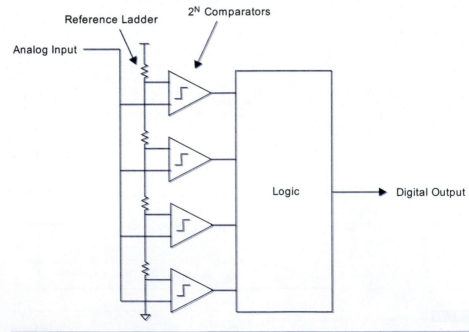

FIG 12.79 Flash ADC architecture

cuit uses one or two inverters [Dingwall79, Dingwall85]. The basic circuit is shown in Figure 12.80(a). The circuit consists of three CMOS switches, a capacitor, and a CMOS inverter, connected as shown. The operation is explained in Figure 12.80(b). In the reset cycle, the inverter is biased by connecting its input to its output. This settles the output at the inverter threshold voltage, as defined in Section 2.5.1. One end of the sampling capacitor is connected to the input of the inverter. In addition, the analog input is connected to the other end of the capacitor. In this configuration, the capacitor assumes a charge $Q = C \cdot (V_a - V_{cm})$, where V_{cm} is the inverter threshold voltage and V_a is the analog input voltage. In the second phase (Figure 12.80(c)), the reference voltage V_{ref} is connected to the capacitor and the charge is maintained. As the inverter is in high gain mode, the difference in analog input and reference voltage is amplified by the inverter. If the gain is high enough, the inverter drives the output voltage to the rails.

If a single stage does not have enough gain, another capacitively coupled gain stage can be added, as shown in Figure 12.81 [Dingwall85]. This figure shows the complete circuit as well as a latch to store the comparator output.

The design decisions are fairly simple with this design. The capacitor size has to be selected. The size of the capacitor in relation to the input size of the inverter determines the overall gain of a single stage. A value of roughly 10x the input inverter gate capacitance is a start. The inverter also has to be sized. Normally, there is no advantage to making the inverter much larger than minimum-sized because its size determines the sampling capacitor and the overall input capacitance of the converter. However, with sub-micron transistors, it is advantageous to lengthen the gate from minimum to flatten the I-V characteristic and reduce g_{ds}. The nMOS and pMOS transistors should be sized to place V_{cm} in the center of V_{ref}, which is commonly V_{DD}. The disadvantages of such a simple circuit compared to differential circuits are that the comparator is sensitive to common mode noise on the input, ground, and V_{DD}. Using V_{DD} as a reference also implies that this supply should be filtered. However, as supply voltages are reduced, the opportunity to build more sophisticated circuits is lessened and careful layout and design can yield good results for this comparator. This is especially true with an insulated SOI substrate or triple-well processes.

The output of the comparators is a thermometer code and the next stage has to decode this to a binary value. A "bubble gate," as shown in Figure 12.82, is used to determine the highest one in the thermometer code. While a simple 2-input gate can be used, the gate shown detects a sequence of 110 in the output. This prevents a spurious one (and zero bubble) from falsely triggering the decoder.

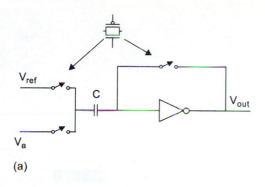

(a)

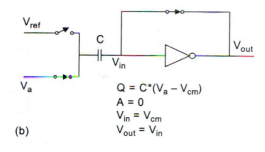

(b)

$$Q = C^*(V_a - V_{cm})$$
$$A = 0$$
$$V_{in} = V_{cm}$$
$$V_{out} = V_{in}$$

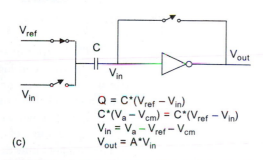

(c)

$$Q = C^*(V_{ref} - V_{in})$$
$$C^*(V_a - V_{cm}) = C^*(V_{ref} - V_{in})$$
$$V_{in} = V_a - V_{ref} - V_{cm}$$
$$V_{out} = A^*V_{in}$$

FIG 12.80 Simple CMOS inverter-based comparator

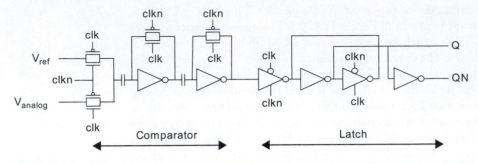

FIG 12.81 Improved gain comparator circuit with output latch

FIG 12.82 Thermometer bubble gate

The output of the bubble gate is then fed to a 1-hot-to-binary encoder. The first example shown in Figure 12.83(a) uses pseudo-nMOS NOR gates to convert to a binary number. The second encoder, shown in Figure 12.83(b), uses multiplexers. Another method is to count the number of 1's in the thermometer code with a cascade of adders.

This completes the design of the flash converter. Although it is the fastest converter, the large number of comparators places a significant load on any circuit driving the ADC. This usually limits the flash converter to fewer than 8 bits. Representative flash converters can be found in [Jiang03, Uyttenhove03, Donovan02, Scholtens02]. Some reduction in the number of comparators can be achieved using interpolation techniques and a technique called *folding*. Representative converters can be found in [Nauta95, Li03].

The *pipeline ADC* essentially trades the high speed and low latency of the flash converter for longer latency and a slightly more complicated (and hence lower speed) design. However, the power dissipation can be much lower than for a flash converter of the same speed. The design is outlined in Figure 12.84(a). The ADC is composed of (ideally) N identical stages for an N-bit converter. Each stage contributes a bit to the overall result. A single stage is shown is Figure 12.84(b). The input V_{in} is presented to an ADC that subtracts or adds a reference voltage (V_{ref}) from the input. The difference is then amplified by

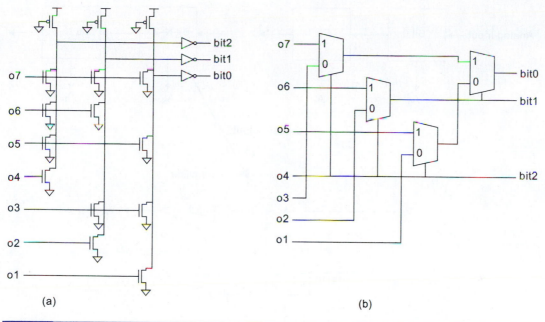

(a)　　　　　　　　　　　　　　　(b)

FIG 12.83 Flash ADC encoder

a factor of two and the process repeated. (This amounts to a distributed successive approximation converter.)

While all of these operations may seem difficult to achieve, a typical stage is shown in Figure 12.85 using CMOS switches, capacitors, and a gain stage.

The gain stage in the pipeline converter has to have enough gain to accurately subtract and perform the multiplication (assuming matching capacitors). The *folded cascode* stage shown in Figure 12.86 is a popular gain stage because it combines good gain into a single stage. This illustrates that the circuit complexity can be moderate. However, the design of the amplifier is beyond the scope of this text and the reader is referred to the literature for further details [Gray01].

Representative pipeline converters and descriptions of their operation can be found in [Lin91, Cho95, Jamal02, Chang03, Poulton03].

One of the most ubiquitous ADC architectures in use today is the *sigma-delta* architecture shown in Figure 12.87. This converter, which was developed in the late 1970s, is ideal for processes where digital circuits are easier to implement than analog circuits [Candy76]. The converter works by subtracting a delayed digitized sample of the analog input from itself. This is passed through a filter, $F(s)$, and then sent to a comparator. In the first-order example shown, this is then applied to a 1-bit DAC, which completes the loop. The digital output of the comparator is passed through a digital filter and the digital code for the analog input retained. Sigma-delta converters are oversampled converters. That is, they operate at a multiple of the required signal frequency. Oversampling ratios of

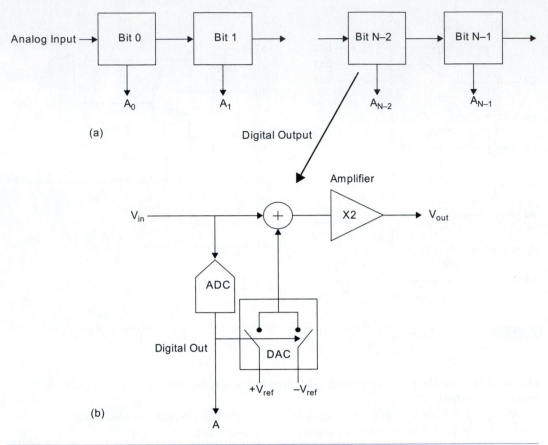

(a)

(b)

FIG 12.84 Pipeline ADC block diagram

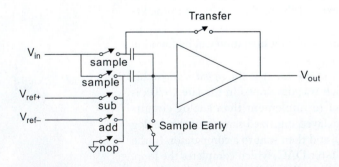

FIG 12.85 Pipeline ADC single stage

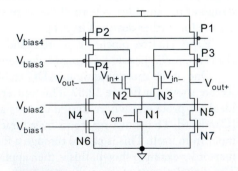

FIG 12.86 CMOS folded cascode op-amp

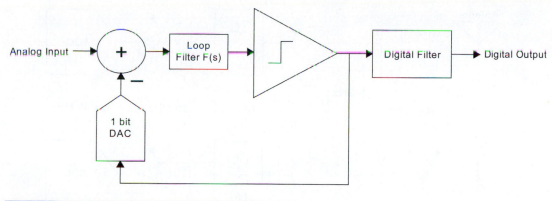

FIG 12.87 Sigma-delta ADC

between 40 to over 1000 are commonly used. As the oversampling ratio increases, so does the ADC precision (of course, at the expense of speed). Sigma-delta ADCs are commonly used in CD players at audio frequencies with resolutions up to 20 bits, but they have also been used at radio frequencies (10 MHz) for moderate precision (8–10 bits). They are popular CMOS converters because the analog components are limited to a few blocks, which can be designed carefully to obtain high performance. Representative examples can be found in [Kappes03, Sauerbrey02, Gupta03].

Apart from the basic ADCs presented, many more architectures have been invented. A single flash converter can be modified to be used twice at half the resolution and half the speed. This is the basis for a two-step flash converter. Finally, any converter can be used in parallel with delayed clocks to form an arbitrarily high-speed converter by interleaving the parallel converters. A CMOS converter implemented from 80 8-bit 250 MHz current-mode pipeline converters demonstrated operation at 20 GHz [Poulton03].

12.6.10 Radio Frequency (RF) Circuits

RF circuits are generally low in device count but high in design effort. While we will illustrate some RF circuits in this section, it must be understood that device sizing and component selection should be done in conjunction with a specialist RF design text [Lee98, Razavi98, Leung02]. To repeat, the point of showing these circuits is to encourage digital designers to explore these allied CMOS design areas.

As an introduction, a typical radio transceiver is shown in Figure 12.88. The receiver is what is called a *direct conversion* or *homodyne* architecture. It takes a 5 GHz carrier and downconverts this to a baseband signal by multiplying with a 5 GHz on-chip oscillator. The resulting signal is filtered by a 10 MHz low-pass filter and then amplified by a variable-gain amplifier. This signal is then fed to a 40 MHz ADC. While this is a simplified receiver, it includes most of the key modules required for RF applications at any frequency, namely, amplifiers (fixed-gain and variable-gain), oscillators, mixers, and filters.

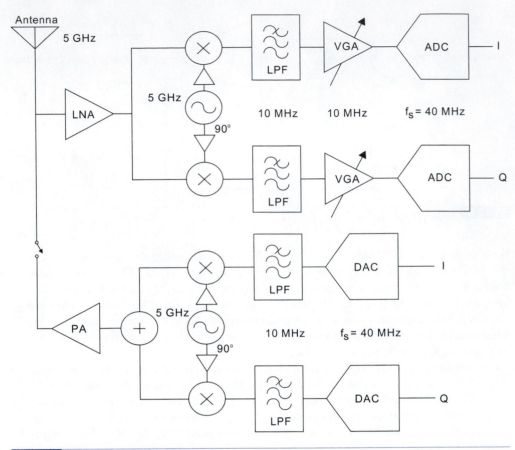

FIG 12.88 Typical CMOS radio transceiver

The transmitter path is roughly the reverse of the receiver. It starts with digital IQ values, which are fed to a DAC. The output is filtered by a low-pass reconstruction filter and then upconverted to 5 GHz. The output of the upmixer is amplified and fed to an antenna via an external transmit/receive switch. The transmitter is simpler than the receiver because of the larger signal amplitudes [vanZeijl02].

The rest of this section will introduce representative examples of each type of circuit introduced in this architecture.

A variety of CMOS RF amplifiers are shown in Figure 12.89 [Lee98]. Figure 12.89(a) shows a simple resistively loaded common source gain stage. It is biased by resistor $R1$ connected to V_{bias}. These circuits can give bandwidths in the GHz regions for submicron processes and should not be overlooked despite their simplicity. The stage is inherently wide-band. Figure 12.89(b) shows a tuned amplifier. Inductor $L1$ resonates with device and load capacitance to provide gain at a particular frequency. The stage is biased similarly to the resistively loaded example. Figure 12.89(c) shows the corresponding dif-

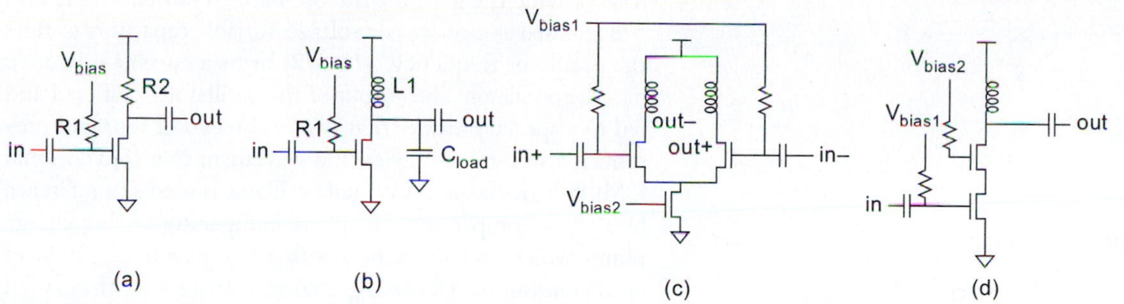

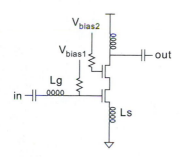

FIG 12.89 CMOS RF amplifiers

ferential stage, while Figure 12.89(d) shows a cascode gain stage. This is normally the starting point for a low-noise amplifier (LNA) that might be used at the front-end of a receiver.

Figure 12.90 shows the cascode RF amplifier with some added inductors. The source and gate inductances are used to tune out the input capacitance and present a resistive (50 Ω usually) load at the input. This stage can also be implemented differentially, which illustrates that while the initial circuit might be simple, extracting top performance from a simple gain stage requires sophisticated design.

Figure 12.91 shows an LC *voltage-controlled oscillator* (VCO) capable of operation well into the microwave regions (> 10 GHz) with modern processes in conjunction with the circuitry required to implement a phase locked loop. The inductors (L1, L2) resonate with the stray capacitance at the drain of M3 and M4 to oscillate at the required frequency. The cross-coupled nMOS transistors (M3,

FIG 12.90 Cascode RF amplifier

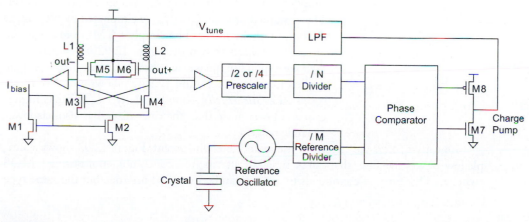

FIG 12.91 LC oscillator in a phased locked loop

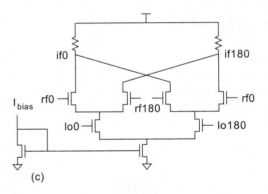

FIG 12.92 CMOS mixers

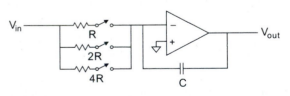

FIG 12.93 Low-pass filter

$M4$) provide the gain for the oscillator. Transistors $M5$ and $M6$ are used as *varactors* or voltage-variable capacitors to tune the oscillator frequency. $M1/M2$ forms a current mirror to bias the oscillator. The output of the oscillator is buffered and fed to a special *prescaler* that divides by two or four and presents a lower-frequency digital waveform to a conventional CMOS logic divider. A crystal oscillator is used as a reference by a phase comparator. The phase comparator feeds a charge pump, which, in conjunction with a low-pass filter, produces an increasing or decreasing analog voltage V_{tune} that is fed back to the LC oscillator. When configured properly with the correct ratios in the VCO divider and the reference divider, the feedback loop stabilizes the LC oscillator and prevents it from drifting with voltage or temperature.

Parameters of interest in an oscillator include the frequency of operation, power dissipation, and phase noise. The latter often determines the usefulness of a given oscillator in a particular system. Phase noise in turn is determined mainly by the gain of the transistors at the frequency of interest and the circuit Q of the inductors. A lot of effort goes into creating high Q inductors to achieve low-phase noise oscillators.

A *mixer* is an analog multiplier that converts one frequency to another. Figure 12.92(a) shows a symbol for a mixer and a typical application where a high radio frequency (RF) signal (say 5 GHz) is converted to a lower intermediate frequency (IF) signal (say 1 GHz) by mixing with a local oscillator (say 4 GHz). The mixer produces the sum and difference of the RF and LO (i.e., 1 GHz and 9 GHz). This is simply the result of multiplying two sine waves together. The unwanted product (9 GHz) is eliminated by filtering at the output of the mixer. The simplest CMOS mixer is the quad FET switch shown in Figure 12.92(b). Signals with their corresponding phases are shown. Correct bias has to be applied to each port. Figure 12.92(c) shows an active mixer based on a *Gilbert cell* [Zhou03, Melly01]. This has higher gain and lower noise than the ring mixer, but also has lower dynamic range.

Low-pass filters are circuits of use in an RF environment. Figure 12.93 shows a simple low-pass active filter that can be tuned by altering resistor combinations using CMOS switches. This form of continuous time filter can be increased in order, but the same type of tuning can be employed. Alternatively, capacitors can be switched and fixed resistors used.

The CMOS switch idea can also be used to build variable-gain amplifiers as shown in Figure 12.94. This design employs binary weighted capacitors.

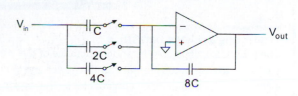

As mentioned previously, decreasing V_{DD} presents challenges for linear circuits, especially amplifiers. Figure 12.95 shows a high-speed op-amp with feed-forward compensation that only uses pMOS loaded differential stages, which work well at low supply voltages [Harrison03]. This is suitable for the filter and active gain control (AGC) applications mentioned previously. As with other circuits included in this section, the circuit is included here to illustrate the point that high-frequency amplifiers are not necessarily complex in circuit terms.

FIG 12.94 Gain controlled amplifier

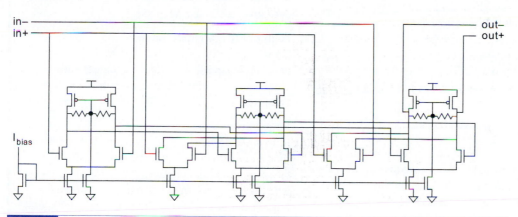

FIG 12.95 A high-speed CMOS op-amp

12.6.11 Analog Summary

This section was included to provide a brief introduction to the type of circuits that a designer might encounter in a mixed-signal analog or RF system-on-chip design. In such a short space, we can hardly do justice to this expansive area. However, we hope that the principles and circuits provided here will provide a springboard to understanding these circuits in more depth by consulting specialist texts and experimenting with circuits through simulation.

12.7 Pitfalls and Fallacies

Neglecting package parasitics

The resistance, capacitance, and inductance of the package have enormous impact on the power and I/O signal integrity of high-speed digital chips. They must be incorporated into modeling.

Using an inadequate power grid

A power grid should use generous amounts of the top two metal layers running in orthogonal directions. A mesh that mostly runs in only one direction is subject to excessive IR drops when many gates on a single wire switch simultaneously. It can also lead to serious inductive problems because of the huge current loops. The power grid should use many narrow wires interdigitated with the signals to provide a low S:R ratio rather than a few wide wires forming large current return loops. The grid should also avoid slots and other discontinuities that might lead to large current loops and high inductance.

Goofing your PLL/DLL

Phase-locked loops are notoriously difficult to design correctly. If poorly designed, they can oscillate at the wrong frequency, fail to acquire lock, or have excessive jitter. Careful circuit design is necessary to ensure they work across process variation and reject power supply noise. If the PLL does not work, testing the rest of the chip can be difficult or impossible. Most successful companies either have an in-house team that specializes in PLLs or they license their loops from a reputable third party.

Top Six Ways to Fool the Masses about Clock Skew

1) **Calculate clock skew without using process variation data**

   Random skew depends entirely on the mismatch of transistors (especially L_e) and wires on a chip. This mismatch varies with distance and layout technique. The process corners model worst-case variation from chip to chip, which can be far greater than between two nearby transistors; this results in unacceptably conservative skew budgets. But reliable data for on-chip variation can be hard to obtain, especially for small ASIC design teams and universities. Unless this data is used, clock skew budgeting is largely a matter of guesswork.

2) **Claim "zero skew"**

   Many papers state that a system has *zero skew* when the writers really mean that it has zero systematic skew. These systems may have significant random skew as well as drift and jitter. The term *zero skew* is deceptive and is best avoided.

3) **Report only systematic skew**

   Many papers also report only the systematic skew. In a well-balanced clock distribution network, systematic skew is often smaller than random skew and jitter.

4) **Ignore jitter**

   Jitter depends on time and space and is difficult to model or estimate. Unsophisticated clocking strategies sometimes ignore jitter. This results in unrealistic skew budgets. In particular, active deskew buffers increase clock distribution delay. Voltage noise on the buffers appears as jitter. Unless the supplies are unusually quiet, the buffers can increase jitter more than they decrease systematic or random skew.

5) **Report measured skew at only two elements over a brief period of time in a quiet environment**

   Measuring skew on a chip is also difficult. Some papers measure clock interarrival times at only two or a few points on the chip for a brief period of time and report those as the skew. As a chip has many clocked elements, you are unlikely to find the worst-case skew by measuring just a few points. Moreover, measurements over a brief time interval are unlikely to capture worst-case jitter. The chip should be exercised through a variety of modes that cause large fluctuations in supply current to cause maximum power supply noise and clock jitter.

6) **Don't report the skew budget used during design** Designers often choose rather conservative clock skew budgets during design because they must en-

sure the design will operate correctly. Reporting a "measured" skew rather than a skew budget will give a smaller number.

12.8   Historical Perspective

Clock distribution is a persistent challenge in VLSI design. The clock often accounts for 30–50% of the chip's dynamic power. As clock periods have decreased, the clock distribution network requires more elaborate design to deliver clocks with skews below 10% of the cycle time. The DEC Alpha series of microprocessors led the clock frequency race through much of the 1990s. In this section, we will trace the evolution of their clock grids. Figure 12.96(a) shows the grids for the Alpha 21064, 21164, and 21264 microprocessors, while Figure 12.96(b) plots the systematic clocks skew across the die [Gronowski98].

In the 200 MHz Alpha 21064 [Dobberpuhl92], the clock grid drove TSPC latches directly without any local clock gaters. The final clock load is 3.5 nF, driven by a 35 cm (!) wide inverter arranged along the center of the chip. Such a wide inverter is built from many smaller inverters ganged in parallel. A binary tree layout is used to distribute the external clock to all of the inverters simultaneously. The clock skew is zero near the center and increases toward the edges. The clock driver generates so much heat that it raises the chip temperature by 30° C nearby. Checking for hold time violations is simply a matter of counting enough gates between latches.

In the 300–433 MHz Alpha 21164 [Bowhill95, Gronowski96], the clock grid directly drove conventional two-phase dynamic transmission gate latches. Two banks of clock buffers located midway between the center and the edges of the die drive the 3.75 nF clock load. The banked drivers also lead to a more even temperature distribution across the die.

The 600 MHz Alpha 21264 [Gieseke97, Gronowski98, Bailey98] used a clock grid that in turn drove an assortment of local clocks through gaters. The grid is divided into four "window panes," each of which is driven from its four edges. Each driver along the edge of a pane uses many inverters in parallel. The clock is distributed to the inverters in a fashion that equalizes the RC delay by tapering the wires appropriately. The systematic skew is close to zero along the edges of the panes and remains relatively small near the center because the wires are short. Some gaters perform clock gating to reduce power consumption. Others generate delayed clocks to fix critical paths. Because the system now has multiple logical clocks with different arrival times, checking for hold time violations requires more careful timing analysis.

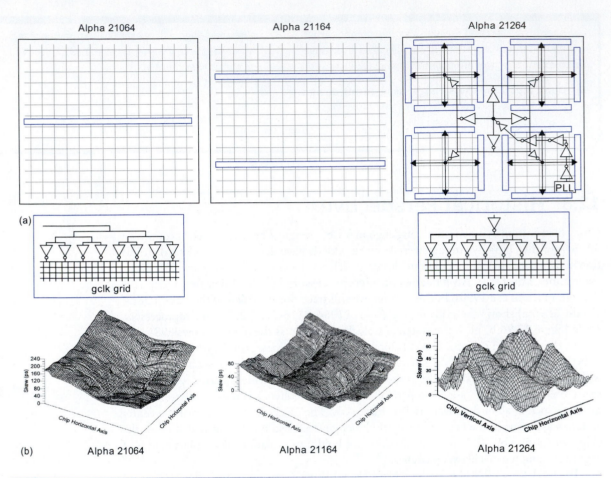

FIG 12.96 Alpha clock grids and systematic skew. (b) © 1998 IEEE.

In the Alpha 21064 and 21164, the global clock distribution network must have low resistance to directly drive such a large load capacitance. This in turn requires an enormous amount of metal wiring dedicated to the clock grid. The wire also has a high capacitance, so the clock grid consumes a large amount of power. For example, the 21164 clock distribution system consumes 20 W, or 40% of total chip power. In the 21264, the local clock gaters have electrical effort that reduces the capacitance seen directly on the global clock. This is beneficial because it reduces the metal usage and capacitance of the global clock grid while simultaneously reducing global clock skew. However, the local gaters introduce additional skew from random variation, drift, and jitter.

Summary

This chapter has surveyed package, power distribution, I/O, clock, and analog subsystem design. While each topic is a book in itself and a specialty design area, the short fat VLSI designer must understand enough about each area to optimize the system as a whole.

Packages connect the chip to the board or module, protect the chip, and are the first link in removing heat. They should offer plenty of connections, low thermal resistance, and low parasitics, while still being inexpensive to manufacture and test. Flip-chip packaging using solder bumps distributed across the die has become popular because of the large number of connections and low inductance.

The power distribution network consists of elements on the chip, package, and board. It must deliver a stable voltage across the chip under fluctuating current demands. Noise is caused by both average and peak current requirements. Multiple bypass capacitors offer low impedance to help filter high-frequency IR and L di/dt noise, but the DC supply resistance must be low enough to deliver the average current. V_{DD} and GND lines should be interdigitated in both directions with signal wires to provide small current return loops and low inductance. The supply wires must also have enough cross-sectional area to avoid electromigration problems. These requirements imply large amounts of metal and bypass capacitance, yet cost constraints dictate no more chip area than necessary.

I/O signals include inputs, outputs, bidirectional signals, and analog signals. The I/O pads must deliver adequate bandwidth to large off-chip capacitances at voltage levels compatible with other chips. They must also protect the core circuitry against overvoltage and electrostatic discharge.

A clocking subsystem includes clock generation, distribution, and gater elements. The clock generator can use a PLL to align the on-chip clock to an external reference for synchronous communication and to perform frequency multiplication. The clock distribution network should send the global clock to all clocked elements with low skew, yet not consume excessive power or area. The gaters perform local clock stopping or can produce multiple phases from the single global clock.

Basic analog building blocks include common source amplifiers, current mirrors, and differential amplifiers. From these, we can construct operational amplifiers, D/A converters, and A/D converters.

Exercises

12.1 A ceramic PGA package with a good heat sink and fan has a thermal resistance to the ambient of 10° C/W. The thermal resistance from the die to the package is 2° C/W. If the package is in a chassis that will never exceed 50° C and the maximum acceptable die temperature is 110° C, how much power can the chip dissipate?

12.2 Explain how an electrostatic discharge event could cause latchup on a CMOS chip.

12.3 Comment on the advantages and disadvantages of H-trees and clock grids. How does the hybrid tree/grid improve on a standard grid?

12.4 Calculate the bias-point and small-signal low-frequency gain of the common source amplifier from Figure 12.46 if $V_t = 0.7$, $\beta = 240\ \mu A/V^2$, and the nMOS output impedance is infinite. Let $V_{BIAS} = 3\ V$, $V_{DD} = 15\ V$, and $R_L = 10\ k\Omega$.

12.5 Prove EQ (12.24).

12.6 Calculate the output impedance of the Wilson current mirror in Figure 12.97.

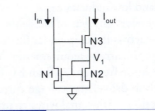

FIG 12.97 Wilson current mirror

12.7 Design a current source that sinks 200 μA, using the process parameters from the example in Section 12.6.4. What is the minimum drain voltage over which your current source operates?

12.8 Find the output impedance of your current source from Exercise 12.7. By what fraction does the current change as the output node changes by 1 V?

12.9 Simulate the operational amplifier of Figure 12.63. Using minimum-size transistors and a 10 kΩ resistor, what is the gain?

12.10 What changes would you make to the amplifier from Exercise 12.9 to increase the gain? What are the tradeoffs involved? What gain can you achieve using reasonable changes?

12.11 Prove EQ (12.31).

12.12 Use SPICE to find the transconductance and output resistance of a minimum-size transistor in your process biased at $V_{gs} = V_{ds} = V_{DD}/2$. What is the $g_m r_o$ product?

12.13 Repeat Exercise 12.12 for a transistor with 2x minimum channel length. How does the product change?

12.14 In Section 12.6.8 on resistor string DACs, it was mentioned that a similar DAC can be implemented with capacitors. Design the architecture of a 4-bit capacitor DAC.

12.15 A bias generator is required to generate 16 steps from 0 to 100 μA to bias an amplifier. Design a CMOS DAC to do this, assuming the presence of a 50 μA reference current.

12.16 Differential circuits provide good noise immunity and have the advantage of dual rail inputs and outputs. Pseudo-differential circuits based on CMOS inverter amplifiers can be implemented by using two signal paths that process each signal, but do not provide for the same level of noise immunity. Design a single stage of a pipeline ADC that uses this style of differential circuit.

12.17 To reduce clock and decoder skew in a current-mode DAC, a latch is often included in the current cell. Design the circuit for such a cell, demonstrating where the latch would be placed. If this is a slave latch, where would the master latch be located?

Verilog

A.1 Introduction

This appendix gives a quick introduction to the Verilog Hardware Description Language (HDL). There are many texts on Verilog ([Smith00, Thomas02, Ciletti99] and others) that provide a more in-depth treatment. The IEEE standard itself is quite readable as well as authoritative [IEEE1364-01]. Many books treat Verilog as a programming language, which is not the best way of viewing it. Verilog is better understood as a shorthand for describing digital hardware. It is best to begin your design process by planning, on paper or in your mind, the hardware you want. (For example, the MIPS processor consists of an FSM controller and a datapath built from registers, adders, multiplexers, etc.) Then, write Verilog that implies that hardware to a synthesis tool. A common error among beginners is to write a program without thinking about the hardware that is implied. If you don't know what hardware you are implying, you are almost certain to get something that you didn't want. Sometimes this means extra latches appearing in your circuit in places you didn't expect. Other times, it means that the circuit is much slower than required or it takes far more gates than it would have if it were more carefully described.

The Verilog language was developed by Gateway Design Automation as a proprietary language for logic simulation in 1984. Gateway was acquired by Cadence in 1989 and Verilog was made an open standard in 1990 under the control of Open Verilog International. The language, with some revisions, became an IEEE standard in 1995 and was updated in 2001. This appendix is consistent with the 2001 standard.

As mentioned in Section 1.8.4, there are two general styles of description: *behavioral* and *structural*. Structural Verilog describes how a module is composed of simpler modules or basic primitives such as gates or transistors. Behavioral Verilog describes how the outputs are computed as functions of the inputs. There are two general types of statements used in behavioral Verilog. *Continuous assignment* statements necessarily imply combinational logic[1] because the output on the left side is a function of the inputs on the right side. *Always* blocks can imply combinational logic or sequential logic, depending on how they are used. It is good practice to partition your design into combinational and sequential components and then write Verilog in such a way that you get what you want. If you

---

[1] Recall that the outputs of *combinational logic* depend only on the present inputs, while outputs of *sequential logic* depend on both past and present inputs. In other words, combinational logic is memoryless, while sequential logic has memory or *state*.

don't know whether a block of logic is combinational or sequential, you are likely to get the wrong thing. A particularly common mistake is to use `always` blocks to model combinational logic, but to accidentally imply latches or flip-flops.

This appendix focuses on a subset of Verilog sufficient to synthesize any hardware function. The language contains many other commands that are beyond the scope of this tutorial.

A.2 Behavioral Modeling with Continuous Assignments

A 32-bit adder is a complex design at the schematic level of representation. It can be constructed from 32 full adder cells, each of which in turn requires about six 2-input gates. Verilog provides a much more compact description. In each of the examples in this appendix, Synplify Pro was used to synthesize the Verilog into hardware. The Verilog code is shown adjacent to the schematic it implies.

```
module adder(input  [31:0] a,
             input  [31:0] b,
             output [31:0] y);

    assign y = a + b;
endmodule
```

A Verilog module is like a cell in a schematic. It begins with a description of the inputs and outputs, which in this case are 32-bit busses.

During simulation, an `assign` statement causes the left side (`y`) to be updated any time the right side (`a/b`) changes. This necessarily implies combinational logic; the output on the left side is a function of the current inputs given on the right side. A 32-bit adder is a good example of combinational logic.

A.2.1 Bitwise Operators

Verilog has a number of *bitwise* operators that act on busses. For example, the following module describes four inverters.

```
module inv(input  [3:0] a,
           output [3:0] y);

    assign y = ~a;
endmodule
```

Similar bitwise operations are available for the other basic logic functions:

```
module gates(input [3:0]   a, b,
             output [3:0] y1, y2, y3, y4, y5);

   /* Five different two-input logic
      gates acting on 4 bit busses */
   assign y1 = a & b;    // AND
   assign y2 = a | b;    // OR
   assign y3 = a ^ b;    // XOR
   assign y4 = ~(a & b); // NAND
   assign y5 = ~(a | b); // NOR
endmodule
```

A.2.2   Comments and White Space

The previous examples showed two styles of comments, just like those used in C or Java. Comments beginning with /* continue, possibly across multiple lines, to the next */. Comments beginning with // continue to the end of the line. It is important to properly comment complex logic so that six months from now, you can understand what you did or so that some poor slob assigned to fix your buggy code will be able to figure it out. ☺

Verilog is not picky about the use of white space. Nevertheless, proper indenting and spacing is helpful to make nontrivial designs readable. Verilog is case-sensitive. Be consistent in your use of capitalization and underscores in signal and module names. Be sparing with the use of underscores because they can increase the risk of carpal tunnel syndrome.

A.2.3   Reduction Operators

Reduction operators imply a multiple-input gate acting on a single bus. For example, the following module describes an 8-input AND gate with inputs a[0], a[1], a[2], ..., a[7].

```
module and8(input   [7:0] a,
            output        y);

   assign y = &a;
endmodule
```

As one would expect, |, ^, ~&, and ~| reduction operators are available for OR, XOR, NAND, and NOR as well. Recall that a multi-bit XOR performs parity, returning true if an odd number of inputs are true.

A.2.4 Other Operators

The conditional operator ?: works like the same operator in C or Java and is useful for describing multiplexers. It is called a *ternary operator* because it takes three inputs. If the first input is nonzero, the result is the expression in the second input. Otherwise, the result is the expression in the third input.

```
module mux2(input  [3:0] d0, d1,
            input        s,
            output [3:0] y);

   assign y = s ? d1 : d0;
endmodule
```

A number of arithmetic functions are supported, including +, -, *, <, >, <=, >=, ==, !=, <<, >>, <<<, >>>, /, and %. Recall from other languages that % is the modulo operator: a%b equals the remainder of a when divided by b. These operations imply a vast amount of hardware. == and != (equality/inequality) on N-bit inputs require N 2-input XNORs to determine equality of each bit and an N-input AND or NAND to combine all the bits, as shown in Figure 10.52. Addition, subtraction, and comparison all require an adder, which is expensive in hardware. Variable logical left and right shifts << and >> and arithmetic left and right shifts <<< and >>> imply a barrel shifter. Multipliers are even more costly. Do not use these statements without contemplating the number of gates you are generating. Moreover, the implementations are not necessarily efficient for your problem.

Some synthesis tools ship with optimized libraries for special functions like adders and multipliers. For example, the Synopsys DesignWare libraries produce reasonably good multipliers. If you do not have a license for the libraries, you'll probably be disappointed with the speed and gate count of a multiplier your synthesis tool produces from when it sees *. Many synthesis tools choke on / and % because these are nontrivial functions to implement in combinational logic.

A.3 Basic Constructs

A.3.1 Internal Signals

Often it is convenient to break a complex calculation into intermediate variables. For example, in a full adder, we sometimes define the propagate signal as the XOR of the two inputs *A* and *B*. The sum from the adder is the XOR of the propagate signal and the carry-in. We can declare the propagate signal using a wire statement, in much the same way we use local variables in a programming language.

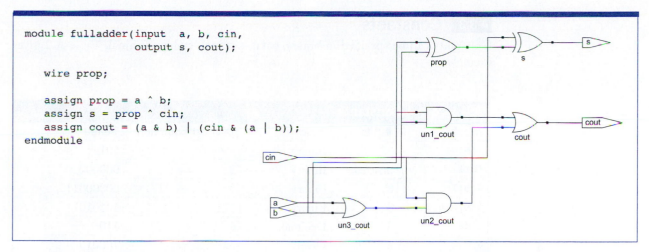

```
module fulladder(input  a, b, cin,
                 output s, cout);

   wire prop;

   assign prop = a ^ b;
   assign s = prop ^ cin;
   assign cout = (a & b) | (cin & (a | b));
endmodule
```

Technically, it is not necessary to declare single-bit wires. However, it is necessary to declare multi-bit busses and is good practice to declare all signals. Some Verilog simulation and synthesis tools give errors that are difficult to decipher when a wire is not declared.

A.3.2 Precedence

Notice that we fully parenthesized the `cout` computation. We could take advantage of operator precedence to use fewer parentheses:

```
assign cout = a&b | cin&(a|b);
```

The operator precedence from highest to lowest is much as you would expect in other languages, as shown in Table A.1. AND has precedence over OR.

Table A.1	Operator Precedence	
Symbol	**Meaning**	**Precedence**
~	NOT	Highest
*, /, %	MUL, DIV, MODULO	
+, -	PLUS, MINUS	
<<, >>,	Logical Left/Right Shift	
<<<, >>>	Arithmetic Left/Right Shift	
<, <=, >, >=	Relative Comparison	
==, !=	Equality Comparison	
&, ~&	AND, NAND	
^, ~^	XOR, XNOR	
|, ~|	OR, NOR	
?:	Conditional	Lowest

A.3.3 Constants

Constants can be specified in binary, octal, decimal, or hexadecimal. Table A.2 gives examples.

Table A.2	Constants			
Number	**# Bits**	**Base**	**Decimal Equivalent**	**Stored**
3'b101	3	Binary	5	101
'b11	unsized	Binary	3	000000..00011
8'b11	8	Binary	3	00000011
8'b1010_1011	8	Binary	171	10101011
3'd6	3	Decimal	6	110
6'o42	6	Octal	34	100010
8'hAB	8	Hexadecimal	171	10101011
42	unsized	Decimal	42	0000...00101010

It is good practice to specify the length of the number in bits, even though the second row shows that this is not strictly necessary. If you don't specify the length, one day you may be surprised when Verilog assumes the constant has additional leading 0's that you didn't intend. Underscores in numbers are ignored and can be helpful in breaking long numbers into more readable chunks. If the base is omitted, the number is assumed to be decimal.

A.3.4 Hierarchy

Nontrivial designs are developed in a hierarchical form, in which complex modules are composed of submodules. For example, a 4-input multiplexer can be constructed from three 2-input multiplexers:

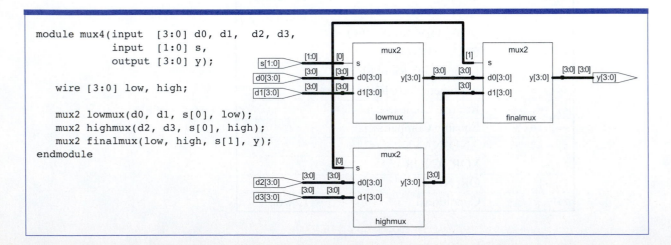

```
module mux4(input   [3:0] d0, d1,   d2, d3,
            input   [1:0] s,
            output  [3:0] y);

    wire [3:0] low, high;

    mux2 lowmux(d0, d1, s[0], low);
    mux2 highmux(d2, d3, s[0], high);
    mux2 finalmux(low, high, s[1], y);
endmodule
```

This is an example of the structural coding style because the mux is built from simpler modules. It is good practice to avoid (or at least minimize) mixing structural and behavioral descriptions within a single module. Generally, simple modules are described behaviorally and larger modules are composed structurally from these building blocks.

A.3.5 Tristates

It is possible to leave a bus floating rather than drive it to 0 or 1. This floating value is called 'z in Verilog. For example, a tristate buffer produces a floating output when the enable is false.

```
module tristate(input  [3:0] a,
                input        en,
                output [3:0] y);

    assign y = en ? a : 4'bz;
endmodule
```

Floating inputs to gates causes undefined outputs, displayed as 'x in Verilog. At start-up, state nodes such as the internal node of flip-flops are also usually initialized to 'x, as we will see later.

We could define a multiplexer using two tristates so that the output is continuously driven by exactly one tristate. This guarantees that there are no floating nodes.

```
module mux2(input  [3:0] d0, d1,
            input        s,
            output [3:0] y);

    tristate t0(d0, ~s, y);
    tristate t1(d1, s, y);
endmodule
```

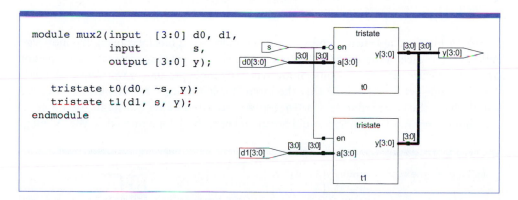

A.3.6 Bit Swizzling

Often it is necessary to operate on parts of a bus or to concatenate (join together) signals to construct busses. In the mux4 example, the least significant bit s[0] of a 2-bit select signal was used for the low and high muxes and the most significant bit s[1] was used for the final mux. Use ranges to select subsets of a bus. For example, an 8-bit wide 2-input mux can be constructed from two 4-bit wide 2-input muxes.

```verilog
module mux2_8(input  [7:0] d0, d1,
             input        s,
             output [7:0] y);

   mux2 lsbmux(d0[3:0], d1[3:0], s, y[3:0]);
   mux2 msbmux(d0[7:4], d1[7:4], s, y[7:4]);
endmodule
```

The {} notation is used to concatenate busses. For example, the following 8 × 8 multiplier produces a 16-bit result, which is placed on the upper and lower 8-bit result busses.

```verilog
module mul(input [7:0] a, b,
           output [7:0] upper, lower);

   assign {upper, lower} = a*b;
endmodule
```

A 16-bit 2's complement number is sign-extended to 32 bits by copying the most significant bit to each of the upper 16 positions. The Verilog syntax concatenates 16 copies of a[15] to the 16-bit a[15:0] bus. Some synthesis tools produce a warning that a is a "feedthrough net." This means that the input "feeds through" to the output. y[15:0] should have the same value as a[15:0], so we did intend a feedthrough. If you get a feedthrough net warning where you did not intend one, check for a mistake in your Verilog.

```verilog
module signextend(input  [15:0] a,
                  output [31:0] y);

   assign y = {{16{a[15]}}, a[15:0]};
endmodule
```

The next statement generates a silly combination of two busses. Don't confuse the 3-bit binary constant `3'b101` with bus `b`. Note that it was important to specify the length of 3 bits in the constant; otherwise many additional 0's might have appeared in the middle of `y`.

```
assign y = {a[2:1], {3{b[0]}}, a[0], 3'b101, b[1:3]};
```

This produces

```
y = a[2] a[1] b[0] b[0] b[0] a[0] 1 0 1 b[1] b[2] b[3]
```

A.3.7   Delays

The delay of a statement can be specified in arbitrary units. For example, the following code defines an inverter with a 42-unit propagation delay. Delays have no impact on synthesis, but can be helpful while debugging simulation waveforms because they make cause and effect more apparent.

```
assign #42 y = ~a;
```

A.4   Behavioral Modeling with `Always` Blocks

`Assign` statements are reevaluated every time any term on the right side changes. Therefore, they must describe combinational logic. `Always` blocks are reevaluated only when signals in the header (called a *sensitivity list*) change. Depending on the form, `always` blocks can imply either sequential or combinational circuits.

A.4.1   Registers

Verilog refers to edge-triggered flip-flops as *registers*. Registers are described with an `always @(posedge clk)` statement:

```
module flop(input          clk,
            input     [3:0] d,
            output reg [3:0] q);

    always @(posedge clk)
        q <= d;
endmodule
```

The body of the `always` statement is only evaluated on the rising (positive) edge of the clock. At this time, the output `q` is copied from the input `d`. The `<=` is called a *nonblocking assignment* and is pronounced *"gets,"* as in *"q gets d."* Think of it as a regular equals sign for now; we'll return to the more subtle points in Section A.4.6. Notice that `<=` is used instead of `assign` inside the `always` block.

All the signals on the left side of assignments in `always` blocks must be declared as `reg`. This is a confusing point for new Verilog users. In this circuit, `q` is also the output. Declaring a signal as `reg` does not mean the signal is actually a register! All it means is that it appears on the left side in an `always` block. We will see examples of combinational signals later that are declared `reg`, but that have no flip-flops.

At startup, the `q` output is initialized to `'x`. Generally, it is good practice to use resettable registers so that on power-up you can put your system in a known state. The reset can be either asynchronous or synchronous, as discussed in Section 7.3.4. Asynchronous resets occur immediately. Synchronous resets only change the output on the rising edge of the clock.

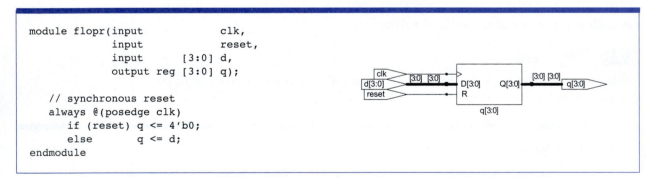

```
module flopr(input           clk,
             input           reset,
             input    [3:0]  d,
             output reg [3:0] q);

    // asynchronous reset
    always @(posedge clk, posedge reset)
        if (reset) q <= 4'b0;
        else       q <= d;
endmodule
```

```
module flopr(input           clk,
             input           reset,
             input    [3:0]  d,
             output reg [3:0] q);

    // synchronous reset
    always @(posedge clk)
        if (reset) q <= 4'b0;
        else       q <= d;
endmodule
```

Note that the asynchronously resettable flop evaluates the always block when either `clk` or `reset` rise so that it immediately responds to `reset`. The synchronously reset flop is not sensitized to `reset` in the @ list, so it waits for the next clock edge before clearing the output.

You can also consider registers with enables that only respond to the clock when the enable is true. The following register with enable and asynchronous reset retains its old value if both `reset` and `en` are false.

```
module flopren(input            clk,
               input            reset,
               input            en,
               input     [3:0]  d,
               output reg [3:0] q);

    // asynchronous reset
    always @(posedge clk, posedge reset)
        if (reset) q <= 4'b0;
        else if (en) q <= d;
endmodule
```

A.4.2 Latches

Always blocks can be used to model transparent latches, also known as D latches When the clock is high, the latch is *transparent* and the data input flows to the output. When the clock is low, the latch goes *opaque* and the output remains constant.

```
module latch(input            clk,
             input     [3:0]  d,
             output reg [3:0] q);

    always @(clk, d)
        if (clk) q <= d;
endmodule
```

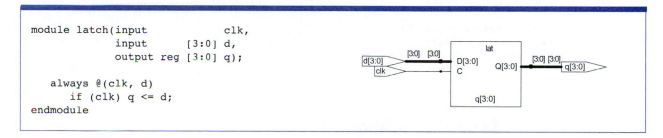

The latch evaluates the always block any time either clk or d change. If clk is high, the output gets the input. Notice that even though q is a latch node, not a register node, it is still declared as reg because it is on the left side of a <= in an always block. Some synthesis tools are primarily intended to target edge-triggered flip-flops and will produce warnings when generating latches.

A.4.3 Counters

Consider two ways of describing a 4-bit counter with asynchronous reset. The first scheme (behavioral) implies a sequential circuit containing both the 4-bit register and an adder. The second scheme (structural) explicitly declares modules for the register and adder.

Either scheme is good for a simple circuit such as a counter. As you develop more complex finite state machines, it is a good idea to separate the next state logic from the registers in your Verilog code. Verilog does not protect you from yourself here and there are many simple errors that lead to circuits unlike those you intended, as will be explored in Section A.9.

```
module counter(input           clk,
               input           reset,
               output reg [3:0] q);

   // counter using always block

   always @(posedge clk)
      if (reset) q <= 4'b0;
      else       q <= q+1;
endmodule
```

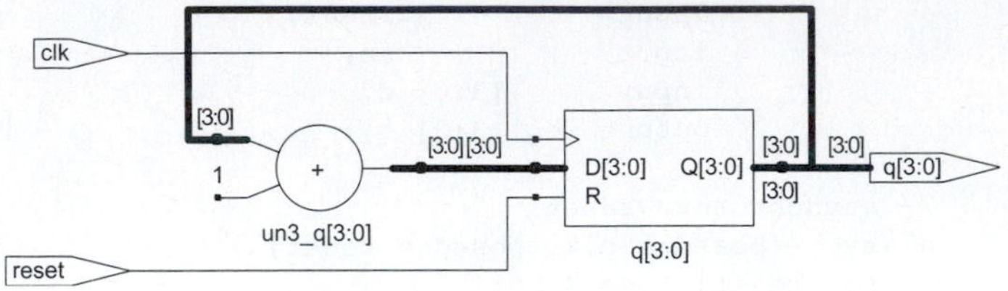

```
module counter(input           clk,
               input           reset,
               output [3:0]    q);

   wire [3:0] nextq;

   // counter using module calls

   flopr qflop(clk, reset, nextq, q);
   adder inc(q, 4'b0001, nextq);
   // assumes a 4-bit adder
endmodule
```

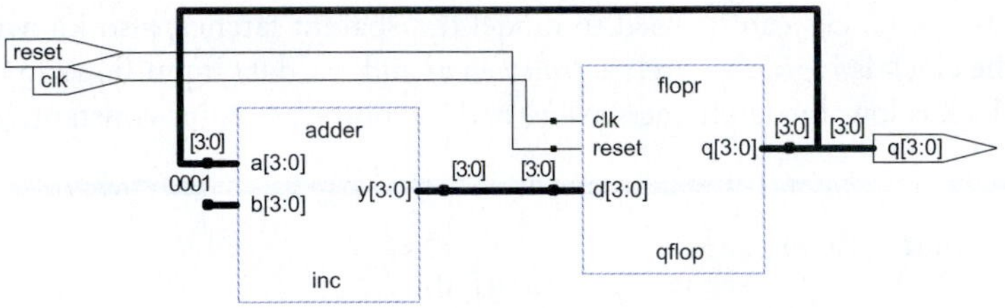

A.4.4   Combinational Logic

Always blocks imply sequential logic when some of the inputs do not appear in the @ stimulus list or might not cause the output to change. For example, in the flop module, d is not in the @ list, so the flop does not immediately respond to changes of d. In the latch, d is in the @ list, but changes in d are ignored unless clk is high. Always blocks can also be used to imply combinational logic if they are written in such a way that the output is reevaluated every time there are changes in any of the inputs. The following code shows how to define a bank of inverters with an always block. Note that y must be declared as reg because it appears on the left side of a <= or = sign in an always block. Nevertheless, y is the output of combinational logic, not a register.

```
module inv(input      [3:0] a,
           output reg [3:0] y);

   always @(*)
      y <= ~a;
endmodule
```

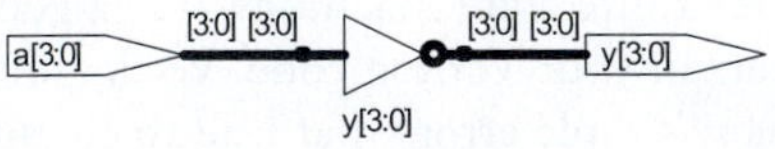

always @(*) evaluates the statements inside the **always** block whenever any of the signals on the right side of <= or = change inside the **always** block. Thus @(*) is a safe way to model combinational logic. In this particular example, @(a) would also have sufficed.

Similarly, the next example defines five banks of different kinds of gates. In this case, an @(a, b) would have been equivalent to @(*). However, @(*) is better because it avoids common mistakes of missing signals in the stimulus list. Also notice that the begin / end construct is necessary because multiple commands appear in the **always** block. This is analogous to { } block structure in C or Java. The begin / end was not needed in the **flopr** example because an **if** / **else** command counts as a single statement.

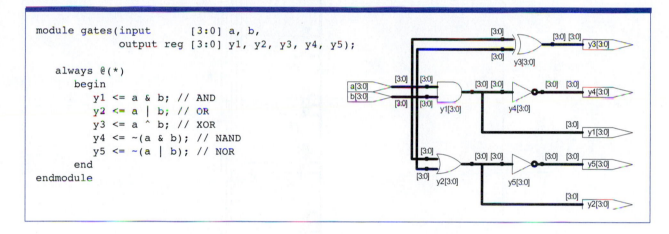

```
module gates(input        [3:0] a, b,
             output reg [3:0] y1, y2, y3, y4, y5);

   always @(*)
      begin
         y1 <= a & b;      // AND
         y2 <= a | b;      // OR
         y3 <= a ^ b;      // XOR
         y4 <= ~(a & b);   // NAND
         y5 <= ~(a | b);   // NOR
      end
endmodule
```

These two examples are poor applications of **always** blocks for modeling combinational logic because they require more lines than the equivalent approach with **assign** statements. Moreover, they pose the risk of inadvertently implying sequential logic (see Section A.9.2). A better application of the **always** block is a decoder, which takes advantage of the **case** statement that can only appear inside an **always** block.

```
module decoder_always(input        [2:0] a,
                      output reg [7:0] y);

   // a 3:8 decoder
   always @(*)
      case (a)
         3'b000:   y <= 8'b00000001;
         3'b001:   y <= 8'b00000010;
         3'b010:   y <= 8'b00000100;
         3'b011:   y <= 8'b00001000;
         3'b100:   y <= 8'b00010000;
         3'b101:   y <= 8'b00100000;
         3'b110:   y <= 8'b01000000;
         3'b111:   y <= 8'b10000000;
      endcase
endmodule
```

continued

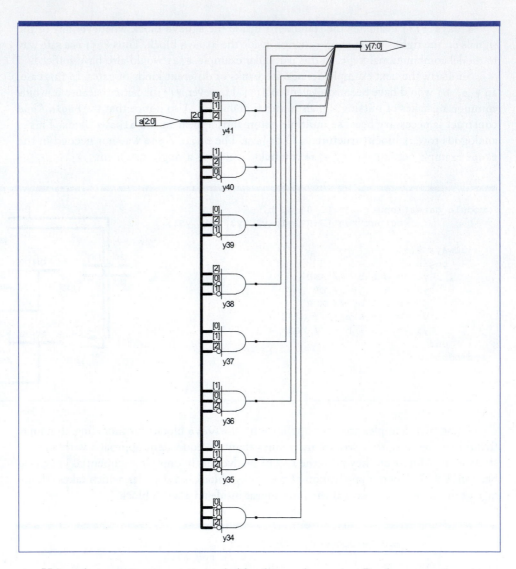

Using the case statement is probably clearer than using Boolean equations in an assign statement:

```
module decoder_assign(input  [2:0] a,
                      output [7:0] y);

  assign y[0] = ~a[0] & ~a[1] & ~a[2];
  assign y[1] =  a[0] & ~a[1] & ~a[2];
  assign y[2] = ~a[0] &  a[1] & ~a[2];
  assign y[3] =  a[0] &  a[1] & ~a[2];
  assign y[4] = ~a[0] & ~a[1] &  a[2];
  assign y[5] =  a[0] & ~a[1] &  a[2];
  assign y[6] = ~a[0] &  a[1] &  a[2];
  assign y[7] =  a[0] &  a[1] &  a[2];
endmodule
```

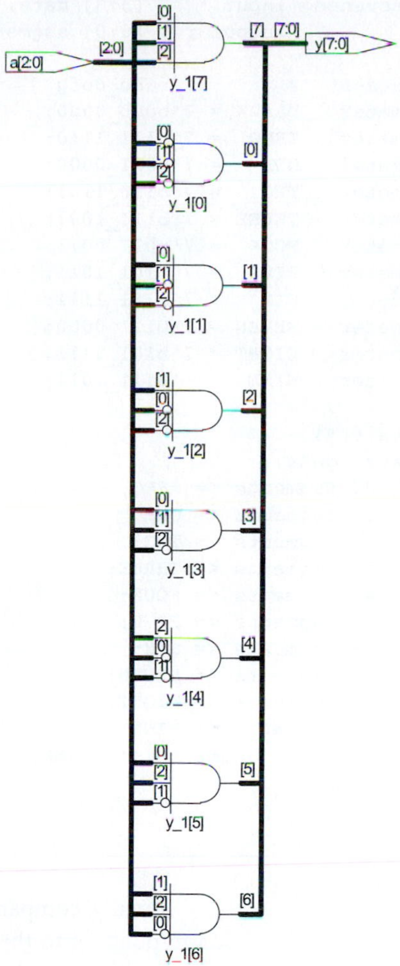

An even better example is the logic for a 7-segment display decoder from [Ciletti99]. The 7-segment display is shown in Figure A.1. The decoder takes a 4-bit number and displays its decimal value on the segments. For example, the number 0111 = 7 should turn on segments a, b, and c. The equivalent logic with assign statements describing the detailed logic for each bit would be tedious. This more abstract approach is faster to write, clearer to read, and can be automatically synthesized down to an efficient logic implementation. This example also illustrates the use of parameters to define constants to make the code more readable. The case statement has a default to display a blank output when the input is outside the range of decimal digits.

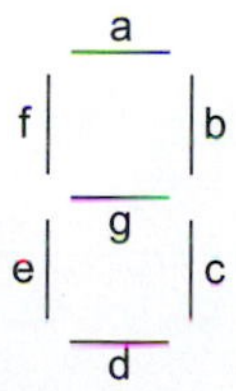

FIG A.1 7-segment display mapping

```
module sevenseg(input        [3:0] data,
                output reg [6:0] segments);

   // Segment #           abc_defg
   parameter   BLANK = 7'b000_0000;
   parameter   ZERO  = 7'b111_1110;
   parameter   ONE   = 7'b011_0000;
   parameter   TWO   = 7'b110_1101;
   parameter   THREE = 7'b111_1001;
   parameter   FOUR  = 7'b011_0011;
   parameter   FIVE  = 7'b101_1011;
   parameter   SIX   = 7'b101_1111;
   parameter   SEVEN = 7'b111_0000;
   parameter   EIGHT = 7'b111_1111;
   parameter   NINE  = 7'b111_1011;

   always @(*)
      case (data)
         0: segments <= ZERO;
         1: segments <= ONE;
         2: segments <= TWO;
         3: segments <= THREE;
         4: segments <= FOUR;
         5: segments <= FIVE;
         6: segments <= SIX;
         7: segments <= SEVEN;
         8: segments <= EIGHT;
         9: segments <= NINE;
         default: segments <= BLANK;
      endcase
endmodule
```

Finally, compare three descriptions of a priority encoder that sets one output true corresponding to the most significant input that is true. The `if` statement can appear in `always` blocks and makes the logic quite readable. The `casez` statement also appears in `always` blocks and allows don't care's in the case logic, indicated with the `?` symbol. The `assign` statements synthesize to the same results, but are arguably less clear to read. Note that $a[3]$ is another example of a feedthrough net because $y[3] = a[3]$. Of these three styles, the `if/else` approach is recommended for describing priority encoders because it is the easiest for most engineers to recognize. `case` statements are best reserved for functions specified by truth tables and `casez` statements should be used for functions specified by truth tables with don't cares. In the first two descriptions, `y` must be declared as a `reg` because it is assigned inside an `always` block.

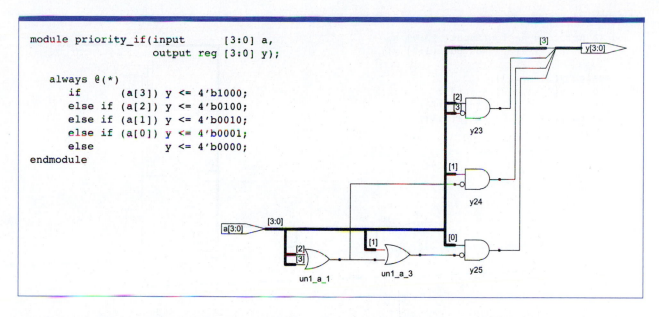

```
module priority_if(input        [3:0] a,
                   output reg [3:0] y);

   always @(*)
      if      (a[3]) y <= 4'b1000;
      else if (a[2]) y <= 4'b0100;
      else if (a[1]) y <= 4'b0010;
      else if (a[0]) y <= 4'b0001;
      else           y <= 4'b0000;
endmodule
```

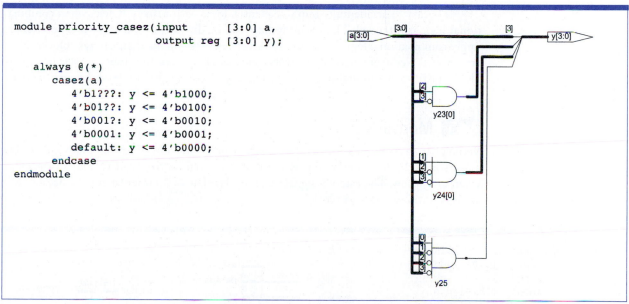

```
module priority_casez(input        [3:0] a,
                      output reg [3:0] y);

   always @(*)
      casez(a)
         4'b1???: y <= 4'b1000;
         4'b01??: y <= 4'b0100;
         4'b001?: y <= 4'b0010;
         4'b0001: y <= 4'b0001;
         default: y <= 4'b0000;
      endcase
endmodule
```

```
module priority_assign(input  [3:0] a,
                       output [3:0] y);

   assign y[3] = a[3];
   assign y[2] = a[2] & ~a[3];
   assign y[1] = a[1] & ~|a[3:2];
   assign y[0] = a[0] & ~|a[3:1];
endmodule
```

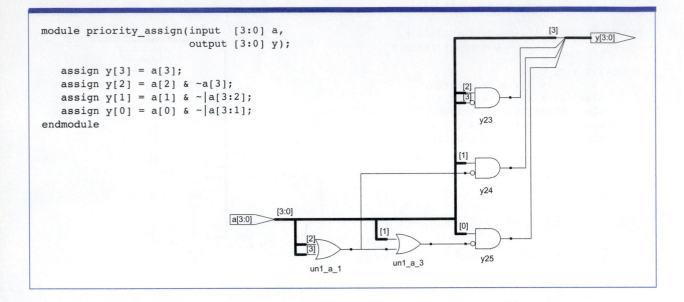

It is easy to accidentally imply sequential logic with `always` blocks when combinational logic is intended. The resulting bugs can be difficult to track down. Therefore, to imply combinational logic, it is safer to use `assign` statements than `always` blocks. Nevertheless, the convenience of constructs such as `if` or `case` that must appear in `always` blocks justifies the modeling style as long as you thoroughly understand what you are doing.

A.4.5 Memories

Verilog has an array construct used to describe memories. The following module describes a 64-word × 16-bit synchronous RAM that is written on the positive edge of the clock when `wrb` is low. The internal signal `mem` is declared as `reg` rather than `wire` because it is assigned inside an `always` block.

```
module ram(input         clk,
           input  [5:0]  addr,
           input         wrb,
           input  [15:0] din,
           output [15:0] dout);

   reg  [15:0] mem[63:0]; // the memory

   always @(posedge clk)
      if (~wrb) mem[addr] <= din;

   assign dout = mem[addr];
endmodule
```

Synthesis tools are often restricted to generating gates from a library and produce poor memory arrays. A specialized memory generator is commonly used instead.

A.4.6   Blocking and Nonblocking Assignment

Verilog supports two types of assignments inside an `always` block. *Blocking assignments* use the = statement. *Nonblocking assignments* use the <= statement. Do not confuse either type with the `assign` statement, which cannot appear inside `always` blocks at all.

A group of blocking assignments inside a `begin/end` block are evaluated sequentially, just as you would expect in a standard programming language. A group of nonblocking assignments are evaluated in parallel; all of the statements are evaluated before any of the left sides are updated. This is what you would expect in hardware because real logic gates all operate independently rather than waiting for the completion of other gates.

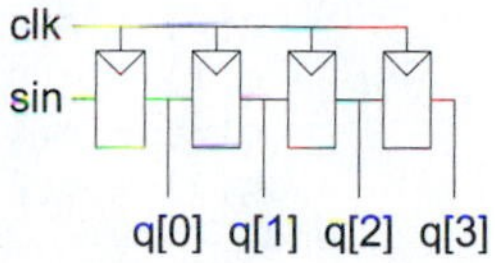

For example, consider two attempts to describe a shift register. On each clock edge, the data at `sin` should be shifted into the first flop, as shown in Figure A.2. The first flop shifts to the second flop. The data in the second flop shifts to the third flop, and so on until the last element drops off the end.

FIG A.2   Intended shift register

```
module shiftreg(input          clk,
                input          sin,
                output reg [3:0] q);

   // This is a correct implementation
   // using nonblocking assignment

   always @(posedge clk)
      begin
         q[0] <= sin; // nonblocking <=
         q[1] <= q[0];
         q[2] <= q[1];
         q[3] <= q[2];
         // even better to write
         // q <= {q[2:0], sin};
      end
endmodule
```

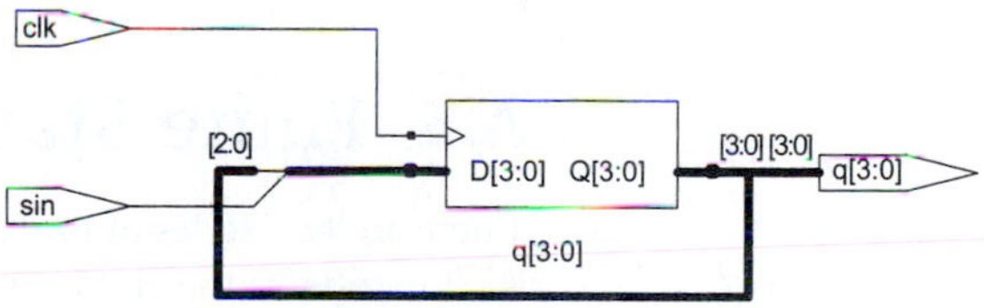

The nonblocking assignments mean that all of the values on the right sides are assigned simultaneously. Therefore, q[1] will get the original value of q[0], not the value of sin that gets loaded into q[0]. This is what we would expect from real hardware. Of course, all of this could be written on one line for brevity.

Blocking assignments are more familiar from traditional programming languages, but they inaccurately model hardware. Consider the same module using blocking assignments. When clk rises, the Verilog says that q[0] should be copied from sin. Then q[1] should be copied from the new value of q[0] and so forth. All four registers immediately get the sin value.

```
module shiftreg(input            clk,
                input            sin,
                output reg [3:0] q);

   // This is a bad implementation
   // using blocking assignment

   always @(posedge clk)
      begin
         q[0] = sin; // blocking =
         q[1] = q[0];
         q[2] = q[1];
         q[3] = q[2];
      end
endmodule
```

The moral of this illustration is to use nonblocking assignments in `always` blocks when modeling sequential logic. Using sufficient ingenuity, such as reversing the orders of the four commands, you could make blocking assignments work correctly, but they offer no advantages and harbor great risks.

Finally, note that each `always` block implies a separate block of logic. Therefore, a given `reg` can be assigned in only one `always` block. Otherwise, two pieces of hardware with shorted outputs will be implied.

A.5 Finite State Machines

There are two styles of finite state machines. In *Mealy machines* (Figure A.3(a)), the output is a function of the current state and inputs. In *Moore machines* (Figure A.3(b)), the output is a function of only the current state.

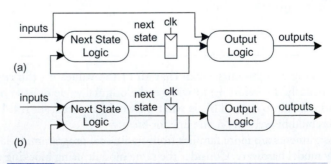

(a)

(b)

FIG A.3 Moore and Mealy machines

FSMs are modeled in Verilog with an `always` block defining the state registers and combinational logic defining the next state and output logic.

Let us first consider a simple finite state machine with one output and no inputs, a divide-by-3 counter. The output should be asserted every three clock cycles. The state transition diagram for a Moore machine is shown in Figure A.4. The output value is labeled in each state because the output is only a function of the state.

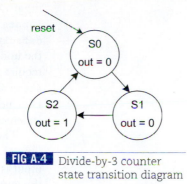

FIG A.4 Divide-by-3 counter state transition diagram

```verilog
module divideby3FSM(input   clk,
                    input   reset,
                    output out);

   reg  [1:0] state, nextstate;

   parameter S0 = 2'b00;
   parameter S1 = 2'b01;
   parameter S2 = 2'b10;

   // State Register
   always @(posedge clk, posedge reset)
      if (reset) state <= S0;
      else       state <= nextstate;

   // Next State Logic
   always @(*)
      case (state)
         S0: nextstate <= S1;
         S1: nextstate <= S2;
         S2: nextstate <= S0;
         default: nextstate <= S0;
      endcase

   // Output Logic
   assign out = (state == S2);
endmodule
```

The FSM model is divided into three portions: the *state register, next state logic,* and *output logic.* The state register logic describes an asynchronously resettable register that resets to an initial state and otherwise advances to the computed next state. Defining states with parameters allows the easy modification of state encodings and makes the code easier to read. The next state logic computes the next state as a function of the current state and inputs; in this example, there are no inputs. A case statement in an always @(state or inputs) block is a convenient way to define the next state. It is important to have a default if not all cases are enumerated; otherwise nextstate would not be assigned in the undefined cases. This implies that nextstate should keep its old value, which would require the existence of latches. Finally, the output logic may be a function of the current state alone in a Moore machine or of the current state and inputs in a Mealy machine. Depending on the complexity of the design, assign statements, if statements, or case statements may be most readable and efficient.

The next example shows a finite state machine with an input a and two outputs. Output x is true when the input is the same now as it was last cycle. Output y is true when the input is the same now as it was for the past two cycles. This is a Mealy machine because the output depends on the current inputs as well as the state. The outputs are labeled on each transition after the input. The state transition diagram is shown in Figure A.5.

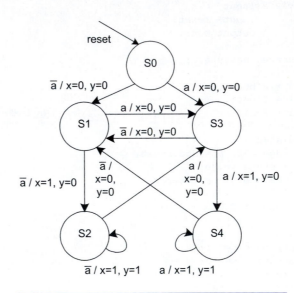

FIG A.5 History FSM state transition diagram

```verilog
module historyFSM(input   clk,
                  input   reset,
                  input   a,
                  output x, y);

   reg  [2:0] state, nextstate;

   parameter S0 = 3'b000;
   parameter S1 = 3'b010;
   parameter S2 = 3'b011;
   parameter S3 = 3'b100;
   parameter S4 = 3'b101;

   // State Register
   always @(posedge clk, posedge reset)
      if (reset) state <= S0;
      else       state <= nextstate;

   // Next State Logic
   always @(*)
      case (state)
         S0: if (a) nextstate <= S3;
             else   nextstate <= S1;
         S1: if (a) nextstate <= S3;
             else   nextstate <= S2;
         S2: if (a) nextstate <= S3;
             else   nextstate <= S2;
         S3: if (a) nextstate <= S4;
             else   nextstate <= S1;
         S4: if (a) nextstate <= S4;
             else   nextstate <= S1;
         default: nextstate <= S0;
      endcase

   // Output Logic
   assign x = (state[1] & ~a) |
              (state[2] & a);
   assign y = (state[1] & state[0] & ~a) |
              (state[2] & state[0] & a);
endmodule
```

continued

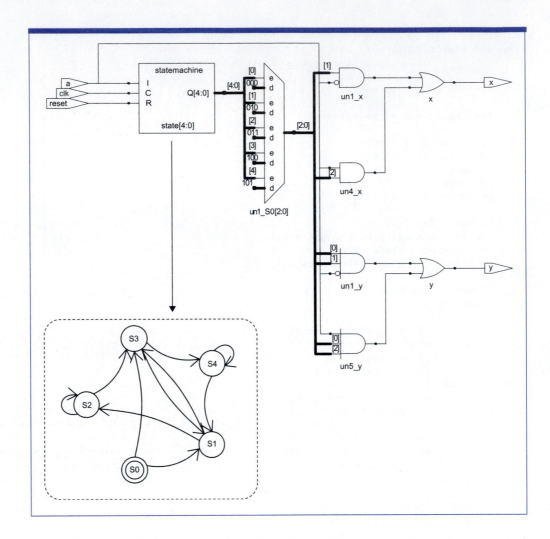

The output logic equations depend on the specific state encoding and were worked out by hand. A more general approach is independent of the encodings and requires less thinking, but more code:

```
// Output Logic
always @(state or a)
    case (state)
        S0:   begin
                  x <= 0; y <= 0;
              end
        S1:   if (a) begin
                  x <= 0; y <= 0;
              end else begin
                  x <= 1; y <= 0;
              end
        S2:   if (a) begin
                  x <= 0; y <= 0;
              end else begin
                  x <= 1; y <= 1;
              end
        S3:   if (a) begin
                  x <= 1; y <= 0;
              end else begin
                  x <= 0; y <= 0;
              end
        S4:   if (a) begin
                  x <= 1; y <= 1;
              end else begin
                  x <= 0; y <= 0;
              end
    endcase
```

You may be tempted to simplify the `case` statement. For example, case S4 could be reduced to:

```
// bad simplification of S4
S4:    if (a) begin
           y <= 1;
       end else begin
           x <= 0; y <= 0;
       end
```

The designer reasons that to get to state S4, we must have passed through state S3 with a high, setting x high. Therefore, the assignment of x is optimized out of S4 when a is high. This is incorrect reasoning. The modified approach implies sequential logic. Specifically, a latch is implied that holds the old value of x when x is not assigned. The latch holds its output under a peculiar set of circumstances; a and the state must be used to compute the latch clock signal. This is undoubtedly not what you wanted, but was easy to inadvertently imply. The moral of this example is that if any signal gets assigned in any branch of an `if` or `case` statement, it must be assigned in all branches lest a latch be implied.

A.6 Parameterized Modules

So far, all of our modules have had fixed-width inputs and outputs. Thus, we have needed separate modules for 4- and 8-bit wide 2-input multiplexers. Verilog permits variable bit widths using *parameterized modules*. For example, we can declare a parameterized bank of inverters with a default of 8 gates as:

```
module inv
   #(parameter width = 8)
    (input  [width-1:0] a,
     output [width-1:0] y);

   assign y = ~a;
endmodule
```

We can adjust the parameter when we instantiate the block. For example, we can build a bank of 12 buffers using a pair of banks of 12 inverters.

```
module buffer
   #(parameter numbits = 12)
    (input  [numbits-1:0] a,
     output [numbits-1:0] y);

   wire [numbits-1:0] x;

   inv #(numbits) i1(a, x);
   inv #(numbits) i2(x, y);
endmodule
```

A.7 Structural Primitives

When coding at the structural level, primitives exist for basic logic gates and transistors. Examples for a full adder carry circuit (majority gate) were given in Section 1.8.4.

Gate primitives include not, and, or, xor, nand, nor, and xnor. The output is declared first; multiple inputs may follow. For example, a 4-input AND gate may be given as

```
and g1(y, a, b, c, d);
```

Transistor primitives include tranif1, tranif0, rtranif1, and rtranif0. tranif1 is an nMOS transistor that turns ON when the gate is '1' while tranif0 is a pMOS transistor. The rtranif primitives are resistive transistors, i.e., weak transistors that can be overcome by a stronger driver. For example, a pseudo-nMOS NOR gate with a weak pull-up is modeled with three transistors. Most synthesis tools map only onto gates, not transistors, so these transistor primitives are only for simulation.

```
module nor2(input   a, b,
            output y);

    tranif1  n1(y, gnd, a);
    tranif1  n2(y, gnd, b);
    rtranif0 p1(y, vdd, gnd);
endmodule
```

The `tranif` devices are bi-directional; that is, the source and drain are symmetric. Verilog also supports unidirectional `nmos` and `pmos` primitives that only allow a signal to flow from the input terminal to the output terminal. Real transistors are inherently bi-directional, so unidirectional models can result in simulation not catching bugs that would exist in real hardware. Therefore, `tranif` primitives are preferred for simulation.

A.8   Test Benches

Verilog models are tested through simulation. For small designs, it may be practical to manually apply inputs to a simulator and visually check for the correct outputs. For larger designs, this procedure is usually automated with a *test bench*.

The following code shows an adder and its associated test bench. The test bench uses nonsynthesizable system calls to read a file, apply the *test vectors* to the *device under test* (DUT), check the results, and report any discrepancies. The `initial` statement defines a block that is executed only on startup of simulation. In it, the `$readmemh` reads a file (in hexadecimal form) into an array in memory. The `testvector.tv` example file shows four test vectors, each consisting of two 32-bit inputs and an expected 32-bit output. As the array is much larger than the number of test vectors, the remaining entries are filled with x's. The next `always` block defines a clock that repeats forever, being low for 50 units of time, then high for 50. On each positive edge of the clock, the next test vector is applied to the inputs and the expected output is saved. The actual output is sampled on the negative edge of the clock to permit some time for it to settle. It is not uncommon for bad logic to generate z and x values. The `!=` and `==` comparison operations return x if either argument has z or x, so they are not reliable. The `!==` and `===` commands check for an exact match, including z or x values. If there is a mismatch, the `$display` command is used to print the discrepancy. Each time `vectornum` is incremented, the test bench checks to see if the test is complete. It terminates testing with the `$finish` command after 100 vectors have been applied or the test vector consists of x's; for our `testvector.tv` file, it will end after the four vectors are applied.

```verilog
module adder(a, b, y);
    input      [31:0]      a, b;
    output     [31:0]      y;

    assign y = a + b;
endmodule

module testbench();
    reg      [31:0]  testvectors[1000:0];
    reg              clk;
    reg      [10:0]  vectornum, errors;
    reg      [31:0]  a, b, expectedy;
    wire     [31:0]  y;

    // instantiate device under test
    adder    dut(a, b, y);

    // read the test vector file and initialize test
    initial
        begin
            $readmemh("testvectors.tv", testvectors);
            vectornum = 0; errors = 0;
        end

    // generate a clock to sequence tests
    always
        begin
            clk = 0; #50; clk = 1; #50;
        end

    // on each clock step, apply next test
    always @(posedge clk)
        begin
            a = testvectors[vectornum*3];
            b = testvectors[vectornum*3 + 1];
            expectedy = testvectors[vectornum*3 + 2];
        end

    // then check for correct results
    always @(negedge clk)
        begin
            vectornum = vectornum + 1;
            if (y !== expectedy) begin
                $display("Inputs were %h, %h", a, b);
                $display("Expected %h but actual %h", expectedy, y);
                errors = errors + 1;
            end
        end
```

```
    // halt at the end of file
    always @(vectornum)
        begin
            if (vectornum == 100 || testvectors[vectornum*3] === 32'bx)
                begin
                    $display("Completed %d tests with %d errors. ",
                             vectornum, errors);
                    $finish;
                end
        end
endmodule
```

testvectors.tv file:
```
00000000
00000000
00000000

00000001
00000000
00000001

ffffffff
00000003
00000002

12345678
12345678
2468acf0
```

A.9   Pitfalls

This section includes a set of style guidelines and examples of a number of bad circuits produced by common Verilog coding errors. The examples include the warnings given by Synopsys Design Compiler and Synplify Pro synthesis tools.

A.9.1   Verilog Style Guidelines

1. Use only nonblocking assignments inside `always` blocks.

2. Define your combinational logic using `assign` statements when practical. Only use `always` blocks to define combinational logic if constructs like `if` or `case` make your logic much clearer or more compact.

3. When modeling combinational logic with an `always` block, if a signal is assigned in any branch of an `if` or `case` statement, it must be assigned in all branches.

4. Include `default` cases in your `case` statements.

5. Partition your design into leaf cells and non-leaf cells. Leaf cells contain behavioral code (`assign` statements or `always` blocks), but do not instantiate other cells. Non-leaf cells contain structural code (i.e., they instantiate other cells, but contain no logic). Minor exceptions to this guideline can be made to keep the code readable.

6. Use parameters to define state names and constants.

7. Properly indent your code, as shown in the examples in this guide.

8. Use comments liberally.

9. Use meaningful signal names. Use `a`, `b`, `c`, … for generic logic gate inputs. Use `x`, `y`, `z` for generic combinational outputs and `q` for a generic state element output. Use descriptive names for nongeneric cells. Do not use `foo`, `bar`, or `baz`!

10. Be consistent in your use of capitalization and underscores.

11. Do not ignore synthesis warnings unless you understand what they mean.

The clock, registers, and latches are common places to introduce bugs. Many FPGA and ASIC design flows use a conservative methodology of supplying a single clock to edge-triggered, asynchronously resettable registers with clock enables. While this will sacrifice performance and cost extra area, it reduces both the number of errors a designer can make and the expense of debugging these errors. The following guidelines are common to such a style.

1. Use only positive edge-triggered registers. Avoid `@(negedge clk)`, SR latches, and transparent latches.

2. Be certain not to inadvertently imply latches. Check synthesis reports for warnings such as

   (Synopsys) *Warning: Latch inferred in design '…' read with 'hdlin_check_no_latch'.*
   (Synplicity) *@W: …Latch generated from always block for signal …*

3. Provide an asynchronous reset to all of your registers with a common signal name.

4. Provide a common clock to all of your registers whenever possible. Avoid gated clocks, which may lead to extra clock skew and hold time failures. Use a clock enable instead.

5. If you get any "Bus Conflict" messages or x's in your simulation, be sure to find their cause and fix the problem.

A.9.2 Incorrect Stimulus List

The following circuit was intended to be a transparent latch, but the `d` input was omitted from the stimulus list. When synthesized, it still produces a transparent latch, but with a warning. If the Verilog is simulated, `q` will change on the rising edge of `clk` but not on a change in `d` while `clk` is stable high. Thus, the circuit will simulate as if it were a flip-flop.

Inconsistencies between simulation and synthesis are a common reason for chips to fail, so correcting these warnings is vital.

(Synopsys) *Warning: Variable 'd' is being read*
   in routine notquitealatch line 5 in file 'notquitealatch.v',
   but does not occur in the timing control of the block which begins there. (HDL-180)
(Synplicity) *@W: notquitealatch.v(5): Incomplete sensitivity list – assuming completeness*
   @W:" notquitealatch.v":5:12:5:15
   @W: notquitealatch.v(6): Referenced variable d is not in sensitivity list
   @W:"c: notquitealatch.v":6:20:6:21

```
module notquitealatch(input              clk,
                      input       [3:0] d,
                      output reg [3:0] q);

    always @(clk) // left out 'or d'
        if (clk) q <= d;
endmodule
```

Similarly, the b input in the following combinational logic was omitted from the stimulus list of the always block. Synthesis tools do generate the intended gates, but give another warning.

(Synopsys) *Warning: Variable 'b' is being read*
   in routine gates line 4 in file 'gates.v',
   but does not occur in the timing control of the block which begins there. (HDL-180)
(Synplicity) *@W: gates_bad.v(4): Incomplete sensitivity list – assuming completeness*
   @W:" gates_bad.v":4:12:4:13
   @W: gates_bad.v(6): Referenced variable b is not in sensitivity list
   @W:"c: gates_bad.v":6:19:6:20

```
module gates(input       [3:0] a, b,
             output reg [3:0] y1, y2, y3, y4, y5);

    always @(a) // missing 'or b'
        begin
            y1 <= a & b; // AND
            y2 <= a | b; // OR
            y3 <= a ^ b; // XOR
            y4 <= ~(a & b); // NAND
            y5 <= ~(a | b); // NOR
        end
endmodule
```

The next example is supposed to model a multiplexer, but the author incorrectly wrote `@(posedge s)` rather than `@(s)`. Because s is by definition high immediately after its positive edge, this circuit actually behaves as a flip-flop that captures d1 on the rising edge of `clk` and ignores d0. Some tools will synthesize the flip-flop, while others will just produce an error message.

(Synopysis) *Error: clock variable s is being used as data. (HDL-175)*
(Synplicity) *@W: badmux.v(1): Input d0 is unused @W:" badmux.v":1:31:1:33*

```
module badmux(input      [3:0] d0, d1,
              input          s,
              output reg [3:0] y);

   always @(posedge s)
      if (s) y <= d1;
      else   y <= d0;

endmodule
```

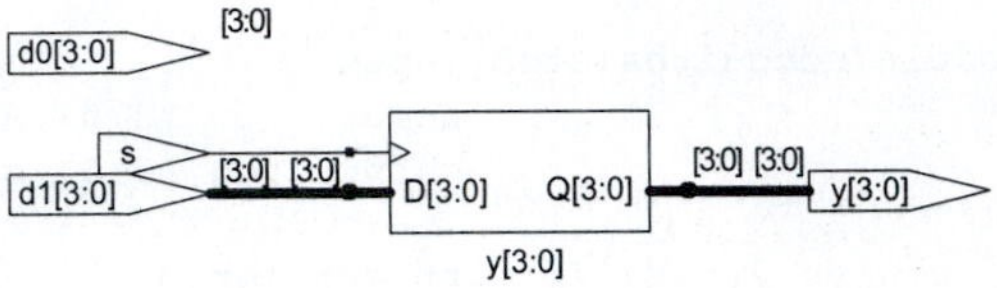

A.9.3 Missing `begin`/`end` Block

In the following example, two variables are supposed to be assigned in the `always` block. The `begin`/`end` block is missing. This is a syntax error.

(Synopsys) *Error: syntax error at or near token '[' (File: notquiteatwobitflop.v Line: 7) (VE-0)*
      Error: Can't read 'verilog' file 'J:/Classes/E155/Fall2000/synopsys/flop2.v'. (UID-59)
(Synplicity) *@E: notquiteatwobitflop.v(7): Expecting : @E:" notquiteatwobitflop.v":7:9:7:9*

```
module notquiteatwobitflop(input          clk,
                           input    [1:0] d,
                           output reg [1:0] q);

   always @(posedge clk)
      q[1] = d[1];
      q[0] = d[0];
endmodule
```

A.9.4 Undefined Outputs

In the next example of a finite state machine, the user intended out1 to be high when the state is 0 and out2 to be high when the state is 1. However, the code never sets the outputs low. Synopsys produces a circuit with an SR latch and a transparent latch that can set the output high, but never resets the output low. Synplicity produces a circuit in which both outputs are hardwired to 1.

```verilog
module FSMbad(input       clk,
              input       a,
              output reg out1, out2);

   reg  state;

   always @(posedge clk)
      if (state == 0) begin
         if (a) state <= 1;
      end else begin
         if (~a) state <= 0;
      end

   always @(*)
      if (state == 0) out1 <= 1;
      else            out2 <= 1;
      // neglect to set out1/out2 to 0
endmodule
```

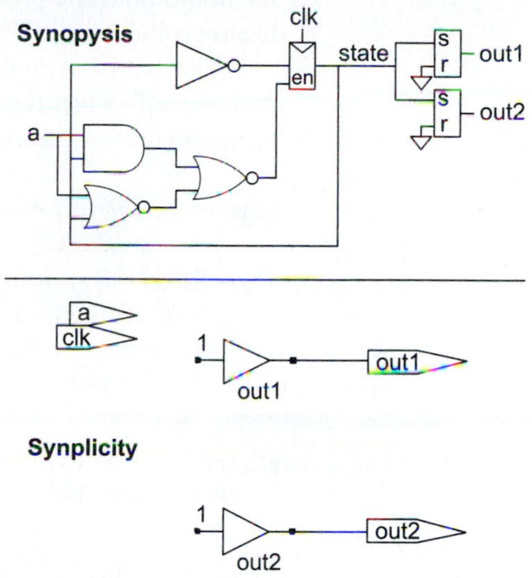

A corrected version of the code produces the desired state machine.

```verilog
module FSMgood(input       clk,
               input       a,
               output reg out1, out2);

   reg  state;

   always @(posedge clk)
      if (state == 0) begin
         if (a) state <= 1;
      end else begin
         if (~a) state <= 0;
      end

   always @(*)
      if (state == 0) begin
         out1 <= 1;
         out2 <= 0;
      end else begin
         out2 <= 1;
         out1 <= 0;
      end
endmodule
```

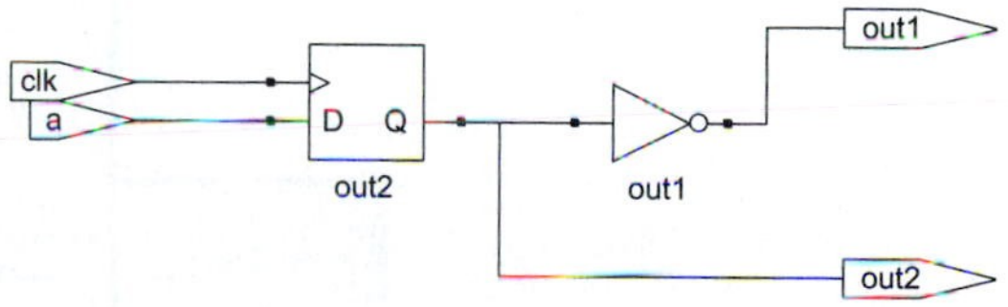

A.9.5 Incomplete Specification of Cases

The next examples show an incomplete specification of input possibilities. The priority encoder fails to check for the possibility of no true inputs. It therefore incorrectly implies latches to hold the previous output when all four inputs are false. Synplicity produces a bizarre circuit with SR latches.

The synthesis tool will warn that a latch or memory device is implied. The astute designer will detect the problem by knowing that a priority encoder should be a combinational circuit and therefore have no memory devices.

(Synopsys) *Inferred memory devices in process in routine priority_always line 4 in file 'priority_always_bad.v'.*

(Synplicity) *@W: priority_always_bad.v(5): Latch generated from always block for signal y[3:0], probably caused by a missing assignment in an if or case stmt @W:" priority_always_bad.v".5:6:5:8*

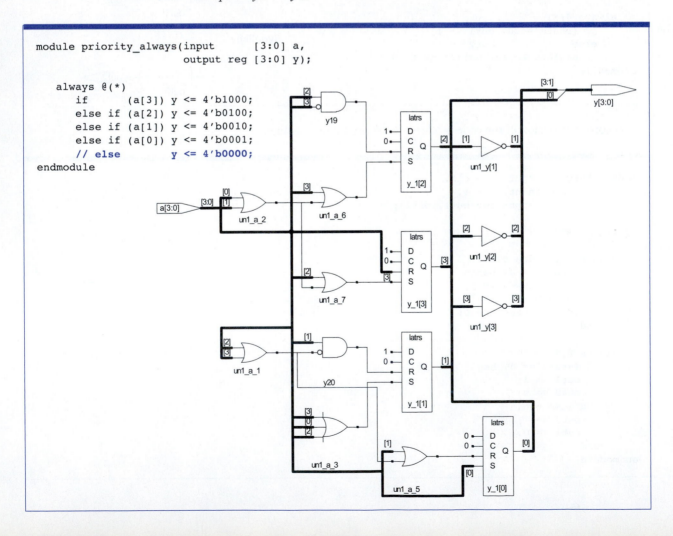

```
module priority_always(input      [3:0] a,
                       output reg [3:0] y);

    always @(*)
        if      (a[3]) y <= 4'b1000;
        else if (a[2]) y <= 4'b0100;
        else if (a[1]) y <= 4'b0010;
        else if (a[0]) y <= 4'b0001;
        // else        y <= 4'b0000;
endmodule
```

The next example of a 7-segment display decoder shows the same type of problem in a case statement.

```verilog
module seven_seg_display_decoder(input      [3:0] data,
                                 output reg [6:0] segments);

    always @(*)
        case (data)
            0: segments <= 7'b000_0000; // ZERO
            1: segments <= 7'b111_1110; // ONE
            2: segments <= 7'b011_0000; // TWO
            3: segments <= 7'b110_1101; // THREE
            4: segments <= 7'b011_0011; // FOUR
            5: segments <= 7'b101_1011; // FIVE
            6: segments <= 7'b101_1111; // SIX
            7: segments <= 7'b111_0000; // SEVEN
            8: segments <= 7'b111_1111; // EIGHT
            9: segments <= 7'b111_1011; // NINE
            // default: segments <= 7'b000_0000;
        endcase
endmodule
```

Similarly, it is a common mistake to forget the default in the next state or output logic of an FSM.

```verilog
module divideby3FSM(input   clk,
                    input   reset,
                    output out);

    reg [1:0] state, nextstate;

    parameter S0 = 2'b00;
    parameter S1 = 2'b01;
    parameter S2 = 2'b10;

    // State Register
    always @(posedge clk, posedge reset)
        if (reset) state <= S0;
        else       state <= nextstate;

    // Next State Logic
    always @(state)
        case (state)
            S0: nextstate <= S1;
            S1: nextstate <= S2;
            S2: nextstate <= S0;
            //default: nextstate <= S0;
        endcase
```

continued

```
                        // Output Logic
                        assign out = (state == S2);
                     endmodule
```

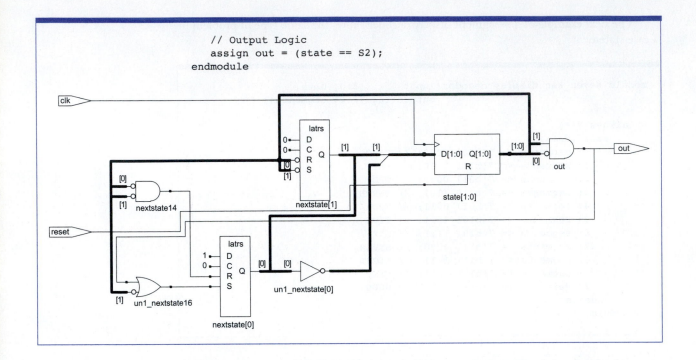

A.9.6 Shorted Outputs

Bad code can sometimes lead to shorted outputs of gates. For example, the tristate drivers in the following multiplexer should have mutually exclusive enable signals, but instead are both active simultaneously and produce a conflict when d0 and d1 are not equal.

Synthesis may not report any errors. However, during simulation, you will observe x's rather than 0's or 1's when the bus is simultaneously being driven high and low. You may also get a "Bus Conflict" warning message.

```
module mux2(input  [3:0] d0, d1,
            input        s,
            output [3:0] y);

   tristate t0(d0, s, y); // wanted ~s
   tristate t1(d1, s, y);
endmodule
```

Another cause of shorted outputs is when a `reg` is assigned in two different `always` blocks. For example, the following code tries to model a register with asynchronous reset and asynchronous set. The first `always` block models the reset and ordinary operation. The second `always` block attempts to incorporate the asynchronous set. Synthesis infers a separate piece of hardware for each `always` block, with a shorted output and may report an error.

(Synopsys) *Error: the net '/ver1/q' has more than one driver*
(Synplicity) *@E: floprs_bad.v(5): Only one always block may assign a given variable q[3:0]*
   @E:"\floprs_bad.v":5:33:5:34

```verilog
module floprsen(input              clk,
                input              reset,
                input              set,
                input     [3:0] d,
                output reg [3:0] q);

   // bad asynchronous set and reset

   always @(posedge clk, posedge reset)
      if (reset) q <= 0;
      else q <= d;

   always @(set)
      if (set) q <= 1;
endmodule
```

A.9.7   Incorrect Use of Nonblocking Assignments

Section A.4.6 recommended using nonblocking assignments in all `always` blocks. This can occasionally get you in trouble, as shown in the following 3-input AND gate.

```verilog
module and3bad(a, b, c, y);
   input              a, b, c;
   output             y;

   reg                y;
   reg                tmp;

   always @(a, b, c)
   begin
      tmp <= a & b;
      y <= tmp & c;
   end
endmodule
```

In this example, suppose that initially, a = 0 and b = c = 1. Therefore, tmp initially is 0. When a rises, the always block is triggered. tmp is given the value 1, but y is in parallel given the value 0 because the new value of tmp has not yet been written. The code can synthesize correctly, but simulate incorrectly without warning. The problem can be avoided by using always @(*) to describe combinational logic.

A.10   Example: MIPS Processor

To illustrate a nontrivial Verilog design, this section lists the Verilog code and test bench for the MIPS processor subset discussed in Chapter 1. The example handles only the LB, SB, ADD, SUB, AND, OR, SLT, BEQ, and J instructions. It uses an 8-bit datapath and only eight registers. Because the instruction is 32 bits wide, it is loaded in four successive fetch cycles across an 8-bit path to external memory.

The test bench initializes a 512-byte memory with instructions and data from a text file. The code exercises each of the instructions. The mipstest.asm assembly language file and memfile.dat text file are shown below. The test bench runs until it observes a memory write. If the value 7 is written to address 5, the code probably executed correctly. If all goes well, the test bench should take 100 cycles (1000 ns) to run.

```
# mipstest.asm
# 9/16/03 David Harris David_Harris@hmc.edu
#
# Test MIPS instructions.  Assumes memory was
# initialized as:
# word 16: 3 — be careful of endianness
# word 17: 5
# word 18: 12

main:     #Assembly Code          effect                    Machine Code
          lb $2, 68($0)           # initialize $2 = 5        80020044
          lb $7, 64($0)           # initialize $7 = 3        80070040
          lb $3, 69($7)           # initialize $3 = 12       80e30045
          or $4, $7, $2           # $4 <= 3 or 5 = 7         00e22025
          and $5, $3, $4          # $5 <= 12 and 7 = 4       00642824
          add $5, $5, $4          # $5 <= 4 + 7 = 11         00a42820
          beq $5, $7, end         # shouldn't be taken       10a70008
          slt $6, $3, $4          # $6 <= 12 < 7 = 0         0064302a
          beq $6, $0, around      # should be taken          10c00001
          lb $5, 0($0)            # shouldn't happen         80050000
around:   slt $6, $7, $2          # $6 <= 3 < 5 = 1          00e2302a
          add $7, $6, $5          # $7 <= 1 + 11 = 12        00c53820
          sub $7, $7, $2          # $7 <= 12 - 5 = 7         00e23822
          j end                   # should be taken          0800000f
          lb $7, 0($0)            # shouldn't happen         80070000
end:      sb $7, 0($2)            # write adr 5 <= 7         a0470000
```

```
memfile.dat
80020044
80070040
80e30045
00e22025
00642824
00a42820
10a70008
0064302a
10c00001
80050000
00e2302a
00c53820
00e23822
0800000f
80070000
a0470000
03000000
05000000
0c000000
```

```verilog
//————————————————————
// mips.v
// Max Yi (byyi@hmc.edu) and David_Harris@hmc.edu 12/9/03
// Model of subset of MIPS processor described in Ch 1
//————————————————————

// top level design for testing
module top #(parameter WIDTH = 8, REGBITS = 3)();

    reg                 clk;
    reg                 reset;
    wire                memread, memwrite;
    wire    [WIDTH-1:0] adr, writedata;
    wire    [WIDTH-1:0] memdata;

    // instantiate devices to be tested
    mips #(WIDTH,REGBITS) dut(clk, reset, memdata, memread, memwrite, adr, writedata);

    // external memory for code and data
    exmemory #(WIDTH) exmem(clk, memwrite, adr, writedata, memdata);

    // initialize test
    initial
        begin
            reset <= 1; # 22; reset <= 0;
        end

    // generate clock to sequence tests
    always
        begin
            clk <= 1; # 5; clk <= 0; # 5;
        end

    always@(negedge clk)
        begin
            if(memwrite)
                if(adr == 5 & writedata == 7)
                    $display("Simulation completely successful");
                else $display("Simulation failed");
        end
endmodule
```

```verilog
// external memory accessed by MIPS
module exmemory #(parameter WIDTH = 8)
                 (input                 clk,
                  input                 memwrite,
                  input      [WIDTH-1:0] adr, writedata,
                  output reg [WIDTH-1:0] memdata);

  reg  [31:0] RAM [(1<<WIDTH-2)-1:0];
  wire [31:0] word;

  initial
    begin
      $readmemh("memfile.dat",RAM);
    end

  // read and write bytes from 32-bit word
  always @(posedge clk)
    if(memwrite)
      case (adr[1:0])
        2'b00: RAM[adr>>2][7:0]   <= writedata;
        2'b01: RAM[adr>>2][15:8]  <= writedata;
        2'b10: RAM[adr>>2][23:16] <= writedata;
        2'b11: RAM[adr>>2][31:24] <= writedata;
      endcase

  assign word = RAM[adr>>2];
  always @(*)
    case (adr[1:0])
      2'b00: memdata <= word[31:24];
      2'b01: memdata <= word[23:16];
      2'b10: memdata <= word[15:8];
      2'b11: memdata <= word[7:0];
    endcase
endmodule

// simplified MIPS processor
module mips #(parameter WIDTH = 8, REGBITS = 3)
             (input                 clk, reset,
              input  [WIDTH-1:0] memdata,
              output                memread, memwrite,
              output [WIDTH-1:0] adr, writedata);

  wire [31:0] instr;
  wire        zero, alusrca, memtoreg, iord, pcen, regwrite, regdst;
  wire [1:0]  aluop,pcsource,alusrcb;
  wire [3:0]  irwrite;
  wire [2:0]  alucont;

  controller  cont(clk, reset, instr[31:26], zero, memread, memwrite,
                   alusrca, memtoreg, iord, pcen, regwrite, regdst,
                   pcsource, alusrcb, aluop, irwrite);
  alucontrol  ac(aluop, instr[5:0], alucont);
  datapath    #(WIDTH, REGBITS)
              dp(clk, reset, memdata, alusrca, memtoreg, iord, pcen,
                 regwrite, regdst, pcsource, alusrcb, irwrite, alucont,
                 zero, instr, adr, writedata);
endmodule

module controller(input clk, reset,
                  input      [5:0] op,
                  input            zero,
                  output reg       memread, memwrite, alusrca, memtoreg, iord,
                  output           pcen,
                  output reg       regwrite, regdst,
                  output reg [1:0] pcsource, alusrcb, aluop,
                  output reg [3:0] irwrite);
```

```verilog
parameter   FETCH1  =  4'b0001;
parameter   FETCH2  =  4'b0010;
parameter   FETCH3  =  4'b0011;
parameter   FETCH4  =  4'b0100;
parameter   DECODE  =  4'b0101;
parameter   MEMADR  =  4'b0110;
parameter   LBRD    =  4'b0111;
parameter   LBWR    =  4'b1000;
parameter   SBWR    =  4'b1001;
parameter   RTYPEEX =  4'b1010;
parameter   RTYPEWR =  4'b1011;
parameter   BEQEX   =  4'b1100;
parameter   JEX     =  4'b1101;

parameter   LB      =  6'b100000;
parameter   SB      =  6'b101000;
parameter   RTYPE   =  6'b0;
parameter   BEQ     =  6'b000100;
parameter   J       =  6'b000010;

reg [3:0] state, nextstate;
reg       pcwrite, pcwritecond;

// state register
always @(posedge clk)
   if(reset) state <= FETCH1;
   else state <= nextstate;

// next state logic
always @(*)
   begin
      case(state)
         FETCH1:  nextstate <= FETCH2;
         FETCH2:  nextstate <= FETCH3;
         FETCH3:  nextstate <= FETCH4;
         FETCH4:  nextstate <= DECODE;
         DECODE:  case(op)
                     LB:      nextstate <= MEMADR;
                     SB:      nextstate <= MEMADR;
                     RTYPE:   nextstate <= RTYPEEX;
                     BEQ:     nextstate <= BEQEX;
                     J:       nextstate <= JEX;
                     default: nextstate <= FETCH1; // should never happen
                  endcase
         MEMADR:  case(op)
                     LB:      nextstate <= LBRD;
                     SB:      nextstate <= SBWR;
                     default: nextstate <= FETCH1; // should never happen
                  endcase
         LBRD:    nextstate <= LBWR;
         LBWR:    nextstate <= FETCH1;
         SBWR:    nextstate <= FETCH1;
         RTYPEEX: nextstate <= RTYPEWR;
         RTYPEWR: nextstate <= FETCH1;
         BEQEX:   nextstate <= FETCH1;
         JEX:     nextstate <= FETCH1;
         default: nextstate <= FETCH1; // should never happen
      endcase
   end

always @(*)
   begin
   // set all outputs to zero, then conditionally assert just the appropriate ones
      irwrite <= 4'b0000;
      pcwrite <= 0; pcwritecond <= 0;
      regwrite <= 0; regdst <= 0;
      memread <= 0; memwrite <= 0;
      alusrca <= 0; alusrcb <= 2'b00; aluop <= 2'b00;
      pcsource <= 2'b00;
      iord <= 0; memtoreg <= 0;
```

```verilog
case(state)
    FETCH1:
        begin
            memread <= 1;
            irwrite <= 4'b1000;
            alusrcb <= 2'b01;
            pcwrite <= 1;
        end
    FETCH2:
        begin
            memread <= 1;
            irwrite <= 4'b0100;
            alusrcb <= 2'b01;
            pcwrite <= 1;
        end
    FETCH3:
        begin
            memread <= 1;
            irwrite <= 4'b0010;
            alusrcb <= 2'b01;
            pcwrite <= 1;
        end
    FETCH4:
        begin
            memread <= 1;
            irwrite <= 4'b0001;
            alusrcb <= 2'b01;
            pcwrite <= 1;
        end
    DECODE: alusrcb <= 2'b11;
    MEMADR:
        begin
            alusrca <= 1;
            alusrcb <= 2'b10;
        end
    LBRD:
        begin
            memread <= 1;
            iord    <= 1;
        end
    LBWR:
        begin
            regwrite <= 1;
            memtoreg <= 1;
        end
    SBWR:
        begin
            memwrite <= 1;
            iord     <= 1;
        end
    RTYPEEX:
        begin
            alusrca <= 1;
            aluop   <= 2'b10;
        end
    RTYPEWR:
        begin
            regdst   <= 1;
            regwrite <= 1;
        end
    BEQEX:
        begin
            alusrca    <= 1;
            aluop      <= 2'b01;
            pcwritecond <= 1;
            pcsource   <= 2'b01;
        end
    JEX:
        begin
            pcwrite  <= 1;
            pcsource <= 2'b10;
        end
```

```
            endcase
        end
    assign pcen = pcwrite | (pcwritecond & zero); // program counter enable
endmodule

module alucontrol(input      [1:0] aluop,
                  input      [5:0] funct,
                  output reg [2:0] alucont);

    always @(*)
        case(aluop)
            2'b00: alucont <= 3'b010;  // add for lb/sb/addi
            2'b01: alucont <= 3'b110;  // sub (for beq)
            default: case(funct)       // R-Type instructions
                    6'b100000: alucont <= 3'b010; // add (for add)
                    6'b100010: alucont <= 3'b110; // subtract (for sub)
                    6'b100100: alucont <= 3'b000; // logical and (for and)
                    6'b100101: alucont <= 3'b001; // logical or (for or)
                    6'b101010: alucont <= 3'b111; // set on less (for slt)
                    default:   alucont <= 3'b101; // should never happen
                endcase
        endcase
endmodule

module datapath #(parameter WIDTH = 8, REGBITS = 3)
                 (input               clk, reset,
                  input  [WIDTH-1:0] memdata,
                  input               alusrca, memtoreg, iord, pcen, regwrite, regdst,
                  input  [1:0]        pcsource, alusrcb,
                  input  [3:0]        irwrite,
                  input  [2:0]        alucont,
                  output              zero,
                  output [31:0]       instr,
                  output [WIDTH-1:0] adr, writedata);

    // the size of the parameters must be changed to match the WIDTH parameter
    parameter CONST_ZERO = 8'b0;
    parameter CONST_ONE =  8'b1;

    wire [REGBITS-1:0] ra1, ra2, wa;
    wire [WIDTH-1:0]   pc, nextpc, md, rd1, rd2, wd, a, src1, src2, aluresult,
                       aluout, constx4;

    // shift left constant field by 2
    assign constx4 = {instr[WIDTH-3:0],2'b00};

    // register file address fields
    assign ra1 = instr[REGBITS+20:21];
    assign ra2 = instr[REGBITS+15:16];
    mux2       #(REGBITS) regmux(instr[REGBITS+15:16], instr[REGBITS+10:11], regdst, wa);

    // independent of bit width, load instruction into four 8-bit registers over four cycles
    flopen     #(8)      ir0(clk, irwrite[0], memdata[7:0], instr[7:0]);
    flopen     #(8)      ir1(clk, irwrite[1], memdata[7:0], instr[15:8]);
    flopen     #(8)      ir2(clk, irwrite[2], memdata[7:0], instr[23:16]);
    flopen     #(8)      ir3(clk, irwrite[3], memdata[7:0], instr[31:24]);

    // datapath
    flopenr    #(WIDTH)  pcreg(clk, reset, pcen, nextpc, pc);
    flop       #(WIDTH)  mdr(clk, memdata, md);
    flop       #(WIDTH)  areg(clk, rd1, a);
    flop       #(WIDTH)  wrd(clk, rd2, writedata);
    flop       #(WIDTH)  res(clk, aluresult, aluout);
    mux2       #(WIDTH)  adrmux(pc, aluout, iord, adr);
    mux2       #(WIDTH)  src1mux(pc, a, alusrca, src1);
    mux4       #(WIDTH)  src2mux(writedata, CONST_ONE, instr[WIDTH-1:0],
                                 constx4, alusrcb, src2);
    mux4       #(WIDTH)  pcmux(aluresult, aluout, constx4, CONST_ZERO, pcsource, nextpc);
    mux2       #(WIDTH)  wdmux(aluout, md, memtoreg, wd);
    regfile    #(WIDTH,REGBITS) rf(clk, regwrite, ra1, ra2, wa, wd, rd1, rd2);
```

```verilog
    alu        #(WIDTH) alunit(src1, src2, alucont, aluresult);
    zerodetect #(WIDTH) zd(aluresult, zero);
endmodule

module alu #(parameter WIDTH = 8)
            (input      [WIDTH-1:0] a, b,
             input      [2:0]       alucont,
             output reg [WIDTH-1:0] result);

    wire    [WIDTH-1:0] b2, sum, slt;

    assign b2 = alucont[2] ? ~b:b;
    assign sum = a + b2 + alucont[2];
    // slt should be 1 if most significant bit of sum is 1
    assign slt = sum[WIDTH-1];

    always@(*)
       case(alucont[1:0])
          2'b00: result <= a & b;
          2'b01: result <= a | b;
          2'b10: result <= sum;
          2'b11: result <= slt;
       endcase
endmodule

module regfile #(parameter WIDTH = 8, REGBITS = 3)
               (input                clk,
                input                regwrite,
                input  [REGBITS-1:0] ra1, ra2, wa,
                input  [WIDTH-1:0]   wd,
                output [WIDTH-1:0]   rd1, rd2);

    reg  [WIDTH-1:0] RAM [(1<<REGBITS)-1:0];

    // three ported register file
    // read two ports combinationally
    // write third port on rising edge of clock
    // register 0 hardwired to 0
    always @(posedge clk)
       if (regwrite) RAM[wa] <= wd;

    assign rd1 = ra1 ? RAM[ra1] : 0;
    assign rd2 = ra2 ? RAM[ra2] : 0;
endmodule

module zerodetect #(parameter WIDTH = 8)
                   (input [WIDTH-1:0] a,
                    output            y);

    assign y = (a==0);
endmodule

module flop #(parameter WIDTH = 8)
            (input                clk,
             input      [WIDTH-1:0] d,
             output reg [WIDTH-1:0] q);

    always @(posedge clk)
       q <= d;
endmodule

module flopen #(parameter WIDTH = 8)
              (input                clk, en,
               input      [WIDTH-1:0] d,
               output reg [WIDTH-1:0] q);

    always @(posedge clk)
       if (en) q <= d;
endmodule
```

```verilog
module flopenr #(parameter WIDTH = 8)
                (input                clk, reset, en,
                 input      [WIDTH-1:0] d,
                 output reg [WIDTH-1:0] q);

   always @(posedge clk)
      if      (reset) q <= 0;
      else if (en)    q <= d;
endmodule

module mux2 #(parameter WIDTH = 8)
             (input  [WIDTH-1:0] d0, d1,
              input             s,
              output [WIDTH-1:0] y);

   assign y = s ? d1 : d0;
endmodule

module mux4 #(parameter WIDTH = 8)
             (input      [WIDTH-1:0] d0, d1, d2, d3,
              input      [1:0]       s,
              output reg [WIDTH-1:0] y);

   always @(*)
      case(s)
         2'b00: y <= d0;
         2'b01: y <= d1;
         2'b10: y <= d2;
         2'b11: y <= d3;
      endcase
endmodule
```

Appendix B

VHDL

B.1 Introduction

This appendix gives a quick introduction to VHDL. VHDL is an acronym for the VHSIC Hardware Description Language. VHSIC is in turn an acronym for the US Department of Defense Very High Speed Integrated Circuits program. [Ashenden01] offers a definitive and readable treatment of the language.

VHDL was originally developed in 1981 by the Department of Defense as a language to describe the structure and function of hardware. The IEEE standardized it in 1987, and a revised version was adopted in 1993, updated in 2000, and updated again in 2002 [IEEE1076-02]. The language was first envisioned for documentation, but quickly was adopted for simulation and synthesis. Compared to Verilog, VHDL is more verbose and cumbersome, as you might expect of a language developed by committee. However, VHDL has some features that are convenient for large team design projects, and government contractors and telecommunications companies use it extensively. Religious wars have raged over which HDL is superior, but both are used too widely for CAD vendors not to support them.

As mentioned in Section 1.8.4, there are two general styles of description: *behavioral* and *structural*. Structural VHDL describes how a module is composed of simpler modules or basic primitives such as gates or transistors. Behavioral VHDL describes how the outputs are computed as functions of the inputs. There are two general types of statements used in behavioral VHDL. *Concurrent signal assignments* imply combinational logic. *Processes* can imply combinational logic or sequential logic, depending how they are used. It is good practice to partition your design into combinational and sequential components and then write VHDL in such a way that you get what you want. If you don't know whether a block of logic is combinational or sequential, you are likely to get the wrong thing.

This appendix focuses on a synthesizable subset of the VHDL language. The language also contains extensive capabilities for system modeling that are beyond the scope of this tutorial.

B.2 Behavioral Modeling with Concurrent Signal Assignments

A 32-bit adder is a complex design at the schematic level of representation. It can be constructed from 32 full adder cells, each of which in turn requires about six 2-input gates. VHDL provides a more compact description:

```
library IEEE;
use IEEE.STD_LOGIC_1164.all;
use IEEE.STD_LOGIC_UNSIGNED.all;

entity adder is
    port(a, b: in  STD_LOGIC_VECTOR(31 downto 0);
         y:    out STD_LOGIC_VECTOR(31 downto 0));
end;

architecture synth of adder is
begin
    y <= a + b;
end;
```

This example has three parts: the *library use clauses*, the *entity declaration*, and the *architecture body*. The library part will be discussed in Section B.3.7. The entity defines the inputs and outputs of the adder *block*, which in this case are 32-bit busses. The architecture describes what the block does. In this case, the output y is computed as the sum of a and b. y <= a + b is called a *concurrent signal assignment* statement because multiple statements could happen in parallel just as multiple logic gates can operate concurrently.

B.2.1 Bitwise Operators

VHDL has a number of *bitwise* operators that act on busses. For example, the following module describes four inverters.

```
library IEEE; use IEEE.STD_LOGIC_1164.all;

entity inv is
    port(a: in  STD_LOGIC_VECTOR(3 downto 0);
         y: out STD_LOGIC_VECTOR(3 downto 0));
end;

architecture synth of inv is
begin
    y <= not a;
end;
```

Similar bitwise operations are available for the other basic logic functions:

```
library IEEE; use IEEE.STD_LOGIC_1164.all;

entity gates is
    port(a, b:                    in  STD_LOGIC_VECTOR(3 downto 0);
           y1, y2, y3, y4, y5: out STD_LOGIC_VECTOR(3 downto 0));
end;

architecture synth of gates is
begin
    -- Five different two-input logic gates acting on 4 bit busses
    y1 <= a and b;
    y2 <= a or b;
    y3 <= a xor b;
    y4 <= a nand b;
    y5 <= a nor b;
end;
```

B.2.2 Comments and White Space

The previous examples showed a comment. Comments begin with -- and continue to the end of the line. Comments spanning multiple lines use -- at the beginning of each line.

VHDL is not picky about the use of white space. Nevertheless, proper indenting and spacing is helpful to make nontrivial designs readable. VHDL is not case-sensitive, but a consistent use of upper and lower case also makes the code readable and more portable to other tools that are case-sensitive.

B.2.3 Other Operators

VHDL defines a number of operators available in concurrent assignment statements including:

Multiplying Operators:	`*, /, mod, rem`
Adding Operators:	`+, -`
Relational Operators:	`=, /=, <, <=, >, >=`
Logical Operators:	`not, and, or, nand, nor, xor, xnor`

= and /= (equality / inequality) on N-bit inputs require N 2-input XNORs to determine equality of each bit and an N-input AND or NAND to combine all the bits. Addition, subtraction, and comparison all require an adder, which is expensive in hardware. Multipliers are even more costly. Do not use these statements without contemplating the number of gates you are generating. Moreover, the implementations are not always particularly efficient for your problem. Division, mod (modulo), and rem (remainder) are only supported when the right operand is a power of two.

Some synthesis tools ship with optimized libraries for special functions like adders and multipliers. For example, the Synopsys DesignWare libraries produce reasonably good multipliers. If you do not have a license for the libraries, you'll probably be disappointed with the speed and gate count of a multiplier your synthesis tool produces from when it sees *.

B.2.4 Conditional Signal Assignment Statements

Conditional signal assignments perform different operations depending on certain conditions. For example, a 2:1 multiplexer can use conditional signal assignment to select one of two 4-bit inputs.

```
library IEEE; use IEEE.STD_LOGIC_1164.all;

entity mux2 is
    port(d0, d1: in  STD_LOGIC_VECTOR(3 downto 0);
         s:      in  STD_LOGIC;
         y:      out STD_LOGIC_VECTOR(3 downto 0));
end;

architecture synth of mux2 is
begin
    y <= d0 when s = '0' else d1;
end;
```

A 4:1 multiplexer can select one of four inputs using multiple else clauses in the conditional signal assignment.

```
library IEEE; use IEEE.STD_LOGIC_1164.all;

entity mux4 is
    port(d0, d1, d2, d3: in  STD_LOGIC_VECTOR(3 downto 0);
         s:              in  STD_LOGIC_VECTOR(1 downto 0);
         y:              out STD_LOGIC_VECTOR(3 downto 0));
end;

architecture synth of mux4 is
begin
    y <= d0 when s = "00" else
         d1 when s = "01" else
         d2 when s = "10" else
         d3;
end;
```

B.2.5 Selected Signal Assignment Statements

Selected signal assignment statements provide a shorthand when selecting from one of several possibilities. They are analogous to using a case statement in place of multiple if/

`else` statements, as will be described in Section B.4.4. The 4:1 multiplexer could have been written with a selected signal assignment as:

```
library IEEE; use IEEE.STD_LOGIC_1164.all;

entity mux4 is
    port(d0, d1, d2, d3: in  STD_LOGIC_VECTOR(3 downto 0);
         s:              in  STD_LOGIC_VECTOR(1 downto 0);
         y:              out STD_LOGIC_VECTOR(3 downto 0));
end;

architecture synth2 of mux4 is
begin
    with s select y <=
        d0 when "00",
        d1 when "01",
        d2 when "10",
        d3 when others;
end;
```

B.3 Basic Constructs

B.3.1 Blocks, Entities, and Architectures

A cell or module in VHDL is called a *block* and is the basic unit of design. It has inputs and outputs specified in an *entity declaration* using the *port* statement. An *architecture body* defines what the block does. VHDL separates the architecture body from the entity declaration to allow multiple bodies for a single block. The syntax of the architecture body is

```
architecture <name> of <block> is
begin
   -- body goes here
end;
```

For example, we have seen body for the `adder` block with an architecture named `synth`. It uses a concurrent signal assignment in a behavioral description of the block. The name of the architecture carries no special meaning to VHDL, but helps the human reader understand that the code is intended to be synthesizable (rather than just for simulation). As the design is refined, you can imagine creating a structural description such as 32 full adders connected in a ripple-carry fashion. This architecture might be named `struct`. While VHDL permits selecting among multiple architectures for a single entity, the syntax is complicated and the simplest approach is to only provide one architecture for each entity.

A side effect of separating architectures from entities is that simple blocks involve substantially more typing in VHDL than in Verilog.

B.3.2 Internal Signals

Often it is convenient to break a complex calculation into intermediate variables. For example, in a full adder, we sometimes define the propagate signal as the XOR of the two inputs a and b. The sum from the adder is the XOR of the propagate signal and the carry-in. We can name the propagate signal using a `signal` statement, in much the same way we use local variables in a programming language.

```
library IEEE; use IEEE.STD_LOGIC_1164.all;

entity fulladder is
    port(a, b, cin:  in STD_LOGIC;
         s, cout:    out STD_LOGIC);
end;

architecture synth of fulladder is
    signal prop: STD_LOGIC;
begin
    prop <= a xor b;
    s <= prop xor cin;
    cout <= (a and b) or (cin and (a or b));
end;
```

B.3.3 Precedence

Notice that we fully parenthesized the `cout` computation. This is necessary in VHDL because the logical operators all have equal precedence. If we had written

```
cout = a and b or cin and (a or b);
```

it could have been interpreted as either

```
cout = ((a and b) or cin) and (a or b));
out = (a and b) or (cin and (a or b)));
```

These are definitely not the same function, so VHDL demands parentheses.

Table B.1 lists operator precedence. All the operators on a particular row have equal precedence and have higher precedence than the row below. Thus, multiplication takes place before addition, as you would expect in most programming languages or in mathematics. On the other hand, AND does not take place before OR, unlike what you would expect in a conventional Boolean equation. Parentheses are used to group expressions when necessary.

Table B.1	Operator Precedence
not	**Highest**
`*, /, mod, rem`	
`+, -, &`	
`=, /=, <, <=, >, >=`	
`=, = =, !=`	
`and, or, nand, nor, xor`	**Lowest**

B.3.4 Hierarchy

Nontrivial designs are developed in a hierarchical form, in which complex blocks are composed of simpler blocks. For example, a 4-input multiplexer can be constructed from three 2-input multiplexers that were defined earlier.

```
library IEEE; use IEEE.STD_LOGIC_1164.all;

entity mux4 is
    port(d0, d1, d2, d3: in  STD_LOGIC_VECTOR(3 downto 0);
         s:              in  STD_LOGIC_VECTOR(1 downto 0);
         y:              out STD_LOGIC_VECTOR(3 downto 0));
end;

architecture struct of mux4 is
    component mux2 port(d0, d1: in  STD_LOGIC_VECTOR(3 downto 0);
                        s:      in  STD_LOGIC;
                        y:      out STD_LOGIC_VECTOR(3 downto 0));
    end component;
    signal low, high: STD_LOGIC_VECTOR(3 downto 0);
begin
    lowmux:   mux2 port map(d0, d1, s(0), low);
    highmux:  mux2 port map(d2, d3, s(0), high);
    finalmux: mux2 port map(low, high, s(1), y);
end;
```

This is an example of the structural coding style because the mux4 is built from simpler blocks. It is good practice to avoid (or at least minimize) mixing structural and behavioral descriptions within a single module. Generally, simple modules are described behaviorally and larger modules are composed structurally from these building blocks.

The architecture must first declare the mux2 ports using the *component declaration* statement. This allows VHDL tools to make sure that the component you wish to use has the same ports as the component that was declared somewhere else in an entity statement, preventing errors caused by changing the entity but not the use. However, it makes VHDL code rather cumbersome.

B.3.5 Bit Swizzling

Often it is necessary to work on parts of a bus or to concatenate (join together) signals to construct busses. The mux4 example showed using the least significant bit s(0) of a 2-bit select signal for the low and high muxes and the most significant bit s(1) for the final mux. Use ranges to select subsets of a bus. For example, an 8-bit wide 2-input mux can be constructed from two 4-bit wide 2-input muxes:

```
library IEEE; use IEEE.STD_LOGIC_1164.all;

entity mux2_8 is
    port(d0, d1: in  STD_LOGIC_VECTOR(7 downto 0);
         s:      in  STD_LOGIC;
         y:      out STD_LOGIC_VECTOR(7 downto 0));
end;
```

```
architecture struct of mux2_8 is
    component mux2 port(d0, d1: in  STD_LOGIC_VECTOR(3 downto 0);
                        s:     in  STD_LOGIC;
                        y:     out STD_LOGIC_VECTOR(3 downto 0));
    end component;
begin

    lsbmux:  mux2 port map(d0(3 downto 0), d1(3 downto 0),
                           s, y(3 downto 0));
    msbhmux: mux2 port map(d0(7 downto 4), d1(7 downto 4),
                           s, y(7 downto 4));
end;
```

The & operator is used to concatenate busses. For example, the following code multiplies a number by four by shifting it left two positions.

```
library IEEE; use IEEE.STD_LOGIC_1164.all;

entity shift2 is
    port(a: in  STD_LOGIC_VECTOR(7 downto 0);
         y: out STD_LOGIC_VECTOR(7 downto 0));
end;

architecture synth of shift2 is
begin
    y <= a(5 downto 0) & "00";
end;
```

Variable left and right shifts can also be done with the SHL and SHR functions:

```
library IEEE;
use IEEE.STD_LOGIC_1164.all;
use IEEE.STD_LOGIC_UNSIGNED.all;

entity shifter is
    port(a:   in  STD_LOGIC_VECTOR(7 downto 0);
         amt: in  STD_LOGIC_VECTOR(2 downto 0);
         y:   out STD_LOGIC_VECTOR(7 downto 0));
end;

architecture synth of shifter is
begin
    y <= SHL(a, amt);
end;
```

B.3.6 Types

VHDL uses a strict data typing system that can be clumsy at times. In logic design, our signals are typically binary digits or binary words. In addition to '0' and '1,' it is convenient to have values such as 'z' (high impedance, representing a tristate buffer with a floating output as described in Section B.3.8), 'u' (uninitialized, representing a flip-flop at startup before reset where the value might be high or low), and 'x' (unknown, representing a node with contention that is driven high by one gate and low by another or the output of a gate whose inputs are 'z')[1]. The STD_LOGIC and STD_LOGIC_VECTOR types represent single- and multiple-bit binary values. Table B.2 shows a truth table for an AND gate using STD_LOGIC types. Notice that the operators are optimistic in that they can sometimes determine the result despite some inputs being unknown. For example '0' and 'z' returns '0' because the output of an AND gate is always '0' if either input is '0.'

Table B.2		AND gate truth table				
		A				
		0	1	z	x	u
B	0	0	0	0	0	0
	1	0	1	x	x	u
	z	0	x	x	x	u
	x	0	x	x	x	u
	u	0	u	u	u	u

Despite its fundamental importance, the STD_LOGIC type is not built into VHDL. Instead, it is part of the IEEE.STD_LOGIC_1164 library defined by the IEEE 1164 standard [IEEE1164-93]. Thus, every file must include the library statements we have seen in previous examples. VHDL does have built-in BIT and BIT_VECTOR types that define only '0' and '1,' but we will avoid these types in this tutorial because STD_LOGIC is better suited for logic simulation with tristates, uninitialized nodes, and contention.

In simulation, it may be necessary to define what to do when an input is 'x' or 'u.' For example, the select signal to a multiplexer might be an illegal or undefined value. The mux4 examples in Sections B.2.4 and B.2.5 default to selecting d3 in such an event.

VHDL also has a *Boolean* type with two values: true and false. Boolean values are created by comparisons (like s = '0') and used in conditional statements such as if or when. Boolean and STD_LOGIC values are not interchangeable despite the temptation to think true means '1' and false means '0.' Thus, the following statements are illegal:

```
y <= d1 when s else d0;
q <= (state = S2);
```

Instead we must write

```
y <= d1 when s = '1' else d0;
q <= '1' when state = S2 else '0';
```

---

[1]Technically, STD_LOGIC also has values of W, L, H, and—for weak unknown, low, and high values and for don't cares, but we will not need them in this tutorial.

While we will not declare any signals to be Boolean, they are automatically implied by comparisons and used by conditional statements.

Similarly, VHDL has an INTEGER type representing a 2's complement integer of at least 32 bits (spanning values $-2^{31} \ldots 2^{31}-1$). Integer values are used as array indices. For example, in the statement

```
lowmux:    mux2 port map(d0, d1, s(0), low);
```

0 is an integer serving as an index to choose one bit of the s signal. We cannot directly index an array with a STD_LOGIC or STD_LOGIC_VECTOR signal. Instead, we must convert the signal to an integer. This is demonstrated in the following 8:1 multiplexer that selects one bit from a vector using a 3-bit index. The conv_integer function is defined in STD_LOGIC_UNSIGNED and performs the conversion from STD_LOGIC_VECTOR to integer for positive (unsigned) values.

```
library IEEE;
use IEEE.STD_LOGIC_1164.all;
use IEEE.STD_LOGIC_UNSIGNED.all;

entity mux8 is
  port(d: in  STD_LOGIC_VECTOR(7 downto 0);
       s: in  STD_LOGIC_VECTOR(2 downto 0);
       y: out STD_LOGIC);
end;

architecture synth of mux8 is
begin
    y <= d(conv_integer(s));
end;
```

VHDL also permits enumeration types. For example, the divide-by-3 finite state machine described in Section B.5 uses three states. We can give the states names using the enumeration type rather than referring to them by binary values.

```
type statetype is (S0, S1, S2);
signal state, nextstate: statetype;
```

B.3.7 Library and Use Clauses

VHDL uses a number of standard libraries that extend the built-in capabilities of the language. As we have seen, the IEEE.STD_LOGIC_1164 library defines the essential STD_LOGIC and STD_LOGIC_VECTOR types. Unfortunately, it does not define basic operations such as addition, comparison, shifts, or conversion to integer for STD_LOGIC_VECTOR data.

Synopsys has developed another freely available library to perform these functions for unsigned numbers. Many other CAD vendors have adopted it for compatibility. The library is typically called STD_LOGIC_UNSIGNED. An equivalent library called STD_LOGIC_SIGNED handles STD_LOGIC_VECTORs representing signed 2's complement numbers. Right-shifts fill the most significant bit with 0's for unsigned numbers and sign-extend for signed numbers.

The syntax to load a library and use all of its functions is:

```
library IEEE;
use IEEE.STD_LOGIC_1164.all;
use IEEE.STD_LOGIC_UNSIGNED.all;
```

B.3.8 Tristates

It is possible to leave a bus floating rather than drive it to '0' or '1.' This floating value is called 'z' in VHDL. For example, a tristate buffer produces a floating output when the enable is '0.'

```
library IEEE; use IEEE.STD_LOGIC_1164.all;

entity tristate is
    port(a:  in  STD_LOGIC_VECTOR(3 downto 0);
         en: in  STD_LOGIC;
         y:  out STD_LOGIC_VECTOR(3 downto 0));
end;


architecture synth of tristate is
begin
    y <= "ZZZZ" when en = '0' else a;
end;
```

Multiple tristates driving different nonfloating values onto a bus causes contention, displayed as 'x' for STD_LOGIC signals. Floating inputs to gates also cause undefined outputs, displayed as 'x.' Thus, a bus driven by multiple tristates should have one and only one driver active at any given time.

We could define a multiplexer using two tristates so that the output is always driven by exactly one tristate. This guarantees that there are no floating nodes.

```
library IEEE; use IEEE.STD_LOGIC_1164.all;

entity mux2 is
    port(d0, d1: in  STD_LOGIC_VECTOR(3 downto 0);
         s:      in  STD_LOGIC;
         y:      out STD_LOGIC_VECTOR(3 downto 0));
end;

architecture struct of mux2 is
    component tristate port(a:  in  STD_LOGIC_VECTOR(3 downto 0);
                            en: in  STD_LOGIC;
                            y:  out STD_LOGIC_VECTOR(3 downto 0));
    end component;
    signal sbar: STD_LOGIC;
begin
    sbar <= not s;
    t0: tristate port map(d0, sbar, y);
    t1: tristate port map(d1, s, y);
end;
```

B.3.9   Delays

The delay of a statement can be specified in arbitrary units. For example, the following code defines an inverter with a 100 ps propagation delay. Delays have no impact on synthesis, but can be helpful while debugging simulation waveforms because they make cause and effect more apparent.

```
y <= not a after 100 ps;
```

B.4   Behavioral Modeling with Process Statements

Concurrent signal assignments are reevaluated every time any term on the right side changes. Therefore, they must describe combinational logic. Process statements are reevaluated only when signals in the header (called a *sensitivity list*) change. Depending on the form, process statements can imply either sequential or combinational circuits.

B.4.1   Flip-flops

Positive edge-triggered flip-flops are described with a process controlled by clk:

```
library IEEE; use IEEE.STD_LOGIC_1164.all;

entity flop is
    port(clk: in  STD_LOGIC;
         d:   in  STD_LOGIC_VECTOR(3 downto 0);
         q:   out STD_LOGIC_VECTOR(3 downto 0));
end;

architecture synth of flop is
begin
    process(clk) begin
        if clk'event and clk = '1' then -- or use "if RISING_EDGE(clk) then"
            q <= d;
        end if;
    end process;
end;
```

VHDL has several idioms for specifying positive edge-triggered flip-flops. In this example, the flip-flop copies d to q when an event takes place on clk and clk has a value of '1.' An event is a change in a signal value, i.e., a rising or falling transition. The RISING_EDGE(clk) syntax is also acceptable.

At startup, the q output is initialized to 'u' by the simulator before the first clock edge arrives. Generally, it is good practice to use resettable registers so that on power-up you can put your system in a known state. The reset can be either asynchronous or synchronous. Asynchronous resets occur immediately. Synchronous resets only change the output on the rising edge of the clock.

```
library IEEE; use IEEE.STD_LOGIC_1164.all;

entity flopr is
    port(clk, reset: in  STD_LOGIC;
         d:              in  STD_LOGIC_VECTOR(3 downto 0);
         q:              out STD_LOGIC_VECTOR(3 downto 0));
end;

architecture synchronous of flopr is
begin
    process(clk) begin
        if clk'event and clk = '1' then
            if reset = '1' then
                q <= "0000";
            else q <= d;
            end if;
        end if;
    end process;
end;

architecture asynchronous of flopr is
begin
    process(clk, reset) begin
        if reset = '1' then
            q <= "0000";
        elsif clk'event and clk = '1' then
            q <= d;
        end if;
    end process;
end;
```

Note that the asynchronously resettable flop evaluates the process statement when either clk or reset change so that it immediately responds to reset. The synchronously reset flop is not sensitized to reset, so it waits for the next clock edge before clearing the output.

You can also consider registers with enables that only respond to the clock when the enable is true. The following register retains its old value if both reset and en are false.

```
library IEEE; use IEEE.STD_LOGIC_1164.all;

entity flopenr is
    port(clk, reset, en: in  STD_LOGIC;
         d:                  in  STD_LOGIC_VECTOR(3 downto 0);
         q:                  out STD_LOGIC_VECTOR(3 downto 0));
end;
```

```
architecture asynchronous of flopenr is -- asynchronous reset
begin
    process(clk, reset) begin
        if reset = '1' then
            q <= "0000";
        elsif clk'event and clk = '1' then
            if en = '1' then
                q <= d;
            end if;
        end if;
    end process;
end;
```

B.4.2 Latches

Transparent latches are also modeled with process statements. When the clock is high, the latch is *transparent* and the data input flows to the output. When the clock is low, the latch goes *opaque* and the output remains constant.

```
library IEEE; use IEEE.STD_LOGIC_1164.all;

entity latch is
    port(clk: in  STD_LOGIC;
         d:   in  STD_LOGIC_VECTOR(3 downto 0);
         q:   out STD_LOGIC_VECTOR(3 downto 0));
end;

architecture synth of latch is
begin
    process(clk, d) begin
        if clk = '1' then q <= d;
        end if;
    end process;
end;
```

The latch evaluates the `process` statement any time either `clk` or `d` change. If `clk` is high, the output gets the input.

B.4.3 Counters

Consider two ways of describing a 4-bit counter with asynchronous reset. The first scheme (behavioral) implies a sequential circuit containing both the 4-bit register and an adder. The second scheme (structural) explicitly declares modules for the register and adder. Either scheme is good for a simple circuit such as a counter. As you develop more complex finite state machines, it is a good idea to separate the next state logic from the registers in your code.

```
library IEEE;
use IEEE.STD_LOGIC_1164.all;
use IEEE.STD_LOGIC_UNSIGNED.all;

entity counter is
    port(clk, reset: in      STD_LOGIC;
           q:              buffer STD_LOGIC_VECTOR(3 downto 0));
end;

architecture synth of counter is
begin
    process(clk) begin
        if clk'event and clk = '1' then
            if reset = '1' then q <= "0000";
            else q <= q + "0001";
            end if;
        end if;
    end process;
end;

architecture struct of counter is
    component flopr port(clk, reset: in  STD_LOGIC;
                           d:              in  STD_LOGIC_VECTOR(3 downto 0);
                           q:              out STD_LOGIC_VECTOR(3 downto 0));
    end component;
    component adder port(a, b:  in  STD_LOGIC_VECTOR(3 downto 0);
                           y:      out STD_LOGIC_VECTOR(3 downto 0));
    end component;
    signal nextq: STD_LOGIC_VECTOR(3 downto 0);
begin
    qflop: flopr port map(clk, reset, nextq, q);
    inc:   adder_4 port map(q, "0001", nextq);
end;
```

Note that q is defined as buffer instead of out. Buffer ports are required when a signal is used as both input and output within a block, as in the case of q <= q + "0001". Also, adder_4 is identical to the adder block defined earlier except that it is 4 bits wide.

B.4.4 Combinational Logic

Process statements imply sequential logic when some of the inputs do not appear in the @ stimulus list or might not cause the output to change. For example, in the flop module, d is not in the @ list, so the flop does not immediately respond to changes of d. In the latch, d is in the @ list, but changes in d are ignored unless clk is high so the latch is also sequential. Processes can also be used to imply combinational logic if they are written in such a way that the process is reevaluated every time there are changes in any of the inputs. The following architecture shows how to define a bank of inverters with a process. Only the architecture is listed; the rest of the code is the same as the other inverter example. The begin and end process statements are required even though the process only contains one assignment.

```
architecture proc of inv is
begin
    process(a) begin
        y <= not a;
    end process;
end;
```

Similarly, the next example defines five banks of different kinds of gates.

```
architecture proc of gates is
begin
    process(a, b) begin
        y1 <= a and b;
        y2 <= a or b;
        y3 <= a xor b;
        y4 <= a nand b;
        y5 <= a nor b;
    end process;
end;
```

Processes can also contain if statements. The following example describes a priority encoder that determines the most significant input bit that is asserted.

```
library IEEE; use IEEE.STD_LOGIC_1164.all;

entity priority is
    port(a: in  STD_LOGIC_VECTOR(3 downto 0);
         y: out STD_LOGIC_VECTOR(3 downto 0));
end;

architecture synth of priority is
begin
    process(a) begin
        if    a(3) = '1' then y <= "1000";
        elsif a(2) = '1' then y <= "0100";
        elsif a(1) = '1' then y <= "0010";
        elsif a(0) = '1' then y <= "0001";
        else                  y <= "0000";
        end if;
    end process;
end;
```

Processes can also contain case statements. For example, a 3:8 decoder could be written using a case statement. This is easier to read than a description of the same decoder using Boolean equations.

```
library IEEE; use IEEE.STD_LOGIC_1164.all;

entity decoder is
    port(a: in  STD_LOGIC_VECTOR(2 downto 0);
         y: out STD_LOGIC_VECTOR(7 downto 0));
end;
```

```
architecture proc_case of decoder is
begin
    process(a) begin
        case a is
            when "000" =>  y <= "00000001";
            when "001" =>  y <= "00000010";
            when "010" =>  y <= "00000100";
            when "011" =>  y <= "00001000";
            when "100" =>  y <= "00010000";
            when "101" =>  y <= "00100000";
            when "110" =>  y <= "01000000";
            when others => y <= "10000000";
        end case;
    end process;
end;


architecture boolean of decoder is
begin
    y(0) <= not a(2) and not a(1) and not a(0);
    y(1) <= not a(2) and not a(1) and a(0);
    y(2) <= not a(2) and a(1) and not a(0);
    y(3) <= not a(2) and a(1) and a(0);
    y(4) <= a(2) and not a(1) and not a(0);
    y(5) <= a(2) and not a(1) and a(0);
    y(6) <= a(2) and a(1) and not a(0);
    y(7) <= a(2) and a(1) and a(0);
end;
```

Another even better application of `case` is the logic for a 7-segment display decoder. The 7-segment display is shown in Figure B.1. The decoder takes a 4-bit number and displays its decimal value on the segments. For example, the number 0111 = 7 should turn on segments a, b, and c. The equivalent logic with concurrent signal assignments describing the detailed logic for each bit would be tedious. This more abstract approach is faster to write, clearer to read, and can be automatically synthesized down to an efficient logic implementation.

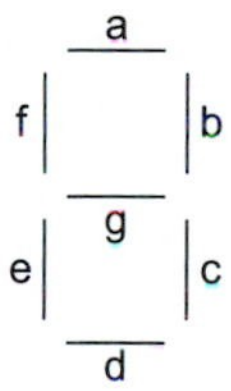

FIG B.1 7-segment display mapping

```
library IEEE; use IEEE.STD_LOGIC_1164.all;

entity seven_seg_decoder is
    port(data:      in  STD_LOGIC_VECTOR(3 downto 0);
         segments: out STD_LOGIC_VECTOR(6 downto 0));
end;

architecture synth of seven_seg_decoder is
--                                    Segment #'s: abcdefg
    constant BLANK: STD_LOGIC_VECTOR(6 downto 0) := "0000000";
    constant ZERO:  STD_LOGIC_VECTOR(6 downto 0) := "1111110";
    constant ONE:   STD_LOGIC_VECTOR(6 downto 0) := "0110000";
    constant TWO:   STD_LOGIC_VECTOR(6 downto 0) := "1101101";
    constant THREE: STD_LOGIC_VECTOR(6 downto 0) := "1111001";
```

```
            constant FOUR:  STD_LOGIC_VECTOR(6 downto 0) := "0110011";
            constant FIVE:  STD_LOGIC_VECTOR(6 downto 0) := "1011011";
            constant SIX:   STD_LOGIC_VECTOR(6 downto 0) := "1011111";
            constant SEVEN: STD_LOGIC_VECTOR(6 downto 0) := "1110000";
            constant EIGHT: STD_LOGIC_VECTOR(6 downto 0) := "1111111";
            constant NINE:  STD_LOGIC_VECTOR(6 downto 0) := "1111011";
begin
    process(data) begin
        case data is
            when "0000" => segments <= ZERO;
            when "0001" => segments <= ONE;
            when "0010" => segments <= TWO;
            when "0011" => segments <= THREE;
            when "0100" => segments <= FOUR;
            when "0101" => segments <= FIVE;
            when "0110" => segments <= SIX;
            when "0111" => segments <= SEVEN;
            when "1000" => segments <= EIGHT;
            when "1001" => segments <= NINE;
            when others => segments <= BLANK;
        end case;
    end process;
end;
```

This example shows the use of *constant declarations* to make the code more readable. The `case` statement has an `others` clause to display a blank output when the input is outside the range of decimal digits.

Overall, concurrent, conditional, and selected signal assignment statements can mimic the effects of processes with sequential assignments, `if` statements, and `case` statements. Moreover, processes can inadvertently imply sequential logic if the output is not assigned a value for all inputs. One reasonable style of VHDL coding is to use processes only for flip-flops and latches.

B.4.5 Memories

VHDL has an array construct used to describe memories. The following module describes a 64-word × 16-bit RAM that is written when `wrb` is low and otherwise read. Synthesis tools are often restricted to generating gates from a library and produce poor memory arrays. A specialized memory generator is commonly used instead.

```
library IEEE;
use IEEE.STD_LOGIC_1164.all;
use IEEE.STD_LOGIC_UNSIGNED.all;

entity memory is
    port(addr: in  STD_LOGIC_VECTOR(5 downto 0);
         wrb:  in  STD_LOGIC;
         din:  in  STD_LOGIC_VECTOR(15 downto 0);
         dout: out STD_LOGIC_VECTOR(15 downto 0));
end;
```

```
architecture synth of memory is
    type ramtype is array (63 downto 0) of STD_LOGIC_VECTOR(15 downto 0);
    signal mem: ramtype;
begin
    process(addr, wrb, din) begin
        if wrb = '0' then mem(conv_integer(addr)) <= din;
        else dout <= mem(conv_integer(addr));
        end if;
    end process;
end;
```

Note that the memory contents are stored in the signal mem. It is declared to be of type ramtype, an array of 64 16-bit words. Arrays are indexed with an integer, but addr is a STD_LOGIC_VECTOR. Hence, the conv_integer function is used to convert between data types. This does not produce any hardware, but keeps the VHDL type system happy.

B.5   Finite State Machines

There are two styles of finite state machines. In *Mealy machines* (Figure B.2(a)), the output is a function of the current state and inputs. In *Moore machines* (Figure B.2(b)), the output is a function of only the current state. FSMs are modeled in VHDL with process defining the *state registers* and combinational logic defining the *next state* and *output logic*.

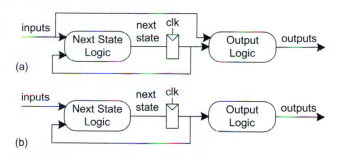

(a)

(b)

FIG B.2   Moore and Mealy machines

Let us first consider a simple finite state machine with one output and no inputs, a divide-by-3 counter. The output should be asserted every three clock cycles. The state transition diagram for a Moore machine is shown in Figure B.3. The output value is labeled in each state because the output is only a function of the state.

```
library IEEE; use IEEE.STD_LOGIC_1164.all;

entity divideby3FSM is
    port(clk, reset: in  STD_LOGIC;
         q:           out STD_LOGIC);
end;

architecture synth of divideby3FSM is
    type statetype is (S0, S1, S2);
    signal state, nextstate: statetype;
begin
    -- state register
    process(clk, reset) begin
```

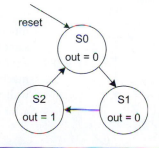

FIG B.3   Divide-by-3 counter state transition diagram

```
              if reset = '1' then state <= S0;
              elsif clk'event and clk = '1' then state <= nextstate;
              end if;
          end process;

          -- next state logic
          nextstate <= S1 when state = S0 else
                       S2 when state = S1 else
                       S0;

          -- output logic
          q <= '1' when state = S2 else '0';
      end;
```

Defining states with a type rather than a binary encoding makes the code easier to read. It also allows the synthesis tool to search for an encoding that is fastest or uses the least area. For example, the synthesis tool might choose s0 = 00, s1 = 01, and s2 = 10 or might choose s0 = 000, s1 = 010, s2 = 100.

The FSM model is divided into three portions: the state register, next state logic, and output logic. The state register logic describes an asynchronously resettable register that resets to an initial state and otherwise advances to the computed next state. The next state logic computes nextstate as a function of state and the inputs; in this example, there are no inputs. The final else is essential to make sure nextstate gets a value even when state has an illegal value; otherwise, latches might be implied. Finally, the output logic may be a function of the current state alone in a Moore machine or of the current state and inputs in a Mealy machine.

It would be tempting to write the output logic as

```
q <= (state = S2);
```

Recall that(state = S2) returns a boolean result, either true or false. q is of type STD_LOGIC. VHDL is picky about types, so the when/else clause is needed to convert Boolean to '1' or '0' logic values.

The next example shows a finite state machine with an input a and two outputs. Output x is true when the input is the same now as it was last cycle. Output y is true when the input is the same now as it was for the past two cycles. This is a Mealy machine because the output depends on the current inputs as well as the state. The outputs are labeled on each transition after the input. The state transition diagram is shown in Figure B.4.

```
library IEEE; use IEEE.STD_LOGIC_1164.all;

entity historyFSM is
    port(clk, reset: in  STD_LOGIC;
```

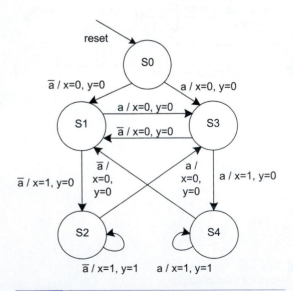

FIG B.4 History FSM state transition diagram

```vhdl
        a:             in  STD_LOGIC;
        x, y:          out STD_LOGIC);
end;

architecture synth of historyFSM is
    type statetype is (S0, S1, S2, S3, S4);
    signal state, nextstate: statetype;
begin
    -- state register
    process(clk, reset) begin
        if reset = '1' then state <= S0;
        elsif clk'event and clk = '1' then state <= nextstate;
        end if;
    end process;

    -- next state logic
    process(state, a) begin
        case state is
            when S0 =>  if a = '1' then nextstate <= S3;
                        else            nextstate <= S1;
                        end if;
            when S1 =>  if a = '1' then nextstate <= S3;
                        else            nextstate <= S2;
                        end if;
            when S2 =>  if a = '1' then nextstate <= S3;
                        else            nextstate <= S2;
                        end if;
            when S3 =>  if a = '1' then nextstate <= S4;
                        else            nextstate <= S1;
                        end if;
            when S4 =>  if a = '1' then nextstate <= S4;
                        else            nextstate <= S1;
                        end if;
            when others =>              nextstate <= S0;
        end case;
    end process;

    -- output logic
    x <= '1' when ((state = S1 or state = s2) and a = '1') or
                  ((state = S3 or state = S4) and a = '1')
            else '1';
    y <= '1' when (state = S2 and a = '1') or (state = S4 and a = '1')
            else '1';
end;
```

B.6 Parameterized Blocks

So far, all of our blocks have had fixed-width inputs and outputs. Thus, we have needed separate entities for 4- and 8-bit wide 2-input multiplexers. VHDL can automatically produce hardware with parameterized bit width. For example, we can declare a parameterized bank of inverters with a default of 8 gates as:

```
library IEEE; use IEEE.STD_LOGIC_1164.all;

entity inv is
    generic(width: integer := 8);
    port(a: in  STD_LOGIC_VECTOR(width-1 downto 0);
         y: out STD_LOGIC_VECTOR(width-1 downto 0));
end;

architecture synth of inv is
begin
    y <= not a;
end;
```

We can adjust the parameter when we instantiate the block. For example, we can build a bank of 12 buffers using a pair of banks of 12 inverters.

```
library IEEE; use IEEE.STD_LOGIC_1164.all;

entity buf is
    generic(numbits: integer := 12);
    port(a: in  STD_LOGIC_VECTOR(numbits-1 downto 0);
         y: out STD_LOGIC_VECTOR(numbits-1 downto 0));
end;

architecture synth of buf is
    component inv generic(width: integer);
                  port(a: in  STD_LOGIC_VECTOR(width-1 downto 0);
                       y: out STD_LOGIC_VECTOR(width-1 downto 0));
    end component;
    signal x: STD_LOGIC_VECTOR(numbits-1 downto 0);
begin
    i1: inv generic map(numbits) port map(a, x);
    i2: inv generic map(numbits) port map(x, y);
end;
```

VHDL also has the ability to generate an adjustable number of gates connected in arbitrary ways. The following code produces an N-input AND gate as a cascade of $N-1$ 2-input AND gates (N = width = 8). It uses the generate command in conjunction with for and if statements to conditionally produce an array of gates. Generate can easily produce large numbers of gates, so use it with care to be sure you intend to imply all the hardware.

```
library IEEE; use IEEE.STD_LOGIC_1164.all;

entity andN is
    generic(width: integer := 8);
    port(a: in  STD_LOGIC_VECTOR(width-1 downto 0);
         y: out STD_LOGIC);
end;
```

```vhdl
architecture synth of andN is
    signal i: integer;
    signal x: STD_LOGIC_VECTOR(width-1 downto 1);
begin
    AllBits: for i in width-1 downto 1 generate
        LowBit: if i = 1 generate
            A1: x(1) <= a(0) and a(1);
        end generate;
        OtherBits: if i /= 1 generate
            Ai: x(i) <= a(i) and x(i-1);
        end generate;
    end generate;
    y <= x(width-1);
end;
```

In a similar example, `generate` can instantiate an array of other blocks. For example, an *N*-bit ripple-carry adder can be built from *N* full adders with the carry signals chained together. Of course, using the + symbol to imply an adder is less effort and gives the synthesis tool more freedom to trade off area and delay as the constraints demand.

```vhdl
library IEEE; use IEEE.STD_LOGIC_1164.all;


entity rippleadder is
    generic(width: integer := 8);
    port(a, b: in  STD_LOGIC_VECTOR(width-1 downto 0);
         s:    out STD_LOGIC_VECTOR(width-1 downto 0));
end;


architecture synth of rippleadder is
    component fulladder port(a, b, cin:  in  STD_LOGIC;
                             s, cout:    out STD_LOGIC);
    end component;
    signal i: integer;
    signal c: STD_LOGIC_VECTOR(width-1 downto 0);
begin
    AllBits: for i in width-1 downto 0 generate
        LowBit: if i = 0 generate
            FA0: fulladder port map(a(0), b(0), '0', s(0), c(0));
        end generate;
        OtherBits: if i /= 0 generate
            FAi: fulladder port map(a(i), b(i), c(i-1), s(i), c(i));
        end generate;
    end generate;
end;
```

B.7 Example: MIPS Processor

To illustrate a nontrivial VHDL design, this section lists the VHDL code and test bench for the MIPS processor subset discussed in Chapter 1. The example handles only the LB, SB, ADD, SUB, AND, OR, SLT, BEQ, and J instructions. It uses an 8-bit datapath and only eight registers. As the instruction is 32 bits wide, it is loaded in four successive fetch cycles across an 8-bit path to external memory.

The test bench initializes a 512-byte memory with instructions and data from a text file. The code exercises each of the instructions. The `mipstest.asm` assembly language file and `memfile.dat` text file are shown in Section A.10. The test bench runs until it observes a memory write. If the value 7 is written to address 5, the code executed correctly. If all goes well, the test bench should take 100 cycles (1000 ns) to run.

The code uses the `conv_std_logic_vector` function that converts an integer into a `STD_LOGIC_VECTOR` of the specified width. This is helpful to create constants of parameterized widths. The function is in the `IEEE.STD_LOGIC_ARITH` library. It also uses the `**` exponentiation operator to determine the number of register file entries given the number of bits of register address.

```vhdl
-----------------------------------------------------------
-- mips.vhd
-- David_Harris@hmc.edu 9/9/03
-- Model of subset of MIPS processor described in Ch 1
-----------------------------------------------------------

-----------------------------------------------------------
-- Entity Declarations
-----------------------------------------------------------

library IEEE; use IEEE.STD_LOGIC_1164.all; use IEEE.STD_LOGIC_UNSIGNED.all;
entity top is -- top-level design for testing
    generic(width:   integer := 8;       -- default 8-bit datapath
            regbits: integer := 3);      -- and 3 bit register addresses (8 regs)
end;

library IEEE; use IEEE.STD_LOGIC_1164.all; use STD.TEXTIO.all;
use IEEE.STD_LOGIC_UNSIGNED.all;  use IEEE.STD_LOGIC_ARITH.all;
entity memory is -- external memory accessed by MIPS
    generic(width: integer);
    port(clk, memwrite:  in STD_LOGIC;
         adr, writedata: in STD_LOGIC_VECTOR(width-1 downto 0);
         memdata:        out STD_LOGIC_VECTOR(width-1 downto 0));
end;

library IEEE; use IEEE.STD_LOGIC_1164.all;
entity mips is -- simplified MIPS processor
    generic(width:   integer := 8;       -- default 8-bit datapath
            regbits: integer := 3);      -- and 3 bit register addresses (8 regs)
    port(clk, reset:         in  STD_LOGIC;
         memdata:            in  STD_LOGIC_VECTOR(width-1 downto 0);
         memread, memwrite:  out STD_LOGIC;
         adr, writedata:     out STD_LOGIC_VECTOR(width-1 downto 0));
end;

library IEEE; use IEEE.STD_LOGIC_1164.all;
entity controller is -- control FSM
    port(clk, reset:                     in  STD_LOGIC;
         op:                             in  STD_LOGIC_VECTOR(5 downto 0);
         zero:                           in  STD_LOGIC;
         memread, memwrite, alusrca, memtoreg,
         iord, pcen, regwrite, regdst:   out STD_LOGIC;
         pcsource, alusrcb, aluop:       out STD_LOGIC_VECTOR(1 downto 0);
         irwrite:                        out STD_LOGIC_VECTOR(3 downto 0));
end;

library IEEE; use IEEE.STD_LOGIC_1164.all;
entity alucontrol is -- ALU control decoder
    port(aluop:  in  STD_LOGIC_VECTOR(1 downto 0);
         funct:  in  STD_LOGIC_VECTOR(5 downto 0);
```

```
                alucont: out STD_LOGIC_VECTOR(2 downto 0));
end;

library IEEE; use IEEE.STD_LOGIC_1164.all; use IEEE.STD_LOGIC_ARITH.all;
entity datapath is  -- MIPS datapath
    generic(width, regbits: integer);
    port(clk, reset:          in  STD_LOGIC;
         memdata:             in  STD_LOGIC_VECTOR(width-1 downto 0);
         alusrca, memtoreg, iord, pcen,
         regwrite, regdst:    in  STD_LOGIC;
         pcsource, alusrcb: in  STD_LOGIC_VECTOR(1 downto 0);
         irwrite:             in  STD_LOGIC_VECTOR(3 downto 0);
         alucont:             in  STD_LOGIC_VECTOR(2 downto 0);
         zero:                out STD_LOGIC;
         instr:               out STD_LOGIC_VECTOR(31 downto 0);
         adr, writedata:      out STD_LOGIC_VECTOR(width-1 downto 0));
end;

library IEEE; use IEEE.STD_LOGIC_1164.all;
use IEEE.STD_LOGIC_ARITH.all; use IEEE.STD_LOGIC_UNSIGNED.all;
entity alu is -- Arithmetic/Logic unit with add/sub, AND, OR, set less than
    generic(width: integer);
    port(a, b:      in  STD_LOGIC_VECTOR(width-1 downto 0);
         alucont: in  STD_LOGIC_VECTOR(2 downto 0);
         result:  out STD_LOGIC_VECTOR(width-1 downto 0));
end;

library IEEE; use IEEE.STD_LOGIC_1164.all;
use IEEE.STD_LOGIC_UNSIGNED.all; use IEEE.STD_LOGIC_ARITH.all;
entity regfile is -- three-port register file of 2**regbits words x width bits
    generic(width, regbits: integer);
    port(clk:          in  STD_LOGIC;
         write:        in  STD_LOGIC;
         ra1, ra2, wa: in  STD_LOGIC_VECTOR(regbits-1 downto 0);
         wd:           in  STD_LOGIC_VECTOR(width-1 downto 0);
         rd1, rd2:     out STD_LOGIC_VECTOR(width-1 downto 0));
end;

library IEEE; use IEEE.STD_LOGIC_1164.all;
entity zerodetect is -- true if all input bits are zero
    generic(width: integer);
    port(a: in  STD_LOGIC_VECTOR(width-1 downto 0);
         y: out STD_LOGIC);
end;

library IEEE; use IEEE.STD_LOGIC_1164.all;
entity flop is -- flip-flop
    generic(width: integer);
    port(clk: in  STD_LOGIC;
         d:   in  STD_LOGIC_VECTOR(width-1 downto 0);
         q:   out STD_LOGIC_VECTOR(width-1 downto 0));
end;

library IEEE; use IEEE.STD_LOGIC_1164.all;
entity flopen is -- flip-flop with enable
    generic(width: integer);
    port(clk, en: in  STD_LOGIC;
         d:       in  STD_LOGIC_VECTOR(width-1 downto 0);
         q:       out STD_LOGIC_VECTOR(width-1 downto 0));
end;

library IEEE; use IEEE.STD_LOGIC_1164.all; use IEEE.STD_LOGIC_ARITH.all;
entity flopenr is -- flip-flop with enable and synchronous reset
    generic(width: integer);
    port(clk, reset, en: in  STD_LOGIC;
         d:                 in  STD_LOGIC_VECTOR(width-1 downto 0);
         q:                 out STD_LOGIC_VECTOR(width-1 downto 0));
end;

library IEEE; use IEEE.STD_LOGIC_1164.all;
entity mux2 is -- two-input multiplexer
    generic(width: integer);
    port(d0, d1: in  STD_LOGIC_VECTOR(width-1 downto 0);
         s:      in  STD_LOGIC;
         y:      out STD_LOGIC_VECTOR(width-1 downto 0));
end;

library IEEE; use IEEE.STD_LOGIC_1164.all;
entity mux4 is  -- four-input multiplexer
    generic(width: integer);
```

```
       port(d0, d1, d2, d3: in  STD_LOGIC_VECTOR(width-1 downto 0);
            s:              in  STD_LOGIC_VECTOR(1 downto 0);
            y:              out STD_LOGIC_VECTOR(width-1 downto 0));
end;

-----------------------------------------------------------
-- Architecture Definitions
-----------------------------------------------------------
architecture test of top is
    component mips generic(width:   integer := 8;   -- default 8-bit datapath
                           regbits: integer := 3);  -- and 3 bit register addresses (8 regs)
        port(clk, reset:       in  STD_LOGIC;
             memdata:          in  STD_LOGIC_VECTOR(width-1 downto 0);
             memread, memwrite: out STD_LOGIC;
             adr, writedata:   out STD_LOGIC_VECTOR(width-1 downto 0));
    end component;
    component memory generic(width: integer);
        port(clk, memwrite: in  STD_LOGIC;
             adr, writedata: in  STD_LOGIC_VECTOR(width-1 downto 0);
             memdata:        out STD_LOGIC_VECTOR(width-1 downto 0));
    end component;
    signal clk, reset, memread, memwrite: STD_LOGIC;
    signal memdata, adr, writedata: STD_LOGIC_VECTOR(width-1 downto 0);
begin
    -- mips being tested
    dut: mips generic map(width, regbits)
             port map(clk, reset, memdata, memread, memwrite, adr, writedata);
    -- external memory for code and data
    exmem: memory generic map(width)
               port map(clk, memwrite, adr, writedata, memdata);

    -- Generate clock with 10 ns period
    process begin
        clk <= '1';
        wait for 5 ns;
        clk <= '0';
        wait for 5 ns;
    end process;

    -- Generate reset for first two clock cycles
    process begin
        reset <= '1';
        wait for 22 ns;
        reset <= '0';
        wait;
    end process;

    -- check that 7 gets written to address 5 at end of program
    process (clk) begin
        if (clk'event and clk = '0' and memwrite = '1') then
            if (conv_integer(adr) = 5 and conv_integer(writedata) = 7) then
                report "Simulation completed successfully";
            else report "Simulation failed.";
            end if;
        end if;
    end process;
end;


architecture synth of memory is
begin
    process is
        file mem_file: text open read_mode is "memfile.dat";
        variable L: line;
        variable ch: character;
        variable index, result: integer;
        type ramtype is array (511 downto 0) of STD_LOGIC_VECTOR(7 downto 0);
        variable mem: ramtype;
    begin
        -- initialize memory from file
        for i in 0 to 511 loop -- set all contents low
            mem(conv_integer(i)) := "00000000";
        end loop;
        index := 0;
        while not endfile(mem_file) loop
            readline(mem_file, L);
            for j in 0 to 3 loop
                result := 0;
                for i in 1 to 2 loop
                    read(L, ch);
                    if '0' <= ch and ch <= '9' then
                        result := result*16 + character'pos(ch)-character'pos('0');
                    elsif 'a' <= ch and ch <= 'f' then
```

```
                           result := result*16 + character'pos(ch)-character'pos('a')+10;
                     else report "Format error on line " & integer'image(index)
                          severity error;
                     end if;
                 end loop;
             mem(index*4+j) := conv_std_logic_vector(result, width);
         end loop;
         index := index + 1;
     end loop;
     -- read or write memory
     loop
         if clk'event and clk = '1' then
             if (memwrite = '1') then mem(conv_integer(adr)) := writedata;
             end if;
         end if;
         memdata <= mem(conv_integer(adr));
         wait on clk, adr;
     end loop;
  end process;
end;

architecture struct of mips is
    component controller
        port(clk, reset:                in  STD_LOGIC;
             op:                        in  STD_LOGIC_VECTOR(5 downto 0);
             zero:                      in  STD_LOGIC;
             memread, memwrite, alusrca, memtoreg,
             iord, pcen, regwrite, regdst: out STD_LOGIC;
             pcsource, alusrcb, aluop:  out STD_LOGIC_VECTOR(1 downto 0);
             irwrite:                   out STD_LOGIC_VECTOR(3 downto 0));
    end component;
    component alucontrol
        port(aluop:     in  STD_LOGIC_VECTOR(1 downto 0);
             funct:     in  STD_LOGIC_VECTOR(5 downto 0);
             alucont:   out STD_LOGIC_VECTOR(2 downto 0));
    end component;
    component datapath generic(width, regbits: integer);
        port(clk, reset:         in  STD_LOGIC;
             memdata:            in  STD_LOGIC_VECTOR(width-1 downto 0);
             alusrca, memtoreg, iord, pcen,
             regwrite, regdst:   in  STD_LOGIC;
             pcsource, alusrcb:  in  STD_LOGIC_VECTOR(1 downto 0);
             irwrite:            in  STD_LOGIC_VECTOR(3 downto 0);
             alucont:            in  STD_LOGIC_VECTOR(2 downto 0);
             zero:               out STD_LOGIC;
             instr:              out STD_LOGIC_VECTOR(31 downto 0);
             adr, writedata:     out STD_LOGIC_VECTOR(width-1 downto 0));
    end component;
    signal instr: STD_LOGIC_VECTOR(31 downto 0);
    signal zero, alusrca, memtoreg, iord, pcen, regwrite, regdst: STD_LOGIC;
    signal aluop, pcsource, alusrcb: STD_LOGIC_VECTOR(1 downto 0);
    signal irwrite: STD_LOGIC_VECTOR(3 downto 0);
    signal alucont: STD_LOGIC_VECTOR(2 downto 0);
begin
    cont: controller port map(clk, reset, instr(31 downto 26), zero,
                              memread, memwrite, alusrca, memtoreg,
                              iord, pcen, regwrite, regdst,
                              pcsource, alusrcb, aluop, irwrite);
    ac: alucontrol port map(aluop, instr(5 downto 0), alucont);
    dp: datapath generic map(width, regbits)
                 port map(clk, reset, memdata, alusrca, memtoreg,
                          iord, pcen, regwrite, regdst,
                          pcsource, alusrcb, irwrite,
                          alucont, zero, instr, adr, writedata);
end;


architecture synth of controller is
    type statetype is (FETCH1, FETCH2, FETCH3, FETCH4, DECODE, MEMADR,
                       LBRD, LBWR, SBWR, RTYPEEX, RTYPEWR, BEQEX, JEX);
        constant LB:    STD_LOGIC_VECTOR(5 downto 0) := "100000";
        constant SB:    STD_LOGIC_VECTOR(5 downto 0) := "101000";
        constant RTYPE: STD_LOGIC_VECTOR(5 downto 0) := "000000";
        constant BEQ:   STD_LOGIC_VECTOR(5 downto 0) := "000100";
        constant J:     STD_LOGIC_VECTOR(5 downto 0) := "000010";
        signal state, nextstate:    statetype;
        signal pcwrite, pcwritecond: STD_LOGIC;
begin
    process (clk) begin -- state register
        if clk'event and clk = '1' then
            if reset = '1' then state <= FETCH1;
```

```
            else state <= nextstate;
            end if;
        end if;
end process;

process (state, op) begin -- next state logic
    case state is
        when FETCH1 => nextstate <= FETCH2;
        when FETCH2 => nextstate <= FETCH3;
        when FETCH3 => nextstate <= FETCH4;
        when FETCH4 => nextstate <= DECODE;
        when DECODE => case op is
                           when LB | SB => nextstate <= MEMADR;
                           when RTYPE => nextstate <= RTYPEEX;
                           when BEQ => nextstate <= BEQEX;
                           when J => nextstate <= JEX;
                           when others => nextstate <= FETCH1; -- should never happen
                       end case;
        when MEMADR => case op is
                           when LB => nextstate <= LBRD;
                           when SB => nextstate <= SBWR;
                           when others => nextstate <= FETCH1; -- should never happen
                       end case;
        when LBRD => nextstate <= LBWR;
        when LBWR => nextstate <= FETCH1;
        when SBWR => nextstate <= FETCH1;
        when RTYPEEX => nextstate <= RTYPEWR;
        when RTYPEWR => nextstate <= FETCH1;
        when BEQEX => nextstate <= FETCH1;
        when JEX => nextstate <= FETCH1;
        when others => nextstate <= FETCH1; -- should never happen
    end case;
end process;

process (state) begin
    -- set all outputs to zero, then conditionally assert just the appropriate ones
    irwrite <= "0000";
    pcwrite <= '0'; pcwritecond <= '0';
    regwrite <= '0';  regdst <= '0';
    memread <= '0'; memwrite <= '0';
    alusrca <= '0'; alusrcb <= "00"; aluop <= "00";
    pcsource <= "00";
    iord <= '0'; memtoreg <= '0';

    case state is
        when FETCH1 => memread <= '1';
                       irwrite <= "1000";
                       alusrcb <= "01";
                       pcwrite <= '1';
        when FETCH2 => memread <= '1';
                       irwrite <= "0100";
                       alusrcb <= "01";
                       pcwrite <= '1';
        when FETCH3 => memread <= '1';
                       irwrite <= "0010";
                       alusrcb <= "01";
                       pcwrite <= '1';
        when FETCH4 => memread <= '1';
                       irwrite <= "0001";
                       alusrcb <= "01";
                       pcwrite <= '1';
        when DECODE => alusrcb <= "11";
        when MEMADR => alusrca <= '1';
                       alusrcb <= "10";
        when LBRD =>   memread <= '1';
                       iord <= '1';
        when LBWR =>   regwrite <= '1';
                       memtoreg <= '1';
        when SBWR =>   memwrite <= '1';
                       iord <= '1';
        when RTYPEEX => alusrca <= '1';
                        aluop <= "10";
        when RTYPEWR => regdst <= '1';
                        regwrite <= '1';
        when BEQEX =>  alusrca <= '1';
                       aluop <= "01";
                       pcwritecond <= '1';
                       pcsource <= "01";
        when JEX =>    pcwrite <= '1';
                       pcsource <= "10";
```

```
            end case;
        end process;

    pcen <= pcwrite or (pcwritecond and zero); -- program counter enable
end;

architecture synth of alucontrol is
begin
    process(aluop, funct) begin
        case aluop is
            when "00" => alucont <= "010"; -- add (for lb/sb/addi)
            when "01" => alucont <= "110"; -- sub (for beq)
            when others => case funct is          -- R-type instructions
                                when "100000" => alucont <= "010"; -- add (for add)
                                when "100010" => alucont <= "110"; -- subtract (for sub)
                                when "100100" => alucont <= "000"; -- logical and (for and)
                                when "100101" => alucont <= "001"; -- logical or (for or)
                                when "101010" => alucont <= "111"; -- set on less (for slt)
                                when others   => alucont <= "---"; -- should never happen
                           end case;
        end case;
    end process;
end;

architecture struct of datapath is
    component alu generic(width: integer);
        port(a, b:    in  STD_LOGIC_VECTOR(width-1 downto 0);
             alucont: in  STD_LOGIC_VECTOR(2 downto 0);
             result:  out STD_LOGIC_VECTOR(width-1 downto 0));
    end component;
    component regfile generic(width, regbits: integer);
        port(clk:        in  STD_LOGIC;
             write:      in  STD_LOGIC;
             ra1, ra2, wa: in STD_LOGIC_VECTOR(regbits-1 downto 0);
             wd:         in  STD_LOGIC_VECTOR(width-1 downto 0);
             rd1, rd2:   out STD_LOGIC_VECTOR(width-1 downto 0));
    end component;
    component zerodetect generic(width: integer);
        port(a: in  STD_LOGIC_VECTOR(width-1 downto 0);
             y: out STD_LOGIC);
    end component;
    component flop generic(width: integer);
        port(clk: in  STD_LOGIC;
             d:   in  STD_LOGIC_VECTOR(width-1 downto 0);
             q:   out STD_LOGIC_VECTOR(width-1 downto 0));
    end component;
    component flopen generic(width: integer);
        port(clk, en: in  STD_LOGIC;
             d:       in  STD_LOGIC_VECTOR(width-1 downto 0);
             q:       out STD_LOGIC_VECTOR(width-1 downto 0));
    end component;
    component flopenr generic(width: integer);
        port(clk, reset, en: in  STD_LOGIC;
             d:              in  STD_LOGIC_VECTOR(width-1 downto 0);
             q:              out STD_LOGIC_VECTOR(width-1 downto 0));
    end component;
    component mux2 generic(width: integer);
        port(d0, d1: in  STD_LOGIC_VECTOR(width-1 downto 0);
             s:      in  STD_LOGIC;
             y:      out STD_LOGIC_VECTOR(width-1 downto 0));
    end component;
    component mux4 generic(width: integer);
        port(d0, d1, d2, d3: in  STD_LOGIC_VECTOR(width-1 downto 0);
             s:              in  STD_LOGIC_VECTOR(1 downto 0);
             y:              out STD_LOGIC_VECTOR(width-1 downto 0));
    end component;
    constant CONST_ONE:  STD_LOGIC_VECTOR(width-1 downto 0) := conv_std_logic_vector(1, width);
    constant CONST_ZERO: STD_LOGIC_VECTOR(width-1 downto 0) := conv_std_logic_vector(0, width);
    signal ra1, ra2, wa: STD_LOGIC_VECTOR(regbits-1 downto 0);
    signal pc, nextpc, md, rd1, rd2, wd, a,
           src1, src2, aluresult, aluout, dp_writedata, constx4: STD_LOGIC_VECTOR(width-1 downto 0);
    signal dp_instr: STD_LOGIC_VECTOR(31 downto 0);

begin
    -- shift left constant field by 2
    constx4 <= dp_instr(width-3 downto 0) & "00";

    -- register file addrss fields
    ra1 <= dp_instr(regbits+20 downto 21);
    ra2 <= dp_instr(regbits+15 downto 16);
```

```
        regmux:  mux2 generic map(regbits) port map(dp_instr(regbits+15 downto 16),
                                         dp_instr(regbits+10 downto 11), regdst, wa);

        -- independent of bit width, load dp_instruction into four 8-bit registers over four cycles
        ir0:     flopen generic map(8) port map(clk, irwrite(0), memdata(7 downto 0), dp_instr(7 downto 0));
        ir1:     flopen generic map(8) port map(clk, irwrite(1), memdata(7 downto 0), dp_instr(15 downto 8));
        ir2:     flopen generic map(8) port map(clk, irwrite(2), memdata(7 downto 0), dp_instr(23 downto 16));
        ir3:     flopen generic map(8) port map(clk, irwrite(3), memdata(7 downto 0), dp_instr(31 downto 24));
        -- datapath
        pcreg:   flopenr generic map(width) port map(clk, reset, pcen, nextpc, pc);
        mdr:     flop generic map(width) port map(clk, memdata, md);
        areg:    flop generic map(width) port map(clk, rd1, a);
        wrd:     flop generic map(width) port map(clk, rd2, dp_writedata);
        res:     flop generic map(width) port map(clk, aluresult, aluout);
        adrmux:  mux2 generic map(width) port map(pc, aluout, iord, adr);
        src1mux: mux2 generic map(width) port map(pc, a, alusrca, src1);
        src2mux: mux4 generic map(width) port map(dp_writedata, CONST_ONE,
                                      dp_instr(width-1 downto 0), constx4, alusrcb, src2);
        pcmux:   mux4 generic map(width) port map(aluresult, aluout, constx4, CONST_ZERO, pcsource, nextpc);
        wdmux:   mux2 generic map(width) port map(aluout, md, memtoreg, wd);
        rf:      regfile generic map(width, regbits) port map(clk, regwrite, ra1, ra2, wa, wd, rd1, rd2);
        alunit:  alu generic map(width) port map(src1, src2, alucont, aluresult);
        zd:      zerodetect generic map(width) port map(aluresult, zero);

    -- drive outputs
    instr <= dp_instr; writedata <= dp_writedata;
end;


architecture synth of alu is
    signal b2, sum, slt: STD_LOGIC_VECTOR(width-1 downto 0);
begin
    b2 <= not b when alucont(2) = '1' else b;
    sum <= a + b2 + alucont(2);
    -- slt should be 1 if most significant bit of sum is 1
    slt <= conv_std_logic_vector(1, width) when sum(width-1) = '1'
         else conv_std_logic_vector(0, width);
    with alucont(1 downto 0) select result <=
        a and b when "00",
        a or b  when "01",
        sum     when "10",
        slt     when others;
end;


architecture synth of regfile is
    type ramtype is array (2**regbits-1 downto 0) of STD_LOGIC_VECTOR(width-1 downto 0);
        signal mem: ramtype;
begin
    -- three-ported register file
    -- read two ports combinationally
    -- write third port on rising edge of clock
    process(clk) begin
        if clk'event and clk = '1' then
            if write = '1' then mem(conv_integer(wa)) <= wd;
            end if;
        end if;
    end process;
    process(ra1, ra2) begin
        if (conv_integer(ra1) = 0) then rd1 <= conv_std_logic_vector(0, width); -- register 0 holds 0
        else rd1 <= mem(conv_integer(ra1));
        end if;
        if (conv_integer(ra2) = 0) then rd2 <= conv_std_logic_vector(0, width);
        else rd2 <= mem(conv_integer(ra2));
        end if;
    end process;
end;


architecture synth of zerodetect is
    signal i: integer;
    signal x: STD_LOGIC_VECTOR(width-1 downto 1);
begin -- N-bit AND of inverted inputs
    AllBits: for i in width-1 downto 1 generate
        LowBit: if i = 1 generate
                    A1: x(1) <= not a(0) and not a(1);
                end generate;
                OtherBits: if i /= 1 generate
                    Ai: x(i) <= not a(i) and x(i-1);
                end generate;
        end generate;
    y <= x(width-1);
end;
```

```vhdl
architecture synth of flop is
begin
    process(clk) begin
        if clk'event and clk = '1' then -- or use "if RISING_EDGE(clk) then"
            q <= d;
        end if;
    end process;
end;

architecture synth of flopen is
begin
    process(clk) begin
        if clk'event and clk = '1' then
            if en = '1' then q <= d;
            end if;
        end if;
    end process;
end;

architecture synchronous of flopenr is
begin
    process(clk) begin
        if clk'event and clk = '1' then
            if reset = '1' then
                q <= CONV_STD_LOGIC_VECTOR(0, width); -- produce a vector of all zeros
            elsif en = '1' then q <= d;
            end if;
        end if;
    end process;
end;

architecture synth of mux2 is
begin
    y <= d0 when s = '0' else d1;
end;

architecture synth of mux4 is
begin
    y <= d0 when s = "00" else
        d1 when s = "01" else
        d2 when s = "10" else
        d3;
end;
```

References

The IEEE Journal of Solid-State Circuits is abbreviated as *JSSC* because it is cited heavily. Most of the references in IEEE publications since 1988 can be obtained from `ieeexplore.ieee.org`.

[Acken83]   J. Acken, "Testing for bridging faults (shorts) in CMOS circuits," *Proc. Design Automation Conf.*, 1983, pp. 717-718.

[Acosta95]   A. Acosta, M. Valencia, A. Barriga, M. Bellido, and J. Huertas, "SODS: A new CMOS differential-type structure," *JSSC*, vol. 30, no. 7, July 1995, pp. 835-838.

[Afghahi90]   M. Afghahi and C. Svensson, "A unified single-phase clocking scheme for VLSI systems," *JSSC*, vol. 25, no. 1, Feb. 1990, pp. 225-233.

[Allam00]   M. Allam, M. Anis, and M. Elmasry, "High-speed dynamic logic styles for scaled-down CMOS and MTCMOS technologies," *Proc. Intl. Symp. Low Power Electronics and Design*, 2000, pp. 155-160.

[Allam01]   M. Allam and M. Elmasry, "Dynamic current mode logic (DyCML): a new low-power high-performance logic style," *JSSC*, vol. 36, no. 3, March 2001, pp. 550-558.

[Alvandpour02] A. Alvandpour, R. Krishnamurthy, K. Soumyanath, and S. Borkar, "A sub-130-nm conditional keeper technique," *JSSC*, vol. 37, no. 5, May 2002, pp. 633-638.

[Amrutur98]   B. Amrutur and M. Horowitz, "A replica technique for wordline and sense control in low-power SRAM's," *JSSC*, vol. 33, no. 8, Aug. 1998, pp. 1208-1219.

[Amrutur00]   B. Amrutur and M. Horowitz, "Speed and power scaling of SRAM's," *JSSC*, vol. 35, no. 2, Feb. 2000, pp. 175-185.

[Amrutur01]   B. Amrutur and M. Horowitz, "Fast low-power decoders for RAMs," *JSSC*, vol. 36, no. 10, Oct. 2001, pp. 1506-1515.

[Anastasakis02] D. Anastaskasis, R. Damiano, H. Ma, and T. Stanion, "A practical and efficient method for compare-point matching," *Proc. Design Automation Conf.*, June 2002, pp. 305-310.

[Anderson02]   F. Anderson, J. Wells, and E. Berta, "The core clock system on the next generation Itanium microprocessor," *Proc. IEEE Intl. Solid-State Circuits Conf.*, Feb. 2002, pp. 146-147, 453.

[Ando80]   H. Ando, "Testing VLSI with random access scan," *Digest of Papers COMPCON 80*, Feb. 1980, pp. 50-52.

[Artisan02]   Artisan Components, *TSMC 0.18μm Process 1.8-Volt SAGE-X Standard Cell Library Databook*, Release 4.0, Feb. 2002.

[Ashenden01]   P. Ashenden, *The Designer's Guide to VHDL*, 2nd ed., San Francisco, CA: Morgan Kaufmann, 2001.

[Ayers03]   D. Ayers, "VLSI Power Delivery," EE371 Lecture Notes, Stanford University, April 29, 2003.

[Baghini02]   M. Baghini and M. Desai, "Impact of technology scaling on metastability performance of CMOS synchronizing latches," *Proc. Intl. Conf. VLSI Design*, 2002, pp. 317-322.

[Bailey98]   D. Bailey and B. Benschneider, "Clocking design and analysis for a 600-MHz Alpha microprocessor," *JSSC*, vol. 33, no. 11, Nov. 1998, pp. 1627-1633.

[Baker97]   K. Baker and J. van Beers, "Shmoo plotting: the black art of IC testing," *IEEE Design and Test of Computers*, vol. 14, no. 3, July-Sept. 1997, pp. 90-97.

[Baker98]   R. Jacob Baker, H. Li, and D. Boyce, *CMOS Circuit Design, Layout, and Simulation*, New York: Wiley-Interscience, 1998.

[Baker02]   J. Baker, *CMOS Mixed-Signal Circuit Design*, Piscataway, NJ: IEEE Press, 2002.

[Bakoglu90]   H. Bakoglu, *Circuits, Interconnections, and Packaging for VLSI*, Reading, MA: Addison-Wesley, 1990.

[Balamurugan01]   G. Balamurugan and N. Shanbhag, "The twin-transistor noise-tolerant dynamic circuit technique," *JSSC*, vol. 36, no. 2, Feb. 2001, pp. 273-280.

[Barke88]   E. Barke, "Line-to-ground capacitance calculation for VLSI: a comparison," *IEEE Trans. Computer-Aided Design*, vol. 7, no. 2, Feb. 1988, pp. 295-298.

[Baugh73]   C. Baugh and B. Wooley, "A two's complement parallel array multiplication algorithm," *IEEE Trans. Computers*, vol. C-22, no. 12, Dec. 1973, pp. 1045-1047.

[Beaumont-Smith99]   A. Beaumont-Smith, N. Burgess, S. Lefrere, and C. Lim, "Reduced latency IEEE floating-point adder architectures," *Proc. IEEE Symp. Computer Arithmetic*, April 1999, pp. 35-42.

[Beaumont-Smith01]   A. Beaumont-Smith and C. Lim, "Parallel prefix adder design," *Proc. IEEE Symp. Computer Arithmetic*, 2001, pp. 218-225.

[Bedrij62]   O. Bedrij, "Carry-select adder," *IRE Trans. Electronic Computers*, vol. 11, June 1962, pp. 340-346.

[Bernstein99]   K. Bernstein, K. Carrig, C. Durham, P. Hansen, D. Hogenmiller, E. Nowak, and N. Roher, *High Speed CMOS Design Styles*, Boston: Kluwer Academic Publishers, 1999.

[Bernstein00]   K. Bernstein and N. Rohrer, *SOI Circuit Design Concepts*, Boston: Kluwer Academic Publishers, 2000.

[Best03]   R. Best, *Phase-Locked Loops: Design, Simulation, and Applications*, 5th ed., McGraw-Hill, 2003.

[Bewick94]   G. Bewick, Fast Multiplication: Algorithms and Implementation, Ph.D. Thesis, Stanford University, CSL-TR-94-617, 1994.

[Black69]   J. Black, "Electromigration—A brief survey and some recent results," *IEEE Trans. Electron Devices*, vol. ED-16, no. 4, April 1969, pp. 338-347.

[Blackburn96]   J. Blackburn, L. Arndt, and E. Swartzlander, "Optimization of spanning tree carry lookahead adders," *Proc. 30th Asilomar Conf. Signals, Systems, and Computers*, vol. 1, 1996, pp. 177-181.

[Booth51]   A. Booth, "A signed binary multiplication technique," *Quarterly J. Mechanics and Applied Mathematics*, vol. IV, part 2, June 1951, pp. 236-240.

[Borkar03]   S. Borkar, T. Kamik, S. Narendra, J. Tschanz, A. Keshavarzi, and V. De, "Parameter variations and impact on circuits and microarchitecture," *Proc. Design Automation Conf.*, 2003, pp. 338-342.

[Bouldin03]   D. Bouldin, A. Miller, and C. Tan, "Teaching custom integrated circuit design and verification," *Proc. Microelectronics Systems Education Conf.*, 2003, pp. 48-49.

[Bowhill95]   W. Bowhill et al., "Circuit implementation of a 300-MHz 64-bit second-generation CMOS Alpha CPU," *Digital Technical Journal*, vol. 7, no. 1, 1995, pp. 100-115.

[Bowman99]   K. Bowman, B. Austin, J. Eble, X. Tang, and J. Meindl, "A physical alpha-power law MOS-FET model," *JSSC*, vol. 34, no. 10, Oct. 1999, pp. 1410-1414.

[Brent82]   R. Brent and H. Kung, "A regular layout for parallel adders," *IEEE Trans. Computers*, vol. C-31, no. 3, March 1982, pp. 260-264.

[Brooks95]   F. Brooks, *The Mythical Man-Month*, Boston: Addison-Wesley, 1995.

[Brown03]   A. Brown, "Fast films," *IEEE Spectrum*, vol. 40, no. 2, Feb. 2003, pp. 36-40.

[Bugeja00]   A. Bugeja and B. Song, "A self-trimming 14-b 100MSample/s CMOS DAC," *JSSC*, vol. 35, no. 12, Dec. 2000, pp. 1841-1852.

[Burks46]   A. Burks, H. Goldstine, and J. von Neumann, *Preliminary discussion of the logical design of an electronic computing instrument, part 1*, vol. 1, Inst. Advanced Study, Princeton, NJ, 1946.

[Burleson98]   W. Burleson, M. Ciesielski, F. Klass, and W. Liu, "Wave-pipelining: a tutorial and research survey," *IEEE Trans. VLSI*, vol. 6, no. 3, Sept. 1998, pp. 464-474.

[Calma84]   Calma Corporation, GDS II Stream Format, July 1984.

[Candy76]   J. Candy, W. Ninke, and B. Wooley, "A per-channel A/D converter having 15-segment μ-255 companding," *IEEE Transactions on Communications*, vol. 24, no. 1, Jan. 1976, pp. 33-42.

[Carr72]   W. Carr and J. Mize, *MOS/LSI Design and Application*, New York: McGraw-Hill, 1972.

[Celik02]   M. Celik, L. Pileggi, and A. Odabasioglu, *IC Interconnect Analysis*, Boston: Kluwer Academic Publishers, 2002.

[Chan90]   P. Chan and M. Schlag, "Analysis and design of CMOS Manchester adders with variable carry-skip," *IEEE Trans. Computers*, vol. 39, no. 8, Aug. 1990, pp. 983-992.

[Chandrakasan01] A. Chandrakasan, W. Bowhill, and F. Fox, ed., *Design of High-Performance Microprocessor Circuits*, Piscataway, NJ: IEEE Press, 2001.

[Chaney73]   T. Chaney and C. Molnar, "Anomalous behavior of synchronizer and arbiter circuits," *IEEE Trans. Computers*, vol. C-22, April 1973, pp. 421-422.

[Chaney83]   T. Chaney, "Measured flip-flop responses to marginal triggering," *IEEE Trans. Computers*, vol. C-32, no. 12, Dec. 1983, pp. 1207-1209.

[Chang03]   D. Chang and U. Moon, "A 1.4-V 25-MS/s pipelined ADC using opamp-reset switching technique," *JSSC*, vol. 38, no. 8, Aug. 2003, pp. 1401-1404.

[Chao89]   H. Chao and C. Johnston, "Behavior analysis of CMOS D flip-flops," *JSSC*, vol. 24, no. 5, Oct. 1989, pp. 1454-1458.

[Chappell91]   T. Chappell, B. Chappell, S. Schuster, J. Allan, S. Klepner, R. Joshi, and R. Franch, "A 2-ns cycle, 3.8-ns access 512-kb CMOS ECL SRAM with a fully pipelined architecture," *JSSC*, vol. 26, no. 11, Nov. 1991, pp. 1577-1585.

[Cheng99]   Y. Cheng and C. Hu, *MOSFET Modeling & BSIM3 User's Guide*, Boston: Kluwer Academic Publishers, 1999.

[Cheng00y]   Y. Cheng, C. Tsai, C. Teng, and S. Kang, *Electrothermal Analysis of VLSI Systems*, Boston: Kluwer Academic Publishers, 2000.

[Chern92]   J. Chern, J. Huang, L. Arledge, P. Li, and P. Yang, "Multilevel metal capacitance models for CAD design synthesis systems," *IEEE Electron Device Letters*, vol. 13, no. 1, Jan. 1992, pp. 32-34.

[Childs84]   R. Childs, J. Crawford, D. House, and R. Noyce, "A processor family for personal computers," *Proc. IEEE*, vol. 72, no. 3, March 1984, pp. 363-376.

[Chillarige03] Y. Chillarige, S. Dubey, S. Sompur, and B. Wong, "A 399ps arithmetic logic unit (ALU) implemented using Propagate (P), Generate (G), and Kill (K) signals in push-pull style for a next generation UltraSparc microprocessor," *Proc. IEEE Custom Integrated Circuits Conf.*, 2003.

[Chinnery02]   D. Chinnery and K. Keutzer, *Closing the Gap Between ASIC and Custom: Tools and techniques for high-performance ASIC design*, Boston: Kluwer Academic Publishers, 2002.

[Cho95]   T. Cho and P. Gray, "A 10 b, 20 Msample/s, 35mW pipeline A/D converter," *JSSC*, vol. 30, no. 3, March 1995, pp. 166-172.

[Choi97]   J. Choi, L. Jang, S. Jung and J. Choi, "Structured design of a 288-tap FIR filter by optimized partial product tree compression," *JSSC*, vol. 32, no. 3, March 1997, pp. 468-476.

[Choudhury97] M. Choudhury and J. Miller, "A 300 MHz CMOS microprocessor with multi-media technology," *Proc. IEEE Intl. Solid-State Circuits Conf.*, 1997, pp. 170-171.

[Chu86]   K. Chu and D. Pulfrey, "Design procedures for diffential cascode voltage switch circuits," *JSSC*, vol. SC-21, no. 6, Dec. 1986, pp. 1082-1087.

[Chu87]   K. Chu and D. Pulfrey, "A comparison of CMOS circuit techniques: differential cascode voltage switch logic versus conventional logic," *JSSC*, vol. SC-22, no. 4, Aug. 1987, pp. 528-532.

[Ciletti99]   M. Ciletti, *Modeling, Synthesis, and Rapid Prototyping with the VERILOG HDL*, Upper Saddle River, NJ: Prentice Hall, 1999.

[Clark02]   L. Clark, S. Demmons, N. Deutscher, and F. Ricci, "Standby power management for a 0.18µm microprocessor," *Proc. Intl. Symp. Low Power Electronics and Design*, Aug. 2002, pp. 7-12.

[Cobbold66]   R. Cobbold, "Temperature effects on M.O.S. transistors," *Electronics Letters*, vol. 2, no. 6, June 1966, pp. 190-192.

[Cobbold70]   R. Cobbold, *Theory and Application of Field Transistors*, New York: Wiley Interscience, 1970.

[Collins01]   P. Collins, M. Arnold, and P. Avouris, "Engineering carbon nanotubes and nanotube circuits using electrical breakdown," *Science*, vol. 292, 27 April 2001, pp. 706-709.

[Colwell95]   R. Colwell and R. Steck, "A 0.6 µm BiCMOS processor with dynamic execution," *Proc. IEEE Solid-State Circuits Conf.*, 1995, pp. 176-177.

[Cortadella92] J. Cortadella and J. Llabería, "Evaluation of A+B=K conditions without carry propagation," *IEEE Trans. Computers*, vol. 41, no. 11, Nov. 1992, pp. 1484-1487.

[Covino97]   J. Covino, "Dynamic CMOS circuit with noise immunity," US Patent 5,650,733, 1997.

[Crews03]   M. Crews and Y. Yuenyongsgool, "Practical design for transferring signals between clock domains," *EDN Magazine*, Feb. 20, 2003, pp. 65-71.

[Curran02]   B. Curran et al., "IBM eServer z900 high-frequency microprocessor technology, circuits, and design methodology," *IBM J. Research and Development*, vol. 46, no. 4/5, July/Sept. 2002, pp. 631-644.

[Dabral98]   S. Dabral and T. Maloney, *Basic ESD and I/O Design*, New York: John Wiley & Sons, 1998.

[Dadda65]   L. Dadda, "Some schemes for parallel multipliers," *Alta Frequenza*, vol. 34, no. 5, May 1965, pp. 349-356.

[Dally98]   W. Dally and J. Poulton, *Digital Systems Engineering*, Cambridge, UK: Cambridge University Press, 1998.

[Davari99]   B. Davari, "CMOS technology: present and future," *Symp. VLSI Circuits Digest Tech. Papers*, 1999, pp. 5-10.

[Dekker90]   R. Dekker, F. Beenker, and L. Thijssen, "A realistic fault model and test algorithms for static random access memories," *IEEE Trans. Computer-Aided Design*, vol. 9, no. 6, June 1990, pp. 567-572.

[Deleganes02] D. Deleganes, J. Douglas, B. Kommandur, and M. Patyra, "Designing a 3GHz, 130nm, Intel Pentium 4 processor," *Symp. VLSI Circuits Digest Tech. Papers*, 2002, pp. 130-133.

[Delgado-Frias00] J. Delgado-Frias and J. Nyathi, "A high-performance encoder with priority lookahead," *IEEE Trans. Circuits and Systems I*, vol. 47, no. 9, Sept. 2000, pp. 1390-1393.

[Dennard68]   R. Dennard, "Field-effect transistor memory," US Patent 3,387,286, 1968.

[Dennard74]   R. Dennard et al., "Design of ion-implanted MOSFET's with very small physical dimensions," *JSSC*, vol. SC-9, no. 5, Oct. 1974, pp. 256-268.

[Dhanesha95]   H. Dhanesha, K. Falakshahi, and M. Horowitz, "Array-of-arrays architecture for parallel floating point multiplication," *Proc. Conf. Advanced Research in VLSI*, 1995, pp. 150-157.

[Dike99]   C. Dike and E. Burton, "Miller and noise effects in a synchronizing flip-flop," *JSSC*, vol. 34, no. 6, June 1999, pp. 849-855.

[Dingwall79]   A. Dingwall, "Monolithic expandable 6 bit 20 MHz CMOS/SOS A/D converter," *JSSC*, vol. SC-14, no. 6, Dec. 1979, pp. 926-932.

[Dingwall85]   A. Dingwall and V. Zazzu, "An 8-MHz CMOS subranging 8-bit A/D converter," *JSSC*, vol. SC-20, no. 6, Dec. 1985, pp. 1138-1143.

[Dobbalaere95] I. Dobbalaere, M. Horowitz, and A. El Gamal, "Regenerative feedback repeaters for programmable interconnect," *JSSC*, vol. 30, no. 11, Nov. 1995, pp. 1246-1253.

[Dobberpuhl92] D. Dobberpuhl et al., "A 200-MHz 64-b dual-issue CMOS microprocessor," *JSSC*, vol. 27, no. 11, Nov. 1992, pp. 1555-1867.

[Dobson95]   J. Dobson and G. Blair, "Fast two's complement VLSI adder design," *Electronics Letters*, vol. 31, no. 20, Sept. 1995, pp. 1721-1722.

[Donnay03]   S. Donnay and G. Gielen, eds., *Substrate Noise Coupling in Mixed-Signal ASICs*, Boston: Kluwer Academic Publishers, 2003.

[Donovan02]   C. Donovan and M. Flynn, "A "digital" 6-bit ADC in 0.25μm CMOS," *JSSC*, vol. 37, no. 3, March 2002, pp. 432-437.

[Doyle91]   B. Doyle, B. Fishbein, and K. Mistry, "NBTI-enhanced hot carrier damage in p-channel MOSFETs," *Proc. Intl. Electron Devices Meeting*, 1991, pp. 529-532A.

[Draper97]   D. Draper et al., "Circuit techniques in a 266-MHz MMX-enabled processor," *JSSC*, vol. 32, no. 11, Nov. 1997, pp. 1650-1664.

[D'Souza96]   G. D'Souza, "Dyanmic logic circuit with reduced charge leakage," US Patent 5,483,181, 1996.

[Edwards93]   B. Edwards, A. Corry, N. Weste and C. Greenberg, "A single-chip video ghost canceller," *JSSC*, vol. 28, no. 3, March 1993, pp. 379-383..

[Eichelberger78] E. Eichelberger and T. Williams, "A logic design structure for LSI testability," *J. Design Automation and Fault Tolerant Computing*, vol. 2, no. 2, May 1978, pp. 165-178.

[Elmore48]   W. Elmore, "The transient response of damped linear networks with particular regard to wideband amplifiers," *J. Applied Physics*, vol. 19, no. 1, Jan. 1948, pp. 55-63.

[Ercegovac04]   M. Ercegovac and T. Lang, *Digital Arithmetic*, San Francisco: Morgan Kaufmann, 2004.

[Estreich82]   D. Estreich and R. Dutton, "Modeling latch-up in CMOS integrated circuits," *IEEE Trans. Computer-Aided Design*, vol. CAD-1, no. 4, Oct. 1982, pp. 157-162.

[Faggin96]   F. Faggin, M. Hoff, S. Mazor, and M. Shima, "The history of the 4004," *IEEE Micro*, vol. 16, no. 6, Dec. 1996, pp. 10-20.

[Fahim02]   A. Fahim and M. Elmasry, "Low-power high-performance arithmetic circuits and architectures," *JSSC*, vol. 37, no. 1, Jan. 2002, pp. 90-94.

[Fetzer02]   E. Fetzer, M. Gibson, A. Klein, N. Calick, C. Zhu, E. Busta, and B. Mohammad, "A fully bypassed six-issue integer datapath and register file on the Itanium-2 microprocessor," *JSSC*, vol. 37, no. 11, Nov. 2002, pp. 1433-1440.

[Flannagan85]   S. Flannagan, "Synchronization reliability in CMOS technology," *JSSC*, vol. SC-20, no. 4, Aug. 1985, pp. 880-882.

[Flynn01]   M. Flynn and S. Oberman, *Advanced Computer Arithmetic Design*, New York: John Wiley & Sons, 2001.

[Foty96]   D. Foty, *MOSFET Modeling with SPICE: Principles and Practices*, Upper Saddle River, NJ: Prentice Hall, 1996.

[Friedman84]   V. Friedman and S. Liu, "Dynamic logic CMOS circuits," *JSSC*, vol. SC-19, no. 2, April 1984, pp. 263-266.

[Frohman69]   D. Frohman-Bentchkowsky and A. Grove, "Conductance of MOS transistors in saturation," *IEEE Trans. Electron Devices*, vol. ED-16, no. 1, Jan. 1969, pp. 108-113.

[Frowerk77]   R. Frowerk, "Signature Analysis: A New Digital Field Service Method," *Hewlett Packard Journal*, May 1977, pp. 2-8.

[Gajski83]   D. Gajski and R. Kuhn, "New VLSI tools," *Computer*, vol. 16, no. 12, Dec. 1983, pp. 11-14.

[Galiay80]   J. Galiay, Y. Crouzet, and M. Verginiault, "Physical versus logical fault models MOS LSI circuits: impact on their testability," *IEEE Trans. Computers*, vol. C-29, no. 6, June 1980, pp. 527-531.

[Gauthier02]   C. Gauthier and B. Amick, "Inductance: Implications and solutions for high-speed digital circuits: the chip electrical interface," *Proc. IEEE Intl. Solid-State Circuits Conf.*, vol.2, 2002, pp. 565-565.

[Geannopoulos98]   G. Geannopoulos and X. Dai, "An adaptive digital deskewing circuit for clock distribution networks," *Proc. IEEE Intl. Solid-State Circuits Conf.*, 1998, pp. 400-401.

[Gelsinger01]   P. Gelsinger, "Microprocessors for the new millennium: challenges, opportunities, and new frontiers," *Proc. IEEE Intl. Solid-State Circuits Conf.*, 2001, pp. 22-25.

[George96]   S. George, A. Ott, and J. Klaus, "Surface chemistry for atomic layer growth," *J. Phys. Chem.*, vol. 100, 1996, pp. 13121-13131.

[Gerosa94]   G. Gerosa et al., "A 2.2 W, 80 MHz superscalar RISC microprocessor," *JSSC*, vol. 29, no. 12, Dec. 1994, pp. 1440-1452.

[Gielis91]   G. Gielis, R. van de Plassche, and J. van Valburg, "A 540-MHz 10-b polar-to-cartesian converter," *JSSC*, vol. 26, no. 11, Nov. 1991, pp. 1645-1650.

[Gieseke97]   B. Gieseke et al., "A 600-MHz superscalar RISC microprocessor with out-of-order execution," *Proc. IEEE Intl. Solid-State Circuits Conf.*, 1997, pp. 176-177, 451.

[Glasser85]   L. Glasser and D. Dobberpuhl, *The Design and Analysis of VLSI Circuits*, Reading, MA: Addison Wesley, 1985.

[Golden99]   M. Golden et al., "A seventh-generation x86 microprocessor," *JSSC*, vol. 34, no. 11, Nov. 1999, pp. 1466-1477.

[Golomb81]   S. Golomb, *Shift Register Sequences*, Revised Edition, Laguna Hills, CA: Aegean Park Press, 1981.

[Gonclaves83]   N. Gonclaves and H. DeMan, "NORA: a racefree dynamic CMOS technique for pipelined logic structures," *JSSC*, vol. SC-18, no. 3, June 1983, pp. 261-266.

[Gonzalez96]   R. Gonzalez and M. Horowitz, "Energy dissipation in general purpose microprocessors," *JSSC*, vol. 31, no. 9, Sept. 1996, pp. 1277-1284.

[Gray53]   F. Gray, "Pulse code communications," US Patent 2,632,058, 1953.

[Gray01]   P. Gray, P. Hurst, S. Lewis, and R. Meyer, *Analysis and Design of Analog Integrated Circuits*, 4th ed., New York: John Wiley & Sons, 2001.

[Grayver98]   E. Grayver and B. Daneshrad, "Direct digital frequency synthesis using a modified CORDIC," *Proc. IEEE Intl. Symp. Circuits and Systems*, May 1998, pp. 241-244.

[Greenhill02]   D. Greenhill, Design for Reliability Tutorial, *Proc. IEEE Intl. Solid-State Circuits Conf.*, 2002.

[Griffin83]   W. Griffin and J. Hiltebeitel, "CMOS 4-way XOR circuit," *IBM Technical Disclosure Bulletin*, vol. 25, no. 11B, April 1983, pp. 6066-6067.

[Gronowski96]   P. Gronowski et al., "A 433-MHz 64-b quad-issue RISC microprocessor," *JSSC*, vol. 31, no. 11, Nov. 1996, pp. 1687-1696.

[Gronowski98]   P. Gronowski, W. Bowhill, R. Preston, M. Gowan, and R. Allmon, "High-performance microprocessor design," *JSSC*, vol. 33, no. 5, May 1998, pp. 676-686.

[Grosspietsch92] K. Grosspietsch, "Associative processors and memories: a survey," *IEEE Micro*, vol. 12, no. 3, June 1992, pp. 12-19.

[Grotjohn86]   T. Grotjohn and B. Hoefflinger, "Sample-set differntial logic (SSDL) for complex high-speed VLSI," *JSSC*, vol. SC-21, no. 2, April 1986, pp. 367-369.

[Gupta03]   S. Gupta and V. Fong, "A 64-MHz clock-rate sigma-delta ADC with 88-dB SNDR and –105-dB IM3 distortion at a 1.5-MHz signal frequency," *JSSC*, vol. 37, no. 12, Dec. 2002, pp. 1653-1661.

[Gutierrez01]   E. Gutierrez, J. Deen, and C. Claeys (eds.), *Low Temperature Electronics: Physics, Devices, Circuits, and Applications*, New York: Academic Press, 2001.

[Gutnik00]   V. Gutnik and A. Chandrakasan, "Active GHz clock network using distributed PLLs," *JSSC*, vol. 35, no. 11, Nov. 2000, pp. 1553-1560.

[Guyot87]   A. Guyot, B. Hochet, and J. Muller, "A way to build efficient carry-skip adders," *IEEE Trans. Computers*, vol. 36, no. 10, Oct. 1987, pp. 1144-1152.

[Guyot97]   A. Guyot and S. Abou-Samra, "Modeling power consumption in arithmetic operators," *Microelectronic Engineering*, vol. 39, 1997, pp. 245-253.

[Hamming50]   R. Hamming, "Error detecting and error correcting codes," *Bell Systems Technical Journal*, vol. 29, pp. 147-160.

[Hamzaoglu02]   F. Hamzaoglu and M. Stan, "Circuit-level techniques to control gate leakage for sub-100nm CMOS," *Proc. Intl. Symp. Low Power Electronics and Design*, 2002, pp. 60-63.

[Han87]   T. Han and D. Carlson, "Fast area-efficient VLSI adders," *Proc. IEEE Symp. Computer Arithmetic*, 1987, pp. 49-56.

[Harame01a]   D. Harame and B. Meyerson, "The early history of IBM's SiGe mixed signal technology," *IEEE Transactions on Electron Devices*, vol. 48, no. 11, Nov. 2001, pp. 2555-2567.

[Harame01b]   D. Harame et al., "Current status and future trends of SiGe BiCMOS technology," *IEEE Transactions on Electron Devices*, vol. 48, no. 11, Nov. 2001, pp. 2575-2594.

[Haring96]   R. Haring et al., "Self-resetting logic register and incrementer," *Symp. VLSI Circuits Digest Tech. Papers*, 1996, pp. 18-19.

[Harrer02]   H. Harrer et al., "First and second-level packaging for the IBM eServer z900," *IBM J. Research and Development*, vol. 46, no. 4/5, July/Sept. 2002, pp. 397-420.

[Harris97]   D. Harris and M. Horowitz, "Skew-tolerant domino circuits," *JSSC*, vol. 32, no. 11, Nov. 1997, pp. 1702-1711.

[Harris99]   D. Harris, M. Horowitz, and D. Liu, "Timing analysis including clock skew," *IEEE Trans. Computer-Aided Design*, vol. 18, no. 11, Nov. 1999, pp. 1608-1618.

[Harris01a]   D. Harris, *Skew-Tolerant Circuit Design*, San Francisco, CA: Morgan Kaufmann, 2001.

[Harris01b]   D. Harris and S. Naffziger, "Statistical clock skew modeling with data delay variations," *IEEE Trans. VLSI*, vol. 9, no. 6, Dec. 2001, pp. 888-898.

[Harris03]   D. Harris, "A taxonomy of prefix networks," *Proc. 37th Asilomar Conf. Signals, Systems, and Computers*, 2003, pp. 2213-2217.

[Harrison03]   J. Harrison and N. Weste, "A 500 MHz CMOS anti-alias filter using feed-forward op-amps with local common-mode feedback," *Proc. IEEE Intl. Solid-State Circuits Conf.*, Feb. 2003, pp. 132-133.

[Hashemian92]   R. Hashemian and C. Chen, "A new parallel technique for design of decrement/increment and two's complement circuits," *Proc. IEEE Midwest Symp. Circuits and Systems*, vol. 2, 1992, pp. 887-890.

[Hashimoto02]   T. Hashimoto et al., "Integration of a 0.13-μm CMOS and a high performance self-aligned SiGe HBT featuring low base resistance," *Proc. Intl. Electron Devices Meeting*, Dec. 2002, pp. 779-782.

[Hatamian86]   M. Hatamian and G. Cash, "A 70-MHz 8-bit x 8-bit parallel pipelined multiplier in 2.5-μm CMOS," *JSSC*, vol. 21, no. 4, Aug. 1986, pp. 505-513.

[Haykin00]   S. Haykin, *Digital Communications*, New York: John Wiley & Sons, 2000.

[Hazucha00]   P. Hazucha, C. Svensson, and S. Wender, "Cosmic-ray soft error rate characterization of a standard 0.6-μm CMOS process," *JSSC*, vol. 35, no. 10, Oct. 2000, pp. 1422-1429.

[Heald93]   R. Heald and J. Holst, "A 6-ns cycle 256 kb cache memory and memory management unit," *JSSC*, vol. 28, no. 11, Nov. 1993, pp. 1078-1083.

[Heald98]   R. Heald et al., "64-Kbyte sum-addressed-memory cache with 1.6-ns cycle and 2.6-ns latency," *JSSC*, vol. 33, no. 11, Nov. 1998, pp. 1682-1689.

[Heald00]   R. Heald et al., "A third-generation SPARC v9 64-b microprocessor," *JSSC*, vol. 35, no. 11, Nov. 2000, pp. 1526-1538.

[Hedenstierna87]   N. Hedenstierna and K. Jeppson, "CMOS circuit speed and buffer optimization," *IEEE Trans. Computer-Aided Design*, vol. CAD-6, no. 2, March 1987, pp. 270-281.

[Heikes94]   C. Heikes, "A 4.5mm^2 multiplier array for a 200MFLOP pipelined coprocessor," *Proc. IEEE Intl. Solid-State Circuits Conf.*, 1994, pp. 290-291.

[Heller84]   L. Heller, W. Griffin, J. Davis and N. Thoma, "Cascode voltage switch logic: a differential CMOS logic family," *Proc. IEEE Intl. Solid-State Circuits Conf.*, 1984, pp. 16-17.

[Hennessy90]   J. Hennessy and D. Patterson, *Computer Architecture: A Quantitative Approach*, San Mateo, CA: Morgan Kaufmann Publishers, Inc. 1990.

[Hess94]   C. Hess and L. Weiland, "Drop in process control checkerboard test structure for efficient online process characterization and defect problem debugging," *Proc. IEEE Int. Conf. Microelectronic Test Structures*, vol. 7, March, 1994, pp. 152-159.

[Hidaka89]   H. Hidaka, K. Fujishima, Y. Matsuda, M. Asakura, and T. Yoshihara, "Twisted bit-line architectures for multi-megabit DRAM's," *JSSC*, vol. 24, no. 1, Feb. 1989, pp. 21-27.

[Hill68]   C. Hill, "Noise margin and noise immunity in logic circuits," *Microelectronics*, vol. 1, April 1968, pp. 16-21.

[Hinton01]   G. Hinton et al., "A 0.18-μm CMOS IA-32 processor with a 4-GHz integer execution unit," *JSSC*, vol. 36, no. 11, Nov. 2001, pp. 1617-1627.

[Hisamoto98]   D. Hisamoto et al., "A folded-channel MOSFET for deep-sub-tenth micron era," *Tech. Digest Intl. Electron Devices Meeting*, San Francisco, Dec. 1998, pp. 1032-1034..

[Ho98]   R. Ho, B. Amrutur, K. Mai, B. Wilburn, T. Mori, and M. Horowitz, "Application of on-chip samplers for test and measurement of integrated circuits," *Symp. VLSI Circuits Digest Tech. Papers*, 1998, pp. 138-139.

[Ho01]   R. Ho, K. Mai, and M. Horowitz, "The future of wires," *Proc. IEEE*, vol. 89, no. 4, April 2001, pp. 490-504.

[Ho03a]   R. Ho, K. Mai, and M. Horowitz, "Efficient on-chip global interconnects," *Symp. VLSI Circuits Digest Tech. Papers*, 2003, pp. 271-274.

[Ho03b]   R. Ho, K. Mai, and M. Horowitz, "Managing wire scaling: a circuit perspective," *Proc. IEEE Interconnect Technology Conf.*, 2003, pp. 177-179.

[Hoeneisen72]   B. Hoeneisen and C. Mead, "Fundamental limitations in Microelectronics-I. MOS technology," *Solid-State Electronics*, vol. 15, 1972, pp. 819-829.

[Hoeschele94]   D. Hoeschele, *Analog-to-Digital and Digital-to-Analog Conversion Techniques*, New York: Wiley-Interscience, 1994.

[Hoppe90]   B. Hoppe, G. Neuendorf, D. Schmitt-Landsiedel, and W. Specks, "Optimization of high-speed CMOS logic circuits with analytical models for signal delay, chip area, and dynamic power dissipation," *IEEE Trans. Computer-Aided Design*, vol. 9, no. 3, March 1990, pp. 236-247.

[Horowitz83]   M. Horowitz and R. Dutton, "Resistance extraction from mask layout data," *IEEE Trans. Computer-Aided Design*, vol. CAD-2, no. 3, July 1983, pp. 145-150.

[Horowitz87]   M. Horowitz et al., "MIPS-X: a 20-MIPS peak, 32-bit microprocessor with on-chip cache," *JSSC*, vol. SC-22, no. 5, Oct. 1987, pp. 790-799.

[Horowitz02]   M. Horowitz, EE371 Course Notes, Stanford University, Spring 2002, www.stanford.edu/class/ee371

[Horstmann89]   J. Horstmann, H. Eichel, and R. Coates, "Metastability behavior of CMOS ASIC flip-flops in theory and test," *JSSC*, vol. 24, no. 1, Feb. 1989, pp. 146-157.

[Hrishikesh02]   M. Hrishikesh et al., "The optimal logic depth per pipeline stage is 6 to 8 FO4 inverter delays," *Proc. Intl. Symp. Computer Architecture*, 2002, pp. 14-24.

[Hsu91]   W. Hsu, B. Sheu, and S. Gowda, "Design of reliable VLSI circuits using simulation techniques," *JSSC*, vol. 26, no. 3, March 1991, pp. 452-457.

[Hsu92]   W. Hsu, B. Sheu, S. Gowda and C. Hwang, "Advanced integrated-circuit reliability simulation including dynamic stress effects," *JSSC*, vol. 27, no. 3, March 1992, pp. 247-257.

[Hu90]   Y. Hu and S. Chen, "GM_Plan: A Gate Matrix Layout Algorithm Based on Artificial Intelligence Planning Techniques," *IEEE Trans. Computer-Aided Design*, vol. 9, no. 8, Aug. 1990, pp. 836-845.

[Hu92]   C. Hu, "IC reliability simulation," *JSSC*, vol. 27, no. 3, March 1992, pp. 241-246.

[Hu95]   C. Hu, K. Rodbell, T. Sullivan, K. Lee, and D. Bouldin, "Electromigration and stress-induced voiding in fine Al and Al-alloy thin-film lines," *IBM J. Research and Development*, vol. 39, no. 4, July 1995, pp. 465-497.

[Huang00]   Z. Huang and M. Ercegovac, "Effect of wire delay on the design of prefix adders in deep-submicron technology," *Proc. 34th Asilomar Conf. Signals, Systems, and Computers*, vol. 2, 2000, pp. 1713-1717.

[Huang02]   C. Huang, J. Wang and Y. Huang, "Design of high-performance CMOS priority encoders and incrementer/decrementers using multilevel lookahead and multilevel folding techniques," *JSSC*, vol. 37, no. 1, Jan. 2002, pp. 63-76.

[Huang03]   X. Huang et al., "Loop-based interconnect modeling and optimization approach for multigigahertz clock network design," *JSSC*, vol. 38, no. 3, March 2003, pp. 457-463.

[Huh98]   Y. Huh, Y. Sung, and S. Kang, "A study of hot-carrier-induced mismatch drift: a reliability issue for VLSI circuits," *JSSC*, vol. 33, no. 6, June 1998, pp. 921-927.

[Huitema03]   E. Huitema et al., "Plastic transistors in active-matrix displays," *Proc. IEEE Intl. Solid-State Circuits Conf.*, Feb. 2003, pp. 380-381.

[Hwang89]   I. Hwang and A. Fisher, "Ultrafast compact 32-bit CMOS adders in multiple-output domino logic," *JSSC*, vol. 24, no. 2, April 1989, pp. 358-369.

[Hwang99a]   W. Hwang, R. Joshi, and W. Henkels, "A 500-MHz, 32-Word x 64-bit, eight-port self-resetting CMOS register file," *JSSC*, vol. 34, no. 1, Jan. 1999, pp. 56-67.

[Hwang99b]   W. Hwang, G. Gristede, P. Sanda, S. Wang, and D. Heidel, "Implementation of a self-resetting CMOS 64-bit parallel adder with enhanced testability," *JSSC*, vol. 34, no. 8, Aug. 1999, pp. 1108-1117.

[Hwang02]   D. Hwang, F. Dengwei, A. Willson, Jr, "A 400-MHz processor for the efficient conversion of rectangular to polar coordinates for digital communications applications," *Symp. VLSI Circuits Digest Tech. Papers*, June 2002, pp. 248-51.

[IEEE1076-02]   *IEEE Standard 1076-2002, VHDL Language Reference Manual.*

[IEEE1149.1-01]   *IEEE Standard 1149.1-2001, Test Access Port and Boundary-Scan Architecture.*

[IEEE1164-93]   *IEEE Standard 1164-1993, Multivalue Logic System for VHDL Model Interoperability (Std_logic_1164).*

[IEEE1364-01]   *IEEE Standard 1364-2001, Verilog Hardware Description Language.*

[Ingino01]   J. Ingino and V. von Kaenel, "A 4-GHz clock system for a high-performance system-on-a-chip design," *JSSC*, vol. 36, no. 11, Nov. 2001, pp. 1693-1698.

[Intel03]   Intel Corporation, *Microprocessor Quick Reference Guide*, http://www.intel.com/pressroom/kits/quickreffam.htm, 2003.

[Ismail99]   Y. Ismail, E. Friedman, and J. Neves, "Figures of merit to characterize the importance of on-chip interconnect," *IEEE Trans. VLSI*, vol. 7, no. 4, Dec. 1999, pp. 442-449.

[Itoh01k]   K. Itoh, *VLSI Memory Chip Design*, Berlin: Springer-Verlag, 2001.

[Itoh01n]   N. Itoh et al., "A 600-MHz 54 x 54-bit multiplier with rectangular-styled Wallace tree," *JSSC*, vol. 36, no. 2, Feb. 2001, pp. 249-257.

[Jamal02]   S. Jamal et al., "A 10-b 120-Msample/s time-interleaved analog-to-digital converter with digital background calibration," *JSSC*, vol. 37, no. 12, Dec. 2002, pp. 1618-1627.

[Jayasumana91] A. Jayasumana, Y. Malaiya, and R. Rajsuman," Design of CMOS circuits for stuck-open fault testability," *JSSC*, vol. 26, no. 1, Jan. 1991, pp. 58-61.

[Jiang03]   X. Jiang, Z. Wang, and F. Chang, "A 2 GS/s 6 b ADC in 0.18 μm CMOS," *Proc. IEEE Intl. Solid-State Circuits Conf.*, Feb. 2003, pp.322-323.

[Ji-ren87]   Y. Ji-ren, I. Karlsson, and C. Svensson, "A true single-phase-clock dynamic CMOS circuit technique," *JSSC*, vol. SC-22, no. 5, Oct. 1987, pp. 899-901.

[Johns96]   D. Johns and K. Martin, *Analog Integrated Circuit Design*, New York: John Wiley & Sons, 1996.

[Johnson88]   M. Johnson, "A symmetric CMOS NOR gate for high-speed applications," *JSSC*, vol. SC-23, no. 5, Oct. 1988, pp. 1233-1236.

[Johnson91]   B. Johnson, T. Quarles, A. Newton, D. Pederson, A. Sangiovanni-Vincentelli, *SPICE3 Version 3e User's Manual*, UC Berkeley, April 1991.

[Johnson93]   H. Johnson and M. Graham, *High-Speed Digital Design: A Handbook of Black Magic*, Upper Saddle River, NJ: Prentice Hall, 1993.

[Josephson02]   D. Josephson, "The manic depression of microprocessor debug," *Proc. Intl. Test Conf.*, 2002, pp. 657-663.

[Juhnke95]   T. Juhnke and H. Klar, "Calculation of the soft error rate of submicron CMOS logic circuits," *JSSC*, vol. 30, no. 7, July 1995, pp. 830-834.

[Jung01]   S. Jung, S. Yoo, K. Kim, and S. Kang, "Skew-tolerant high-speed (STHS) domino logic," *Proc. IEEE Intl. Symp. Circuits and Systems*, 2001, pp. 154-157.

[Kamon94]   M. Kamon, J. Tsuk, and J. White, "FASTHENRY: a multipole-accelerated 3-D inductance extraction program," *IEEE Trans. Microwave Theory and Techniques*, vol. 42, no. 9, Sept. 1994, pp. 1750-1758.

[Kang03]   S. Kang and Y. Leblebici, *CMOS Digital Integrated Circuits*, 3rd ed., Boston: McGraw Hill, 2003.

[Kantabutra91] V. Kantabutra, "Designing optimum carry-skip adders," *Proc. IEEE Symp. Computer Arithmetic*, 1991, pp. 146-153.

[Kantabutra93] V. Kantabutra, "A recursive carry-lookahead / carry-select hybrid adder," *IEEE Trans. Computers*, vol. 42, no. 12, Dec. 1993, pp. 1495-1499.

[Kao01]   J. Kao and A. Chandrasakan, "MTCMOS sequential circuits," *Proc. 27th European Solid-State Circuits Conf.*, 2001, pp. 332-335.

[Kappes03]   M. Kappes, "A 2.2-mW CMOS bandpass continuous-time delta-sigma ADC with 68 dB of dynamic range and 1-MHz bandwidth for wireless applications," *JSSC*, vol. 38, no. 7, July 2003, pp. 1098-1104.

[Karnik01]   T. Karnik, B. Bloechel, K. Soumyanath, V. De, and S. Borkar, "Scaling trends of cosmic rays induced soft errors in static latches beyond 0.18m," *Symp. VLSI Circuits Digest Tech. Papers*, 2001, pp. 61-62.

[Keeth01]   B. Keeth and J. Baker, *DRAM Circuit Design: A Tutorial*, Piscataway, NJ: IEEE Press, 2001.

[Keshavarzi01]   A. Keshavarzi et al., "Effectivness of reverse body bias for leakage control in scaled dual Vt CMOS ICs," *Proc. Intl. Symp. Low Power Electronics and Design*, 2001, pp. 207-212.

[Keyes70]   R. Keyes, E. Harris, and K. Konnerth, "The role of low temperatures in the operation of logic circuitry," *Proc. IEEE*, vol. 58, no. 12, Dec. 1970, pp. 1914-1932.

[Kielkowski95]   R. Kielkowski, SPICE: *Practical Device Modeling*, Boston: McGraw-Hill, 1995.

[Kilburn59]   T. Kilburn, D. Edwards, and D. Aspinall, "Parallel addition in a digital computer - a new fast carry," *Proc. IEE*, vol. 106B, 1959, pp. 460-464.

[Kim02]   Y. Kim et al., "50 nm gate length logic technology with 9-layer Cu interconnects for 90 nm node SoC applications," *Proc. Intl. Electron Devices Meeting*, 2002, p. 69.

[Kio01]   S. Kio, L. McMurchie, and C. Sechen, "Application of output prediction logic to differential CMOS," *Proc. IEEE Computer Society Workshop on VLSI*, 2001, pp. 57-65.

[Klass99]   F. Klass et al., "A new family of semidynamic and dynamic flip-flops with embedded logic for high-performance processors," *JSSC*, vol. 34, no. 5, May 1999, pp. 712-716.

[Klaus98]   J. Klaus, A. Ott, A. Dillon, and S. George, "Atomic layer controlled growth of Si_3N_4 films using sequential surface reactions," *Surf. Sci.*, vol. 418, 1998, pp. L14-L19.

[Knebel98]   D. Knebel et al., "Diagnosis and characterization of timing-related defects by time-dependent light emission," *IEEE Intl. Test Conf.*, 1998, pp. 733-739.

[Knowles01]   S. Knowles, "A family of adders," *Proc. IEEE Symp. Computer Arithmetic*, 2001, pp. 277-284.

[Koenemann79]   B. Koenemann, J. Mucha, and G. Zwiehoff, "Built-in logic block observation techniques," *Proc. Intl. Test Conf.*, Oct. 1979, pp. 37-41.

[Kogge73]   P. Kogge and H. Stone, "A parallel algorithm for the efficient solution of a general class of recurrence equations," *IEEE Trans. Computers*, vol. C-22, no. 8, Aug. 1973, pp. 786-793.

[Kong01w]   W. Kong, R. Venkatraman, R. Castagnetti, F. Duan, and S. Ramesh, "High-density and high-performance 6T-SRAM for system-on-chip in 130 nm CMOS technology," *Tech. Digest Symp. VLSI Technology*, 2001, pp. 105-106.

[Koren02]   I. Koren, *Computer Arithmetic Algorithms*, 2nd ed., Natick, Mass.: A.K. Peters, 2002.

[Kovacs98]   G. Kovacs, *Micromachined Transducers Sourcebook*, Boston: McGraw-Hill, 1998.

[Kozu96]   S. Kozu et al., "A 100 MHz 0.4W RISC processor with 200 MHz multiply-adder, using pulse-register technique," *Proc. IEEE Intl. Solid-State Circuits Conf.*, 1996, pp. 140-141.

[Krambeck82]   R. Krambeck, C. Lee, and H. Law, "High speed compact circuits with CMOS," *JSSC*, vol. SC-17, no. 3, June 1982, pp. 614-619.

[Kumar94]   R. Kumar, "ACMOS: an adaptive CMOS high performance logic," *Electronics Letters*, vol. 30, no. 6, March 1994, pp. 483-484.

[Kumar01]   R. Kumar, "Interconnect and noise immunity design for the Pentium 4 processor," *Intel Technology Journal*, vol. 5, no. 1, Q1 2001, pp. 1-12.

[Kuo01]   J. Kuo and S. Lin, *Low-Voltage SOI CMOS VLSI Devices and Circuits*, New York: Wiley Interscience, 2001.

[Kurd01]   N. Kurd, J. Barkatullah, R. Dizon, T. Fletcher, and P. Madland, "A multigigahertz clocking scheme for the Pentium 4 microprocessor," *JSSC*, vol. 36, no. 11, Nov. 2001, pp. 1647-1653.

[Kuroda96]   T. Kuroda et al., "A 0.9-V, 150-MHz, 10-mW, 4 mm^2, 2-D discrete cosine transform core processor with variable threshold-voltage (VT) scheme," *JSSC*, vol. 31, no. 11, Nov. 1996, pp. 1770-1779.

[Ladner80]   R. Ladner and M. Fischer, "Parallel prefix computation," *J. ACM*, vol. 27, no. 4, Oct. 1980, pp. 831-838.

[Lai97]   F. Lai and W. Hwang, "Design and implementation of differential cascode voltage switch with pass-gate (DCVSPG) logic for high-performance digital systems," *JSSC*, vol. 32, no. 4, April 1997, pp. 563-573.

[Lakshmanan01] A. Lakshmanan and R. Sridhar, "Input controlled refresh for noise tolerant dynamic circuits," *Proc. 14th IEEE Intl. ASIC/SOC Conf.*, 2001, pp. 129-133.

[Larsson94]   P. Larsson and C. Svensson, "Impact of clock slope on true single phase clocked (TSPC) CMOS circuits," *JSSC*, vol. 29, no. 6, June 1994, pp. 723-726.

[Larsson97]   P. Larsson, "Parasitic resistance in an MOS transistor used as on-chip decoupling capacitance," *JSSC*, vol. 32, no. 4, April 1997, pp. 574-576.

[Lasserre99]   F. Lasserre et al., "Laser beam backside probing of CMOS integrated circuits," *Microelectronics and Reliability*, June 1999, vol. 39, no. 6, pp. 957-961.

[Leblebici96]   Y. Leblebici, "Design considerations for CMOS digital circuits with improved hot-carrier reliability," *JSSC*, vol. 31, no. 7, July 1996, pp. 1014-1024.

[Lee02]   S. Lee and H. Yoo, "Race logic architecture (RALA): a novel logic concept using the race scheme of input variables," *JSSC*, vol. 37, no. 2, Feb. 2002, pp. 191-201.

[Lee04]   T. Lee, *The Design of CMOS Radio-Frequency Integrated Circuits*, 2nd ed., Cambridge: Cambridge University Press, 2004.

[Lee86]   C. Lee and E. Szeto, "Zipper CMOS," *IEEE Circuits and Systems Magazine*, May 1986, pp. 10-16.

[Lee92]   K. Lee and M. Breuer, "Design and test rules for CMOS circuits to facilitate IDDQ testing of bridging faults," *IEEE Trans. On CAD of Integrated circuits*, vol. 11, no. 5, May 1992, pp. 659-670.

[Lee98]   M. Lee, "A multilevel parasitic interconnect capacitance modeling and extraction for reliable VLSI on-chip clock delay evaluation," *JSSC*, vol. 33, no. 4, April 1998, pp. 657-661.

[Lehman61]   M. Lehman and N. Burla, "Skip technique for high-speed carry-propagation in binary arithmetic units," *IRE Trans. Electronic Computers*, vol. 10, Dec. 1961, pp. 691-698.

[Leighton92]   F. Leighton, *Introduction to Parallel Algorithms and Architectures: Arrays; Trees; Hypercubes*, San Francisco: Morgan Kaufmann, 1992.

[Leung02]   B. Leung, *VLSI for Wireless Communication*, Upper Saddle River, NJ: Prentice Hall, 2002.

[Lev95]   L. Lev et al., "A 64-b microprocessor with multimedia support," *JSSC*, vol. 30, no. 11, Nov. 1995, pp. 1227-1238.

[Li03]   Y. Li and E. Sanchez-Sinencio, "A wide input bandwidth 7-bit 300-Msample/s folding and current-mode interpolating ADC," *JSSC*, vol. 38, no. 8, Aug. 2003, pp.1405-1410.

[Liew90]   B. Liew, N. Cheung, and C. Hu, "Projecting interconnect electromigration lifetime for arbitrary current waveforms," *IEEE Trans. Electron Devices*, vol. 37, no. 5, May 1990, pp. 1343-1351.

[Lim72]   R. Lim, "A barrel switch design," *Computer Design*, Aug. 1972, pp. 76-78.

[Lin83]   S. Lin and D. Costello, *Error Control Coding: Fundamentals and Applications*, Upper Saddle River, NJ: Prentice Hall, 1983.

[Lin91]   Y. Lin, B. Kim and P. Gray, "A 13-b 2.5-MHz self-calibrated pipelined A/D converter in 3-μm CMOS," *JSSC*, vol. 26, no. 4, April 1991, pp. 628-636.

[Ling81]   H. Ling, "High-speed binary adder," *IBM J. Research and Development*, vol. 25, no. 3, May 1981, pp. 156-166.

[Liu01]   X. Liu, C. Lee, C. Zhou, and J. Han, "Carbon nanotube field-effect inverters," *Appl. Phys. Letters*, vol. 79, no. 20, Nov. 2001, pp. 3329-3331

[Lohstroh79] J. Lohstroh, "Static and dynamic noise margins of logic circuits," *JSSC*, vol. SC-14, no. 3, June 1979, pp. 591-598.

[Lohstroh83] J. Lohstroh, E. Seevinck, and J. de Groot, "Worst-case static noise margin criteria for logic circuits and their mathematical equivalence," *JSSC*, vol. SC-18, no. 6, Dec. 1983, pp. 803-807.

[Lovett98]   S. Lovett, M. Welten, A. Mathewson, and B. Mason, "Optimizing MOS transistor mismatch," *JSSC*, vol. 33, no. 1, Jan. 1998, pp. 147-150.

[Lu88b]   S. Lu, "Implementation of iterative networks with CMOS differential logic," *JSSC*, vol. 23, no. 4, Aug. 1988, pp. 1013-1017.

[Lu91]   S. Lu and M. Ercegovac, "Evaluation of two-summand adders implemented in ECDL CMOS differential logic," *JSSC*, vol. 26, no. 8, Aug. 1991, pp. 1152-1160.

[Lu93]   F. Lu, H. Samueli, J.Yuan, C. Svensson, "A 700 MHz 24-b pipelined accumulator in 1.2-μm CMOS for application as a numerically controlled oscillator," *JSSC*, vol. 28, no. 8, Aug 1993, pp. 878-886.

[Lu93b]   F. Lu and H. Samueli, "A 200-MHz CMOS pipelined multiplier-accumulator using a quasi-domino dynamic full-adder cell design," *JSSC*, vol. 28, no. 2, Feb. 1993, pp. 123-132.

[Lynch91]   T. Lynch and E. Swartzlander, "The redundant cell adder," *Proc. IEEE Symp. Computer Arithmetic*, 1991, pp. 165-170.

[Lynch92]   T. Lynch and E. Swartzlander, "A spanning tree carry lookahead adder," *IEEE Trans. Computers*, vol. 41, no. 8, Aug. 1992, pp. 931-939.

[Lyon87]   R. Lyon and R. Schediwy, "CMOS static memory with a new four-transistor memory cell," *Proc. Advanced Research in VLSI*, March 1987, pp. 111-132

[Ma98]   T. Ma, "Making silicon nitride film a viable gate dielectric," *IEEE Trans. Electron Devices*, vol. 45, no. 3, March 1998, pp. 680-690.

[MacSorley61] O. MacSorley, "High-speed arithmetic in binary computers," *Proc. IRE*, vol. 49, part 1, Jan. 1961, pp. 67-91.

[Mahalingam85] M. Mahalingam, "Thermal management in semiconductor device packages," *Proc. IEEE Custom Integrated Circuits Conf.*, 1985, pp. 46-49.

[Maier97]   C. Maier et al., "A 533-MHz BiCMOS superscalar RISC microprocessor," *JSSC*, vol. 32, no. 11, Nov. 1997, pp. 1625-1634.

[Majerski67] S. Majerski, "On determination of optimal distributions of carry skips in adders," *IEEE Trans. Electronic Computers*, vol. EC-16, no. 1, 1967, pp. 45-58.

[Maluf00]   N. Maluf, *An Introduction to Microelectromechanical Systems Engineering*, Norwood, MA: ArtechHouse, 2000.

[Maneatis96]   J. Maneatis, "Low-jitter process-independent DLL and PLL based on self-biased techniques," *JSSC*, vol. 31, no. 11, Nov. 1996, pp. 1723-1732.

[Maneatis03]   J. Maneatis, I. McClatchie, J. Maxey, and M. Shankaradas, "Self-biased high-bandwidth low-jitter 1-to-4096 multiplier clock generator PLL," *JSSC*, vol. 38, no. 11, Nov. 2003, pp. 1795-1803.

[Mathew03]   S. Mathew, M. Anders, R. Krishnamurthy, and S. Borkar, "A 4-GHz 130-nm address generation unit with 32-bit sparse-tree adder core," *JSSC*, vol. 38, no. 5, May 2003, pp. 689-695.

[Matsui94]   M. Matsui et al., "A 200 MHz 13 mm^2 2-D DCT macrocell using sense-amplifier pipeline flip-flop scheme," *JSSC*, vol. 29, no. 12, Dec. 1994, pp. 1482-1490.

[May79]   T. May and M. Woods, "Alpha-particle-induced soft errors in dynamic memories," *IEEE Trans. Electron Devices*, vol. ED-26, no. 1, Jan. 1979, pp. 2-9.

[McMurchie00]   L. McMurchie, S. Kio, G. Yee, T. Thorp, and C. Sechen, "Output prediction logic: a high-performance CMOS design technique," *Proc. Intl. Conf. Computer Design*, 2000, pp. 247-254.

[Mead80]   C. Mead and L. Conway, *Introduction to VLSI Systems*, Reading, MA: Addison-Wesley, 1980.

[Mehta99]   G. Mehta, D. Harris, and D. Singh, "Pulsed Domino Latches," US Patent 5,880,608, 1999.

[Meier99]   N. Meier, T. Marieb, P. Flinn, R. Gleixner, and J. Bravman, "In-situ studies of electromigration voiding in passivated copper interconnects," *AIP Conf. Proc. 491*, Fifth Intl. Workshop on Stress-Induced Phenomena in Metallization, June 1999, p. 180.

[Meijs84]   N. van der Meijs, J. Fokkema, "VLSI circuit reconstruction from mask topology," *Integration, The VLSI Journal, vol. 2*, no. 2, June 1984, pp. 85-119.

[Melly01]   T. Melly, A. Porret, C. Enz, and E. Vittoz, "An analysis of flicker noise rejection in low-power and low-voltage CMOS mixers," *JSSC*, vol. 36, no. 1, Jan. 2001, pp. 102-109.

[Merchant01]   S. Merchant, S. Kang, M. Sanganeria, B. van Schravendijk, and T. Mountsier, "Copper interconnects for semiconductor devices," *JOM: Journal of the Minerals, Metals, and Materials Society*, vol. 53, no. 6, June 2001, pp. 43-48.

[Messerschmitt90]   D. Messerschmitt, "Synchronization in digital system design," *IEEE J. Selected Areas Communications*, vol. 8, no. 8, Oct. 1990, pp. 1404-1419.

[Mezhiba03]   A. Mezhiba and E. Friedman, *Power Distribution Networks in High Speed Integrated Circuits*, Boston: Kluwer Academic Publishers, 2003.

[Miller00]   R. Miller et al., "The development of 157nm small field and mid-field microsteppers," *Proc. SPIE*, vol. 4000, Optical Microlithography XIII, Christopher J. Progler ed., 2000, pp. 567-578.

[Miyatake01]   H. Miyatake, M. Tanaka, and Y. Mori, "A design for high-speed low-power CMOS fully parallel content-addressable memory macros," *JSSC*, vol. 36, no. 6, June 2001, pp. 956-968.

[Mizuno94]   T. Mizuno, J. Okumtura, and A. Toriumi, "Experimental study of threshold voltage fluctuation due to statistical variation of chanel dopant number in MOSFET's," *IEEE Trans. Electron Devices*, vol. 41, no. 11, Nov. 1994, pp. 2216-2221.

[Moazzami90]   R. Moazzami and C. Hu, "Projecting gate oxide reliability and optimizing reliability screens," *IEEE Trans. Electron Devices*, vol. 37, no. 7, July 1990, pp. 1643-1650.

[Montanaro96]   J. Montanaro et al., "A 160-MHz, 32-b, 0.5-W CMOS RISC microprocessor," *JSSC*, vol. 31, no. 11, Nov. 1996, pp. 1703-1714.

[Moore65]   G. Moore, "Cramming more components onto integrated circuits," *Electronics*, vol. 38, no. 8, April 1965.

[Moore03]   G. Moore, "No exponential is forever: but 'forever' can be delayed!" *Proc. IEEE Intl. Solid-State Circuits Conf.*, 2003, pp. 1-19.

[Morgan59]   C. Morgan and D. Jarvis, "Transistor logic using current switching routing techniques and its application to a fast carry-propagation adder, *Proc. IEE*, vol. 106B, 1959, pp. 467-468.

[Morrison61]   P. Morrison and E. Morrison, eds., *Charles Babbage: On the Principles and Development of the Calculator*, New York: Dover, 1961.

[Mortezapour00] S. Mortezapour and E. Lee, "A 1-V, 8-bit successive approximation ADC in standard CMOS process," *JSSC*, vol. 35, no. 4, April 2000, pp. 642-646.

[Morton99]   S. Morton, "On-chip inductance issues in multiconductor systems," *Proc. Design Automation Conf.*, 1999, pp. 921-926.

[Morton02]   S. Morton, "Inductance: Implications and solutions for high-speed digital circuits," *Proc. IEEE Intl. Solid-State Circuits Conf.*, vol. 2 , Feb 2002, pp. 554-557.

[Mou90]   Z. Mou and F. Jutand, "A class of close-to-optimum adder trees allowing regular and compact layout," *Proc. IEEE Intl. Conf. Computer Design*, 1990, pp. 251-254.

[Mule02]   A. Mule, E. Glytsis, T. Gaylord, and J. Meindl, "Electrical and optical clock distribution networks for gigascale microprocessors," *IEEE Trans. VLSI*, vol. 10, no. 5, Oct. 2002, pp. 582-594.

[Muller03]   R. Muller, T. Kamins, and M. Chan, *Device Electronics for Integrated Circuits*, 3rd ed., New York: John Wiley & Sons, 2003.

[Murabayashi96] F. Murabayashi et al., "2.5 V CMOS circuit techniques for a 200 MHz superscalar RISC processor," *JSSC*, vol. 31, no. 7, July 1996, pp. 972-980.

[Murphy80]   B. Murphy, "Unified field-effect transistor theory including velocity saturation," *JSSC*, vol. SC-15, no. 3, June 1980, pp. 325-327.

[Mutoh95]   S. Mutoh et al., "1-V power supply high-speed digital circuit technology with multithreshold-voltage CMOS," *JSSC*, vol. 30, no. 8, Aug. 1995, pp. 847-854.

[Nabors92]   K. Nabors, S. Kim, and J. White, "Fast capacitance extraction of general three-dimensional structures," *IEEE Trans. Microwave Theory and Techniques*, vol. 40, no. 7, July 1992, pp. 1496-1506.

[Nadig77]   H. Nadig, "Signature analysis—concepts, examples and guidelines," *Hewlett Packard Journal*, vol. 28, no. 9, May 1977, pp. 15-21.

[Naffziger96]   S. Naffziger, "A subnanosecond 0.5μm 64b adder design," *Proc. IEEE Intl. Solid-State Circuits Conf.*, 1996, pp. 362-363.

[Naffziger98]   S. Naffziger, "High speed addition using Ling's equations and dynamic CMOS logic," US Patent 5,719,803, 1998.

[Naffziger02]   S. Naffziger, G. Colon-Bonet, T. Fischer, R. Riedlinger, T. Sullivan, and T. Grutkowski, "The implemetnation of the Itanium 2 microprocessor," *JSSC*, vol. 37, no. 11, Nov. 2002, pp. 1448-1460.

[Nagel75]   L. Nagel, *SPICE2: a computer program to simulate semiconductor circuits*, Memo ERL-M520, Dept. of Electrical Engineering and Computer Science, University of California at Berkeley, May 9, 1975.

[Nair78]   R. Nair, S. Thatte, and J. Abraham, "Efficient algorithms for testing semiconductor random-access memories," *IEEE Trans. Computers*, vol. C-27, no. 6, June 1978, pp. 572-576.

[Nalamalpu02]   A. Nalamalpu, S. Srinivasan, and W. Burleson, "Boosters for driving long onchip interconnects—design issues, interconnect synthesis, and comparison with repeaters," *IEEE Trans. Computer–Aided Design*, vol. 21, no. 1, Jan. 2002, pp. 50-62.

[Nambu98]   H. Nambu et al., "A 1.8-ns access, 550-MHz, 4.5-Mb CMOS SRAM," *JSSC*, vol. 33, no. 11, Nov. 1998, pp. 1650-1658.

[Narayanan96]   V. Narayanan, B. Chappell, and B. Fleischer, "Static timing analysis for self-resetting circuits," *Proc. Intl. Conf. Computer–Aided Design*, 1996, pp. 119-126.

[Narendra99]   S. Narendra, D. Antoniadis, and V. De, "Impact of using adaptive body bias to compensate die-to-die Vt variation on within-die Vt variation," *Proc. Intl. Symp. Low Power Electronics and Design*, 1999, pp. 229-232.

[Narendra01]   S. Narendra, S. Borkar, V. De, D. Antoniadis, and A. Chandrakasan, "Scaling of stack effect and its application for leakage reduction," *Proc. Intl. Symp. Low Power Electronics and Design*, 2001, pp. 195-200.

[Narendra03]   S. Narendra, A. Keshavarzi, B. Bloechel, S. Borkar, and V. De, "Forward body bias for microprocessors in 130-nm technology generation and beyond," *JSSC*, vol. 38, no. 5, May 2003, pp. 696-701.

[Nauta95]   B. Nauta and A. Venes, "A 70-MS/s 110-mW 8-b CMOS folding and interpolating A/D converter," *JSSC*, vol. 30, no. 12, Dec. 1995, pp. 1302-1308.

[Ng96]   P. Ng, P. Balsara, and D. Steiss, "Performance of CMOS differential circuits," *JSSC*, vol. 31, no. 6, June 1996, pp. 841-846.

[Nikolí00]   B. Nikolí, V. Oklobdíja, V. Stojanoví, W. Jia, J. Chiu, and M. Leung, "Improved sense-amplifier-based flip-flop: design and measurements," *JSSC*, vol. 35, no. 6, June 2000, pp. 876-884.

[Noice83]   D. Noice, *A clocking discipline for two–phase digital integrated circuits*, Stanford University Technical Report, Jan. 1983.

[Nowak02]   E. Nowak, "Maintaining the benefits of CMOS scaling when scaling bogs down," *IBM J. Research and Development*, vol. 46, no. 2/3, March/May 2002, pp. 169-180

[Nowka98]   K. Nowka and T. Galambos, "Circuit design techniques for a gigahertz integer microprocessor," *Proc. Intl. Conf. Computer Design*, 1998, pp. 11-16.

[Ohkubo95]   N. Ohkubo et al., "A 4.4 ns CMOS 54 x 54-b multiplier using pass-transistor multiplexer," *JSSC*, vol. 30, no. 3, March 1995, pp. 251-257.

[Oklobdíja85]   V. Oklobdíja and E. Barnes, "Some optimal schemes for ALU implementation in VLSI technology," Proc. IEEE Symp. Comp. Arithmetic, 1985.

[Oklobdíja86]   V. Oklobdíja and R. Montoye, "Design-performance trade-offs in CMOS-domino logic," *JSSC*, vol. SC-21, no. 2, April 1986, pp. 304-309.

[Ortiz-Conde02]   A. Ortiz-Conde, F. Sánchez, J. Liou, A. Cerdeira, M. Estrada, and Y. Yue, "A review of recent MOSFET threshold voltage extraction methods," *Microelectronics Reliability*, vol. 42, 2002, pp. 583-596.

[Oshawa87]   T. Oshawa et al., "A 60-ns 4-Mbit CMOS DRAM with built-in self-test function," *JSSC*, vol. SC-22, no. 5, Oct. 1987, pp. 663-668.

[Paik96]   W. Paik, H. Ki, and S. Kim, "Push-pull pass-transistor logic family for low voltage and low power," *Proc. 22nd European Solid–State Circuits Conf.*, 1996, pp. 116-119.

[Parameswar96]   A. Parameswar, H. Hara, and T. Sakurai, "A swing restored pass-transistor logic-based multi-ply and accumulate circuit for multimedia applications," *JSSC*, vol. 31, no. 6, June 1996, pp. 804-809.

[Paraskevopoulos87]   D. Paraskevopoulos and C. Fey, "Studies in LSI technology economics III: design schedules for application-specific integrated circuits," *JSSC*, vol. SC-22, no. 2, April 1987, pp. 223-229.

[Parhami00]   B. Parhami, *Computer Arithmetic Algorithms and Hardware Designs*, New York: Oxford University Press, 2000.

[Parihar01]   S. Parihar et al., "A high density 0.10 μm CMOS technology using low K dielectric and copper interconnect," *Proc. Intl. Electron Devices Meeting*, 2001, pp. 11.4.1-11.4.4.

[Park99]   J. Park, J. Lee, and W. Kim, "Current sensing differential logic: a CMOS logic for high reliability and flexibility," *JSSC*, vol. 34, no. 6, June 1999, pp. 904-908.

[Parker03]   K. Parker, *The Boundary-Scan Handbook*, Boston: Kluwer Academmic Publishers, 2003.

[Partovi94]   H. Partovi and D. Draper, "A regenerative push-pull differential logic family," *Proc. IEEE Intl. Solid-State Circuits Conf.*, 1994, pp. 294-295.

[Partovi96]   H. Partovi et al., "Flow-through latch and edge-triggered flip-flop hybrid elements," *Proc. IEEE Intl. Solid-State Circuits Conf.*, 1996, pp. 138-139.

[Pasternak87]   J. Pasternak, A. Shubat, and C. Salama, "CMOS differential pass-transistor logic design," *JSSC*, vol. SC-22, no. 2, April 1987, pp. 216-222.

[Pasternak91]   J. Pasternak and C. Salama, "Design of submicrometer CMOS differential pass-transistor logic circuits," *JSSC*, vol. 26, no. 9, Sept. 1991, pp. 1249-1258.

[Patterson04]   D. Patterson and J. Hennessy, *Computer Organization and Design*, 3rd ed., San Francisco, CA: Morgan Kaufmann, 2004.

[Paul02]   B. Paul and K. Roy. Testing cross-talk induced delay faults in static CMOS circuit through dynamic timing analysis. *Proc. Intl. Test Conf.*, Oct. 2002, pp. 384-390.

[Pelgrom89]   M. Pelgrom, A. Duinmaijer, and A. Welbers, "Matching properties of MOS transistors," *JSSC*, vol. 24, no. 5, Oct. 1989, pp. 1433-1440.

[Peng02]   C. Peng et al., "A 90 nm generation copper dual damascene technology with ALD TaN barrier," *Tech. Digest Intl. Electron Devices Meeting*, Dec. 2002, pp. 603-606.

[Penney72]   W. Penney and L. Lau, *MOS Integrated Circuits*, New York: Van Nostrand Reinhold, 1972.

[Petegem94]   W. van Petegem, B. Geeraerts, W. Sansen, and B. Graindourze, "Electrothermal simulation and design of integrated circuits," *JSSC*, vol. 29, no. 2, Feb. 1994, pp. 143-146.

[Pfennings85]   L. Pfennings, W. Mol, J. Bastiens, and J. van Dijk, "Differential split-level CMOS logic for subnanosecond speeds," *JSSC*, vol. SC-20, no. 5, Oct. 1985, pp. 1050-1055.

[Pihl98]   J. Pihl, "Single-ended swing restoring pass transistor cells for logic synthesis and optimization," *Proc. IEEE Intl. Symp. Circuits and Systems*, vol. 2, 1998, pp. 41-44.

[Piña02]   C. Piña, "Evolution of the MOSIS VLSI educational program," *Proc. Electronic Design, Test, and Applications Workshop*, 2002, pp. 187-191.

[Poulton03]   K. Poulton et al., "A 20Gs/s 8 b ADC with a 1 MB Memory in 0.18 μm CMOS," *Proc. IEEE Solid-State Circuits Conf.*, Feb. 2003, pp. 318-319.

[Pretorius86]   J. Pretorius, A. Shubat, and A. Salama, "Latched domino CMOS logic," *JSSC*, vol. SC-21, no. 4, Aug. 1986, pp. 514-522.

[Price95]   D. Price, "Pentium FDIV flaw—lessons learned," *IEEE Micro*, vol. 15, no. 2, April 1995, pp. 86-88.

[Prince99]   B. Prince, *High Performance Memories: New Architecture DRAMs and SRAMs—Evolution and Function*, New York: John Wiley & Sons, 1999.

[Proebsting91]   R. Proebsting, "Speed enhancement technique for CMOS circuits," US Patent 4,985,643, 1991.

[Promizter01]   G. Promitzer, "12-bit low-power fully differential switched capacitor noncalibrating successive approximation ADC with 1MS/s," *JSSC*, vol. 36, no. 7, July 2001, pp. 1138-1143.

[Quader94]   K. Quader, E. Minami, W. Huang, P. Ko, and C. Hu, "Hot-carrier-reliability design guidelines for CMOS logic circuits," *JSSC*, vol. 29, no. 3, March 1994, pp. 253-262.

[Rabaey03]   J. Rabaey, A. Chandrakasan, and B. Nikolic, *Digital Integrated Circuits*, 2nd Ed., Upper Saddle River, NJ: Prentice Hall, 2003.

[Razavi98]   B. Razavi, *RF Microelectronics*, Upper Saddle River, NJ: Prentice Hall, 1998.

[Reddy02]   V. Reddy et al., "Impact of negative bias temperature instability on digital circuit reliability," *Proc. 40th IEEE Intl. Reliability Physics Symp.*, 2002, pp. 248-254.

[Restle98]   P. Restle and A. Deutsch, "Designing the best clock distribution network," *Symp. VLSI Circuits Digest Tech. Papers*, 1998, pp. 2-5.

[Restle01]   P. Restle et al., "A clock distribution network for microprocessors," *JSSC*, vol. 36, no. 5, May 2001, pp. 792-799.

[Riordan97]   M. Riordan and L. Hoddeson, *Crystal Fire: The Invention of the Transistor and the Birth of the Information Age*, New York: W. W. Norton & Co, 1998.

[Rodgers89]   B. Rodgers and C. Thurber, "A Monolithic ±5½-digit BiMOS A/D converter," *JSSC*, vol. 24, no. 3, June 1989, pp. 617-626.

[Rotella02]   F. Rotella, V. Blaschke, and D. Howard, "A broad-band scalable lumped-element inductor model using analytic expressions to incorporate skin effect, substrate loss, and proximity effect," *Tech. Digest Intl. Electron Devices Meeting*, Dec. 2002, pp. 471-474.

[Ruehli73]   A. Ruehli and P. Brennan, "Efficient capacitance calculations for three-dimensional multiconductor systems," *IEEE Trans. Microwave Theory and Techniques*, vol. MTT-21, No. 2, Feb. 1973, pp. 76-82.

[Rusu00]   S. Rusu and G. Singer, "The first IA-64 microprocessor," *JSSC*, vol. 35, no. 11, Nov. 2000, pp. 1539-1544.

[Rusu03]   S. Rusu, J. Stinson, S. Tam, J. Leung, H. Muljono, and B. Cherkauer, "A 1.5-GHz 130-nm Itanium 2 processor with 6-MB on-die L3 cache," *JSSC*, vol. 38, no. 11, Nov. 2003, pp. 1887-1895.

[Ryan01]   P. Ryan et al., "A single chip PHY COFDM modem for IEEE 802.11a with integrated ADCs and DACs," *Proc. IEEE Intl. Solid-State Circuits Conf.*, 2001, pp. 338-339, 463.

[Rzepka98]   S. Rzepka, K. Banerjee, E. Meusel, and C. Hu, "Characterization of self-heating in advanced VLSI interconnect lines based on thermal finite element simulation," *IEEE Trans. Components, Packaging, and Manufacturing Technology* - Part A, vol. 21, no. 3, Sept. 1998, pp. 406-411.

[Sah64]   C. T. Sah, "Characteristics of the Metal-Oxide-Semiconductor Transistors," *IEEE Trans. Electron Devices*, ED-11, July 1964, pp. 324-345.

[Saint02]   C. Saint and J. Saint, *IC Mask Design: Essential Layout Techniques*, New York: McGraw-Hill, 2002.

[Sakurai83]   T. Sakurai, "Approximation of wiring delay in MOSFET LSI," *JSSC*, vol. SC-18, no.4, Aug. 1983, pp. 418-426.

[Sakurai86]   T. Sakurai, K. Nogami, M. Kakumu, and T. Iizuka, "Hot-carrier generation in submicrometer VLSI environment," *JSSC*, vol. SC-21, no. 1, Feb. 1986, pp. 187-192.

[Sakurai90]   T. Sakurai and R. Newton, "Alpha-Power Law MOSFET Model and its Applications to CMOS Inverter Delay and Other Formulas," *JSSC*, vol. 25, no. 2, April 1990, pp. 584-594.

[Samavati98]   H. Samavati, A. Hajimiri, A. Shahani, G. Nazzerbakht, and T. Lee, "Fractal capacitors," *JSSC*, vol. 33, no. 12, Dec. 1998, pp. 2035-2041.

[Santoro89]   M. Santoro, Design and Clocking of VLSI Multipliers, Ph.D. Thesis, Stanford University, CSL-TR-89-397, 1989.

[Sauerbrey02]   J. Sauerbrey, T. Tille, D. Schmitt-Landsiedel and R. Thewes, "A 0.7-V MOSFET-only switched-opamp sigma delta modulator in standard digital CMOS technology," *JSSC*, vol. 37, no. 12, Dec. 2002, pp. 1662-1669.

[Sauerbrey03]   J. Sauerbrey, D. Schmitt-Landsiedel and R. Thewes, "A 0.5V 1-μW successive approximation ADC," *JSSC*, vol. 38, no. 7, July 2003, pp. 1261-1265.

[Schellenberg03] F. Schellenberg, "A little light magic," *IEEE Spectrum*, vol. 40, no. 9, Sept. 2003, pp. 34-39.

[Schmitt38]   O.H. Schmitt, "A thermionic trigger," *J. Scientific Instruments*, vol. 15, Jan. 1938, pp. 24-26.

[Scholtens02]   P. Scholtens and M. Vertregt, "A 6-b 1.6-Gsample/s flash ADC in 0.18 μm CMOS using averaging termination," *JSSC*, vol. 37, no. 12, Dec. 2002, pp. 1599-1609.

[Schultz90]   K. Schultz, R. Francis and K. Smith, "Ganged CMOS: trading standby power for speed," *JSSC*, vol. SC-25, no. 3, June 1990, pp. 870-873.

[Schultz95]   K. Schultz and P. Gulak, "Architectures for large-capacity CAMs," *Integration*, vol. 18, no. 2-3, 1995, pp. 151-171.

[Schutten03]   R. Schutten, T. Fitzbatrick, "Design for verification—blueprint for productivity and product quality," Synopsys white paper, 2003.

[Seeds67]   R. Seeds, "Yield and cost analysis of bipolar LSI," *Intl. Electron Device Meeting*, Oct. 1967.

[Shahidi02]   G. Shahidi, "SOI technology for the GHz era," *IBM J. Research and Development*, vol. 46, no. 2/3, March/May 2002, pp. 121-131.

[Sharma00]   `http://www.free-ip.com/cordic/`

[She02]   M. She et al., "JVD silicon nitride as tunnel dielectric in p-channel flash memory," *IEEE Electron Device Letters*, vol. 23, no. 2, Feb. 2002, pp.   91-93.

[Shepard99]   K. Shepard, V. Narayanan, and R. Rose, "Harmony: static noise analysis of deep submicron digital integrated circuits," *IEEE Trans. Computer-Aided Design*, vol. 18, no. 8, Aug. 1999, pp. 1132-1150.

[Sheu87a]   B. Sheu and P. Ko, "Measurement and modeling of short-channel MOS transistor gate capacitances," *JSSC*, vol. SC-22, no. 3, June 1987, pp. 464-472.

[Sheu87b]   B. Sheu, D. Scharfetter, P. Ko, and M. Jeng, "BSIM: Berkeley short-channel IGFET model for MOS transistors," *JSSC*, vol. SC-22, no. 4, Aug. 1987, pp. 558-566.

[Shichman68]   H. Shichman and D. Hodges, "Modeling and simulation of insulated-gate field-effect transistor switching circuits," *JSSC*, vol. SC-3, no. 3, Sept. 1968, pp. 285-289.

[Shockley52]   W. Shockley, "A unipolar 'field-effect' transistor," *Proc. IRE*, vol. 40, 1952, pp. 1365-1376.

[Shoji82]   M. Shoji, "Electrical design of BELLMAC-32a microprocessor," *Proc. IEEE Intl. Conf. Circuits and Computers, Sept. 1982*, pp. 112-115.

[Shoji85]   M. Shoji, "FET scaling in domino CMOS gates," *JSSC*, vol. SC-20, no. 5, Oct. 1985, pp. 1067-1071.

[Shoji86]   M. Shoji, "Elimination of process-dependent clock skew in CMOS VLSI," *JSSC*, vol. SC-21, no. 5, Oct. 1986, pp. 875-880.

[SIA97]   Semiconductor Industry Association, *International Technology Roadmap for Semiconductors*, 1997.

[SIA02]   Semiconductor Industry Association, *International Technology Roadmap for Semiconductors*, 2002 Update, public.itrs.net.

[Silberman98]   J. Silberman et al., "A 1.0-GHz single-issue 64-bit PowerPC integer microprocessor," *JSSC*, vol. 33, no. 11, Nov. 1998, pp. 1600-1608.

[Simon92]   T. Simon, "A fast static CMOS NOR gate," *Proc. Advanced Research in VLSI and Parallel Systems*, 1992, pp. 180-192.

[Sklansky60]   J. Sklansky "Conditional-sum addition logic," *IRE Trans. Electronic Computers*, vol. EC-9, June 1960, pp. 226-231.

[Smith00]   D. Smith and P. Franzon, *Verilog Styles for Synthesis of Digital Circuits*, Upper Saddle River, NJ: Prentice Hall, 2000.

[Soeleman01]   H. Soeleman, K. Roy, and B. Paul, "Robust subthreshold logic for ultra-low power operation," *IEEE Trans. VLSI*, vol. 9, no. 1, Feb. 2001, pp. 90-99.

[Solomatnikov00]   A. Solomatnikov, D. Somasekhar, K. Roy, and C. Koh, "Skewed CMOS: noise-immune high-performance low-power static circuit family," *Proc. IEEE Intl. Conf. Computer Design*, 2000, pp. 241-246.

[Somasekhar96]   D. Somasekhar and K. Roy, "Differential current switch logic: a low power DCVS logic family," *JSSC*, vol. 31, no. 7, July 1996, pp. 981-991.

[Somasekhar98]   D. Somasekhar and K. Roy, "LVDCSL: a high fan-in, high-performance, low-voltage differential current switch logic family," *IEEE Trans. VLSI*, vol. 6, no. 4, Dec. 1998, pp. 573-577.

[Somasekhar00]   D. Somasekhar, S. Choi, K. Roy, Y. Ye, and V. De, "Dynamic noise analysis in precharge-evaluate circuits," *Proc. Design Automation Conf.*, 2000, pp. 243-246.

[Song96]   M. Song, G. Kang, S. Kim, and B. Kang, "Design methodology for high speed and low power digital circuits with energy economized pass-transistor logic (EEPL)," *Proc. 22nd European Solid-State Circuits Conf.*, 1996, pp. 120-123.

[Song01]   S. Song et al., "On the gate oxide scaling of high performance CMOS transistors," *Proc. Intl. Electron Devices Meeting*, 2001, pp. 3.2.1-3.2.4.

[Sowiati01]   T. Sowlati, V. Vathulya, and D. Leenaerts, "High density capacitance structures in submicron CMOS for low power RF applications," *Proc. Intl. Symp. Low Power Electronics and Design*, Aug. 2001, pp. 243-246.

[Sparsø01]   J. Sparsø and S. Furber, eds., *Principles of Asynchronous Circuit Design: A Systems Perspective*, Boston: Kluwer Academic Publishers, 2001.

[Srinivas92]   H. Srinivas and K. Parhi, "A fast VLSI adder architecture," *JSSC*, vol. 27, no. 5, May 1992, pp. 761-767.

[Stinson03]   J. Stinson and S. Rusu, "A 1.5 GHz third generation Itanium processor," *Proc. Design Automation Conf.*, 2003, pp. 706-709.

[Stojanovic99]   V. Stojanovic and V. Oklobdíja, "Comparative analysis of master-slave latches and flip-flops for high-performance and low-power systems," *JSSC*, vol. 34, no. 4, April 1999, pp. 536-548.

[Stroud02]   C. Stroud, *A Designer's Guide to Built-in Self-Test*, Boston: Kluwer Academic Publishers, 2002.

[Sun87]   J. Sun, Y. Taur, R. Dennard, and S. Klepner, "Submicrometer-channel CMOS for low-temperature operation," *IEEE Trans. Electron Devices*, vol. ED-34, no. 1, Jan. 1987, pp. 19-26.

[Sutherland99]   I. Sutherland, B. Sproull, and D. Harris, *Logical Effort: Designing Fast CMOS Circuits*, San Francisco, CA: Morgan Kaufmann, 1999.

[Suzuki73]   Y. Suzuki, K. Odagawa and T. Abe, "Clocked CMOS calculator circuitry," *JSSC*, vol. SC-8, no. 6, Dec. 1973, pp. 462-469.

[Suzuki93]   M. Suzuki, N. Ohkubo, T. Shinbo, T. Yamanaka, A. Shimizu, K. Sasaki, and Y. Nakagome, "A 1.5-ns 32-b CMOS ALU in double pass-transistor logic," *JSSC*, vol. 28, no. 11, Nov. 1993, pp. 1145-1151.

[Sweeney02]   P. Sweeney, *Error Control Coding: From Theory to Practice*, New York: John Wiley & Sons, 2002.

[Sylvester98]   D. Sylvester and K. Keutzer, "Getting to the bottom of deep submicron," *Proc. IEEE/ACM Intl. Conf. Computer-Aided Design*, 1998, pp. 203-211.

[Tam00]   S. Tam, S. Rusu, U. Desai, R. Kim, J. Zhang, and I. Young, "Clock generation and distribution for the first IA-64 microprocessor," *JSSC*, vol. 35, no. 11, Nov. 2000, pp. 1545-1552.

[Tam04]   S. Tam, R. Limaye, and U. Desai, "Clock generation and distribution for the 130-nm Itanium 2 processor with 6-MB on-die L3 cache," *JSSC*, vol. 39, no. 4, Apr. 2004.

[Tharakan92]   G. Tharakan and S. Kang, "A new design of a fast barrel switch network," *JSSC*, vol. 27, no. 2, Feb. 1992, pp. 217-221.

[Thomas02]   D. Thomas and P. Moorby, *The Verilog Hardware Description Language*, 5th Ed. Boston: Kluwer Academic Publishers, 2002.

[Thorp99]   T. Thorp, G. Yee, and C. Sechen, "Design and synthesis of monotonic circuits," *Proc. IEEE Intl. Conf. Computer Design*, 1999, pp. 569-572.

[Tiilikainen01]   M. Tiilikainen, "A 14-bit 1.8-V 20-mW 1-mm^2 CMOS DAC," *JSSC*, vol. 36, no. 7, July 2001, pp. 1144-1147.

[Timko80]   M. Timko and P. Holloway, "Circuit techniques for achieving high speed-high resolution A/D conversion," *JSSC*, vol. SC-15, no. 6, Dec. 1980, pp. 1040-1051.

[Timmermann94]   D. Timmermann, B. Rix, H. Hahn and B. Hosticka, "A CMOS floating-point vector-arithmetic unit," *JSSC*, vol. 29, no. 5, Sept. 1994, pp. 634-639.

[Tobias95]   P. Tobias and D. Trindade, *Applied Reliability*, 2nd ed., New York: Van Nostrand Reinhold, 1995.

[Troutman86]   R. Troutman, *Latchup in CMOS Technology: The Problem and its Cure*, Boston: Kluwer Academic Publishers, 1986.

[Tschanz02]   J. Tschanz et al., "Adaptive body bias for reducing impacts of die-to-die and within-die parameter variations on microprocessor frequency and leakage," *JSSC*, vol. 37, no. 11, Nov. 2002, pp. 1396-1402.

[Tsividis99]   Y. Tsividis, *Operation and Modeling of the MOS Transistor*, 2nd ed., Boston: McGraw-Hill, 1999.

[Tyagi93]   A. Tyagi, "A reduced-area scheme for carry-select adders," *JSSC*, vol. 42, no. 10, Oct. 1993, pp. 1163-1170.

[Uehara81]   T. Uehara and W. vanCleemput, "Optimal layout of CMOS functional arrays," *IEEE Trans. Computers*, vol. C-30, no. 5, May 1981, pp. 305-312.

[Unger86]   S. Unger and C. Tan, "Clocking schemes for high-speed digital systems," *IEEE Trans. Computers*, vol. 35, no. 10, Oct. 1986, pp. 880-895.

[Usami94]   K. Usami and M. Horowitz, "Clustered voltage scaling technique for low-power design," *Proc. Intl. Symp. Low Power Electronics*, 1994, pp. 3-8.

[Uyttenhove03]   K. Uyttenhove and M. Steyaert, "A 1.8-V 6-bit 1.3 GHz flash ADC in 0.25μm CMOS," *JSSC*, vol. 38, no. 7, July 2003, pp. 1115-1122.

[Vadasz66]   L. Vadasz and A. Grove, "Temperature dependence of MOS transistor characteristics below saturation," *IEEE. Trans. Electron Devices*, vol. ED-13, no. 13, 1966, pp. 863-866.

[Vadasz69]   L. Vadasz, A. Grove, T. Rowe, and G. Moore, "Silicon-gate technology," *IEEE Spectrum*, vol. 6, no. 10, Oct. 1969, pp. 28-35.

[van Berkel99]   C. van Berkel and C. Molnar, "Beware the three-way arbiter," *JSSC*, vol. 34, no. 6, June 1999, pp. 840-848.

[Vangal02]   S. Vangal et al., "5-GHz 32-bit integer execution core in 130-nm dual-V_T CMOS," *JSSC*, vol. 37, no. 11, Nov. 2002, pp. 1421-1432.

[vanZeijl02]   P. van Zeijl et al., "A Bluetooth radio in 0.18μm CMOS," *JSSC*, vol. 37, no. 12, Dec. 2002, pp. 1679-1687.

[Veendrick80]   H. Veendrick, "The behavior of flip-flops used as synchronizers and prediction of their failure rate," *JSSC*, vol. SC-15, no. 2, April 1980, pp. 169-176.

[Veendrick84]   H. Veendrick, "Short-circuit dissipation of static CMOS circuitry and its impact on the design of buffer circuits," *JSSC*, vol. SC-19, no. 4, Aug. 1984, pp. 468-473.

[Vittal99]   A. Vittal et al., "Crosstalk in VLSI interconnections," *IEEE Trans. Computer-Aided Design*, vol. 18, no. 12, Dec. 1999, pp. 1817-1824.

[Volder59]   J. Volder, "The CORDIC trigonometric computing technique," *IRE Trans. Electronic Computers*, vol. EC-8, no. 3, Sept. 1959, pp-330-334.

[Vollertsen99]   R. Vollertsen, "Burn-in," *IEEE Integrated Reliability Workshop Final Report*, 1999, pp. 167-173.

[Wadell01]   B. Wadell, *Transmission Line Design Handbook*, Norwood, MA: Artech House, 1991.

[Wakerly00]   J. Wakerly, *Digital Design Principles and Practices*, 3rd ed., Upper Sadde River, NJ: Prentice Hall, 2000.

[Wallace64]   C. Wallace, "A suggestion for a fast multiplier," *IEEE Trans. Electronic Computers*, Feb. 1964, pp. 14-17.

[Wang86]   L. Wang and E. McCluskey, "Complete feedback shift register design for built-in self test," *Proc. Design Automation Conf.*, Nov. 1986, pp. 56-59.

[Wang89]   J. Wang, C. Wu, and M. Tsai, "CMOS nonthreshold logic (NTL) and cascode nonthreshold logic (CNTL) for high-speed applications," *JSSC*, vol. 24, no. 3, June 1989, pp. 779-786.

[Wang93]   Z. Wang, G. Jullien, W. Miller, and J. Wang, "New concepts for the design of carry-looka-head adders," *Proc. IEEE Intl. Symp. Circuits and Systems*, 1993, vol. 3, pp. 1837-1840.

[Wang94]   J. Wang, S. Fang, and W. Feng, "New efficient designs for XOR and XNOR functions on the transistor level," *JSSC*, vol. 29, no. 7, July 1994, pp. 780-786.

[Wang97]   Z. Wang, G. Jullien, W. Miller, J. Wang, and S. Bizzan, "Fast adders using enhanced multiple-output domino logic," *JSSC*, vol. 32, no. 2, Feb. 1997, pp. 206-214.

[Wang00j]   J. Wang and C. Huang, "High-speed and low-power CMOS priority encoders," *JSSC*, vol. 35, no. 10, Oct. 2000, pp. 1511-1514.

[Wang00l]   L. Wang and N. Shanbhag, "An energy-efficient noise-tolerant dynamic circuit technique," *IEEE Trans. Circuits and Systems II*, vol. 47, no. 11, Nov. 2000, pp. 1300-1306.

[Wang01]   J. Wang, C. Chang, and C. Yeh, "Analysis and design of high-speed and low-power CMOS PLAs," *JSSC*, vol. 36, no. 8, Aug. 2001, pp. 1250-1262.

[Wanlass63]   F. Wanlass and C. Sah, "Nanowatt logic using field effect metal-oxide semiconductor triodes," *Proc. IEEE Intl. Solid-State Circuits Conf.*, 1963, pp. 32-33.

[Webb97]   C. Webb et al., "A 400-MHz S/390 microprocessor," *JSSC*, vol. 32, no. 11, Nov. 1997, pp. 1665-1675.

[Wei98]   L. Wei, Z. Chen, M. Johnson, K. Roy, and V. De, "Design and optimization of low voltage high performance dual threshold CMOS circuits," *Proc. Design Automation Conf.*, 1998, pp. 489-494.

[Weinberger58] A. Weinberger and J. Smith, "A logic for high-speed addition," System Design of Digital Computer at the National Bureau of Standards: Methods for High-Speed Addition and Multiplication, National Bureau of Standards, Circular 591, Section 1, Feb. 1958, pp. 3-12.

[Weinberger81] A. Weinberger, "4-2 carry-save adder module," *IBM Technical Disclosure Bulletin*, vol. 23, no. 8, Jan. 1981, pp. 3811-3814.

[Weiss02]   D. Weiss, J. Wuu, and V. Chin, "The on-chip 3-MB subarray-based third-level cache on an Itanium microprocessor," *JSSC*, vol. 37, no. 11, Nov. 2002, pp. 1523-1529.

[Weste93]   N. Weste and K. Eshraghian, *Principles of CMOS VLSI Design*, 2nd ed., Reading, MA: Addison-Wesley, 1993.

[Wicht01]   B. Wicht, S. Paul, and D. Schmitt-Landsiedel, "Analysis and compensation of the bitline multiplexer in SRAM current sense amplifiers," *JSSC*, vol. 36, no. 11, Nov. 2001, pp. 1745-1755.

[Williams83]   T. Williams and K. Parker, "Design for Testability—A Survey," *Proc. IEEE*, vol. 71, no. 1, Jan. 1983, pp. 98-112.

[Williams86]   T. Williams, "Design for testability," *Proc. NATO Advanced Study Inst. Computer Design Aids for VLSI Circuits*, (P. Antognetti et al. ed.), NATO AIS Series, 1986, Martinus Nijhoff Publishers, pp. 359-416.

[Williams91]   T. Williams and M. Horowitz, "A zero-overhead self-timed 160-ns 54-b CMOS divider," *JSSC*, vol. 26, no. 11, Nov. 1991, pp. 1651-1661.

[Wing82]   O. Wing, "Automated gate matrix layout," *Proc. IEEE Intl. Symp. Circuits and Systems*, vol. 2, 1982, pp. 681-685.

[Wolf00]   S. Wolf and R. Tauber, *Silicon Processsing for the VLSI Era*, 2nd ed., Sunset Beach, CA: Lattice Press, 2000.

[Wong02]   H. Wong, "Beyond the conventional transistor," *IBM Journal of Research and Development*, vol. 46, no. 2/3, March/May 2002, pp. 133-168.

[Wood01]   J. Wood, T. Edwards, and S. Lipa, "Rotary traveling-wave oscillator arrays: a new clock technology," *JSSC*, vol. 36, no. 11, Nov. 2001, pp. 1654-1665.

[Wu87]   C. Wu, J. Wang and M. Tsai, "The analysis and design of CMOS multidrain logic and stacked multidrain logic," *JSSC*, vol. SC-22, no. 1, Feb. 1987, pp. 47-56.

[Wu91]   C. Wu and K. Cheng, "Latched CMOS differential logic (LCDL) for complex high-speed VLSI," *JSSC*, vol. 26, no. 9, Sept. 1991, pp. 1324-1328.

[Wurtz93]   L. Wurtz, "An efficient scaling procedure for domino CMOS logic," *JSSC*, vol. 28, no. 9, Sept. 1993, pp. 979-982.

[Yamada95]   H. Yamada, T. Hotta, T. Nishiyama, F. Murabayashi, T. Yamauchi, and H. Sawamoto, "A 13.3 ns double-precision floating-point ALU and multiplier," *Proc. Intl. Conf. Computer Design*, 1995, pp. 466-470.

[Yang98]   S. Yang et al., "A high performance 180 nm generation logic technology," *Tech. Digest Intl. Electron Device Meeting*, Dec. 1998, pp. 197-200.

[Yano90]   K. Yano, T. Yamanaka, T. Nishida, M. Saito, K. Shimohigashi, and A. Shimizu, "A 3.8-ns 16 x 16-b multiplier using complementary pass-transistor logic," *JSSC*, vol. 25, no. 2, April 1990, pp. 388-395.

[Yano96]   K. Yano, Y. Sasaki, K. Rikino, and K. Seki, "Top-down pass-transistor logic design," *JSSC*, vol. 31, no. 6, June 1996, pp. 792-803.

[Ye98]   Y. Ye, S. Borkar, and V. De, "A new technique for standby leakage reduction in high-performance circuits," *Symp. VLSI Circuits Digest Tech. Papers*, 1998, pp. 40-41.

[Ye00]   Y. Ye, J. Tschanz, S. Narendra, S. Borkar, M. Stan, and V. De, "Comparative delay, noise and energy of high-performance domino adders with stack node preconditioning (SNP)," *Symp. VLSI Circuits Digest Tech. Papers*, 2000, pp. 188-191.

[Yee00]   G. Yee and C. Sechen, "Clock-delayed dominio for dynamic circuit design," *IEEE Trans. VLSI*, vol. 8, no. 4, Aug. 2000, pp. 425-430.

[Yoon02]   J. Yoon et al., "CMOS-compatible surface-micromachined suspended-spiral inductors for multi-GHz silicon RF Ics" *IEEE Electron Device Letters*, vol. 23, no. 10, Oct. 2002, pp. 591-593.

[Yoshimoto83]   M. Yoshimoto et al., "A divided word-line structure in the static RAM and its application to a 64K full CMOS RAM," *JSSC*, vol. SC-18, no. 5, Oct. 1983, pp. 479-485.

[Young00]   K. Young et al., "A 0.13 μm CMOS technology with 193 nm lithography and Cu/Low-k for high performance applications," *Proc. Intl. Electron Devices Meeting*, 2000, pp. 563-566.

[Yuan82]   C. Yuan and T. Trick, "A simple formula for the estimation of the capacitance of two-dimensional interconnects in VLSI circuits," *IEEE Electron Device Letters*, vol. EDL-3, Dec. 1982, pp. 391-393.

[Yuan89]   J. Yuan and C. Svensson, "High-speed CMOS circuit technique," *JSSC*, vol. 24, no. 1, Feb. 1989, pp. 62-70.

[Zhou99]   X. Zhou, K. Lim, and D. Lim, "A simple and unambiguous definition of threshold voltage and its implications in deep-submicron MOS device modeling," *IEEE Trans. Electron Devices*, vol. 46, no. 4, April 1999, pp. 807-809.

[Zhou03]   Y. Zhou and J. Yuan, "A 10-bit wide-band CMOS direct digital RF amplitude modulator," *JSSC*, vol. 38, no. 7, July 2003, pp. 1182-1188.

[Zhuang92]   N. Zhuang and H. Wu, "A new design of the CMOS full adder," *JSSC*, vol. 27, no. 5, May 1992, pp. 840-844.

[Ziegler02]   J. Ziegler, *Ion-Implantation—Science and Technology, 2002 Edition*, IIT Press, 2002.

[Ziegler96]   J. Ziegler, "Terrestrial cosmic rays," *IBM J. Research and Development*, vol. 40, no. 1, Jan. 1996, pp. 19-39.

[Zimmermann96] R. Zimmermann, "Non-heuristic optimization and synthesis of parallel-prefix adders," *Proc. Intl. Workshop on Logic and Architecture Synthesis*, Dec. 1996, pp. 123-132.

[Zimmermann97a] R. Zimmermann and W. Fichtner, "Low-power logic styles: CMOS versus pass-transistor logic," *JSSC*, vol. 32, no. 7, July 1997, pp. 1079-1090.

[Zimmermann97b] R. Zimmermann, *Binary Adder Architectures for Cell-Based VLSI and their Synthesis*, ETH Dissertation 12480, Swiss Federal Institute of Technology, 1997.

[Zuras86]   D. Zuras and W. McAllister, "Balanced delay trees and combinatorial division in VLSI," *JSSC*, vol. SC-21, no. 5, Oct. 1986, pp. 814-819.

Index